Vasily E. Tarasov, Valentina V. Tarasova
Economic Dynamics with Memory

Fractional Calculus in Applied Sciences and Engineering

Volume 8

Vasily E. Tarasov, Valentina V. Tarasova

Economic Dynamics with Memory

Fractional Calculus Approach

DE GRUYTER

Mathematics Subject Classification 2010
91B02, 91B55, 26A33, 34A08, 47G20

Authors

Prof. Dr. Vasily E. Tarasov
Lomonosov Moscow State University
Skobeltsyn Inst. of Nuclear Physics
Leninskie Gory 1-2, GSP-1
119991 Moscow
Russian Federation
tarasov@theory.sinp.msu.ru

Dr. Valentina V. Tarasova
Lomonosov Moscow State University
Faculty of Economics
Leninskie Gory 1-46, GSP-1
119991 Moscow
Russian Federation
v.v.tarasova@mail.ru

ISBN 978-3-11-062460-1
e-ISBN (PDF) 978-3-11-062745-9
e-ISBN (EPUB) 978-3-11-062481-6
ISSN 2509-7210

Library of Congress Control Number: 2020950073

Bibliographic information published by the Deutsche Nationalbibliothek
The Deutsche Nationalbibliothek lists this publication in the Deutsche Nationalbibliografie; detailed bibliographic data are available on the Internet at http://dnb.dnb.de.

Typesetting: VTeX UAB, Lithuania
Printing and binding: CPI books GmbH, Leck

www.degruyter.com

Preface

The proposed book outlines the basic concepts, models and principles of Economics with Memory as one of the new modern branches of economic science. Economics with Memory is a science of the behavior and interaction of economic agents with memory, about economic processes with memory and nonlocality over time. The presence of memory in the economic process means that this process depends on the history of changes of the process in the past during a finite time interval. Fractional calculus is used as the basic mathematical tool to take into account memory and nonlocality in time.

For the first time, the importance of long-range time dependence in economic data was recognized in 1966 by Clive W. J. Granger, who received the Nobel memorial prize in economic sciences in 2003 "for methods of analyzing economic time series with common trends (cointegration)." In fact, the phenomenon of memory in modern economics was first discovered by Granger. Then the first economic models with memory were proposed in the work of Clive W. J. Granger and Roselyne Joyeux in 1980. These models are called the fractional ARIMA models, which generalize autoregressive integrated moving average (ARIMA) models by using noninteger values of orders of the differencing and integrating.

Now, it is becoming more and more obvious that when describing the behavior of economic agents, we must take into account that agents may have memory. The description of economic processes should take into account that the behavior of economic agents may depend on the history of previous changes in economy. To describe this type of behavior, we need new economic concepts and notions that allow us to take into account the presence of memory in economic processes. New economic models and methods are needed to take into account that economic agents may remember the changes of economic indicators and factors in the past. The presence of memory can change the behavior of agents and the decisions they make. As a result, to describe economic processes with memory, we cannot use the standard tools of differential and difference equations of integer orders. In fact, the differential equations describe only such economic processes, in which agents actually have amnesia. In other words, economic models, which use only derivatives of integer orders, can be applied only when economic agents forget the history of changes in economic indicators and factors during an infinitely small time interval. Differential equations of integer orders describe an economy consisting of economic agents with amnesia. It is becoming clear that the use of such equations in models with continuous time holds back the development of the modern economics.

At the present time, a revolution is taking place in modern economics, which can be called the "Memory Revolution." This revolution is intended to include in the modern economic theory new economic concepts and notions that allow us to take into account the presence of memory in economic processes. The main mathematical tool de-

https://doi.org/10.1515/9783110627459-201

signed to "cure amnesia" in economics is the theory of derivatives and integrals of noninteger order (the fractional calculus), fractional differential and difference equations. Note that the fractional calculus as a theory of integrals, derivatives and differences of arbitrary (integer and noninteger) orders includes standard calculus as a special case. Derivatives and integrals of arbitrary (integer and noninteger) orders are well-known in mathematics more than 300 years. In theory of economic processes with memory, fractional calculus should replace standard calculus of integrals, derivatives and differences of integer orders.

To describe economic processes with long memory Clive W. J. Granger and Roselyne Joyeux in 1980 proposed the difference operators that were called the fractional differencing and fractional integrating. Approach, which is based on the discrete operators proposed by Granger and Joyeux, is now the most common among economists. In fact, these operators are the Grunwald–Letnikov fractional differences, which are well known in mathematics for more than 150 years. In modern mathematics, there are various types of fractional integral and differential operators that were proposed by such well-known mathematicians as Riemann, Liouville, Grunwald, Letnikov, Sonine, Marchaud, Weyl, Riesz, Kober, Erdelyi and other scientists. Obviously, the restriction of mathematical tools only to the Grunwald–Letnikov fractional differences significantly reduces the possibilities for studying economic processes with memory and nonlocality in time. Now is the time to use all the tools of fractional calculus in the modern economics, and not be restricted only by the discrete Grunwald–Letnikov operators.

The "Memory Revolution" has led to the emergence of a new branch of economics, which can be called "Economics with Memory." The mathematical language of this branch of the economics is fractional calculus, which allows us to formulate new concepts and principles. However, this does not mean that the Economics with Memory can be considered as part of applied mathematics or mathematical economics. Economics with Memory focuses on the behavior and interactions of economic agents with memory and economic processes with memory and nonlocality in time. Economics with Memory should be a science based on new economic concepts, notions, principles, rules and models to take into account a memory. The modern revolutionary situation in economic theory means the importance of developing new economic concepts, notions, models and principles. This is especially important due to the fact that the Economics with Memory is now only being formed as new branch of science.

We can assume that Economics with Memory can include, as a special case, the standard economics. The standard economic model, which does not take into account the fading memory, can be considered as a special case of the economic model with memory, when the parameters of memory fading take integer values. This statement is based on the fact that fractional differential equations of economic models with memory include differential equations of integer orders, which describe standard models, as a special case. The situation is similar for generalized economic concepts. For example, a marginal value with power-law memory contains the standard marginal and average values as special cases corresponding to the fading parameter equal to zero and one, respectively.

In writing this book, an ambitious task was set to create and describe the foundation of the Economics with Memory as a new branch of the economics. For this purpose, generalizations of basic economic concepts, models and principles of economic dynamics were proposed. It is obvious that such an ambitious task cannot be completely solved within the framework of one book, despite its volume. Many questions, concepts, notions, models and parts of economics remained beyond the scope of this book. However, the authors hope that the proposed book will lead to new publications of articles and books on Economics with Memory. In this case, we will consider that the proposed book has achieved its purpose. We hope that the Economics with Memory will become an actively developing branch of economics. Perhaps in the future, standard economics will become a part of the Economics with Memory, in the same way as classical physics became a part of quantum and relativistic physics.

In this book, we proposed the construction of the Economics with Memory as a new branch of economics, and we offer new notions, concepts, effects, phenomena, principles and models, which take into account the features of economic processes with memory. This book is based on our works written in the period from 2016 to 2020.

The following economic concepts and notions were proposed between 2016 and 2020: the marginal values of noninteger order with memory and nonlocality; the economic multiplier with memory; the economic accelerator with memory; the exact discretization of economic accelerators and multipliers based on exact fractional differences; the accelerator with memory and periodic sharp bursts; the duality of the multiplier with memory and the accelerator with memory; the accelerators and multipliers with memory and distributed lag; the elasticity of fractional order for processes with memory and nonlocality; the measures of risk aversion with nonlocality and with memory; the warranted (technological) rate of growth with memory; the nonlocal methods of deterministic factor analysis; the productivity with fatigue and memory; the chronological memory ordering; and some others.

Using these concepts, the following economic models were proposed: the natural growth model with memory; the growth model with constant pace and memory; the Harrod–Domar model with memory; the Keynes model with memory; the dynamic Leontief (intersectoral) model with memory; the model of dynamics of fixed assets (or capital stock) with memory; the model of price dynamics with memory; the logistic growth model with memory; the model of logistic growth with memory and periodic sharp splashes (kicks); the time-dependent dynamic intersectoral model with memory; the Phillips model with memory and distributed lag; the Harrod–Domar growth model with memory and distributed lag; the dynamic Keynesian model with memory and distributed lag; the model of productivity with fatigue and memory; the Solow–Swan model with memory; the Kaldor-type model of business cycles with memory; the Evans model with memory; the Cagan model with memory; the Lucas models with memory; and some other economic models.

Some of these generalized economic concepts and models for processes with memory are described in the proposed book on economic dynamics with memory.

We hope that the book will be useful to anyone interested in modern economics and applied mathematics. Preliminary knowledge of fractional calculus for reading the book is not expected. However, knowledge of standard calculus and the introductory course of economics are necessary. Chapters of the advanced parts of the book may be skipped by economists on their first reading.

In conclusion of this preface, the authors want to thank the editor-in-chief of the book series, Professor Changpin Li, for supporting us in writing this book. The authors also express their gratitude to the members of the Editorial Board, Professors Virginia Kiryakova, Francesco Mainardi, Dragan Spasic, Bruce Ian Henry and YangQuan Chen.

Vasily E. Tarasov
Valentina V. Tarasova
Moscow, 9 May 2020

Contents

Part II: Concepts of economics with memory

Part V: **Advanced models: distributed lag and memory**

Part VI: **Advanced models: discrete time approach**

Introduction: economics with memory

The birth of modern economics occurred almost simultaneously with the appearance of new economic concepts, which began to be actively used in various economic models. "Marginal revolution" and "Keynesian revolution" in economics introduced fundamental economic concepts, including the concepts of "marginal value," "economic multiplier," "economic accelerator," "elasticity" and many others. These concepts are based on the use of mathematical tools, which were not previously used in the economics. The most important mathematical tools, which are become actively used in description of economic processes, are the theory of derivatives and integrals of integer orders, the theory of differential and difference equations. These mathematical tools allowed economists to build economic models in a mathematical form and on their basis to describe a wide range of economic processes and phenomena. "Marginal revolution" and "Keynesian revolution" led to the use of standard mathematical calculus, which is theory of the derivatives and integrals of integer orders, and the differential and difference equations of integer orders.

As a result of the past marginal and Keynesian revolutions, the economic models with continuous and discrete time began to be described by differential equations with derivatives of integer orders and equations with finite differences of integer orders. However, these tools have a number of shortcomings that lead to the incompleteness of descriptions of economic processes. It is known that the derivatives of integer orders are determined by the properties of a differentiable function only in an infinitely small neighborhood of the point, in which these derivatives are considered. As a result, differential equations with derivatives of integer orders, which are used in economic models, cannot describe processes with memory and nonlocality in time. In fact, such equations describe only economic processes, in which all economic agents have complete amnesia and they do not remember the past events in the economy. Obviously, this assumption about the absence of memory among economic agents is a strong restriction for economic models. Therefore, standard models have drawbacks, since they cannot take into account the important aspects of economic processes and phenomena.

At the present moment, a new revolution, which can be called "Memory revolution," is actually taking place in modern economics. This revolution is intended to "cure amnesia" of modern economic theory, which is caused by the use of differential and integral operators of integer orders. In economics, it becomes obvious that when describing the behavior of economic agents, we must take into account that agents may have memory. The description of economic processes should take into account that the behavior of economic agents may depend on the history of previous changes in economy. To describe this type of behavior, we need new economic concepts and notions that allow us to take into account the presence of memory in economic processes. New economic models and methods are needed to take into account that economic

https://doi.org/10.1515/9783110627459-202

agents may remember the changes of economic indicators and factors in the past. The presence of memory changes the behavior of agents and influences the decisions they make. As a result, to describe economic processes with memory, we cannot use the standard tools of differential (or difference) equations of integer orders. In fact, these equations describe only such economic processes, in which agents actually have amnesia. In other words, economic models, which use only derivatives of integer orders, can be applied when economic agents forget the history of changes of economic indicators and factors during an infinitesimally small period of time. It is becoming clear that this restriction holds back the development of economic sciences.

The “Memory revolution” is intended to include in the modern economic theory new economic concepts and notions that allow us to take into account the presence of memory in economic processes. The main mathematical tool designed to “cure amnesia” in economics is the theory of derivatives and integrals of noninteger order (the fractional calculus), fractional differential and difference equations [335, 202, 308, 200, 90, 164, 165, 119]. Fractional calculus as a theory of integrals, derivatives, and finite differences of arbitrary (integer and noninteger) orders includes standard calculus as a special case. Derivatives and integrals of arbitrary (integer and noninteger) orders are well known in mathematics more than 300. There are different types of fractional integral and differential operators that have been proposed by such well-known mathematicians as Riemann, Liouville, Grunwald, Letnikov, Sonine, Marchaud, Weyl, Riesz, Kober, Erdelyi and other scientists. In modern economics, fractional calculus is intended to replace standard calculus of integrals, derivatives and differences of integer orders.

The “Memory revolution” has led to the emergence of a new economic science, which can be called “Economics with Memory.” The mathematical language of this science is fractional calculus, which allows us to formulate new concepts and principles. However, this does not mean that the economics with memory can be considered as part of applied mathematics or mathematical economics. Economics with memory is the science of the behavior of the behavior and interactions of economic agents with memory and economic processes with memory and nonlocality in time. Economics with Memory should be based on new economic concepts, notions, principles and models that allow us to describe economic processes with memory. The modern revolutionary situation in economic theory means that now an important task is the formulation of new economic concepts, concepts, models and principles. This is especially important due to the fact that the economics with memory is now only being formed as new branch of science.

The beginning of the “Memory revolution” in economics may be associated with the works, which were published in 1966 and 1980 by Clive W. J. Granger [147, 148, 152, 155, 153, 151]. Note that Granger received the Nobel memorial prize in economic sciences in 2003 “for methods of analyzing economic time series with common trends (cointegration)” [479]. For the first time, the importance of long-range time dependence in economic data was recognized by Clive W.J. Granger in his article [148] in

1966 (see also [147, 152, 155, 29]). Granger showed that a number of spectral densities, which are estimated from economic time series, have a similar form. We can state that the phenomenon of long memory in modern economics was discovered by Granger. Then, to describe economic processes with memory Granger and Joyeux [155] in 1980 proposed the fractional ARIMA models, which are also called ARFIMA(p, d, q). The fractional ARIMA(p, d, q) models are generalization of ARIMA(p, d, q) model from a positive integer order d to noninteger (positive and negative) orders d [29, 30]. To generalize ARIMA models, Granger and Joyeux [155] (see also [153, 180]) proposed the so-called fractional differencing and integrating for discrete time case (see books [29, 295, 30, 481, 231, 485], and reviews [151, 21, 303, 23, 135]). The fractional difference operators of Granger and Joyeux were proposed and then began to be used in economics up to the present time without any connection with the fractional calculus and the well-known fractional differences of noninteger orders. In fact, these fractional differencing and integrating are the well-known Grunwald–Letnikov fractional differences, which were suggested in 1867 and 1868 in works [157, 222]. Then the Grunwald–Letnikov fractional differences are actively used in the fractional calculus [335, 308, 200] and began to apply in physics and other sciences. We should also note that in the continuous limit the Grunwald–Letnikov fractional differences of positive orders can give the Grunwald–Letnikov, Marchaud and Liouville fractional derivatives [335].

Approach, which is based on the discrete operators proposed by Granger and Joyeux is the most common among economists [21, 303, 23, 135]. The approach based on discrete operators of Granger–Joyeux, more precisely on the Grunwald–Letnikov fractional differences, is restricted by only one type of fractional finite differences. Unfortunately, this approach is used without an explicit connection with the modern mathematics and the development of fractional calculus in the last 200 years. It should be emphasized that the Granger–Joyeux approach to economics with memory is restricted by models with discrete time and application of the Grunwald–Letnikov fractional differences. Obviously, the restriction of mathematical tools to only to the discrete Grunwald–Letnikov operators significantly reduces the possibilities to describe processes with memory and nonlocality in time. In recent years, attempts have been made to suggest a different approach to the description of processes with memory in economics and finance. A brief description of the history of attempts to describe economic and finance processes with memory is given in work [385] (see also book [254, pp. 5–32]).

We can state that the use of fractional calculus in economic models will significantly expand the possibilities of describing processes with memory and will lead to new results. We propose to use all methods of modern fractional calculus to describe economic processes with memory, and to generalize the basic economic notions, concepts, models and principles to form a new economic science, the Economics with Memory.

In this book, fractional calculus is used to formulate the concept of memory itself for economic processes, and to define basic concepts, models and principles for description of economic processes with memory and nonlocality in time.

From our subjective point of view, the formation of new basic economic concepts and notions of new science "Economics with Memory" began with a proposal of generalizations of the basic economic concepts and notions at the beginning of 2016, when the concept of elasticity for economic processes with memory was proposed in works [418] and [430, 419]. Then, in 2016, we proposed such new concepts as the marginal values with memory [424, 416, 425], accelerator with memory and multiplier with memory [412, 417], the nonlocal measures of risk aversion [428] and nonlocal deterministic factor analysis [414]. In 2016, these concepts are used in generalizations of some standard economic models [420, 421, 426, 423, 427, 422] that describe the dynamics of economic processes with memory.

During 2016–2020, we use the fractional calculus to define basic concepts of Economics with Memory, and to describe the dynamics of economic processes with memory. Some of these generalized economic concepts and models for processes with memory are described in the proposed book on economics with memory.

It should be noted that formal replacements of derivatives of integer order by fractional derivatives in standard differential equations, which describe economic processes, and solutions of the obtained fractional differential equations were considered in papers published before 2016. However, these papers were purely mathematical works, in which generalizations of economic concepts and notions were not proposed. In these works, fractional differential equations were not usually derived, since a formal replacement of integer-order derivatives by fractional derivatives cannot be recognized as a derivation of the equations. Formulations of economic conclusions and interpretations from the obtained solutions are not usually suggested in these papers. Examples of incorrectness and errors in such generalizations are given in [386], (see also book [254, pp. 43–92]). Review article [386] formulates five principles of the "fractional-dynamic" generalization of standard dynamic models. We can state that in the works with formal fractional generalizations of standard economic equations, the Principles of Derivability and Interpretability were neglected [386]. Let us give a brief formulation of the Principles of Derivability and Interpretability.

Derivability Principle: It is not enough to generalize the differential equations describing the dynamic model. It is necessary to generalize the whole scheme of obtaining (all steps of derivation) these equations from the basic principles, concepts and assumptions. In this sequential derivation of the equations, we should take into account the nonstandard characteristic properties of fractional derivatives and integrals. If necessary, generalizations of the notions, concepts and methods, which are used in this derivation, should also be obtained.

Interpretability Principle: The subject (physical, economic) interpretation of the mathematical results, including solutions and their properties, should be obtained.

Differences and, first of all qualitative differences from the results obtained for the standard model, should be described.

Note that the problems and difficulties arising in the construction of generalized dynamic models with memory for the standard economic models by using the fractional calculus are described in [386] with details. The most important purpose of the "Memory revolution" is the inclusion of memory and time nonlocality into the economic theory, into the basic economic concepts and methods. The economics should be extended and generalized such that it takes into account the memory and time nonlocality. Generalizations of standard economic models should be constructed only on this conceptual basis. The most important purpose of studying such generalizations is the search and formulation of qualitatively new effects and phenomena caused by memory in the behavior of economic processes.

Let us list some generalizations of economic concepts and generalizations of economic models that have already been proposed in recent years.

The generalizations of some basic economic notions are described in the proposed book on Economics with memory. The list of these new notions and concepts primarily include the following:

1. the marginal value of noninteger order [424, 416, 425, 439, 446] with memory and nonlocality;
2. the economic multiplier with memory [412, 397];
3. the economic accelerator with memory [412, 397];
4. the exact discretization of economic accelerators and multipliers [441, 442, 431, 432] based on exact fractional differences;
5. the accelerator with memory and periodic sharp bursts [417, 436, 444];
6. the duality of the multiplier with memory and the accelerator with memory [412, 397];
7. the accelerators and multipliers with memory and distributed lag [405, 404, 402];
8. the elasticity of fractional order [418, 430, 429, 419] for processes with memory and nonlocality;
9. the measures of risk aversion with nonlocality and with memory [428, 449];
10. the warranted (technological) rate of growth with memory [450, 447, 445, 453, 393, 387];
11. the nonlocal methods of deterministic factor analysis [414, 452];
12. the productivity with fatigue and memory [454];
13. the chronological memory ordering [433, 398];
14. and some other.

The use of these notions and concepts makes it possible for us to generalize some classical economic models, which are proposed by the following well-known economists:

1. Henry Roy F. Harrod [170, 171, 172] and Evsey D. Domar [92, 93];
2. John M. Keynes [191, 193, 194, 192];
3. Wassily W. Leontief [219, 220, 354, 355];

4. Alban W.H. Phillips [305, 306];
5. Roy G. D. Allen [8, 9, 10, 11, 12];
6. Robert M. Solow [350, 351] and Trevor W. Swan [357];
7. Nicholas Kaldor [187, 188];
8. Griffith C. Evans [116, 117];
9. Phillip D. Cagan [43];
10. Robert E. Lucas Jr. [239, 238, 237, 235, 236];
11. and other scientists.

Economic models with continuous time are usually described by differential equations with derivatives of integer orders. In the standard economic models, the memory effects and memory fading are neglected. At the same time, it is obvious that it is necessary to take into account the influence of memory effects on economic processes. We should take into account the memory, since the assumption of a lack of memory among economic agents, that is, the assumption that economic agents have lost their memory and have amnesia is a very strong restriction on economic models.

The presence of memory in the economic process means that the behavior of the process depends not only on the variables and parameters of this process at the present time, but also on the history of changes in these variables and parameters on a finite time interval. The concept of memory is very important for an adequate description of real economic processes.

From a mathematical point of view, the neglect of memory effects in standard economic models is due to the used equations with derivatives of integer orders, which are determined by the properties of the function in an infinitely small neighborhood of the considered time.

A very effective and powerful tool for describing memory effects is fractional calculus of derivatives and integrals of noninteger orders [335, 202, 308, 200, 90, 164, 165], which have a long history of more than 300 years [221, 328, 330, 329, 203, 467], [469, 470, 489, 464]. We should note that the characteristic property of fractional derivatives of noninteger order is the violations of standard rules and properties that are fulfilled for derivatives of integer order [367, 287, 377, 376, 382]. These nonstandard mathematical properties allow us to describe nonstandard processes and phenomena associated with nonlocality and memory [386, 406, 385]. On the other hand, these nonstandard properties lead to difficulties [386] in sequential constructing generalizations of standard models. In article [386] (see also book [254, pp. 43–92]), we show how problems arise when building generalizations of standard dynamic models of economics.

The need to take into account the non-standard properties of fractional derivatives and integrals leads us to the following [386]:

First of all, it is not enough to generalize the differential equations describing the standard dynamic model. It is necessary to generalize the whole scheme of obtaining (all steps of derivation) these equations from the basic principles, concepts and assumptions (see the Derivability Principle in [386]). Therefore, it is necessary to consider the possibility of generalizing all stages of obtaining equations of standard economic models.

Second, in this sequential derivation of the equations, we should take into account the nonstandard characteristic properties of fractional derivatives and integrals. This leads us to the fact that it is necessary to take into account the violation of standard product and chain rules, the violation of semigroup properties for the repeated action of fractional and integer derivatives.

Third, for standard models, there is a set of various generalizations, due to the existence of various types of fractional operators and violation of s-equivalence for fractional differential equations (see the Multiplicity Principle in [386]).

Because of this, in this book, we restrict ourselves mainly to such generalizations of standard models in which there are exact analytical solutions of models equations.

In this book, we propose some economic models, in which effects of fading memory are taken into account. We describe the possible difficulties and ambiguities that arise in the generalization of standard economic models. We derive the fractional differential equations of the proposed economic models with memory and, then we find solutions of these equations. Using these solutions, we formulate some principles of economic dynamics with fading memory. For generalized models with memory, we describe differences from standard models that arise due to the presence of memory in the processes.

In this book, we will describe various economic models with fading memory, which are generalizations of the classical models. For example, the following economic models were proposed.

1. the natural growth model with memory [420, 440];
2. the growth model with constant pace and memory [447, 438];
3. the Harrod–Domar model with memory [421, 426] and [450, 445, 453, 393];
4. the Keynes model with memory [423, 427, 422] and [402, 407];
5. the dynamic Leontief (intersectoral) model with memory [437, 451, 433, 398];
6. the model of dynamics of fixed assets (or capital stock) with memory [447, 438];
7. the model of price dynamics with memory [438];
8. the logistic growth model with memory [386] (see also [403]);
9. the model of logistic growth with memory and periodic sharp splashes (kicks) [444];
10. the time-dependent dynamic intersectoral model with memory [433, 398];
11. the Phillips model with memory and distributed lag [405];
12. the Harrod–Domar growth model with memory and distributed lag [404];
13. the dynamic Keynesian model with memory and distributed lag [402, 407];
14. the model of productivity with fatigue and memory [454];
15. the Solow–Swan model with memory [386, 383];
16. the Kaldor-type model of business cycles with memory [386];
17. the Evans model with memory [390];

18. the Cagan model with memory [388];
19. the Lucas models with memory [391, 389, 381];
20. and some other economic models.

We should note that the standard concepts and models, which do not take into account the fading memory, can be considered as special cases of the proposed generalized concepts and models with memory, when the parameter of memory fading takes integer values. For example, fractional differential equations of economic models with memory include differential equations of integer orders that describe standard models. The situation is similar for generalized economic concepts. For example, marginal value with power-law memory contains the standard marginal and average values as special cases corresponding to the fading parameter equal to zero and one, respectively.

In this book, we proposed new concepts, models and principles for Economics with Memory. Qualitatively new effects due to the presence of memory in the economic process are described. The most important element in the construction of the Economics with memory as a branch of economics are the formation of new notions, concepts, effects, phenomena, principles and methods, which are specific only to this branch.

In writing this book, an ambitious task was set to create and describe the foundation of the Economics with Memory as a new branch of the economics. For this purpose, generalizations of basic economic concepts, models and principles of economic theory were proposed. It is obvious that such an ambitious task cannot be completely solved within the framework of one book, despite its volume. Many questions, concepts, notions, models, methods and even sections of economics remained beyond the scope of this book. However, the authors hope that the proposed book will lead to new publications of articles and books on Economics with Memory, which will fill in the missing elements. In this case, we will consider that the proposed book has achieved its purpose. We hope that the "Economy with Memory" will become in the near future an independent and actively developing new branch of economics.

Part I: **Concept of memory**

1 Concept of memory in economics

In this chapter, we discuss the concept of memory and its properties from the point of view of its possible application in economics. We also describe some general restrictions that can be imposed on the properties of memory. These restrictions include the following principles: (A) The Causality Principle; (B) The Principle of Nonlocality in Time; (C) The Linear Superposition Principle; (D) The Principle of Memory Fading (the Fading Principle); (E) The Principle of Nonaging Memory (the Principle of Memory Homogeneity on Time); (F) The Principle of Memory Reversibility (the Principle of Memory Recovery). Examples of operators and functions are proposed for describing different types of memory. This chapter is based on articles [450, 448, 399, 386, 406].

1.1 Introduction

Concept of memory is actively used not only in psychology, biology or computer science, but also in physics and mechanics (e. g., see books [88, 320, 321, 230, 322], [7, 245, 363, 15] and [167, 168]). Currently, processes with memory have begun to be studied in economics and finance (e. g., see books [29, 295, 30, 481, 231, 485] and [385, 254, 392]).

Memory can be considered as a characteristic (property) of processes, which describes the dependence of the process at a given time on the states of this process in the past. The economic process with memory assumes awareness of economic agents about the change history of this process in the past, and assumes the impact of this information on the behavior of agents at the current time. In such processes, which is described by a set of state variables $X(t)$, the behavior of economic agents is based not only on information on the state of the process $\{t, X(t)\}$ at a given moment of time t. The presence of memory in the process assumes that economic agents also use information on about the process states $\{\tau, X(\tau)\}$ at time instants $\tau \in [t_0, t]$.

In economic processes, the presence of memory means the existence of a variable, which depends not only on the values of another variable (or variables) at present, but also on their values at previous points in time. The effect of memory is related to the fact that the same change in a variable $X(t)$ can lead to another change in the corresponding dependent variable $Y(t)$. This can lead to the multivalued dependencies of these variables. The multivalued dependencies are caused by the fact that the economic agents remember previous changes of these variables and, therefore, can react differently despite repeating the situation. As a result, an identical change of the variable $X(t)$ may lead to a different type of behavior of dependent variable(s) $Y(t)$.

In this chapter, we discuss a concept of memory and some mathematical methods, which allow us to describe economic processes with memory primarily in the framework of the continuous time approach. However, the discrete approach will also be briefly reviewed. This chapter is based on articles [450, 448, 399, 386, 406]. In first section, we discuss the definition of the concept of memory. Second section deals with

https://doi.org/10.1515/9783110627459-001

the general characteristics, properties and principles of memory, which impose restrictions on the description of the memory. The next section provides examples of memory based on the use of fractional calculus. Then we briefly discuss the relationship between the continuous and discrete forms of the mathematical description of memory.

1.2 Definition of memory

The variables, which are used in economic models can be divided into two types. At first, the exogenous variables (factors) are considered as independent (autonomous) variables, which are external to the model. Second, the endogenous variables (indicators) are internal variables of the model, which are formed inside the model. These variables are described as variables that depend on the independent variables (factors). Thus endogenous variable can be considered as a response (reaction) to the action, which is described by the exogenous variable. Usually, the terms exogenous and endogenous variables are used in macroeconomic models. However, these terms are relatively universal and can be used for a wide class of models in various branches of economic theory and other sciences.

Let us consider the case, when variables, which describe the states of economic processes, are real single-valued functions of continuous time t.

A general formulation of the economic process with memory can be given the following form. In the economic process with memory, there exists at least one variable $Y(t)$ at the time t, which depends on the history of the change of another variable(s) $X(\tau)$ at $\tau \in (t_0, t]$. Instead of the given verbal formulation, we can use the following symbolic expression

$$Y(t) = \mathcal{F}_{t_0}^{t}(X(\tau)). \tag{1.1}$$

In equation (1.1), the symbol $\mathcal{F}_{t_0}^{t}$ denotes a certain method that allows us to find the value of $Y(t)$ for any time t, if it is known $X(\tau)$ for $\tau \in (t_0, t]$.

As the initial moment of time t_0, for example, we can consider zero or minus infinity ($t_0 = 0$ or $t_0 = -\infty$). Minus infinity means that the economic process is considered starting from a very distant past. Actually any economic process exists only during a finite time interval.

Remark 1.1. We can say that $\mathcal{F}_{t_0}^{t}$ is an operator, which is a mapping from one space of functions to another. In continuum mechanics and physics $\mathcal{F}_{t_0}^{t}$ is called a functional, which transforms each history of changes of $X(\tau)$ for $\tau \in (t_0, t]$ into the appropriate history of changes of $Y(\tau)$ with $\tau \in (t_0, t]$. In mathematics, a functional is a map from a space of functions into its underlying field of scalars (numbers). For this reason, we will use the mathematical term “operator.”

The operator $\mathcal{F}_{t_0}^t$ is said to be a linear operator, if the condition

$$\mathcal{F}_{t_0}^t(aX_1(\tau) + bX_2(\tau)) = a\mathcal{F}_{t_0}^t(X_1(\tau)) + b\mathcal{F}_{t_0}^t(X_2(\tau)) \tag{1.2}$$

is satisfied for all a and b that is arbitrary real numbers, i. e., $a, b \in \mathbb{R}$.

For a wide range of processes, we can use operators $\mathcal{F}_{t_0}^t$ in the form of linear integral and integro-differential operators. For example, we can use the integral operators, which are defined by the expression

$$\mathcal{F}_{t_0}^t(X(\tau)) = \int_{t_0}^{t} M(t,\tau)X(\tau)\, d\tau. \tag{1.3}$$

The integro-differential operators can be considered as compositions of integral operator (1.3) and differential operators. For example,

$$\mathcal{F}_{t_0}^t(X^{(n)}(\tau)) = \int_{t_0}^{t} M(t,\tau)X^{(n)}(\tau)\, d\tau, \tag{1.4}$$

where $X^{(n)}(\tau) = d^n X(\tau)/d\tau^n$ is the derivative of the integer order $n > 0$. Operator (1.3) is sometimes called the Volterra operator. The function $M(t,\tau)$, which is the kernel of the integral operator (1.3), is called the memory function. In this case, we can say that the memory is described by the Volterra operator $\mathcal{F}_{t_0}^t$, or by the memory function $M(t,\tau)$.

Remark 1.2. Expression (1.3) has a sense, if the integral (1.3) exists. Therefore, the variable $X(\tau)$ is not necessarily a continuous function of time and the memory function can have an integrable singularity of some kind.

Let us give the following definition of the concept of memory that we will use in economics.

Definition 1.1. In economic process, memory is the property of the process, for which there exists at least one variable $X(t)$, and another variable $Y(t)$ depending on it, and this dependence is described by the linear integral equation

$$Y(t) = \int_{t_0}^{t} M(t,\tau)X(\tau)\, d\tau, \tag{1.5}$$

where $M(t,\tau)$ is a function, which is called the memory function.

In equations (1.5), we can call $X(t)$ as an impact variable (impact), $Y(t)$ as a response variable (response), and the function $M(t,\tau)$ as a linear response function.

It is obvious that not every equation of the form (1.5) can be used to describe the memory in the economic processes. In the next sections, we will consider restrictions on expression (1.5) and examples of memory function.

As a impact variable $X(\tau)$ of equation (1.5), we can consider the integer derivative $X^{(n)}(\tau) = d^n X(\tau)/d\tau^n$ of $X(\tau)$ with respect to time τ, where n is a nonnegative integer number. In this case, we can use the equation

$$Y(t) = \int_{t_0}^{t} M(t,\tau) X^{(n)}(\tau)\, d\tau. \tag{1.6}$$

An integral of integer order n of the variable $X(\tau)$ can also be considered instead of the variable $X(t)$ in equation (1.5).

Remark 1.3. Equation (1.5) can be interpreted as an equation of economic multiplier with memory, which is characterize by the function $M(t,\tau)$. Equation (1.6) with $n = 1$ can be interpreted as an equation of economic accelerator with memory, which is characterized by the function $M(t,\tau)$. Note that the concept of the multiplier with memory and the accelerator with memory were proposed in articles [412, 397].

Remark 1.4. We can use generalized linear operators $\mathcal{F}_0^t$, which is characterized by the function $M(t,\tau)$. For example, we can use equation (1.1) in the form of the integro-differential equations

$$Y(t) = \frac{d^n}{dt^n} \int_{t_0}^{t} M(t,\tau) X(\tau)\, d\tau, \tag{1.7}$$

and

$$Y(t) = \frac{d^{n-k}}{dt^{n-k}} \int_{t_0}^{t} M(t,\tau) X^{(k)}(\tau)\, d\tau, \tag{1.8}$$

where $k = 0, 1, \ldots, n$ and n is a nonnegative integer number.

Remark 1.5. In nonlinear cases, we should use equations that take into account nonlinearity. For example, instead of equation (1.5), we can use the equation

$$Y(t) = f(\mathcal{F}_{t_0}^t(X(\tau))), \tag{1.9}$$

or

$$Y(t) = \int_{t_0}^{t} M(t,\tau) f(X(\tau))\, d\tau, \tag{1.10}$$

where $f(X)$ is some nonlinear function. In a more general case, we can take into account a finite number of derivatives of integer orders by using the equation

$$Y(t) = \int_{t_0}^{t} M(t,\tau) f(X(\tau), X^{(1)}(\tau), \ldots, X^{(n)}(\tau))\, d\tau. \tag{1.11}$$

To take into account the dependence of $Y(t)$ on the history of change not only $X(\tau)$ on interval (t_0, t), but also the history of change of $Y(\tau)$ for $\tau \in (t_0, t)$, we can apply equation (1.1) in the form

$$Y(t) = \int_{t_0}^{t} M(t,\tau) f(X(\tau), Y(\tau))\, d\tau. \tag{1.12}$$

One can also consider the dependencies not only on the change of $X(\tau)$ and $Y(\tau)$, but also on a finite number of their derivatives. However, in this book, we will not strive for the maximum generality of the mathematical description. Therefore, only the simplest nonlinear equations that take into account memory will be considered.

Keeping in mind a possibility of applying the concept of memory in economics, in the next section we describe some general restrictions that can be imposed on the structure and properties of the memory function. These restrictions primarily include the following principles:
(A) The Causality Principle;
(B) The Principle of Nonlocality in Time;
(C) The Linear Superposition Principle;
(D) The Principle of Memory Fading (the Fading Principle);
(E) The Principle of Nonaging Memory (the Principle of Memory Homogeneity on Time);
(F) The Principle of Memory Reversibility (the Principle of Memory Recovery).

To simplify the mathematical description, we will not consider in detail the topological properties of the function spaces to which the functions $X(t)$ and $Y(t)$ belong, and the topological properties of operators $\mathcal{F}_{t_0}^{t}$.

1.3 General principles of memory

1.3.1 Principle of causality in time domain

The separation of variables into impact and response is based on the causal relationship: the impact variable is the cause (the action), and the response variable is the consequence. In the general case, the causal relationship assumes the influence of some economic processes on others.

The history of the concept of causality in economics is described by Hoover in the article "Causality in economics and econometrics" [178] (see also [176, 177]).

An important concept of causality was proposed by Granger in 1969 [149] (see also [150, 154]). The main assumption of Granger was that the future cannot be the cause of the present or the past. This means that the cause must at least precede the effect. However, the fact of precedence does not say anything about the existence of a causal relationship. Note the importance of the fact that the previous values of the "cause" should have a perceptible influence on the future values of the "effect" and, moreover, the past values of the "effect" did not have a significant effect on the future values of the "cause." The concept of the Granger causality is used in econometrics and time series analysis to formalize the notion of a cause-effect relationship between time series [149, 150, 154]. However, the Granger causality is necessary, but not sufficient, for a cause-effect relationship.

Clive Granger [150, 154, p. 330] proposed the following definition of causality that takes into account the memory effect in the form of the past histories: a variable $X(t)$ causes $Y(t)$, if the probability of $Y(t)$ conditional on its own past history and the past history of $X(t)$ does not equal the probability of $Y(t)$ conditional on its own past history alone. In the economic literature, this definition is called the Granger-causality, which is applied in econometrics. The Granger-causality is formulated for the time domain.

The concept of causality often appears in discussion of the interpretation of a correlation coefficient or regression, since an observed relationship does not allow us to say anything about the causal relationship between the variables.

Unfortunately, is no generally accepted procedure for testing for causality, partially, because of a lack of the definition of this concept that can be universal and can be accepted by all.

Let us explain what the causality principle leads to when we use fractional calculus to describe economic processes with memory. There are left-side and right-side fractional derivatives. We will use only left-sided derivatives, since the economic process at time t depends only on the state changes of this process in the past, that is, for $\tau < t$. The right-sided fractional derivative is determined by the values $\tau > t$, therefore, cannot describe memory.

Let us describe the causality for economic processes with memory in the time domain.

Suppose that a variable $Y(t)$ at the time t depends on another variable $X(\tau)$ at all other times $\tau \leq t$ by the linear relationship

$$Y(t) = \int_{-\infty}^{+\infty} R(t,\tau)X(\tau)\,d\tau. \tag{1.13}$$

The function $R(t,\tau)$ can be interpreted as a response function. In linear case, the response function describes how some time-dependent property $Y(t)$ of an economic process responds to an impulse impact $X(t)$. For $R(t,\tau) = m\delta(t-\tau)$, equation (1.13) gives $Y(t) = mX(t)$ that can be interpreted as standard multiplier equation, where m is the multiplier coefficient.

The principle of "strict causality" implies the following statement "no output can occur before the input." Therefore, the response function $R(t,\tau)$ must be zero for $t < \tau$ since a process cannot react to the impact before it is applied.

Strict causality, which means that the past can determine the present, but the future cannot, is expressed as $R(t,\tau) = 0$ for $\tau > t$. This allows us to represent the response function in the form

$$R(t,\tau) = H(t-\tau)M(t,\tau), \tag{1.14}$$

where $H(t-\tau)$ is the Heaviside step function that is defined as

$$H(t-\tau) = \begin{cases} 0 & \text{for } t < \tau \\ 1 & \text{for } t > \tau. \end{cases} \tag{1.15}$$

The appearance of the Heaviside step function in the definition of the response function reminds us that the response variable $Y(t)$ can only depend on the past values of the impact variable $X(\tau)$, i. e., $X(\tau)$ can be considered as an impact on the process and $Y(t)$ can be considered as the response of the process to this impact. Therefore, equation (1.13) gives a causal relationship between the variables $Y(t)$ and $X(\tau)$ in the form

$$Y(t) = \int_{-\infty}^{t} M(t,\tau)X(\tau)\,d\tau. \tag{1.16}$$

Note that the memory function $M(t,\tau)$ must be zero for $t < \tau$ since a process cannot react to the impact before it is applied.

Remark 1.6. The causality principle in frequency domain is described in article [399].

1.3.2 Principle of nonlocality in time

Memory is one of the types of nonlocality in time. Therefore, the nonlocality in time is an important property of processes with memory. This inalienable property of memory can be formulated in the form of the following principle.

Principle 1.1 (Principle of nonlocality in time). *If equation* (1.1) *can be represented in the form of differential equation of finite integer order in time domain, then equation* (1.1) *describes processes without memory.*

Let us formulate this principle of nonlocality in the symbolic form. If equation (1.1) in the form

$$Y(t) = \mathcal{F}_{t_0}^{t}(X(\tau)) \tag{1.17}$$

can be represented in the form of the differential equation

$$\mathit{Diff.Eq.}[t, X(t), X^{(1)}(t), \ldots, X^{(n)}(t), Y^{(1)}(t), \ldots, Y^{(m)}(t)] = 0, \tag{1.18}$$

which contains only integer-order derivatives $X^{(k)}(t)$, $(k = 0, 1, \ldots, n < \infty)$, and $Y^{(l)}(t)$, $(l = 0, 1, \ldots, m < \infty)$, then equation (1.17) cannot describe processes with memory. Equation (1.18) assumes that all these integer-order derivatives must exist. Note that differential equation (1.18) should not depend on the initial values such as the initial time t_0, the initial values of the variables $X(t_0)$, $Y(t_0)$ and the initial values of the derivatives $X^{(k)}(t_0)$, $(k = 0, 1, \ldots, n < \infty)$, and $Y^{(l)}(t_0)$, $(l = 0, 1, \ldots, m < \infty)$.

This principle can be formulated for equation (1.5), which is a special form of equation (1.17). In this case, the principle of nonlocality actually imposes restrictions on the memory function $M(t, \tau)$ of equation (1.5).

Example 1.1. Let us consider equation (1.5) with the function $M(t, \tau)$ in the form

$$M(t, \tau) = \lambda \exp\{-\lambda(t - \tau)\}, \tag{1.19}$$

where $\lambda > 0$. Then equation (1.5) with function (1.19) and $t_0 = 0$ takes the form

$$Y(t) = \lambda \int_0^t \exp\{-\lambda(t - \tau)\} X(\tau)\, d\tau. \tag{1.20}$$

Using the equality,

$$\frac{d}{dt} \int_0^t f(\tau)\, d\tau = f(t), \tag{1.21}$$

equation (1.20) can be represented in the form

$$\frac{dY(t)}{dt} = -\lambda(Y(t) - X(t)). \tag{1.22}$$

This representation is well known, and it is actively used for accelerators with the exponentially distributed lag (for details, see [10, pp. 26–27]).

The Principle of Nonlocality in Time is based on the well-known fact that the derivatives of integer orders are determined by properties of differentiable functions only in infinitely small neighborhood of the considered point. As a result, the processes or systems, which are described by the differential equations with finite number of integer-order derivatives, are local and cannot be considered as nonlocal. This means that memory as a special type of nonlocality in time cannot be described by these equations.

We should emphasize that the presence of nonlocality in time does not mean that we have a memory. The nonlocality in time may be due, for example, to a continuously distributed lag (time delay) and depreciation, or to continuously distributed time scaling [400, 406].

1.3.3 Principle of linear superposition

The linear superposition principle states that total value of Y is a linear sum of the values of the indicator Y, which have arisen in the processes at the corresponding instants of time. This principle was given by Ludwig Boltzmann in 1874 and 1876 [35, 36, 38, 39, 37] for physical model of isotropic viscoelastic media that are actually media with memory.

Boltzmann postulated that the response value $Y(t)$ accumulated by time t is equal to the sum of the response values caused by individual ΔX effects.

Due to the presence of memory in a process disturbed by the action, the response does not disappear when the effect ceases. A consequence of this fact is that a change in the variable Y depends not only on the actual value of the variable X at the same time, but also on previous changes in X (on the history of changes in X) that have occurred with the process in the past. In the linear behavior mode, responses to various disturbances overlap each other. Suppose that the effects of X_k (constant in magnitude) arise (act) at time t_k, respectively, $(k = 1, \dots, n)$.

An example of one of the possible impact-response histories is schematically shown Figure 1.1 for the case

$$Y(t) = M(t, t_1)\Delta X(t_1) + M(t, t_2)\Delta X(t_2) + M(t, t_3)\Delta X(t_3). \tag{1.23}$$

In more general case, the response $Y(t)$ is given by the sum

$$Y(t) = \sum_{k=1}^{N} M(t, t_k)\Delta X(t_k). \tag{1.24}$$

The Boltzmann superposition principle can be expressed in a general form by the equation

$$Y(t) = \sum_{t_k=-\infty}^{t_k=t} M(t, t_k)\Delta X(t_k). \tag{1.25}$$

Equation (1.25) as one of possible forms of the Boltzmann superposition principle, states that the influence of the history of process changes with memory is linearly additive.

Boltzmann postulated that expression (1.25) is valid for all small-enough step sizes $\Delta X(t_k)$ (or $\Delta t_k = t_{k+1} - t_k$).

If the impact $X(t)$ varies with time in very small steps so that $X = X(t)$ can be assumed to be a continuous differentiable function of time, then equation (1.25) can be written in the form

$$Y(t) = \int_{-\infty}^{t} M(t, \tau)\, dX(\tau) = \int_{-\infty}^{t} M(t, \tau)X^{(1)}(\tau)\, d\tau \tag{1.26}$$

for continuous time case.

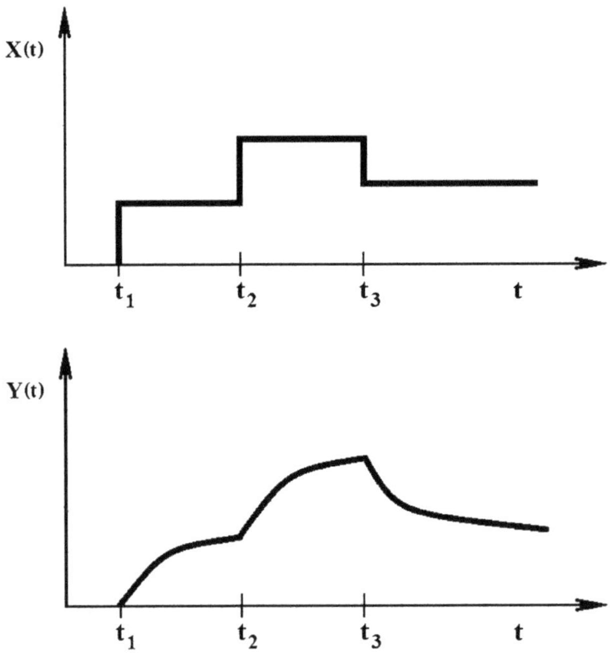

Figure 1.1: An example of one of the possible impact-response histories is schematically shown on this Figures for the case (1.23). The function $X(t)$ describes an impact and the function $Y(t)$ describes a response on this impact.

1.3.4 Principle of memory fading

The first mathematical formulation of the principle of memory fading has been proposed by Ludwig Boltzmann in 1874 and 1876 [35, 36, 38, 39, 37] for physical model of isotropic viscoelastic media that are actually media with memory. The memory fading principle states that the increasing of the time interval leads to a decrease in the corresponding contribution to the variable Y at time t. Then this principle is used in the works of Vito Volterra in 1928 and 1930 [496, 495, 497, 255, 498]. A more complete definition of the principle of fading memory has been suggested in [499, 79, 78, 77, 337] for materials and continuous media with memory. Such detailed mathematical formulations of the principle of memory fading are more complicated and too formalized than those required for application in economic theory. For our purposes, when we restrict ourselves to the consideration of memory described by expressions of the form (1.5), we will use a somewhat simplified formulation of this principle.

(I) The first approach. To describe the properties of the memory function and the fading memory, we consider a variable $X(\tau)$, which is different from zero on a finite time interval ($X(\tau) \neq 0$ for $\tau \in [0, T]$), and which is zero outside this interval ($X(\tau) = 0$ for $t > T$). In this case, the variable $X(\tau)$ is represented through the Heaviside step function

$$H(T-\tau) = \begin{cases} 0 & \text{for } T-\tau < 0 \\ 1 & \text{for } T-\tau > 0. \end{cases} \tag{1.27}$$

Substituting $X(\tau) = X_0(\tau)H(T-\tau)$ into equation (1.5) with $t_0 = 0$ and $t \in [T, \infty)$ instead of $X(\tau)$, we get the equation

$$Y(t) = \int_0^T M(t,\tau)X_0(\tau)\,d\tau, \tag{1.28}$$

in which the upper limit is T instead of t. From equation (1.28), we see that for any given time $t > T$ there is no impact ($X(t) = 0$ for $t > T$), but the response is different from zero ($Y(t) \neq 0$ for $t > T$). This means that the memory about the impact, which existed during the time $\tau \in [0, T]$, is stored in the economic process. In other words, the economic process saves the history of changes of the variable $X(t)$.

By the mean value theorem, there is a value $\xi \in [0, T]$ such that equation (1.28) can be represented in the form

$$Y(t) = M(t,\xi)X_0(\xi)T, \tag{1.29}$$

where $t > T > \xi > 0$.

(II) The second approach. We can consider a variable $X(\tau)$, which is expressed through the Dirac delta-function $\delta(\tau - \xi)$, where $\xi \in (0, T)$ and use the equation

$$\int_0^t f(\tau)\delta(\tau-\xi)\,d\tau = f(\xi), \tag{1.30}$$

in the form

$$\int_0^t M(t,\tau)\delta(\tau-\xi)\,d\tau = M(t,\xi). \tag{1.31}$$

Substitution of $X(\tau) = X_0(\tau)\delta(\tau - \xi)$ into equation (1.5) with $t_0 = 0$ gives the response $Y(t)$ for $t > \xi$ in the form

$$Y(t) = M(t,\xi)X_0(\xi). \tag{1.32}$$

As a result, we can see that the behavior of the variable $Y(t)$, which is considered as a response on the impact (perturbation) $X(\tau)$, is determined by the behavior of the memory function $M(t, \xi)$, i. e., by the function $M(t, \tau)$ with fixed constant time $\tau = \xi$. The behavior of the memory function $M(t, \tau)$ at infinite increase of t, i. e., $t \to \infty$, and fixed τ determines the type of behavior of the economic process with memory.

Statement 1.1. *The economic dynamics of processes with memory essentially depends on the behavior of the memory function $M(t,\tau)$ at infinity $t \to \infty$ with fixed $\tau < t$.*

Let us consider three types of behavior of memory function at infinity.

Type 1: Tends to zero. If the memory function tends to zero ($M(t,\tau) \to 0$) at $t \to \infty$ and fixed τ, then the economic process completely forgotten the impact, which it has been subjected in the last time. In this case, the economic process, which is described by equation (1.5), is reversible (is repeated) in a sense. In other words, the memory effects did not lead to irreversible changes of the economic process and economic agents, since the memory about the impact has not been preserved forever.

Type 2: Tends to finite value. If the memory function $M(t,\tau)$ tends to a finite limit at $t \to \infty$, the impact of the variable $X(t)$ on the economic process leads to the irreversible consequences in the sense that the memory of the impact is preserved forever.

Type 3: Tends to infinity. Unbounded increase of the memory function $M(t,\tau)$ at $t \to \infty$ characterizes an unstable economic process with memory. This memory function cannot be used to describe stable economic processes. However, this type of functions can be applied in the models, which take into account the economic crises and economic shocks, when we can expect a manifestation of instability phenomena.

Using the Type 1, in the formulation of basic models of economic processes with memory, we can consider such memory functions that satisfy the following principle of fading memory.

Principle 1.2 (Principle of fading memory). *In economic process, operator* (1.3) *describes fading memory, if the memory function satisfies the condition $M(t,\tau) \to 0$ at $t \to \infty$ with fixed τ.*

Using the symbolic representation of the memory, we can write the principle of fading memory in the form

$$\lim_{t\to\infty} \mathcal{F}_0^t(\delta(\tau - T)) = 0. \tag{1.33}$$

Definition 1.2. The memory will be called the memory with power-law fading if there is a parameter $\beta > -1$ such that the following limit is a finite constant for fixed τ:

$$\lim_{t\to\infty} t^{-\beta} M(t,\tau) = M_\infty, \tag{1.34}$$

where $0 < |M_\infty| < \infty$.

For example, the memory with power-law fading can be described by the memory function

$$M(t,\tau) = M_\infty (t - \tau)^{\alpha-1}, \tag{1.35}$$

where $\alpha = \beta + 1 > 0$ is a parameter that characterize the power-law fading, $t > \tau$, and M_∞ is a nonzero real number. Note that function (1.35) coincides with the kernel of the

Riemann–Liouville fractional integral of order $\alpha > 0$ up to a numerical factor [335, 308, 200]. In this example, integral operator (1.3) and equation (1.5) are characterized by the following types of behavior of memory function at infinity. For $0 < \alpha < 1$ memory function (1.35) tends to zero ($M(t,\tau) \to 0$) at $t \to \infty$ and fixed τ. For $\alpha > 1$, we have unbounded increase of the memory function $M(t,\tau)$ at $t \to \infty$ and fixed τ.

1.3.5 Monotonous memory fading and significant events

Note that in modern physics and mechanics the concept of fading memory assumes a set of stronger restrictions on memory function. For example, it is often assumed that the memory function tends to zero monotonically with increasing t.

Let us explain this restriction in more detail. If some variable $Y(t)$ is a linear function of the previous history of changes of variable $X(t)$, then the principle of fading memory states that the variable $Y(t)$ strongly depends on the changes in the recent changes of $X(t)$, rather than on more distant changes $X(t)$. In other words, the dependence of $Y(t)$ on the variable $X(\tau)$ at the previous times ($\tau < t$) is determined by using a memory function, which provides a continuously decreasing dependence on past events as they moving away from the considered points in time. In order to dependence of $Y(t)$ on the variable $X(\tau)$ satisfies the principle of fading memory, it is sufficient to assume that the memory function $M(t,\tau)$ has been continuously decreasing function of time, i. e., $M(t_2,\xi) \le M(t_1,\xi)$ for all t_1 and t_2 such that $t_2 > t_1 > \xi$.

The principle of fading memory assumes that it is less probable to expect of strengthening of the memory with respect to the more distant events. However, in certain economic processes, we should take into account that the economic agents can remember sharp and significant changes of the variable $X(\tau)$, despite the fact that these changes were in the more distant past in comparison with the more weak changes in the nearest past. For this reason, it is acceptable to use the memory function, which is not monotonic decrease.

1.3.6 Principle of nonaging memory

The homogeneity of time means that equations, which describe process change over time, are invariant under the shift $t \to t + s$. Homogeneity of time means translational invariance with respect to time variable. Using the symbolic representation, we can write the property of memory homogeneity on time in the form

$$\mathcal{F}_{t_0+s}^{t+s}(X(\tau)) = \mathcal{F}_{t_0}^{t}(X(\tau)). \tag{1.36}$$

Economic processes, for which condition (1.36) is satisfied for all $t > 0$, can be interpreted as processes with the nonaging memory.

The homogeneity of time means that the flow of economic processes with memory in the same conditions, but at different times of their observation, is identical. The same conditions for the processes means, in particular, the same "process age." If there are no allocated moments of time and the description of the economic process with memory does not depend on the initial moment of time, then the memory function $M(t,\tau)$ satisfies the condition

$$M(t+s,\tau+s) = M(t,\tau). \tag{1.37}$$

Equation (1.37) can be represented in a simpler form if the memory function is continuously differentiable. Differentiating equation (1.37) with respect to the parameter s, and then assuming $s = 0$, we obtain the equation

$$\frac{\partial M(t,\tau)}{\partial t} + \frac{\partial M(t,\tau)}{\partial \tau} = 0. \tag{1.38}$$

The general solution of equation (1.38) is an arbitrary function of $t-\tau$, that is,

$$M(t,\tau) = M(t-\tau). \tag{1.39}$$

In this case, the memory that satisfies condition (1.39) can be interpreted as the nonaging memory.

As a result, we can formulate the following principle.

Principle 1.3 (Principle of nonaging memory). *In economic process, memory, which is described by equation* (1.5), *is nonaging memory (homogeneous on time) if the memory function satisfies the condition* $M(t,\tau) = M(t-\tau)$ *for all* τ *and* $t > \tau$.

This principle can be also called the Principle of Memory Homogeneity on Time.

For $t_0 = 0$, equation (1.5) takes the form

$$Y(t) = \int_0^t M(t-\tau)X(\tau)\,d\tau. \tag{1.40}$$

Using expression (1.39), and replacing the integration variable ($\tau \to t-\tau$) allows us to write equation (1.40) in the form

$$Y(t) = \int_0^t M(\tau)X(t-\tau)\,d\tau. \tag{1.41}$$

For $t_0 = -\infty$, equation (1.5) takes the form

$$Y(t) = \int_{-\infty}^t M(t-\tau)X(\tau)\,d\tau. \tag{1.42}$$

Using equation (1.39) and changing the integration variable, equation (1.42) can be written as

$$Y(t) = \int_0^\infty M(\tau)X(t-\tau)\,d\tau. \tag{1.43}$$

Remark 1.7. Equation (1.5) with $t_0 = 0$ is mathematically interpreted as the Laplace convolution of functions (e. g., see equation 1.4.10 in [200, p. 19]), which is also called the Duhamel convolution (e. g., see equation 2.1.1 in [202, p. 66]). In this case, the short notation $Y = M * X$ is used in mathematics. Note that the convolution operation is the commutative ($Y * X = X * Y$) and associative ($X * (Y * Z) = (X * Y) * Z$) operation. A consequence of the commutativity of this operation was the possibility of using two representations of (1.5) that are given by equations (1.40) and (1.41).

Memory functions (1.39) are important for description of economic processes, when the properties of processes do not depend on the choice of the time origin. For these processes, we can always choose the beginning of the reference time at $t_0 = 0$.

If the properties of economic process with memory do not depend on some selected points in time, and thus, do not depend on the choice of the reference time, the memory function will depend only on the difference $t - \tau$. Here, $t - \tau$ is the time that separates the current point of time from the memorized event. Thus, the memory function for such processes can be represented in form (1.39).

In general, the memory function cannot be represented in the form (1.39). In economic processes, the memory about the event can depend not only on the time $(t-\tau)$, but also on the states of the economic process in moments of time τ. The initial moment in time can be considered the moment the process was created. If this time is considered as the reference point of time $t = 0$, then the time moment $t > 0$ can be interpreted as the age of process, and $t = 0$ can be interpreted as the time of the birth of process. In the case $M(t,\tau) = M(t-\tau)$, the behavior of the process with memory depends only on age, and not on the date of birth.

1.3.7 Changing type of behavior at infinity

In this subsection, we consider equation (1.6) with $t_0 = 0$ and $M(t,\tau) = M(t-\tau)$, where the variable $X(\tau)$ is represented by the Dirac delta-function $\delta(\tau - T)$ and the Heaviside step function $H(T-\tau)$.

Let us consider the impact variable in the form of the Dirac delta-function, without using the assumption that the memory function has the form $M(t,\tau) = M(t-\tau)$. We first consider equation (1.6) with t_0 and $X(\tau)$, which is represented by the Dirac delta-function $\delta(\tau - \xi)$. Substitution of $X(\tau) = \delta(\tau - \xi)$ into equation (1.6) with $t_0 = 0$ and integer $n \geq 1$ gives the response $Y(t)$ for $t > \xi$ in the form

$$Y(t) = (-1)^n \left(\frac{\partial^n M(t,\tau)}{\partial \tau^n}\right)_{\tau=\xi}. \tag{1.44}$$

Note that expression (1.44) can be obtained in a slightly different way. We can consider equation (1.5) with $t_0 = 0$ and $X(\tau)$, which is represented by derivative of integer order $n \geq 1$ of the Dirac delta-function $\delta(\tau - \xi)$.

Let us assume that the memory function has the form $M(t, \tau) = M(t - \tau)$. Substitution of the variable $X(\tau) = \delta^{(n)}(\tau - \xi)$ into (1.5) with $t_0 = 0$ gives (1.44). Substitution of $X(\tau) = \delta(\tau - \xi)$ into equation (1.6) with $t_0 = 0$ and $M(t, \tau) = M(t - \tau)$ gives the response $Y(t)$ for $t > \xi$ in the form

$$Y(t) = \frac{\partial^n M(t, \xi)}{\partial t^n}, \tag{1.45}$$

where we use

$$\frac{\partial M(t, \tau)}{\partial t} = -\frac{\partial M(t, \tau)}{\partial \tau}. \tag{1.46}$$

Note that we can consider equation (1.5) with t_0 and $X(\tau)$, which is represented by derivative of integer order $n \neq 1$ of the Dirac delta-function $\delta(\tau - \xi)$. Substitution the variable $X(\tau) = \delta^{(n)}(\tau - \xi)$ into (1.5) with $t_0 = 0$ and $M(t, \tau) = M(t - \tau)$ gives (1.45).

From equation (1.45), we can see that the behavior of the variable $Y(t)$, which is considered as a response on the impact $X(\tau)$, is determined by the behavior of the n-th derivatives of the memory function $M^{(n)}(t, \xi)$ with respect to time t with fixed ξ.

As a result, if the behavior of memory functions at infinity has the form $M(t, \tau) \sim t^k$, $(k \in \mathbb{N})$, then we have Type 3 "Tends to Infinity" for operator (1.5). For operator (1.6) with $t_0 = 0$, we have the behavior of the variable $Y(t)$ corresponds to Type 3 for $n < k$, Type 2 for $n = k$, and Type 1 "Tends to Zero" for $n > k$.

These results allow us to formulate the following statement.

Statement 1.2. *For nonaging memory (for memory, which is homogeneous in time), the use of integro-differential operator* (1.4), *and equations* (1.6), (1.7), (1.8) *with* $t_0 = 0$ *can change the type of behavior at infinity* ($t \to \infty$) *compared to the use of integral operator* (1.3).

Let us consider the impact variable in the form of the Heaviside step function, without using the assumption that the memory function has the form $M(t, \tau) = M(t - \tau)$. Similarly, substituting $X(\tau) = H(\xi - \tau)$ into equation (1.6) with $t_0 = 0$ and integer $n \geq 1$, we get

$$Y(t) = (-1)^{n-1} \left(\frac{\partial^{n-1} M(t, \tau)}{\partial \tau^{n-1}} \right)_{\tau = \xi}. \tag{1.47}$$

Here, we used that the first derivative of the Heaviside step function is the Dirac delta-function, where the derivative is considered in a generalized sense, i. e., on the space of test functions.

Let us assume that the memory function has the form $M(t,\tau) = M(t-\tau)$. Then, substitution of $X(\tau) = H(\xi - \tau)$ into equation (1.6) with $t_0 = 0$ gives

$$Y(t) = \frac{\partial^{n-1} M(t,\xi)}{\partial t^{n-1}}. \tag{1.48}$$

Therefore, the behavior of the variable $Y(t)$, which is a response on the impact $X(\tau) = H(\xi - \tau)$, is determined by the behavior of the n-th derivatives of the memory function $M^{(n-1)}(t,\xi)$ with respect to time t with fixed ξ.

As a result, the behavior of the variable $Y(t)$ corresponds to Type 3 "Tends to Infinity" for $n-1 < k$, Type 2 "Tends to Finite Value" for $n-1 = k$, and Type 1 "Tends to Zero" for $n-1 > k$ in the case of operator (1.6) with $t_0 = 0$, if $M(t,\tau) \sim t^k$ at $t \to \infty$.

1.3.8 Principle of memory reversibility

Let us consider the case, when $X(t)$ and $Y(t)$ are variables that are connected by equation (1.5). Until now we assumed that the function $X(t)$ is considered as an impact variable, and the function $Y(t)$ is interpreted as a response to the impact. In the general case, the equation $Y(t) = \mathcal{F}_0^t(X(\tau))$ is irreversible, that is, does not exist an inverse operator G_0^t, for which $X(t) = G_0^t(Y(\tau))$, where $t > 0$.

Principle 1.4 (Principle of memory reversibility (memory recovery)). *The memory, which is described by the equation $Y(t) = \mathcal{F}_0^t(X(\tau))$ with the Volterra operator* (1.3), *is reversible if there is an operator G_0^t such that $X(t) = G_0^t(Y(\tau))$ for all $t > 0$ and for wide class of functions $X(t)$.*

In equation (1.5), we can consider $Y(t)$ as a given function, and the variable $X(t)$ can be considered as the unknown function. In this case, equation (1.5) is called the linear Volterra equation of the first kind. Existence of solutions of this equation, i. e., the reversibility of equation (1.5), is being studied in the theory of integral equations (e. g., see [311] and references therein).

It should be noted that the question of the reversibility of equations (1.5) and (1.6) is connected with the principle of duality of accelerator with memory and multiplier with memory, which was proposed in [412, 397].

Equations (1.5) and (1.6) can be associated with the so-called inverse problems of dynamics [129, 130, 356, 99], which is the economic dynamics in our case. For example, if we know the memory function $M(t,\tau)$ of an economic process and a variable $Y(t)$, then we can consider the problem of finding (restoration) of the impact function $X(t)$. This type of problems includes a finding of solutions for the equations of macroeconomic growth models with memory.

Another inverse problem is related to econometrics of processes with memory. The known functions of the impact variable $X(t)$ and the response variable $Y(t)$ can be used to determine the memory function $M(t,\tau)$ of the considered economic process

[399]. This problem is discussed in [399] as the problem of determining the parameters of memory fading. Note that this problem has a great importance for the study of properties of memory in the real economic processes.

1.4 Examples of memory and nonlocality in time

1.4.1 Absence of memory: total amnesia and instant amnesia

For processes without memory, the memory functions are expressed in terms of the Dirac delta-function in the form

$$M(t,\tau) = M(t-\tau) = m(\tau)\delta(t-\tau), \tag{1.49}$$

where $\delta(t-\tau)$ is the Dirac delta-function. Substitution of (1.49) into equation (1.5) gives

$$Y(t) = m(t)X(t). \tag{1.50}$$

The absence of memory means that the economic indicator $Y(t)$ is determined by a factor $X(\tau)$, only at $\tau = t$. Equation (1.50) means the instantaneous forgetting of the history of the factor changes.

Definition 1.3. The process without memory (the absence of memory) connects the sequence of subsequent states of the economic process to the previous state only through the current state for each time t. This case can be called the total amnesia (total memory loss) or complete amnesia (complete loss of memory).

In macroeconomics equation, (1.50) can be interpreted as equation of standard economic multiplier, which describes process without memory and time delay (lag).

As an impact variable $X(\tau)$, we can consider the integer derivative $X^{(n)}(\tau) = d^nX(\tau)/d\tau^n$ of the factor $X(\tau)$ with respect to time τ, where n is an positive integer number. In this case, we can use equation (1.6). Then equation (1.6) with memory function (1.49) gives the equation

$$Y(t) = m(t)X^{(n)}(t). \tag{1.51}$$

In macroeconomics, equations (1.51) with $n = 1$ is the equation of the standard accelerator without memory and lag, where $m(t)$ is the accelerator coefficient. For $n = 2$, equations (1.51) can be interpreter as the "second-order accelerator (SOA)" for continuous time models (e. g., see [82, 174, 175], and [146, pp. 13–14]).

Remark 1.8. It should be noted that we cannot consider the multiplication of the Dirac delta-functions since these functions are the generalized functions, which are treated as functionals on a space of test functions. Therefore, we cannot substitute expression (1.49) and the factor $X(\tau) = \delta(\tau - T)$ into equations (1.5) and (1.6) simultaneously.

It is known that the derivatives $X^{(n)}(t)$ of the integer order n are determined by the values of $X(\tau)$ in the infinitesimal neighborhood of the time point $\tau = t$. In case (1.51), the forgetting occurs in an infinitely small time interval. In fact, the equations with derivatives of integer orders describe a process, in which all economic agents have infinitely fast amnesia. This case can be called the "instant amnesia." In other words, the economic models, which use only integer-order derivatives, describes processes, in which economic agents forget the history of changes of the economic factors during an infinitely small time interval (in an infinitesimal small neighborhood of the current time) immediately disappears, as if it did not exist. As a result, the differential equations with derivatives of integer orders with respect to time cannot describe the memory effect.

If the memory function $M(t,\tau) = M(t-\tau)$ is described as $M(t) = m\delta(t-T)$, then equation (1.5) can be written as an equation of multiplier with is a fixed-time delay T, where the time-constant of the lag T is a given positive real number: $Y(t) = mX(t-T)$ [10, p. 23].

1.4.2 Distributed lag or memory

Let us consider a finite fixed time interval $[t_0, t_1]$ (or infinite time interval if $t_1 = \infty$), in which the factor $X(t)$ influences on the indicator $Y(t)$ in form (1.1). Using the symbolic representation, we can consider the condition in the form

$$\mathcal{F}_{t_0}^{t_1}(1) = 1. \tag{1.52}$$

Equation (1.52) can be called the normalization condition. For the case of dependence (1.5) with the homogeneity on time $M(t,\tau) = M(t-\tau)$, condition (1.52) takes the form

$$\int_{t_0}^{t_1} M(\tau)\, d\tau = 1. \tag{1.53}$$

Condition (1.52) gives the restriction on the memory function. The function $M(t,\tau) = M(t-\tau)$, which satisfies normalization condition (1.53) is often called the weighting function [10, p. 26]. Mathematically, this function can be interpreted as the probability density function (pdf) of a distribution on integral $[t_0, t_1]$. Economically, this function can be interpreted as the distribution density of the delay time. This type of functions $M(t)$ is often used to describe economic processes with continuously distributed lag [10, p. 25]. The existence of the time delay (lag) is connected with the fact that the processes take place with a finite speed, and the change of the economic factor does not lead to instant changes of indicator that depends on it.

Statement 1.3. *If normalization condition* (1.53) *of function* $M(t)$ *holds, the economic process that is described by equation* (1.5) *goes through all the states without any losses.*

It should be emphasized that the time delay is due to the finiteness of the speed of the processes, and not the presence of memory in the process or system. The use of the term "memory" for the processes of this type can lead to confusion.

The time delay (lag) is caused by finite speeds of processes, i. e., the change of impact variable does not lead to instant changes of response variable. Therefore, the distributed lag (time delay) cannot be considered as a memory in processes. For example, the change in the value of electromagnetic field at the point A of observation is delayed with respect to the change in the sources of the field located at the point B at the time $\tau = |AB|/c$, where c is the speed of propagation of disturbances, and $|AB|$ is the distance between points A and B. It is known that the processes of propagation of the electromagnetic field in a vacuum do not mean the presence of memory in this process.

In the simplest form the lag is described by the translation operator [335, pp. 95–96], which is also called the shift operator and is defined by the equation

$$(\mathbf{T}_\tau X)(t) = X(t-\tau). \tag{1.54}$$

The operator $\mathbf{T}_\tau$, where $\tau \in \mathbb{R}_+$, maps a function $X(t)$ defined on $\mathbb{R}$ to its translation $X(t-\tau)$ on the fixed value $\tau > 0$. The translation operator allows us to describe the lag. The parameter $\tau > 0$ is positive constant that characterizes the fixed-time delay.

In a more general form, the translation operator can be described for a continuously distributed lag.

Definition 1.4. The translation operator with continuously distributed lag is described by the equation

$$(\mathbf{T}_M X)(t) = \int_{t_0}^{t_1} M_T(\tau) T_\tau X(t)\, d\tau = \int_{t_0}^{t_1} M_T(\tau) X(t-\tau)\, d\tau, \tag{1.55}$$

where the weighting function $M(\tau)$ satisfies the condition of nonnegativity and the normalization condition

$$M_T(\tau) \geq 0, \quad \int_{t_0}^{t_1} M_T(\tau)\, d\tau = 1. \tag{1.56}$$

Here, we assume that $X(t)$ and $M_T(t)$ are piecewise continuous functions on $\mathbb{R}$ and the integral

$$\int_{t_0}^{t_1} M_T(\tau) \big|X(t-\tau)\big|\, d\tau \tag{1.57}$$

converges.

Note that the translation operator with fixed-time lag with period $h > 0$ is a particular case of (1.55), in which the weighting function is used in the form $M_T(\tau) = \delta(\tau - h)$, in the framework of distribution theory.

Operator (1.55) can be also called the operator of the continuously distributed translation or the operator of the continuously distributed lag. Note that such operator is actively used in economics to describe macroeconomic growth models with continuously distributed lag [10]. In this case, the dependences of impact variable $X(t)$ and response variable $Y(t)$ are considered in the form

$$Y(t) = (\mathbf{T}_M X)(t), \quad Y(t) = \int_{t_0}^{t_1} M_T(\tau) X(t-\tau)\, d\tau, \tag{1.58}$$

where the upper limit t_1 is fixed. For example, the lower and upper limits can be equal to zero and infinity, respectively ($t_0 = 0$ and $t_1 = \infty$).

To describe the exponential distribution of delay time [10], we should use the probability density function in the form

$$M(\tau) = \begin{cases} \lambda \exp(-\lambda\tau) & \text{for } \tau > 0, \\ 0 & \text{for } \tau \leq 0, \end{cases} \tag{1.59}$$

where $\lambda > 0$ is positive constant that is often called the rate parameter. This distribution describes many phenomena such as waiting time for an insurance event; time of receipt of the order for the enterprise; time between visits by shop or supermarkets; duration of telephone conversations; the service life of parts and components in the complex product.

Note that the exponential distribution describes the time between events in a Poisson point process, i. e., a process in which events occur continuously and independently at a constant average rate. It should be emphasized that the exponential distribution is the continuous analogue of the geometric distribution [10], which has the key property of being memoryless. The lack of memory here is associated with the fact that for this distribution the number of past "failures" does not affect the number of future "failures."

Using equation (1.59), the exponentially distributed lag operator is defined by the equation

$$(\mathbf{T}_{\exp} X)(t) = \int_0^\infty \lambda \exp(-\lambda\tau) X(t-\tau)\, d\tau, \tag{1.60}$$

and equation (1.58) takes the form

$$Y(t) = \lambda \int_0^\infty \exp(-\lambda\tau) X(t-\tau)\, d\tau. \tag{1.61}$$

In economics, the continuously distributed lag has been considered starting with the works of Michal A. Kalecki [189] and Alban W. H. Phillips [305, 306]. The continuous uniform distribution of delay time is considered by M. A. Kalecki in 1935 [189] (see also Sections 8.4 of [10, pp. 251–254]) for dynamic models of business cycles. The continuously distributed lag with the exponential distribution of delay time is considered by A. W. H. Phillips [305, 306] in 1954. In the Phillips' papers, generalizations of the economic concepts of accelerator and multiplier were proposed by taking into account the exponentially distributed lag. The operators with continuously distributed lag were considered by Roy G. D. Allen [8], in 1956 (see also [10, pp. 23–29]). Logistic equation with continuously distributed lag for exponential and gamma distributions of delay time and its application in economics was proposed in [403]. Currently, economic models with delay are actively used to describe the processes in the economy.

In the general case, the distribution of delay time can be described by other continuous probability distributions, not just uniform nd exponential distributions. The time delay is caused by finite speeds of economic processes. Therefore, the distributed lag (time delay) is not interpreted as a memory in economics.

Remark 1.9. Note that economic models with the simultaneous action of distributed lag and memory were proposed in articles [404, 402, 407, 405]. Equations of these economic models use the fractional derivatives and integrals with distributed lag, which are proposed in article [400].

1.4.3 Memory with power-law fading

To describe the economic processes with fading memory, we can use the equations with derivatives and integrals of noninteger orders [335, 308, 200, 165], which are actually special cases of integro-differential operators. To describe the memory with power-law fading, we can use the memory function in the form

$$M(t,\tau) = M_{\alpha}(t-\tau) = \frac{1}{\Gamma(\alpha)} \frac{m(\alpha)}{(t-\tau)^{1-\alpha}}, \tag{1.62}$$

where $\Gamma(\alpha)$ is the gamma function, $\alpha > 0$ is a parameter that characterize the power-law fading, $t > \tau$, and $m(\alpha)$ is a positive real number, which can depend on α, in general. Substitution of expression (1.62) into equation (1.5) gives the fractional integral equation of the order $\alpha > 0$ in the form

$$Y(t) = m(\alpha)(I^{\alpha}_{\mathrm{RL};0+}X)(t), \tag{1.63}$$

where $I^{\alpha}_{\mathrm{RL};0+}$ is called the left-sided Riemann–Liouville integral of the order $\alpha > 0$ [200, pp. 69–70] with respect to time variable.

Definition 1.5. The left-sided Riemann–Liouville fractional integral is defined by the equation

$$(I^{\alpha}_{\mathrm{RL};0+}X)(t) = \frac{1}{\Gamma(\alpha)} \int\limits_{0}^{t} \frac{X(\tau)\,d\tau}{(t-\tau)^{1-\alpha}}, \tag{1.64}$$

where $\Gamma(\alpha)$ is the gamma function and $t > 0$. In equation (1.64), the function $X(t)$ is assumed to satisfy the condition $X(t) \in L_1(0,T)$, which implies that $X(\tau)$ is a Lebesgue integrable function and the inequality

$$\int\limits_{0}^{T} |X(\tau)|\,d\tau < \infty \tag{1.65}$$

holds.

Remark 1.10. The Riemann–Liouville integral (1.64) is a generalization of the standard integration [335]. Note that the Riemann–Liouville integration (1.64) of the order $\alpha = 1$ gives the standard integration of the first order,

$$(I^{1}_{\mathrm{RL};0+}X)(t) = \int\limits_{0}^{t} X(\tau)\,d\tau. \tag{1.66}$$

For $\alpha = 2$, we get the standard integration of the second order,

$$(I^{2}_{\mathrm{RL};0+}X)(t) = \int\limits_{0}^{t}\left(\int\limits_{0}^{\tau_1} X(\tau_2)\,d\tau_2\right)d\tau_1 = \int\limits_{0}^{t} d\tau_1 \int\limits_{0}^{\tau_1} d\tau_2 X(\tau_2). \tag{1.67}$$

For $\alpha = n \in \mathbb{N}$, we get the standard integration of the integer order n in the form

$$(I^{n}_{\mathrm{RL};0+}X)(t) = \int\limits_{0}^{t} d\tau_1 \int\limits_{0}^{\tau_1} d\tau_2 \cdots \int\limits_{0}^{\tau_{n-1}} d\tau_n X(\tau_n) = \frac{1}{(n-1)!} \int\limits_{0}^{t} (t-\tau)^{n-1} X(\tau)\,d\tau. \tag{1.68}$$

Remark 1.11. In economics, equation (1.63) can be interpreted as multiplier with power-law memory, which is proposed in [412, 397], and the parameter $m(\alpha)$ is the coefficient of this multiplier. Equation (1.63) with $\alpha = 1$ takes the form

$$Y(t) = m(1) \int\limits_{0}^{t} X(\tau)\,d\tau. \tag{1.69}$$

The action of the derivative of first order on equation (1.69) gives the equation of the standard accelerator $X(t) = aY^{(1)}(t)$, where $a = 1/m(1)$ is the accelerator coefficient.

Let consider impact variable $X(\tau)$, which is represented through the Heaviside step function $H(T-\tau)$, such that $X(\tau) = X_0 = \text{const}$ for $\tau \in [0, T]$ and $X(\tau) = 0$ for $\tau \in [T, \infty)$. Substitution of $X(\tau) = X_0 H(T - \tau)$ into equation (1.63) with $t \in [T, \infty)$, we get

$$Y(t) = \frac{m(\alpha) X_0}{\Gamma(\alpha + 1)} (t^\alpha - (t - T)^\alpha), \tag{1.70}$$

where we use $\alpha\Gamma(\alpha) = \Gamma(\alpha + 1)$. We see that for $t > T$, when there is no impact ($X(t) = 0$ for $t > T$), the response is different from zero ($Y(t) \neq 0$ for $t > T$) and it has the power-law form.

Let us consider the impact variable $X(t)$, which is represented through the Dirac delta-function $\delta(\tau - T)$, such that $X(\tau) = X_0(\tau)\delta(\tau - T)$ for $T \in (0, t)$. Then equation (1.63) gives the response variable for $t > T$ in the form

$$Y(t) = \frac{m(\alpha)}{\Gamma(\alpha)} X_0(T)(t - T)^{\alpha - 1}, \tag{1.71}$$

where $t > T$ and $|X_0(T)| < \infty$. As a result, the behavior of the response $Y(t)$ on the impact $X(\tau)$ has the power-law type. Therefore, the memory, which is defined by memory function (1.62), can be interpreted as memory with power-law fading. We can see that the response tends to zero ($Y(t) \to 0$) for $0 < \alpha < 1$ at $t \to \infty$, and tends to infinity ($|Y(t)| \to \infty$) for $\alpha > 1$ at $t \to \infty$.

Remark 1.12. Let us consider the impact variable $X(\tau)$, which is perturbation in the form of the periodic sequence of delta-function-type pulses (kicks) following with period T:

$$X(\tau) = x(\tau) \sum_{k=1}^{\infty} \delta\left(\frac{\tau}{T} - k\right), \tag{1.72}$$

where $m(\alpha)$ is an amplitude of the pulses, and $X_0(\tau)$ is some function. Here, we assume that $x(\tau)$ is continuous at all points $t = kT$. Then the response $Y(t)$ for t: $NT < t < (N + 1)T$ takes the form

$$Y(t) = \frac{m(\alpha)}{\Gamma(\alpha)} \sum_{k=0}^{n} (t - kT)^{\alpha - 1} x(kT). \tag{1.73}$$

Using the notation,

$$Y_n = Y(NT - 0) = \lim_{\varepsilon \to 0+} Y(NT - \varepsilon), \tag{1.74}$$

$$X_k = x(kT - 0) = \lim_{\varepsilon \to 0+} x(kT - \varepsilon), \tag{1.75}$$

we get the discrete map with memory

$$Y_{n+1} = \frac{m(\alpha) t^{\alpha - 1}}{\Gamma(\alpha)} \sum_{k=0}^{n} (N + 1 - k)^{\alpha - 1} X_k. \tag{1.76}$$

Using equation (1.76) with replacement $N + 1$ by N, and subtraction the result from equation (1.76), we get the discrete map with memory in the form

$$Y_{n+1} = Y_n + \frac{m(\alpha)t^{\alpha-1}}{\Gamma(\alpha)} X_n + \frac{m(\alpha)t^{\alpha-1}}{\Gamma(\alpha)} \sum_{k=0}^{n-1} V_\alpha(N-k) X_k, \tag{1.77}$$

where $V_\alpha(z)$ is defined by $V_\alpha(z) = (z+1)^{\alpha-1} - z^{\alpha-1}$. In economics, equation (1.77) can be interpreted as a discrete multiplier with power-law memory [417, 436]. The concepts of discrete economic multipliers with power-law memory have been proposed in articles [441, 442, 431, 432] and [417, 436].

Let us consider the integer derivative $X^{(n)}(\tau) = d^n X(\tau)/d\tau^n$ as an impact variable. In this case, we can use the power-law memory function is the form

$$M(t,\tau) = M_{n-\alpha}(t-\tau) = \frac{1}{\Gamma(n-\alpha)} \frac{a(\alpha)}{(t-\tau)^{\alpha-n+1}}, \tag{1.78}$$

where $n = [\alpha] + 1$ for noninteger values of $\alpha > 0$, and $n = [\alpha]$ for integer values of $\alpha > 0$. Then we get the equation

$$Y(t) = a(\alpha)(D^\alpha_{C;0+} X)(t), \tag{1.79}$$

where $D^\alpha_{C;0+}$ is the left-sided Caputo fractional derivative of the order $\alpha \geq 0$ with respect to variable t [200, p. 92].

Definition 1.6. The Caputo fractional derivative is defined by the equation

$$(D^\alpha_{C;0+} X)(t) = \frac{1}{\Gamma(n-\alpha)} \int_0^t \frac{X^{(n)}(\tau)\, d\tau}{(t-\tau)^{\alpha-n+1}}, \tag{1.80}$$

where $\Gamma(\alpha)$ is the gamma function, $t \in [0, T]$, and $X^{(n)}(\tau)$ is the derivative of the integer order n with respect to τ. It is assumed that $X(\tau) \in AC^n[0, T]$, i. e., the function $X(\tau)$ has integer-order derivatives up to $(n-1)$-th order, which are continuous functions on the interval $[0, T]$, and the derivative $X^{(n)}(\tau)$ is Lebesgue summable on the interval $[0, T]$.

In economics, equation (1.79) can be interpreted as equation of an accelerator with memory [417, 436] with the power-law fading parameter $\alpha \geq 0$, where $a(\alpha)$ is the acceleration coefficient. The concepts of the accelerators with memory have been proposed in [441, 442, 431, 432] and [417, 436].

Let us consider equation (1.79), where the factor $X(\tau)$ is represented by the Dirac delta-function $\delta(\tau - T)$. Substitution of $X(\tau) = X_0 \delta(\tau - T)$ with $X_0 = \text{const}$ into equation (1.79) gives the response $Y(t)$ for $t > T$ in the form

$$Y(t) = \frac{\partial^n M_{n-\alpha}(t-T)}{\partial t^n} X_0 = \frac{a(\alpha) X_0}{\Gamma(-\alpha)} (t-T)^{-\alpha-1} \tag{1.81}$$

for noninteger values of the order $\alpha > 0$. As a result, the behavior of the response $Y(t)$ tends to zero ($Y(t) \to 0$) for $\alpha > 0$ at $t \to \infty$.

We can consider equation (1.79) for the factor $X(\tau)$ that is represented by the Heaviside step function $H(T-\tau)$. Substitution of $X(\tau) = X_0 H(T-\tau)$ into equation (1.79) gives

$$Y(t) = \frac{\partial^{n-1} M_{n-\alpha}(t-T)}{\partial t^{n-1}} X_0 = \frac{a(\alpha) X_0}{\Gamma(-\alpha)} (t-T)^{-\alpha}, \tag{1.82}$$

where α is noninteger positive parameter. As a result, the behavior of the variable $Y(t)$, which is a response on the impact $X(\tau)$, has the power-law type. We can see that the response $Y(t)$ tends to zero ($Y(t) \to 0$) for $\alpha > 0$ at $t \to \infty$.

1.4.4 Memory with multiparameter power-law fading

In the previous sections, we considered examples of simple forms of memory, in which the power-law fading is characterized by one parameter α. In economic models, we can take into account the presence of different types of memory fading, which characterize the different types of economic agents. In this case, to describe fading memory, we can use the memory function $M(t,\tau)$ as a sum of functions (1.62) with different parameters of memory fading. For example, we can use the following two-parameter description of the fading memory

$$M(t,\tau) = M_{\alpha,\beta}(t-\tau) = \frac{1}{\Gamma(1-\alpha)} \frac{m_\alpha}{(t-\tau)^\alpha} + \frac{1}{\Gamma(1-\beta)} \frac{m_\beta}{(t-\tau)^\beta}, \tag{1.83}$$

where $\alpha > 0$, $\beta > 0$, $t > \tau$ and m_α, m_β are numerical coefficients. Substitution of expression (1.83) into equation (1.5) gives

$$Y(t) = m_\alpha (I^\alpha_{\mathrm{RL};0+} X)(t) + m_\beta (I^\beta_{\mathrm{RL};0+} X)(t). \tag{1.84}$$

In order to take into account N different parameters of memory fading, we can use the equation

$$Y(t) = \sum_{k=1}^{n} m_k (I^{\alpha_k}_{\mathrm{RL};0+} X)(t). \tag{1.85}$$

Similarly, we can consider the generalization of the accelerator with memory in the form

$$Y(t) = a_\alpha (D^\alpha_{C;0+} X)(t) + a_\beta (D^\beta_{C;0+} X)(t). \tag{1.86}$$

In order to take into account N different parameters of memory fading, we can use the equation

$$Y(t) = \sum_{k=1}^{n} a_k (D^{\alpha_k}_{C;0+} X)(t). \tag{1.87}$$

Using the memory with multiparameter power-law fading, we can consider economic models to describe the memory effects in economy with different type of economic agents.

1.4.5 Memory with variable fading

In some economic processes with memory, the parameter α of memory fading can be changed during the time, i. e., $\alpha = \alpha(t)$. In this case, we can consider the memory function (1.62) in the form

$$M(t,\tau) = M_{\alpha(t)}(t-\tau) = \frac{1}{\Gamma(\alpha(t))} \frac{m}{(t-\tau)^{1-\alpha(t)}}, \tag{1.88}$$

where the variable order $\alpha(t) \geq 0$. Substituting (1.88) into equation (1.5), we obtain the integral equation in the form

$$Y(t) = m(I^{\alpha(t)}_{RL;0+}X)(t), \tag{1.89}$$

where $I^{\alpha(t)}_{RL;0+}$ is the Riemann–Liouville fractional integral of variable order. If $\alpha(t) = \alpha$, then equation (1.89) gives (1.63). Equation (1.89) can be interpreted as equation of economic multiplier with variable fading memory.

Using $X^{(n)}(\tau) = d^nX(\tau)/d\tau^n$ with $n = [\alpha] + 1$ instead of $X(\tau)$, and the memory function $M(t,\tau) = M_{n-\alpha(t)}(t-\tau)$ instead of $M(t,\tau) = M_{\alpha(t)}(t-\tau)$, we get the equation of accelerator of variable order

$$Y(t) = a(D^{\alpha(t)}_{C;0+}X)(t), \tag{1.90}$$

where $D^{\alpha(t)}_{C;0+}$ is the left-sided Caputo fractional derivative of the variable order $\alpha(t) \geq 0$. For $\alpha(t) = \alpha$, equation (1.90) gives (1.79). Equation (1.90) can be interpreted as equation of economic accelerator with variable fading memory.

As a result, to describe processes with variable memory fading, we should use fractional derivatives and integrals of variable orders (e. g., see [234, 14, 455]).

1.4.6 Complete (perfect, ideal) memory

Let us assume that the factor $X(t)$ influences on the indicator $Y(t)$ such that equation (1.1) with $t_0 = 0$ holds. Let us consider the condition

$$\mathcal{F}^t_0(1) = 1 \quad \forall t > 0. \tag{1.91}$$

Note that this condition is not equivalent to the normalization condition (1.91), where t_0 and t_1 are fixed values.

Economic processes, for which condition (1.91) is satisfied for all $t > 0$, can be interpreted as processes with unit preserving memory. For the linear operators $\mathcal{F}^t_0$, this condition means that constant impact variable does not change. In other words, the memory remembers the unchanged variable infinitely long. For the Volterra operator, condition (1.91) leads to the following restriction on the memory function:

$$\int_0^t M(t,\tau)\,d\tau = 1, \tag{1.92}$$

where we assume that condition (1.92) holds for all $t > 0$. Condition (1.92) means that equation (1.5) should give $Y(t) = 1$ for $X(t) = 1$ for all $t > 0$.

If condition (1.91) or (1.92) holds, the economic process goes through all the states without any losses. In this case, we say that the memory function describes a unit preserving memory or the *complete (perfect, ideal) memory*.

If the memory is nonaging memory ($M(t,\tau) = M(t-\tau)$), then condition (1.92) can be used in the form

$$\int_0^t M(\tau)\,d\tau = 1 \tag{1.93}$$

that should holds for all $t > 0$. This requirement means the absence of memory.

As a special case of the function $M(t,\tau)$, we can consider the memory function in the form

$$M(t,\tau) = A(t)M_0(t-\tau), \tag{1.94}$$

where $A(t)$ and $M_0(t)$ are some functions. If the function has the form

$$A(t) = \left\{\int_0^t M_0(\tau)\,d\tau\right\}^{-1}, \tag{1.95}$$

then the memory function (1.94) can be interpreted as a complete (perfect, ideal) memory in some cases. An example of such a function is given in the next subsection.

1.4.7 Memory with generalized power-law fading

Memory can be characterized by more complex fading behavior than those discussed in the previous subsections. In this case, the memory should be described by more complex memory function. In order to describe the economic dynamics with more complex memory, we can use the generalized fractional calculus [202, 203]. Let us give some examples of the generalized memory functions $M(t,\tau)$ by using the kernels of the generalized fractional operators.

However, in this case, the question arises as to whether the generalized data operator can describe the memory or it describes other economic effects, such as time scaling. For illustration, we consider the function

$$M(t,\tau) = \frac{t^{-\alpha-\eta}}{\Gamma(\alpha)} \frac{m\tau^{\eta}}{(t-\tau)^{1-\alpha}}, \tag{1.96}$$

where η is the real number and $\alpha > 0$. Substituting (1.96) into equation (1.5), we obtain the integral equation of the order $\alpha > 0$ in the form

$$Y(t) = m(I^{\alpha}_{K;0+;\eta}X)(t), \tag{1.97}$$

where $I^{\alpha}_{K;0+;\eta}$ is the left-sided Kober fractional integral of the order $\alpha > 0$ with respect to time variable [200, p. 106], [335, pp. 322–325]. This integral operator was proposed as a generalization of the Riemann–Liouville fractional integral (1.64).

Definition 1.7. The Kober fractional integral is defined by the expression

$$(I^{\alpha}_{K;0+;\eta}X)(t) = \frac{t^{-\alpha-\eta}}{\Gamma(\alpha)} \int_0^t \tau^{\eta}(t-\tau)^{\alpha-1}X(\tau)\,d\tau, \tag{1.98}$$

where $\alpha > 0$ is the order of integration and $\eta \in \mathbb{R}$.

For the functions $X(t) \in L_p(\mathbb{R}_+)$, where $1 \le p < \infty$, and $\eta > (1-p)/p$, operator (1.98) is bounded [335, p. 323].

Remark 1.13. For $\eta = 0$, operator (1.98) can be expressed through the Riemann–Liouville integration (1.64) by the expression

$$(I^{\alpha}_{K;0+;1}X)(t) = t^{-\alpha}(I^{\alpha}_{RL;0+}X)(t). \tag{1.99}$$

Using expression 2.2.4.8 of [316, p. 296], [315] in the form

$$\int_0^t \tau^{\eta}(t-\tau)^{\alpha-1}\,d\tau = t^{\alpha+\eta}B(\eta+1,\alpha), \tag{1.100}$$

where $B(\alpha,\beta)$ is the beta function, we get

$$(I^{\alpha}_{EK;0+;\eta}1)(t) = \frac{1}{\Gamma(\alpha)}B(\eta+1,\alpha) = \frac{\Gamma(\eta+1)}{\Gamma(\eta+\alpha+1)}. \tag{1.101}$$

This allows us to assume that this operator can describe a memory that stores information about constant impact values without changes. However, it is not so simple.

Making the change of variable by $x = \tau/t$, the Kober operator (1.98) can be represented in the form

$$(I^{\alpha}_{K;0+;\eta}X)(t) = \frac{1}{\Gamma(\alpha)} \int_0^1 x^{\eta}(1-x)^{\alpha-1}X(xt)\,dx. \tag{1.102}$$

Expression (1.102) gives a possibility to use the probability density function (pdf) of the beta-distribution [411, 124, 214, 488] up to a constant factor (1.101), in the form

$$f_{\alpha;\beta}(x) = \frac{1}{B(\alpha,\beta)} x^{\alpha-1}(1-x)^{\beta-1} \tag{1.103}$$

for $x \in [0,1]$ and $f_{\alpha;\beta}(x) = 0$ if $x \notin [0,1]$, where $B(\alpha,\beta)$ is the beta function. Using equation (1.103), the Kober fractional integral is represented by equation

$$(I^{\alpha}_{K;0+;\eta}X)(t) = \frac{\Gamma(\eta+\alpha+1)}{\Gamma(\eta+1)} \int_0^1 f_{\eta+1;\alpha}(x) X(xt)\, dx. \tag{1.104}$$

We should emphasize that equation (1.104) contains $X(xt)$ instead of $X(t)$. Therefore, we can consider the variable $x > 0$ as a random variable that describes time scaling (dilation), which has the gamma distribution. To describe the change of scale (dilation), we can use the operator S_x [335, pp. 95–96], [200, p. 11] such that

$$(S_x X)(t) = X(xt), \tag{1.105}$$

where $x > 0$. It is known that the dilation of Euclidean geometric figures changes the size, when the shape is not changed. In physics and economics, the dilation is the change of scale of objects and processes. Using the scaling operator (1.105), the Kober fractional integral is represented by the equation

$$(I^{\alpha}_{K;0+;\eta}X)(t) = \frac{\Gamma(\eta+\alpha+1)}{\Gamma(\eta+1)} \int_0^1 f_{\eta+1;\alpha}(x)(S_x X)(t)\, dx. \tag{1.106}$$

Equation (1.106) allows us to propose the probabilistic interpretation of the Kober fractional integrals [400, 401]. Operator (1.106) can be interpreted as an expected value of a random variable $x > 0$ that describes the time scaling and has the beta distribution up to numerical factor (1.101).

As a result, expression (1.106) allows us to state that the Kober operator (1.98) can be interpreted as a continuously distributed dilation operator, in which the scaling variable has the beta distribution up to a constant factor (1.101).

A similar probabilistic interpretation can be proposed for some other generalized fractional integrals and derivatives. For example, for the Erdelyi–Kober fractional integral [400, 401, 406].

The Erdelyi–Kober fractional integration [335, pp. 322–325] (see also [241, 204, 201, 252, 253]) is defined by the function

$$M(t,\tau) = M^{\alpha,\eta}_{\sigma}(t,\tau) = \frac{\sigma t^{-\sigma(\alpha+\eta)}}{\Gamma(\alpha)} \frac{m\tau^{\sigma(\eta+1)-1}}{(t^{\sigma}-\tau^{\sigma})^{1-\alpha}}, \tag{1.107}$$

where $\sigma > 0$, η is the real number and $\alpha > 0$. For $\sigma = 1$, expression (1.107) takes form (1.96). Substituting (1.107) into equation (1.5), we obtain the integral equation in the form

$$Y(t) = m(I^{\alpha}_{EK;0+;\sigma,\eta}X)(t), \tag{1.108}$$

where $I^{\alpha}_{EK;0+;\sigma,\eta}$ is the left-sided Erdelyi–Kober fractional integral of the order $\alpha > 0$ with respect to time variable [200, p. 105] (see also [241, p. 251], where $\beta = \sigma$, $\gamma = \eta$, $\delta = \alpha$). For $\sigma = 1$, equation (1.108) takes the form of the Kober fractional integral (1.97).

Remark 1.14. For $\eta = 0$ and $\sigma = 1$, equation (1.108) can be expressed through the Riemann–Liouville fractional integral (1.64) by the equation $(I^{\alpha}_{EK;0+;1,0}X)(t) = t^{-\alpha}(I^{\alpha}_{RL;0+}X)(t)$.

In economics, equations (1.97) and (1.108) can be interpreted as generalized multiplier with distributed time scaling [400, 401, 406, 252, 253].

Let us consider equation (1.108) for unit impact variable $X(t) = 1$. Using equation (1.108) with $X(t) = 1$ and equation 2.2.4.8 of [316, p. 296] in the form

$$\int_0^t \tau^{\sigma(\eta+1)-1}(t^{\sigma} - \tau^{\sigma})^{\alpha-1}\, d\tau = \sigma^{-1} t^{\sigma(\alpha+\eta)} B(\eta + 1, \alpha), \tag{1.109}$$

where $B(x, y) = \Gamma(x)\Gamma(y)/\Gamma(x + y)$ is the beta function (the Euler integral of first kind), we get

$$Y(t) = m(I^{\alpha}_{EK;0+;\sigma,\eta}1)(t) = \frac{m}{\Gamma(\alpha)} B(\eta + 1, \alpha) = \frac{m\Gamma(\eta + 1)}{\Gamma(\eta + \alpha + 1)}. \tag{1.110}$$

As a result, we have a question about the kernels of Kober and Erdelyi–Kober fractional integrals. Can the kernels of these integrals be interpreted as memory functions, which describe a unit preserving memory up to a constant factor? The term "unit preserving memory" means that processes with this memory, which is described by equation (1.108), can remember a constant value of impact variables during the infinitely long period of time. Note that the memory (1.108) cannot be considered as nonaging memory, which is homogeneous on time $M(t, \tau) = M(t - \tau)$, in the general case.

Remark 1.15. The Caputo modification of the Erdelyi–Kober fractional derivative has been suggested by Yuri F. Luchko and Juan J. Trujillo [241, p. 260] and then it has been generalized by Virginia Kiryakova and Yuri F. Luchko in [204].

1.4.8 Memory with distributed fading

In economic processes with memory, the parameter α of memory fading can be distributed on interval $[\alpha_1, \alpha_2]$, where the distribution is described by a weighting function $\rho(\alpha)$. In this case, we should consider the memory function in the form, which depends on the weighting function $\rho(\alpha)$ and the interval $[\alpha_1, \alpha_2]$.

For the memory with power-law fading of distributed order, we can use the memory function in the form

$$M(t-\tau) = M^{[\alpha_1,\alpha_2]}_{\rho(\alpha)}(t-\tau) = \int_{\alpha_1}^{\alpha_2} \frac{\rho(\alpha)}{\Gamma(\alpha)} \frac{m}{(t-\tau)^{1-\alpha}} d\alpha, \tag{1.111}$$

where $\alpha_2 > \alpha_1 \geq 0$ and the weight function $\rho(\alpha)$ satisfies the normalization condition

$$\int_{\alpha_1}^{\alpha_2} \rho(\alpha)\, d\alpha = 1. \tag{1.112}$$

Substitution of (1.111) into equation (1.5) gives the integral equation

$$Y(t) = m(I^{[\alpha_1,\alpha_2]}_{RL;0+} X)(t), \tag{1.113}$$

where $I^{[\alpha_1,\alpha_2]}_{RL;0+}$ is the Riemann–Liouville fractional integral of distributed order. In economics, equation (1.113) can be interpreted as the equation of the economic multiplier with distributed fading memory [412, 397].

Using the left-sided Riemann–Liouville integral of the order $\alpha > 0$, which is defined by equation (1.64), the fractional integral of distributed order can be defined by the equation

$$(I^{[\alpha_1,\alpha_2]}_{0+} X)(t) = \int_{\alpha_1}^{\alpha_2} \rho(\alpha)(I^{\alpha}_{RL;0+} X)(t)\, d\alpha, \tag{1.114}$$

where $\rho(\alpha)$ satisfies the normalization condition (1.112) and $\alpha_2 > \alpha_1 \geq 0$. In equation (1.114), the integration with respect to time and the integration with respect to order can be permuted for a wide class of functions $X(\tau)$. As a result, equation (1.114) can be represented in the form

$$(I^{[\alpha_1,\alpha_2]}_{RL;0+} X)(t) = \int_0^t M^{[\alpha_1,\alpha_2]}_{\rho(\alpha)}(t-\tau) X(\tau)\, d\tau, \tag{1.115}$$

where the kernel $M^{[\alpha_1,\alpha_2]}_{\rho(\alpha)}(t-\tau)$ is defined by expression (1.111) with $m = 1$.

The concept of the integration and differentiation of distributed order was first proposed by Caputo in [45] and then developed in [45, 19, 20, 46, 234]. Derivatives and integrals of distributed order are discussed in book [184].

In equation (1.113), we can use the integer derivative $X^{(n)}(\tau) = d^n X(\tau)/d\tau^n$ as the impact variable. If $[\alpha_1] = [\alpha_2]$, we can use the memory function in the form

$$M^{[n-\alpha_2,n-\alpha_1]}_{\rho(n-\alpha)}(t-\tau) = a \int_{\alpha_1}^{\alpha_2} \frac{\rho(\alpha)(t-\tau)^{n-\alpha-1}}{\Gamma(n-\alpha)} d\alpha = a \int_{n-\alpha_2}^{n-\alpha_1} \frac{\rho(n-\alpha)(t-\tau)^{\alpha-1}}{\Gamma(\alpha)} d\alpha, \tag{1.116}$$

where $n = [\alpha_1] + 1 = [\alpha_2] + 1$. Substituting (1.116) into equation (1.6), where we use $X^{(n)}(\tau) = d^n X(\tau)/d\tau^n$, we obtain the fractional differential equation in the form

$$Y(t) = a(D^{[\alpha_1,\alpha_2]}_{C;0+} X)(t), \tag{1.117}$$

where $D^{[\alpha_1,\alpha_2]}_{C;0+}$ is the Caputo fractional derivative of distributed order. In economics, equation (1.117) can be interpreted as the equation of the economic accelerator with distributed fading memory [412, 397].

Using the left-sided Caputo derivative of the order $\alpha > 0$, which is defined by equation (1.80), we can define the fractional derivative of distributed order by the equation

$$(D^{[\alpha_1,\alpha_2]}_{C;0+} X)(t) = \int_{\alpha_1}^{\alpha_2} \rho(\alpha)(D^{\alpha}_{C;0+} X)(t)\, d\alpha, \tag{1.118}$$

where $n = [\alpha_1] + 1 = [\alpha_2] + 1$. Equation (1.118) can be represented in the form

$$(D^{[\alpha_1,\alpha_2]}_{C;0+} X)(t) = \int_0^t M^{[n-\alpha_2,n-\alpha_1]}_{\rho(n-\alpha)}(t-\tau) X^{(n)}(\tau)\, d\tau, \tag{1.119}$$

where the kernel $M^{[n-\alpha_2,n-\alpha_1]}_{\rho(n-\alpha)}(t-\tau)$ is defined by expression (1.116) with $a = 1$.

Fractional derivatives of distributed order can be generalized for the case $[\alpha_1] < [\alpha_2]$.

In the simplest case, we can use the continuous uniform distribution (CUD) that is defined by the expression

$$\rho(\alpha) = \frac{1}{\alpha_2 - \alpha_1} \tag{1.120}$$

for the case $\alpha \in [\alpha_1, \alpha_2]$ with $\alpha_2 - \alpha_1 > 0$, and $\rho(\alpha) = 0$ for other cases.

The fractional derivative of uniform distributed order can be expressed through the derivative of continual order that are suggested in [450, 412, 397]. For the distribution function (1.120), the memory function has the form

$$M^{[\alpha_1,\alpha_2]}_{\mathrm{CUD}}(t) = \frac{m}{\alpha_2 - \alpha_1} \int_{\alpha_1}^{\alpha_2} \frac{t^{\alpha-1}}{\Gamma(\alpha)}\, d\alpha. \tag{1.121}$$

In the general case, we can consider different distribution functions. These functions describe distributions of the parameter of memory fading on the set of economic agents. This is important for the economics, since various types of economic agents may have different parameters of memory fading.

In our opinion, the truncated normal (or Gaussian) distribution of fading memory parameter with the probability density function $\rho(\alpha, \sigma, \mu)$ can be of great interest for economic models. This type of memory function and corresponding fractional derivatives and integrals are proposed in [450, 448], [450, 412, 397]. The processes with memory, which is small deviation from classical case of amnesia, can be represented by the Gaussian distribution functions with integer mean and infinitesimally small variance. It is known that the delta-function can be considered as a limit of a family of the Gaussian functions, when the variance becomes smaller ($\rho(\alpha, \sigma, \mu) \to \delta(\alpha - \mu)$ at $\sigma > 0$). The processes without memory are described by the Dirac delta-function $\rho(\alpha) = \delta(\alpha - n)$. Therefore, the Gaussian distribution functions with integer mean and infinitesimally small variance can be used to describe economic processes with memory, which is distributed around the classical case.

1.5 Memory in economics: discrete time approach

In economics, memory can be described in discrete time approach. To describe economic processes with memory, Clive W. J. Granger and Roselyne Joyeux proposed [155] the difference operators that were called the fractional differencing and integrating. These Granger–Joyeux operators were suggested without any connection with the fractional calculus [396, 434, 443, 435]. In fact, the Granger–Joyeux operators are the Grunwald–Letnikov fractional differences, which are well known in mathematics for more than 150 years [157, 222]. However, these operators are currently used without any connection with a branch of mathematics that is called the fractional calculus. Equations of ARIMA(p, d, q) and ARFIMA(p, d, q) models, which are proposed by Granger, Joyeux and Hosking [155, 180], should be considered as equations with the Grunwald–Letnikov differences of order d.

We should note that continuous analogues of the Grunwald–Letnikov fractional difference are the Grunwald–Letnikov fractional derivatives of noninteger order [335, 200]. In continuous time approach, the economic processes with the memory, which is characterized by the power law, can be described by the fractional differential equations with the Grunwald–Letnikov fractional derivatives. Note that the Gurnwald–Letnikov fractional derivatives coincide with the Marchaud and Liouville fractional derivatives for a wide class of functions (see Theorem 20.4 in [335, p. 382], and Section 5.4 in [335, pp. 109–111]).

1.5.1 Long memory in discrete time approach

Let us consider the relation of the impact $X(t)$ and the response $Y(t)$ in the form

$$Y(t) = \int_0^\infty M(\tau) X(t - \tau)\, d\tau, \tag{1.122}$$

where the function $M(\tau)$ is expressed through the sum of the Dirac delta-functions with fixed coefficients in the form

$$M(\tau) = \sum_{k=0}^{\infty} a_k \delta(\tau - kT). \tag{1.123}$$

Here, T is positive time-constant, and a_k ($k \in \mathbb{N}$) are fixed coefficients.

Substitution of expression (1.123) into equation (1.122) gives

$$\begin{aligned} Y(t) &= \sum_{k=0}^{\infty} a_k \int_0^{\infty} \delta(\tau - kT) X(t-\tau)\, d\tau = \sum_{k=0}^{\infty} a_k X(t - kT) \\ &= a_0 X(t) + a_1 X(t-T) + a_2 X(t-2T) + \cdots, \end{aligned} \tag{1.124}$$

where $T > 0$ and $a_k \in \mathbb{R}$. Here, we use the following property of the delta-function:

$$\int_0^{\infty} \delta(\tau - \tau_0) X(\tau)\, dx = X(\tau_0), \tag{1.125}$$

which holds for all continuous compactly supported functions $X(\tau)$ on the positive semiaxis $(0, \infty)$, where $\tau_0 \in (0, \infty)$.

Let us consider some special cases of equation (1.124). The following particular cases are well known in economics.

(1) If $a_0 \neq 0$ and $a_k = 0$ for all $k = 1, 2, 3, \ldots$, then equation (1.124) has the form

$$Y(t) = a_0 X(t) \tag{1.126}$$

that describes the relation of the impact and response without memory and lag (time delay).

(2) If $a_0 = 0$, $a_1 \neq 0$ and $a_k = 0$ for all $k = 2, 3, 4, \ldots$, then equation (1.124) has the form

$$Y(t) = a_1 X(t - T). \tag{1.127}$$

In this case, the relation of the impact and response is realized with lag that is fixed time delay [10, pp. 23–24]. The response variable depends on only one previous value of the impact variable $X(t)$.

(3) If $a_0 = 0$, $a_k \neq 0$ for all $k = 1, 2, 3, \ldots$, and

$$\sum_{k=0}^{\infty} a_k = a, \tag{1.128}$$

where $0 < |a| < \infty$, then equation (1.124) has the form

$$Y(t) = \sum_{k=1}^{\infty} a_k X(t - kT) \tag{1.129}$$

that describes the distributed lag [10, p. 24]. The coefficients a_k is interpreted as the weighting coefficients. In this case, the response variable depends not on one prior value of impact variable $X(t)$, but on a whole sequence of such values.

(4) If $a_0 = 0$, $a_k = a(1-r)r^{k-1}$ for all $k = 1, 2, 3, \dots$, where $0 < r < 1$, then equation (1.124) describes the well-known case of the distributed lag, which is called geometric lag. In this lag, the weighting coefficients decrease in given fixed ratio $r \in (0, 1)$ in an infinite geometric series [10, p. 24]. This type of discrete distributed lag is often used in economics.

(5) Let us consider the coefficients a_k in the form

$$a_k = (-1)^k \binom{\alpha}{k}, \tag{1.130}$$

where $\binom{\alpha}{k}$ are the generalized binomial coefficients [200, p. 26] that are defined by the equation

$$\binom{\alpha}{0} = 1, \quad \binom{\alpha}{k} = \frac{\alpha(\alpha-1)\dots(\alpha-k+1)}{k!}, \tag{1.131}$$

where $k \in \mathbb{N}$ and $\alpha \in \mathbb{R}$. In particular, when $\alpha = n \in \mathbb{N}$, we have

$$\binom{n}{k} = \frac{n!}{k!(n-k)!} \quad (n \geq k \geq 0), \quad \binom{n}{k} = 0, \quad (k > n \geq 0). \tag{1.132}$$

If $(-\alpha) \neq 1, 2, 3, \dots$, then the generalized binomial coefficients can be represented through the gamma function

$$\binom{\alpha}{k} = \frac{\Gamma(\alpha+1)}{k!\Gamma(\alpha-k+1)}. \tag{1.133}$$

Note that $\Gamma(k+1) = k!$ for $k \in \mathbb{N}$.

In the case (1.130), the right-hand side of equation (1.124) is the Grunwals–Letnikov fractional difference of order α [200, p. 121] that are defined by the equation

$$(\Delta_T^{\alpha} X)(t) = \sum_{k=0}^{\infty} (-1)^k \binom{\alpha}{k} X(t-kT). \tag{1.134}$$

For integer values of $\alpha = n$, the Grunwals–Letnikov fractional differences are the standard differences of integer order. For example, if $\alpha = 1$, then (1.134) takes the form

$$(\Delta_T^1 X)(t) = X(t) - X(t-T), \tag{1.135}$$

and for $\alpha = 0$ we have

$$(\Delta_T^0 X)(t) = X(t). \tag{1.136}$$

As a result, equation (1.124) with coefficients (1.130) can be written in the form

$$Y(t) = (\Delta_T^{\alpha} X)(t), \tag{1.137}$$

where the operators Δ_T^{α} is defined by equation (1.134). Equation (1.137) describes the relation of the impact and the response with memory if the parameter α takes noninteger values.

Remark 1.16. Let us consider $t = nT$, where $n \in \mathbb{Z}$. Using the notation,

$$X_k = X(kT), \quad Y_n = Y(nT), \tag{1.138}$$

where $k, n \in \mathbb{Z}$, equation (1.124) takes the form

$$Y_n = \sum_{k=0}^{\infty} a_k X_{n-k}. \tag{1.139}$$

In economics, the designations X_t, Y_t (or x_t, y_t) are often used, in which it is assumed that t takes integer values ($t \in \mathbb{Z}$), and $T = 1$.

Remark 1.17. If $M(t,\tau)$ cannot be represented as $M(t-\tau)$ in the continuous time approach, then in the discrete time approach we can consider the expression

$$Y_n = \sum_{k=0}^{\infty} a_{n,k} X_{n-k} \tag{1.140}$$

as a generalization of equation (1.139).

1.5.2 Definition of process with memory for discrete time

In discrete time approach, economic processes with memory are actively studied in recent decades (e. g., see [29, 295, 30, 481, 231, 485, 21]). Reviews of econometric articles on long memory were suggested in [21, 481, 23]. The mathematical statistics for the long-memory processes has been described in detail by [29, 30]. For discrete real-valued time series, there are several ways of defining long memory which are formulated for the time and the frequency domains.

In time domain, economic processes with long memory are described by the correlations $\rho(\tau)$ that are asymptotically (at $\tau \to \infty$) equal to a constant c_ρ times $\tau^{-\alpha}$ for some $\alpha \in (0,1)$. These processes are often called process with long-range dependence or processes with slowly decaying (or long-range) correlations.

The following definition [29, p. 42] of stationary process with long memory (or long-range dependence) is usually used.

Definition 1.8. Let X_t be an economic stationary stochastic process for which there exist a real number $\alpha \in (0,1)$ and a constant c_ρ such that

$$\rho(\tau) \sim c_\rho \tau^{-\alpha} \quad (\tau \to \infty), \tag{1.141}$$

where $\rho(\tau) = \operatorname{cov}(X_t, X_{t+\tau})$ is the autocovariance function (ACF) at lag τ. In equation (1.141), the symbol $\sim$ means that the ratio of the left- and right-hand sides is finite, when τ tends to infinity, i. e., we have

$$\lim_{\tau \to \infty} \frac{\rho(\tau)}{\tau^{-\alpha}} = c_\rho, \tag{1.142}$$

where $0 < |c_\rho| < \infty$. Then X_t is called a stationary process with long memory.

The knowledge of covariance (autocovariance function) is equivalent to knowledge of a spectral density. For this reason, the second method for describing processes with memory is also used. In the frequency domain, economic processes with long memory are described by the spectral density $S(\omega)$, which asymptotically (at $\omega \to \infty$) has a pole at zero that is equal to a constant c_S times $\tau^{-\beta}$ for some $\beta \in (0,1)$.

Definition 1.9. Let X_t be an economic stationary stochastic process for which there exist a real number $\beta \in (0,1)$ and a constant c_S such that

$$S(\omega) \sim c_S \omega^{-\beta} \quad (\omega \to 0), \tag{1.143}$$

where $S(\omega) = F\{\rho(\tau)\}$ is the spectral density, and the symbol F denotes the Fourier transform. In equation (1.143), the symbol $\sim$ means that the ratio of the left- and right-hand sides is finite, when ω tends to zero

$$\lim_{\omega \to 0} \frac{S(\omega)}{\omega^{-\beta}} = c_S, \tag{1.144}$$

where $0 < |c_S| < \infty$. Then X_t is called a stationary process with long memory.

Sometimes instead of two parameters α and β, we can use one parameter d, such that $-\alpha = 2d - 1$ and $-\beta = -2d$ (for details, see [30, p. 32]).

It is important to note that these definitions are asymptotic definitions. For example, (1.142) tells us only about the behavior of the correlations as the lag τ tends to infinity. Note that the restriction to low frequencies close to zero, which is used in equation (1.143), leads to insensitivity of this definition to different short term shocks.

There are different estimators of the memory fading parameter for economic processes with long memory. One of the most actively used methods is semiparametric estimation, which is based on the spectral density function in the neighborhood of zero according to condition (1.143). This estimator is based on the information included in a periodogram, but for very low frequencies. Note that this restriction of only low frequencies leads to insensitivity of these estimators with respect to different short term shocks.

1.5.3 Granger–Joyeux fractional differencing

In economics, the long memory was first described by using so-called fractional differencing and integrating in the paper of C. W. J. Granger and R. Joyeux [155] in the framework of the discrete time stochastic process (see also [180, 21, 303, 23, 135]). Granger and Joyeux, and Hosking [155, 180] independently proposed the so-called autoregressive fractional integrated moving average model (ARFIMA model). Note that C. W. J. Granger received the Nobel memorial prize in economic sciences in 2003 "for methods of analyzing economic time series with common trends (cointegration)" [479]. We say that $\{y_t, t = 1, 2, \dots, T\}$ is an ARFIMA(0, d, 0) model, if we have the following equation of discrete time stochastic process

$$(1 - L)^d y_t = \varepsilon_t, \tag{1.145}$$

where L is the translation operator (also known as shift operator). The operator L is called the lag operator in economics and it is defined by the equation

$$L y_t = y_{t-1}. \tag{1.146}$$

In equation (1.146), the parameter d is the order of the fractional differencing (integration), which need not be an integer order, y_t is the stochastic process, and ε_t is independent and identically distributed (i. i. d.) white noise process of random variables with the mean $E(\varepsilon_t) = 0$ and the variance $V(\varepsilon_t) = \sigma_\varepsilon^2$.

The expression $(1 - L)^d$ can be defined by the series expansion [335, p. 371] in the form

$$(1 - L)^d = \sum_{m=0}^{\infty} (-1)^m \binom{d}{m} L^m, \tag{1.147}$$

where $\binom{d}{m}$ are the generalized binomial coefficients [335, p. 14] that are defined by the equation

$$\binom{d}{m} = \frac{\Gamma(d+1)}{\Gamma(d-m+1)\Gamma(m+1)}, \tag{1.148}$$

where $\Gamma(z)$ is the Euler gamma function. Equation (1.147) can be written in the form

$$(1 - L)^d = 1 - dL - \frac{1}{2} d(1-d) L^2 - \frac{1}{6} d(1-d)(2-d) L^3 - \cdots. \tag{1.149}$$

Using equation 1.48 of [335, p. 14], the binomial coefficients (1.148) can be written as

$$\binom{d}{m} = \frac{(-1)^{m-1} d \Gamma(m-d)}{\Gamma(1-d)\Gamma(m+1)}. \tag{1.150}$$

Using equation (1.150) and $\Gamma(z+1) = z\Gamma(z)$, equation (1.147) can be represented in the form

$$(1-L)^d = \sum_{m=0}^{\infty} \frac{\Gamma(m-d)}{\Gamma(-d)\Gamma(m+1)} L^m, \tag{1.151}$$

which is usually used in the econometric papers on long memory and time series.

It should be noted that the operator

$$\Delta^d = (1-L)^d \tag{1.152}$$

is the difference of fractional (integer or noninteger) order, which is called the Grunwald–Letnikov fractional difference of order d with the unit step $T = 1$ [335, 308, 200].

Definition 1.10. The Grunwald–Letnikov fractional difference Δ_T^α of order α with the step T is defined by the equation

$$\Delta_T^\alpha Y(t) = (1-L_T)^\alpha Y(t) = \sum_{m=0}^{\infty} (-1)^m \binom{\alpha}{m} Y(t-mT), \tag{1.153}$$

where $L_T Y(t) = Y(t-T)$ is fixed-time delay and the time-constant T is any given positive value. The series (1.153) converges absolutely and uniformly for each $\alpha > 0$ and for every bounded function $Y(t)$.

Remark 1.18. It should be noted that the Grunwald–Letnikov fractional difference (1.153) may converge for $\alpha < 0$, if $Y(t)$ has a "good" decrease at infinity [335, p. 372]. For example, we can use the functions $Y(t)$ such that $|Y(t)| \leq c(1+|t|)^{-\mu}$, where $\mu > |\alpha|$. This allows us to use (1.153) as a discrete fractional integration in the nonperiodic case.

In economics, the memory was first related to fractional difference operators by Granger and Joyeux [155] in the framework of the discrete time approach [29, 295, 30, 481, 231, 485]. The proposed Granger–Joyeux fractional differences, which were called the fractional differencing and integrating in [155, 180], are not directly connected in these works to the fractional calculus and the well-known finite differences of noninteger orders [335, 202, 308, 200, 90, 164, 165, 119]. We proved [396, 434, 443, 435] that the fractional differencing and integrating, which are proposed by Granger, Joyeux and Hosking in [155, 180], are the well-known Grunwald–Letnikov fractional differences. Note that these Grunwald–Letnikov differences were proposed over 150 years ago [157, 222].

Therefore, we can formulate the following statement.

Statement 1.4. *The Granger–Joyeux fractional differencing and integrating* (1.152) *of the ordered are the well-known Grunwald-Letnikov fractional differences* Δ_T^α *of order* $\alpha = d$ *with the step* $T = 1$.

As a result, equation (1.145) is the fractional difference equation. Equation (1.145) can be generalized for the continuous time case by using the fractional difference equation

$$\Delta_T^\alpha Y(t) = E(t). \tag{1.154}$$

Equations of ARFIMA models are the fractional difference equations with the Grunwald–Letnikov fractional differences (1.153) of order $\alpha = d$.

Fractional calculus, which is a branch of modern mathematics and has existed for more than 300 years, was not even mentioned in works of Granger, Joyeux, Hosking and in subsequent works in economics.

Obviously, the use of only one type of fractional differential and difference operators in the description of economic processes with memory is illogical self-restriction in the work about economy.

Remark 1.19. Note that the Grunwald–Letnikov fractional differences, and thus the Granger–Joye fractional differencing gives fractional derivatives in the continuous limit. This situation is similar to the fact that a finite difference of integer order in the continuous limit gives an integer derivative.

1.5.4 Continuous limit of fractional differencing

To describe memory in continuous time approach, we can use fractional-order derivatives instead of fractional-order differences. The Grunwald–Letnikov fractional difference (1.153) allows us to define the fractional-order derivatives [335, p. 373] as the limit

$$(D^\alpha_{\mathrm{GL};\pm} Y)(t) = \lim_{T \to +0} \frac{1}{T^\alpha} \Delta^\alpha_{\pm T} Y(t). \tag{1.155}$$

The operator $(D^\alpha_{\mathrm{GL};\pm} Y)$ is called the Grunwald–Letnikov fractional derivative of order α [335, 308, 200].

Using the fractional-order derivatives (1.155), equation (1.154) can be generalized for the continuous time case. Memory processes with continuous time can be described by the equation

$$(D^\alpha_{\mathrm{GL};+} Y)(t) = E(t). \tag{1.156}$$

Equation (1.156) is the fractional differential equation with derivatives of order α [308, 200]. Equation (1.156) is the continuous analog of the fractional difference equation (1.154) with the Grunwald–Letnikov fractional differences.

As a result, the continuous time description of the economic processes with the memory can be based on the fractional derivatives and integrals, and the fractional differential equations [335, 308, 200]. This mathematical tool allows us to get new fractional dynamical models for economic processes with memory.

Remark 1.20. Note that the Gurnwald–Letnikov derivatives (1.155) coincide with the Marchaud fractional derivatives (see Theorems 20.2 and 20.4 of [335]). The Grunwald–Letnikov and Marchaud derivatives have the same domain of definition. The Marchaud fractional derivatives coincide with the Liouville fractional derivatives for a wide class of functions [335, pp. 110–111]. As a result, the Gurnwald–Letnikov fractional derivatives coincide with the Marchaud and Liouville fractional derivatives for a wide class of functions.

Remark 1.21. We should note that the Fourier transform of the Liouville fractional integral and derivative (and Grunwald–Letnikov fractional integral and derivative for a wide class of functions) has the power-law form [200, p. 90], and [335, p. 137]. The Fourier transform of the Liouville fractional integral is

$$F\{I^{\alpha}_{L;\pm}Y(t)\}(\omega) = (\mp i\omega)^{-\alpha}F\{Y(t)\}(\omega). \tag{1.157}$$

The Fourier transform of the Liouville fractional derivative is represented by the expression

$$F\{D^{\alpha}_{L;\pm}Y(t)\}(\omega) = (\mp i\omega)^{\alpha}F\{Y(t)\}(\omega). \tag{1.158}$$

As a result, we can see that the Fourier transforms of the Liouville fractional derivatives and integrals is described by the power law exactly. We emphasize that the fractional differencing (1.147), (1.152) and the fractional-order difference (1.153) satisfy a power law only asymptotically at $\omega \to 0$ [335, p. 373].

1.5.5 Power-law memory and exact fractional differences

It is important to emphasize that definition of long memory in frequency domain are asymptotic definitions. It tells us only about the behavior of the spectral density at the frequency ω tends to zero. We can consider an economic process X_t with discrete or continuous time such that its spectral density function $S(\omega)$ satisfies the power-law exactly for all frequencies holds for all $\omega > 0$. In this case, the process X_t exhibits a long memory of the power-law form. The spectral density function of such economic process is exactly the power-law function.

It is known that in the nonperiodic case the Fourier transform F of $\Delta^{\alpha}_{T}y(t)$ is given [335, p. 373], by the formula

$$F\{\Delta^{\alpha}_{T}y(t)\}(\omega) = (1 - \exp(i\omega T))^{\alpha}F\{y(t)\}(\omega). \tag{1.159}$$

As a result, the Grunwald–Letnikov fractional difference Δ^{α}_{T} of order α cannot correctly describe the processes with of power-law memory. The fractional differences (1.153) cannot be considered as an exact discrete (difference) analog of the Liouville and Gurnwald–Letnikov fractional derivatives and integrals [335, 308, 200], since the

Fourier transform of the Grunwald–Letnikov fractional differences are not the power law, i. e., we have

$$F\{\Delta_T^{\alpha} y(t)\}(\omega) \neq (i\omega T)^{\alpha} F\{y(t)\}(\omega). \tag{1.160}$$

In order to have equality in (1.160), we can use the exact fractional differences that are suggested in article [378], and then considered in papers [441, 442, 431, 432]. The kernel $K_{\alpha}(m)$ of the exact fractional differences $\Delta_{T,\text{exact}}^{\alpha}$ is expressed by the generalized hypergeometric functions $F_{1,2}(a;;b,c;;z)$ instead the gamma functions in (1.148).

Note that Hosking used the hypergeometric functions to describe two-parameter ARIMA(p, d, q) processes [180, p. 172]. These processes are most conveniently expressed in terms of the hypergeometric functions [180].

We should emphasize that the Fourier transform F of the exact fractional differences (1.162) with kernel (1.163) has the power-law form exactly, i. e., we have the equality

$$F\{\Delta_{T,\text{exact}}^{\alpha} y(t)\}(\omega) = (i\omega T)^{\alpha} F\{y(t)\}(\omega) \tag{1.161}$$

holds in contrast to the fractional differencing (1.147), (1.151) and fractional difference (1.153), where we have inequality (1.160). Equation (1.161) means that the spectral density function $S_y(\omega)$ satisfies the power law exactly.

The exact fractional difference can be considered as an exact discrete analog of the Liouville fractional derivatives and integrals [441, 442, 431, 432], since these operators have the same power-law behavior (1.157), (1.158) and (1.161).

For $\alpha < 0$ ($\alpha \in (-1, 0)$), the exact fractional differences define the discrete fractional integration [378].

The exact discretization of the derivatives of integer and noninteger orders [378, 371] and the corresponding exact finite differences [378] were initially proposed in [359, 363, 370, 370, 370, 372, 373] as derivatives on lattices. In economics, they were used in [441, 442, 431, 432]. Derivatives of noninteger order and the corresponding fractional finite differences allow us to describe economic processes with power-law long memory.

The exact fractional differences of integer orders and the standard derivatives of integer orders have the same algebraic properties [378, 371] in contrast to the standard finite-differences integer order [431].

As a result, we can conclude that in the discrete-time approach, the memory with power-law fading should be described by the exact fractional-order differences, which demonstrate the power law (1.161). Fractional differencing (1.147), (1.151), which are the Grunwald–Letnikov fractional differences (1.153), cannot give exact results for the long memory with power-law fading, since these discrete operators satisfy inequality (1.160). These discrete operators lead us to insensitivity of these mathematical tools with respect to different short term shocks, since the Fourier transform of these difference operators satisfy the power law in the neighborhood of zero only. The correct description of the discrete time long memory with power law should be based on the exact fractional differences that are suggested in papers [378] and [441, 442, 431, 432].

1.5.6 Definition of exact fractional differences

The exact fractional difference is defined [378] by the equation

$$\Delta^{\alpha}_{T,\mathrm{exact}} y(t) = \sum_{m=-\infty}^{\infty} K_\alpha(m) y(t - mT). \tag{1.162}$$

For $\alpha < 0$, equation (1.162) with kernel (1.163) defines the discrete fractional integration [378].

The kernel of the exact fractional differences is represented by the equation

$$K_\alpha(m) = \cos\left(\frac{\pi\alpha}{2}\right) K^+_\alpha(m) + \sin\left(\frac{\pi\alpha}{2}\right) K^-_\alpha(m), \tag{1.163}$$

where

$$K^+_\alpha(m) = \frac{\pi^\alpha}{\alpha+1} F_{1,2}\left(\frac{\alpha+1}{2}; \frac{1}{2}, \frac{\alpha+3}{2}; -\frac{\pi^2 m^2}{4}\right), \quad (\alpha > -1), \tag{1.164}$$

$$K^-_\alpha(m) = -\frac{\pi^\alpha m}{\alpha+2} F_{1,2}\left(\frac{\alpha+2}{2};; \frac{3}{2}, \frac{\alpha+4}{2};; -\frac{\pi^2 m^2}{4}\right), \quad (\alpha > -2). \tag{1.165}$$

The generalized hypergeometric function $F_{1,2}(a;; b, c; z)$ is defined as

$$F_{1,2}(a;; b, c;; z) = \sum_{k=0}^{\infty} \frac{\Gamma(a+k)\Gamma(b)\Gamma(c)}{\Gamma(a)\Gamma(b+k)\Gamma(c+k)} \frac{z^k}{k!}. \tag{1.166}$$

Using equation (1.166), kernel (1.163) can be represented in the form

$$K_\alpha(m) = \sum_{k=0}^{\infty} \frac{(-1)^k \pi^{2k+\alpha+\frac{1}{2}} m^{2k}}{2^{2k} k! \Gamma(k+\frac{1}{2})} \left(\frac{\cos(\frac{\pi\alpha}{2})}{\alpha+2k+1} - \frac{\pi m \sin(\frac{\pi\alpha}{2})}{(\alpha+2k+2)(2k+1)}\right). \tag{1.167}$$

For the arbitrary positive integer order $\alpha = n$, the kernel $K_\alpha(m)$ of the exact difference can be represented by the equation

$$K_n(m) = \sum_{k=0}^{[\frac{n+1}{2}]+1} \frac{(-1)^{m+k} n! \pi^{n-2k-1}}{(n-2k)! m^{2k+2}} \left((n-2k)\cos\left(\frac{\pi n}{2}\right) + \pi m \sin\left(\frac{\pi n}{2}\right)\right) \tag{1.168}$$

for $m \neq 0$, and for $m = 0$ the kernel is written by the expression

$$K_n(0) = \frac{\pi^n}{n+1} \cos\left(\frac{\pi n}{2}\right). \tag{1.169}$$

The exact finite difference of the first order ($\alpha = 1$) is defined by the equation

$$\Delta^1_{T,\text{exact}} Y(t) = \sum_{m=1}^{\infty} \frac{(-1)^m}{m} (Y(t - Tm) - Y(t + Tm)), \quad (1.170)$$

where the sum implies the Cesaro or Poisson-Abel summation [378]. An important characteristic property of the exact finite difference (1.170) is the Leibniz rule (the product rule) in the form

$$\Delta^1_{T,\text{exact}}(X(t)Y(t)) = (\Delta^1_{T,\text{exact}}X(t))Y(t) + X(t)(\Delta^1_{T,\text{exact}}Y(t)), \quad (1.171)$$

which is satisfied for all $X(t)$, $Y(t)$ from the space of entire functions. Exact finite difference of second and next integer orders can be derived by the recurrence formulas

$$\Delta^{k+1}_{T,\text{exact}} Y(t) = \Delta^1_{T,\text{exact}}(\Delta^k_{T,\text{exact}} Y(t)). \quad (1.172)$$

Equations of discrete macroeconomic models, which are used the exact finite differences, are exact discrete analogs of differential equations of models with continuous time for a wide class of solutions (e. g., see [378] and [432]).

1.6 Methods of describing processes with memory

The basic methods for describing processes with memory can be conditionally divided into the following three main approaches:

(a) The IE/IDE approach based on integral equations and integro-differential equations;
(b) The MS/TSA approach based on mathematical statistics and time series analysis;
(c) The FC approach based on fractional calculus.

These methods and corresponding models can also be divided into two groups with the discrete time and continuous time.

Let us briefly discuss these basic approaches.

1.6.1 Integral equations and integro-differential equations

The first description of processes with memory and nonlocality in time was given by Ludwig Boltzmann in 1874 and 1876 [35, 36, 38, 39, 37]. He proposed physical model of isotropic viscoelastic media that are actually media with memory. Boltzmann used the term aftereffect (“Nachwirkung”). Boltzmann assumed that the stress at time t depends on the strains not only at the present time t, but also on the history of the process at $\tau < t$. He also proposed two principles: the linear superposition principle and the memory fading principle. Boltzmann proposed the integral equation to describe the dynamics of the isotropic viscoelastic media, whose behavior is interpreted as memory effects.

The Boltzmann theory has been significantly developed by Vito Volterra in 1928 and 1930 [496, 495, 497, 255, 498] in the form of the heredity concept and its application. Volterra used the terms hereditary and heredity ("ereditari," "hereditaires"). He formulated the principles, which has been called by him the general laws of heredity (see Section 148 of [498]). Volterra made a significant contribution to the development of the theory of integral equations, which can be used as one of the tools to describe processes with memory.

If the integral equations also contain derivatives of unknown functions, then they are called integro-differential equations. In modern physics, the integral and integro-differential equations are used to describe processes with memory in physics and mechanics in the framework of the models with continuous time. Note that the general theory of the integral (and integro-differential) equations is used very rarely in the economic and financial models with memory and continuous time.

A wide class of integral and integro-differential equations (e. g., the equations with kernels of power-law type) refers to fractional calculus that studies equations with derivatives and integrals of noninteger orders. The equations with fractional integrals and derivatives of noninteger orders can be considered as a special type of the integral and integro-differential equations.

We also should note that the fractional calculus is actively expanding to include some parts of the theory of integral and integro-differential operators and equations. For example, in recent decades, generalized fractional calculus [202, 203], general fractional calculus [209, 210, 211, 347], fractional calculus of a distributed order [19, 20, 234, 184] and fractional calculus with a distributed lag [400, 403, 405, 404, 402] have appeared (see also [467, 470]). Therefore, the fractional calculus and fractional differential equation contains most of the methods of IE/IDE approach to describe processes with memory and nonlocality in time.

1.6.2 Mathematical statistics and time series analysis

Long memory effects were known before the development of adequate methods of mathematical statistics. Scientists have empirically observed the long-range time dependencies, for which the correlations between observations decay to zero more slowly than it can be expected from independent data or data resulting from classical Markov and ARMA models [29, 295, 30]. An intuitive interpretation of the relationship between events, which grow very slowly over time, is that this process has a long memory. In development of the mathematical statistics, the processes with memory were introduced to the theory and the mathematical tools for describing such processes were created. In general, there are two main types of statistical methods for analyzing the behavior of time series. The first type is more related to the time-domain analysis such as the correlation analysis. The second type is more connected with an analysis in the frequency domain such as the spectral analysis.

Processes with long memory are characterized by slowly decaying autocorrelations or by a spectral density, which has a pole at the origin. This property significantly changes the statistical behavior of estimates and predictions. As a consequence, many of the theoretical results and approaches, which are used for analyzing classical time series of the Markov and ARMA processes, cannot be used to describe processes with memory (e. g., see [29, 295, 30]).

For the first time, the importance of long-range time dependence in economic data was recognized by Clive W. J. Granger in his technical report [147] at 1964, and then in his article [148] in 1966 on "The typical spectral shape of an economic variable" (see also [152]). Granger showed that a number of spectral densities, which are estimated from economic time series, have a similar form. Then, to describe economic processes with memory Granger and Joyeux in 1980, and, independently of them, Hosking in 1981 proposed the fractional ARIMA models [155, 180], [153, pp. 321–337], which are also called ARFIMA(p, d, q). The fractional ARIMA models greatly improved the methods of description of processes with long memory in the framework of statistical approach. Note that Clive W. J. Granger received the Nobel memorial prize in economic sciences in 2003 "for methods of analyzing economic time series with common trends (cointegration)" [479].

The fractional ARIMA models are generalization of the classical ARMA and ARIMA models, which are the most popular types of linear models in time series analysis, because of their simplicity and flexibility. The fractional ARIMA(p, d, q) models mathematically can be considered as generalization of ARIMA(p, d, q) model from a positive integer order d to noninteger (positive and negative) orders d. To generalize ARMA and ARIMA models Granger, Joyeux and Hosking [155, 180] proposed the so-called fractional differencing and integrating in the framework of the discrete time models (see also [21, 303, 23, 135]). These fractional differencing and integrating were proposed in [155, 180] and then began to be used up to the present time without any connection with the fractional calculus and the well-known fractional differences of noninteger orders.

We demonstrate [396, 434] that these fractional differencing and integrating are the well-known Grunwald–Letnikov fractional differences, which have been suggested in 1867 and 1868, i. e., 150 years ago [157, 222]. These fractional differences of noninteger orders are actively used in the fractional calculus [335, pp. 371–388], [308, pp. 43–62], [200, pp. 121–123]. Note that in the continuous limit these fractional differences of positive orders give the Grunwald–Letnikov fractional derivatives that coincide with the Marchaud fractional derivatives of nonintegral order (see Theorems 4.2 and 4.4 of [335, pp. 380, 382]). Moreover, these fractional derivatives have the same domain of definition [335]. In addition the Marchaud fractional derivatives (and, therefore, the Grunwald–Letnikov fractional derivatives) coincide with the Liouville fractional derivatives for wide class of functions (e. g., see Section 5.4 of [335, pp. 109–112]).

The restriction in economics to one type of fractional finite differences, the neglect of fractional derivatives and integrals, significantly reduces the possibility of describing processes with memory in the economy. We proposed to use the modern fractional calculus, fractional differential and difference equations to significantly expand the possibilities of describing economic processes with memory and nonlocality in time.

1.6.3 Fractional calculus

To describe memory, it is possible to use the theory of equations with integrals and derivatives of noninteger order [335, 202, 308, 200, 90, 164, 165]. There are different types of fractional integral, differential and difference operators that are suggested by the well-known mathematicians Riemann, Liouville, Grunwald, Letnikov, Sonine, Marchaud, Weyl, Riesz, Kober, Erdelyi and other scientists. The fractional derivatives have a set of nonstandard properties [367, 287, 377, 376, 382, 338, 81, 406]. For example, such properties include a violation of the standard Leibniz rule [367, 81], which describes the derivative of the product of functions. Another important example is the violation of standard chain rules [367, 81], which described the derivative of the composition of functions. It should be emphasized that the violation of the standard form of the Leibniz rule is a characteristic property of derivatives of noninteger orders [367, 81]. These nonstandard properties of the fractional derivatives complicate the use of fractional calculus in the formulation of new concepts and the construction of models of economic processes with memory [386] (see also book [254, pp. 43–92]). However, thanks to these nonstandard properties, we can describe the nonstandard characteristic properties of processes with memory and nonlocality in time.

A description of the concept of memory itself using fractional calculus is discussed in [450, 448, 399, 380].

At present time, the fractional integro-differential equations have become actively used in continuous time models of physics to describe wide class of processes with memory [363, 167, 168].

A brief history of the application of fractional calculus to describe economic and financial processes is given in review [385] (see also book [254, pp. 5–32]). Note that most of the first works were devoted to financial processes and not economics. The basic economic concepts and principles for economic processes with memory were not considered in these works.

From our point of view, the formulation of new basic economic concepts and notions of Economics with Memory began with a proposal of generalizations of the basic economic concepts and notions at the beginning of 2016, when the concept of elasticity for economic processes with memory was proposed [418] and [430, 419]. Then in 2016 the following economic concepts were proposed: the marginal values with memory [424, 416, 425], the concept of accelerator and multiplier with memory [412, 417], the nonlocal measures of risk aversion [428], and nonlocal deterministic factor analysis [414]. In 2016, these concepts are used in generalizations of some standard economic models [420, 421, 426, 423, 427, 422] that describe the dynamics of economic processes with memory.

As a result, during 2016–2020 the following basic economic concepts were proposed: the marginal value of noninteger order [424, 416, 425, 439, 446] with memory and nonlocality; the economic multiplier with memory [412, 397]; the economic accelerator with memory [412, 397]; the exact discretization of economic accelerators and multipliers [441, 442, 431, 432]; based on exact fractional differences; the accelerator with memory and periodic sharp bursts [417, 436, 444]; the duality of the multiplier with memory and the accelerator with memory [412, 397]; the accelerators and multipliers with memory and distributed lag [405, 404, 402]; the elasticity of fractional order [418, 430, 429, 419] for processes with memory and nonlocality; the measures of risk aversion with nonlocality and with memory [428, 449]; the warranted (technological) rate of growth with memory [450, 447, 445, 453, 393, 387]; the nonlocal methods of deterministic factor analysis [414, 452]; the productivity with fatigue and memory [454]; the chronological memory ordering [433, 398]; and some other.

Using these concepts, the following economic models were proposed: the natural growth model with memory [420, 440]; the growth model with constant pace and memory [447, 438]; the Harrod–Domar model with memory [421, 426] and [450, 445, 453, 393]; the Keynes model with memory [423, 427, 422] and [402, 407]; the dynamic Leontief (intersectoral) model with memory [437, 451, 433, 398]; the model of dynamics of fixed assets (or capital stock) with memory [447, 438]; the model of price dynamics with memory [438]; the logistic growth model with memory [386] (see also [403]); the model of logistic growth with memory and periodic sharp splashes (kicks) [444]; the time-dependent dynamic intersectoral model with memory [433, 398]; the Phillips model with memory and distributed lag [405]; the Harrod–Domar growth model with memory and distributed lag [404]; the dynamic Keynesian model with memory and distributed lag [402, 407]; the model of productivity with fatigue and memory [454]; the Solow–Swan model with memory [386, 383]; the Kaldor-type model of business cycles with memory [386]; the Evans model with memory [390]; the Cagan model with memory [43]; the Lucas models with memory [391, 389, 381]; and some other economic models.

Some of these generalized economic concepts and models for processes with memory are described in the following chapters of this proposed book on economic dynamics with memory.

1.7 Conclusion

The memory effects are caused by the fact that economic agents remember the history of changes of variables that characterize the economic process. The agents can take into account these changes in making economic decisions. The continuous time description of the economic processes with the fading memory can be based on the fractional calculus, the fractional differential and difference equations. In economics, we proved that the inclusion of memory effects into the economic models can lead to qualitatively new results at the same the other parameters. In constructions of appropriate economic models, we should take into account a possible dependence of economic processes on memory effects predict the dynamics of economic processes the dynamics of economic processes.

Part II: **Concepts of economics with memory**

2 Concepts of marginal values with memory

In this chapter, we consider the generalized concepts of marginal values of economic indicator, which are proposed in works [424, 416, 425, 439, 446]. The proposed concepts take into account the dependence of economic processes of state change on a finite time interval. The proposed concept of marginal value with memory (or hereditary value of economic indicator) contains the standard concepts of average and marginal values of indicators as special cases, when the parameter of memory fading is equal to zero and one, respectively. This chapter is based on articles [424, 416, 425, 439, 446].

2.1 Introduction

The standard concept of marginal value includes but is not limited to the following notions:
- the marginal utility (MU);
- the marginal rate of substitution (MRS);
- the marginal benefit (MB);
- the marginal cost (MC);
- the marginal product (MP);
- the marginal product of labor (MPL);
- the marginal product of capital (MPK);
- the marginal rate of transformation (MRT);
- the marginal revenue product (MRP);
- the marginal propensity to save (MPS);
- the marginal propensity to consume (MPC);
- the marginal tax rate (MTR).

The generalizations of standard marginal values were proposed in articles [424, 416, 425, 439, 446]. These generalizations take into account the memory, which is described as a dependence of economic processes on changes in the past in a finite time interval. For generalization of standard concepts of marginal and average values, we use a mathematical tool of derivative of arbitrary (noninteger and integer) order. The standard average value and marginal value are special cases of the proposed concept, when the parameter of memory fading is equal to zero and one, respectively. The proposed concept allows us to use a wide range of characteristics of economic processes that are intermediate between the average and marginal values.

2.2 Economic behavior of consumers and presence of memory

In this chapter, we discuss the concept of marginal utility and methods for describing economic processes that take into account the dependence of the present state of

https://doi.org/10.1515/9783110627459-002

the process on all previous states on a finite time interval. The necessity of taking into account the memory of consumers in the models of economic behavior of consumers is shown. Using the notion of a derivative of a noninteger order, generalizations of marginal utility are defined, which make it possible to describe the behavior of economic processes with memory.

In the seventies of the XIX century, marginalist economic theory arose. In the center of this theory was the consumer with his needs [263, 265, 264, 491]. The subjective motivation of the economic behavior of the consumers was adopted as the starting point of this theory [263, 265, 264, 491].

To describe this behavior, the economics began to use the concept of marginal values and mathematical apparatus of integer-order derivatives.

The use of derivatives of only integer orders means that the presence of memory in consumers is neglected in describing the process. In other words, in the standard marginalist economic theory all consumers are assumed to be suffering from complete amnesia. If we assume that economic agents and consumers have a memory of the past of the economic process, in which they participate, then it is necessary to take into account all previous states of this process. Mathematically, this means that to describe such a process, it is not enough to know the initial values of the state parameters and the finite number of derivatives of the integer order in time for these parameters. To describe the processes with memory, it is necessary to know the history of changes in the state of this process during the previous finite period of time. This is due to the fact that the current state of the economic process with memory depends on the previous history of changes in this process.

The economic behavior of consumers in many economic processes can be determined by the presence of a memory. In this case, the economic analysis should use a generalization of marginal indicators, which takes into account memory.

In fact, the marginalistic approach to the description of economic processes implicitly assumes the presence of memory among economic agents and consumers. To confirm this, we can refer to the work "Principles of Economics" [263, 265, 264] of 1871, written by Carl Menger, who is one of the founders of marginalism. In this work, Menger developed a theory of value and marginal utility.

In his book [263, 265, 264], Menger defined the concept of value as follows:

"Value is thus the importance that individual goods or quantities of goods attain for us because we are conscious of being dependent on command of them for the satisfaction of our needs," [265, p. 115] and [264, p. 125].

It is obvious that the awareness of the existence of a good implies that the conscious consumer remembers the presence or absence of this good when forming its economic behavior. Menger explained the following.

"Value is therefore nothing inherent in goods, no property of them, but merely the importance that we first attribute to the satisfaction of our needs, that is, to our lives and well-being, and in consequence carry over to economic goods as the exclusive causes of the satisfaction of our needs," [265, p. 116] and [264, p. 128].

He further writes: *"Value is thus nothing inherent in goods, no property of them, nor an independent thing existing by itself. It is a judgment economizing men make about the importance of the goods at their disposal for the maintenance of their lives and well-being. Hence value does not exist outside the consciousness of men,"* [265, pp. 120–121] and [264, p. 132].

As a result, value is a judgment about the good that exists in the mind of the consumer. At the same time, the very fact of the existence of any judgment in consciousness for a certain finite period of time presupposes the presence of memory.

According to Menger: *"The value an economizing individual attributes to a good is equal to the importance of the particular satisfaction that depends on his command of the good,"* [265, p. 146] and [264, p. 158].

As a result, we can formulate the following basic principle of economic theory [424, p. 109].

Principle 2.1 (Principle of nonamnesia consumer). *Economic theory should assume that the consumer ("economizing individual") must remember whether he satisfied his need earlier or not. The buyer should remember that he recently bought the goods. He can remember which products (or services) he bought earlier, in what quantity and at what price. Whether the buyer will buy new products or not may be determined by what the price of this product was previously, and how it has changed over the past period of time. Consumer behavior will also be determined by how the quantity of goods already purchased (acquired) by him has changed, and how the quantity of goods and services offered to him has changed. Moreover, the consumer himself and his judgments about the value of certain goods may change.*

Therefore, in the description of the economic behavior of consumers, it is necessary to take into account the memory effects. The behavior models of consumers with amnesia do not fully reflect real economic processes. Placing the consumer with his needs in the center of marginal economic theory, it is implied that this consumer has a memory, and his behavior is determined not only by his state at a given time. However, the use of standard derivatives of the integer orders in the definition of marginal utility allows one to take into account only an infinitely small neighborhood of a given point in time, and thereby neglect the memory effect.

As a result, we can formulate the following statement.

Statement 2.1. *In the center of the marginal economic theory, there should not be a consumer with amnesia, but a consumer with memory. It is necessary to take into account that the consumer can remember changes on a certain finite interval of time, both the goods themselves and their value, and the results of satisfying their needs with these goods.*

In view of the above, we can formulate new concept of "consumer with memory" ("economizing individual with memory").

Definition 2.1. Consumer with memory is a consumer that is characterized by the following properties:

1. **Consumer remembers.** A consumer with memory remembers whether he satisfied his needs earlier or not, remembers what products or services he bought earlier, in what quantity and at what price. A consumer with memory remembers how the price of products has changed over a certain period of time, how has the number of products and services already purchased by him changed during this time interval. The consumer remembers how his judgments about the value and cost of products and services changed.
2. **Consumer forgets.** The consumer's memory is characterized by fading. A consumer with memory gradually over time forgets the history of purchases of products and services, their cost and price, and the value that these products and services had for the consumer in the past.
3. **Memory and memory fading affects consumer behavior.** A consumer with memory makes decisions about purchasing a product or service at the present moment in time, based not only on the value, cost, quality and other parameters of these products and services at the moment (at the time of purchase), but also taking into account the history of changes in these parameters in the past for a certain period of time. At the same time, the past is not fully taken into account due to the fading of memory. A consumer with a memory may have poor memories of ordinary purchases of products and services made in the distant past.

Remark 2.1. It is known that Menger was negative about the use of mathematical methods in utility theory, insisting that the function of economics is to investigate entities, not quantitative relationships and dependencies in economic phenomena. We can assume that this was largely due to the fact that the mathematical apparatus used at that time did not allow us to take into account the effects of memory, and many believed that memory effects relate only to psychology. Note that the first mathematical description of physical system with memory has been given by Ludwig Boltzmann in 1874 and 1876, that is, 3 and 5 years after the publication of the Menger's book in 1871. In economics, the first mathematical description of the memory effects will be formulated almost a century later.

2.3 Consumer behavior, utility and memory effect

In modern economic theory, consumer behavior models and the notion of utility are used. The main attention is focused on the study of models of consumer demand, analysis of supply, market research and pricing at the microeconomic level [490]. The theory of consumer behavior used some assumptions. First of all, it is assumed that the consumer always behaves rationally, i. e., the consumer seeks to get the maximum utility for himself, while "utility" is a subjective concept. The most important characteristic of utility lies in the fact that utility has no physical or material existence, since utility exists in the mind of the consumer. In this place, we can assume that the consumer minds have a memory.

Statement 2.2. *The theory of consumer behavior should assume the utility exists in the consumer mind that has a memory. This theory should be based on the concept of a consumer with memory.*

The utility of a good (product or service) lies in its ability to satisfy any need of the consumer. In this case, the consumer must remember that his need is satisfied. A consumer with amnesia does not remember this. If we assume that each consumer may have its own utility rating, which does not have to coincide with the average, then we can use memory with a distributed fading parameter.

One of the first to express ideas, which then became an important part of the marginal utility theory, was the German economist Hermann H. Gossen [144, 145]. He proposed a law on saturation of needs (the first law of Gossen). According to this law, with satisfaction of the need for any good, its value decreases or as the quantity of goods increases, its utility decreases. In other words, the first law of Gossen states that with consistent consumption, the utility of each subsequent unit of product is lower than the previous one. The first law of Gossen was generalized for marginal utility, and was formulated as a law of decreasing marginal utility. This law reflects the relationship between the amount of good consumed and the degree of satisfaction with the consumption of each additional unit. In this place, we can state the following.

Statement 2.3. *To describe behavior of consumer with memory, we should generalize the standard concept of marginal utility to takes into account a memory.*

The law of diminishing marginal utility is that with an increase in the consumption of a good (with a constant consumption of all the others goods), the overall utility, which is received by the consumer, increases, but increases more and more slowly. Mathematically, this is described by the following properties of the utility function: the first derivative of the utility function in terms of the quantity of a given good is positive, and the second derivative is negative. In other words, the law of diminishing marginal utility states that the utility function is increasing and convex up. The property of diminishing marginal utility is mathematically represented by the concavity of the utility function.

Statement 2.4. *We can say that the first derivative is positive, and the second derivative of this function is negative, only if the utility function is single-valued and twice differentiable. In general, it is impossible to assume the uniqueness of the functions describing the utility.*

As a result, we can state the following.

Statement 2.5. *For a correct description of consumer behavior with memory, a generalized concept of marginal utility should be applied.*

As with any law, the law of decreasing marginal utility has its limitations on applicability of the standard theory of consumer behavior.

- The first limitations of the law of diminishing marginal utility are largely due to the fact that the utility function can be unique only on a short time interval. Among the restrictions on the applicability of the law is the absence of changes in the consumer. This means that there should not be a change in tastes, habits, customs, preferences and incomes of the consumer. Changing one of these factors may change the utility of the product, and the consumption function will become ambiguous. This ambiguity can be largely due to the presence of memory in the consumer. It is this ambiguity that leads to the need to generalize marginal utility.
- The second limitation is the smallness of the time interval. All units of the product must be consumed within a short period of time. If there were some time intervals between the consumption of a product or service, then the consumption of subsequent units of the product can give satisfaction equivalent to the satisfaction obtained with the consumption of the previous units of the product. At small time intervals, the memory of past events can sometimes be ignored. The increase in the time interval leads to the need to take into account the influence of memory and memory fading on consumer behavior.
- The third limitation is also the absence of changes in the consumed goods or services. This means that goods and services should not be changed in price, quality, prevalence and availability in time and space.
- The fourth important limitation is price consistency. Product prices should remain the same for the period of time under consideration. The consumer can refuse to purchase or increase purchases only because of changes in the price of a product or service, and not because the assessment of the utility of the product has changed.
- The fifth important limitation of the standard concept is the absence of memory in consumers, that is, all consumers have lost their memory and have amnesia.

It is obvious that in real economic processes the simultaneous fulfillment of these five conditions is rarely encountered.

The fact that the utility function can be ambiguous, i. e., it is not a single-valued function, is largely due to the presence of memory among consumers. In addition, the decrease in marginal utility over a short time interval is also largely due to the presence of memory in the consumer. The consumer has a memory and remembers that he purchased this product (product or service) in the past period of time. Therefore, taking into account the effects of memory in modern economic theory, using consumer behavior models and the concept of utility, is necessary for a more adequate description of economic processes. For this, we can use a new type of marginal utility, which can be called the marginal utility with memory (the hereditary marginal utility). The hereditary (the presence of memory) means that this indicator takes into account not only the present state of the economic process and the infinitely close previous state, but also all previous states in which the process was located.

The hereditary marginal value (marginal value with memory) should take into account that the state of the economic process at the present time depends on the state of this process in the past. This dependence is largely due to the memory of economic agents, that the subjects remember the changes in the economic process over a certain period of time.

2.4 Standard average and marginal values and ambiguity

One of the most important tasks of economic theory is the study and description of economic processes. For this, different economic concepts are used, including the concepts of the averages and marginal values.

The concept of marginal value allows the use of standard mathematical calculus, which includes the theory of derivatives of integer orders, to describe changes in economic processes. Usually, the marginal value is defined as the first-order derivative of the function of a certain index with respect to its determining factor. The standard marginal value shows the increase in the corresponding indicator per unit increase in its determining factor.

In 2016, we proposed a new concept [424, 416, 425] (see also articles [439, 446, 323]) that expands the range of tools of economic analysis by using derivatives of non-integer order to describe economic processes with memory.

Statement 2.6. *To define a generalized indicator, it is not enough to formally replace the derivative of the first order by the derivative of the fractional (non-integral) order, since memory effects can be associated with ambiguity in the dependence of the studied indicator on the determining factor.*

In order to explain this statement [386], we give standard definitions of the marginal and average values that do not take into account the effects of memory.

In the study of economic processes, as a rule, the average and marginal values are calculated for the various economic indicators, which are presented as functions of some factors.

Let a single-valued function $Y = Y(X)$ be given, which describes the dependence of the economic indicator Y on some factor X. A single-valued function is a function that for each point in the domain has a unique value in the range. The average value (AY_X) of the indicator Y is defined as the ratio of the function $Y = Y(X)$ to the corresponding value of the factor X, i. e., we have the equality

$$AY_X = \frac{Y(X)}{X}, \tag{2.1}$$

where $X \neq 0$.

To define the standard marginal value for the function $Y(X)$, this function must be differentiable. A differentiable function $Y(X)$ of one real variable X is a function whose derivative exists at each point in its domain.

If the given function $Y = Y(X)$ is single-valued and differentiable, then the marginal value (MY_X) of the indicator Y is determined as the first derivative of the function $Y = Y(X)$ with respect to the factor X in the form

$$MY_X = \frac{dY(X)}{dX}. \tag{2.2}$$

The most important condition for the applicability of equation (2.2), which determines the marginal value, is the assumption that the indicator Y can be represented as a single-valued function of the factor X. In general, this assumption is not satisfied [424, 416, 425, 415] (see also [439, 446, 323]), and the dependence of Y on X is ambiguous (multivalued), that is, several different values of Y can correspond to the same value of X.

Let us give example of the multi-valued dependency of an indicator Y on a factor X, using results of our article [419] (see also [439, 446]). This dependency describes the volume of exchange trading Y in millions of dollars as a function of the weighted average rate of X dollar to ruble in 2015, when interpolation by fourth-degree polynomials is used [419] (see also [439, 446]). Let us consider the indicator and the factor as functions of time t that are given by the equations

$$X(t) = 72.65 - 0.579t + 5.383\,10^{-3}t^2 - 1.471\,10^{-5}t^3 + 8.185\,10^{-9}t^4 \tag{2.3}$$

$$Y(t) = 1687 - 23.69t + 0.5103t^2 - 3.491\,10^{-3}t^3 + 7.496\,10^{-6}t^4. \tag{2.4}$$

Equations (2.3) and (2.4) define the parametric dependence of Y on X, which is represented graphically in Figure 2.1.

From graph presented in Figure 2.1, it is clear that the value of X may correspond to more than one value of Y in different points of the X-axis.

In many cases, the indicator Y cannot be represented as a single-valued function of the factor X. If the dependence of the indicator Y on the factor X is not single-valued function, we cannot use equations (2.1) and (2.2) to calculate the average and marginal values of indicator. However, it is possible to avoid this problem in the economic analysis. In many cases, the indicator and the factor can be considered as single-valued functions of time. Therefore, to define the marginal and average values of the indicators, we can use parametric dependence of an indicator Y on a factor X in the form of the single-valued functions $Y = Y(t)$ and $X = X(t)$, where the parameter t is time.

If the dependence of the indicator Y on the factor X is multivalued, then equation (2.1) cannot be applied. In this case, we should use the single-valued functions $Y = Y(t)$ and $X = X(t)$ to define the average and marginal values at time t (the T-indicators or Temporal-indicators).

As a result, the average and marginal values of indicators for t can be defined by the following definition.

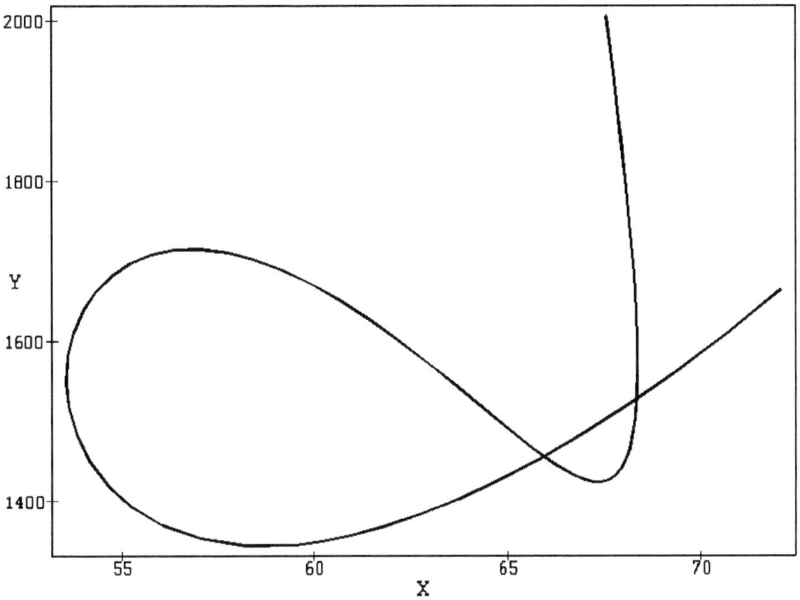

Figure 2.1: The dependence of Y on X, which are defined by equations (2.3) and (2.4) for $t \in [0, 200]$. The value $X(t)$ is plotted along the X-axis, and the value $Y(t)$ is plotted along the Y-axis.

Definition 2.2. Let $Y = Y(t)$ and $X = X(t)$ are single-valued differentiable functions, which describe the parametric dependence of the economic indicator Y on some factor X. Then the standard average and marginal values at time t (T-indicators, Temporal-indicators) are defined by the equations

$$AY_X(t) = \frac{Y(t)}{X(t)}, \tag{2.5}$$

$$MY_X(t) = \frac{dY(t)/dt}{dX(t)/dt}, \tag{2.6}$$

where $X(t) \neq 0$ and $dX(t)/dt \neq 0$.

If the dependence of $Y(t)$ on $X(t)$ can be represented as a single-valued differentiable function $Y = Y(X)$, by excluding the time parameter t, then equations (2.2) and (2.6) will be equivalent by the standard chain rule

$$\frac{dY(X(t))}{dt} = \left(\frac{dY(X)}{dX}\right)_{X=X(t)} \frac{dX(t)}{dt}. \tag{2.7}$$

In other words, if the function $X = X(t)$ is invertible in a neighborhood of the point t, then

$$MY_X(t) = \frac{dY(t)/dt}{dX(t)/dt} = \frac{dY(X)}{dX} = MY_X. \tag{2.8}$$

In this case, equations (2.1) and (2.5) will also be equivalent

$$AY_X(t) = \frac{Y(t)}{X(t)} = \frac{Y(X)}{X} = AY_X. \tag{2.9}$$

In order for equations (2.1) and (2.5) to be satisfied, it is sufficient that the function $X = X(t)$ be reversible. In this case, equations (2.2) and (2.6) will also be equivalent.

In the general case, equalities (2.8) and (2.9) are not fulfilled and the dependence of $Y(t)$ on $X(t)$ is not representable as a single-valued differentiable function $Y = Y(X)$. Since the indicator and factor can be considered as unambiguous (single-valued) functions of time, the definition of the marginal value by equation (2.6) is more general.

In the differential calculus, the variable Y as function of argument X is called the given parametrically if both of these variables are functions of a third variable t. In our case, the third variable is time t. We emphasize that formula (2.6) is a standard definition of the parametric derivative of the first order, if the function $X = X(t)$ has an inverse function in a neighborhood of t and the functions $X = X(t)$ and $Y = Y(t)$ have the first derivatives. Moreover, equation (2.6) can be considered as a generalization of the parametric derivative of the indicator $Y = Y(t)$ by the factor $X = X(t)$ at time t, if $dX(t)/dt \neq 0$. As a result, equations (2.5) and (2.6) can be used for parametric dependences given by equations (2.3), (2.4). Expressions (2.1) and (2.2) cannot be used for these dependencies.

2.5 Ambiguity and memory effects

It should be noted that one of the reasons for the violation of single-valued property of indicators is the presence of memory among economic agents and processes [424, 416, 425, 439, 446, 415]. Economic agents may remember previous notable changes of the indicator $Y(t)$ and the factor $X(t)$. In this case, at repeated similar changes the agents can react on these changes in a different way than it did before. As a result, the indicator will be different, despite similar changes in the factor. This leads to the fact that the same value of the factor corresponds to a number of different values of the indicator, which is described as a many-valued function $Y = Y(X)$.

Memory leads to the fact that the marginal values of an economic indicator at the time t can depend on all changes of the variables $Y(t)$ and $X(t)$ on a finite time interval preceding the considered moment of time t. The standard average and marginal values of indicator depend only on the given time t and its infinitesimal neighborhood. Therefore, the standard definitions of these values are applicable only on the condition that all economic agents have full amnesia. It is obvious that this assumption is not always possible to use in economic analysis.

Because of this, it is necessary to generalize the concept of marginal value to take into account the effects of memory in economic processes. Let us emphasize that using equation (2.6) does not solve the problem, because it uses derivatives of integer order, which are determined by the behavior of the functions $X = X(t)$ and $Y = Y(t)$ in an infinitely small neighborhood of the time instant t, which means instant forgetting (amnesia) of all the changes that were made before. Mathematically, this problem caused by the use of derivatives of integer orders.

To describe the processes with memory, mathematical tools should allow us to take into account previous changes of variables $Y(t)$ and $X(t)$ in a finite time interval preceding the considered moment t.

First, we give a definition of marginal value with memory in a general form by using the operators introduced earlier to describe the effects of memory.

Definition 2.3. Let the variables $X = X(t)$ and $Y = Y(t)$ be single-valued differentiable m and n times, respectively, functions with respect to time variable $t \in [t_i; t_f]$. The generalized $\mathcal{F}$-marginal value $MY^{(n,m)}_{\mathcal{F},X}([t_i,t])$ of $Y = Y(t)$ with respect to $X = X(t)$ for processes with memory is defined by the equation

$$MY^{(n,m)}_{\mathcal{F},X}([t_i,t]) = \frac{\mathcal{F}^t_{t_i}(Y^{(n)}(\tau))}{\mathcal{F}^t_{t_i}(X^{(m)}(\tau))}, \tag{2.10}$$

where $\tau \in [t_i, t]$, and $t_i < t < t_f$, $\mathcal{F}^t_{t_i}(X^{(m)}(\tau)) \neq 0$.

If the memory can be described by the memory functions, then this definition will take the following form.

Definition 2.4. Let the variables $X = X(t)$ and $Y = Y(t)$ be single-valued differentiable m and n times, respectively, functions with respect to time variable $t \in [t_i; t_f]$ and the memory is described by memory functions $M_x(t,\tau)$ and $M_y(t,\tau)$ for X and Y, respectively. Then the generalized M-marginal value $MY^{(n,m)}_{M,X}([t_i,t])$ of $Y = Y(t)$ with respect to $X = X(t)$ for processes with memory is defined by the equation

$$MY^{(n,m)}_{M,X}([t_i,t]) = \frac{\int_{t_i}^t M_y(t,\tau)Y^{(n)}(\tau)\,d\tau}{\int_{t_i}^t M_x(t,\tau)X^{(m)}(\tau)\,d\tau}, \tag{2.11}$$

where $\tau \in [t_i, t]$, and $t_i < t < t_f$, the expression in the denominator of the fraction is assumed to be nonzero.

The proposed generalizations of marginal value are formal in a sense. The following are examples for memory with power-law fading. For this, we need the Caputo fractional derivatives and their properties. In the next subsection, we describe these operators and some of their properties.

2.6 Caputo fractional derivative and its properties

As a mathematical tool for generalizing the marginal values of economic indicators, we can use the theory of fractional derivatives, derivatives of noninteger order [335,

202, 308, 200, 90, 164, 165]. Note that fractional derivatives are a generalization of the concept of derivatives of integer orders, since the fractional derivatives with integer values of the orders coincide with the standard integer-order derivatives. There are various types of derivatives of noninteger orders. For simplification, we will the Caputo fractional derivative. One of the distinguishing features of this derivative is that its action on the constant function gives zero. The use of the Caputo derivative produces zero value of marginal value of noninteger order for the constant functions of the corresponding indicator. There are left-side and right-side fractional derivatives. We will use only left-sided derivatives, since the economic process at time t depends only on the state changes of this process in the past, that is, for $\tau < t$. The right-sided fractional derivative at time t is determined by the values of function $X(\tau)$ for $\tau > t$.

Let us give the definition of the Caputo fractional derivative [200, 90].

Definition 2.5. The left-sided Caputo fractional derivative of order $\alpha \geq 0$ is defined by the equation

$$(D^{\alpha}_{C;t_i+}X)(t) = \frac{1}{\Gamma(n-\alpha)} \int_{t_i}^{t} \frac{X^{(n)}(\tau)\, d\tau}{(t-\tau)^{\alpha-n+1}}, \tag{2.12}$$

where $\Gamma(\alpha)$ is the gamma function, $t > 0$, and $X^{(n)}(\tau)$ is standard derivative of integer order n of the function $X(\tau)$ in time τ. Here, $n = [\alpha] + 1$ for noninteger values of α and $n = \alpha$ for integer α. Operator (2.12) exists if $X(\tau) \in AC^n[t_i, t_f]$, i. e., the function $X(\tau)$ has integer-order derivatives up to $(n-1)$-th order, which are continuous functions on the interval $[t_i, t_f]$, and the derivative $X^{(n)}(\tau)$ is Lebesgue summable on the interval $[t_i, t_f]$.

Remark 2.2. For positive integer values $\alpha = n$, the Caputo derivative coincides (see Theorem 2.1 in [200, p. 92]) with the standard derivative of integer order n, i. e., we have the equality

$$(D^{n}_{C;0+}X)(t) = \frac{d^n X(t)}{dt^n}, \quad (D^{0}_{C;0+}X)(t) = X(t). \tag{2.13}$$

To calculate the Caputo derivatives of the order $\alpha \geq 0$ and $t_i = 0$ for a power function, we can use the following equations (see Example 3.1 of book [90, p. 49] or see Property 2.16 in [200, p. 95], where $\beta - 1 - \alpha$ should be used instead of $\beta - 1$ on the right sides of equation (2.4.28)) that have the form

$$(D^{\alpha}_{C;0+}\tau^{\beta})(t) = \frac{\Gamma(\beta+1)}{\Gamma(\beta-\alpha+1)} t^{\beta-\alpha}, \quad (t \geq 0, \beta - \alpha \geq 0), \tag{2.14}$$

$$(D^{\alpha}_{C;0+}\tau^{k})(t) = 0, \quad (k = 0, 1, \ldots, n-1), \tag{2.15}$$

where $n = [\alpha] + 1$ for noninteger values of α and $n = \alpha$ for integer α. In the special case, we have

$$(D^{\alpha}_{C;0+}\tau^{\alpha})(t) = \Gamma(\alpha+1), \quad (D^{\alpha}_{C;0+}1)(t) = 0. \tag{2.16}$$

Note that the exponential function is not invariant with respect to the action of the left-sided Caputo derivative of noninteger order [200, p. 98]. The generalization of the exponential function is the Mittag–Leffler function $E_\alpha[\lambda\tau^\alpha]$ [143], for which the property $E_1[\lambda t] = e^{\lambda t}$ is valid for any real values of λ. The Mittag–Leffler function $E_\alpha[\lambda t^\alpha]$ is invariant (see equation (2.4.58) in [200, p. 98]) with respect to the action of the left-sided Caputo derivative of order $\alpha > 0$, i. e., we have the equality

$$(D^{\alpha}_{C;0+} E_\alpha[\lambda\tau^\alpha])(t) = \lambda E_\alpha[\lambda t^\alpha], \tag{2.17}$$

where $\lambda \in \mathbb{C}$ or $\lambda \in \mathbb{R}$ (see Lemma 2.23 in [200, p. 98]).

To describe processes with memory, we can also use parametric derivatives of noninteger order, which are usually called the fractional derivatives of function with respect to another function. The parametric Riemann–Liouville derivatives are considered in Section 18.2 of book [335] and in Section 2.5 of monograph [200, pp. 99–105]. The Caputo parametric fractional derivative was proposed in Definition 3 of article [418, p. 224]. Then the properties of this derivative were described in article [13].

Definition 2.6. Let $X(\tau)$ be an increasing positive monotonic function having a continuous derivative $X^{(1)}(t)$. Then the Caputo parametric fractional derivative of the order $\alpha \geq 0$ is defined by the equation

$$(D^{\alpha}_{X(t)} Y)(t) = \frac{1}{\Gamma(n-\alpha)} \int_0^t d\tau \frac{X^{(1)}(\tau)}{(X(t) - X(\tau))^{\alpha+1-n}} \left(\frac{1}{X^{(1)}(\tau)} \frac{d}{d\tau} \right)^n Y(\tau), \tag{2.18}$$

where $n - 1 < \alpha \leq n$ and $0 < \tau < t$.

If $X(t) = t$, then equation (2.18) takes the form

$$(D^{\alpha}_{X(t)} Y)(t) = (D^{\alpha}_{C;0+} Y)(t). \tag{2.19}$$

For the parametric Caputo derivative (2.18) of the function $Y(t) = (X(t) - X(0))^\beta$, we have the equation

$$(D^{\alpha}_{X(t)} Y)(t) = \frac{\Gamma(\beta+1)}{\Gamma(\beta-\alpha+1)} (X(t) - X(0))^{\beta-\alpha}, \tag{2.20}$$

where $\beta - \alpha > 0$.

2.7 Generalization of concepts of average and marginal values

2.7.1 One-parameter marginal values with memory

Using the left-sided Caputo derivative, we can define a generalization of the concepts of the marginal value (2.6) and the average value (2.5), which allows us to take into account the power-law memory. The proposed generalized indicator can be called the marginal value with memory or the hereditary marginal value. Hereditarity refers to the dependence on changes in $Y(t)$ and $X(t)$ in a finite time interval preceding the considered moment. This dependence occurs when the behavior of economic agents is determined by the presence of the memory of changes in $Y(t)$ and $X(t)$. As a result, we can give the following definition [416].

Definition 2.7. Let the economic indicator $Y = Y(t)$ and its determining factor $X(t)$ are functions of time $t > 0$, for which there is the left-sided Caputo derivative (2.12) of the order $\alpha \geq 0$ and $t_i = 0$. Then one-parameter the marginal value with memory $MY_X(\alpha, [0,t])$ (the hereditary T-indicator $HY_X(\alpha, [0,t])$ of order $\alpha \geq 0$, the marginal value of order α) at time t is defined by the equation

$$MY_X(\alpha, [0,t]) = \frac{(D^{\alpha}_{C;0+}Y)(t)}{(D^{\alpha}_{C;0+}X)(t)}, \tag{2.21}$$

where $(D^{\alpha}_{C;0+}X)(t) \neq 0$, and $D^{\alpha}_{C;0+}$ is the left-sided Caputo fractional derivative (2.12) of the order $\alpha \geq 0$, which characterizes the memory fading. Here, it is assumed that $X(t), Y(t) \in AC^n[t_i, t_f]$, where $n - 1 < \alpha \leq n$ and $t_i = 0$.

Remark 2.3. Note that due to the violation of the standard chain rule (see Section 2.7.3 of [308, pp. 97–98] and [377, 81]) for derivatives of noninteger orders α, equation (2.21) cannot be represented as a parametric fractional derivative of the function $Y(t)$ with respect to another function $X(t)$ even if there is an inverse function for $X = X(t)$.

The standard marginal value (2.6) of the economic indicator takes into account changes of the indicator and factor in an infinitely small neighborhood of time t. The proposed concept of marginal value with memory (2.21) allows us to describe the dependence of economic processes on all changes of $Y(t)$ and $X(t)$ on a finite time interval, and not only in an infinitely small neighborhood of time t. This allows us to apply the suggested concept to describe economic processes with memory. The proposed marginal value with memory makes it possible to take into account in economic analysis the fact that consumers (and economic agents) remember what the values of the indicator and factor were earlier.

Statement 2.7. *The concept of the marginal value with memory $MY_X(\alpha, [0,t])$ (the hereditary T-indicator $HY_X(\alpha, [0,t])$ of order $\alpha \geq 0$) which is defined by equation* (2.21), *includes, as special cases, the standard concepts of the average value and the marginal value, which are defined by equations* (2.5) *and* (2.6). *Mathematically, this is expressed i the form of the identities*

$$MY_X(0, [0,t]) = AY_X(t), \quad MY_X(1, [0,t]) = MY_X(t). \tag{2.22}$$

Proof. To prove this statement, we consider special cases of the proposed general marginal value (2.21), when the single-valued function $Y = Y(X)$ exists.

If $\alpha = 0$, then using the equality $(D^0_{C;0+}Y)(t) = Y(t)$ [200, p. 92], hereditary T-indicator (2.21) gives the standard average value

$$MY_X(0,[0,t]) = \frac{(D^0_{C;0+}Y)(t)}{(D^0_{C;0+}X)(t)} = \frac{Y(t)}{X(t)} = AY_X(t). \tag{2.23}$$

If $\alpha = 1$, then using the equality $(D^1_{C;0+}Y)(t) = dY(t)/dt$ [200, p. 92], hereditary T-indicator (2.21) gives the standard marginal value

$$MY_X(1,[0,t]) = \frac{(D^1_{C;0+}Y)(t)}{(D^1_{C;0+}X)(t)} = \frac{dY(t)/dt}{dX(t)/dt} = MY_X(t), \tag{2.24}$$

which is defined by the parametric derivative of the first order for the indicator $Y(t)$ by the factor $X(t)$. □

As a result, average value (2.5) and marginal value (2.6) are special cases of the hereditary T-indicator (2.21) of the order $\alpha \geq 0$.

Statement 2.8. *Hereditary T-indicator* (2.21) *allows us to consider not only the standard average and marginal characteristics of economic processes, but also characteristics that are intermediate* ($0 < \alpha < 1$) *between the standard average and marginal values. The proposed marginal value* (2.21) *with memory (the marginal value of order* α*) includes the whole range of intermediate characteristics from the average value to the marginal value.*

Note that equation (2.21) can be applied to the parametric dependence of Y on X, given by equations (2.3) and (2.4). To do this, it is enough to apply equations (2.14) and (2.15), which allow us to calculate the Caputo derivatives of order $\alpha \geq 0$ for power-law functions.

2.7.2 Two-parameter marginal values with memory

In more general cases, to describe economic processes with memory, we can take into account that memory fading parameters of the indicator $Y(t)$ and the factor $X(t)$ do not coincide. In this case, two-parameter marginal value should be used.

Definition 2.8. Let the economic indicator $Y(t)$ and its determining factor $X(t)$ be functions of time t, for which there are the left-sided Caputo derivatives (2.12) of orders $\alpha \geq 0$ and $\beta \geq 0$, respectively. Then the two-parameter marginal value with memory (the two-parameter hereditary T-indicator) $HY_X(\alpha,\beta,[0,t])$ for the time t is defined by the equation

$$HY_X(\alpha,\beta,[0,t]) = \frac{(D^\alpha_{C;0+}Y)(t)}{(D^\beta_{C;0+}X)(t)}, \tag{2.25}$$

where α is the memory fading parameter of the indicator $Y(t)$, and β is the memory fading parameter of the factor $X(t)$ and $(D^\beta_{C;0+}X)(t) \neq 0$. Here, it is assumed that $Y(t) \in$

$AC^n[t_i, t_f]$, where $n - 1 < \alpha \leq n$, and $X(t) \in AC^m[t_i, t_f]$, where $m - 1 < \beta \leq m$ and $t_i = 0$.

In case of equality of parameters $\alpha = \beta$, equation (2.25) gives expression (2.21), that is,

$$HY_X(\alpha, \alpha, [0, t]) = MY_X(\alpha, [0, t]). \tag{2.26}$$

The one-parameter marginal value (2.21) can be used only if the fading of memory about change in the indicator $Y(t)$ and the factor $X(t)$ are the same.

2.7.3 General marginal values through parametric derivative

The generalized marginal values can be also defined by the using the parametric Caputo fractional derivative. If $Y(t)$ can be represented as a single-valued function of $X(t)$, then we can define the marginal value with memory by using the parametric Caputo fractional derivatives (2.18), which is the fractional derivatives of the function $Y(t)$ with respect to the function $X(t)$.

Definition 2.9. Let $Y(t)$ be a function that can be represented as a single-valued function of $X(t)$, which is an increasing positive monotonic function having a continuous derivative $X^{(1)}(t)$. Then the marginal value of the order $\alpha \geq 0$ can be defined by the equation

$$MY_X(\alpha, t) = (D^{\alpha}_{X(t)} Y)(t), \tag{2.27}$$

where $D^{\alpha}_{X(t)}$ is the Caputo parametric derivative (2.18).

Remark 2.4. In the case $Y(t) = X(t)$ and $Y(t) = X(t) - X(0)$, using equation (2.20) for expressions (2.27) and $(D^{\alpha}_{X(t)} X(0))(t) = 0$, we get

$$MX_X(\alpha, t) = \frac{1}{\Gamma(2 - \alpha)} (X(t) - X(0))^{1-\alpha}. \tag{2.28}$$

In order to the marginal value of noninteger order gives one at $Y(t) = X(t)$, i. e., $MX_X(\alpha, t) = 1$, we can define the marginal value by the equation

$$MY_X(\alpha, t) = \frac{(D^{\alpha}_{X(t)} Y)(t)}{(D^{\alpha}_{X(t)} X)(t)} \tag{2.29}$$

instead of equation (2.27).

In definition (2.27), we use the fractional derivative of a function with respect to another function, which can be considered fractional analog of the standard parametric derivative of first order. However, it should be emphasized that this fractional derivative exists only under the condition that the function is an increasing, positively monotonic function.

As a result, the "parametric" marginal values (2.27) and (2.29) are essentially restricted in application. This restriction is caused by the following conditions: the function $Y = Y(t)$ must be representable as a single-valued function of $X = X(t)$; the function $X(t)$ must be positive and monotonically increasing. These conditions are rarely implemented in the economic processes.

Therefore, to describe the properties of economic processes with memory, we should use two-parameter and one parameter marginal values with memory, which are defined by equations (2.25) and (2.21).

2.8 Examples of calculations of generalized marginal values

Let us give illustrative examples of calculation of the average, marginal and hereditary values of an economic indicator for the simple case. We first give an example of calculating the standard average and marginal values.

Let the economic indicator $Y = Y(t)$ and the factor $X(t)$ be described by the functions

$$X(t) = a_0 t, \quad Y(t) = a_1 t + a_2 t^2 - a_3 t^3, \tag{2.30}$$

where $a_0 = 5$, $a_1 = 175$, $a_2 = 10$ and $a_3 = 6.25$. We can calculate the average and marginal values of the indicator Y in the absence of memory for a moment of time equal to $t = 2$, which corresponds to a factor X with value of 10 units ($X(2) = 10$).

The function of the average value of the indicator has the form

$$AY_X(t) = \frac{Y(t)}{X(t)} = \frac{a_1}{a_0} + \frac{a_2}{a_0} t - \frac{a_3}{a_0} t^2 = \frac{a_1}{a_0} + \frac{a_2}{(a_0)^2} X - \frac{a_3}{a_0^3} X^2. \tag{2.31}$$

For $t = 2$, the average value (2.31) is equal to

$$AY_X(2) = 175/5 + (10/5)2 - (6.25/5)2^2 = 34. \tag{2.32}$$

The function of the marginal value of the indicator Y is given by the equation

$$MY_X(t) = \frac{dY(t)/dt}{dX(t)/dt} = \frac{a_1}{a_0} + \frac{2a_2}{a_0} t - \frac{3a_3}{a_0} t^2 = \frac{a_1}{a_0} + \frac{2a_2}{(a_0)^2} X - \frac{3a_3}{(a_0)^3} X^2. \tag{2.33}$$

Substituting $t = 2$, we find that standard marginal value (2.33) is equal to

$$MY_X(2) = 175/5 + 2(10/5)2 - 3(6.25/5)2^2 = 28. \tag{2.34}$$

As a result, for $t = 2$ (the value of the factor X equal to 10), while neglecting the memory effect, we get the average value equal to 34 units, and the marginal value equal to 28 units.

We now consider examples of calculating the generalized (hereditary) T-indicators (2.21) and (2.25) for the economic process with memory. We first give a calculation of the hereditary value (2.21) of order $\alpha \in [0,1]$. For simplicity, let us use functions (2.30), which describe the dependence of the indicator and factor on time. Using the linearity of the Caputo derivative, we get

$$(D^{\alpha}_{C;0+}X)(t) = a_0(D^{\alpha}_{C;0+}\tau)(t), \tag{2.35}$$

$$(D^{\alpha}_{C;0+}Y)(t) = a_1(D^{\alpha}_{C;0+}\tau)(t) + a_2(D^{\alpha}_{C;0+}\tau^2)(t) - a_3(D^{\alpha}_{C;0+}\tau^3)(t). \tag{2.36}$$

Then, applying equation (2.14) of the derivative for the power function, we obtain the expression

$$(D^{\alpha}_{C;0+}X)(t) = \frac{a_0\Gamma(2)}{\Gamma(2-\alpha)}t^{1-\alpha}, \tag{2.37}$$

$$(D^{\alpha}_{C;0+}Y)(t) = \frac{a_1\Gamma(2)}{\Gamma(2-\alpha)}t^{1-\alpha} + \frac{a_2\Gamma(3)}{\Gamma(3-\alpha)}t^{2-\alpha} - \frac{a_3\Gamma(4)}{\Gamma(4-\alpha)}t^{3-\alpha}. \tag{2.38}$$

Substituting (2.37) and (2.38) into equation (2.21) and using $\Gamma(n+1) = n!$, we get

$$MY_X(\alpha,[0,t]) = \frac{(D^{\alpha}_{C;0+}Y)(t)}{(D^{\alpha}_{C;0+}X)(t)} = \frac{a_1}{a_0} + \frac{2a_2\Gamma(2-\alpha)}{a_0\Gamma(3-\alpha)}t - \frac{6a_3\Gamma(2-\alpha)}{a_0\Gamma(4-\alpha)}t^2. \tag{2.39}$$

As a result, the one-parameter T-indicator (2.21) is equal to

$$MY_X(\alpha,[0,t]) = \frac{a_1}{a_0} + \frac{2a_2}{a_0(2-\alpha)}t - \frac{6a_3}{a_0(2-\alpha)(3-\alpha)}t^2. \tag{2.40}$$

Here, we use the equality $\Gamma(z+1) = z\Gamma(z)$ in the form

$$\Gamma(4-\alpha) = (3-\alpha)\Gamma(2-\alpha)\Gamma(2-\alpha), \quad \Gamma(3-\alpha) = (3-\alpha)\Gamma(2-\alpha), \tag{2.41}$$

and $\Gamma(4) = 3! = 6$, $\Gamma(2) = 1! = 1$.

In Table 2.1, we present the numerical values for the one-parameter marginal values with memory (2.40), which are calculated by using equation (2.25) for different values of the memory fading parameter $\alpha \in (0,1)$.

Table 2.1: Values of one-parameter hereditary T-indicator (2.40) for different values of memory fading parameter α at $t = 2$.

α	1.0	0.9	0.8	0.7	0.6	0.4	0.2	0.1	0
$MY_X(\alpha,[0,2])$	28.0	29.3	30.0	31.1	31.8	32.8	33.5	33.7	34.0

Using Table 2.1, we see that $MY_X(0,[0,t]) = AY_X(t)$ and $MY_X(1,[0,t]) = MY_X(t)$. As a result, for our example, the values of the one-parameter marginal value lie in the interval $28 \leq MY_X(0,[0,t]) \leq 34$, i. e., we have

$$AY_X(t) \leq MY_X(\alpha,[0,t]) \leq MY_X(t), \tag{2.42}$$

if $\alpha \in [0,1]$. We see that the marginal value with memory (the marginal value of order $\alpha \in (0,1)$) includes the whole range of intermediate characteristics from the average value to the marginal value.

Similarly, we can calculate the hereditary T-indicator (2.25), and get the expression

$$HY_X(\alpha,\beta,[0,t]) = \frac{a_1\Gamma(2-\beta)}{a_0\Gamma(2-\alpha)}t^{\beta-\alpha}+ \tag{2.43}$$

$$\frac{2a_2\Gamma(2-\beta)}{a_0\Gamma(3-\alpha)}t^{\beta-\alpha+1} - \frac{6a_3\Gamma(2-\beta)}{a_0\Gamma(4-\alpha)}t^{\beta-\alpha+2}. \tag{2.44}$$

For $\beta = \alpha$, expression (2.43) takes the form (2.40), that is,

$$HY_X(\alpha,\alpha,[0,t]) = MY_X(\alpha,[0,t]). \tag{2.45}$$

In Table 2.2, we present the numerical values for two-parameter marginal values with memory (2.43), which are calculated by using equation (2.25) for different values of the memory fading parameters $\alpha, \beta \in (0,1)$.

Table 2.2: Values of two-parameter hereditary T-indicator (2.43) for different values of memory fading parameters α and β at $t = 2$.

$\alpha \backslash \beta$	0	0.1	0.3	0.5	0.7	0.9	1.0
0	34.0	35.1	38.0	42.6	49.6	60.4	68.0
0.1	32.8	33.7	36.6	41.1	47.8	58.2	65.5
0.3	29.7	30.6	33.2	37.2	43.2	52.6	59.3
0.5	25.8	26.6	28.9	32.3	37.6	45.8	51.6
0.7	21.3	22.0	23.9	26.8	31.1	37.9	42.7
0.9	16.5	17.0	18.4	20.7	24.1	29.3	33.0
1.0	14.0	14.4	15.7	17.6	20.4	24.9	28.0

Using Table 2.2, we see that

$$HY_X(0,0,[0,t]) = AY_X(t), \quad HY_X(1,1,[0,t]) = MY_X(t). \tag{2.46}$$

Let us note that, the two-parameter marginal values do not lie in the range from 28 to 34. In other words, the values of the two-parameter marginal value with memory $HY_X(\alpha,\beta,[0,t])$ do not lie between the values of the standard average value and standard marginal value. In our example, we have the inequality

$$14 < HY_X(\alpha,\beta,[0,t]) < 68 \tag{2.47}$$

for $\alpha, \beta \in [0,1]$. This is due to the fact that when the memory fading parameters tend to different integer values, we get two quantities that describe the ratio of the rates of change ($Y^{(1)}$, $X^{(1)}$ to the values ($X(t)$, $Y(t)$) themselves in the form

$$HY_X(1,0,[0,t]) = \frac{Y^{(1)}(t)}{X(t)}, \quad HY_X(0,1,[0,t]) = \frac{Y(t)}{X^{(1)}(t)}, \tag{2.48}$$

if $X(t) \neq 0$ and $X^{(1)}(t) \neq 0$.

Based on Tables 2.1 and 2.2, we can make the following statement.

Statement 2.9. *Taking into account the memory effects can significantly change the marginality of economic processes (as in large and smaller side) even if the other parameters of these processes remain unchanged.*

Plots of two-parameter hereditary T-indicator $HY_X(\alpha, \beta, [0,t])$, given by equation (2.43), is shown in Figures 2.2 and 2.3 for $t = 1$ and $t = 2$, respectively.

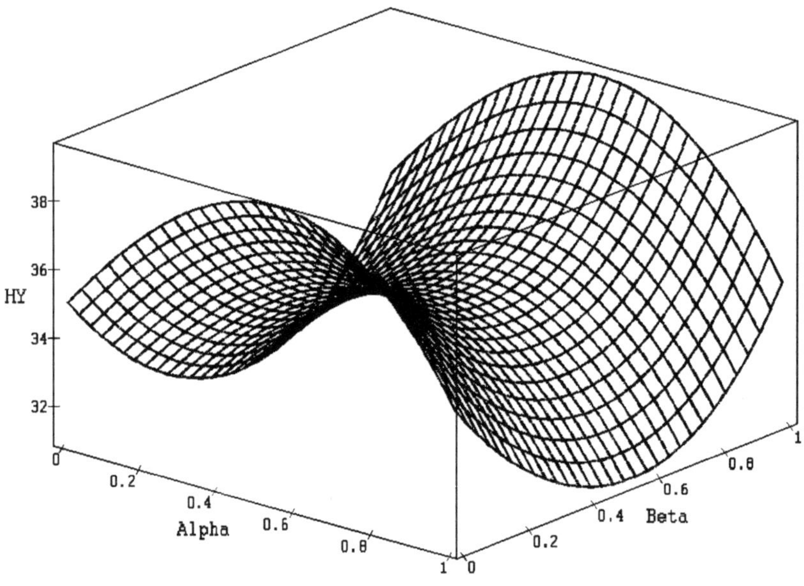

Figure 2.2: Dependence of the two-parameter T-indicator $HY_X(\alpha, \beta, [0,t])$ on the parameters α and β of the fading of the memory of changes in $Y(t)$ and $X(t)$ for $t = 1$.

The tables and graphs show that the generalized T-indicators depend on the fading parameters of the memory about changes of X and Y in the economic process.

2.9 Conclusion

In the conclusion, we can formulate the following statements:

1. Accounting for the presence of memory among consumers is an important condition for the adequacy of the description of economic processes.

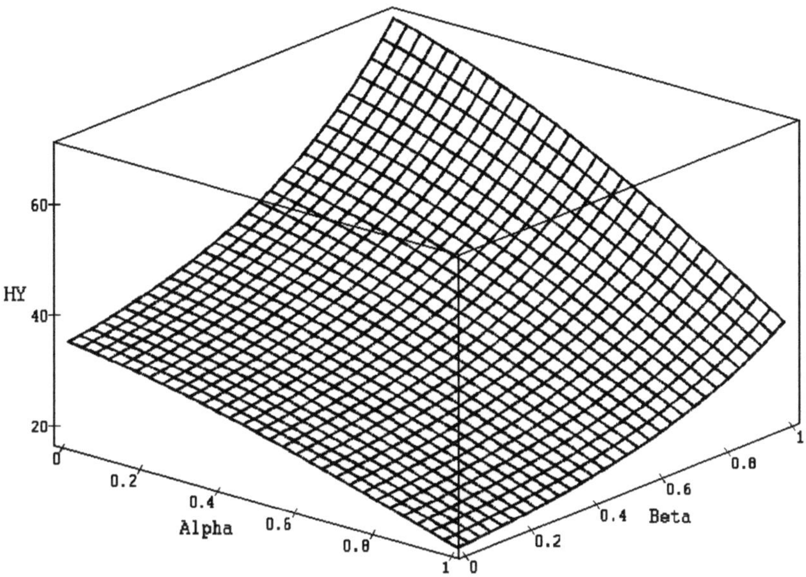

Figure 2.3: Dependence of the two-parameter T-indicator $HY_X(\alpha, \beta, [0, t])$ on the parameters α and β of the fading of the memory of changes in $Y(t)$ and $X(t)$ for $t = 2$.

2. As a tool of economic analysis for processes with memory, we can use the proposed marginal values with memory and the hereditary T-indicators.
3. The standard average and marginal values are special cases of the proposed marginal value with memory, which correspond to zero and one for the parameters of memory fading.
4. The calculations show that the proposed generalized economic T-indicators may significantly depend on the presence of memory and memory fading parameters.
5. The use of generalized (hereditary) indicators (the marginal values with memory) allows us to quantitatively describe the characteristics of economic processes that are intermediate between averages and marginal values.

3 Marginal values of noninteger order in economic analysis

In this chapter, we consider generalizations of the marginal values of economic indicators, which are suggested in [424, 416, 425, 439, 446] to take into account the changes of this indicator on a finite interval of the determining factor [425]. This generalized concept can be used to take into account all values of the indicator at all points of the interval, and not only in an infinitesimal neighborhood of the factor value. We proposed [425, 414, 452] to use new tools that expand the possibilities of economic analysis. The proposed analytical tools allow us to take into account the properties of economic processes with nonlocality in the parameter space of variables and temporal nonlocality (memory). The application of the concept of marginal values of noninteger order in economic analysis is described. This chapter is based on articles [425] and [416, 425, 414, 452].

3.1 Introduction

The standard marginal values of economic indicators, which are used in the economic analysis, are defined in terms of derivatives of integer (first) order of the indicator functions with respect to determining factor. The use of derivatives of integer order really means that the standard marginal values are only local characteristics of economic processes. This is due to the fact that these derivatives of integer orders are determined by the properties indicator function only in the infinitely small neighborhood of the considered value of the factor. In 2016 [424, 416, 425], generalizations of the marginal values of economic indicators are proposed to take into account the changes of this indicator on a finite interval of the determining factor. The proposed approach allows us to take into account all values of the indicator at all points of the interval, and not only in an infinitesimal neighborhood of the factor value. New types of marginal values are defined by using mathematical tools of derivatives and integrals of arbitrary (integer and noninteger) orders.

In article [425] (see also [414, 452]), we prove that an economic analysis, which is based on the marginal values of noninteger order, can give more accurate results than an approach that is based on standard marginal values. It is shown that the use of marginal values of noninteger orders can give exact description of the growth of economic indicators that are described by the functions of power-law type. The formulas that allow us to calculate the total value by the marginal values for noninteger order are suggested.

One of the most important goals and objectives of economic analysis is the study and description of economic processes by using the total and marginal values for economic indicators. In the economic analysis, the marginal utility (MU), the marginal

https://doi.org/10.1515/9783110627459-003

product (MP), the marginal revenue (MR), the marginal cost (MC), the marginal demand (MD) and others are considered as important marginal values. The marginal values show the change (increase) of the corresponding economic indicator per unit growth of factor, on which the indicator under consideration depends. The marginal value of the indicator is usually defined as a derivative of the first order of the corresponding function with respect to the factor, on which the indicator depends. The total value is often understood as a single-valued function $Y = Y(X)$, which describes the dependence of the economic indicator Y on some factor X. In economic analysis, for example, the total utility (TU), the total product (TP), the total cost (TC), the total revenue (TR), the total demand (MD) and others are considered as important total values. The total value can be calculated as a first-order integral of the marginal value. The use of the concept of a marginal value in economic analysis allowed the use of a mathematical calculus that includes differential and integral operators of integer order.

The standard marginal values, which are defined as first-order derivatives with respect to the factor determining them, characterize the economic situation only locally. This locality property is based on the fact that these derivatives are defined by the properties of the function only in an infinitely small neighborhood of the factor value. To describe economic processes, in which the current state of the process depends not only on the infinitely small neighborhood of this state, a generalization of the concepts and methods of economic analysis is necessary. In particular, it is important to have a generalization of the concept of marginal values of economic indicators, which allow us to take into account the change of this indicator on a finite interval of the value of the determining factor. We should have methods that allow us to take into account all the values of the indicator in all points of a finite (or infinite) interval, and not only in an infinitely small neighborhood of the considered value. Such a generalization of the methods of economic analysis is possible only under the condition of the use of an adequate mathematical tools that allow us to describe nonlocality (and memory) in the space of factors (and parameters) characterizing the economic process.

3.2 Standard marginal value in economic analysis

First, we give definitions of standard marginal value. Let the function $Y = Y(X)$ be a single-valued and differentiable function in a neighborhood X_0. Then the marginal value of the economic indicator Y is defined as a first-order derivative of the function $Y = Y(X)$ with respect to the factor X at the point X_0,

$$MY_X = \left(\frac{dY(X)}{dX}\right)_{X=X_0}. \tag{3.1}$$

Marginal value (3.1) is defined by the first-order derivative $Y^{(1)}_X(X) = dY(X)/dX$, that is, the properties of the function $Y(X)$ in the small neighborhood of the point X_0

in the space of factors. The function $Y = Y(X)$ is often called the total value of the indicator. The marginal value (3.1) characterizes the change in the indicator Y caused by a factor X increase by one, provided that all other factors are constant.

Let us briefly describe how the standard marginal value (3.1) allows us to calculate the changes in the economic indicator Y, when the factor X changes. The mathematical basis for this description is the Taylor formula. For the function $Y = Y(X)$, the Taylor formula can be written as

$$Y(X) = Y(X_0) + Y_X^{(1)}(X_0)\Delta X + R_2(X), \tag{3.2}$$

where $Y_X^{(1)}(X_0)$ is the value of the first-order derivative of the function $Y(X)$ with respect to X at the point X_0; ΔX is the absolute increase (increment) of the corresponding factor ($\Delta X = X - X_0$); $R_2(X)$ is residual term. In the Peano form (asymptotic form), the residual term is $R_2(X) = o(\Delta X)$, i. e., it is an infinitely small value (an infinitesimal function) in the neighborhood of the point X_0.

The total increment is defined by the equation

$$\Delta Y = Y(X_1) - Y(X_0), \tag{3.3}$$

which describes the absolute change of the indicator Y, when the factor changes from X_0 to X_1. Neglecting the infinitesimal value of $R_2(X)$ in equation (3.2), the total increment can be written in terms of the increment of the factor ΔX as

$$\Delta Y_X = Y_X^{(1)}(X_0)\Delta X = MY_{X_0}\Delta X, \tag{3.4}$$

where X_0 of the factor X, on which the indicator Y depends; X_1 is the actual value of this factor; $\Delta X = X_1 - X_0$ is the absolute changes (deviations) of the factor X. At the same time, the residual term $R_2(X)$, which is neglected, is interpreted as a method error that is defined by the expression

$$\delta = \Delta Y - \Delta Y_X, \tag{3.5}$$

where ΔY is defined by equation (3.3), and ΔY_X is defined by equation (3.4).

3.3 Concept of marginal value of noninteger order

In works [425, 414, 452], we proposed to use new tools that will expand the possibilities of economic analysis. The proposed analytical tools allow us to take into account the properties of economic processes with nonlocality in the parameter space of (endogenous and exogenous) variables and temporal nonlocality (memory). For this, we use differential and integral calculus of noninteger order.

In economic analysis, the description of change (3.4) of the economic indicator Y with changes in the factor X is based on the concept of the marginal value (3.1) and on the Taylor formula (3.2).

To generalize methods of economic analysis, we proposed [425] to generalize the marginal value by using fractional derivatives of noninteger order [424, 416]. We also use a generalization of the Taylor formulas for the fractional derivatives. We will use a generalization of the Taylor series for the Caputo fractional derivative, which was proposed in [286]. In general, other generalizations of the Taylor series based on other types of fractional derivatives can be used.

Let assume that $f(x)$ is a single-valued function and $f(x) \in C^k[a,b]$, i. e., $f(x)$ has continuous derivatives of all orders up to and including k (the zero-order derivative is the function itself).

Definition 3.1. Let (a,b) be an open set on the real line and a function $f(x)$ defined on that set with real values. The function $f(x)$ is said to be of (differentiability) class $C^k[a,b]$, i. e., $f(x) \in C^k[a,b]$, if the derivatives of the integer orders $f^{(1)}(x), f^{(2)}(x), \dots, f^{(k)}(x)$ exist and are continuous (the continuity is implied by differentiability for all the derivatives except for $f^{(k)}(x)$).

It is known [286, p. 289] that the function $f(x) \in C^N[a,b]$ for $x \geq a$ can be expanded into the Taylor series with the Caputo derivatives of order $0 < \alpha \leq 1$ in the form

$$f(x) = \sum_{k=0}^{N-1} \frac{((D^{\alpha}_{C;a+})^k f)(a)}{\Gamma(k\alpha + 1)} (x-a)^{k\alpha} + R_{N\alpha}(x, a+), \tag{3.6}$$

where we can use $x = X(t)$ and $a = X(0)$. Here, $D^{\alpha}_{C;a+}$ is the Caputo derivative of order $0 < \alpha \leq 1$ that is defined by the equation

$$(D^{\alpha}_{C;a+} f)(x) = \frac{1}{\Gamma(1-\alpha)} \int_a^x \frac{f^{(1)}(\xi)\, d\xi}{(x-\xi)^{\alpha}}, \tag{3.7}$$

where $\Gamma(\alpha)$ is a gamma function, $a < x < b$. Operator (3.7) exists if $f(x) \in AC^n[a,b]$ (see Theorem 2.1 in [200, p. 92]), i. e., the function $f(x)$ has integer-order derivatives up to $(n-1)$-th order, which are continuous functions on the interval $[a,b]$, and the derivative $f^{(n)}(x)$ is Lebesgue summable on the interval $[a,b]$.

In equation (3.6), $R_{N\alpha}(x, a+)$ is the residual term that can be represented as

$$R_{N\alpha}(x, a+) = \frac{((D^{\alpha}_{C;a+})^N f)(\xi_+)}{\Gamma(N\alpha + 1)} (x-a)^{N\alpha}, \tag{3.8}$$

where $a \leq \xi_+ \leq x$.

For $N = 2$, equation (3.6) has the form

$$f(x) = f(a) + \frac{(D^{\alpha}_{C;a+} f)(a)}{\Gamma(\alpha + 1)} (x-a)^{\alpha} + R_{2\alpha}(x, a+). \tag{3.9}$$

Remark 3.1. If $x < a$, then the function $f(x)$ cannot be expanded into a Taylor series (3.6) with the Caputo derivatives (3.7) of the noninteger order α. In addition, a disadvantage of equations (3.6) and (3.9) is the coincidence of the initial and final values of a variable in the expressions of the Caputo derivative in equation (3.6). This leads to restrictions on the applicability of the formulas (3.6) and (3.9) in economic analysis, narrowing the interval characterized by the marginal value of the indicator to the neighborhood of the point $x = a$.

In order to consider a finite interval in the marginal values of noninteger order, we propose a generalization of the Taylor formula (3.9) for the case, when the Caputo derivative is calculated for an arbitrary point $x_0 \geq a$. Consideration of the Taylor series (3.9) for the function and for the Caputo derivatives themselves at the point $x = x_0$ allows us to obtain the required formula

$$f(x) = f(x_0) + \frac{(D^{\alpha}_{C;a+}f)(x_0)}{\Gamma(\alpha+1)} \Delta_{\alpha}x + R_{2\alpha}(x), \tag{3.10}$$

where $x_0 \geq a$, $x \geq a$, $R_{2\alpha}(x)$ is the new remainder term, and we use the notation

$$\Delta_{\alpha}x = (x-a)^{\alpha} - (x_0 - a)^{\alpha}. \tag{3.11}$$

Note that the new the remainder term

$$R_{2\alpha}(x) = R_{2\alpha}(x, a+) - R_{2\alpha}(x_0, a+) \tag{3.12}$$

cannot be considered as an infinitely small value (an infinitesimal function) in the neighborhood of a point, in the general case.

For simplicity, we consider only the case with the starting point $a = 0$. This condition does not give strong restrictions, since many factors can be described by positive real numbers. In this case, equation (3.10) is written as

$$Y(X) = Y(X_0) + \frac{(D^{\alpha}_{C;0+}Y)(X_0)}{\Gamma(\alpha+1)} \Delta_{\alpha}X + R_{2\alpha}(X), \tag{3.13}$$

where we use $X = X(t) \geq 0$ and $X_0 = X(0) \geq 0$. Note that $\Delta_{\alpha}X = X^{\alpha} - X_0^{\alpha}$ can take both positive and negative values. For $\alpha = 1$, the Taylor formula (3.13) gives standard equation (3.2).

Neglecting the term $R_{2\alpha}(X)$ in equation (3.10) and using $X = X_1$, the increment $\Delta Y = Y(X_1) - Y(X_0)$ of Y is written through the increment of the factor X in the form

$$\Delta Y_{X,\alpha} = \frac{(D^{\alpha}_{C;0+}Y)(X_0)}{\Gamma(\alpha+1)} \Delta_{\alpha}X, \tag{3.14}$$

where X_0 is the basic value of the factor X, which affects the index Y; X_1 is the actual value of this factor X; $\Delta_{\alpha}X = X_1^{\alpha} - X_0^{\alpha}$ is a generalized (power) change (deviation) of the factor X. Using equation (3.14), we can give the following definition of the marginal value of an arbitrary (integer and noninteger) order.

Definition 3.2. Let $Y = Y(X)$ be a single-valued function that describe the dependence of the indicator Y on the factor X. We assume that function $Y = Y(X)$ has integer-order derivatives up to $(n-1)$-th order, which are continuous functions on the interval $[0, X_1]$, and the derivative $Y^{(n)}(X)$ is Lebesgue summable on the interval $[0, X_1]$. Then the marginal value of order $\alpha \in [0, 1]$ for the indicator Y on the factor X on the interval $[0, X_1]$ is defined by the equation

$$MY_X(\alpha) = (D^{\alpha}_{C;0+} Y)(X), \tag{3.15}$$

where $0 \leq X \leq X_1$.

In equation (3.15), we can use the analytic continuation for the parameter α from the unit interval $[0, 1]$ to the positive real semi-axis. As a result, the marginal value $MY_X(\alpha)$ can be considered for any nonnegative orders $\alpha \geq 0$.

Let us consider special cases of marginal value (3.15) of order α.

For $\alpha = 1$, using equation (3.15) and $\Gamma(2) = 1$, we get that equation (3.15) takes standard form (3.1), we have the equality

$$MY_X(1) = (D^{1}_{C;0+} Y)(X_0) = \left(\frac{dY(X)}{dX}\right)_{X=X_0} = MY_X. \tag{3.16}$$

In other words, the marginal value of the first order ($\alpha = 1$) is the standard marginal value.

For $\alpha = 0$, using the property $(D^{0}_{C;a+} f)(x) = f(x)$ [335, p. 92] and $\Gamma(1) = 1$, we get that the marginal value of the zero order is equal to the total value, i. e., we have the equality

$$MY_X(0) = (D^{0}_{C;0+} Y)(X) = Y(X). \tag{3.17}$$

For $0 < \alpha < 1$, the marginal value $MY_{X_0}(\alpha) = (D^{\alpha}_{C;0+} Y)(X_0)$ is expressed by the equations

$$MY_{X_0}(\alpha) = \frac{1}{\Gamma(1-\alpha)} \int\limits_0^{X_0} \frac{Y^{(1)}(x)\, dx}{(X-x)^{\alpha}}. \tag{3.18}$$

Substituting (3.18) into (3.14), we get the desired formula

$$\Delta Y_{X,\alpha} = \frac{MY_{X_0}(\alpha)}{\Gamma(\alpha+1)} \Delta_{\alpha} X, \tag{3.19}$$

where $\Delta_{\alpha} X = X_1^{\alpha} - X_0^{\alpha}$ is generalized (power) change (deviation) of factor X. For $\alpha = 1$, equation (3.19) reduces to equation (3.4). As a result, the marginal values (3.15) of the order α allow us to describe the increment of the economic indicator $\Delta Y = Y(X_1) - Y(X_0)$ by changing the factor X.

It is known that the Caputo derivative of order $\alpha > 0$ of a constant function is equal to zero. Therefore, for the constant function of the indicator ($Y(X) = \text{const}$), we obtain the zero marginal value ($MY_X(\alpha) = 0$) for any orders $\alpha > 0$, and, as a result, zero increment of the indicator $\Delta Y = 0$.

3.4 Example of calculating of generalized marginal values

Power laws and functions of the power type play an important role in economic theory and finance [125, 126]. However, it should be noted that formulas for standard marginal values (3.4) without the remainder term cannot give exact results for power functions with noninteger exponents. In article [425], we proved that the marginal values of non-integer order in the form (3.19) (even without the remainder term) can give exact values. To prove this statement, we use the equation for the Caputo fractional derivatives of the power-law function in the form

$$D^{\alpha}_{C;a+}(x-a)^{\beta} = \frac{\Gamma(\beta+1)}{\Gamma(\beta-\alpha+1)}(x-a)^{\beta-\alpha}, \tag{3.20}$$

$$D^{\alpha}_{C;a+}(x-a)^{k} = 0, \tag{3.21}$$

where $n-1<\alpha<n$, $\beta>n-1$, and $k=0,1,\dots,n-1$. In particular, we have $D^{\alpha}_{C;a+}1=0$ and $D^{\alpha}_{C;a+}(x-a)^{\alpha}=\Gamma(\alpha+1)$.

Let us consider an illustrative example for calculation of the marginal value of the order α. We can consider economic indicator Y in the form $Y(X)=AX^{\beta}+B$, where A and B are some constants, and β is the exponent. Using equations (3.20) and (3.21) for the function $Y(X)=AX^{\beta}+B$, we get

$$(D^{\alpha}_{C;0+}Y)(X) = \frac{A\Gamma(\beta+1)}{\Gamma(\beta-\alpha+1)}X^{\beta-\alpha}. \tag{3.22}$$

As a result, marginal value (3.15) of the order of $\alpha>0$ has the form

$$MY_{X_0}(\alpha) = \frac{A\Gamma(\beta+1)}{\Gamma(\beta-\alpha+1)}X_0^{\beta-\alpha}, \tag{3.23}$$

and the change in the indicator Y, which is calculated by equation (3.19), has the form

$$\Delta Y_{X,\alpha} = \frac{A\Gamma(\beta+1)}{\Gamma(\beta-\alpha+1)\Gamma(\alpha+1)}X_0^{\beta-\alpha}\Delta_{\alpha}X, \tag{3.24}$$

where

$$\Delta_{\alpha}X = X_1^{\alpha}-X_0^{\alpha}. \tag{3.25}$$

If the order of the Caputo derivative is equal to the exponent of the function ($\alpha=\beta$), then equation (3.22) gives the expression

$$(D^{\alpha}_{C;0+}Y)(X) = A\Gamma(\alpha+1). \tag{3.26}$$

In this case ($\alpha=\beta$), equation (3.23) gives $MY_X(\alpha)=A\Gamma(\alpha+1)$, and we leads to the following statement.

Statement 3.1. *The marginal value of the order $\alpha > 0$, which is equal to the exponent of the power function, is a constant value.*

As a result, using equation (3.24), we see that for power function $Y(X) = AX^{\beta} + B$, the increase of the indicator is directly proportional to the change $\Delta_{\alpha}X$ of the factor X:

$$\Delta Y_{X,\alpha} = A\Delta_{\alpha}X. \tag{3.27}$$

Absolute increase $\Delta Y = Y(X_1) - Y(X_0)$ of the indicator Y, which is defined by the function $Y(X) = AX^{\beta}$, is equal to

$$\Delta Y = AX_1^{\beta} - AX_0^{\beta}. \tag{3.28}$$

Let us compare the change in the indicator Y, which is given by expression (3.24), and the absolute increase in the indicator (3.28). The difference of these values determines the method error δ_{α}, based on indicators of noninteger order (3.15). In general, a method error arises because of discarding the remainder term and is defined by the expression

$$\delta_{\alpha} = \Delta Y - \Delta Y_{X,\alpha}, \tag{3.29}$$

where ΔY is defined by equation (3.28), and $\Delta Y_{X,\alpha}$ is defined by expression (3.24),

For the power function $Y(X) = AX^{\beta} + B$, the substitution of equations (3.33) and (3.35) in (3.36) gives the expression

$$\delta_{\alpha} = AX_1^{\beta} - AX_0^{\beta} - \frac{A\Gamma(\beta+1)X_0^{\beta-\alpha}(X_1^{\alpha} - X_0^{\alpha})}{\Gamma(\beta-\alpha+1)\Gamma(\alpha+1)}. \tag{3.30}$$

For the case $\alpha = \beta$, equation (3.30) takes the form

$$\delta_{\alpha=\beta} = AX_1^{\beta} - AX_0^{\beta} - A(X_1^{\alpha} - X_0^{\alpha}) = 0. \tag{3.31}$$

As a result, we obtain the following statement.

Statement 3.2. *For power functions the marginal value of noninteger order can describe the increase with absolute accuracy, i. e., with zero error, $\delta_{\alpha} = 0$.*

Let us now compare the results obtained with the standard method based on equations (3.1) and (3.4). If the order of the Caputo derivative is equal to one ($\alpha = 1$), then equation (3.15) will define the standard marginal value (3.1). Therefore, equation (3.23) with $\alpha = 1$ defined the standard marginal value for the power function $Y(X) = AX^{\beta} + B$ in the form

$$MY_{X_0} = MY_{X_0}(1) = A\beta X_0^{\beta-1}. \tag{3.32}$$

Equation (3.24) with $\alpha = 1$ defines the standard change of the indicator Y, which is represented by equation (3.4), in the form

$$\Delta Y_X = A\beta X_0^{\beta-1}(X_1 - X_0). \tag{3.33}$$

The method error $\delta = \delta_1 = \Delta Y - \Delta Y_X$ of the standard method, which is based on marginal values (3.1), is equal to

$$\delta = AX_1^{\beta} - AX_0^{\beta} - A\beta X_0^{\beta-1}(X_1 - X_0). \tag{3.34}$$

We can see that error (3.34) is nonzero ($\delta \neq 0$).

As a result, we proved that an economic analysis based on marginal values of noninteger order can produce more accurate results compared to an approach based on standard marginal values. Therefore, we can conclude that the use of marginal values of noninteger order can give a more adequate description than standard marginal values for a wide class of functions of economic indicators, including power-law functions.

3.5 From marginal value of noninteger order to total value

To find the total value of the marginal value is used the antiderivative and the integral operators of integer orders. The mathematical basis of the method of calculating the total value is the fundamental (basic) theorem of standard mathematical calculus and the standard Newton–Leibniz formula

$$\int_a^b f_x^{(1)}(x)\, dx = f(b) - f(a). \tag{3.35}$$

Let us give an example of standard formulas that describe the relationship between the total value and the marginal value of the indicator. For the function of one variable $Y = Y(X)$, the Newton–Leibniz formula, which allows us to calculate the influence of the factor X on the indicator Y, has the form

$$\Delta Y = \int_{X_0}^{X_1} Y_X^{(1)}(X)\, dX = \int_{X_0}^{X_1} Y_\xi^{(1)}(\xi)\, d\xi. \tag{3.36}$$

The total value can be considered as antiderivative of the standard marginal value. Every continuous function $Y(X)$ has an antiderivative, and the antiderivative can be found by the definite integral with variable upper boundary. Therefore, to find the total value of the standard marginal value, we can use the equation

$$Y(X) = \int_0^X MY_x\, dx + Y(0) \tag{3.37}$$

up to a constant $Y(0)$.

Let us consider the use of integrals of noninteger order in economic analysis to find the total value for the marginal value of noninteger order. In [360], it was proved that the inverse operation for the Caputo derivative of order α is the Riemann–Liouville integration of the same order. Let us give definition of the Riemann–Liouville integration [335, pp. 41–42].

Definition 3.3. Let the function $f(x)$ satisfy the condition $f(x) \in L_1(a,b)$, which implies that $f(x)$ is a Lebesgue integrable function on the interval (a,b) and the inequality

$$\int_a^b |f(\xi)|\, d\xi < \infty, \tag{3.38}$$

holds. Then the Riemann–Liouville integral of the order $\alpha \geq 0$ with respect to the variable x is defined by the equation

$$(I^{\alpha}_{\mathrm{RL};a+}f)(x) = \frac{1}{\Gamma(\alpha)} \int_a^x \frac{f(\xi)\, d\xi}{(x-\xi)^{1-\alpha}}, \tag{3.39}$$

where $\Gamma(\alpha)$ is the gamma function, and $a < x < b$.

The Riemann–Liouville integral (3.39) is a generalization of standard integration of the integer order n. Note that the Riemann–Liouville integral (3.39) for the order $\alpha = 1$ gives the standard first-order integral

$$(I^{1}_{\mathrm{RL};a+}f)(x) = \int_a^x f(\xi)\, d\xi. \tag{3.40}$$

Thus, the first-order integration, which is used in the formula for the total value, is a special case of the integral of the order $\alpha \geq 0$.

Equation (3.37) is based on the Newton–Leibniz formula and the fundamental theorem of mathematical calculus. To generalize equation (3.37), it is necessary to use a generalization of the fundamental theorem and Newton–Leibniz formula for the case of derivatives of noninteger orders. The fundamental theorem of differential and integral calculi of noninteger order was formulated in article [360], and in monographs [363, pp. 263–267]. Some additional aspects of this theorem are discussed in [156, 240]. For noninteger order $\alpha > 0$, the following generalized Newton–Leibniz formula holds

$$(I^{\alpha}_{\mathrm{RL};a+}D^{\alpha}_{\mathrm{C};a+}f)(b) = f(b) - f(a) - V(a,b), \tag{3.41}$$

where

$$V(a,b) = \sum_{k=1}^{n-1} \frac{f^{(n)}(a)}{k!}(b-a)^k \tag{3.42}$$

and $n-1 \le \alpha < n$. Note that due to the property

$$(I^{1}_{\mathrm{RL};a+} D^{1}_{C;a+} f)(b) = \int_{a}^{b} f^{(1)}_{x}(x)\, dx, \tag{3.43}$$

equation (3.41) with $\alpha = 1$ gives the standard Newton–Leibniz formula (3.35).

For $0 \le \alpha \le 1$, equation (3.41) can be written in the form

$$(I^{\alpha}_{\mathrm{RL};0+} D^{\alpha}_{C;0+} Y)(X) = Y(X) - Y(0). \tag{3.44}$$

As a result, we obtain equation that allows to find the total value by the marginal value of $MY_X(\alpha)$ of noninteger order α. This equation has the form

$$(I^{\alpha}_{\mathrm{RL};0+} MY_X(\alpha))(X) = Y(X) - Y(0). \tag{3.45}$$

To apply this formula in economic analysis for power functions with the exponent $\beta > 0$, we can use the Riemann–Liouville integral of the order $\alpha \ge 0$ [200, p. 71] of the power-law functions, that has the form

$$I^{\alpha}_{\mathrm{RL};a+} (x-a)^{\beta} = \frac{\Gamma(\beta+1)}{\Gamma(\beta+\alpha+1)} (x-a)^{\beta+\alpha}. \tag{3.46}$$

For $\alpha = 1$, equations (3.44)–(3.46) takes the standard form with the integrals of the first order.

In addition to equation (3.46) for power-law functions, there is a set of integration formulas for non-integer order given in Table 9.1 in the monograph [335, pp. 140-141]. For analytical calculations of the total value of the marginal value of noninteger order, the use of these table integrals is required.

3.6 Conclusion

The proposed concept of the marginal value of noninteger order allows us to expand an application of economic analysis. The use of marginal values of noninteger order, for a wide class of functions (including functions of a power type), can give more accurate results compared to the standard approach. In addition, the marginal values of non-integer order make it possible to take into account the effects of nonlocality and memory in economic analysis. In application of marginal values of noninteger orders, we should take into account that the standard product and chain rules for differentiation of the product is not satisfied for the derivatives of noninteger order [377, 81]. To calculate the total values of the marginal values of noninteger order, we can use the table integrals given in book [335, pp. 140–141].

4 Deterministic factor analysis of processes with memory

In this chapter, we proposed basic concepts and methods that allow us to take into account the effects of memory and nonlocality in deterministic factor analysis. We generalize deterministic factor analysis by using the fractional calculus as a mathematical tool. The suggested methods give a quantitative description of the influence of individual factors on the change of the effective economic indicator for processes with memory and nonlocality. In works [425, 414, 452], we proposed generalized differential and integral methods for the deterministic factor analysis of economic processes with memory. It has been shown that these methods, which are based on the integro-differentiation of noninteger order, can give more accurate results than the standard methods of factor analysis, which are based on differentiation and integration of integer orders. This chapter is based on works [414, 452].

4.1 Introduction

The economic analysis actively uses methods of the deterministic factor analysis, which gives characteristics of influence of factors on change of the effective indicator. The main methods of the standard deterministic factor analysis are the differential method and the integral method. These methods are based on the mathematical theory of derivatives and integrals of integer orders.

In works [425, 414, 452], we proposes generalizations of these standard differential and integral methods of factor analysis. The suggested methods are based on the theory of fractional derivatives and integrals of noninteger orders.

Let us briefly describe the standard method of factor analysis, which is based on the differential calculus. The mathematical basis of this method is the derivatives (differentiation) of integer orders and the Taylor series.

Let $z = f(x, y)$ be a function of two independent real variables x and y. The Taylor formula for the function $z = f(x, y)$ can be written in the form

$$f(x, y) = f(x_0, y_0) + f_x^{(1)}(x_0, y_0)\Delta x + f_y^{(1)}(x_0, y_0)\Delta y + R_2(x, y), \tag{4.1}$$

where

$$f_x^{(1)}(x_0, y_0) = \left(\frac{\partial f(x, y_0)}{\partial x}\right)_{x=x_0}, \quad f_y^{(1)}(x_0, y_0) = \left(\frac{\partial f(x_0, y)}{\partial y}\right)_{y=y_0} \tag{4.2}$$

are the values of the partial derivatives of the first order with respect to x and y at the point (x_0, y_0). The expressions Δx and Δy denote factor increments of the corresponding variables ($\Delta x = x - x_0$, $\Delta y = y - y_0$); $R_2(x, y)$ is the remainder term, which is of an infinitesimal value in the neighborhood of the point (x_0, y_0).

https://doi.org/10.1515/9783110627459-004

Let us consider increment of the function $z = f(x,y)$. Total increment Δz of a function $z = f(x,y)$ of the variables x and y (arguments) is the increment of the function, when all the arguments undergo, as a rule, nonzero increments

$$\Delta z = f(x + \Delta x, y + \Delta y) - f(x,y). \tag{4.3}$$

Neglecting the remainder term $R_2(x,y)$, the total increment of the function $\Delta z = f(x_1,y_1) - f(x_0,y_0)$ can be written by the increment of the factors $\Delta x = x_1 - x_0$ and $\Delta y = y_1 - y_0$ in the form

$$\Delta z \approx f_x^{(1)}(x_0,y_0)\Delta x + f_y^{(1)}(x_0,y_0)\Delta y. \tag{4.4}$$

The influence of the factor x on the generalizing indicator z is calculated by the equation

$$\Delta z_x = f_x^{(1)}(x_0,y_0)\Delta x, \tag{4.5}$$

and the influence of the factor y is described by the equation

$$\Delta z_y = f_y^{(1)}(x_0,y_0)\Delta y, \tag{4.6}$$

where x_0, y_0 are the basic (planned) values of the factors x and y that have influence on the effective indicator; x_1, y_1 are the actual values of these factors; $\Delta x = x_1 - x_0$ and $\Delta y = y_1 - y_0$ are the absolute changes (deviations) of the factors x and y.

In the differential calculus method, it is assumed that the total increment of the functions Δz is decomposed into the terms Δz_x and Δz_y. The values of each of these terms are calculated as the product of the corresponding partial derivative and the increment of the variable (factor). In this method, the remainder term $R_2(x,y)$, which is neglected, is interpreted as an error of the differential calculus method. The neglecting of an the remainder term is one of the disadvantages of this method, since for economic calculations the exact balance of the change of the effective indicator and the algebraic sum of the influence of all factors are often required.

In the standard approach to deterministic factor analysis, the differential method uses the derivatives of integer orders. It is known that the derivatives of integer orders are determined by the properties of the differentiable function only in an infinitesimal neighborhood of the considered point. As a result, the differential equations (4.4)–(4.6) with derivatives of integer orders cannot describe processes with memory and nonlocality.

In works [414, 452], we proposed the generalizations of the deterministic factor analysis methods, which are based on the fractional integration and differentiation of arbitrary (noninteger) orders. For simplification, we will use the Caputo fractional derivatives. The main distinctive feature of these derivatives is that their effect on the constant function gives zero. This leads us to the zero effect of a constant factor on the indicator. Let us give the definitions of the Caputo derivative.

Definition 4.1. The left-sided Caputo derivatives of order $\alpha \geq 0$ on the interval $[a, b]$ are defined by the equation

$$(D^{\alpha}_{C;a+}f)(x) = \frac{1}{\Gamma(n-\alpha)} \int_{a}^{x} \frac{f^{(n)}(\xi)\, d\xi}{(x-\xi)^{\alpha-n+1}}, \tag{4.7}$$

where $\Gamma(\alpha)$ is the gamma function, $a < x < b$, $n-1 < \alpha \leq n$, and $f^{(n)}(\xi)$ is the derivative of the integer order n of the function $f(\xi)$ with respect to the variable ξ. Operator (4.7) exists if $f(\xi) \in AC^{n}[a, b]$, i. e., the function $f(\xi)$ has integer-order derivatives up to $(n-1)$-th order, which are continuous functions on the interval $[a, b]$, and the derivative $f^{(n)}(\xi)$ is Lebesgue summable on the interval $[a, b]$.

For positive integer values $\alpha = n$, the Caputo derivatives coincide [200, p. 92] with the standard derivative of the integer order n:

$$(D^{n}_{C;a+}f)(x) = f^{(n)}(x). \tag{4.8}$$

Therefore, the standard derivatives are special cases of the fractional derivatives.

In general, indicators and factors can be considered as single-valued functions of time. If the parametric dependence of indicator $z(t)$ on factors $x(t)$ and $y(t)$ can be represented by a single-valued function, then we can use the parametric fractional derivatives and integrals. The Riemann–Liouville parametric fractional derivative has been considered in Section 18.2 of [335, pp. 325–329] and Section 2.5 of [200]. The Caputo parametric fractional derivatives are proposed in [360], and then described in [13].

Definition 4.2. The left-sided Caputo parametric fractional derivative of order $\alpha > 0$ is defined by the equation

$$(D^{\alpha,x}_{C;a+}f)(t) = \frac{1}{\Gamma(n-\alpha)} \int_{a}^{t} d\tau \frac{x^{(1)}(\tau)}{(x(t)-x(\tau))^{\alpha+1-n}} \left(\frac{1}{x^{(1)}(\tau)} \frac{d}{d\tau}\right)^{n} f(\tau), \tag{4.9}$$

where $a < t < b$, $n-1 < \alpha \leq n$, and $x(\tau)$ is a increasing and positive monotonic function having a continuous derivative $x^{(1)}(\tau) = dx(\tau)/d\tau$. The derivative (4.9) is also called the Caputo fractional derivatives of function $f(t)$ by a function $x(t)$ of the order $\alpha > 0$.

In works [414, 452], we propose a generalization of the deterministic factor analysis. This generalization can be used to describe properties of economic indicators, which are represented by power-law functions of factors of processes with nonlocality and memory.

4.2 Differential method of noninteger order: single variable

The standard differential method of deterministic factor analysis is based on the Taylor formula. We proposed to use a generalization of the standard Taylor formula for the Caputo fractional derivative to generalize this method by using the derivatives of noninteger orders.

Let us consider the generalization of the Taylor series, which was proposed in [286].

Statement 4.1. *The function $f(x)$ with $x \in [a,b]$ can be expanded by using the generalized Taylor series with the left-sided Caputo derivatives of order $0 < \alpha \le 1$ in the form*

$$f(x) = \sum_{k=0}^{N-1} \frac{((D^{\alpha}_{C;a+})^k f)(a)}{\Gamma(k\alpha+1)} (x-a)^{k\alpha} + R_{N\alpha}(x,a+), \tag{4.10}$$

where $R_{N\alpha}(x,a+)$ is the remainder term, which can be represented in the form

$$R_{N\alpha}(x,a+) = \frac{((D^{\alpha}_{C;a+})^N f)(\xi_+)}{\Gamma(N\alpha+1)} (x-a)^{N\alpha}, \tag{4.11}$$

where $a \le \xi_+ \le x$. The generalized Taylor series (4.10), is applicable to the functions $f(x)$ that satisfy the condition

$$((D^{\alpha}_{C;a+})^k f)(x) \in C[a,b], \tag{4.12}$$

for $k = 0, 1, \ldots, N$.

Remark 4.1. Note that equation (4.10) cannot be applied for the case $x < a$, if the parameter α is not an integer [414, 452]. In this case, we can use the right-sided Caputo derivative of the order $\alpha > 0$ [200, p. 95].

For simplicity, we will consider deterministic factor analysis only for cases $x > a$, $X > X_0$ ($X(t) > X(0)$ for all $t > 0$). We also will assume that $a = 0$, $X(0) = 0$, i. e., $x = X(t) > 0$ for all $t > 0$.

For example, series (4.10) with $N = 2$ have the form

$$f(x) = f(a) + \frac{(D^{\alpha}_{C;a+} f)(a)}{\Gamma(\alpha+1)} (x-a)^{\alpha} + R_{2\alpha}(x,a+), \tag{4.13}$$

where $x \in [a,b]$.

The deterministic factor analysis of processes with memory should take into account that the indicators and factors can be described as single-valued functions of time. If the parametric dependence of the indicator $z(t)$ on the factors $x(t)$ and $y(t)$ can be represented by a single-valued function, then we can use the fractional Taylor formula with the parametric fractional derivatives. Using Theorem 18 of [335, p. 474], we can write the fractional Taylor formula with the Caputo parametric fractional derivative as

$$f(x(t)) = \sum_{k=0}^{N-1} \frac{((D^{\alpha,x}_{C;a+})^k f)(a)}{\Gamma(k\alpha+1)} (x(t)-x(a))^{k\alpha} + R_{N\alpha}(x(t),a+), \tag{4.14}$$

where $R_{N\alpha}(x(t), a+)$ is the remainder terms, which can be represented in the form

$$R_{N\alpha}(x(t), a+) = \frac{((D^{\alpha,x}_{C;a+})^N f)(\xi_+)}{\Gamma(N\alpha + 1)} (x(t) - x(a))^{N\alpha}, \tag{4.15}$$

where we assume that $x(t) \geq x(a)$ for all $t \in [a, b]$

For equations (4.10), (4.13), (4.14) there is an additional problem, which is caused by the coincidence of the initial and final values in the derivatives $(D^{\alpha}_{C;a+}f)(a)$ and $(D^{\alpha}_{C;a-}f)(a)$. It is important to have the generalized Taylor series at an arbitrary point x_0, which does not coincide with the initial point of fractional derivative $D^{\alpha}_{C;a+}$, i. e., $x_0 \neq a$. In this case, we can avoid some restrictions on a possible application of equations (4.10), (4.13), to calculating the influence of factors on the change in the effective indicator. To solve this problem, we propose a generalization of the Taylor formulas (4.10) to the case when the Caputo derivative is considered at an arbitrary point $x_0 \geq a$.

Consideration of the Taylor series (4.13) for the function $f(x)$ and for the Caputo derivatives $(D^{\alpha}_{C;a+}f)(x)$ at the point $x = x_0$ allows us to obtain the equation

$$f(x) = f(x_0) + \frac{(D^{\alpha}_{C;a+}f)(x_0)}{\Gamma(\alpha + 1)} \Delta_{\alpha} x + R_{2\alpha}(x, x_0, a+), \tag{4.16}$$

where $x_0 \geq a$, $x \geq a$, and the term $R_{2\alpha}(x, x_0, a+)$ is equal to the difference between $R_{2\alpha}(x, a-)$ and $R_{2\alpha}(x_0, a-)$, in the form

$$R_{2\alpha}(x, x_0, a+) = R_{2\alpha}(x, a-) - R_{2\alpha}(x_0, a-). \tag{4.17}$$

In equation (4.16), we use the notation

$$\Delta_{\alpha} x = (x - a)^{\alpha} - (x_0 - a)^{\alpha}. \tag{4.18}$$

For equations (4.14) and (4.15) with the parametric fractional derivatives, we have

$$f(x(t)) = f(x(t_0)) + \frac{(D^{\alpha,x}_{C;a+}f)(a)}{\Gamma(\alpha + 1)} \Delta_{\alpha} x + R_{2\alpha}(x(t), x(t_0), a+), \tag{4.19}$$

where we use

$$\Delta_{\alpha} x = (x(t) - x(a))^{\alpha} - (x(t_0) - x(a))^{\alpha}. \tag{4.20}$$

For simplicity, we will consider only the case $a = 0$ and $x(a) = 0$. Neglecting the term $R_{2\alpha}(x, x_0, a+)$, equation (4.16) takes the form the form

$$f(x) - f(x_0) \approx \frac{(D^{\alpha}_{C;0+}f)(x_0)}{\Gamma(\alpha + 1)} \Delta_{\alpha} x, \tag{4.21}$$

where $x > 0$, $x_0 > 0$, and

$$\Delta_\alpha x = x^\alpha - x_0^\alpha. \tag{4.22}$$

Neglecting the term $R_{2\alpha}(x(t), x(a), a+)$, equation (4.19) takes the form

$$f(x(t)) - f(x(t_0)) \approx \frac{(D^{\alpha,x}_{C;0+}f)(t_0)}{\Gamma(\alpha+1)} \Delta_\alpha x + R_{2\alpha}(x(t), x(t_0), 0+), \tag{4.23}$$

where $t > 0$, $t_0 > 0$, $x(t) \geq 0$, $x(t_0) \geq 0$, $x(0) = 0$, and

$$\Delta_\alpha x = (x(t))^\alpha - (x(t_0))^\alpha. \tag{4.24}$$

Note that term (4.17) cannot be considered as a small value, in the general case.

4.3 Differential method of noninteger order: two variables

For a function of two variables $z = f(x, y)$, we have the formula

$$f(x, y) - f(x_0, y_0) \approx \frac{(D^\alpha_{0+;x}f)(x_0, y_0)}{\Gamma(\alpha+1)} \Delta_\alpha x + \frac{(D^\beta_{0+;y}f)(x_0, y_0)}{\Gamma(\beta+1)} \Delta_\beta y. \tag{4.25}$$

As a result, the total increment of the function

$$\Delta z = f(x_1, y_1) - f(x_0, y_0) \tag{4.26}$$

is written in terms of the increments $\Delta_\alpha x$ and $\Delta_\beta y$ of the factors x and y in the form

$$\Delta z \approx \frac{(D^\alpha_{0+;x}f)(x_0, y_0)}{\Gamma(\alpha+1)} \Delta_\alpha x + \frac{(D^\beta_{0+;y}f)(x_0, y_0)}{\Gamma(\beta+1)} \Delta_\beta y, \tag{4.27}$$

where x_0, y_0 are the basic (planned) values of the factors x and y that have influence on effective indicator z; the values x_1, y_1 describe the actual values of these factors. The expressions

$$\Delta_\alpha x = x_1^\alpha - x_0^\alpha, \quad \Delta_\beta y = y_1^\alpha - y_0^\beta \tag{4.28}$$

can be interpreted as generalized absolute changes (deviations) of the factors x and y.

For $\alpha = \beta = 1$, expression (4.25) gives standard equation (4.4).

As a result, the influence of the factor x on the indicator z will be calculated by the equation

$$\Delta z_{x,\alpha} = \frac{(D^\alpha_{0+;x}f)(x_0, y_0)}{\Gamma(\alpha+1)} \Delta_\alpha x, \tag{4.29}$$

and the influence of the factor y can be calculated by the equation

$$\Delta z_{y,\beta} = \frac{(D^{\beta}_{0+;y}f)(x_0, y_0)}{\Gamma(\beta + 1)} \Delta_{\beta} y. \tag{4.30}$$

For $\alpha = \beta = 1$, equations (4.29) and (4.30), which describe the influence of the factors x and y, take standard form (4.5) and (4.6), respectively.

It is known that the Caputo fractional derivative of a constant function is equal to zero. As a result, for the case $z = f(x, y) = \text{const}$, we get a zero effect of both factors $\Delta z_{x,\alpha} = \Delta z_{y,\beta} = 0$. In addition, for the indicator, which does not depend on one of the factors, the effect of this factor will be zero.

4.4 Comparison with standard differential method

The standard Taylor's equation (4.1) without remainder term cannot give exact results for power functions with noninteger exponents. The Taylor equation (4.25) with derivatives of noninteger order even without remainder term is a more accurate tool for approximating the power functions.

If we use the standard Taylor equation for the expansion of power functions in a neighborhood of the point $x_0 > 0$, then we get an infinite power series, in general. For the same power function, the expression of the generalized Taylor equation (4.19) can be obtained by using the equations for the left-sided Caputo derivative of the power-law function

$$D^{\alpha}_{C;a+}(x - a)^{\beta} = \frac{\Gamma(\beta + 1)}{\Gamma(\beta - \alpha + 1)}(x - a)^{\beta - \alpha}, \tag{4.31}$$

where $n - 1 < \alpha < n$, $\beta > n - 1$, and

$$D^{\alpha}_{C;a+}(x - a)^{k} = 0, \quad (k = 0, 1, \ldots, n - 1). \tag{4.32}$$

If the indicators are given as parametric function of factors, then we can use the parametric left-sided Caputo derivative. For example, we can use [13, p. 464] the equation

$$D^{\alpha,x}_{C;a+}(x(t) - x(a))^{\beta} = \frac{\Gamma(\beta + 1)}{\Gamma(\beta - \alpha + 1)}(x(t) - x(a))^{\beta - \alpha}, \tag{4.33}$$

where $\alpha > 0$, and $\beta > -1$.

To compare the standard method and the proposed generalization, we will consider a power-law production function in an illustrative numerical example. An example of a power function of two real variables we take the Cobb–Douglas production function.

The production function $F(K, L)$ for a single good with two factors shows the maximum amount of the good that can be produced using alternative combination of capital (K) and labor (L) [283, p. 268].

Obviously, when describing processes with memory, we should consider a dynamic production function in the general case. In the dynamic function, all variables are functions of time ($K = K(t)$, $L = L(t)$), and an explicit dependence on the time of the function itself and its parameters is possible.

As an illustrative numerical example, we will consider the Cobb–Douglas production function with parameters and notation that were used in the original work of Cobb and Douglas [76]. The Cobb–Douglas function was proposed by Charles W. Cobb and Paul H. Douglas in the paper "A theory of production" in 1928 [76, p. 151] for the US industry for the period 1899–1922. The initial form of this function was given [76, p. 151] by the expression

$$P(L, C) = 1.01L^{3/4}C^{1/4}. \tag{4.34}$$

The Cobb–Douglas production function $P = P(L, C)$ describes the dependence of the volume of production P (total production) on the labor costs L (labor input) and capital costs C (capital input). In this case, the production is the indicator ($z = P$) and the variables L and C are the factors ($x = L$, $y = C$). The Cobb–Douglas function has the form

$$P(L, C) = AL^{a}C^{b}, \tag{4.35}$$

where A, a and b are all nonnegative constants. Here, $A > 0$ is the aggregate productivity of the factors, and a and b are interpreted as capital and labor elasticities, respectively. The values of the constant $0 \le a < 1$ and $0 \le b < 1$ are determined by the available technologies. In initial form of this function $A = 1.01$, $a = 0.75$ and $b = 0.25$ (see [76, p. 151]).

The total increment of the Cobb–Douglas production function is defined by the equations

$$\Delta P = P(L_1, C_1) - P(L_0, C_0), \tag{4.36}$$

where L_0, C_0 are the basic (planned) values of the labor L and capital C that have influence on production P; the values L_1, C_1 describe the actual values of labor and capital.

Let us consider the total increment of the production as a sum of two decomposed influence of the labor and capital

$$\Delta P \approx \Delta P_{L,\alpha} + \Delta P_{C,\beta} = \frac{(D^{\alpha}_{0+;L}P)(L_0, C_0)}{\Gamma(\alpha+1)}\Delta_{\alpha}L + \frac{(D^{\beta}_{0+;C}P)(L_0, C_0)}{\Gamma(\beta+1)}\Delta_{\beta}C. \tag{4.37}$$

Let us calculate these two independent terms. Using equation (4.31) for the Cobb–Douglas function $P(L, C) = AL^{a}C^{b}$, we obtain

$$(D^{\alpha}_{0+;L}P)(L, C) = \frac{A\Gamma(a+1)}{\Gamma(a-\alpha+1)}L^{a-\alpha}C^{b}, \tag{4.38}$$

$$(D^{\beta}_{0+;C}P)(L,C) = \frac{A\Gamma(b+1)}{\Gamma(b-\beta+1)}L^{a}C^{b-\beta}. \quad (4.39)$$

If the orders of the Caputo derivatives are taken equal to the exponents of the Cobb–Douglas function ($\alpha = a, \beta = b$), then equations (4.38) and (4.39) give the expressions

$$(D^{\alpha}_{0+;L}P)(L,C) = A\Gamma(\alpha+1)C^{\beta}, \quad (4.40)$$

$$(D^{\beta}_{0+;C}P)(L,C) = AL^{\alpha}\Gamma(\beta+1). \quad (4.41)$$

If the orders of the Caputo derivatives are equal to one ($\alpha = \beta = 1$), then equations (4.38) and (4.39) give the standard expressions.

As a result of the influence of the factors L and C on the indicator P, which is calculated by equations (4.29) and (4.30), have the form

$$\Delta P_{L,\alpha} = \frac{(D^{\alpha}_{0+;L}P)(L_0,C_0)}{\Gamma(\alpha+1)}\Delta_{\alpha}L = \frac{A\Gamma(a+1)}{\Gamma(a-\alpha+1)\Gamma(\alpha+1)}L_0^{a-\alpha}C_0^{b}\Delta_{\alpha}L, \quad (4.42)$$

$$\Delta P_{C,\beta} = \frac{(D^{\beta}_{0+;C}P)(L_0,C_0)}{\Gamma(\beta+1)}\Delta_{\beta}C = \frac{A\Gamma(b+1)}{\Gamma(b-\beta+1)\Gamma(\beta+1)}L_0^{a}C_0^{b-\beta}\Delta_{\beta}C, \quad (4.43)$$

where $\Delta_{\alpha}L = L_1^{\alpha} - L_0^{\alpha}$ and $\Delta_{\beta}C = C_1^{\alpha} - C_0^{\beta}$.

If we consider the orders of the derivatives $\alpha = a$ and $\beta = b$, then equations (4.42) and (4.43) can be written in the form

$$\Delta P_{L,a} = AC_0^{b}\Delta_{a}L, \quad \Delta P_{C,b} = AL_0^{a}\Delta_{b}C. \quad (4.44)$$

As a numerical example, we consider the production function in the form $P(L,C) = 1.01L^{3/4}C^{1/4}$, that is, $A = 1.01$, $a = 0.75$ and $b = 0.25$. For simplicity, the basis (planned) and actual values of the factors are chosen in the form of quaternary powers: $L_0 = 6.1^4$, $L_1 = 6.2^4$ and $C_0 = 0.2^4$, $C_1 = 0.5^4$. In this case, equations (4.42) and (4.43) lead to the following values:

$$\Delta P_{L,\alpha=a} = AC_0^{b}\Delta_{a}L = 2.2920940 \approx 2.29, \quad (4.45)$$

$$\Delta P_{C,\beta=b} = AL_0^{a}\Delta_{b}C = 68.775243 \approx 68.8. \quad (4.46)$$

The total influence of the factors L and C is

$$\Delta P \approx \Delta P_{L,\alpha} + \Delta P_{C,\beta} = 71.067337 \approx 71.1. \quad (4.47)$$

In this case, the increment $\Delta P = P(L_1,C_1) - P(L_0,C_0)$ of the effective indicator P, which is given by the production function $P(L,C) = AL^{a}C^{b}$, has the form

$$\Delta P = P(L_1,C_1) - P(L_0,C_0) = AL_1^{a}C_1^{b} - AL_0^{a}C_0^{b} = 74.505478 \approx 74.5. \quad (4.48)$$

The total influence of factors, which are calculated by the method of differentiation of fractional order, differs from the actual change ΔP approximately by 4.6 %. The additional increase of the increment of production from the interaction of factors is determined by the expression

$$\delta_{\alpha,\beta} = \Delta P - (\Delta P_{L,\alpha} + \Delta P_{C,\beta}) = 3.438141 \approx 3.44. \tag{4.49}$$

Let us now compare the results of the proposed method with results of the standard differential method. If the orders of the Caputo derivatives in equations (4.42) and (4.43) are equal to one ($\alpha = \beta = 1$), then we obtain standard expressions of the standard differential method given by equations (4.5) and (4.6). For the standard approach with the Cobb–Douglas production function $P(L,C) = 1.01L^{3/4}C^{1/4}$, the influence of the factors L and C is described by the values

$$\Delta P_{L,1} = AaL_0^{a-1}C_0^b\Delta L = 2.310983484 \approx 2.31, \tag{4.50}$$

$$\Delta P_{C,1} = AbL_0^aC_0^{b-1}\Delta C = 436.2929478 \approx 436. \tag{4.51}$$

The corresponding total influence of the factors L and C is given by the values

$$\Delta P_{L,1} + \Delta P_{C,1} = 438.6039313 \approx 439. \tag{4.52}$$

The total influence of factors calculated by the standard differential method differs from the real change in the effective index ΔP approximately by 4.9 times, i. e., by 390 %. We emphasize that the difference of the proposed method is 4.6 %, that is, it differs by almost 85 times.

The standard expression of the additional change in the effective index from the interaction of factors is given by

$$\delta = \delta_{1,1} = \Delta P - (\Delta P_{L,1} + \Delta P_{C,1}) = -364.0984533 \approx -364. \tag{4.53}$$

Note that all the first numerical values of equations (4.45)–(4.53) are absolutely exact values like $5^4 = 625$. Approximate values of these numbers are given for simplification.

From a comparison of the values of $\delta_{1,1}$, and the expressions $\delta_{\alpha,\beta}$, it can be seen that the error of the standard method is greater than the error of the proposed methods by more than 100 times

$$\left|\frac{\delta_{1,1}}{\delta_{\alpha,\beta}}\right| \approx 106. \tag{4.54}$$

As a result, the proposed method is a hundred times better than the standard, in the considered example.

As a result, we can state that the proposed differential method of deterministic factor analysis, which is based on fractional differential calculus, can give more accurate results in comparison with the standard method.

4.5 Integral method of arbitrary (noninteger) order

The mathematical basis of the standard integral method is the first-order integration, the fundamental theorem of standard mathematical calculus, and the classical Newton–Leibniz equation

$$\int_a^b f_x^{(1)}(x)\,dx = f(b) - f(a). \tag{4.55}$$

The integral method is one of the most common methods of factor analysis, which allows us to decompose the overall increase in the effective index by factor increments.

Let us give an example of standard equations that describe the relationship between the increment of a function and the increment of factor characteristics. For simplicity, we consider the function of two real variables $z = f(x, y)$. The equation of the integral method, which makes it possible to calculate the influence of the factor x on the resultant indicator, has the form

$$\Delta z_x = \int_{x_0}^{x_1} f_x^{(1)}(x, y)\,dx. \tag{4.56}$$

The influence of factor y is calculated by the equation

$$\Delta z_y = \int_{y_0}^{y_1} f_y^{(1)}(x, y)\,dy. \tag{4.57}$$

It is obvious that as integration variables we can consider $x = \xi$ and $y = \eta$. The integral method allows us to obtain accurate estimations of factor influences and does not imply the separation of factors into quantitative and qualitative ones.

In works [414, 452], we investigate the possibility of applying the integrals of noninteger orders in the deterministic economic factor analysis. As was shown in [360] and [363, pp. 241–264], the inverse operation for the Caputo derivative is the Riemann–Liouville integration of the same order.

Let us give the definition of this integration [200, pp. 69–70].

Definition 4.3. The left-sided Riemann–Liouville integral of the order $\alpha \geq 0$ on the interval $[a, b]$ is defined by the equation

$$(I^{\alpha}_{\mathrm{RL};a+} f)(x) = \frac{1}{\Gamma(\alpha)} \int_a^x \frac{f(\xi)}{(x - \xi)^{1-\alpha}}\,d\xi, \tag{4.58}$$

where $\Gamma(\alpha)$ is the gamma function, $a < x < b$. The function $f(\xi)$ is assumed to satisfy the condition $f(\xi) \in L_1(a, b)$, which implies that $f(\xi)$ is a Lebesgue integrable function on the interval (a, b) and the inequality

$$\int_a^b |f(\xi)|\, d\xi < \infty \tag{4.59}$$

holds.

The Riemann–Liouville integration (4.58) is a generalization of the standard n-fold integration [335, 308, 200]. Note that the Riemann–Liouville integral (4.58) of the order $\alpha = 1$ is equal to the standard integral of the first order

$$(I^1_{\mathrm{RL};a+} f)(x) = \int_a^x f(\xi)\, d\xi. \tag{4.60}$$

Thus, the integration, which is used in the basic equations of the standard integral method of factor analysis, can be considered as a special case of the integration of fractional order.

If the parametric dependence of indicator $z(t)$ on factors $x(t)$ and $y(t)$ can be represented by a single-valued function, then we can use the parametric derivatives and integrals. The parametric fractional integrals are defined [200, pp. 99–100] in the following expression.

Definition 4.4. The left-sided Riemann–Liouville parametric fractional integral of a function f with respect to another function x of the order $\alpha > 0$ on $[a, b]$ is defined by the equation

$$(I^{\alpha,x}_{\mathrm{RL};a+} f)(t) = \frac{1}{\Gamma(\alpha)} \int_a^t \frac{x^{(1)}(\tau)}{(x(t) - x(\tau))^{1-\alpha}} f(\tau)\, d\tau, \tag{4.61}$$

where $x(t) > x(a)$, $a < t < b$, $\alpha > 0$, and $x(\tau)$ is a increasing and positive monotonic function that has a continuous derivative $x^{(1)}(\tau) = dx(\tau)/d\tau$.

Since the integral method is based on the Newton–Leibniz formula and the fundamental theorem of standard calculus, we should use fundamental theorem of fractional calculus and the Newton–Leibniz formula to the case of fractional (noninteger) operators. The fundamental theorem of the theory of integro-differentiation of a fractional (noninteger) order was formulated in [360] and [363, pp. 247–248]. Some additional aspects of this theorem are discussed in [156, 240]. We give a generalization of the Newton–Leibniz formula to the case of integrals and derivatives of noninteger order. For the left-sided operators, the following generalized Newton–Leibniz formula is satisfied

$$(I^{\alpha}_{\mathrm{RL};a+} D^{\alpha}_{\mathrm{C};a+} f)(b) = f(b) - f(a) - \sum_{k=1}^{n-1} \frac{f^{(k)}(a)}{k!} (b - a)^k, \tag{4.62}$$

where $n - 1 < \alpha < n$.

Using the expressions,

$$(I^1_{\mathrm{RL};a+} D^1_{C;a+} f)(b) = \int_a^b f^{(1)}_x(x)\, dx, \tag{4.63}$$

we see that equation (4.62) with $\alpha = 1$ gives the standard Newton–Leibniz formula (4.55).

To apply the method of fractional integration of noninteger order in deterministic factor analysis, we can use the equation for the left-sided Riemann–Liouville integral of order $\alpha \geq 0$ for the power function [335, p. 71] that has the form

$$I^{\alpha}_{\mathrm{RL};a+}(x-a)^{\beta} = \frac{\Gamma(\beta+1)}{\Gamma(\beta+\alpha+1)}(x-a)^{\beta+\alpha}, \quad (\beta > 0). \tag{4.64}$$

The use of the fractional calculus makes it possible to obtain more accurate results for the influence of factors in comparison with the differential method of noninteger order. This is due to the fact that the additional increase in the resultant indicator, which arises from the interaction of factors, is distributed among them in equal proportions.

As an example, we give equation for power-law function of two variables, which is analogous to the Cobb–Douglas function considered above. For the function $P(L, C) = AL^{\alpha}C^{\beta}$ with $\alpha > 0$ and $\beta > 0$, similar to the standard deterministic factor analysis [345], we have

$$\delta_{\alpha,\beta} = \Delta P - \Delta P_{x,\alpha} - \Delta P_{y,\beta} = A\Delta_{\alpha}L\Delta_{\beta}C, \tag{4.65}$$

where $\Delta P = P(L_1, C_1) - P(L_0, C_0)$, and $\Delta P_{L,\alpha}$, $\Delta P_{C,\beta}$ are defined by equations (4.29) and (4.30). Then using $\Delta P = \Delta P_{L,\alpha} + \Delta P_{C,\beta}$ and

$$\Delta P_{L,\alpha} = \frac{(D^{\alpha}_{0+;L}P)(L_0, C_0)}{\Gamma(\alpha+1)}\Delta_{\alpha}L + \frac{1}{2}\delta_{\alpha,\beta}, \tag{4.66}$$

$$\Delta P_{C,\beta} = \frac{(D^{\beta}_{0+;C}P)(L_0, C_0)}{\Gamma(\beta+1)}\Delta_{\beta}C + \frac{1}{2}\delta_{\alpha,\beta}, \tag{4.67}$$

we obtain the expressions

$$\Delta P_{L,\alpha} = AC^{\beta}_0\Delta_{\alpha}L + \frac{1}{2}A\Delta_{\alpha}L\Delta_{\beta}C = \frac{1}{2}A(C^{\beta}_1 + C^{\beta}_0)\Delta_{\alpha}x, \tag{4.68}$$

$$\Delta P_{C,\beta} = AL^{\alpha}_0\Delta_{\beta}C + \frac{1}{2}A\Delta_{\alpha}L\Delta_{\beta}C = \frac{1}{2}A(L^{\alpha}_1 + L^{\alpha}_0)\Delta_{\beta}C. \tag{4.69}$$

For $\alpha = \beta = 1$, equations (4.65)–(4.69) give the standard equations for the integral method for a multiplicative model with the function $P(L, C) = ALC$ [345, pp. 67, 323].

For the production function $P(L,C) = 1.01L^{3/4}C^{1/4}$, the planned and actual values of the factors $L_0 = 6.1^4$, $L_1 = 6.2^4$, $C_0 = 0.2^4$, $C_1 = 0.5^4$, equations (4.68)–(4.69) gives the values

$$\Delta P_{L,\alpha=a} = AC_0^b \Delta_a L + \frac{A}{2}\Delta_\alpha L \Delta_\beta C = 4.0111645 \approx 4.0, \tag{4.70}$$

$$\Delta P_{C,\beta=b} = AL_0^a \Delta_b C + \frac{A}{2}\Delta_\alpha L \Delta_\beta C = 70.4943135 \approx 70. \tag{4.71}$$

As a result, we get

$$\Delta P_{L,a} + \Delta P_{C,b} = \Delta P = 74.5054780 \approx 74. \tag{4.72}$$

As a result, the sum of influences of the labor ($\Delta P_{L,a}$) and capital ($\Delta P_{C,b}$) is exactly equal to the total increment of production (ΔP).

In addition to (4.64), there is a set of the integration formulas of a noninteger order that are given in Tables 9.1–9.2 of book [335, pp. 173–174]. To apply the proposed generalization of the integral method, it is required to use these tables of formulas that allow us to develop final working formulas for the most common types of factor dependencies and make this method more accessible.

4.6 Conclusion

The proposed integral and differential methods, which are based on the fractional calculus as a theory of integrals and derivatives of noninteger orders expand the scope of the deterministic factor analysis. The differential method of noninteger order can give more accurate results in comparison with the standard method, which is used the derivatives of integer orders, for a wide class of functions including the power functions. In addition, the integral and differential methods of non-integer order make it possible to take into account the effects of memory and nonlocality in economic processes.

The proposed fractional differential and integral methods of the deterministic factor analysis can be used in the analysis of economic or financial processes, in which indicators are power-law functions of factors. These methods can be used to study the processes, which are described by power laws (e. g., the Cobb–Douglas production function). The suggested methods allow us to more accurately describe the total influence of factors compared with the standard methods.

The advantage of the proposed methods is especially clear for processes described by power-law functions and functions, which are representable by series with noninteger powers.

In application of suggested methods of the deterministic factor analysis, it should be taken into account that derivatives of noninteger order have a number of nonstandard properties, including violation of the standard product rule and the chain rule [367, 377, 376, 382, 338, 81].

5 Elasticity for processes with memory

In this chapter, we describe generalizations of concept of point elasticity for processes with memory by using fractional derivatives of noninteger orders. The proposed concepts of elasticity of $Y(t)$ with respect to $X(t)$ takes into account the dependence on changes of these variables at previous points in time. As an example, we also define generalizations of point price elasticity of demand to the case of processes with power-law memory. In these generalizations, we take into account dependence of demand not only from price at current time, but also all changes of price and demand on some time interval. For simplification, we will assume that there is one parameter, which characterizes memory fading. The properties of the proposed fractional elasticities for processes with memory and examples of calculations of these elasticities are presented. This chapter is based on articles [418, 430].

5.1 Introduction

One of the important economic concepts is the concept of elasticity. In microeconomics, the following concepts are often used:

- the price elasticity of demand;
- the income elasticity of demand;
- the price elasticity of supply;
- the elasticity of substitution.

Elasticity shows a relative change of an economic indicator under influence of change of an economic factor on which it depends, when remaining factors acting on it are constant. Memory effects are ignored in the standard concept of elasticity. For example, the standard point-price elasticity of demand is defined by the equation

$$E(Q;P) = EQ_P = \frac{P}{Q}\frac{dQ}{dP}, \tag{5.1}$$

where Q is the quantity demanded and P is the price of a good.

The most important condition for the applicability of equation (5.1) is the assumption that the indicator $Y = Q$ can be represented as a single-valued function of the factor $X = P$. In general, this assumption is not satisfied [429, 419], and the dependence of Q on P is multivalued, that is, several different values of Q can correspond to the same value of P.

If the dependence of the indicator $Y = Q$ on the factor $X = P$ is not single-valued function, we cannot use equation (5.1) to calculate the price elasticity of demand. To avoid this problem, we can use the fact that the indicator $Y = Q$ and the factor $X = P$ can be considered as singlee-valued functions of time. In this case, to define point elasticity, we can use parametric dependence of an indicator Y on a factor X in the

https://doi.org/10.1515/9783110627459-005

form of the single-valued functions $Y = Y(t)$ and $X = X(t)$, where the parameter t is time.

As a result, we can define the standard point-price elasticity of demand by the equation

$$E(Q(t); P(t); t) = \frac{P(t)}{Q(t)} \frac{dQ(t)/dt}{dP(t)/dt}, \tag{5.2}$$

where $Q(t) \neq 0$ and $dP(t)/dt \neq 0$. Equation (5.2) assumes that the elasticity depends only on the current price at t and price at infinitesimal neighborhood of point t.

If the dependence of $Q(t)$ on $P(t)$ can be represented as a single-valued differentiable function $Q = Q(P)$, by excluding the time parameter t, then equation (5.2) will be equivalent to standard equation (5.2) by the standard chain rule

$$\frac{dQ(P(t))}{dt} = \left(\frac{dQ(P)}{dP}\right)_{P=P(t)} \frac{dP(t)}{dt}. \tag{5.3}$$

In other words, if the function $P = P(t)$ is invertible in a neighborhood of the point t, then

$$E(Q(t); P(t); t) = \frac{P(t)}{Q(t)} \frac{dQ(t)/dt}{dP(t)/dt} = \frac{P}{Q(P)} \frac{dQ(P)}{dP} = E(Q; P). \tag{5.4}$$

Equations (5.2) and (5.1) are equivalent, if the function $P = P(t)$ is reversible.

Note that equation (5.2) can be used if the indicator $Y = Q$ cannot be represented as a single-valued function of the factor $X = P$, when equation (5.1) cannot be applied.

Example 5.1. The ambiguity of the demand function $Q = Q(P)$ arises when, for the same price $P(t_1) = P(t_2)$, the demand $Q(t_1)$ and $Q(t_2)$ takes different values. In other words, despite the same price at different points in time, the demand at these points in time differs while other parameters of the model remain unchanged. Mathematically, this property can be written in the form: there are such $t_1, t_2 \in [t_i, t_f]$, that $Q(t_1) \neq Q(t_2)$ for $P(t_1) = P(t_2)$. As a second illustrative example, let us assume that the price at equal time intervals $[t_1, t_2]$ and $[t_3, t_4]$ behaves the same (the graphs $P(t)$ are shifted in these areas coincide), when the behavior of demand $Q(t)$ at these intervals can differ significantly. Mathematically, this can be written as the inequality $Q(t) \neq Q(t + T)$, when $P(t) = P(t + T)$ for $t \in [t_1, t_2]$, where $T = t_2 - t_1 = t_4 - t_3$ is the duration of these time intervals.

As a result, we can state that equation (5.2) is more general than equation (5.1). In addition to take into account a memory effect, we should include the time variable into equations.

In general, we must take into account that demand may depend on all changes of prices over a finite time interval. This is due to the fact that behavior of buyers can be determined by the presence of a memory of previous price changes. We can say that expression (5.2) can be used only if all buyers have a total amnesia.

In 2016, we proposed [418, 430] a generalization of the point elasticity by using fractional derivatives to remove amnesia of buyers in the concept of elasticity. Fractional derivatives of noninteger orders allow us to take into account a fading memory. Therefore, the proposed fractional generalization of point-price elasticity of demand cannot be considered as a point economic indicator. A point indicator is local in time, and it cannot describe such nonlocal property as memory. We proposed the concept of elasticity, which takes into account a memory, depends on finite interval of time and, therefore, is nonlocal in time. To simplify, we will assume that there is only one parameter α, which characterizes memory fading.

It should be emphasized that the standard properties of the elasticity are not satisfied for elasticity that takes into account a fading memory. This is due to the fact that the fractional derivatives of noninteger orders have a set of nonstandard properties [367, 377, 376, 382, 81]. For example, such properties include a violation of the standard product rule (the Leibniz rule) and a violation of standard chain rules [367, 377, 376, 382, 81].

It should be emphasized that the violations of these rules can be considered as characteristic properties of fractional derivatives of noninteger orders. The nonstandard properties of fractional derivatives lead to the violation of the properties of the standard elasticity (elasticity without memory), which is defined by equations (5.1) and (5.2).

5.2 Concept of elasticity with memory

Elasticity is a measure of the proportional change of an economic variable in response to a change in another. Elasticity shows the relative change in an economic indicator under the influence of changes in the economic factors on which it depends, at constant remaining factors affecting it.

Elasticity can be considered as a measure of the sensitivity of one variable (e. g., supply or demand) to another change (e. g., price, income), showing how many percent the first indicator will change, when the second changes by one percent.

In this section, we define generalizations of point elasticity of Y with respect to X by taking into account a memory. Let us consider the following type of generalizations of the elasticity: In general, economic indicators and factors can be considered as single-valued functions of time $t \in [t_i; t_f]$. The absence of memory (amnesia) and lag means that the state of economic process at time point t is determined only by the states of process in infinitely small neighborhood of this point. The presence of a memory means that the value of $Y(t)$ depends on values of $X(\tau)$ and $Y(\tau)$ at all points $\tau \in [t_i; t]$.

First, we give a definition of elasticity with memory in a general form by using the operators introduced earlier to describe the effects of memory.

Definition 5.1. Let the variables $X = X(t)$ and $Y = Y(t)$ be single-valued differentiable m and n times, respectively, functions with respect to time variable $t \in [t_i; t_f]$. The generalized $\mathcal{F}$-elasticity $E_{\mathcal{F}}^{(n,m)}(Y(t); X(t); [t_i, t])$ of $Y = Y(t)$ with respect to $X = X(t)$ for processes with memory is defined by the equation

$$E_{\mathcal{F}}^{(n,m)}(Y(t); X(t); [t_i, t]) = \frac{X(t)}{Y(t)} \frac{\mathcal{F}_{t_i}^{t}(Y^{(n)}(\tau))}{\mathcal{F}_{t_i}^{t}(X^{(m)}(\tau))}, \tag{5.5}$$

where $t \in [t_i, t_f]$, $\mathcal{F}_{t_i}^{t}(X^{(m)}(\tau)) \neq 0$, $Y(t) \neq 0$.

Definition 5.2. Let the variables $X = X(t)$ and $Y = Y(t)$ be single-valued differentiable m and n times, respectively, functions with respect to time variable $t \in [t_i; t_f]$, and the memory is described by memory functions $M_x(t,\tau)$ and $M_y(t,\tau)$ for X and Y, respectively. Then the generalized M-elasticity $E_M^{(n,m)}(Y(t); X(t); [t_i, t])$ of $Y = Y(t)$ with respect to $X = X(t)$ for processes with memory, which is described by memory functions $M_x(t,\tau)$ and $M_y(t,\tau)$, is defined by the equation

$$E_M^{(n,m)}(Y(t); X(t); [t_i, t]) = \frac{X(t)}{Y(t)} \frac{\int_{t_i}^{t} M_y(t,\tau) Y^{(n)}(\tau)\, d\tau}{\int_{t_i}^{t} M_x(t,\tau) X^{(m)}(\tau)\, d\tau}, \tag{5.6}$$

where $t \in [t_i, t_f]$, the expressions in the denominators of the fractions are assumed to be nonzero.

For simplification, we will assume that there is one parameter α, which characterizes the power-law memory fading. In this case, to generalize the concept of elasticity for economic processes with memory, we can use the theory of fractional derivatives of noninteger order. Note that the fractional derivatives with positive integer values of the orders α coincide with the standard integer-order derivatives. There are various types of derivatives of noninteger orders. We will use the Caputo fractional derivatives of the orders $\alpha \geq 0$. One of the distinguishing features of this derivative is that its action on the constant function gives zero. The order α of the Caputo fractional derivative is interpreted as the memory fading parameter.

Definition 5.3. Let the variables $X = X(t)$ and $Y = Y(t)$ be single-valued functions of time variable, which have the Caputo fractional derivatives of the orders α_x and α_y for $t \in [t_i; t_f]$, respectively. The fractional T-elasticity $E_\alpha(Y(t); X(t); [t_i, t])$ of the order $\alpha = \{\alpha_y, \alpha_x\}$ of $Y = Y(t)$ with respect to $X = X(t)$ is defined by the equation

$$E_\alpha(Y(t); X(t); [t_i, t]) = \frac{X(t)}{Y(t)} \frac{{}_{t_i}^{C}D_t^{\alpha_y}[\tau] Y(\tau)}{{}_{t_i}^{C}D_t^{\alpha_x}[\tau] X(\tau)}, \tag{5.7}$$

where α_x and α_y are parameters of memory fading with respect to $X = X(t)$ and $Y = Y(t)$, respectively, ${}_{t_i}^{C}D_t^{\alpha}[\tau]$ is the left-sided Caputo fractional derivative of order α that is defined by the equation

$$ {}_{t_i}^{C}D_t^{\alpha}[\tau]f(\tau) = \frac{1}{\Gamma(n-\alpha)} \int_{t_i}^{t} (t-\tau)^{n-\alpha-1} f^{(n)}(\tau)\, d\tau, \tag{5.8} $$

with $t \in [t_i, t_f]$ and $n-1 < \alpha \leq n$. Here, it is assumed that $Y(t) \in AC^n[t_i, t_f]$, where $n-1 < \alpha_y \leq n$, and $X(t) \in AC^m[t_i, t_f]$, where $m-1 < \alpha_x \leq m$.

The fractional T-elasticity $E_\alpha(Y(t); X(t); [t_i, t])$ is a measure of sensitivity of the variable $Y(t)$ in response to a change in another variable $X(t)$ for economic processes with power-law memory. The proposed concept of elasticity with memory takes into account the dependences on the current point in time, but also on changes of the variables $Y(\tau)$ and $X(\tau)$ in the past for $\tau \in [t_i; t]$. The parameter $\alpha = \{\alpha_y, \alpha_x\}$ characterizes the fading of the memory about changes of the variables $Y(t)$ and $X(t)$ in past. In the special case, we can consider fractional T-elasticity with two coincident parameters of memory fading, i. e., with $\alpha_y = \alpha_x = \alpha$.

Let us give examples of generalizations of point-price elasticity of demand for economic processes with memory. These generalizations take into account demand and price not only at the current time, but also changes of these variables in a finite time interval.

Example 5.2. Let demand $Q = Q(t)$ and price $P = P(t)$ be single-valued functions of time variable t, which have the Caputo fractional derivatives of the order $\alpha > 0$ on the interval $[t_i, t_f]$. The elasticity with memory $E_\alpha(Q(t); P(t); [t_i, t])$ of demand $Q(t)$ with respect to price $P(t)$ (the price elasticity of demand with memory) is defined by the equation

$$ E_\alpha(Q(t); P(t); [t_i, t]) = \frac{P(t)}{Q(t)} \frac{{}_{t_i}^{C}D_t^{\alpha}[\tau]Q(\tau)}{{}_{t_i}^{C}D_t^{\alpha}[\tau]P(\tau)}, \tag{5.9} $$

where $\tau \in [t_i, t]$, and $t_i < t < t_f$, and the denominators are not equal to zero. Here we assume that $\alpha_Q = \alpha_P = \alpha$. In general, we can consider $\alpha_Q \neq \alpha_P$.

The fractional T-elasticity (5.9) describes a price elasticity of demand for the processes with memory. This type of elasticity depends not only on demand $Q(t)$ and price $P(t)$ at the current time t, but also on changes in demand $Q(\tau)$ and price $P(\tau)$ for $\tau \in [t_i; t]$. The order α is interpreted as the parameter that characterizes the memory fading for Q and P.

Example 5.3. Analogously to generalization of the price elasticities of demand, we can generalize of other types of elasticity. For example, we can give definition of income elasticity of demand for processes with memory. Using the demand function $Q = Q(\tau)$ and income function $I = I(\tau)$ that are defined on the time interval $[t_i; t_f]$, the elasticity with memory (fractional T-elasticity of order α) $E_{\alpha}(Q(t);I(t);[t_i,t])$ can be defined by the equation

$$E_{\alpha}(Q(t);I(t);[t_i,t]) = \frac{I(t)}{Q(t)} \frac{{}_{t_i}^{C}D_t^{\alpha}[\tau]Q(\tau)}{{}_{t_i}^{C}D_t^{\alpha}[\tau]I(\tau)}, \tag{5.10}$$

where the denominators are not equal to zero.

Remark 5.1. Note that we use the Caputo fractional derivatives in order to fractional elasticity of constant demand will be equal to zero, that allows us to correspond it to a perfectly inelastic demand.

Let us consider a generalization of the Log-elasticity of Y with respect to X.

Using the equation,

$$\frac{d\ln(X(t))}{dt} = \frac{1}{X(t)}\frac{dX(t)}{dt}, \tag{5.11}$$

the standard (point) elasticity of Y with respect to X can be represented as a derivative of $F(t) = \ln(Y(t))$ by $G(t) = \ln(X(t))$ in the form

$$E(F(t), G(t), t) = \frac{dF(t)/dt}{dG(t)/dt} = \frac{d\ln(Y(t))/dt}{d\ln(X(t))/dt}. \tag{5.12}$$

Remark 5.2. Equation (5.12) can be interpreted as a marginal value of the indicator $F(t)$ with respect to the factor $G(t)$.

Using the concept of the marginal value with memory, expression (5.12) allows us to define the new fractional generalization of elasticity in the following form:

$$E_{\alpha}(Y(t);X(t);[t_i,t]) = MF_G(\alpha_y,\alpha_x,[t_i,t]) = \frac{{}_{t_i}^{C}D_t^{\alpha_y}[\tau]F(\tau)}{{}_{t_i}^{C}D_t^{\alpha_x}[\tau]G(\tau)}, \tag{5.13}$$

where $F(t) = \ln(Y(t))$ and $G(t) = \ln(X(t))$. Expression (5.13) allows us to formulate the definition in the following form.

Definition 5.4. Let the variables $X = X(t)$ and $Y = Y(t)$ be single-valued functions of time variable, for which the functions $F(t) = \ln(Y(t))$ and $G(t) = \ln(X(t))$ have the Caputo fractional derivatives of the orders α_x and α_y for $t \in [t_i;t_f]$, respectively. Then the fractional Log-elasticity $E_{\alpha,\mathrm{Log}}(Y(t);X(t);[t_i,t])$ of the order $\alpha = \{\alpha_y,\alpha_x\}$ (the elasticity with memory of $Y = Y(t)$ with respect to $X = X(t)$) is defined by the equation

$$E_{\alpha,\mathrm{Log}}(Y(t);X(t);[t_i,t]) = \frac{{}_{t_i}^{C}D_t^{\alpha_y}[\tau]\ln\{Y(\tau)\}}{{}_{t_i}^{C}D_t^{\alpha_x}[\tau]\ln\{X(\tau)\}}, \tag{5.14}$$

where $t \in [t_i,t_f]$, and the parameters α_x and α_y characterize the fading of memory for $X = X(t)$ and $Y = Y(t)$, respectively.

Remark 5.3. Using that

$${}_{t_i}^{C}D_t^1[\tau]f(\tau) = \frac{df(t)}{dt}, \tag{5.15}$$

the fractional elasticity (5.7) with $\alpha = 1$ takes the standard form

$$E_1(Y(t); X(t); [t_i, t]) = \frac{X(t)}{Y(t)} \frac{dY(t)/dt}{dX(t)/dt}. \tag{5.16}$$

Equation (5.16) describes the standard point elasticity. This means that in the case $\alpha = 1$, the suggested elasticities correspond to the economic processes without memory. The fractional elasticity of the order $\alpha = 1$ takes the form of standard elasticity that can be used to describe only economic processes without memory, i. e., in the framework of economics with total amnesia of agents. We should emphasize that the standard point elasticity of Y with respect to X describes economic processes, in which all economic agents have lost their memory and have amnesia.

Remark 5.4. Expression (5.12) also allows us to define the corresponding fractional generalization by using the fractional derivative of function $F(t) = \ln(Y(t))$ with respect to the function $G(t) = \ln(X(t))$ (see Section 18.2 of [335] and Section 2.5 of [200]), which are also called the parametric fractional derivatives. The fractional Param-elasticity $E_{\alpha,\text{Param}}(Y(t); X(t); [t_i, t])$ of order α at $t \in [t_i; t_f]$ is defined by the equation

$$E_{\alpha,\text{Param}}(Y(t); X(t); [t_i, t]) = \frac{1}{\Gamma(n-\alpha)} \int_{t_i}^{t} d\tau G^{(1)}(\tau) \frac{F(\tau)}{(G(t) - G(\tau))^{\alpha+1-n}} \left(\frac{1}{G^{(1)}(\tau)} \frac{d}{d\tau} \right)^n f(\tau), \tag{5.17}$$

where $t > t_i$, $n - 1 \le \alpha \le n$, $F(t) = \ln(Y(t))$ and $G(t) = \ln(X(t))$. However, the scope of application of this concept of "parametric" elasticity (5.17) is strongly restricted by the conditions for the function $G(\tau)$, which should be an increasing positive monotonic function.

5.3 Properties of elasticity with memory

Let us give some basic properties of proposed elasticities with memory (5.7) (the fractional T-elasticity) with $\alpha_y = \alpha_x = \alpha$ that is defined by the equation

$$E_\alpha(Y(t); X(t); [t_i, t]) = \frac{X(t)}{Y(t)} \frac{{}_{t_i}^{C}D_t^\alpha[\tau]Y(\tau)}{{}_{t_i}^{C}D_t^\alpha[\tau]X(\tau)}. \tag{5.18}$$

For simplification, we describe these properties for the case $t_i = 0$.

1. For $\alpha_x = \alpha_y = \alpha$, the elasticity with memory is a dimensionless quantity,

$$E_\alpha(\lambda Y(t); X(t); [0,t]) = E_\alpha(Y(t); X(t); [0,t]), \tag{5.19}$$

$$E_\alpha(Y(t); \lambda X(t); [0,t]) = E_\alpha(Y(t); X(t); [0,t]). \tag{5.20}$$

   These equations mean that the elasticity with memory does not depend on units of the economic indicator Y and the economic factor X.

2. The elasticity with memory of inverse function is inverse

$$E_\alpha(X(t); Y(t); [0,t]) = \frac{1}{E_\alpha(Y(t); X(t); [0,t])}. \tag{5.21}$$

3. In general, the elasticity with memory for the product of two functions, which depend on the same factor, does not equal to the sum of elasticities

$$E_\alpha(Y_1(t) \cdot Y_2(t); X(t); [0,t]) \neq E_\alpha(Y_1(t); X(t); [0,t]) + E_\alpha(Y_2(t); X(t); [0,t]) \tag{5.22}$$

   for $\alpha \neq 1$. This inequality becomes an equality for $\alpha = 1$. Inequality (5.22) caused by the violation of the standard product rule (the Leibniz rule) for the fractional derivatives of noninteger orders [367, 376, 338, 81].

4. The elasticity with memory for the sum of two functions, which depend on the same argument, is given by the equation

$$E_\alpha(Y_1(t) + Y_2(t); X(t); [0,t]) = \frac{Y_1(t)}{Y_1 + Y_2} E_\alpha(Y_1(t); X(t); [0,t]) + \frac{Y_2(t)}{Y_1 + Y_2} E_\alpha(Y_2(t); X(t); [0,t]). \tag{5.23}$$

5. The elasticity with memory for the power-law functions $Y(t) = \lambda_y t^\beta$ and $X(t) = \lambda_x t^\gamma$ takes the constant value

$$E_\alpha(\lambda_y t^\beta; \lambda_x t^\gamma; [0,t]) = \frac{\Gamma(\beta+1)\Gamma(\gamma-\alpha+1)}{\Gamma(\beta-\alpha+1)\Gamma(\gamma+1)}, \tag{5.24}$$

   where $n-1 < \alpha < n$ and $\beta, \gamma > n-1$ with $n \in \mathbb{N}$.

6. The elasticity with memory of the constant function $Y(t) = \text{const} \neq 0$ is given by the equation

$$E_\alpha(\text{const}; X(t); [0,t]) = 0, \tag{5.25}$$

   if ${}_{t_i}^{C}D_t^\alpha[\tau]X(\tau) \neq 0$.

7. The elasticity with memory with $t_i = 0$ for the Mittag–Leffler functions $E_\alpha[z]$ [143] is described by the equation

$$E_\alpha(E_\alpha[\lambda_y t^\alpha]; E_\alpha[\lambda_x t^\alpha]; [0,t]) = \frac{\lambda_y}{\lambda_x}, \tag{5.26}$$

where $\alpha > 0$, $\lambda_x \neq 0$, and we use equation 2.4.58 of [200, p. 98]. Note that $E_1[z] = e^z$ (see equation 1.8.2 in [200, p. 40]). Therefore, property (5.26) generalizes the property of the standard elasticity for exponential function.

8. The elasticity with memory with $t_i = -\infty$ for the exponential functions is described by the expression

$$E_\alpha(e^{-\lambda_y t}; e^{-\lambda_x t}; (-\infty, t]) = \left(\frac{\lambda_y}{\lambda_x}\right)^\alpha, \tag{5.27}$$

where $\lambda_x, \lambda_y > 0$, and we use equation 2.4.57 of [200, p. 98].

9. For fractional derivatives of noninteger orders, the standard chain rule is not satisfied in general (see equation (2.209) in Section 2.7.3 of [308, p. 98]). Therefore, the elasticity with memory (5.7) and the fractional Log-elasticity (5.14) should be considered as independent characteristics of economic processes with memory.

Note that these properties are directly derived from the properties of the Caputo fractional derivative [200, 90] and the definition of the elasticity with memory (5.18).

5.4 Elasticity through marginal value with memory

It is known that the standard elasticity $E(Y;X) = EY_X$ of the indicator Y by the factor X, for which there is a single-valued function $Y = Y(X)$, can be defined through the standard average and marginal values by the equation

$$EY_X = \frac{MY_X}{AY_X} = \frac{X}{Y(X)} \frac{dY(X)}{dX}, \tag{5.28}$$

where

$$AY_X = \frac{Y(X)}{X}, \quad MY_X = \frac{dY(X)}{dX}. \tag{5.29}$$

Let us assume that $Y = Y(t)$ and $X = X(t)$ are the single-valued functions of time variable t. In the case of parametric dependences of an indicator Y on a factor X, the standard elasticity can also be define through the standard average and marginal values by the equation

$$E(Y(t); X(t); t) = \frac{MY_X(t)}{AY_X(t)}, \tag{5.30}$$

where

$$AY_X(t) = \frac{Y(t)}{X(t)}, \quad MY_X(t) = \frac{dY(t)/dt}{dX(t)/dt}, \tag{5.31}$$

and $X(t), Y(t) \neq 0$, $dX(t)/dt \neq 0$.

The standard average and marginal values are special cases of the marginal value with memory, which are proposed in 2016 [424, 416] (see also [425, 439, 446]). Using the one-parameter marginal values with memory, expression (5.30) can be represented in the form

$$E(Y(t);X(t);t)=\frac{MY_X(1,[0,t])}{MY_X(0,[0,t])}, \tag{5.32}$$

since $MY_X(0,[0,t])=AY_X(t)$ and $MY_X(1,[0,t])=MY_X(t)$.

Using the one-parameter marginal values with memory, we can define the two-parameter generalized elasticity with memory.

Definition 5.5. The two-parameter generalized elasticity of Y over X is defined by the equation

$$EY_X(\alpha,\beta,[0,t])=\frac{MY_X(\alpha,[0,t])}{MY_X(\beta,[0,t])}, \tag{5.33}$$

where $\alpha \geq 0$ and $\beta \geq 0$ are parameters of memory fading, and $MY_X(\alpha,[0,t])$, $MY_X(\beta,[0,t]) \neq 0$ are one-parameter marginal values with memory (2.21).

The elasticity (5.33) with memory shows the relative change in Y as a function of the change in factor X for various memory fading parameters. Note that the concept of elasticity of noninteger order was proposed in [418, 430]. For $\alpha = 1$, $\beta = 0$ and the condition of existence of a single-valued function $Y = Y(X)$, equation (5.33) gives expression (5.28), and we have the equality

$$EY_X(1,0,[0,t])=EY_X, \tag{5.34}$$

where $EY_X = E(Y;X)$ is standard elasticity (5.28).

Using the representation (5.33) of elasticity with memory through marginal values with memory, it is easy to verify that the following properties:

$$EY_X(\alpha,\beta,[0,t])=\frac{1}{EY_X(\beta,\alpha,[0,t])}, \tag{5.35}$$

$$EY_X(0,1,t)=\frac{1}{EX_Y(1,0,[0,t])}=\frac{1}{EY_X}=EX_Y, \tag{5.36}$$

$$EY_X(\alpha,\alpha,[0,t])=1. \tag{5.37}$$

For descriptive reasons, we write properties (5.34) and (5.36) in the form

$$EY_X(1,0,[0,t])=E(Y;X), \quad EY_X(0,1,[0,t])=E(X;Y). \tag{5.38}$$

From the given equations, we can see the difference between property (5.35) of two-parameter generalized elasticity with memory (5.33) and property (5.21) of one-parameter elasticity with memory (5.7).

We should emphasize that two-parameter and one-parameter elasticities with memory are nonlocal characteristics of economic processes for noninteger values of memory fading parameters, since these characteristics depend on the behavior of variables on a finite time interval $[t_i, t]$, where we use took $t_i = 0$ for convenience.

5.5 Generalized marginal rate of substitution with memory

Generalizing equation (5.28) to the case of several factors X_m with its memory fading parameters α_m, where $m = 1, 2, \dots, k$, we can define the marginal rate of substitution (MRS) with memory.

Definition 5.6. Let $Y(t)$ and $X_m = X_m(\tau)$, where $m = 1, 2, \dots, k$, be single-valued functions of time variable $t \in [t_i; t_f]$, which have the Caputo fractional derivatives of order α_m for $t \in [t_i; t_f]$, respectively. Then the marginal rate of substitution (MRS) with memory of one factor X_i by another X_j for economic processes with memory, where memory fading parameters are $\alpha_m > 0$, $m = 1, 2, \dots, k$, is defined by the equation

$$\mathrm{MRS}_{ij}(\alpha_i, \alpha_j,) = \frac{MY_{X_i}(\alpha_i, [0, t])}{MY_{X_j}(\alpha_j, [0, t])}, \tag{5.39}$$

where $i, j = 1, 2, \dots, k$, and it is assumed that the denominator is not equal to zero.

In a simplified case, all the parameters of memory fading can be considered the same ($\alpha_m = \alpha$ for all $m = 1, 2, \dots, k$).

It should be noted that in the standard interpretation, the marginal rate of substitution (MRS) is considered as a rate, at which the consumer wants to replace (or refuse) a certain quantity of one product in exchange for another, while maintaining the same level of utility. In this case, the marginal rate of substitution with memory of good or service Y for good or service X should be considered as a ration of the marginal utility $\mathrm{MU}_X(\alpha, [0, t])$ with memory about $X(t)$ to the marginal utility $\mathrm{MU}_Y(\beta, [0, t])$ with memory about $Y(t)$. This leads to the expression

$$\mathrm{MRS}_{XY}(\alpha, \beta, [0, t]) = \frac{\mathrm{MU}_X(\alpha, [0, t])}{\mathrm{MU}_Y(\beta, [0, t])}. \tag{5.40}$$

For $\alpha = \beta = 1$, equation (5.40) takes the standard expression

$$\mathrm{MRS}_{XY}(1, 1, [0, t]) = \frac{\mathrm{MU}_X(1, [0, t])}{\mathrm{MU}_Y(1, [0, t])} = \frac{U_t^{(1)}}{X_t^{(1)}} : \frac{U_t^{(1)}}{Y_t^{(1)}} = \frac{Y_t^{(1)}}{X_t^{(1)}} = \frac{dY}{dX}, \tag{5.41}$$

if the single-valued function $Y = Y(X)$ exists.

In the interpretation, which is given by equation (5.39), the marginal rate of substitution with memory can be considered as a generalization of the proposed concept of elasticity with memory.

Statement 5.1. *The concept of two-parameter elasticity with power-law memory* $EY_X(\alpha, \beta, [0, t])$ *is a special case of the the marginal rate of substitution (MRS) with memory* $\mathrm{MRS}_{YX}(\alpha, \beta, [0, t])$.

Proof. Let us prove this statement by the following simple example. For $k = 2$ and $X_1 = Y, X_2 = X, \alpha_1 = \alpha, \alpha_2 = \beta$, and using the definitions of the elasticity with memory (5.33) and the marginal rate of substitution (MRS) with memory (5.39), we get the equality

$$\mathrm{MRS}_{YX}(\alpha, \beta, [0, t]) = EY_X(\alpha, \beta, [0, t]). \tag{5.42}$$

As a result, the elasticity with memory (5.33) is a special case of the marginal rate of substitution (MRS) with memory (5.39). □

Note that an interpretation, which is given by this statement and equation (5.42), is not applicable for standard concepts in the framework of economics without memory.

5.6 Nonlocal elasticity of noninteger order

In this section, instead of fractional T-elasticity, we consider a different type of elasticity, which will be called fractional X-elasticity. This elasticity will take into account nonlocality in the following form. The presence of a memory can also mean that the values of $Y(X_0)$ actually depends not only on X_0, but it also a depends on X from intervals $[X_i; X_0)$ and $(X_0; X_f]$. Here, we should assume that indicator Y can be represented as single-valued function of the factor X on the interval $[X_i; X_0)$ or $(X_0; X_f]$. Here, we should assume that the time parameter can be excluded to have a single-valued function $Y = Y(X)$ for processes with memory.

Definition 5.7. Let us consider an economic indicator $Y = Y(X)$ as a single-valued function of an economic factor $X \in [X_i; X_f]$ that has integer-order derivatives up to $(n-1)$-th order, which are continuous functions on the interval $[X_i; X_f]$, and the derivative $Y^{(n)}(X)$ is Lebesgue summable on the interval $[X_i; X_f]$. The left-sided and right-sided fractional X-elasticities $E_{\alpha,l}(Y(X); [X_i, X])$ and $E_{\alpha,r}(Y(X); [X, X_f])$ of Y with respect to $X \in [X_i; X_f]$ with the order $\alpha > 0$ are defined by the equations

$$E_{\alpha,l}(Y(X); [X_i, X]) = \frac{X^\alpha}{Y(X)} {}_{X_i}^{C}D_X^\alpha[x] Y(x), \tag{5.43}$$

$$E_{\alpha,r}(Y(X); [X, X_f]) = \frac{X^\alpha}{Y(X)} {}_{X}^{C}D_{X_f}^\alpha[x] Y(x), \tag{5.44}$$

where $X > 0$, $Y(X) \neq 0$, the values $X_i = X_{\min}$ and $X_f = X_{\max}$ are initial and final points of the investigated interval of the economic factor $X \in [X_i; X_f]$. Here ${}_{X_i}^{C}D_X^\alpha[x]$ is the left-sided Caputo derivative and ${}_{X}^{C}D_{X_f}^\alpha[x]$ is the right-sided Caputo derivative of order $\alpha > 0$, that are defined by the equations

$$ {}_{X_i}^{C}D_{X}^{\alpha}[x]Y(x) = \frac{1}{\Gamma(n-\alpha)} \int_{X_i}^{X} (X-x)^{n-\alpha-1} Y^{(n)}(x) v dx, \tag{5.45} $$

$$ {}_{X}^{C}D_{X_f}^{\alpha}[x]Y(x) = \frac{(-1)^n}{\Gamma(n-\alpha)} \int_{X}^{X_f} (x-X)^{n-\alpha-1} Y^{(n)}(x)\, dx, \tag{5.46} $$

where $X_i < X < X_f$, and we use $n = [\alpha] + 1$ for noninteger values of α, and $n = \alpha$ for integer α.

The fractional X-elasticity describes a dependence of Y not only on the current value of X, but also all values x on finite interval $[X_i, X_f]$.

Let us give some properties of the proposed left-sided and right-sided fractional X-elasticities (5.43) and (5.44).

1. For the case $\alpha = 1$, the fractional elasticities that are defined by equation (5.43) take the standard forms

$$ E_{1,l}(Y(X); [X_i, X]) = \frac{X}{Y(X)} \left(\frac{dY(x)}{dx} \right)_{x=X}, \tag{5.47} $$

$$ E_{1,r}(Y(X); [X, X_f]) = \frac{X}{Y(X)} \left(\frac{dY(x)}{dx} \right)_{x=X}, \tag{5.48} $$

   where the elasticity does not depend on $x \neq X$. Equations (5.47), (5.48) describe the standard point elasticity. This means that the suggested elasticities with $\alpha = 1$ correspond to the economic processes without memory and nonlocality.
2. In general, the fractional X-elasticities of inverse functions are not inverse

$$ E_{\alpha,l}(Y(X); [X_i, X]) \neq \frac{1}{E_{\alpha}(X(Y); [Y_i, Y])}, \tag{5.49} $$

$$ E_{\alpha,r}(Y(X); [X, X_f]) \neq \frac{1}{E_{\alpha}(X(Y); [Y, Y_f])} \tag{5.50} $$

   for $ane1$, where $Y_i = Y(X_i)$ and $Y_f = Y(X_f)$. For $\alpha = 1$, these inequalities become equalities.
3. The fractional elasticity of the power-law function is the constant

$$ E_{\alpha,l}(X^{\beta}; [0, X]) = \frac{\Gamma(\beta+1)}{\Gamma(\beta-\alpha+1)}, \tag{5.51} $$

   where $n - 1 < \alpha < n$ with $n \in \mathbb{N}$, and $\beta > n - 1$.
4. The right-sided fractional elasticity of the exponential function is given by the equation

$$ E_{\alpha,r}(e^{-\lambda X}; [X; \infty)) = (\lambda X)^{\alpha}, \tag{5.52} $$

   where $\lambda > 0$, $X > 0$.

5. The left-sided fractional elasticity of the linear function is given by the equation

$$E_{\alpha,l}(a_0 + a_1 X; [0, X]) = \frac{1}{\Gamma(2-\alpha)} \frac{a_1 X}{a_0 + a_1 X}. \tag{5.53}$$

6. For fractional derivatives of noninteger orders, the standard chain rule is not satisfied, in general (see equation (2.209) in Section 2.7.3 of [308, p. 98]). Therefore, we have the inequalities

$$E_{\alpha,l}(Y(X); [X_i, X]) \neq E_\alpha(Y(t); X(t); [t_i, t]), \tag{5.54}$$

$$E_{\alpha,r}(Y(X); [X, X_f]) \neq E_\alpha(Y(t); X(t); [t, t_f]) \tag{5.55}$$

for noninteger values of the order $\alpha > 0$. As a result, the fractional X-elasticities and the fractional T-elasticity should be considered as independent characteristics of economic processes with memory.

7. The fractional elasticities of constant values $Y(t) = \text{const}$ are equal to zero.

$$E_{\alpha,l}(\text{const}; [X_i, X]) = 0, \quad E_{\alpha,r}(\text{const}; [X, X_f]) = 0, \tag{5.56}$$

that corresponds to perfectly inelastic value $Y(t)$ (e. g., perfectly inelastic demand).

Example 5.4. Suppose that there exists a single-valued demand function $Q = Q(P)$ of price $P \in [P_l; P_h]$. The left-sided and right-sided fractional P-elasticities $E_{\alpha,l}(Q(P); [P_l, P])$ and $E_{\alpha,r}(Q(P); [P, P_h])$ of order α at $P \in [P_l; P_h]$ of demand $Q = Q(P)$ is defined by the equations

$$E_{\alpha,l}(Q(P); [P_l, P]) = \frac{P^\alpha}{Q(P)} \, {}_{P_l}^{C}D_P^\alpha[p]Q(p), \tag{5.57}$$

$$E_{\alpha,r}(Q(P); [P, P_h]) = \frac{P^\alpha}{Q(P)} \, {}_{P}^{C}D_{P_h}^\alpha[p]Q(p), \tag{5.58}$$

where $P_l = P_{\min}$ is a lowest price and $P_h = P_{\max}$ is the highest price; ${}_{P_i}^{C}D_P^\alpha[p]$ and ${}_{P}^{C}D_{P_h}^\alpha[p]$ are the left-sided and right-sided Caputo derivatives of the order $\alpha > 0$.

The fractional P-elasticities (5.57) and (5.58) can be interpreted as an elasticity of demand for processes with price memory. Elasticity (5.57) takes into account a "memory of low prices." The "memory of high prices" is taken into account by elasticity (5.58). These elasticities depend of demand Q not only on the current price P, but also on the changes of price p in the given ranges $p \in [P_l, P]$ and $p \in [P, P_h]$, respectively.

Remark 5.5. In the definition of the fractional elasticities, we use the Caputo fractional derivatives. It is caused by the fact that the Caputo fractional derivative of a constant is equal to zero. This property leads us to zero values of fractional P-elasticities for constant demand. Contrary to the Caputo derivative, the Riemann–Liouville fractional derivatives of a constant is not equal to zero (see equation (2.1.20) in book [200, p. 70]), and gives

$$ {}_0^{RL}D_P^{\alpha}[p]Q_0 = \frac{P^{-\alpha}}{\Gamma(1-\alpha)}Q_0, \tag{5.59}$$

where $Q_0 = \text{const}$. Therefore, the fractional P-elasticities, which are defined by using the Riemann–Liouville derivatives, cannot be considered as a perfectly inelastic demand for the constant demand functions. For example, the corresponding left-sided fractional P-elasticity of the constant demand $Q(P) = Q_0 = \text{const}$ is the constant

$$ {}_{RL}E_{\alpha,l}(Q(P);[0,P]) = \frac{P^{\alpha}}{Q(P)}\,{}_0^{RL}D_P^{\alpha}[p]Q(p) = \frac{1}{\Gamma(1-\alpha)}, \tag{5.60}$$

where ${}_0^{RL}D_P^{\alpha}[p]$ is the left-sided Riemann–Liouville derivative [200, pp. 69–70].

We can see that the P-elasticities of the constant demand Q_0 is constant that does not depend on Q_0 and P.

It should be noted that the fractional P-elasticities and the fractional T-elasticity should be considered as independent characteristics of the economic processes with memory. This fact is based on the violation of the standard chain rule for fractional derivatives of noninteger orders.

Let us give numerical examples for standard elasticity $E(Y;X) = (Y/X)(dY/dX)$, and fractional X-elasticity with memory (5.51) for the power-law function $Y(X) = X^{\beta}$.

The first examples, we give for the memory fading parameter equal to $\alpha = 0.6$, i. e., $[\alpha] = 0$ and then $\beta > 0$.

(1) If $\beta = 0.3$, then

$$E(Y;X) < E_{\alpha,l}(Y(X);[0,X]) < 1, \tag{5.61}$$

where

$$E(Y;X) = 0.3, \quad E_{\alpha,l}(Y(X);[0,X]) \approx 0.6913963323. \tag{5.62}$$

(2) If $\beta = 0.9$, then

$$E(Y;X) < 1 < E_{\alpha,l}(Y(X);[0,X]), \tag{5.63}$$

where

$$E(Y;X) = 0.9, \quad E_{\alpha,l}(Y(X);[0,X]) \approx 1.071640373. \tag{5.64}$$

(3) If $\beta = 1.1$, then

$$1 < E(Y;X) < E_{\alpha,l}(Y(X);[0,X]), \tag{5.65}$$

where

$$E(Y;X) = 1.1, \quad E_{\alpha,l}(Y(X);[0,X]) \approx 1.180832828. \tag{5.66}$$

(4) If $\beta = 1.9$, then

$$1 < E_{\alpha,l}(Y(X);[0,X]) < E(Y;X), \tag{5.67}$$

where

$$E(Y;X) = 1.9, \quad E_{\alpha,l}(Y(X);[0,X]) \approx 1.566243623. \tag{5.68}$$

The second examples, we give for the memory fading parameter equal to $\alpha = 1.7$, i. e., $[\alpha] = 1$ and then $\beta > 1$.

(5) If $\beta = 1.2 > 1$, then

$$E_{\alpha,l}(Y(X);[0,X]) < 1 < E(Y;X), \tag{5.69}$$

where

$$E(Y;X) = 1.2, \quad E_{\alpha,l}(Y(X);[0,X]) \approx 0.6216254885. \tag{5.70}$$

(6) If $\beta = 1.5$, then

$$1 < E_{\alpha,l}(Y(X);[0,X]) < E(Y;X), \tag{5.71}$$

where

$$E(Y;X) = 1.5, \quad E_{\alpha,l}(Y(X);[0,X]) \approx 1.141819670. \tag{5.72}$$

(7) If $\beta = 3.5$, then

$$1 < E(Y;X) < E_{\alpha,l}(Y(X);[0,X]), \tag{5.73}$$

where

$$E(Y;X) = 3.5, \quad E_{\alpha,l}(Y(X);[0,X]) \approx 6.938140360. \tag{5.74}$$

Obviously, taking into account the memory effects, we can get other inequalities compared to the standard case of no memory ($\alpha = 1$). These examples allow us to formulate the following statement.

Statement 5.2. *Memory effects can qualitatively change the elasticity of economic processes such that*

$$E(Q,P) < 1, \quad \Longrightarrow \quad E_{\alpha,l}(Q(P);[0,P]) > 1, \tag{5.75}$$

$$E(Q,P) > 1, \quad \Longrightarrow \quad E_{\alpha,l}(Q(P);[0,P]) < 1, \tag{5.76}$$

where $E(Q;P)$ is the standard elasticity, which describes economic process without memory, and $E_{\alpha,l}(Q(P);[0,P])$ is the elasticity of processes with memory.

Based on these numerical illustrative examples and this statement, we can offer the following principle.

Principle 5.1 (Principle of changing elasticity by memory). *Taking into account the memory effects can significantly and qualitatively change the elasticity of economic processes even if the other parameters of these processes remain unchanged.*

As a result, we can conclude that neglecting the memory effects can lead to qualitatively different results. The use of the assumption of the absence of memory can lead to erroneous conclusions in economics.

5.7 Examples of calculations of elasticities with memory

In this section, we give simple illustrative examples of calculations of fractional elasticities with memory. Let us consider the demand $Q = Q(P)$ as a single-valued function of price that is described by the equation

$$Q(P) = a_0 + a_1 P + a_2 P^2, \tag{5.77}$$

where P is the unit price, and $Q(P)$ is the quantity demanded, when the price is P.

Equation (5.77) is considered as a demand function for a product.

The standard point-price elasticity of demand is defined by the equation

$$E(Q;P) = \frac{P}{Q(P)} \frac{dQ(P)}{dP}. \tag{5.78}$$

To find this point elasticity $E(Q;P)$ of demand (5.77), we obtain the first derivative of function (5.77) in the form

$$\frac{dQ(P)}{dP} = a_1 + 2a_2 P. \tag{5.79}$$

As a result, the standard (point-price) elasticity of demand is

$$E(Q;P) = \frac{P}{Q(P)} \frac{dQ(P)}{dP} = \frac{a_1 P + 2a_2 P^2}{a_0 + a_1 P + a_2 P^2}. \tag{5.80}$$

Let us give examples of calculations of fractional elasticity with memory for demand (5.77).

Example 5.5. Let us consider the fractional P-elasticity (5.57) with $P_l = 0$ and $\alpha \in (0;1)$. Using the Caputo fractional derivative of power function

$${}_0^C D_P^\alpha [p] p^\beta = \frac{\Gamma(\beta+1)}{\Gamma(\beta-\alpha+1)} P^{\beta-\alpha}, \quad (P > 0, n-1 < \alpha < n, \beta > n-1). \tag{5.81}$$

We should emphasize that equation (5.81) is not applicable at least for negative powers β if the order α of the derivative is noninteger. Using equation (5.81) for demand function (5.77), we obtain the expression

$$ {}_0^C D_P^\alpha[p]Q(p) = {}_0^C D_P^\alpha[p]a_0 + a_1 {}_0^C D_P^\alpha[p]p + a_2 {}_0^C D_P^\alpha[p]p^2 = a_1 \frac{\Gamma(2)}{\Gamma(2-\alpha)} P^{1-\alpha} + a_2 \frac{\Gamma(3)}{\Gamma(3-\alpha)} P^{2-\alpha}. \tag{5.82} $$

Substitution of (5.82) into (5.57) gives the left-sided fractional P-elasticity in the form

$$ \begin{aligned} E_{\alpha,l}(Q(P);[0,P]) &= \frac{(P)^\alpha}{Q(P)} {}_0^C D_P^\alpha[p]Q(p) = \frac{a_1 \frac{1}{\Gamma(2-\alpha)} P + a_2 \frac{2}{\Gamma(3-\alpha)} P^2}{a_0 + a_1 P + a_2 P^2} \\ &= \frac{1}{\Gamma(2-\alpha)} \frac{a_1 P + a_2 \frac{2}{2-\alpha} P^2}{a_0 + a_1 P + a_2 P^2}, \end{aligned} \tag{5.83} $$

where we use $\Gamma(n+1) = n!$ for $n \in \mathbb{N}$ and $\Gamma(z+1) = z\Gamma(z)$.

For $\alpha = 1$, expression (5.83) takes the standard form of equation (5.80).

Example 5.6. Let us consider the demand and price functions in the form

$$ Q(t) = q_0 + q_1 t + q_2 t^2, \tag{5.84} $$

$$ P(t) = P_0 t. \tag{5.85} $$

It is obvious that substitution of (5.85) into (5.84) gives expression (5.77) with parameters

$$ a_0 = q_0, \quad a_1 = \frac{q_1}{P_0}, \quad a_2 = \frac{q_1}{P_0^2}. \tag{5.86} $$

We see that $Q(t)$ can be represented as single-valued function of P. Let us consider fractional T-elasticity (5.9) with $t_i = 0$ and $\alpha \in (0;1)$. Using the equation of the Caputo derivative of power-law functions in the form

$$ {}_0^C D_t^\alpha[\tau]\tau^\beta = \frac{\Gamma(\beta+1)}{\Gamma(\beta-\alpha+1)} t^{\beta-\alpha}, \quad (t > 0, n-1 < \alpha < n, \beta > n-1), \tag{5.87} $$

we get the expressions

$$ {}_0^C D_t^\alpha[\tau]Q(\tau) = q_1 \frac{\Gamma(2)}{\Gamma(2-\alpha)} t^{1-\alpha} + q_2 \frac{\Gamma(3)}{\Gamma(3-\alpha)} t^{2-\alpha}, \tag{5.88} $$

and

$$ {}_0^C D_t^\alpha[\tau]P(\tau) = P_0 \frac{\Gamma(2)}{\Gamma(2-\alpha)} t^{1-\alpha}. \tag{5.89} $$

Substitution of (5.88) and (5.89) into (5.9) gives the fractional T-elasticity

$$E_\alpha(Q(t);P(t);[0,t]) = \frac{P(t)\;{}_0^CD_t^\alpha[\tau]Q(\tau)}{Q(t)\;{}_0^CD_t^\alpha[\tau]P(\tau)} = \frac{P_0 t}{q_0+q_1t+q_2t^2}\,\frac{q_1\frac{1}{\Gamma(2-\alpha)}t^{1-\alpha}+q_2\frac{2}{\Gamma(3-\alpha)}t^{2-\alpha}}{P_0\frac{1}{\Gamma(2-\alpha)}t^{1-\alpha}}$$
$$= \frac{P_0 t}{q_0+q_1t+q_2t^2}\,\frac{q_1t^{1-\alpha}+q_2\frac{2}{(2-\alpha)}t^{2-\alpha}}{P_0t^{1-\alpha}} = \frac{q_1t+q_2\frac{2}{(2-\alpha)}t^2}{q_0+q_1t+q_2t^2}, \tag{5.90}$$

where we use $\Gamma(z+1) = z\Gamma(z)$ and equations (5.85), (5.86). As a result, we get

$$E_\alpha(Q(t);P(t);[0,t]) = \frac{a_1P+a_2\frac{2}{2-\alpha}P^2}{a_0+a_1P+a_2P^2}. \tag{5.91}$$

Because we have chosen simple equations of (5.84) and (5.85), expressions (5.91) and (5.83) differ only by the factor $\Gamma(2-\alpha)$ such that

$$E_{\alpha,l}(Q(P);[0,P]) = \frac{1}{\Gamma(2-\alpha)}E_\alpha(Q(t);P(t);[0,t]). \tag{5.92}$$

In the general case, the fractional price T-elasticity with memory and the fractional P-elasticity can be distinguished not only by a factor. These generalizations of standard elasticity should be considered as separate independent characteristics of economic processes.

For $\alpha = 1$, expressions (5.91) and (5.83) take standard form (5.80) that describes the elasticity without memory.

Example 5.7. Let impact and response variables be given in the form of single-valued functions of time given in the form

$$Y(t) = at^\beta, \tag{5.93}$$
$$X(t) = bt^\gamma, \tag{5.94}$$

where $\beta > 0$ and $\gamma > 0$. Substituting (5.94) into equation (5.93), we get

$$Y(X) = \frac{a}{b^{\beta/\gamma}}X^{\beta/\gamma}. \tag{5.95}$$

We see that Y can be represented as single-valued function of X, i. e., $Y = Y(X)$. The standard (point-price) elasticity has the form

$$E(Y;X) = \frac{X}{Y(X)}\frac{dY(X)}{dX} = \frac{d\ln Y(X)}{d\ln X} = \frac{\beta}{\gamma}. \tag{5.96}$$

The fractional Log-elasticity (5.14) is given by the equation

$$E_{\alpha,\mathrm{Log}}(Y(t);X(t);[0,t]) = \frac{{}_0^CD_t^\alpha[\tau]\ln\{Y(\tau)\}}{{}_0^CD_t^\alpha[\tau]\ln\{X(\tau)\}} = \frac{{}_0^CD_t^\alpha[\tau]\ln\{a\}+\beta\,{}_0^CD_t^\alpha[\tau]\ln\{\tau\}}{{}_0^CD_t^\alpha[\tau]\ln\{b\}+\gamma\,{}_0^CD_t^\alpha[\tau]\ln\{\tau\}} = \frac{\beta}{\gamma}, \tag{5.97}$$

where we used

$$\ln\{Y(t)\} = \ln\{a\} + \beta \ln\{t\}, \quad \ln\{X(t)\} = \ln\{b\} + \gamma \ln\{t\}, \tag{5.98}$$

and ${}_0^C D_t^\alpha[\tau]\,\mathrm{const} = 0$ [200, 90].

Note that equation (5.97) can be used for positive and negative values of β and γ.

The fractional left-sided X-elasticity for the function $Y = Y(X)$ in form (5.95) is given by the equation

$$E_{\alpha,l}(Y(X); [0,X]) = \frac{X^\alpha}{Y(X)} {}_0^C D_X^\alpha[x] Y(x) = \frac{\Gamma(\beta/\gamma + 1)}{\Gamma(\beta/\gamma + 1 - \alpha)}. \tag{5.99}$$

Here, we assume that $\beta/\gamma > [\alpha]$, where $[\alpha]$ is the integer part of the memory fading parameter $\alpha > 0$. This condition is due to mathematical restrictions for the Caputo fractional derivatives of the power-law function. Note that equation (5.87) is not applicable at least for negative powers β if the order α of the derivative is noninteger. This equation assumes that the power of function is greater than the integer part of the noninteger value of the order of the Caputo fractional derivative, i. e., $\beta > [\alpha]$ in (5.87).

The fractional T-elasticity for variables (5.93) and (5.94) is written in the form

$$E_\alpha(Y(t); X(t); [0,t]) = \frac{X(t)\ {}_0^C D_t^\alpha[\tau] Y(\tau)}{Y(t)\ {}_0^C D_t^\alpha[\tau] X(\tau)} = \frac{\Gamma(\beta + 1)\Gamma(\gamma + 1 - \alpha)}{\Gamma(\gamma + 1)\Gamma(\beta + 1 - \alpha)}. \tag{5.100}$$

Note that equation (5.100) can be used only for positive values of β and γ such that $\beta/\gamma > [\alpha]$. This condition is due to mathematical restrictions on the use of equation (5.81) for the Caputo fractional derivatives. This condition distinguishes the X and T-elasticities from the fractional Log-elasticity.

For $\alpha = 1$, we get

$$E_{\alpha,\mathrm{Log}}(Y(t); X(t); [0,t]) = E_\alpha(Y(t); X(t); [0,t]) = E_{\alpha,l}(Y(X); [0,X]) = E(Y; X), \tag{5.101}$$

since $\Gamma(z + 1) = z\Gamma(z)$, where $E(Y; X)$ is the standard elasticity.

It is easy to see that (5.97), (5.99) (5.100) with $\alpha = 1$ gives expression (5.96) of the standard elasticity without memory by using the equality $\Gamma(z + 1) = z\Gamma(z)$.

We should emphasize that expressions of Log-elasticity with memory (5.97), fractional X-elasticity (5.99) and T-elasticity with memory (5.100) are different for $\alpha \neq 1$. In the general case, these generalizations of standard elasticity should be considered as separate independent characteristics of economic processes with memory.

5.8 Conclusion

We proposed generalizations of point elasticity that allow us to take into account fading memory. The standard point elasticity does not reflect the important element of actual behavior of economic agents. We can state that standard point elasticity can be used only if all economic agents have a total amnesia. In general, we should take into account that indicators and factors can depend on all changes during a finite interval of time, since behavior of economic agents can be determined by the memory about previous changes in economy. As a mathematical tool, we proposed to use theory of derivatives of noninteger orders. The proposed approach to the description of actual behavior of economic agents with memory can give more correct description of elasticity in economics. We state that that neglecting the memory effects can lead to qualitatively different results. The use of the assumption of the absence of memory can lead to erroneous conclusions.

6 Multiplier for processes with memory

In this chapter, generalizations of the important economic concept are proposed. The concept of the multiplier with memory and its properties are described. This chapter is based on articles [412, 397]. The concept of multiplier with memory is also discussed in works [441, 442, 431, 432] and [417, 436].

6.1 Introduction

Macroeconomics studies the functioning of national, regional and global economic systems as a whole. The founder of modern macroeconomic theory is John Maynard Keynes, who published his monograph "The general theory of employment, interest and money" in 1936 [191, 193, 194, 192]. The "Keynesian Revolution" and the birth of modern macroeconomics occurred almost simultaneously with the appearance of concepts of multiplier and accelerator, which were actively used in various macroeconomic models. The "Marginal Revolution" and the "Keynesian Revolution" introduced fundamental economic concepts into economic theory, which are mathematically based on the use of the derivatives and integrals of integer orders, the differential and finite difference equations.

The concepts of multiplier and accelerator are fundamental to modern macroeconomics. The standard concepts of multiplier and accelerator are restricted due to the fact that they do not take into account the effects of memory. In the standard approach, the memory effects and phenomena are neglected. In 2016, we proposed [412, 397] generalizations of the concepts of the multiplier and accelerator, which takes into account the effects of fading memory. We derive equations of the multiplier with memory and the accelerator with memory, which take into account the changes of economic variables on a finite time interval. The proposed generalization includes the standard concepts of the accelerator and the multiplier as special cases. In addition, these generalizations provide a range of intermediate characteristics to take into account the memory effects in macroeconomic models. We prove [412, 397] the duality of the concepts of the multiplier with memory and the accelerator with memory.

6.2 Concept of standard multiplier

Usually, the description of economic processes is realized by using models [10, 11]. In these models, two types of variables are used: the exogenous variables (factors) and the endogenous variables (indicators). The exogenous variables are values that are defined as external for a given model, i. e., that are formed outside the model. These variables are treated as independent (autonomous) variables. The endogenous

https://doi.org/10.1515/9783110627459-006

variables are internal variables of the model, i. e., the variables that are formed within the model. These variables are dependent quantities, behavior of which is determined by the independent variables.

In macroeconomics, the multiplier describes how much a variable, that characterizes states of an economic process, changes in response to a change of some exogenous or endogenous variables. The multipliers show how much the final indicators will change with the change of various economic factors.

The concept of an economic multiplier was first proposed by Richard F. Kahn in the 1931 paper "The Relation of Home Investment to Unemployment" [185]. Using the employment multiplier (the Kahn's multiplier), he proved that government spending on organizing of the public works not only leads to increased employment, but also leads to an increase in consumer demand, and this, in turn, contributes to the growth of production and employment in other sectors of the economy. The increase in aggregate expenditure (e. g., government spending for "public works") can lead to the increase of output and income.

In 1936, John Keynes published the book 'The General Theory of Employment, Interest and Money," [191, 193, 194, 192], which has led to the Keynesian revolution in economic analysis and the birth of modern macroeconomic theory. In this work, Keynes developed the concept of an economic multiplier and multiplicative effects. He proposed the concepts of investment multiplier, saving and consumption multipliers. An investment multiplier (the Keynes multiplier) describes how much the income increases with increasing investment per unit. The multiplier of accumulation (or consumption) shows how much the accumulation (consumption) increases with an increase in income per unit.

In models with continuous time, the indicators and factors (exogenous and endogenous variables, impact and response) are usually described by continuously differentiable functions of time t. Let us consider a variable $Y(t)$, which depends on exogenous (or endogenous) variable $X(t)$. If this dependence is linear, then the multiplier equation is written in the form

$$Y(t) = mX(t) + b, \tag{6.1}$$

where m and b are some constants. The number m is called the multiplier coefficient of the indicator $Y(t)$, and the free term b is interpreted as a part of this indicator, which does not depend on the factor $X(t)$ [10, p. 25]. In linear dependence, the increase of the factor $X(t)$ by the value $\Delta X(t)$ generates the increase of the indicator $Y(t)$ by the quantity $\Delta Y(t)$, which is described by the direct proportionality $\Delta Y(t) = m\Delta X(t)$. If $Y(t)$ and $X(t)$ are continuously differentiable functions, the coefficient m is equal to the marginal value, i. e., $m = MY_X(t) = Y^{(1)}(t)/X^{(1)}(t)$, where $Y^{(1)}(t)$ and $X^{(1)}(t)$ are the derivatives of the first order, and $X^{(1)}(t) \neq 0$. If the function $X(t)$ is invertible in a neighborhood of t, then there is a single-valued function $Y = Y(X)$, and $m = MY_X = dY(X)/dX$.

6.3 Consumption and investment multipliers

Let us consider the standard consumption and investment multipliers without memory. For example, the income $Y(t)$ can be divided into two parts: the consumption $C(t)$ and the saving $S(t)$. The consumption multiplier determines the dependence of the consumption increment $\Delta C(t)$ through the income increment $\Delta Y(t)$ by the equation

$$\Delta C(t) = m_1 \Delta Y(t), \tag{6.2}$$

where m_1 is the consumption multiplier coefficient. The coefficient m_1 is equal to the marginal propensity to consume

$$m_1 = MC_Y = \frac{dC(Y)}{dY} = \frac{C^{(1)}(t)}{Y^{(1)}(t)}, \tag{6.3}$$

where $Y^{(1)}(t) \neq 0$, $C^{(1)}(t) = \frac{dC(t)}{dt}$, and $Y^{(1)}(t) = \frac{dY(t)}{dt}$. In the case of the linear dependence, we have the equation

$$C(t) = m_1 Y(t) + b, \tag{6.4}$$

where the free term b interpreted as part of consumption that is independent of income. The coefficient b allows us to describe negative saving at low income levels [10, p. 46].

The equation of the investment multiplier

$$\Delta Y(t) = m_2 \Delta I(t) \tag{6.5}$$

describes the increment of income $\Delta Y(t)$, which is caused by the increment of investment $\Delta I(t)$, where m_2 is the investment multiplier coefficient. According to Keynes, this coefficient is equal to the inverse value of the marginal propensity to save ($m = 1/MS_Y$, where $MS_Y = 1 - MC_Y$). If the total amount of investments is increased, then the income increases by a certain amount, which is greater than the increase of investment. The concept of an investment multiplier allows us to describe the change in national income when changing government spending.

Macroeconomics also uses a more general concept such an exogenous spending multiplier, which characterizes the change of the national income with a single change of the autonomous exogenous expenditures.

6.4 Multiplier effect and memory

The effect of the multiplier of exogenous expenditures is the effect of the change (increase or decrease) of any component of autonomous expenditures, which leads to a change in the national income by an amount that is greater than the initial change of the expenditure. The autonomous expenditures are expenditures that are considered as independent of the level of income. The mechanism, which leads to the multiplier effect, is caused by the fact that the initial increase in exogenous expenditures may lead to an increase of the consumer spending and, therefore, to an increase of the incomes of other consumers, and this leads to a further increase in consumption, etc. It leads to a general increase of the national income more than the initial value of expenditure. The effect of increasing of the endogenous variable is called the multiplier effect, when the multiplier coefficient is greater than one.

The mechanism of the appearance of the multiplier effect for investment is the following. The initial increase of the investment leads to an increase of the companies incomes. These companies, in turn, increase production and employment. This leads to additional demand for materials and other goods. The increase of the level of employment and income of the company employees leads to an increase in their consumption, and, as a result, to increased production and increased income in industries that create these goods and services. As a result, there is a chain of reaction of the income growth and the increased production, which covers an increasing number of sectors of the economy.

The memory effect can significantly influence the multiplier effect mechanism. Let us explain the action of the memory effect on the multiplier mechanism.

Statement 6.1 (Memory effect for economic multiplier). *Economic agents can remember previous negative (or positive) changes in the economy, and as a consequence, in case of a repetition of the situation, the agents can behave in a different way. For example, the agents can increase or decrease the share of income channeled to saving and accumulation. In general, the propensity to save is related to the presence of the memory of economic agents, since it can depend on changes in income and consumption at previous points in time.*

The process of appearance of the multiplier effect is often compared with the process of formation of circles on water after a primary disturbance caused by a stone thrown into the water. An analogy with the formation of circles and waves in the liquid medium can be continued. The appearance of the multiplier effects for economic processes with memory in a sense is analogous to the propagation of the initial impact (disturbances) and waves in continuous media with a memory (e. g., see [230, 245]).

6.5 Lag effects and memory effect

Usually economic models with continuous time are described by equations that contain only derivatives of positive integer orders with respect to time. It is known that such derivatives are determined by the properties of differentiable functions only in an infinitesimal neighborhood of the considered time point. Therefore, the differential equations of integer orders actually describe only the instantaneous reaction of economic agents without memory. Such equations do not take into account that economic agents have a memory of the history of changes in indicators and factors on a time interval. It can be said that the standard equations of the multiplier and accelerator depend only on infinitesimal neighborhood of the considered point of time. Therefore, these equations are applicable only if all economic agents have total amnesia. Obviously, this condition leads to simplification of economic models, and it cannot always be used to describe economy. This restriction is mathematically conditioned by the use of mathematical tools based on the derivatives of integer orders.

Interrelations between economic agents, as a rule, are not instantaneous, i. e., they are not local in time. Between the causes and consequences, between the primary impact and its effect (e. g., between investment and income generation), there is a period of time, called a time lag (delay). In real economic processes, temporary lags are present in most interconnections. For example, in the modeling of economic dynamics, it is important to take into account an investment lag, which is caused by the finite period from the design of an object to its commissioning at full capacity. In many economic processes, the lag is not a strictly defined quantity, but it is distributed in time. The economic models, which assume the absence of the time lags, are a strong simplification of reality.

To take into account the time lag in macroeconomic models with continuous time, we can use the differential equations with derivatives of integer orders and retarded arguments (deviating arguments) [10, 112]. This delay can be caused by the presence of inertia in some economic agents or the finite speed of economic processes.

In order to accelerate the car to high speed, some finite time is required. This time is due to the presence of internality in the body. We emphasize that in physics this delay is not interpreted as a memory. The fact that goods cannot be instantly delivered from one place to another is determined by the finiteness of the vehicle speed. This delay is not interpreted as a memory.

As a result, economic models with delay do not take into account the memory effects. The distributed delay is not equivalent to the existence of memory in economic agents. It is necessary to distinguish the effects of lag (time delay) and fading memory. The effect of the time lag is related to the fact that all economic processes are realized with finite rates (or inertia), i. e., the change of the economic factor does not lead to an instantaneous change of the indicator.

Let us note some important properties of the memory effect.

(1) The memory effect means that the state of the economy depends not only on the values of the economic indicators and factors at a given time, but also on their values at previous points in time. In general, the state of the economy at present time depends on the state changes at previous times. The memory effect is associated with the presence of memory in economic agents.
(2) The memory effect is due to the fact that the behavior of economic agents depends not only on current economic indicators, but also on how these indicators have

changed in the past. The memory effect can be manifested in the fact that the repetition of similar changes of the economic factor, gives the changes of the economic indicator by another way.

(3) The memory effect leads to the multivalued dependencies of the indicator on the factor. Multivalued dependencies of the indicator on the factor are caused by the following fact. The economic agents remember the previous changes of this factor and the reaction of the considered indicator. Therefore, these agents can react differently to the same changes of the factor. As a result, despite the similar changes of the factor, the behavior of the indicator will be different.

Necessity to take into account memory in the economic process is caused by the fact that agents, which participate in this process, have a memory about the past of the process. As a result, to describe the economic process, we should take into account not only the current state and infinitely close to it previous states, but all the previous states of the considered process on a finite time interval. Mathematically, this means that to describe such a process it is not enough to use the state variables and their derivatives of integer orders with respect to time. In other words, the economic process with memory depends on the previous history of the process change.

In modern mathematics, in addition to derivatives and integrals of integer orders, the different types of derivatives and integrals of noninteger (fractional) orders are known. To describe memory effects in economics, we can use equations with derivatives of noninteger orders, which are actually integro-differential operators, and fractional integrals of noninteger orders.

6.6 Concept of multiplier: from lag to memory

For simplicity, we will consider the multipliers that are described by linear equations with continuous time. Let the variables $X(t)$ and $Y(t)$ be single-valued functions of time t that describe the impact and response variables, respectively. In absence of the lag, the linear multiplier equation has the form

$$Y(t) = mX(t) + b, \tag{6.6}$$

where m and b are some numerical coefficients. The number m is called the multiplier coefficient, and the free term b is interpreted as a part that does not depend on the variable $X(t)$ [10, p. 46]. Equation (6.6) connects the sequence of subsequent states of the economic process with the previous states, through a single current state at each instant of time t. Therefore, equation (6.6) describes economic processes without memory.

In the simplest case of lag with fixed-time delay, the multiplier equation [10, p. 25] has the form

$$Y(t) = mX(t - T) + b, \tag{6.7}$$

where T is a positive time-constant that characterizes the fixed-time delay. A more general case of linear multiplier with distributed lag, which has a fixed-duration T, is written in the form

$$Y(t) = m \sum_{n=0}^{\infty} K_n(T)X(t - nT) + b, \tag{6.8}$$

where it is assumed that the coefficients $K_n(T)$, which is called the weighting coefficients, satisfy the normalization condition

$$\sum_{n=0}^{\infty} K_n(T) = 1. \tag{6.9}$$

Multiplier with continuously distributed lag [10], which is a continuous analog of formula (6.8), is described by the equation

$$Y(t) = m \int_0^{\infty} K(\tau)X(t - \tau)\, d\tau + b, \tag{6.10}$$

where $K(\tau)$ is a weighting function [10, pp. 23–24] that satisfies the normalization condition

$$\int_0^{\infty} K(\tau)\, d\tau = 1. \tag{6.11}$$

The nonnegative function $K(\tau)$ that satisfies condition (6.11) is the probability density function. Equation (6.11) means that the economic process, which is described by equation (6.10), passes through all states without any losses. Therefore, we can say that equation (6.10) corresponds to a complete (perfect, ideal) memory [363, c. 394–395]. However, the use of the term "memory" leads to incorrect interpretation and incorrect explanations of economic processes.

In the general case, normalization condition (6.11) is not necessary. If condition (6.11) is used, the coefficient m of equation (6.10) is the multiplier coefficient. If equality (6.11) is not used as a condition on the function $K(\tau)$, then we can use the function $M(\tau) = mK(\tau)$ that plays the role of the generalized multiplier coefficient, and the coefficient m can be removed from equation (6.10). Equation (6.10) is usually used to describe the distributed delay, and the function $K(\tau)$, characterizes the distribution of the delay time. Using the function $M(\tau) = mK(\tau)$ and the change of variable $(t - \tau \to \tau)$, equation (6.10) can be rewritten in the form

$$Y(t) = \int_{-\infty}^{t} M(t - \tau)X(\tau)\, d\tau + b. \tag{6.12}$$

In equation (6.12), the variable $Y(t)$ depends not only on the values of $X(\tau)$ at one previous moment of time, as in equation (6.7) for time delay with a fixed duration. In equation (6.12), the indicator $Y(t)$ depends on the values of the factor $X(\tau)$ at all past instants on the infinite interval $(-\infty, t)$. The use of infinite limit in the integral of equation (6.12) means that we take into account the values of $X(\tau)$ from an infinite distant past. To describe the economic processes, we do not need to consider the infinitely distant past. In many economic models, we can choose the initial time instant equal to zero in order to consider economic dynamics on a finite time interval of the positive semi-axis. For this, we assume the function $M(\tau)$ to be zero for $\tau < 0$. In this case, equation (6.12), can be written in the form

$$Y(t) = \int_0^t M(t-\tau)X(\tau)\,d\tau + b. \tag{6.13}$$

Let us note three special cases of equation (6.13) that are not related to the memory effect:
1. If the function $M(\tau)$ is expressed in terms of the Dirac delta-function by the equation $M(t-\tau) = m\delta(t-\tau)$, then equation (6.13) takes the form of equation (6.6), which describes the standard multiplier without lag.
2. If $M(t-\tau) = m\delta(t-T-\tau)$, then equation (6.13) takes the form of equation (6.7) that describes the multiplier with fixed-time delay [10, p. 25].
3. If $M(\tau) = mK(\tau)$, where $K(\tau)$ satisfies the normalization condition, then we get multiplier equation for the distributed lag.

As a result, equation (6.13) can be considered as a generalization of the linear equation of the standard multiplier without lag and with distributed lag. The function $M(t-\tau)$ can be interpreted as a generalization of the multiplier coefficient.

6.7 Generalized multiplier with memory

To describe economic processes with memory, it should be taken into account that the indicator $Y(t)$ at time t depends on the changes $X(\tau)$ on a finite time interval $0 \le \tau \le t$. The main reason to take into account the memory effects in economic processes is the fact that the economic agents remember the previous changes of the factor $X(t)$ and their influence on changes of the indicator $Y(t)$. Using the proposed concept of the multiplier with memory that is proposed in [412, 397], we can propose the following definition.

Definition 6.1. The generalized multiplier with memory is the dependence of the (response) variable $Y(t)$ at the time $t \ge 0$ on the history of the change of the (impact) variable $X(\tau)$ on a finite time interval $0 \le \tau \le t$ such that

$$Y(t) = \mathcal{M}_0^t(X(\tau)), \tag{6.14}$$

where $\mathcal{M}_0^t$ is the operator that transforms each history of changes of $X(\tau)$ for $\tau \in [0, t]$ into the corresponding history of changes of $Y(\tau)$ with $\tau \in [0, t]$.

Let us give the integral form of the generalized multiplier with memory.

Definition 6.2. The generalized multiplier with memory (6.14) is called the linear generalized multiplier with memory function $M(t, \tau)$ if multiplier equation (6.14) can be represented in the form

$$Y(t) = \int_0^t M(t, \tau) X(\tau)\, d\tau, \tag{6.15}$$

where $M(t, \tau)$ is the memory function, and we assume that integral (6.15) exists.

For the memory function $M(t, \tau)$, which is represented by the Dirac delta-function

$$M(t, \tau) = m\delta(t - \tau), \tag{6.16}$$

equation (6.15) has the form of the standard multiplier equation $Y(t) = mX(t)$. For the function,

$$M(t, \tau) = m\delta(t - \tau - T), \tag{6.17}$$

where T is the time constant, equation (6.15) describes the linear multiplier with fixed-time delay (6.7), where $b = 0$.

Remark 6.1. Substitution of the expression

$$M(t, \tau) = M_0(t, \tau) + b(\tau)\delta(t - \tau), \tag{6.18}$$

where $\delta(t - \tau)$ is the Dirac delta-function and $\tau \in [0, t]$, into equation (6.15) gives equation of the linear multiplier with memory

$$Y(t) = \int_0^t M_0(t, \tau) X(\tau)\, d\tau + b(t). \tag{6.19}$$

Therefore, in many cases, we can consider equation (6.15) instead of (6.19).

6.8 Multiplier with fading memory

We can consider the variable $X(\tau)$ in the form of the Dirac delta-function $\delta(\tau - T)$, i. e., $X(\tau) = m\delta(\tau - T)$, where $T > 0$ is a fixed time. Using the impact $X(\tau) = m\delta(\tau - T)$, we can consider the response $Y(t)$, which is described by equation of the linear economic multiplier with memory. In general, this response

$$Y(t) = m\mathcal{M}_0^t(\delta(\tau - T)) \tag{6.20}$$

is not equal to zero for $t > T$. This allows us to formulate the concept of multiplier with fading memory.

Definition 6.3. The multiplier with fading memory is the generalized multiplier (6.14), in which the operator $\mathcal{M}_0^t$ satisfies the condition

$$\lim_{t \to \infty} \mathcal{M}_0^t(\delta(\tau - T)) = 0, \tag{6.21}$$

where $T > 0$.

The linear multiplier, which is described by equation (6.15), is the multiplier with fading memory, if the memory function satisfies the condition $M(t, \tau) \to 0$ at $t \to \infty$ with fixed τ.

We can consider the variable $X(\tau)$, which is expressed through the Dirac delta-function in the form

$$X(\tau) = x(\tau)\delta(\tau - T). \tag{6.22}$$

Then equation (6.15) of the linear multiplier with memory function takes the form

$$Y(t) = M(t, T)x(T) \tag{6.23}$$

for $t > T$. As a result, the behavior of the response variable $Y(t)$ on the impact $X(\tau)$ is determined by the behavior of the memory function $M(t, \tau)$ with fixed constant τ. If the memory function tends to zero ($M(t, \tau) \to 0$) at $t \to \infty$ with fixed τ, then the economic process completely forgotten the impact, which has been subjected in the last time. In this case, the effect of the multiplier with memory (6.15) is repeatable, i. e., the effects of multiplier with memory did not lead to irreversible changes of the economic process and economic agents, since the memory about the impact has not been preserved forever.

6.9 Multiplier with nonaging memory

The time homogeneity means that the multiplier effect is invariant under the shift $t \to t + s$. This property can be interpreted as the absence of aging in the multiplier effect. Therefore, we can talk about a generalized multiplier with non-aging memory.

Definition 6.4. The multiplier with nonaging memory is the generalized multiplier (6.14), which satisfies the condition

$$\mathcal{M}_{0+s}^{t+s}(X(\tau)) = \mathcal{M}_0^t(X(\tau)) \tag{6.24}$$

for all variables $t > 0$ and $s > 0$.

Homogeneity of time means that there are no designated points in time and the description of the economic process does not depend on the initial time point. For the multiplier (6.15) with the memory function $M(t,\tau)$, which describes nonaging memory, is represented by the condition

$$M(t+s,\tau+s) = M(t,\tau). \tag{6.25}$$

Condition (6.25) leads to the expression $M(t,\tau) = M(t-\tau)$.

Definition 6.5. The linear multiplier with nonaging memory is the multiplier (6.15) for which the memory function satisfies the condition

$$M(t,\tau) = M(t-\tau) \tag{6.26}$$

for all τ and $t \geq \tau$.

If the effect of multiplier with memory does not depend on the choice of the reference time, then it can be described by using the multiplier with nonaging memory, where the memory function $M(t,\tau)$ depends only on the time interval $t-\tau$.

Remark 6.2. Mathematically, the linear multiplier with nonaging memory is represented as the Laplace (Duhamel) convolution of functions,

$$Y(t) = M(t) * X(t) = \int_0^t M(t-\tau)X(\tau)\,d\tau = \int_0^t M(\tau)X(t-\tau)\,d\tau. \tag{6.27}$$

This type of multipliers describes the economic processes that do not depend on the choice of the time origin. For these processes, we can always choose the beginning of the reference time at $t = 0$.

6.10 Multiplier with power-law memory

Let us give a definition of the multiplier with fading memory, which fading has the power-law form.

Definition 6.6. Multiplier (6.14) with fading memory is called the multiplier with power-law memory, which is denoted as $Y(t) = \mathcal{M}_0^{t;\alpha}(X(\tau))$, if there is parameter $\alpha > 0$ and finite constant $m = m(\alpha)$ such that

$$\lim_{t\to\infty} t^{1-\alpha}\mathcal{M}_0^{t;\alpha}(\delta(\tau-T)) = m(\alpha) \tag{6.28}$$

for fixed values of $T > 0$ and $\tau > T$.

Equation (6.28) means that we should have the response $Y(t) = O(t^{\alpha-1})$ on the impact $X(\tau) = \delta(\tau - T)$ in the multipliers with power-law memory.

If the multiplier is described by equation (6.15), then this definition can be rewritten in the following form.

Definition 6.7. Linear multiplier with memory function (6.15) is called the multiplier with power-law memory, if there are the parameter $\alpha > 0$ and finite constant $m(\alpha)$ such that the limit

$$\lim_{t\to\infty} t^{1-\alpha} M(t,\tau) = m(\alpha) \tag{6.29}$$

is finite for all fixed $\tau < t$.

In order to have a correct definition of the economic multiplier with power-law memory, we should require the implementation of the following correspondence principle.

Principle 6.1 (Correspondence principle for multiplier with memory). *If the memory fading parameter tends to zero ($\alpha \to 0$), then the equation of the linear multiplier with power-law memory*

$$Y(t) = \mathcal{M}_0^{t;\alpha}(X(\tau)) \tag{6.30}$$

should take the form of equation of the standard multiplier without memory, $Y(t) = mX(t)$, where m is the multiplier coefficient.

The correspondence principle for multiplier states that linear multiplier with power-law memory (6.30) should satisfy the condition

$$\lim_{\alpha\to 0} \mathcal{M}_0^{t;\alpha}(X(\tau)) = mX(t) \tag{6.31}$$

for all $t \geq 0$, where $m = m(1)$.

In the simple case of the power-law behavior of the memory fading, we can use the memory function $M(t,\tau)$ in the form $M(t-\tau) = c(\beta)(t-\tau)^{\beta}$, where $c(\beta) > 0$ and $\beta > -1$ are constants. For simplification, we will use the parameters $\alpha = \beta + 1$ and $m(\alpha) = c(\alpha - 1)\Gamma(\alpha)$. In this case, the memory function can be given by the expression

$$M(t-\tau) = \frac{1}{\Gamma(\alpha)} \frac{m(\alpha)}{(t-\tau)^{1-\alpha}}, \tag{6.32}$$

where $\alpha > 0$, $t > \tau$ and $m(\alpha)$ is the numerical coefficient. Substituting expression (6.32) into equation (6.15), we get that multiplier equation (6.15) can be written in the form

$$Y(t) = m(\alpha)(I^{\alpha}_{RL;0+}X)(t), \tag{6.33}$$

where $I^{\alpha}_{RL;0+}$ is the left-sided Riemann–Liouville fractional integral of the order $\alpha > 0$ with respect to the time variable t [335, p. 33].

Definition 6.8. The left-sided Riemann–Liouville fractional integral of the order $\alpha > 0$ is defined by the equation

$$(I^{\alpha}_{RL;0+}X)(t) = \frac{1}{\Gamma(\alpha)} \int_0^t (t-\tau)^{\alpha-1} X(\tau)\, d\tau, \tag{6.34}$$

where $\Gamma(\alpha)$ is the gamma function, $0 < t < T$. Fractional integral (6.34) is defined for functions $X(\tau) \in L_1(0,T)$, which implies that $X(\tau)$ is a Lebesgue integrable function on $(0,T)$, for which the condition $\int_0^T |X(\tau)|\, d\tau < \infty$ holds.

Correspondence principle for multiplier with memory

The Riemann–Liouville integral (6.34) is a generalization of the standard n-th integration.

Using equation (6.33), we can define a simple multiplier with power-law memory, whose fading parameter is $\alpha \geq 0$.

Definition 6.9. The linear multiplier with simple power-law memory, which is characterized by the fading parameter $\alpha > 0$, is described by the equation

$$Y(t) = m(\alpha)(I^{\alpha}_{\mathrm{RL};0+}X)(t), \tag{6.35}$$

where $I^{\alpha}_{\mathrm{RL};0+}$ is the left-sided Riemann–Liouville fractional integral (6.34) of the order $\alpha > 0$ with respect to time, and $m(\alpha)$ is the coefficient of the multiplier with memory.

Remark 6.3. We note that, due to strict inequality for the order $\alpha > 0$ of the integral (6.34), equation (6.35) cannot be considered for $\alpha = 0$. Therefore, equation (6.35) does not include the standard multiplier equation, as a special case. To include the standard equation, it is necessary to extend the definition of the Riemann–Liouville integral (6.34) by using the equation $(I^{0}_{\mathrm{RL};0+}X)(t) = X(t)$. In this case, equation (6.35) of the multiplier with power-law memory includes the standard equation of multiplier.

Remark 6.4. We note that the Riemann–Liouville integral (6.34) for the order $\alpha = 1$ gives the standard integral of the first order

$$(I^{1}_{\mathrm{RL};0+}X)(t) = \int_0^t X(\tau)\, d\tau. \tag{6.36}$$

For positive integer values $\alpha = n$, the fractional Riemann–Liouville integral (6.34) has the form of the standard n-th integral

$$(I^{n}_{\mathrm{RL};0+}X)(t) = \int_0^t dt_1 \int_0^{t_1} dt_2 \cdots \int_0^{t_{n-1}} dt_n X(t_n) = \frac{1}{(n-1)!}\int_0^t (t-\tau)^{n-1} X(\tau)\, d\tau. \tag{6.37}$$

Remark 6.5. For $\alpha = 1$, multiplier equation (6.35) takes the form

$$Y(t) = m(1)\int_0^t X(\tau)\, d\tau. \tag{6.38}$$

Note that differentiation of equation (6.38) with respect to the variable t and the use of the fundamental theorem of calculus, gives

$$\frac{d}{dt}Y(t) = m(1)X(t). \tag{6.39}$$

Equation (6.39) can be written in the form

$$X(t) = a\frac{d}{dt}Y(t), \tag{6.40}$$

where $a = 1/m(1)$. Equation (6.40) can be interpreted as the standard accelerator equation, where a is considered as the accelerator coefficient. As a result, we can state that mathematically the equation of the multiplier with memory includes equation of the standard accelerators as a special case.

6.11 Multiplier with distributed time scaling

In order to describe the economic dynamics with distributed time scaling [400, 401], we can use the generalized fractional calculus [202, 203]. For example, we can consider the memory function by the expression

$$M(t,\tau) = M_{\sigma}^{\alpha,\eta}(t,\tau) = \frac{\sigma t^{-\sigma(\alpha+\eta)}}{\Gamma(\alpha)}\frac{m(\alpha)\tau^{\sigma(\eta+1)-1}}{(t^{\sigma}-\tau^{\sigma})^{1-\alpha}}, \tag{6.41}$$

where $\sigma > 0$, η is the real number and $\alpha > 0$. Substitution of expression (6.41) into equation (6.15) gives the equation of the multiplier with distributed time scaling in the form

$$Y(t) = m(\alpha)(I^{\alpha}_{\mathrm{EK};0+;\sigma,\eta}X)(t), \tag{6.42}$$

where $I^{\alpha}_{\mathrm{EK};0+;\sigma,\eta}$ is the left-sided Erdelyi–Kober fractional integral of the order $\alpha > 0$ with respect to time variable [335, 200].

If $\sigma = 1$, then the Erdelyi–Kober fractional integral (6.42) gives the Kober fractional integral (see [200, pp. 105–106]). The Erdelyi–Kober fractional integration $(I^{\alpha}_{\mathrm{EK},a+;\sigma,\eta}X)(t)$ is a linear map of the space C_{μ} into itself (see Theorem 2.2 of [241, p. 253]) if $\mu \geq -\sigma(\eta+1)$. The space of functions C_{μ} consists of all functions $X(t)$ with $t > 0$, that can be represented in the form $X(t) = t^{p}f(t)$ with $p > \mu$ and $f(t) \in C[0,\infty)$.

Economic multiplier (6.42) can be called the linear multiplier with distributed time scaling. A general approach to describe processes with distributed scaling is suggested in work [400].

Multiplier (6.42) with the impact variable $X(\tau) = \delta(\tau - T)$ gives the expression for the response variable in the form

$$\begin{aligned} Y(t) &= \int_{0}^{t} M_{\sigma}^{\alpha,\eta}(t,\tau)\delta(T-\tau)\,d\tau = \frac{\sigma m(\alpha)T^{\sigma(\eta+1)-1}}{\Gamma(\alpha)}t^{-\sigma(\alpha+\eta)}\left(t^{\sigma}-T^{\sigma}\right)^{\alpha-1} \\ &= \frac{\sigma m(\alpha)T^{\sigma(\eta+1)-1}}{\Gamma(\alpha)}t^{-\sigma(\eta+1)}\left(1-\left(\frac{T}{t}\right)^{\sigma}\right)^{\alpha-1}. \end{aligned} \tag{6.43}$$

This expressions can also be interpreted as the multiplier with fading memory of power-law type if $\sigma(\eta+1) > 0$.

6.12 Superposition principle for multipliers with memory

In this section, we will consider the simultaneous (parallel) action of multipliers with memory. In the equations of multipliers with power-law memory, which are discussed above, the power-law fading is characterized by one parameter $\alpha \geq 0$ only. In economic models, we can take into account the presence of different values of memory fading, which characterize the different types of economic agents.

For example, the multiplier with fading memory can be considered as a sum of N multipliers with one-parameter memory fading that have different values of the fading parameters α_k, $k = 1, \ldots, N$. The equation of the linear multiplier with N-parameter power-law memory can be defined in the form

$$Y(t) = \sum_{k=1}^{N} m_k(\alpha_k)(I^{\alpha_k}_{RL;0+}X)(t), \tag{6.44}$$

where $\alpha_k > 0$, $(k = 1, 2, \ldots, N)$, and $m_k(\alpha)$, $(k = 1, 2, \ldots, N)$ are coefficients of multiplier.

In general, we can propose the principle of superposition of multipliers with memory that describes the simultaneous (parallel) action of these multipliers with memory. Let us assume that the linear multipliers with memory can be characterized by the memory functions $M_k(t, \tau)$, $k = 1, \ldots, N$ and described by the equations

$$Y_k(t) = \int_0^t M_k(t, \tau)X(\tau)\, d\tau. \tag{6.45}$$

The proposed principle states that the simultaneous (parallel) action of several linear multipliers with memory is equivalent to the action of a multiplier, whose memory function is equal to the sum of the memory function of these multipliers.

As a result, we have the following principle of superposition for multipliers with memory.

Principle 6.2 (Principle of superposition of multipliers with general memory). *The multiplier effect, which is created by the simultaneous (parallel) action of finite number of linear multipliers with memory, is equivalent to the action of a linear multiplier, in which memory function is the sum of the memory functions of these multipliers.*

This principle states that the simultaneous action of linear multipliers is characterized by the fact that the action of each of these multipliers does not affect (does not change) the action of other linear multipliers. For linear multipliers with memory the suggested superposition principle means the additivity of the memory functions.

Let us describe the principle of superposition of multipliers with memory in symbolic form. The simultaneous (parallel) action of the linear multipliers that are described by the equations

$$Y_1(t) = \int_0^t M_1(t, \tau)X(\tau)\, d\tau, \tag{6.46}$$

$$Y_2(t) = \int_0^t M_2(t,\tau)X(\tau)\,d\tau \tag{6.47}$$

is equivalent to the action of the multiplier with memory,

$$Y(t) = \int_0^t M(t,\tau)X(\tau)\,d\tau, \tag{6.48}$$

in which the memory function is the sum of the memory functions of these multipliers

$$M(t,\tau) = M_1(t,\tau) + M_2(t,\tau). \tag{6.49}$$

In equations (6.46) and (6.47), the indicators $Y_1(t)$ and $Y_2(t)$ describe the contributions of two groups of economic agents that are characterized by two different memory functions. The action of N multipliers (6.45) with the memory functions $M_k(t,\tau)$, $(k = 1,\dots,N)$, is equivalent to the action of the linear multiplier (6.48) with the memory function

$$M(t,\tau) = \sum_{k=1}^{N} M_k(t,\tau). \tag{6.50}$$

Example 6.1. For two multipliers with memory, which are described by the equations

$$Y_1(t) = m_1(\alpha)(I^{\alpha}_{\mathrm{RL};0+}X)(t), \tag{6.51}$$
$$Y_2(t) = m_2(\alpha)(I^{\alpha}_{\mathrm{RL};0+}X)(t), \tag{6.52}$$

with the same fading parameters, the superposition principle gives the equality

$$Y(t) = (m_1(\alpha) + m_2(\alpha))(I^{\alpha}_{\mathrm{RL};0+}X)(t), \tag{6.53}$$

which is valid for all $\alpha > 0$ and for any $X(t) \in L_1(0,T)$. As a result, we have that the parallel action of multipliers, for which the power-law memory fading parameters are the same, is equivalent to the action of one multiplier, the coefficient of which is equal to the sum of the coefficients of these multipliers.

Example 6.2. Let us now consider the case of noncoinciding values of the power-law memory fading parameters α_k, where $k = 1,2,\dots,N$. According to the proposed principle of superposition, the action of N linear multipliers with memory that are described by the equations

$$Y_k(t) = m_k(\alpha_k)(I^{\alpha_k}_{\mathrm{RL};0+}X)(t), \tag{6.54}$$

with the memory functions

$$M_k(t,\tau) = \frac{m_k(\alpha_k)}{\Gamma(\alpha_k)}(t-\tau)^{\alpha_k-1}, \tag{6.55}$$

where $\alpha_k > 0$, is equivalent to the action of the multiplier with the memory function

$$M(t,\tau) = \sum_{k=1}^{N} \frac{m_k(\alpha_k)}{\Gamma(\alpha_k)} (t-\tau)^{\alpha_k - 1}. \tag{6.56}$$

The use of memory with multiparameter power-law fading, which is described by the sum of memory function, allows us to build economic models that describe the effects of memory for the economy with different type of economic agents.

6.13 Multiplier with distributed power-law memory

Equation (6.44) can be represented in the form

$$Y(t) = m(\alpha) \sum_{k=1}^{N} \rho_k(\alpha) (I^{\alpha_k}_{\mathrm{RL};0+} X)(t), \tag{6.57}$$

where $m(\alpha)$ is the multiplier coefficient

$$m(\alpha) = \sum_{k=1}^{N} m_k(\alpha_k), \tag{6.58}$$

and $\rho_k(\alpha)$ is the discrete probability density function

$$\rho_k(\alpha) = \frac{m_k(\alpha_k)}{m(\alpha)}, \quad \sum_{k=1}^{N} \rho_k(\alpha) = 1. \tag{6.59}$$

In this case, equations (6.57) and (6.44) can be interpreted as a multiplier with discretely distributed memory fading. The memory fading parameter α is considered as a random variable, and $\rho_k(\alpha)$ describe the probability density function.

In general, the parameter α of memory fading can be continuously distributed on the interval $[\alpha_1, \alpha_2]$, where the distribution is described by a weighting function $\rho(\alpha)$, which is a probability density function. The function $\rho(\alpha)$ describes distribution of the memory fading parameter on a set of economic agents. This is important for the economics with memory, since various types of economic agents may have different parameters of memory fading. In this case, we should consider the multipliers with memory function, which depends on the weighting function $\rho(\alpha)$, and the interval $[\alpha_1, \alpha_2]$.

In this case, equation of the multiplier with continuously distributed power-law fading, which is a generalization of equation (6.57), takes the form

$$Y(t) = m([\alpha_1, \alpha_2]) \int_{\alpha_1}^{\alpha_2} \rho(\alpha) (I^{\alpha}_{\mathrm{RL};0+} X)(t) \, d\alpha, \tag{6.60}$$

where $m([\alpha_1, \alpha_2])$ is the coefficient of the multiplier with distributed memory, $I^{\alpha}_{RL;0+}$ is the left-sided Riemann–Liouville integral of the order $\alpha \in [\alpha_1, \alpha_2]$. Here, $\rho(\alpha)$ is the nonnegative weighting function that satisfies the normalization condition

$$\int_{\alpha_1}^{\alpha_2} \rho(\alpha)\, d\alpha = 1. \tag{6.61}$$

Note that the integral in equation (6.60) is the fractional integral of distributed order [45, 19, 20, 234, 184] that is defined by the equation

$$(I^{[\alpha_1,\alpha_2]}_{RL;0+} X)(t) = \int_{\alpha_1}^{\alpha_2} \rho(\alpha)(I^{\alpha}_{RL;0+} X)(t)\, d\alpha, \tag{6.62}$$

where $\rho(\alpha)$ satisfies the normalization condition (6.61) and $\alpha_2 > \alpha_1 \geq 0$. In equation (6.62), the integration with respect to time and the integration with respect to order can be permuted for a wide class of functions $X(\tau)$. As a result, equation (6.62) can be represented in the form

$$(I^{[\alpha_1,\alpha_2]}_{RL;0+} X)(t) = \int_{0}^{t} M^{[\alpha_1,\alpha_2]}_{\rho(\alpha)}(t-\tau) X(\tau)\, d\tau, \tag{6.63}$$

where the kernel $M^{[\alpha_1,\alpha_2]}_{\rho(\alpha)}(t-\tau)$ is defined by the expression

$$M^{[\alpha_1,\alpha_2]}_{\rho(\alpha)}(t-\tau) = \int_{\alpha_1}^{\alpha_2} \frac{\rho(\alpha)}{\Gamma(\alpha)} (t-\tau)^{\alpha-1}\, d\alpha. \tag{6.64}$$

Function (6.64) can be interpreted as new memory function, which can be called the memory function with distributed memory fading.

The multiplier with distributed power-law fading (6.60) can be represented in the form

$$Y(t) = m([\alpha_1, \alpha_2])(I^{[\alpha_1,\alpha_2]}_{RL;0+} X)(t), \tag{6.65}$$

where $I^{[\alpha_1,\alpha_2]}_{RL;0+}$ is the Riemann–Liouville fractional integral of distributed order (6.63) with the kernel (6.64).

6.14 Multiplier with uniform distributed memory fading

In the simplest case, we can use the continuous uniform distribution (CUD) of the memory fading parameter, for which the weighting function $\rho(\alpha)$ is defined by the expression

$$\rho(\alpha) = \begin{cases} \frac{1}{\alpha_2 - \alpha_1} & \text{for } \alpha = [\alpha_1, \alpha_2] \\ 0 & \text{for } \alpha = (-\infty, \alpha_1) \cup (\alpha_2, \infty). \end{cases} \tag{6.66}$$

For distribution function (6.66), memory function (6.64) has the form

$$M_{\mathrm{CUD}}^{[\alpha_1,\alpha_2]}(t-\tau) = m([\alpha_1,\alpha_2])W(\alpha_1,\alpha_2,t-\tau), \tag{6.67}$$

where we use the function

$$W(\alpha,\beta,t) = \frac{1}{(\beta-\alpha)}\int_{\alpha}^{\beta}\frac{t^{\xi-1}\,d\xi}{\Gamma(\xi)}. \tag{6.68}$$

The fractional integral of uniform distributed order is defined [450] by the equation

$$(I_N^{[\alpha,\beta]}X)(t) = \frac{1}{\beta-\alpha}\int_{\alpha}^{\beta}(I_{\mathrm{RL};a+}^{\xi}X)(t)\,d\xi = \int_0^t W(\alpha,\beta,t-\tau)X(\tau)\,d\tau, \tag{6.69}$$

where $\beta > \alpha > 0$. In article [450], this integral is called the Nakhushev fractional integral.

As a result, we can define the multiplier with uniform distributed memory fading (the UDMF multiplier) by the equation

$$Y(t) = m([\alpha,\beta])(I_N^{[\alpha,\beta]}X)(t), \tag{6.70}$$

where $m([\alpha,\beta])$ is the multiplier coefficient, which depends on the interval $[\alpha,\beta]$, in general.

Remark 6.6. The fractional integrals of the uniform distributed order can be expressed thought the so-called continual fractional integrals, which have been suggested by Adam M. Nakhushev [278, 279]. The fractional operators, which are inverse to the continual fractional integrals, was proposed by Arsen V. Pskhu [317, 318]. Note that Adam M. Nakhushev [278, 279] also proposed the continual fractional derivatives. The fractional integration, which is inversed to the continual fractional derivative has been proposed by Arsen V. Pskhu [317, 318].

The fractional integral, which is inverse to the fractional derivatives of uniform distributed orders, can be defined by the equation

$$(I_P^{[\alpha,\beta]}X)(t) = (\alpha-\beta)\int_0^t E_{\beta-\alpha,\beta}[(t-\tau)^{\beta-\alpha}](t-\tau)^{\beta-1}X(\tau)\,d\tau, \tag{6.71}$$

where $\beta > \alpha > 0$ and $E_{\alpha,\beta}[z]$ is the Mittag–Leffler function that is defined by the expression

$$E_{\alpha,\beta}[z] = \sum_{k=0}^{\infty}\frac{z^k}{\Gamma(\alpha k+\beta)}. \tag{6.72}$$

In article [450], this integral (6.71) is called the Pskhu fractional integral. Fractional integrals (6.71) can be interpreted as fractional integrals with nonsingular kernel. As a result, we can define the multiplier with nonsingular distributed memory (the NSDM multiplier) by the equation

$$Y(t) = m([\alpha,\beta])(I_P^{[\alpha,\beta]}X)(t), \tag{6.73}$$

where $m([\alpha,\beta])$ is the multiplier coefficient.

6.15 Sequential action of multipliers with memory

The previous section considered the simultaneous (parallel) action of the multipliers with memory. Let us describe sequential action of the linear multipliers with power-law memory. This property can be presented in the form of the principle of superposition of multipliers with power-law memory.

Principle 6.3 (Principle of superposition of multipliers with power-law memory). *The multiplier effect, which is created at a given moment by sequential action of several linear multipliers with power-law memory, is equivalent to the action of a linear multiplier whose memory fading parameter is the sum of the fading parameters of these multipliers.*

Symbolically, this principle can be written in the following form: The sequential action of two linear multipliers with power-law memory, which are characterized by memory fading parameters α, β and described by the equations

$$Y(t) = \mathcal{M}_0^{t;\alpha}(X(\tau)), \tag{6.74}$$

$$Z(t) = \mathcal{M}_0^{t;\beta}(Y(\tau)), \tag{6.75}$$

is equivalent to the action of the multiplier with power-law memory that is characterized by the fading parameter equal to the sum $\alpha+\beta$ in the form

$$Z(t) = \mathcal{M}_0^{t;\alpha+\beta}(X(\tau)) \tag{6.76}$$

for all $\alpha > 0$ and $\beta > 0$.

Let us formulate the principle of superposition for the special case of multipliers, when the memory is nonaging memory, i. e., all memory functions can be represented as $M_k(t,\tau) = M_k(t-\tau)$, $k = 1, 2$.

Principle 6.4 (Principle of sequential action of multipliers with nonaging memory). *The multiplier effect, which is created at a given moment by sequential action of several linear multipliers with nonaging memory, is equivalent to the action of a linear multiplier whose memory function is represented by the Laplace convolution of memory functions of these multipliers.*

To represent this principle in symbolic form, we will consider two linear multipliers with nonaging memory that are characterized by the memory functions $M_1(t-\tau)$, $M_2(t-\tau)$, and described by the equations

$$Y(t) = M_1(t) * X(t) = \int_0^t M_1(t-\tau)X(\tau)\,d\tau, \tag{6.77}$$

$$Z(t) = M_2(t) * Y(t) = \int_0^t M_2(t-\tau)Y(\tau)\,d\tau, \tag{6.78}$$

where $*$ is the Laplace (Duhamel) convolution of functions. The principle of sequential action of multiplier with memory states that the sequential action of linear multipliers (6.77) and (6.78) is equivalent to the action of the following multiplier with memory:

$$Z(t) = \int_0^t M(t-\tau)X(\tau)\,d\tau \tag{6.79}$$

that is characterized by the memory function $M(t-\tau)$, which is the Laplace convolution of the memory functions

$$M(t) = M_2(t) * M_1(t) = \int_0^t M_2(t-\tau)M_1(\tau)\,d\tau = \int_0^t M_2(\tau)M_1(t-\tau)\,d\tau. \tag{6.80}$$

Mathematically, the proposed principle is based on the associativity of the Laplace convolution

$$M_2(t) * \big(M_1(t) * X(t)\big) = \big(M_2(t) * M_1(t)\big) * X(t). \tag{6.81}$$

Therefore, for multipliers (6.77) and (6.78), the principle of sequential action of multipliers with memory is proved by the equalities

$$Z(t) = M_2(t) * Y(t) = M_2(t) * \big(M_1(t) * X(t)\big) = \big(M_2(t) * M_1(t)\big) * X(t) = M(t) * X(t), \tag{6.82}$$

where $M(t)$ is defined by equation (6.80).

Let us give examples of the principle of sequential action of linear multipliers with memory.

Example 6.3. Let us consider two linear multipliers with power-law memory, which are described by the Riemann–Liouville fractional integrals in the form

$$Y(t) = m_1(\alpha)(I^{\alpha}_{\mathrm{RL};0+}X)(t), \tag{6.83}$$

$$Z(t) = m_2(\beta)(I^{\beta}_{\mathrm{RL};0+}Y)(t). \tag{6.84}$$

The suggested superposition principle states the additivity of the fading parameters and the multiplicativity of the multiplier coefficients. This means that the superposition principle gives the equality

$$Z(t) = m_1(\alpha)m_2(\beta)(I^{\alpha+\beta}_{\mathrm{RL};0+}X)(t) \tag{6.85}$$

for all $\alpha > 0$ and $\beta > 0$. The proof of this property is based on Theorem 2.5 of [335, p. 46], which states that the relation

$$I^{\alpha}_{\mathrm{RL};a+}(I^{\beta}_{\mathrm{RL};a+}X)(t) = (I^{\alpha+\beta}_{\mathrm{RL};a+}X)(t) \tag{6.86}$$

is valid for any $X(t) \in L_1(a,b)$, if $\alpha > 0$ and $\beta > 0$.

As a result, we get the following statement.

Statement 6.2. *The sequential action of two linear multipliers* (6.83) *and* (6.84) *with power-law memory, which is characterized by memory fading parameters α, β and the coefficients $m_1(\alpha)$, $m_2(\beta)$, is equivalent to the action of the multiplier with power-law memory that is characterized by the fading parameter equal to the sum $\alpha + \beta$ and the coefficient equal to the product $m_1(\alpha)m_2(\beta)$.*

Example 6.4. If the linear multipliers are described by the Liouville fractional integrals (see [335, p. 95] and [200, pp. 87–90]), we can also use the principle of sequential action of multipliers with memory, which can be represented by the equation

$$\mathcal{M}^{t;\alpha}_{-\infty}(\mathcal{M}^{t;\beta}_{-\infty}(X(\tau))) = \mathcal{M}^{t;\alpha+\beta}_{-\infty}(X(\tau)) \tag{6.87}$$

for all $\alpha > 0$, $\beta > 0$ and $X(t) \in L_p(-\infty, +\infty)$, where $\alpha + \beta < 1/p$ and $p \geq 1$. The proof is based on the properties of the Liouville fractional integrals that is presented by the relation

$$(I^{\alpha}_{L;+}I^{\beta}_{L;+}X)(t) = (I^{\alpha+\beta}_{L;+}X)(t), \tag{6.88}$$

which is valid for any $X(t) \in L_p(-\infty, +\infty)$ if $\alpha > 0$ and $\beta > 0$, $\alpha + \beta < 1/p$, $p \geq 1$ (see Lemma 2.19 of [200, p. 89]).

Example 6.5. Let us consider the linear multipliers with distributed scaling that are described by the equations

$$Y(t) = m_1(\alpha)(I^{\alpha}_{\mathrm{EK},a+;\sigma,\eta+\alpha}X)(t), \tag{6.89}$$

$$Z(t) = m_2(\beta)(I^{\beta}_{\mathrm{EK},a+;\sigma,\eta}Y)(t) \tag{6.90}$$

with the left-sided Erdelyi–Kober fractional integrals. The principle of sequential action of multipliers with memory states that action of multipliers (6.89) and (6.90) is equivalent to the action of the linear multiplier

$$Z(t) = m_1(\alpha)m_2(\beta)(I^{\alpha+\beta}_{\mathrm{EK},a+;\sigma,\eta}X)(t) \tag{6.91}$$

for all $\alpha > 0$ and $\beta > 0$. The proof of this statement is based on the property of the left-sided Erdelyi–Kober fractional integral that has the form of the equality

$$I^{\alpha}_{\mathrm{EK},a+;\sigma,\eta}(I^{\beta}_{\mathrm{EK},a+;\sigma,\eta+\alpha}X)(t) = (I^{\alpha+\beta}_{\mathrm{EK},a+;\sigma,\eta}X)(t), \tag{6.92}$$

which is valid for any $X(t) \in L_p(a,b)$ if $\alpha > 0$ and $\beta > 0$, $p \geq 1$ (see Lemma 2.29 of [200, p. 107]).

We can formulate the principle of sequential action of finite numbers of multipliers with memory in the form of the following principle.

Principle 6.5 (Principle of sequential action of multipliers with power-law memory). *The sequential action of several linear multipliers with power-law memory at a given time is equivalent to the action of a linear multiplier with memory, for which the memory fading is equal to the sum of the fading parameters and the coefficient is equal to the product of multiplier coefficients that characterize these multipliers individually.*

In general, this superposition principle means that the effect of the impact on the economy of two linear multipliers does not change, when the third multiplier with the same type of memory.

We can generalize the superposition principle from the simple multipliers to the multipliers with multiparameter memory.

Principle 6.6 (Principle of sequential action of multipliers with multi-parameter memory). *The sequential action of two linear multipliers with multi-parameter memory is equivalent to the action of one multiplier with multi-parameter memory, whose fading parameters are equal to all possible sums of two corresponding fading parameters, and the coefficients are equal to all possible products of two corresponding multiplier coefficients.*

This principle can be written symbolically in the following form. The sequential action of the multipliers with multiparameter memory

$$Y(t) = \sum_{k=1}^{N} m_k(\alpha_k)(I^{\alpha_k}_{\mathrm{RL};a+}X)(t), \tag{6.93}$$

$$Z(t) = \sum_{l=1}^{N} m_l(\alpha_l)(I^{\alpha_l}_{\mathrm{RL};a+}Y)(t), \tag{6.94}$$

are equivalent to the action of the multiplier

$$Z(t) = \sum_{k,l=1}^{N} m_{kl}(\alpha_k, \alpha_l)(I^{\alpha_{kl}}_{\mathrm{RL};a+}X)(t), \tag{6.95}$$

where

$$m_{kl}(\alpha_k, \alpha_l) = m_k(\alpha_k)m_l(\alpha_l), \tag{6.96}$$

and

$$\alpha_{kl} = \alpha_k + \alpha_l. \tag{6.97}$$

As an example of application of the principle of sequential action of multipliers with memory, we can consider macroeconomic model with three type of memory, where production-income memory, the income-spending memory and the spending-production memory are taken into account. This model can be considered as a generalization the macroeconomic model with three lags (the production-income lag, the income-spending lag, the spending-production lag), which has been described by Allen (see Section 2.9, in [10, pp. 55–58]).

6.16 Principles of permutability of multipliers with memory

As a consequence of the principle of sequential action of multipliers with memory, we can propose the following statement.

Principle 6.7 (Principle of permutability of multipliers with memory). *The actions of two multipliers with power-law memory are permutable, i. e., the property*

$$\mathcal{M}^{t;\alpha}_{0}(\mathcal{M}^{t;\beta}_{0}(X(\tau))) = \mathcal{M}^{t;\beta}_{0}(\mathcal{M}^{t;\alpha}_{0}(X(\tau))) \tag{6.98}$$

is satisfied if $\alpha > 0$ *and* $\beta > 0$.

The Principle of Permutability (commutativity) states that the effect of the linear multiplier with power-law memory does not depend on the sequence of actions of the multipliers. Mathematically, this principle is based on the semigroup property of fractional integration. For example, the left-sided Riemann–Liouville fractional integrals satisfy the relation

$$(I^{\alpha}_{\mathrm{RL};a+}I^{\beta}_{\mathrm{RL};a+}X)(t) = (I^{\beta}_{\mathrm{RL};a+}I^{\alpha}_{\mathrm{RL};a+}X)(t) \tag{6.99}$$

for any $X(t) \in L_1(a,b)$ and $\alpha > 0$, $\beta > 0$.

If the memory fading parameters α and β are known, then such a permutation may not make sense from an economic point of view. However, if it is impossible to allocate the parameters of memory fading for each sequential action of the multipliers, and only the general fading parameter of this action is known, then the following principle should be used as a consequence of the permutability principle.

Principle 6.8 (Principle of inseparability of memory fading parameters). *If the actions of two multipliers with power-law memory cannot be separated, then it is impossible to separate the corresponding memory fading parameters. The mathematical form of this principle is given by the equality*

$$\mathcal{M}_0^{t;\alpha}(X(\tau)) = \mathcal{M}_0^{t;\alpha_1}(\mathcal{M}_0^{t;\alpha_2}(X(\tau))) = \mathcal{M}_0^{t;\alpha_3}(\mathcal{M}_0^{t;\alpha_4}(X(\tau))), \tag{6.100}$$

which holds for all $\alpha_1 > 0$, $\alpha_2 > 0$ *and* $\alpha_3 > 0$, $\alpha_4 > 0$ *such that*

$$\alpha_1 + \alpha_2 = \alpha, \quad \alpha_3 + \alpha_4 = \alpha. \tag{6.101}$$

This principle complicates econometric studies of the memory effects in the economy. Inseparability means that observing only the joint action of multipliers with memory, it is impossible to determine the fading parameters of each of them.

6.17 Conclusion

In this chapter, we proposed the concept of multiplier with memory, which is a generalization of the standard concept of the multiplier to the case of economic processes with memory. Equations of the standard multiplier are special cases of the equation of multiplier with memory, whose memory function is represented by the Dirac delta-function.

The proposed concept can be used to take into account the fact that economic agents have a memory of the previous changes of the factor and indicator on a finite time interval, and this affects their behavior and decision-making in the economy.

The proposed concepts of the multiplier with memory allow us to build macroeconomic models that take into account the wide class of memory effects by using the mathematical tools of derivatives and integrals of noninteger orders.

In addition to the proposed concept of multiplier with memory, we proposed the concept of accelerator with memory in articles [412, 397]. This concept includes the wide spectrum of new dependencies, which are intermediate between the standard accelerators and the standard multipliers.

7 Accelerator for processes with memory

In this chapter, we consider generalization of the standard concept of economic accelerator that is used in the macroeconomics. This chapter is based on our papers [412, 397], where we proposed generalizations of the concepts of the accelerator and the multiplier to describe economic processes with memory.

7.1 Introduction

In macroeconomic models, various simplifying assumptions are used to describe the economic processes. Because of this, many economic models have disadvantages, since they do not take into account important aspects of economic phenomena and processes. Some of these disadvantages are partially caused by limitations of the used mathematical tools.

Mathematically, the concept of the standard accelerator is based on the theory of derivatives of integer orders and differential equations of the integer orders. It is known that the derivatives of positive integer orders are determined by the properties of the differentiable function only in an infinitesimal neighborhood of the considered point. As a result, differential equations of integer orders with respect to time cannot be used to describe processes with memory. In fact, such equations describe only economic processes, in which all economic agents remember only the infinitesimal past and the amnesia occurs infinitely fast. Differential equations of integer orders describe the behavior of only those economic agents that have lost their memory and have amnesia. It is obvious that the assumption of the absence of memory in economic agents is a strong restriction for economic models.

This chapter is based on our articles [412, 397], where we proposed generalizations of the concepts of the accelerator and the multiplier, which allow us to take into account the fading memory.

7.2 Concept of standard accelerator

In addition to the concept of multiplier, the economic analysis uses the concept of accelerator (see [75], [10, pp. 62–64], [10, p. 73], [12, p. 81]). In macroeconomics, the accelerator describes an indicator as dependent on the rate of change of the factor and not on the level of the factor. The accelerator characterizes how much the indicator changes in response to an increase in the rate of change of the factor by one.

The concept of an accelerator is actually a mathematical expression of the principle of acceleration. According to this principle, to maintain the investment at the same level is necessary to keep a constant rate of growth in the national income. The preservation of income at the same level leads to a reduction in investment to zero.

https://doi.org/10.1515/9783110627459-007

The acceleration principle states a direct relationship between the rate of output of an economy and the level of net investment (see [10, pp. 60–64]).

To explain the principle of acceleration, Allen writes: “If the flow of output is constant (like a velocity), then capital stock is constant and net investment zero. If the flow of output changes (like an acceleration), then the required capital stock also changes and there is (positive or negative) net induced investment. This is the acceleration principle—an increasing flow of output calls for a greater stock of capital and induces investment” [10, p. 61].

For macroeconomic models with continuous time, the equation of the standard accelerator without memory and lag can be written (see [10, p. 62]) in the form

$$Y(t) = F\left(\frac{dX(t)}{dt}\right), \tag{7.1}$$

where F is a function, $dX(t)/dt$ is the rate of the impact variable (e. g., the income), $Y(t)$ is the response variable (e. g., the induced investment). The linear accelerator equation, which expresses the dependence of the indicator $Y(t)$ on the rate of factor $X(t)$, can be given in the form

$$Y(t) = a\frac{dX(t)}{dt}, \tag{7.2}$$

where a is the acceleration coefficient that indicates the power of accelerator (see [10, p. 73]).

Equation (7.2) allows us to interpret the variable $X(t)$ as an impact, and the variable $Y(t)$ as a response. The left and right sides of equation (7.2) are written for the same time. Therefore, an increase or decrease in the rate of change of the factor $X(t)$ leads to an instantaneous change in the indicator $Y(t)$. Obviously, this is a strong restriction for macroeconomic models. In real economic processes, a change of the rate of the factor $X(t)$ at time t cannot immediately lead to a change of the indicator $Y(t)$.

We note that equation (7.2) can be written in the form

$$\frac{dX(t)}{dt} = AY(t), \tag{7.3}$$

where $A = 1/a$. Since the left and right sides of equations (7.2) and (7.3) are written for the same time, equation (7.3) can be interpreted in the opposite direction (inverse interpretation) in comparison with the interpretation of equation (7.2). In equation (7.3), we can consider $Y(t)$ as an impact variable (the factor) and $X^{(1)}(t)$ as a response variable (the indicator). In this case, equation (7.3) describes how much the change of the rate of the indicator $X(t)$ in response of change of the factor $Y(t)$.

The separation of variables into impact and response ones is based on the causal relationship: the impact variable is the cause (the action), and the response variable is the consequence. In the general case, the causal relationship assumes a finite rate of process and mutual influence of economic processes.

Let us give an example of accelerators and their equations. We will designate the national income as a function $Y(t)$ of time t, and the induced investment through $I(t)$. In macroeconomic models, it is assumed that the relationship between induced investment and the growth rate of income is given by the direct proportionality (the accelerator equation) in the form

$$I(t) = a\frac{dY(t)}{dt}, \tag{7.4}$$

where the coefficient a is interpreted as the capital-intensity coefficient for income growth [10, p. 91]. The positive constant a is also called the investment coefficient indicating the power of the accelerator (see [10, p. 62]). This accelerator coefficient shows how much the investment will increase if the current rate of change of national income increases by one [10, pp. 62–63].

Equation (7.4) assumes that the investment changes instantly with a change of rate of income. Moreover, equation (7.4) assumes the absence of the memory about the previous investments and the results of these investments. Equation (7.4) means that new investments do not depend on what investments were made in the past and do not depend on the history of changes in growth rate of income.

Equation (7.4) can be written in the form

$$\frac{dY(t)}{dt} = AI(t), \tag{7.5}$$

where $A = 1/a$ is interpreted as an incremental capital-output ratio or as a marginal productivity of capital. Equation (7.5) assumes that the rate of growth of income changes instantly with the growth of investment, and does not depend on investments made at previous time points. Obviously, these assumptions do not reflect reality and may lead to inadequate conclusions and conclusions.

7.3 Effect of financial accelerator

The concept of a financial accelerator was introduced in [31] to explain the large fluctuations in aggregate economic activity, which arises from seemingly small shocks. It was shown that these phenomena indicate the existence of an accelerator mechanism. The mechanism of the financial accelerator is determined by the relationship between the economy and financial markets, which is based on the need of firms for external financing to invest in business. The ability of firms to borrow significantly depends on the market value of their assets. Lenders, as a rule, require borrowers to guarantee their ability to repay, often in the form of collateralized assets. Therefore, the fall in asset prices deteriorates the balance of firms and the value of assets. As a result, the ability of firms to borrow is deteriorating, which negatively affects their investments, and this, in turn, leads to a decrease in economic activity and a further decline in asset prices. This chain reaction of changes is called a financial accelerator. This financial feedback loop (the credit cycle), which, starting with a small change in the financial markets, can lead to big changes in the economy.

Note that memory effects can weaken and strengthen the effect of the financial accelerator. For example, if economic agents remember that previous changes in economic activity (caused by small shocks), which were insignificant and short-term, then the memory effect may lead to a more rapid decay in the fluctuation of economic activity. On the other hand, if agents remember that previous fluctuations, which are generated by shocks, were large, prolonged, and significantly worsened the economic situation, then repetition of even smaller shocks could lead to more significant changes in the economy by action of agents with memory, than agents without memory.

7.4 Generalized accelerator with memory

To generalize the concept of the economic accelerator to processes with memory, we should take into account that the response variable $Y(t)$ at time t depends on the changes of the impact variable $X(\tau)$ and its integer derivatives $X^{(k)}(\tau)$, $(k = 1, 2, \ldots, n)$ on a finite time interval $0 \leq \tau \leq t$. In economic processes, the memory effects are caused with the fact that economic agents remember the history of previous changes of the impact variable $X(t)$ and the influence on the changes of the response variable $Y(t)$. In the economic process with memory, there exists at least one variable $Y(t)$ at the time t, which depends on the history of the change of $X(\tau)$ and its integer derivatives $X^{(k)}(\tau)$, $(k = 1, 2, \ldots, n)$ at $\tau \in (0, t)$. For this reason, it is impossible to describe the processes with memory by standard accelerator equation (7.1) with the first derivative $X^{(1)}(t)$ at $\tau = t$. As a result, we propose the following definition of the generalized accelerator with memory.

Definition 7.1. The generalized accelerator with memory

$$Y(t) = \mathcal{A}_0^{t,[0,n]}(X(\tau)) \tag{7.6}$$

is the dependence of a response variable $Y(t)$ at the time $t \geq 0$ on the histories of the changes of the impact variable $X(\tau)$ and their derivatives of integer orders up to the order n on a finite time interval $0 \leq \tau \leq t$. Here, $\mathcal{A}_0^{t,[0,n]}$ is an operator that transforms each history of changes of $X(\tau)$ and $X^{(k)}(\tau)$, $k = 1, \ldots, n$ for $\tau \in [0, t]$ into the appropriate history of changes of $Y(\tau)$ with $\tau \in [0, t]$.

We can also use the generalized accelerator with memory in the form

$$Y(t) = \mathcal{A}_0^{t}(X^{(n)}(\tau)), \tag{7.7}$$

where $X^{(n)}(\tau) = d^n X(\tau)/d\tau^n$ is the derivative of $X(\tau)$ of the integer positive order n with respect to time $\tau \in [0, t]$, and $\mathcal{A}_0^t$ is an operator that transforms each history of changes of $X^{(n)}(\tau)$ for $\tau \in [0,t]$ into the appropriate history of changes of $Y(\tau)$ with $\tau \in [0,t]$.

Remark 7.1. Equation (7.7) can be interpreted as a multiplier equation, in which the impact variable is the derivative $X^{(n)}(\tau)$ of integer orders n and the response variable is $Y(t)$. The multiplier with memory can be considered as a special case of the suggested concept of the accelerator with memory, if we include the case $n = 0$ by using $X^{(0)}(\tau) = X(\tau)$ in equation (7.7).

The generalized accelerator with memory (7.6) is called linear, if the condition

$$\mathcal{A}_0^t(aX_1(\tau) + bX_2(\tau)) = a\mathcal{A}_0^t(X_1(\tau)) + b\mathcal{A}_0^t(X_2(\tau)) \tag{7.8}$$

holds for all a and b that are real numbers.

The accelerator with memory (7.7) is linear, if the equality

$$\mathcal{A}_0^t(aX_1^{(n)}(\tau) + bX_2^{(n)}(\tau)) = a\mathcal{A}_0^t(X_1^{(n)}(\tau)) + b\mathcal{A}_0^t(X_2^{(n)}(\tau)) \tag{7.9}$$

holds for all real numbers a and b.

We can consider the linear accelerators with memory, in which the memory is described by the memory function $M(t,\tau)$. It allows us to define the linear generalized accelerator with memory of the special kind.

Definition 7.2. The generalized accelerator (7.7) with memory is called the linear generalized accelerator with memory function $M(t,\tau)$, if accelerator equation (7.7) can be represented in the form

$$Y(t) = \int_0^t M(t,\tau)X^{(n)}(\tau)\,d\tau, \tag{7.10}$$

where n is positive integer number and the function $M(t,\tau)$ describes the memory. Here, we assume that integral (7.10) exists.

If the function $M(t,\tau)$ is expressed in terms of the Dirac delta-function $M(t,\tau) = a\delta(t-\tau)$, then equation (7.10) with $n = 1$ takes the form of the standard accelerator equation $Y(t) = aX^{(1)}(t)$.

Remark 7.2. We can see that the generalized accelerator with memory (7.10), can be mathematically interpreted as a generalized multiplier with memory, which connects the response variable $Y(t)$ and the impact variable $X^{(n)}(\tau)$ with $\tau \in [0,t]$.

Remark 7.3. Note that the linear generalized accelerator with memory, which is characterized by the memory function $M(t,\tau)$, can also be defined by the equations

$$Y(t) = \frac{d^n}{dt^n}\int_0^t M(t,\tau)X(\tau)\,d\tau, \tag{7.11}$$

$$Y(t) = \frac{d^{n-k}}{dt^{n-k}} \int_0^t M(t,\tau) X^{(k)}(\tau)\, d\tau, \tag{7.12}$$

where $n \geq k$. The generalized accelerator with memory can also be described by the equation

$$Y(t) = D^{\alpha-\beta}_{C;0+} \int_0^t M(t,\tau)(D^{\beta}_{C;0+}X)(\tau)\, d\tau, \tag{7.13}$$

where $D^{\alpha-\beta}_{C;0+}$ and $D^{\beta}_{C;0+}$ with $\alpha \geq \beta > 0$ are the Caputo fractional derivatives.

Let us consider the linear accelerator, which is described by equation (7.10) with the impact variable $X(\tau) = \delta(\tau - \xi)$. The substitution of $X(\tau) = \delta(\tau - \xi)$ into equation (7.10) with integer $n \geq 1$ gives

$$Y(t) = \int_0^t M(t,\tau)\delta^{(n)}(\tau - \xi)\, d\tau, \tag{7.14}$$

where $0 < \xi < t$. Using the properties of the delta-function, we get the response $Y(t)$ for $t > \xi$ in the form

$$Y(t) = (-1)^n \left(\frac{\partial^n M(t,\tau)}{\partial \tau^n} \right)_{\tau=\xi}. \tag{7.15}$$

As a result, accelerator (7.10) is the accelerator with fading memory if the response variable (7.15) tends to zero ($Y(t) \to 0$) at $t \to \infty$. Using equation (7.15), we can give the following definition.

Definition 7.3. Generalized accelerator (7.10) is called the accelerator with fading memory if the condition

$$\lim_{t\to\infty} \left(\frac{\partial^n M(t,\tau)}{\partial \tau^n} \right)_{\tau=\xi} = 0 \tag{7.16}$$

is satisfied for $\xi \in (0,t)$.

Let us consider equation (7.14) with $M(t,\tau) = M(t-\tau)$, when the impact variable $X(\tau)$ is represented by the Dirac delta-function $\delta(\tau - \xi)$. Substitution of $X(\tau) = \delta(\tau - \xi)$ into equation (7.10) gives the response $Y(t)$ for $t > \xi > 0$ in the form

$$Y(t) = \frac{\partial^n M(t,\xi)}{\partial t^n}, \tag{7.17}$$

where we use

$$\frac{\partial M(t,\tau)}{\partial t} = -\frac{\partial M(t,\tau)}{\partial \tau}. \tag{7.18}$$

As a result, condition (7.16), which defines the accelerator with fading memory, can be written in the form

$$\lim_{t\to\infty} \frac{\partial^n M(t,\xi)}{\partial t^n} = 0. \tag{7.19}$$

Note that the concepts of the accelerators with memory for the discrete time approach have been suggested in articles [441, 442, 431, 432, 417, 436].

7.5 Accelerator with power-law memory

Let us give a definition of the accelerators with memory, which fading has the power-law form.

Definition 7.4. The accelerator with fading memory

$$Y(t) = \mathcal{A}_0^t(X^{(n)}(\tau)), \tag{7.20}$$

where

$$\mathcal{A}_0^t(X^{(n)}(\tau)) = \int_0^t M(t,\tau) X^{(n)}(\tau)\, d\tau \tag{7.21}$$

is called the accelerator with power-law memory if there is the parameter $\beta > 0$, and a finite nonzero constant $c = c(\beta)$ such that the limit

$$\lim_{t\to\infty} t^{\beta} M(t,\tau) = c(\beta) \tag{7.22}$$

exists for fixed $\tau \in (0,t)$. In this case, we will use the notation

$$\mathcal{A}_0^t(X^{(n)}(\tau)) = \mathcal{A}_{0;\beta}^t(X^{(n)}(\tau)). \tag{7.23}$$

Example 7.1. The linear accelerator with simple power-law memory, which is characterized by the fading parameter $\alpha > 0$, can be described by the equation

$$Y(t) = a(\alpha)(D^{\alpha}_{C;0+}X)(t), \tag{7.24}$$

where $D^{\alpha}_{C;0+}$ is the left-sided Caputo fractional derivative of the order $\alpha > 0$ with respect to time, and $a(\alpha)$ is the coefficient of the accelerator with memory, $n = [\alpha] + 1$ for noninteger values of $\alpha > 0$, and $n = \alpha$ for integer values of α.

The memory function of accelerator (7.24) has the form

$$M(t,\tau) = M_{n-\alpha}(t-\tau) = \frac{a(\alpha)}{\Gamma(n-\alpha)} \frac{1}{(t-\tau)^{\alpha-n+1}}, \tag{7.25}$$

where $t > \tau$. Therefore, we have

$$\lim_{t\to\infty} t^{\alpha-n+1} M(t,\tau) = \frac{a(\alpha)}{\Gamma(n-\alpha)}, \tag{7.26}$$

and the parameters β and $c(\beta)$ are equal to

$$\beta = \alpha - n + 1, \quad c(\beta) = \frac{a(\alpha)}{\Gamma(n-\alpha)}. \tag{7.27}$$

In order to have a correct definition of accelerator (7.20) with power-law memory, we should require the implementation of the following correspondence principle.

Principle 7.1 (Correspondence principle for accelerator with memory). *For accelerator (7.20) with power-law memory, for which condition (7.22) holds, a memory fading parameter $\alpha \geq 0$ must exist, such that $\beta = \beta(\alpha, n)$ and the following condition is satisfied. If the fading parameter $\alpha \geq 0$ tends to one ($\alpha \to 1$), then equation (7.20) should take the form of the standard accelerator without memory, $Y(t) = aX^{(1)}(t)$ for all $t \geq 0$, where $a = a(1)$ is the accelerator coefficient. In this case, accelerator (7.20) will be denoted as*

$$\mathcal{A}_0^{t;\alpha}(X^{(n)}(\tau)) = \mathcal{A}_{0;\beta(\alpha,n)}^{t}(X^{(n)}(\tau)), \tag{7.28}$$

where $n \in \mathbb{N}$, $\tau \in (0,t)$.

This principle means that the linear accelerator with power-law memory (7.20) should satisfy the condition

$$\lim_{\alpha\to 1} \mathcal{A}_0^{t;\alpha}(X^{(n)}(\tau)) = a(1)X^{(1)}(t) \tag{7.29}$$

for all $t \geq 0$, where $a = a(1)$.

Example 7.2. For accelerator (7.24), condition (7.29) is satisfied due to the property of the Caputo fractional derivative

$$(D^1_{C;0+}X)(t) = \frac{dX(t)}{dt}. \tag{7.30}$$

This correspondence principle for accelerator can be formulated with the stronger condition for $\mathcal{A}_0^{t;\alpha}(X^{(n)}(\tau))$: For this linear accelerator with a power-law memory, there is an integer parameter $m = m(\alpha, n)$, for which the condition

$$\lim_{\alpha\to m} \mathcal{A}_0^{t;\alpha}(X^{(n)}(\tau)) = a(m)X^{(m)}(t) \tag{7.31}$$

holds for all $t \geq 0$.

7.6 Accelerator with simplest power-law memory

In the simple case, the accelerator with the power-law behavior of the memory fading, can be described by the memory function $M(t,\tau)$ in the form

$$M(t,\tau) = c(\beta)(t-\tau)^{-\beta}, \tag{7.32}$$

where $c(\beta) > 0$ and $0 < \beta < 1$ are constants. For simplification, we will use the parameters α such that conditions (7.27), i. e., $\beta = \alpha - n + 1$, $n-1 < \alpha < n$ and $c(\beta) = a(\alpha)/\Gamma(n-\alpha)$. In this case, memory function (7.32) takes the form

$$M(t,\tau) = M_{n-\alpha}(t-\tau) = \frac{1}{\Gamma(n-\alpha)} \frac{a(\alpha)}{(t-\tau)^{\alpha-n+1}}, \tag{7.33}$$

where $n-1 < \alpha < n$, $n \in \mathbb{N}$, $t > \tau$ and $a = a(\alpha)$ is the numerical coefficient. Substitution of expression (7.33) into equation (7.20), (7.21), gives the accelerator equation in the form

$$Y(t) = a(\alpha)(D^{\alpha}_{C;0+}X)(t), \tag{7.34}$$

where $D^{\alpha}_{C;0+}$ is the left-sided Caputo fractional derivative of the order $\alpha \geq 0$ of the function $X(t)$. Let us give the definition of the Caputo fractional derivative (for details, see Theorem 2.1 in [200, p. 92]).

Definition 7.5. The left-sided Caputo fractional derivative of the order $\alpha \geq 0$ is defined by the equation

$$(D^{\alpha}_{C;0+}X)(t) = (I^{n-\alpha}_{RL;0+}X^{(n)})(t) = \frac{1}{\Gamma(n-\alpha)} \int_0^t \frac{X^{(n)}(\tau)\, d\tau}{(t-\tau)^{\alpha-n+1}}, \tag{7.35}$$

where $n-1 < \alpha \leq n$, $\Gamma(\alpha)$ is the gamma function, $0 < t < T$, and $X^{(n)}(\tau)$ is the derivative of the integer order $n = [\alpha] + 1$ (and $n = \alpha$ for integer values of α) of the function $X(\tau)$ with respect to the time variable τ, and $I^{n-\alpha}_{RL;0+}$ is the left-sided Riemann–Liouville fractional integral of the order $(n-\alpha) > 0$. In equation (7.35), it is assumed that $X(\tau) \in AC^n[0,T]$, i. e., function $X(\tau)$ has integer-order derivatives up to $(n-1)$-th order, which are continuous functions on the interval $[0,T]$, and the derivative $X^{(n)}(\tau)$ is Lebesgue summable on the interval $[0,T]$.

Remark 7.4. Equation (7.34) describes the equation of economic accelerator with power-law fading memory, where α is the fading parameter, and $a = a(\alpha)$ is a positive constant indicating the power of this accelerator. As a result, we obtain the following definition.

Definition 7.6. The equation of linear accelerator with simple power-law memory, which is characterized by the fading parameter $\alpha \geq 0$, has the form

$$Y(t) = a(\alpha)(D^{\alpha}_{C;0+}X)(t), \tag{7.36}$$

where $D^{\alpha}_{C;0+}$ is the Caputo fractional derivative (7.35) of the order $\alpha \geq 0$ with respect to time, and $a(\alpha)$ is the coefficient of the accelerator with memory. The noninteger values of the parameter $\alpha \geq 0$ are interpreted as the memory fading parameters.

Note that for the integer value of the order $\alpha \in \mathbb{N}$, the Caputo derivative coincides (see Theorem 2.1 in [200, pp. 92–93] and [308, p. 79]) with the standard integer-order derivative.

Statement 7.1. *For the integer values of the orders α, the Caputo derivative coincides with the standard integer-order derivatives*

$$(D^{n}_{C;0+}X)(t) = X^{(n)}(t), \tag{7.37}$$

where $n \in \mathbb{N}$, and for $n = 0$ we have

$$(D^{0}_{C;0+}X)(t) = X(t). \tag{7.38}$$

Proof. The proof of this statement follows from Theorem 2.1 of book [200, pp. 92–93]. □

Statement 7.1 shows that the Caputo derivatives include the standard derivatives of integer orders, as a special case.

As a result, the linear accelerator with memory, which is described by equation (7.36), can be considered as a generalization of the standard equations of accelerator and standard multiplier. These standard concepts are included into proposed concept of the accelerator with memory as special cases.

Corollary 7.1. *For the integer value of $\alpha = 1$, equation of linear accelerator with power-law fading memory* (7.36) *takes the form of the standard accelerator without memory*

$$Y(t) = aX^{(1)}(t), \tag{7.39}$$

where $a = a(1)$.

Corollary 7.2. *For zero value of the parameter $\alpha = 0$, equation of linear accelerator with power-law fading memory* (7.36) *takes the form of the standard multiplier without memory*

$$Y(t) = mX(t), \tag{7.40}$$

where $m = a(0)$.

We must emphasize that the proposed concept of the accelerator with memory, which is described by (7.36), includes not only the standard concept of the accelerator as special case, but it also includes the standard concept of the economic multiplier.

Corollary 7.3. *For noninteger value of the parameter α, equation (7.36) can be interpreted as the equation of a linear accelerator with power-law fading memory, where the order α of fractional derivative is interpreted as the memory fading parameter.*

As a result, equation (7.36) can be considered as a generalization of the standard accelerator and multiplier equations. Equations of the standard accelerator and multiplier are included in equation (7.36) as special cases. In addition to this inclusion, the suggested concept of the accelerator with memory includes the wide spectrum of intermediate dependences of the indicator on the factor. Equation (7.36) allows us to take into account the fact that the economic process depends on the history of previous changes of factors and indicators on the finite time interval.

As an example of the linear accelerator with power-law memory that is represented in the form

$$Y(t) = \frac{d^n}{dt^n} \int_0^t M(t,\tau) X(\tau)\, d\tau, \tag{7.41}$$

we can consider the accelerator that is described by the equation

$$Y(t) = a(\alpha)(D^{\alpha}_{\mathrm{RL};0+}X)(t), \tag{7.42}$$

where $D^{\alpha}_{\mathrm{RL};0+}$ is the left-sided Riemann–Liouville fractional derivative of order $\alpha \geq 0$ of the function $X(t)$. Let us give the definition of the Riemann–Liouville fractional derivative (for details, see [200, p. 70] and [335, p. 37]).

Definition 7.7. The left-sided Riemann–Liouville fractional derivative of the order $\alpha \geq 0$ is defined by the equation

$$(D^{\alpha}_{\mathrm{RL};0+}X)(t) = \frac{d^n}{dt^n}(I^{n-\alpha}_{\mathrm{RL};0+}X)(t) = \frac{1}{\Gamma(n-\alpha)} \frac{d^n}{dt^n} \int_0^t \frac{X(\tau)\, d\tau}{(t-\tau)^{\alpha-n+1}}, \tag{7.43}$$

where $n-1 \leq \alpha < n$, $\Gamma(\alpha)$ is the gamma function, $0 < \tau < t < T$. A sufficient condition of the existence of fractional derivatives (7.43) is $X(t) \in AC^n[0,T]$. The space $AC^n[0,T]$ consists of functions $X(t)$, which have continuous derivatives up to order $n-1$ on $[0,T]$ and function $X^{(n-1)}(t)$ is absolutely continuous on the interval $[0,T]$.

If $X(t) \in AC^n[0,T]$, then fractional derivative (7.43) exists almost everywhere on the interval $[0,T]$, where $n-1 \leq \alpha < n$ (see Lemma 2.2 in book [200, p. 73]). The space $AC^n[0,T]$ is defined in [200, p. 2].

We note that the Riemann–Liouville derivatives of orders $\alpha = 1$ and $\alpha = 0$ give the expressions $(D^{1}_{\mathrm{RL};0+}X)(t) = X^{(1)}(t)$ and $(D^{0}_{\mathrm{RL};0+}X)(t) = X(t)$, respectively (see equation (2.1.7) in [200, p. 70]).

Remark 7.5. Let us consider the function

$$X(t) = \sum_{k=1}^{n} X_k t^{\alpha-k}, \tag{7.44}$$

where X_k are arbitrary constants and $n = [\alpha] + 1$. Function (7.44) plays the same role for the Riemann–Liouville derivative of orders α (see [335, pp. 36–37] and [200, p. 72]) as a constant function for the first derivative and as the functions

$$X(t) = \sum_{k=1}^{n} X_k t^{n-k} \tag{7.45}$$

for the derivative of positive integer order n.

For positive integer values $\alpha = n$ and zero value $\alpha = n = 0$, equation (7.36) takes the form

$$Y(t) = a(n) \frac{d^n X(t)}{dt^n}, \tag{7.46}$$

or $Y(t) = mX(t)$ for $n = 0$, where $m = a(0)$. As a result, the concept of an accelerator with memory and parameter of memory fading $\alpha \geq 0$ includes, as special cases, the equations of "higher-order accelerators" (7.46), which are usually called the linear dynamic elements of order n. For $\alpha = n = 1$, equation (7.36) coincides with equation (7.2), that is, (7.36) include the standard accelerator equation, as a special case.

Remark 7.6. One of the important differences between the accelerator described by the Caputo derivative and the accelerator described by the Riemann–Liouville derivative is the zero value of the response ($Y(t) = 0$) to a constant value of the impact $X(t) = \text{const}$ for noninteger values of α.

For example, if $X(t) = X_0 = \text{const}$, then accelerator equation (7.36) gives,

$$Y(t) = a(\alpha)(D^{\alpha}_{\mathrm{RL};0+} X)(t) = \frac{a(\alpha) X_0}{\Gamma(1-\alpha)} t^{-\alpha}, \tag{7.47}$$

where $\alpha \geq 0$. For details, see equations (2.1.17) and (2.1.20) of Property 2.1 in [200, p. 71].

7.7 Accelerator with distributed time scaling

In order to describe more complex economic dynamics, we can use the generalized fractional calculus [202, 203]. For example, we can consider the accelerator with distributed time scaling.

Example 7.3. For the linear accelerator, we can use the function

$$M^{n-\alpha,\eta+\alpha}_{EKD;\sigma}(t,\tau) = \frac{\sigma t^{-\sigma(n+\eta)}}{\Gamma(n-\alpha)} \frac{a(\alpha)\tau^{\sigma(\eta+\alpha+1)-1}}{(t^{\sigma} - \tau^{\sigma})^{\alpha-n+1}}. \tag{7.48}$$

As a result, the linear accelerator with distributed time scaling is defined by the equation

$$Y(t) = a(\alpha)(D^{\alpha}_{\mathrm{EKC};a+;\sigma,\eta}X)(t), \tag{7.49}$$

where $D^{\alpha}_{\mathrm{EKC};0+;\sigma,\eta}$ is the Caputo modification of the Erdelyi–Kober fractional derivative of the order $\alpha > 0$, which was proposed in [241, p. 260].

Definition 7.8. The Caputo modification of the Erdelyi–Kober fractional derivative is defined by the equation

$$\begin{aligned}(D^{\alpha}_{\mathrm{EKC},a+;\sigma,\eta}X)(t) &= I^{n-\alpha}_{\mathrm{EKC},a+;\sigma,\eta+\alpha} \prod_{k=1}^{n}\left(\frac{\tau}{\sigma}\frac{d}{d\tau} + \eta + k\right)X(\tau) \\ &= \int_a^t M^{n-\alpha,\eta+\alpha}_{EKD;\sigma}(t,\tau) \prod_{k=1}^{n}\left(\frac{\tau}{\sigma}\frac{d}{d\tau} + \eta + k\right)X(\tau)\, d\tau \\ &= \frac{\sigma t^{-\sigma(n+\eta)}}{\Gamma(n-\alpha)} \int_a^t \frac{\tau^{\sigma(\eta+\alpha+1)-1}}{(t^{\sigma}-\tau^{\sigma})^{\alpha-n+1}} \prod_{k=1}^{n}\left(\frac{\tau}{\sigma}\frac{d}{d\tau} + \eta + k\right)X(\tau)\, d\tau,\end{aligned} \tag{7.50}$$

where $n - 1 < \alpha \leq n$, $\sigma > 0$ and $\delta \in \mathbb{C}$.

Example 7.4. Another type of accelerator with distributed time scaling can be defined by the equation

$$Y(t) = a(\alpha)(D^{\alpha}_{\mathrm{EK},a+;\sigma,\eta}X)(t), \tag{7.51}$$

where $D^{\alpha}_{0+;\sigma,\eta}$ is the left-sided Erdelyi–Kober fractional derivative of the Riemann–Liouville type of the order $\alpha > 0$ (see equation (2.6.29) in [200, p. 108]) that are defined by the equation

$$(D^{\alpha}_{\mathrm{EK},a+;\sigma,\eta}X)(t) = t^{-\sigma\eta}\left(\frac{1}{\sigma t^{\sigma-1}}\frac{d}{dt}\right)^{n} t^{\sigma(n+\eta)}(I^{n-\alpha}_{\mathrm{EK},a+;\sigma,\eta+\alpha}X)(t), \tag{7.52}$$

where $n - 1 < \alpha \leq n$. Here $I^{\alpha}_{\mathrm{EK},a+;\sigma,\eta}$ is the left-sided Erdelyi–Kober fractional integral of the order $\alpha > 0$ with respect to time variable [200, p. 105] or [335, pp. 322–325], which is defined by the equation

$$(I^{\alpha}_{\mathrm{EK},a+;\sigma,\eta}X)(t) = \frac{\sigma t^{-\sigma(\alpha+\eta)}}{\Gamma(\alpha)} \int_a^t \frac{\tau^{\sigma(\eta+1)-1}}{(t^{\sigma}-\tau^{\sigma})^{1-\alpha}} X(\tau)\, d\tau, \tag{7.53}$$

where $\sigma > 0$, η is the real number and $\alpha > 0$. For $\sigma = 1$, equation (7.52) takes the form of the Kober fractional integral.

In paper [241] (see equation 19 in [241, p. 254], where $\delta = \alpha$, $\beta = \sigma$, $\gamma = \eta$), the Erdelyi–Kober fractional derivative of the Riemann–Liouville type is defined as

$$(D^{\alpha}_{\mathrm{EK};a+;\sigma,\eta}X)(t) = \prod_{k=1}^{n}\left(\frac{t}{\sigma}\frac{d}{dt} + \eta + k\right)(I^{n-\alpha}_{\mathrm{EK},a+;\sigma,\eta+\alpha}X)(t), \tag{7.54}$$

where $n - 1 < \alpha \leq n$.

Note that both forms of representation of the Erdelyi–Kober fractional derivative, which are given by equations (7.52) and (7.54), are equivalent.

Remark 7.7. In some economic processes with memory, the parameter α of memory fading can be changed during the time, i. e., $\alpha = \alpha(t)$. The accelerator with variable power-law memory can be described by the equation

$$Y(t) = a(\alpha(t))(D^{\alpha(t)}_{C;0+}X)(t), \tag{7.55}$$

where $D^{\alpha(t)}_{C;0+}$ is the left-sided Caputo fractional derivative of the variable order $\alpha(t) \geq 0$. For $\alpha(t) = \alpha$ equation (7.55) gives (7.36).

7.8 Accelerator with distributed memory fading

Let us consider at the beginning of equation of accelerator with discrete distributed memory fading. This accelerator can be represented in the form

$$Y(t) = a(\alpha) \sum_{k=1}^{N} \rho_k(\alpha)(D^{\alpha_k}_{C;0+}X)(t), \tag{7.56}$$

where $a(\alpha)$ is the accelerator coefficient

$$a(\alpha) = \sum_{k=1}^{N} a_k(\alpha_k), \tag{7.57}$$

and $\rho_k(\alpha)$ is the discrete probability density function

$$\rho_k(\alpha) = \frac{a_k(\alpha_k)}{a(\alpha)} : \quad \sum_{k=1}^{N} \rho_k(\alpha) = 1. \tag{7.58}$$

In this case, equation (7.56) can be interpreted as accelerator with discretely distributed memory fading. The memory fading parameter α is considered as a random variable.

In general, the parameter α of memory fading can be continuously distributed on the interval $[\alpha_1, \alpha_2]$, where the distribution is described by a weighting function $\rho(\alpha)$.

The function $\rho(\alpha)$ describes distribution of the parameter of memory fading on the set of economic agents. This is important for the economics, since various types of economic agents may have different parameters of memory fading. In this case, we should consider the accelerators with memory, which depend on the weighting function $\rho(\alpha)$ and the interval $[\alpha_1, \alpha_2]$.

Let us consider an example of the accelerators with distributed memory fading. In the simple case, we can use the continuous uniform distribution (CUD) that is defined by the expression

$$\rho(\alpha) = \frac{1}{\alpha_2 - \alpha_1} \tag{7.59}$$

for the case $\alpha \in [\alpha_1, \alpha_2]$ with $\alpha_2 - \alpha_1 > 0$, and $\rho(\alpha) = 0$ for other cases.

The fractional integrals and derivatives with uniform distributed order can be expressed thought the continual fractional integrals and derivatives, which were suggested by Adam M. Nakhushev [278, 279]. The fractional operators, which are inversed to the continual fractional integrals and derivatives, were proposed by Arsen V. Pskhu [317, 318].

Using the continual fractional integrals and derivatives, we define the integral and derivatives of uniform distributed order [450]. The fractional integral and derivative of the uniform distributed orders are defined by the equations

$$(I_N^{[\alpha,\beta]}X)(t) = \frac{1}{\beta-\alpha}\int_{\alpha}^{\beta}(I_{\mathrm{RL};0+}^{\xi}X)(t)\,d\xi, \tag{7.60}$$

$$(D_N^{[\alpha,\beta]}X)(t) = \frac{1}{\beta-\alpha}\int_{\alpha}^{\beta}(D_{\mathrm{RL};0+}^{\xi}X)(t)\,d\xi, \tag{7.61}$$

where $\beta > \alpha > 0$. The integral (7.60) and derivative (7.61) are called the Nakhushev fractional integral and derivative in [450], respectively. Note that the Nakhushev fractional derivatives cannot be considered as left-inverse operators for the Nakhushev fractional integrals. The corresponding inverse operators are called the Pskhu fractional integrals and derivatives in [450].

Example 7.5 (The multiplier with uniform distributed memory fading). Using equation 5.1.7 of [318, p. 136], we can represent the Nakhushev fractional integral in the form

$$(I_N^{[\alpha,\beta]}X)(t) = \int_0^t W(\alpha,\beta,t-\tau)X(\tau)\,d\tau, \tag{7.62}$$

that is, $(I_N^{[\alpha,\beta]}X)(t) = W(\alpha,\beta,t-\tau) * X(\tau)$, where we use the function

$$W(\alpha,\beta,t) = \frac{1}{(\beta-\alpha)t}\int_{\alpha}^{\beta}\frac{t^{\xi}\,d\xi}{\Gamma(\xi)}. \tag{7.63}$$

This operator allows us to define the multiplier with uniform distributed memory fading (the UDMF multiplier) by the equation

$$Y(t) = m([\alpha,\beta])(I_N^{[\alpha,\beta]}X)(t), \tag{7.64}$$

where $I_N^{[\alpha,\beta]}$ is defined by expression (7.62).

Example 7.6 (The accelerator with uniform distributed memory fading). Using equation 5.1.26 of [318, p. 143], the Nakhushev fractional derivative can be written in the form

$$(D_N^{[\alpha,\beta]}X)(t) = D^n(W(n-\alpha,n-\beta,t-\tau) * X(\tau)), \tag{7.65}$$

where $\beta > \alpha > 0$. Expression (7.65) can be rewritten as

$$(D_N^{[\alpha,\beta]}X)(t) = \left(\frac{d}{dx}\right)^n \int_0^t W(n-\alpha, n-\beta, t-\tau)X(\tau)\, d\tau, \tag{7.66}$$

where $\beta > \alpha > 0$. Using the Nakhushev fractional derivative (7.66), the accelerator with uniform distributed memory fading (the UDMF accelerator) can be defined by the equation

$$Y(t) = a([\alpha,\beta])(D_N^{[\alpha,\beta]}X)(t), \tag{7.67}$$

where $a([\alpha,\beta])$ is the accelerator coefficient, which depends on the interval $[\alpha,\beta]$.

Example 7.7 (The multiplier with nonsingular distributed memory). Using equation (5.1.7) of [318, p. 136], the Pskhu fractional integral can be defined by the expression

$$(I_P^{[\alpha,\beta]}X)(t) = (\beta-\alpha)(\tau^{\beta-1}E_{\beta-\alpha,\beta}[\tau^{\beta-\alpha}]) * X(t-\tau), \tag{7.68}$$

where $\beta > \alpha > 0$, which can be written in the form

$$(I_P^{[\alpha,\beta]}X)(t) = (\alpha-\beta)\int_0^t (t-\tau)^{\beta-1}E_{\beta-\alpha,\beta}[(t-\tau)^{\beta-\alpha}]X(\tau)\, d\tau, \tag{7.69}$$

where $\beta > \alpha > 0$. This integral operator allows us to define the multiplier with nonsingular distributed memory (the NSDM multiplier) by the equation

$$Y(t) = m([\alpha,\beta])(I_P^{[\alpha,\beta]}X)(t), \tag{7.70}$$

where $I_P^{[\alpha,\beta]}$ is defined by equation (7.69).

Example 7.8 (The accelerator with nonsingular distributed memory). Using equation 5.1.7 of [318, p. 136], we can define the Pskhu fractional derivative as

$$(D_P^{[\alpha,\beta]}X)(t) = -(\beta-\alpha)D^n((\tau^{-\alpha}E^{n-1}_{\beta-\alpha,1-\alpha}[\tau^{\beta-\alpha}]) * X(t-\tau)), \tag{7.71}$$

where $\beta > \alpha > 0$ and $E^{\mu}_{\alpha,\beta}[z]$ is defined by the equation

$$E^{\mu}_{\alpha,\beta}[z] = \frac{\partial}{\partial\mu}(z^{\mu}E_{\alpha,\beta+\mu}[z]). \tag{7.72}$$

As a result, we have

$$(D_P^{[\alpha,\beta]}X)(t) = (\alpha-\beta)\left(\frac{d}{dx}\right)^n \int_0^t E^{n-1}_{\beta-\alpha,1-\alpha}[(t-\tau)^{\beta-\alpha}](t-\tau)^{-\alpha}X(\tau)\, d\tau, \tag{7.73}$$

where $\beta > \alpha > 0$. Using this fractional derivative, the accelerator with nonsingular distributed memory (the NSDM accelerator) can be defined by the equation

$$Y(t) = a([\alpha,\beta])(D_P^{[\alpha,\beta]}X)(t), \tag{7.74}$$

where $D_P^{[\alpha,\beta]}$ is defined by expression (7.73).

Note that the Nakhushev fractional derivatives cannot be considered as inverse operators for the Nakhushev fractional integration. The Pskhu fractional derivatives are inverse to the Nakhushev fractional integration and the Pskhu fractional integrals are inverse to the Nakhushev fractional derivatives.

7.9 Superposition principle for accelerators with memory

In simple linear accelerator with power-law memory, the power-law fading is characterized by one parameter α only. In general, the different types of economic agents can be characterized by different types of memory fading. In economics, it is important to take into account the presence of different values of memory fading, which characterize the different types of economic agents.

For example, the accelerator with fading memory can be considered as a sum of N accelerators with one-parameter memory fading that have different values α_k, $k = 1,\dots,N$. The equation of the linear accelerator with N-parameter power-law memory can be used in the form

$$Y(t) = \sum_{k=1}^{N} a_k(\alpha_k)(D_{C;0+}^{\alpha_k}X)(t), \tag{7.75}$$

where $n_k - 1 < \alpha_k < n_k$, $(k = 1,2,\dots,N)$, and $a_k(\alpha)$, $(k = 1,2,\dots,N)$ are numerical coefficients. Note that integer values $\alpha_k = n_k$ correspond to groups of economic agents who have no memory.

In the case, when the accelerators $\mathcal{A}_0^t(X^{(n_k)})$ have $n_k = n$ for all $k = 1,\dots,N$, we can formulate the superposition principle that describes the simultaneous (parallel) action of these accelerators with memory. Let us assume that the linear accelerators with memory can be characterized by the memory functions $M_k(t,\tau)$, $k = 1,\dots,N$ and it can be described by the equations

$$Y_k(t) = \int_0^t M_k(t,\tau)X^{(n)}(\tau)\,d\tau. \tag{7.76}$$

Note that here we assume that all parameters n_k are equal to n, i. e., $n_k = n$ for all $k = 1,\dots,N$. The proposed principle states that the simultaneous (parallel) action of several linear accelerators with memory is equivalent to the action of a accelerator, whose memory function is equal to the sum of the memory function of these accelerators.

Principle 7.2 (Principle of superposition of accelerators with memory). *The accelerator effect, which is created by the simultaneous (parallel) action of finite number of linear accelerators* (7.76) *with memory is equivalent to the action of a linear accelerator, in which memory function is the sum of the memory functions of these accelerators.*

This principle means that the simultaneous action of linear accelerators, is characterized by the fact that the action of each of these accelerators does not affect (does not change) the action of other linear accelerators. For linear accelerators with memory, the suggested superposition principle states the additivity of the memory functions.

Let us describe the superposition principle for accelerators in symbolic form. The simultaneous (parallel) action of the linear accelerators that are described by the equations

$$Y_1(t) = \int_0^t M_1(t,\tau) X^{(n)}(\tau)\, d\tau, \tag{7.77}$$

$$Y_2(t) = \int_0^t M_2(t,\tau) X^{(n)}(\tau)\, d\tau \tag{7.78}$$

is equivalent to the action of the accelerator with memory,

$$Y(t) = \int_0^t M(t,\tau) X^{(n)}(\tau)\, d\tau, \tag{7.79}$$

in which memory function is the sum of the memory functions of these accelerators in the form

$$M(t,\tau) = M_1(t,\tau) + M_2(t,\tau). \tag{7.80}$$

The action of N accelerators (7.76) with the memory functions $M_k(t,\tau)$, ($k = 1,\dots,N$), is equivalent to the action of the linear accelerator (7.79) with the memory function

$$M(t,\tau) = \sum_{k=1}^{N} M_k(t,\tau). \tag{7.81}$$

Example 7.9. For two accelerators (7.76) with power-law memory, which are described by the equations

$$Y_1(t) = a_1(\alpha)(D^{\alpha}_{C;0+}X)(t), \tag{7.82}$$

$$Y_2(t) = a_2(\alpha)(D^{\alpha}_{C;0+}X)(t), \tag{7.83}$$

the superposition principle gives the equality

$$Y(t) = (a_1(\alpha) + a_2(\alpha))(D^{\alpha}_{C;0+}X)(t), \tag{7.84}$$

which is valid for all $\alpha > 0$ and for any $X(t) \in L_1(0,T)$. As a result, we have that the parallel action of accelerators, for which the power-law memory fading parameters are the same, is equivalent to the action of one accelerator, with the same fading parameter the coefficient of which is equal to the sum of the coefficients of these accelerators.

Example 7.10. According to the principle of superposition, the action of N linear accelerators (7.76) with power-law memory

$$Y_k(t) = a_k(\alpha_k)(D^{\alpha_k}_{C;0+}X)(t), \tag{7.85}$$

for which the memory fading parameters $\alpha_k \in (n-1, n)$ for all $k = 1, \dots, N$, and the memory functions

$$M_k(t,\tau) = \frac{a_k(\alpha_k)}{\Gamma(n-\alpha_k)}(t-\tau)^{n-\alpha_k-1}, \tag{7.86}$$

where $\alpha_k > 0$, is equivalent to the action of the accelerator with the memory function

$$M(t,\tau) = \sum_{k=1}^{N} \frac{a_k(\alpha_k)}{\Gamma(n-\alpha_k)}(t-\tau)^{n-\alpha_k-1}. \tag{7.87}$$

As a result, we can formulate the following superposition principle.

Principle 7.3 (Principle of superposition of accelerators with power-law memory). *The simultaneous (parallel) action of finite number of linear accelerators* (7.75) *with memory, for which $n-1 < \alpha_k < n$ for all $k = 1, \dots, N$, is equivalent to the action of a linear accelerator, in which memory function is the sum of the memory functions of these accelerators.*

Memory with multiparameter power-law fading, which is described by the sum of memory functions, can be used in economic models to describe economic agents with various "memory fading parameters."

7.10 Sequential actions of accelerators with memory

Let us compare the sequential actions of two accelerators with power-law memory, for which fading parameters are equal to $\alpha > 0$ and $\beta > 0$, and the action of the accelerator with fading parameter equals to $\alpha + \beta$.

Principle 7.4 (Principle of violation of sequential actions for accelerators with memory). *In the general case, the sequential action of the accelerators with power-law memory cannot be equivalent to the action of an accelerator with power-law memory.*

Symbolically, this principle can be written in the following form. The action of two accelerators with power-law memory

$$Y(t) = \mathcal{A}_0^{t;\alpha}(X(\tau)), \tag{7.88}$$

$$Z(t) = \mathcal{A}_0^{t;\beta}(Y(\tau)) \tag{7.89}$$

with noninteger values of $\alpha > 0$ and $\beta > 0$ is not equivalent to the action of the accelerator with memory

$$Z(t) = \mathcal{A}_0^{t;\alpha+\beta}(X(\tau)). \tag{7.90}$$

As a result, we have the inequality

$$\mathcal{A}_0^{t;\alpha}(\mathcal{A}_0^{t;\beta}(X(\tau))) \neq \mathcal{A}_0^{t;\alpha+\beta}(X(\tau)) \tag{7.91}$$

for $\alpha > 0$ and $\beta > 0$. The principle of violation of superposition of the sequential actions of accelerators with memory means that in general we have the inequality (7.91).

Example 7.11. The sequential action of the accelerators with power-law memory

$$Y(t) = a(\alpha)(D^{\alpha}_{\mathrm{RL};0+}X)(t), \tag{7.92}$$

$$Z(t) = a(\beta)(D^{\beta}_{\mathrm{RL};0+}Y)(t) \tag{7.93}$$

is not equivalent to the action of the accelerator with memory

$$Z(t) = a(\alpha,\beta)(D^{\alpha+\beta}_{\mathrm{RL};0+}X)(t), \tag{7.94}$$

where $a(\alpha,\beta) = a(\alpha)a(\beta)$ for noninteger values $\alpha > 0$ and $\beta > 0$.

In the case of the accelerators with memory, which is described by the Riemann–Liouville fractional derivatives, the violation of superposition is based on the following property of the Riemann–Liouville fractional derivatives (see Property 2.4 of [200, p. 75]).

Statement 7.2. *Let $\alpha > 0$ and $\beta > 0$ be such that $n-1 < \alpha < n$, $m-1 < \beta < m$ and $\alpha + \beta < n$ and let $X(t) \in L_1(a,b)$ with $p \geq 1$, and $(I^{m-\alpha}_{\mathrm{RL};a+}X)(t) \in AC^m[a,b]$. Then the following equation holds:*

$$(D^{\alpha}_{\mathrm{RL};a+}D^{\beta}_{\mathrm{RL};a+}X)(t) = (D^{\alpha+\beta}_{\mathrm{RL};a+}X)(t) - \sum_{k=1}^{m} \frac{(t-a)^{-\alpha-k}}{\Gamma(1-k-\alpha)}(D^{\beta-k}_{\mathrm{RL};a+}X)(a+). \tag{7.95}$$

This statement is proved as Property 2.4 in [200, p. 75].

Equation (7.95) means that, in general, the following inequality holds:

$$(D^{\alpha}_{\mathrm{RL};0+}D^{\beta}_{\mathrm{RL};0+}X)(t) \neq (D^{\alpha+\beta}_{\mathrm{RL};0+}X)(t) \tag{7.96}$$

for noninteger values of $\alpha > 0$ and $\beta > 0$.

For the Caputo fractional derivative, we have the inequality

$$(D^{\alpha}_{C;0+} D^{\beta}_{C;0+} X)(t) \neq (D^{\alpha+\beta}_{C;0+} X)(t). \tag{7.97}$$

The same properties we have for the other type of fractional derivatives of non-integer orders.

The principle of violation of superposition leads to the following consequence that state the nonpermutation (the noncommutativity) of actions of accelerators with memory.

Principle 7.5 (Principle of nonpermutation of accelerator actions). *In the general case, the action of accelerators with memory depends on their action sequences. Symbolically, this principle can be written in the form of the inequality*

$$\mathcal{A}^{t;\alpha}_{0}(\mathcal{A}^{t;\beta}_{0}(X(\tau))) \neq \mathcal{A}^{t;\beta}_{0}(\mathcal{A}^{t;\alpha}_{0}(X(\tau))) \tag{7.98}$$

for $\alpha > 0$ and $\beta > 0$.

For accelerators (7.92) and (7.93), this principle describes a dependence of the accelerator effect on sequence in the case of power-law memory. This fact is based on the inequality that describes noncommutativity of actions of fractional derivatives. For Riemann–Liouville derivatives, we have

$$(D^{\alpha}_{RL;0+} D^{\beta}_{RL;0+} X)(t) \neq (D^{\beta}_{RL;0+} D^{\alpha}_{RL;0+} X)(t) \tag{7.99}$$

for noninteger values of $\alpha > 0$ and $\beta > 0$. For the Caputo fractional derivative, we also have the inequality

$$(D^{\alpha}_{C;0+} D^{\beta}_{C;0+} X)(t) \neq (D^{\beta}_{C;0+} D^{\alpha}_{C;0+} X)(t). \tag{7.100}$$

The same properties are satisfied for other type of fractional derivatives of non-integer orders.

Remark 7.8. We emphasize that, in contrast to multipliers with memory, for accelerators with memory the principle of superposition of the sequential actions of accelerators with memory are not performed.

7.11 Superposition of accelerators with and without memory

In works [412, 397], the principle of superposition of sequential actions of an accelerator with fading memory and a standard accelerator without memory was formulated.

Principle 7.6 (Principle of superposition for accelerators with memory and without memory). *In the general case, the action of the accelerators with power-law memory, which has the fading parameter $\alpha > 0$, and the standard accelerator without memory can be equivalent to the action of an accelerator with power-law memory, fading parameter of which is equal to $\alpha + 1$.*

This superposition principle for accelerators means that we have the equality

$$\mathcal{A}_0^{t;\alpha}(\mathcal{A}_0^{t;\beta}(X(\tau))) = \mathcal{A}_0^{t;\alpha+\beta}(X(\tau)), \tag{7.101}$$

if at least one of the parameters $\alpha > 0$ or $\beta > 0$ takes the positive integer value. Implementation of the superposition for the left-sided or right-sided actions of the standard accelerator depends on the type of memory and type of the fractional derivatives that describe accelerator with memory. Let us give examples to illustrate this principle.

Example 7.12. If the fractional derivatives $(D^{\alpha}_{\mathrm{RL};0+}X)(t)$ and $(D^{\alpha+k}_{\mathrm{RL};0+}X)(t)$ exist, where $\alpha \geq 0$ and $k \in \mathbb{N}$, then the following equality is satisfied

$$D^k D^{\alpha}_{\mathrm{RL};0+}X(t) = D^{\alpha+k}_{\mathrm{RL};0+}X(t). \tag{7.102}$$

For details, see books [308, 200] (Property 2.3 of [200, p. 74] and equation 2.144 of [308, p. 81]). Property (7.102) leads to the superposition for the left-sided action of the standard accelerator.

Example 7.13. If the fractional derivatives $(D^{\alpha}_{C;0+}X)(t)$ and $(D^{\alpha+k}_{C;0+}X)(t)$ exist, where $\alpha > 0$ and k is positive integer number, then we have the equality

$$D^{\alpha}_{C;0+}D^k X(t) = D^{\alpha+k}_{C;0+}X(t). \tag{7.103}$$

In particular, if $k = 1$, then we have $D^{\alpha}_{C;0+}D^1 X(t) = D^{\alpha+1}_{C;0+}X(t)$. For details, see book [308] (equation (2.143) of [308, p. 81]). Property (7.103) leads to the superposition for the right-sided action of the standard accelerator.

Example 7.14. If the Liouville fractional derivatives $(D^{\alpha}_{L;+}X)(t)$ and $(D^{\alpha+k}_{L;+}X)(t)$ exist, where $\alpha > 0$ and $k \in \mathbb{N}$, then we have the equality in the form

$$D^k D^{\alpha}_{L;+}X(t) = D^{\alpha+k}_{L;+}X(t). \tag{7.104}$$

For details, see Property 2.13 of [200, p. 89].

Remark 7.9. For equalities (7.102), (7.103) and (7.104), the left and right positions of the integer-order derivative are important. For other relative positions of the derivatives, the principle of superposition 7.6 does not hold for noninteger values of α.

7.12 Accelerator with memory through standard accelerators

Let us assume that $X(t)$ is a function, for which the Caputo fractional derivative of the order $\alpha > 0$ exist together with the Riemann–Liouville fractional derivatives. Using equation 2.4.6 of [200, p. 91], the Caputo fractional derivative can be expressed though the Riemann–Liouville fractional derivative by the equation

$$(D^{\alpha}_{C;a+}X)(t) = (D^{\alpha}_{\mathrm{RL};a+}X)(t) - \sum_{k=0}^{n-1} \frac{X^{(k)}(a+)}{\Gamma(k-\alpha+1)}(t-a)^{k-\alpha}, \tag{7.105}$$

where $n = [\alpha] + 1$ for noninteger values of $\alpha > 0$. In particular, when $\alpha \in (0,1)$ and $a = 0$, we have

$$(D^{\alpha}_{C;0+}X)(t) = (D^{\alpha}_{RL;0+}X)(t) - \frac{X(0)}{\Gamma(1-\alpha)} t^{-\alpha}. \tag{7.106}$$

Let us assume that $X(t)$ is an analytic function in an interval $(0,T)$, i.e., it is a function, which is expandable into a power series in this interval. Then Lemma 15.3 of [335, p. 278] states that the Riemann–Liouville fractional derivative of the order $\alpha > 0$ can be represented by the infinite series in this interval by the equation

$$(D^{\alpha}_{RL;0+}X)(t) = \sum_{k=0}^{\infty} \binom{\alpha}{k} \frac{t^{k-\alpha}}{\Gamma(k-\alpha+1)} \frac{d^k X(t)}{dt^k}, \tag{7.107}$$

where $\binom{\alpha}{k}$ are the generalized binomial coefficients (see [200, pp. 26–27], [335, p. 14]), that are defined by equation (1.131). Substitution of (7.107) into (7.105) gives the representation of the Caputo fractional derivative in the form

$$\begin{aligned}(D^{\alpha}_{C;0+}X)(t) = &\sum_{k=0}^{\infty} \binom{\alpha}{k} \frac{t^{k-\alpha}}{\Gamma(k-\alpha+1)} \frac{d^k X(t)}{dt^k} \\ &- \sum_{k=0}^{n-1} \frac{X^{(k)}(0+)}{\Gamma(k-\alpha+1)} (t-a)^{k-\alpha}.\end{aligned} \tag{7.108}$$

This representation allows us to propose the following statement.

Statement 7.3. *The economic accelerator with power-law memory, which is described by the Caputo fractional derivative*

$$Y(t) = a(\alpha)(D^{\alpha}_{C;0+}X)(t), \tag{7.109}$$

can be interpreted as a parallel action of the standard linear multiplier

$$Y_0(t) = m(t)X(t) + b(t), \tag{7.110}$$

and an infinite number of standard accelerators of all positive integer orders

$$Y_k(t) = a_k(t) \frac{d^k X(t)}{dt^k}, \tag{7.111}$$

where $k \in \mathbb{N}$, and

$$m(t) = \frac{a(\alpha)}{\Gamma(1-\alpha)} t^{-\alpha}, \tag{7.112}$$

$$b(t) = -a(\alpha) \sum_{k=0}^{n-1} \frac{X^{(k)}(0)}{\Gamma(k-\alpha+1)} t^{k-\alpha}, \tag{7.113}$$

$$a_k(t) = \frac{a(\alpha)}{\Gamma(k-\alpha+1)} t^{k-\alpha}. \tag{7.114}$$

We can also note that the economic accelerator with power-law memory can be derived (and interpreted) as a reconstruction [379] from infinite sequence of standard accelerators of integer orders, where the reconstruction is considered with respect to memory fading parameter. It is analogous to a possibility of a complete reconstruction of the continuous function (or signal) at a discrete reference. This interpretation is connected with the geometric interpretation of fractional derivatives that was proposed in works [375, 181].

7.13 Chain rule and product rule for accelerator with memory

The standard chain rule is used, for example, for the function $C = C(Y(t))$, which describes the dependence of consumption on income [10]. For the function $C = C(Y(t))$, this rule is represented by the equation

$$\frac{dC(Y(t))}{dt} = \left(\frac{\partial C(Y)}{\partial Y}\right)_{Y=Y(t)} \frac{dY(t)}{dt}. \tag{7.115}$$

It is known that the standard chain rule can be used for the standard linear accelerators without memory. Let us consider the standard accelerator equation

$$Y(t) = a\frac{dX(t)}{dt}. \tag{7.116}$$

If we consider the variable $X(t)$ as the composition of two functions $X(t) = X_1(X_2(t))$, then we have the equality

$$Y(t) = a\frac{dX_1(X_2(t))}{dt} = a\frac{\partial X_1}{\partial X_2}\frac{dX_2(t)}{dt} = \frac{\partial X_1}{\partial X_2}Y_2(t) \tag{7.117}$$

for wide class of functions $X_1(t)$ and $X_2(t)$, where we use $Y_2(t) = aX_2^{(1)}(t)$.

It should be noted that the standard chain rule does not hold for derivatives of noninteger order (see [308, pp. 97–98], and [377]). As a result, analysis of the nonlinear relationships between $X(t)$ and $Y(t)$, as well as the composition of functions is significantly more complicated.

It is known that the standard product rule (the rule of differentiation of the product of functions) can be used for standard linear accelerator without memory. Let us consider the standard accelerator equations

$$Y_1(t) = a\frac{dX_1(t)}{dt}, \quad Y_2(t) = a\frac{dX_2(t)}{dt}, \tag{7.118}$$

and the variable $X(t) = X_1(t)X_2(t)$. The action of the standard accelerator satisfies the product rule

$$Y_{12} = a\frac{d}{dt}(X_1(t)X_2(t)) = a\frac{dX_1(t)}{dt}X_2(t) + aX_1(t)\frac{dX_2(t)}{dt} = Y_1(t)X_2(t) + X_1(t)Y_2(t) \tag{7.119}$$

for wide class of functions $X_1(t)$ and $X_2(t)$.

For the accelerators with power-law memory the standard product rule (7.119) is violated since the standard product rule is violated [367, 376, 81] for the fractional derivatives of noninteger orders

$$(D^{\alpha}_{C;0+}(X_1X_2))(t) \neq (D^{\alpha}_{C;0+}X_1)(t)X_2(t) + X_1(t)(D^{\alpha}_{C;0+}X_2)(t). \tag{7.120}$$

These two properties (the violation of the standard chain and product rules) should be taken into account in applications of the concepts of accelerator with power-law memory.

We should emphasize that in constructing economic models with memory, it is not enough to generalize the differential equations that describe the standard economic model. It is necessary to generalize all stages of the derivation of these equations from the basic principles, concepts and assumptions. In this sequential derivation of the equations, the nonstandard properties of fractional derivatives and integrals should be taken into account. The violation of the standard rules for the fractional derivatives, significantly restricts the possibilities of constructing economic models with memory as generalizations of standard models. Examples of incorrectness and errors in constructing such generalizations are given in the review article [386] (see also book [254, pp. 43–92]).

7.14 Conclusion

We proposed the concepts of multiplier and accelerator with memory. These concepts are generalizations of the standard concepts of the accelerator and the multiplier to described economic processes with memory. The equations of the standard accelerator and multiplier are the special cases of the equations of accelerator with memory, when $\alpha = 1$ and $\alpha = 0$, respectively. In addition to the standard accelerator and multiplier, the proposed concept of accelerator with memory includes the wide spectrum of intermediate dependencies that lie between the standard accelerator and the standard multiplier, when $\alpha \in (0, 1)$. The proposed new concepts can also be used when $\alpha > 1$, including the "higher-order accelerators" as special cases.

The proposed concepts can be used to take into account the behavior of economic agents who remember previous changes of the factor and indicator over a finite time interval.

The proposed concepts of the accelerator with memory and the multiplier with memory allow us to build macroeconomic models that take into account the wide class of memory effects, and to obtain solutions for the corresponding equations, which contain derivatives and integrals of noninteger orders, by using the mathematical tools of fractional calculus.

An important property of the suggested new concepts is the duality of the accelerator with memory and the multiplier with memory. The concept of duality is described in the next chapter.

8 Duality of multipliers and accelerators with memory

In this chapter, we consider duality of the accelerator with memory and the multiplier with memory. The duality of the proposed new concepts has no analogues for standard accelerator and the multiplier without memory. This chapter is based on articles [412, 397], where concepts of the accelerator and multiplier with memory, and the duality of these concepts are proposed.

8.1 General duality principle

Let us consider two variables $X(t)$ and $Y(t)$, that are functions of time. The function $X(t)$ is considered as the impact variable. The function $Y(t)$ is interpreted as a response (reaction) to the impact. In the general case, the equation of the generalized multiplier with memory

$$Y(t) = \mathcal{M}_0^t(X(\tau)) \tag{8.1}$$

is not reversible. In the general case, knowing the function $Y(t)$ on the interval $(0, T)$, it is not always possible to determine uniquely the function $X(t)$ from equation (8.1). This irreversibility means that there is no inverse operator $\mathcal{B}_0^t$, such that

$$X(t) = \mathcal{B}_0^t(Y(\tau)) \tag{8.2}$$

for all $t \in (0, T)$, if equation (8.2) is satisfied.

In some cases, the inverse operator $\mathcal{B}_0^t$ exists. This operator is inversed from the left to multiplier equation (8.1), if the equality

$$\mathcal{B}_0^t(\mathcal{M}_0^t(X(\tau))) = X(t) \tag{8.3}$$

holds for a wide class of functions $X(t)$ and for any point in time from a finite interval $(0, T)$. If condition (8.3) is realized, then equation (8.2) can be interpreted as an equation of accelerator with memory. In this case, the accelerator with memory (8.2) will be called dual to the multiplier with memory (8.1). Mathematically, equation (8.3) means that the operator $\mathcal{B}_0^t$ is the left inverse for the operator $\mathcal{M}_0^t$.

The fact that the operator $\mathcal{B}_0^t$ is left inverse does not mean that this operator is right inverse. In the general case, the operator $\mathcal{B}_0^t$, which is left inverse to $\mathcal{M}_0^t$, cannot be considered as a right inverse to the operator $\mathcal{M}_0^t$, i. e., we have the inequality

$$\mathcal{M}_0^t(\mathcal{B}_0^t(Y(\tau))) \neq Y(t) \tag{8.4}$$

in general. This inequality means that substitution of (8.2) into equation (8.1) does not result in the identity.

https://doi.org/10.1515/9783110627459-008

Example 8.1. Let us consider the linear multiplier with simple power-law memory. Equation of multiplier with memory (8.1), which has the form

$$Y(t) = m(\alpha)(I^{\alpha}_{\mathrm{RL};0+}X)(t), \tag{8.5}$$

is reversible such that the dual (inverse) equation has the form

$$X(t) = a(\alpha)(D^{\alpha}_{\mathrm{RL};0+}Y)(t), \tag{8.6}$$

where $a(\alpha) = 1/m(\alpha)$. Equation (8.6) can be considered as an equation of accelerator with memory. Here, $D^{\alpha}_{\mathrm{RL};0+}$ and $I^{\alpha}_{\mathrm{RL};0+}$ are the left-sided Riemann–Liouville fractional derivative and integral of the order $\alpha > 0$, respectively [335, 200]. This statement is based on the property

$$(D^{\alpha}_{\mathrm{RL};a+}I^{\alpha}_{\mathrm{RL};+}X)(t) = X(t), \tag{8.7}$$

which is proved in [335, 200] (see Theorem 2.4 of [335, pp. 44-45], and Lemma 2.4 of [200, p. 74]) for $X(t) \in L_1(a,b)$ and $X(t) \in L_p(a,b)$ with $p \geq 1$, and $t \in (a,b)$. Equation (8.7) is an example of equality (8.3).

In general, we have the inequality

$$(I^{\alpha}_{\mathrm{RL};0+}D^{\alpha}_{\mathrm{RL};0+}Y)(t) \neq Y(t). \tag{8.8}$$

This statement is based on the property

$$\begin{aligned}(I^{\alpha}_{\mathrm{RL};0+}&D^{\alpha}_{\mathrm{RL};0+}Y)(t)\\ &= Y(t) - \sum_{k=0}^{n-1} \frac{t^{\alpha k-1}}{\Gamma(\alpha-k)}(D^{n-k-1}I^{n-\alpha}_{\mathrm{RL};0+}Y)(0),\end{aligned} \tag{8.9}$$

which is proved in [335, 200] (see Theorem 2.4 of [335, pp. 44–45], and Lemma 2.5 of [200, pp. 74-75]) for the function $Y(t) \in L_1(0,T)$ has a summable fractional derivative $(D^{\alpha}_{\mathrm{RL};0+}Y)(t)$, i. e., $(I^{n-\alpha}_{\mathrm{RL};0+}Y)(t) \in AC^n[0,T]$. Equation (8.8) is an example of inequality (8.4), where $n = [\alpha] + 1$.

Let us consider the generalized accelerator with memory in the form

$$Y(t) = \mathcal{A}^t_0(X^{(n)}(\tau)). \tag{8.10}$$

In the general case, equation (8.10) is not reversible and there is no the left inverse operator $\mathcal{H}^t_{n;0}$ such that

$$X(t) = \mathcal{H}^t_{n;0}(Y(\tau)) + b_{X;0+}(t), \tag{8.11}$$

where $b_{X;0+}(t)$ is a function of time t that is defined by some properties of the function $X(t)$ and the integer derivatives $X^{(k)}(\tau)$ of the orders $k = 1, 2, \ldots, n$ at zero ($t \to 0+$).

It should be noted that in some cases, the left inverse operator $\mathcal{H}^t_{n;0}$ exists and equation (8.11), which is inverse to accelerator equation (8.10), can be interpreted as an equation of multiplier with memory. Symbolically, this can be written in the form of the equality

$$\mathcal{H}^t_{n;0}(\mathcal{A}^t_0(X^{(n)}(\tau))) = X(t) - b_{X;0+}(t) \tag{8.12}$$

that holds for a wide class of functions $X(t)$. Here $b_{X;0+}(t)$ means an additional term that depends on t, $X(t)$ and their derivatives at $t \to 0+$. Note that in many cases, we also have the equality

$$\mathcal{A}^t_0((\mathcal{H}^t_{n;0}(Y(\tau)) + b_{X;0+}(t))^{(n)}) = Y(t). \tag{8.13}$$

Equation (8.13) means that substitution of (8.11) into equation (8.10) gives the identity.

Example 8.2. Let us consider the linear accelerator with simple power-law memory. The equation of multiplier with memory (8.1), which has the form

$$Y(t) = a(\alpha)(D^\alpha_{C;0+}X)(t), \tag{8.14}$$

is reversible such that the dual (inverse) equation has the form

$$X(t) = m(\alpha)(I^\alpha_{RL;0+}Y)(t) + b_X(t), \tag{8.15}$$

where $m(\alpha) = 1/a(\alpha)$, $n - 1 < \alpha < n$, and

$$b_X(t) = \sum_{k=0}^{n-1} \frac{t^k}{\Gamma(k+1)} X^{(k)}(0). \tag{8.16}$$

Equation (8.15) can be considered as a solution of fractional differential equation (8.14), if the function $Y(t)$ is given. In economics, equation (8.15) can be interpreted as a linear multiplier with simple power-law memory. In this case, we have the equality

$$a(\alpha)(D^\alpha_{C;0+}(m(\alpha)(I^\alpha_{RL;0+}Y)(\tau) + b(\tau)))(t) = Y(t) \tag{8.17}$$

for wide class of functions $Y(\tau)$. Equation (8.17) is an example of equation (8.13).

Unlike standard accelerator and multiplier concepts, the concepts of accelerator and multiplier with memory are dual to each other. The duality of these concepts is expressed in the fact that the accelerator with memory corresponds to a multiplier with memory, and vice versa. As a result, when constructing economic models, memory effects can be described by one of these two dual concepts. Mathematically, duality is manifested as follows. The principle of duality of the linear accelerator with memory and the linear multiplier with memory states a possibility to represent equation of the multiplier as an equation of the accelerator together with a possibility to represent the equation of accelerator with memory an equation of a multiplier with memory.

Let us formulate a general duality principle, which states a relationship between the concepts of the accelerator with memory and the multiplier with memory.

Principle 8.1 (General duality principle). *The duality of the concepts of accelerator and multiplier with memory is expressed in the fact that the accelerator with memory corresponds to a multiplier with memory, and vice versa. Mathematically, the duality of the linear accelerator and multiplier with memory is described as a possibility to represent equation of the multiplier by equations of the accelerator together with a possibility to represent this accelerator equation by equation of a multiplier with memory.*
[**M** → **A**] *The linear equation of multiplier with memory is a reversible such that the dual (inverse) equation is an equation of accelerator with memory. This accelerator with memory is left inverse to the multiplier with memory.*
[**A** → **M**] *The linear equation of accelerator with memory is reversible such that the dual (inverse) equation can be interpreted as the equations of the multiplier with memory. In the transition from one concept to a dual concept, the impact variable should be interpreted as a response variable, and vice versa.*

In the transformations between concepts of the accelerator and multiplier with memory, the impact and response variables, which are described by the functions with continuous time, change the roles. In other words, when describing memory effects, the transition from multiplier with memory to accelerator with memory is accompanied by the transformation of the impact variables into response variables, and conversely.

Different examples and it proofs will be given in the next sections.

Remark 8.1. Let us consider the linear generalized multiplier with memory of the special kind that is described by the equation

$$Y(t) = \int_0^t M(t,\tau)X(\tau)\,d\tau, \tag{8.18}$$

where $M(t,\tau)$ is the memory function. In equation (8.18), we can consider $Y(t)$ as a given function, and the variable $X(t)$ can be considered as the unknown function. In this case, equation (8.18) is called the linear Volterra equation of the first kind. Existence of solutions of equation (8.18), i. e., the reversibility of equation (8.18), is studied in the theory of integral equations (e. g., see [311] and references therein).

Remark 8.2. Let us consider the linear generalized accelerator with memory that is described by the equation

$$Y(t) = \int_0^t M(t,\tau)X^{(n)}(\tau)\,d\tau, \tag{8.19}$$

where $M(t,\tau)$ is the memory function. In equation (8.19), we can consider $Y(t)$ as a given function, and the variable $X(t)$ can be considered as the unknown function. In this case, equation (8.19) can be associated with the following inverse problems. If we know the memory function $M(t,\tau)$ of the economic process and response variable $Y(t)$, then we can put the problem of finding (the restoration) of the impact variables $X(t)$ for this process. This type of problems includes finding solutions of the fractional differential equations of macroeconomic models of growth with memory.

In the next section, we consider the duality principle for accelerator and multiplier with simple power-law memory with details.

8.2 Duality principle for simple power-law memory

8.2.1 From multiplier with memory to accelerator with memory

Let us consider now the multiplier with simple power-law memory, which is described by the equation with Riemann–Liouville fractional integral. We can use the property, which states that the Caputo derivative is the inverse operation to the Riemann–Liouville integral. This property leads to the important statement about the relationship between the concepts of the multiplier and the accelerator with memory.

Principle 8.2 (Principle of duality of multiplier and accelerator with power-law memory: part 1).
[**M** → **A**]: *Let $X(\tau)$ be the function of the impact variable (factor), which is continuous on the finite interval $[0,T]$, i. e., $X(\tau) \in C[0,T]$ (or $X(\tau) \in L_{\infty}(0,T)$). Then the response variable $Y(t)$, which is described by the linear equation of multiplier with memory*

$$Y(t) = m(\alpha)(I^{\alpha}_{\mathrm{RL};0+}X)(t), \tag{8.20}$$

where $\alpha > 0$, and $m(\alpha)$ is the multiplier coefficient, satisfies the linear equation of accelerator with memory

$$X(t) = a(\alpha)(D^{\alpha}_{C;0+}Y)(t), \tag{8.21}$$

where $a(\alpha) = 1/m(\alpha)$ is the accelerator coefficient. In the duality concepts, the impact and response variables change the roles.

Proof. To prove the statement of this principle, we consider the action of the Caputo derivative of order $\alpha > 0$ on equation (8.20). This action gives the expression

$$(D^{\alpha}_{C;0+}Y)(t) = m(\alpha)(D^{\alpha}_{C;0+}I^{\alpha}_{\mathrm{RL};0+}X)(t). \tag{8.22}$$

Let us use Lemma 2.21 of [200, p. 95], which states that the Caputo fractional derivative is the inverse operation to the Riemann–Liouville fractional integral. According to this Lemma, if $X(t) \in C[0,T]$ or $X(t) \in L_{\infty}(0,T)$, then the identity

$$(D^{\alpha}_{C;0+}I^{\alpha}_{\mathrm{RL};0+}X)(t) = X(t) \tag{8.23}$$

is satisfied for all $\alpha > 0$, where $D^{\alpha}_{C;0+}$ is the left-sided Caputo fractional derivative of the order $\alpha > 0$, and $I^{\alpha}_{RL;0+}$ is the left-sided Riemann–Liouville fractional integral of the order $\alpha > 0$.

Using equation (8.23) for the right-hand side of equation (8.22), we obtain

$$(D^{\alpha}_{C;0+}Y)(t) = m(\alpha)X(t). \tag{8.24}$$

Using the coefficient $a(\alpha) = 1/m(\alpha)$, equation (8.24) is written in form (8.21), which describes an accelerator with the same memory fading parameter $\alpha > 0$. □

As a result, we proved that the equations of accelerator with power-law memory of noninteger fading parameter $\alpha > 0$ can be derived from the equation of multiplier with memory of same fading parameter α. In equation (8.20), the variable $X(t)$ is impact and $Y(t)$ is response variable. In equation (8.21), the variable $Y(t)$ is impact and $X(t)$ is response variable.

8.2.2 From accelerator with memory to multiplier with memory

Let us consider the accelerator with simple power-law memory, which is described equation with the Caputo fractional derivative.

Principle 8.3 (Principle of duality of multiplier and accelerator with power-law memory: part 2).
[**A** → **M**]: *Let $X(\tau)$ be the function of the impact variable, which has integer-order derivatives up to $(n-1)$-th order that are continuous functions on the interval $[0, T]$, and the derivative $X^{(n)}(\tau)$ is Lebesgue summable on the interval $[0, T]$, i. e., $X(\tau) \in AC^n[0, T]$. Then the response variable $Y(t)$, which is described by the equation of linear accelerator with memory*

$$Y(t) = a(\alpha)(D^{\alpha}_{C;0+}X)(t), \tag{8.25}$$

where $a(\alpha)$ is the accelerator coefficient and $0 \leq n-1 < \alpha < n$, satisfies the linear equation of the multiplier with memory

$$X(t) = m(\alpha)(I^{\alpha}_{RL;0+}Y)(t) + b_X(t), \tag{8.26}$$

where $m(\alpha) = 1/a(\alpha)$ is the multiplier coefficient, and

$$b_X(t) = \sum_{k=0}^{n-1} X^{(k)}(0)\frac{t^k}{\Gamma(k+1)}. \tag{8.27}$$

In the duality concepts, the impact and response variables change the roles.

Proof. To prove the statement of this principle, we can consider the actions of the Riemann–Liouville integral on equation (8.25). This action gives the equation

$$(I^{\alpha}_{\mathrm{RL};0+}Y)(t) = a(\alpha)(I^{\alpha}_{\mathrm{RL};0+}D^{\alpha}_{C;0+}X)(t). \tag{8.28}$$

If the function $X(t)$ is continuously differentiable n times, then using Lemma 2.2 of [200, p. 95], we have the equality

$$(I^{\alpha}_{\mathrm{RL};0+}D^{\alpha}_{C;0+}X)(t) = X(t) - \sum_{k=0}^{n-1} X^{(k)}(0)\frac{t^k}{\Gamma(k+1)}, \tag{8.29}$$

where $n-1 < \alpha < n$. Substituting equation (8.29) into equation (8.28), we obtain

$$(I^{\alpha}_{\mathrm{RL};0+}Y)(t) = a(\alpha)X(t) - a(\alpha)\sum_{k=0}^{n-1} X^{(k)}(0)\frac{t^k}{\Gamma(k+1)}. \tag{8.30}$$

Using $m(\alpha) = 1/a(\alpha)$, equation (8.30) gives (8.26). □

It should be noted that the substitution of (8.26) and (8.27) into equation (8.25) gives the identity.

Remark 8.3. Equation (8.24) describes a linear multiplier with memory. As a result, the equation of the linear accelerator with memory can be represented as the linear equation of the multiplier with memory, in which the indicator $Y(t)$ and factor $X(t)$ change interpretations with each other and $m(\alpha) = 1/a(\alpha)$.

Remark 8.4. For $0 < \alpha < 1$, the free term $b(t) = X(0)$. As a result, if $X(0) = 0$ and $0 < \alpha < 1$, then we get the equation of the multiplier with memory in the form $X(t) = m(\alpha)(I^{\alpha}_{\mathrm{RL};0+}Y)(t)$. For $0 \le n-1 < \alpha < n$, we get the same equation, if $X^{(k)}(0) = 0$ for all $k = 0, 1, \dots n-1$.

Remark 8.5. In equation (8.28), the variable $X(t)$ is impact and $Y(t)$ is response variable. In equation (8.29) the variable $Y(t)$ is impact and $X(t)$ is response variable. In the transformations between these equations, the impact and response variables change the roles (interpretations). As a result, we can formulate the following principle.

Principle 8.4 (Principle of interpretation of dual variables). *In describing memory effects, the transition from multiplier with memory to accelerator with memory is accompanied by the fact that impact variables should be considered as response, and conversely. Accelerator coefficient with memory is defined as the inverse value of multiplier coefficient with memory. If the coefficient of the multiplier with memory is greater than one, then the coefficient of the dual accelerator will be less than one, and conversely.*

As a result, the relationship between equations (8.20) and (8.21), and also between equations (8.25) and (8.26), actually means a duality of the concepts of accelerator and multiplier with memory for the case of power-law memory fading.

8.2.3 Formulation of duality for simple power-law memory

Part 1 $[M \to A]$ and Part 2 $[A \to M]$ of the principles of duality of multiplier and accelerator can be combined into one principle of duality. This principle describes the mutual duality of the concepts of accelerator and multiplier with simple power-law memory.

Let us unite in one principle what has been formulated and proved in the two principles proposed above. In this principle, we will consider the accelerator with power-law memory, which is described by the Caputo fractional derivative, and the multiplier with memory that is described by the Riemann–Liouville integral. The concepts of the multiplier and accelerator with simple power-law memory are dual to each other. Let us formulate the principle of duality for these concepts.

Principle 8.5 (First principle of duality of multiplier and accelerator with power-law memory).
[**M** $\to$ **A**] ***From multiplier to accelerator:*** *The multiplier with power-law memory, which is described by the equation*

$$Y(t) = m(\alpha)(I^{\alpha}_{\mathrm{RL};0+}X)(t), \tag{8.31}$$

can be represented as the accelerator with memory

$$X(t) = a(\alpha)(D^{\alpha}_{\mathrm{C};0+}Y)(t), \tag{8.32}$$

where $a(\alpha) = 1/m(\alpha)$ is the coefficient of the dual accelerator, and the impact and response variables change the roles.
[**A** $\to$ **M**] ***From accelerator to multiplier:*** *The accelerator with power-law memory, which is described by the equation*

$$Y(t) = a(\alpha)(D^{\alpha}_{\mathrm{C};0+}X)(t), \tag{8.33}$$

can be represented as the multiplier with memory

$$X(t) = m(\alpha)(I^{\alpha}_{\mathrm{RL};0+}Y)(t) + b_X(t), \tag{8.34}$$

where $m(\alpha) = 1/a(\alpha)$ is the coefficient of the dual multiplier, and

$$b_X(t) = \sum_{k=0}^{n-1} p^C_k(0)t^k \tag{8.35}$$

with $p^C_k(0) = X^{(k)}(0)/k!$. In the duality concepts, the impact and response variables change the roles.

Proof. The proof of the duality of the accelerator and multiplier with power-law memory is based on the Lemma 2.21 and Lemma 2.22 of [200, pp. 95–96]. □

Lemma 2.21 states the following [200, p. 95]. If $\alpha > 0$, then the equality

$$(D^{\alpha}_{C;a+} I^{\alpha}_{RL;a+} X)(t) = X(t) \tag{8.36}$$

is valid for any function $X(t) \in L_{\infty}(a, b)$ or $X(t) \in C[a, b]$. Using equation (8.36), the action of the Caputo fractional derivative on equation (8.31) gives equation (8.32).

Lemma 2.22 states the following [200, p. 96]. Let $\alpha > 0$, $n = [\alpha]+1$ for non-integer α and $n = \alpha$ for integer α. If $X(t) \in AC^n[a, b]$ or $X(t) \in C^n[a, b]$, then we have the equality

$$(I^{\alpha}_{RL;a+} D^{\alpha}_{C;a+} X)(t) = X(t) - \sum_{k=0}^{n-1} p^C_k(a)(t-a)^k, \tag{8.37}$$

where

$$p^C_k(a) = \frac{X^{(k)}(a)}{k!}. \tag{8.38}$$

In particular, for $0 < \alpha < 1$, $(n = 1)$ the equation

$$(I^{\alpha}_{RL;a+} D^{\alpha}_{C;a+} X)(t) = X(t) - X(a) \tag{8.39}$$

is valid if $X(t) \in AC[a, b]$ or $X(t) \in C[a, b]$. Using equation (8.37), the action of the Riemann–Liouville fractional integration on equation (8.33) gives equation (8.34). □

8.3 Principle of decreasing of fading for multiplier with memory

The duality of the accelerator and multiplier with memory has been described for the case of equal fading parameters, i. e., the multiplier $\mathcal{M}^{t;\alpha}_0(X(\tau))$ and the accelerator $\mathcal{A}^{t;\beta}_0(X(\tau))$, when $\alpha = \beta$. Let us consider a combination of actions of an accelerator with memory and a multiplier with memory, when the fading parameters are different.

Principle 8.6 (Principle of decreasing of fading parameter for multiplier with memory). *The sequential action of the accelerator with power-law memory and the multiplier with power-law memory changes the multiplier effect by decreasing the parameter of memory fading. Symbolically, this principle can be written in the form of the equation*

$$\mathcal{A}^{t;\beta}_0(\mathcal{M}^{t;\alpha}_0(X(\tau))) = \mathcal{M}^{t;\alpha-\beta}_0(X(\tau)), \tag{8.40}$$

which is satisfied for memory fading parameters α and β such that $\alpha > \beta > 0$. The fading parameter of the new multiplier is equal to the difference between the fading parameters of the multiplier and the accelerator.

Equality (8.40) means the following. The action of the multiplier with power-law memory

$$Y(t) = \mathcal{M}^{t;\alpha}_0(X(\tau)), \tag{8.41}$$

and the accelerator with power-law memory

$$Z(t) = \mathcal{A}_0^{t;\beta}(Y(\tau)) \tag{8.42}$$

is equivalent to the action of the multiplier with memory

$$Z(t) = \mathcal{M}_0^{t;\alpha-\beta}(X(\tau)), \tag{8.43}$$

if the memory fading parameter α of the multiplier is greater than the memory fading parameter β of the accelerator ($\alpha > \beta > 0$).

Example 8.3. The sequential action of the multiplier with power-law memory

$$Y(t) = m(\alpha)(I^{\alpha}_{\mathrm{RL};0+}X)(t) \tag{8.44}$$

and the accelerator with power-law memory

$$Z(t) = a(\beta)(D^{\beta}_{\mathrm{RL};0+}Y)(t) \tag{8.45}$$

is equivalent to the action of the multiplier with memory

$$Z(t) = m(\alpha,\beta)(I^{\alpha-\beta}_{\mathrm{RL};0+}X)(t), \tag{8.46}$$

where $m(\alpha,\beta) = m(\alpha)a(\beta)$ if $\alpha > \beta > 0$.

The proof of this statement is based on Property 2.2 of [200, p. 74] or Theorem 2.5 of [335, p. 46]. If $\alpha > \beta > 0$ and $X(t) \in L_p(a,b)$ with $p \geq 1$, then the equality

$$(D^{\beta}_{\mathrm{RL};a+}I^{\alpha}_{\mathrm{RL};a+}X)(t) = (I^{\alpha-\beta}_{\mathrm{RL};a+}X)(t) \tag{8.47}$$

is valid (see Property 2.2 of [200, p. 74]) almost everywhere on $[a,b]$. In particular, if $0 < \beta < 1$ and $\alpha = 1$, then we have

$$(D^{\beta}_{\mathrm{RL};a+}I^{1}_{\mathrm{RL};a+}X)(t) = (I^{1-\beta}_{\mathrm{RL};a+}X)(t). \tag{8.48}$$

As a result, the fading parameter of the final multiplier is equal to the difference between the fading parameters of the initial multiplier and the accelerator. The coefficient of the new multiplier is equal to the product of the multiplier and accelerator coefficients.

Example 8.4. It should be noted that this principle can also be used for the standard accelerators without memory. In this case, the fading parameter of the standard accelerator is equal to one or an integer number. The proof of this statement is based on the equation 2.1.33 of Property 2.2 in [200, p. 74]. For example, when $\beta = 1$ and $\alpha > 1$, we have the equality

$$(D^{1}I^{\alpha}_{\mathrm{RL};a+}X)(t) = (I^{\alpha-1}_{\mathrm{RL};a+}X)(t), \tag{8.49}$$

which is valid for $\alpha > 1$.

Example 8.5. If the memory of multiplier and accelerator is describes by the Liouville fractional integral and derivative, this principle is based on Property 2.12 of [200, p. 89], which states that the equality

$$(D^{\beta}_{L;+} I^{\alpha}_{L;+} X)(t) = (I^{\alpha-\beta}_{L;+} X)(t) \tag{8.50}$$

holds for $X(t) \in L_1(-\infty, +\infty)$.

8.4 Examples of duality

8.4.1 Second duality for simplest power-law memory

In the previous section, we consider accelerator that is described by the Caputo fractional derivative. Let us consider the accelerator with power-law memory, which is described by the Riemann–Liouville fractional derivative instead of the Caputo fractional derivative. The multiplier with memory will be described by the equation, with the Riemann–Liouville integral. The multiplier and accelerator with simple power-law memory are dual concepts. Let us formulate the principle of duality for these concepts.

Principle 8.7 (Second principle of duality of multiplier and accelerator with power-law memory).
[**M** → **A**]: *The multiplier with power-law memory, which is described by the equation*

$$Y(t) = m(\alpha)(I^{\alpha}_{\mathrm{RL};0+} X)(t), \tag{8.51}$$

can be considered as the accelerator with memory

$$X(t) = a(\alpha)(D^{\alpha}_{\mathrm{RL};0+} Y)(t), \tag{8.52}$$

where $a(\alpha) = 1/m(\alpha)$ is the coefficient of the dual accelerator and the impact and response variables change the roles.
[**A** → **M**]: *The accelerator with power-law memory, which is described by the equation*

$$Y(t) = a(\alpha)(D^{\alpha}_{\mathrm{RL};0+} X)(t), \tag{8.53}$$

can be considered as the multiplier with memory

$$X(t) = m(\alpha)(I^{\alpha}_{\mathrm{RL};0+} Y)(t) + b_X(t), \tag{8.54}$$

where $m(\alpha) = 1/a(\alpha)$ is the coefficient of the dual multiplier, and

$$b_X(t) = \sum_{k=0}^{n-1} p^{\mathrm{RL}}_k(0) t^{\alpha-k-1} \tag{8.55}$$

with

$$p_k^{\mathrm{RL}}(0) = \frac{1}{\Gamma(\alpha - k)} (D^{n-k-1} I_{\mathrm{RL};0+}^{n-\alpha} X)(0). \tag{8.56}$$

In the duality concepts, the impact and response variables change the roles.

Proof. The proof of the duality of the accelerator and multiplier with power-law memory is based on Theorem 2.4 of [335, pp. 44–45], (and Lemmas 2.4–2.5 of [200, pp. 74–75]). The Riemann–Liouville fractional derivative is left inverse operator for the Riemann–Liouville integration.

Using Theorem 2.4 of [335, pp. 44–45], we have that the equality

$$(D_{\mathrm{RL};a+}^{\alpha} I_{\mathrm{RL};a+}^{\alpha} X)(t) = X(t) \tag{8.57}$$

is valid for any summable function $X(t) \in L_1(a, b)$, where $D_{\mathrm{RL};a+}^{\alpha}$ is the left-sided Riemann–Liouville fractional derivative of the order $\alpha > 0$, and $I_{\mathrm{RL};a+}^{\alpha}$ is the left-sided Riemann–Liouville fractional integral of the order $\alpha > 0$ with respect to the variable t. If $X(t) \in L_p(a, b)$ with $p \geq 1$, then according to Lemma 2.4 of [200, p. 74] equality (8.57) holds almost everywhere on $[a, b]$. Using equation (8.57), the action of the Riemann–Liouville fractional derivative on equation (8.51) gives equation (8.52).

If the function $X(t) \in L_1(a, b)$ has a summable fractional derivative $(D_{\mathrm{RL};a+}^{\alpha} X)(t)$, i. e., $(I_{\mathrm{RL};a+}^{n-\alpha} X)(t) \in AC^n[a, b]$, then (see Theorem 2.4 of [335, pp. 44–45], and Lemma 2.5 of [200, pp. 74–75]), we have the equality

$$(I_{\mathrm{RL};a+}^{\alpha} D_{\mathrm{RL};a+}^{\alpha} X)(t) = X(t) - \sum_{k=0}^{n-1} p_k^{\mathrm{RL}}(a)(t-a)^{\alpha k-1}, \tag{8.58}$$

where $n = [\alpha] + 1$ and

$$p_k^{\mathrm{RL}}(a) = \frac{1}{\Gamma(\alpha - k)} (D^{n-k-1} I_{\mathrm{RL};a+}^{n-\alpha} X)(a). \tag{8.59}$$

In particular, we have

$$(I_{\mathrm{RL};0+}^{\alpha} D_{\mathrm{RL};0+}^{\alpha} X)(t) = X(t) - \frac{t^{\alpha-1}}{\Gamma(\alpha)} (I_{\mathrm{RL};0+}^{1-\alpha} X)(0) \tag{8.60}$$

for $0 < \alpha < 1 (n = 1)$. Using equation (8.58), the action of the Riemann–Liouville fractional integration on equation (8.53) gives equation (8.54). □

Remark 8.6. The difference between the second principle of duality of multiplier and accelerator with power-law memory from the first principle is the use of the Caputo fractional derivative instead of the Riemann–Liouville fractional derivative.

8.4.2 Duality for memory and parametric fractional derivative

The parametric fractional Caputo derivative was suggested in [418] (see equation (23) of Definition 3 of paper [418, p. 224]). The properties of this derivative are described in [13, p. 467].

The duality of the accelerator and multiplier with parametric memory is based on Theorems 4 and 5 of [13, pp. 466–467]. Let us give the statements of these theorems.

If $\alpha > 0$, then (see Theorem 5 in [13, p. 467]) the equality

$$(D^{\alpha,X}_{C;a+} I^{\alpha,X}_{RL;a+} X)(t) = X(t) \tag{8.61}$$

is valid for any function $X(t) \in C^1[a, b]$.

If $X(t) \in C^n[a, b]$, then (see Theorem 4 in [13, p. 466]) we have the equality

$$(I^{\alpha,X}_{RL;a+} D^{\alpha,X}_{C;a+} X)(t) = X(t) - \sum_{k=0}^{n-1} \frac{X^{(k)}(a)}{k!} (X(t) - X(a))^k, \tag{8.62}$$

where $\alpha > 0$, $n = [\alpha] + 1$ for noninteger α and $n = \alpha$ for integer α.

8.4.3 Duality for distributed time scaling

Let us consider two examples of duality principle, when accelerator is described by the Erdelyi–Kober fractional derivatives of Caputo and Riemann–Liouville types.

Example 8.6. Let us consider the accelerator with distributed time scaling, which is described by the equation with the Erdelyi–Kober fractional derivative of the Riemann–Liouville type, and the multiplier with distributed time scaling that is described by the Erdelyi–Kober integral. These multiplier and accelerator are dual concepts. The principle of duality for these concepts can be formulated in the following form.

Principle 8.8 (First principle of duality of multiplier and accelerator with distributed time scaling).
[**M** → **A**]: *The multiplier with distributed time scaling, which is described by the equation*

$$Y(t) = m(\alpha)(I^{\alpha}_{EK,0+;\sigma,\eta} X)(t), \tag{8.63}$$

can be described as the accelerator with distributed time scaling

$$X(t) = a(\alpha)(D^{\alpha}_{EK,0+;\sigma,\eta} Y)(t), \tag{8.64}$$

where $a(\alpha) = 1/m(\alpha)$ is the coefficient of the dual accelerator. The impact and response variables change the roles.

[**A** → **M**]: *The accelerator with distributed time scaling, which is described by the equation*

$$Y(t) = a(\alpha)(D^{\alpha}_{\mathrm{EK},0+;\sigma,\eta}X)(t), \tag{8.65}$$

can be described as the multiplier with distributed time scaling

$$X(t) = m(\alpha)(I^{\alpha}_{\mathrm{EK},0+;\sigma,\eta}Y)(t) + b_X(t), \tag{8.66}$$

where $m(\alpha) = 1/a(\alpha)$ is the coefficient of the dual multiplier,

$$b_X(t) = \sum_{k=0}^{n-1} p_k^{\mathrm{EK}}(0)t^{-\sigma(\eta+1+k)}. \tag{8.67}$$

In the duality, the impact and response variables change the roles.

Proof. The proof of this principle of duality of the accelerator and multiplier, which is described by the Erdelyi–Kober fractional integrals and derivatives, is based on Theorem 3.1 of [241, pp. 255–258], and equation 21 of [241, p. 254], or Property 2.23 of [200, p. 109].

The Erdelyi–Kober fractional derivative provides operation inverse to the Erdelyi–Kober integration from the left. We can use Property 2.23 of [200, p. 109] and equation 21 of [241, p. 254]. If $\alpha > 0$ and $a \geq 0$, then the equality

$$(D^{\alpha}_{\mathrm{EK},a+;\sigma,\eta}I^{\alpha}_{\mathrm{EK},a+;\sigma,\eta}X)(t) = X(t) \tag{8.68}$$

is valid for all $X(t) \in C_{\mu}$, where $\mu \geq -\sigma(\eta+1)$. Using equation (8.68), the action of the Erdelyi–Kober fractional derivative on equation (8.63) gives equation (8.64).

Let $n-1 < \alpha \leq n$, $\mu \geq -\sigma(\eta+1)$ and $X(t) \in C^n_{\mu}$, where the space of functions C^n_{μ} consists of all functions $X(t)$ with $t > 0$, that can be represented in the form $X(t) = t^p f(t)$ with $p > \mu$ and $f(t) \in C^n([0,\infty))$. Then using Theorem 3.1 of [241, pp. 255–258], the following relation between t the Erdelyi–Kober fractional derivative and integral of the order $\alpha > 0$ holds true

$$(I^{\alpha}_{\mathrm{EK},0+;\sigma,\eta}D^{\alpha}_{\mathrm{EK},0+;\sigma,\eta}X)(t) = X(t) - \sum_{k=0}^{n-1} p_k^{\mathrm{EK}}(0)t^{-\sigma(\eta+1+k)}, \tag{8.69}$$

where $p_k^{\mathrm{EK}}(0)$ are defined by the equation

$$\begin{aligned} p_k^{\mathrm{EK}}(0) = {} & \frac{\Gamma(n-k)}{\Gamma(\alpha-k)} \lim_{t\to 0} t^{\sigma(\eta+1+k)} \\ & \times \prod_{j=k+1}^{n-1}\left(\frac{\tau}{\sigma}\frac{d}{d\tau} + \eta + j + 1\right)(I^{n-\alpha}_{\mathrm{EK},0+;\sigma,\eta+\alpha}X)(t). \end{aligned} \tag{8.70}$$

The equations of Theorem 3.1 of [241, pp. 255–258], are similar to the corresponding equations for the Riemann–Liouville fractional integral and derivative

$$(I^{\alpha}_{\mathrm{RL};0+} D^{\alpha}_{\mathrm{RL};0+} X)(t) = X(t) - \sum_{k=1}^{n} p_k^{\mathrm{RL}}(0) t^{\alpha-k}, \tag{8.71}$$

where $p_k^{\mathrm{RL}}(0)$ are defined by the equation

$$p_k^{\mathrm{RL}}(0) = \frac{t^{\alpha-k}}{\Gamma(\alpha-k+1)} \lim_{t\to 0} D^{\alpha-k}_{\mathrm{RL};0+} X(t). \tag{8.72}$$

Using equation (8.69), the action of the Erdelyi–Kober fractional integration on equation (8.65) gives equation (8.66) that describes the linear generalized multiplier. □

Example 8.7. Let us consider the accelerator with distributed time scaling, which is described by the Erdelyi–Kober fractional derivative of the Caputo type, and the multiplier with distributed time scaling that is described by the Erdelyi–Kober integral. These equations of multiplier and accelerator with distributed time scaling describe dual concepts. The duality principle for these concepts can be formulated in the following form.

Principle 8.9 (Second principle of duality of multiplier and accelerator with distributed time scaling).
[**M** → **A**]: *The multiplier with distributed time scaling, which is described by the equation*

$$Y(t) = m(\alpha)(I^{\alpha}_{\mathrm{EK},0+;\sigma,\eta} X)(t), \tag{8.73}$$

can be described as the accelerator with memory

$$X(t) = a(\alpha)(D^{\alpha}_{\mathrm{EK},0+;\sigma,\eta} Y)(t), \tag{8.74}$$

where $a(\alpha) = 1/m(\alpha)$ *is the coefficient of the dual accelerator and the impact and response variables change the roles.*
[**A** → **M**]: *The accelerator with distributed time scaling, which is described by the equation*

$$Y(t) = a(\alpha)(D^{\alpha}_{\mathrm{EKC},0+;\sigma,\eta} X)(t), \tag{8.75}$$

can be described as the multiplier with distributed time scaling

$$X(t) = m(\alpha)(I^{\alpha}_{\mathrm{EK},0+;\sigma,\eta} Y)(t) + b_X(t), \tag{8.76}$$

where $m(\alpha) = 1/a(\alpha)$ *is the coefficient of the dual multiplier,*

$$b_X(t) = \sum_{k=0}^{n-1} p_k^{\mathrm{EKC}}(0) t^{-\sigma(\eta+1+k)}. \tag{8.77}$$

In the duality concepts, the impact and response variables change the roles.

Proof. For this case, when the accelerator with memory is described by the Caputo-type modification of the Erdelyi–Kober fractional derivative, the duality is based on Theorem 4.1 of [241, pp. 260–261], and Theorem 4.3 of [241, pp. 263–264].

The Caputo-type modification of the Erdelyi–Kober fractional derivative is a left-inverse operator to the Erdelyi–Kober fractional integral. This property is valid for the functions from the functional space C_μ, where $\mu \geq -\sigma(\eta+1)$. If $\alpha > 0$ and $0 \leq a < b$, then using Theorem 4.3 of [241, pp. 263–264], we have

$$(D^{\alpha}_{\mathrm{EKC},0+;\sigma,\eta} I^{\alpha}_{\mathrm{EK},0+;\sigma,\eta} X)(t) = X(t) \tag{8.78}$$

for all $X(t) \in C_\mu$, where $\mu \geq -\sigma(\eta+1)$. Using equation (8.78), the action of the Erdelyi–Kober fractional derivative on equation (8.73) gives equation (8.74).

Let $n-1 < \alpha \leq n$, $\mu \geq -\sigma(\alpha+\eta+1)$ and $X(t) \in C^n_\mu$, where the space of functions C^n_μ consists of all functions $X(t)$, $t > 0$, that can be represented in the form $X(t) = t^p f(t)$ with $p > \mu$ and $f(t) \in C^n([0,\infty)$. Then using Theorem 4.1 of [241, pp. 260–261], the following relation between the Caputo-type modification of the Erdelyi–Kober fractional derivative and the Erdelyi–Kober fractional integral of the order $\alpha > 0$ holds true

$$(I^{\alpha}_{\mathrm{EK},0+;\sigma,\eta} D^{\alpha}_{\mathrm{EKC},0+;\sigma,\eta} X)(t) = X(t) - \sum_{k=0}^{n-1} p_k^{\mathrm{EKC}}(0) t^{-\sigma(\eta+1+k)}, \tag{8.79}$$

where p_k are defined by the equation

$$p_k^{\mathrm{EKC}}(0) = \lim_{t\to 0} t^{\sigma(\eta+1+k)} \prod_{j=k+1}^{n-1} \left(\frac{t}{\sigma}\frac{d}{dt} + \eta + j + 1\right) X(t). \tag{8.80}$$

Using equation (8.79), the action of the Erdelyi–Kober fractional derivative on equation (8.75) gives equation (8.76). □

Note that equations (8.78) and (8.79) are similar to the corresponding equations for the Riemann–Liouville fractional integral and the Caputo fractional derivative. We emphasize that the constants p_k depend only on the ordinary derivatives of the function $X(t)$ with some power weights and do not depend on the limit values of the fractional integrals at the point $t = 0$.

8.4.4 Duality for distributed memory fading

Let us consider the multiplier with uniform distributed memory fading (the UDMF multiplier) that is described by the equation with the Nakhushev fractional integral [450, 278, 279].

Principle 8.10 (Principle of duality of UDN multiplier and NSDM accelerator).
[**M** → **A**]: *For the concept of the multiplier with uniform distributed memory fading*

$$Y(t) = m([\alpha,\beta])(I_N^{[\alpha,\beta]} X)(t), \tag{8.81}$$

there is a dual concept of the accelerator with nonsingular distributed memory, which is described by the equation

$$X(t) = a([\alpha,\beta])(D_P^{[\alpha,\beta]}Y)(t), \tag{8.82}$$

where $D_P^{[\alpha,\beta]}$ *is the Pskhu fractional derivative of the nonsingular distributed order.*

[**A** → **M**]: *For the concept of the accelerator with nonsingular distributed memory fading*

$$Y(t) = a([\alpha,\beta])(D_P^{[\alpha,\beta]}X)(t), \tag{8.83}$$

there is a dual concept of the multiplier with uniform distributed memory that is described by the equation

$$X(t) = m([\alpha,\beta])(I_N^{[\alpha,\beta]})Y(t) + b_X(t), \tag{8.84}$$

where

$$b_X(t) = \sum_{k=1}^{n} W(\alpha-k,\beta-k,t)(D_P^{[\alpha-k,\beta-k]}X)(0+). \tag{8.85}$$

In the dual concepts, the impact and response variables change the roles.

Proof. The proof of this duality is based on Theorem 1 of [317], or Theorem 5.1.1 of [318, pp. 137–140], which can be written by the following statement. Let $\beta > \alpha > 0$ and $n-1 \le \alpha < n$. If $X(t) \in L[0,a]$, then

$$(D_P^{[\alpha,\beta]}I_N^{[\alpha,\beta]}X)(t) = X(t). \tag{8.86}$$

Using equation (8.86), the action of the Pskhu fractional derivative on equation (8.81) gives equation (8.82).

If $X(t) \in C_E^{n-1}$, then

$$(I_N^{[\alpha,\beta]}D_P^{[\alpha,\beta]}X)(t) = X(t) - \sum_{k=1}^{n} W(\alpha-k,\beta-k,t)(D_P^{[\alpha-k,\beta-k]}X)(0+), \tag{8.87}$$

where we use the notation

$$(D_P^{[\alpha,\beta]}X)(0+) = \lim_{t\to 0}(D_P^{[\alpha,\beta]}X)(t), \tag{8.88}$$

and the function

$$W(\alpha,\beta,t) = \frac{1}{(\beta-\alpha)t}\int_{\alpha}^{\beta}\frac{t^{\xi}\,d\xi}{\Gamma(\xi)}. \tag{8.89}$$

The function space C^k_E means the following class of functions

$$X(t) \in C^k_*[(0,a); -t^{n-\alpha-1}E_{\beta-\alpha}[t^{\beta-\alpha}; n-\alpha]], \tag{8.90}$$

where n is defined by the conditions $n-1 < \alpha \leq n$ and $1 \leq k \leq n$ [318, p. 137]. We denote by $C^n_*[(0,a); f(t)]$ the space of functions $X(t) \in L[0,a]$ such that $(X(t) * f(t)) \in C^n[0,a]$ and $D^n(X(t) * f(t)) \in AC[0,a]$, where $AC[0,a]$ is the space of functions, which are absolutely continuous on $[0,a]$, and the symbol $*$ denoted the convolution

$$(f * g)(t) = \int_0^t f(\tau)g(t-\tau)\,d\tau = \int_0^t f(t-\tau)g(\tau)\,d\tau. \tag{8.91}$$

Using equation (8.87), the action of the Pskhu fractional integral on equation (8.83) gives equation (8.84). □

The accelerator with uniform distributed memory fading (the UDMF accelerator) can be defined by the equation with the Nakhushev fractional derivative [450, 278, 279].

Principle 8.11 (Principle of duality of UDMF accelerator and NSDM multiplier).
[**A** → **M**]: *For the concept of accelerator*

$$Y(t) = a([\alpha,\beta])(D_N^{[\alpha,\beta]}X)(t), \tag{8.92}$$

there is a dual multiplier, which is the multiplier with nonsingular distributed memory fading (the NSDM multiplier) that is described by the equation

$$X(t) = m([\alpha,\beta])(I_P^{[\alpha,\beta]}Y)(t) + b_X(t), \tag{8.93}$$

where $I_P^{[\alpha,\beta]}$ is the Pskhu fractional derivative of the nonsingular distributed order,

$$b_X(t) = \sum_{k=1}^{n} p^P_k(0)t^{\beta-k}(\alpha-\beta)E_{\beta-\alpha}[t^{\beta-\alpha}; \beta-k+1], \tag{8.94}$$

and $p^P_k(0)$ is defined by the expression $p^P_k(0) = (D_N^{[\alpha-k,\beta-k]}X)(0+)$.
[**M** → **A**]: *At the same time, for the NSDM multiplier that is described by equation*

$$Y(t) = m([\alpha,\beta])(I_P^{[\alpha,\beta]}X)(t), \tag{8.95}$$

there is a dual concept of the accelerator with uniform distributed memory fading

$$X(t) = a([\alpha,\beta])(D_N^{[\alpha,\beta]}Y)(t). \tag{8.96}$$

The impact and response variables change the roles.

Proof. The proof of this duality is based on Theorem 3 of [317] or Theorem 5.1.3 [318, pp. 142–143], which can be rewritten by the following statement. Let $\beta > \alpha$ and $0 < n - 1 \leq \beta < n$, where n is positive integer number. If $X(t) \in L[0, a]$, then

$$(D_N^{[\alpha,\beta]} I_P^{[\alpha,\beta]} X)(t) = X(t). \tag{8.97}$$

If $X(t) \in C_\nu^{n-1}$, then

$$(I_P^{[\alpha,\beta]} D_N^{[\alpha,\beta]} X)(t) \\ = X(t) - \sum_{k=1}^{n} p_k^p(0) t^{\beta-k} (\alpha - \beta) E_{\beta-\alpha}[t^{\beta-\alpha}; \beta - k + 1], \tag{8.98}$$

where

$$p_k^p(0) = (D_N^{[\alpha-k,\beta-k]} X)(0+). \tag{8.99}$$

The function space C_ν^k means the following class of functions:

$$X(t) \in C_*^k[(0, a); W(\alpha-, \beta - n, t)], \tag{8.100}$$

where n is defined by the conditions $n - 1 < \beta \leq n$ and $1 \leq k \leq n$ [318, p. 142]. □

8.5 Conclusion

In this Chapter, we describe duality of the accelerator with memory and the multiplier with memory, which were proposed in [412, 397]. This concept of duality has no analogues for the standard accelerator and multiplier without memory. The concepts of accelerator and multiplier with memory are dual to each other in a sense. Duality means that an accelerator with memory can be represented as a multiplier with memory, and vice versa. As a result, in economic models, the memory can be taken into account by using one of these two dual concepts. Mathematically, the principle of duality states a possibility to represent equation of the multiplier with memory in the form of equation of the accelerator with memory, and, vice versa, a possibility to represent the equation of accelerator with memory in the form of equation of a multiplier with memory. In these transformations between equations of the accelerator and multiplier with memory, the impact and response variables change the roles (interpretations). The representation of multiplier with memory in the form of accelerator with memory is accompanied by the transformation of the impact variables into response variables, and conversely.

Part III: **Linear models of economics with memory**

9 Model of natural growth with memory

In chapters of this part, we proposed macroeconomic models, in which memory effects with power-law fading are taken into account. We derive the fractional differential equations that describe macroeconomic models of growth with power-law memory and then find solutions of these equations. Using properties of these solutions, we formulate principles of economic dynamics with one-parameter fading memory. We demonstrate that the memory effects can significantly change the economic growth rate and dominant parameters, which determine the growth.

In this chapter, we describe new economic model, which is a generalization of the economic model of natural growth by taking into account the power-law memory. This chapter is based on articles [420, 440], we proposed a generalization of the standard model of natural growth.

9.1 Introduction

The standard natural growth model is one of the simplest economic growth models with continuous time [493, 494]. However, this model does not take into account memory effects. In articles [420, 440], we proposed a generalization of the economic model of natural growth by taking into account the power-law memory. The memory means the dependence of the process not only on the current state of the process, but also on the history of changes of this process in the past. We propose equations that take into account the memory with one-parameter power-law fading. The proposed equations use the Caputo fractional derivatives to take into account for power-law memory. Solutions of these equations, which contain fractional derivatives, are suggested. We proved that the growth and decline (downturn) of output depend on the memory effects. This chapter provides examples of the impact of power-law fading memory on economic growth. For example, the memory effect can lead to decrease of output instead of its growth, which is predicted by the standard model without memory. Memory effect can lead to increase of output, rather than decrease, which is predicted by the standard model without memory. A more detailed analysis based on the asymptotic behavior of the solutions will be given in the next chapter on the generalization of the Harrod–Domar model. This chapter is based on articles [420, 440].

9.2 Model of natural growth without memory

Let us describe the economic model of natural growth that does not take into account the effects of the lag and memory effects. Let $Y(t)$ be a function that describes the value of output at time t, and $I(t)$ be a function that describes the net investment, i. e., the investment that is used to expand production.

https://doi.org/10.1515/9783110627459-009

The standard model of natural growth without memory, it is used the following assumptions.

(1) We use the assumption of the unsaturated market, which implies that all manufactured products are sold.

(2) We also assume that the volume of sales is not large and, therefore, does not affect the price of the goods. Therefore, the price $P > 0$ is assumed to be constant

$$P(t) = P = \text{const.} \tag{9.1}$$

(3) In the model of natural growth is assumed that the rate of change of output $(dY(t)/dt)$ are directly proportional to the value of the net investment. Therefore, we can use the accelerator equation

$$\frac{dY(t)}{dt} = \frac{1}{v} I(t), \tag{9.2}$$

where v is a positive constant, that is called the investment coefficient indicating the power of the accelerator [10, p. 62], $1/v$ is the marginal productivity of capital (rate of acceleration), and $dY(t)/dt$ is the first-order derivative of the function $Y(t)$ with respect to time.

(4) Assuming that the value of the net investment is a fixed part of the profit, which is equal to the difference between the income of $PY(t)$ and the costs $C(t)$, we get the equation

$$I(t) = m(PY(t) - C(t)), \tag{9.3}$$

where m is the rate of net investment $(0 < m < 1)$, i. e., the share of profit, which is spent on the net investment.

(5) We assume that the costs $C(t)$ are linearly dependent on the output $Y(t)$, such that we have the equation

$$C(t) = aY(t) + b, \tag{9.4}$$

where a is the marginal costs, and b is the independent costs, i. e., the part of the cost, which does not depend on the value of output.

Substituting expressions (9.3) and (9.4) into equation (9.2), we obtain

$$\frac{dY(t)}{dt} - \frac{m(P-a)}{v} Y(t) = -\frac{mb}{v}. \tag{9.5}$$

Differential equation (9.5) describes the standard economic model of natural growth without memory and lag.

The general solution of differential equation (9.5) has the form

$$Y(t) = \frac{b}{(P-a)} + c \exp\left(\frac{m(P-a)}{v} t\right), \tag{9.6}$$

where c is a constant. Using the initial value $Y(0)$ of function (9.6) at $t = 0$, we obtain $c = Y(0) - b/(P - a)$. As a result, we have the solution

$$Y(t) = \frac{b}{(P-a)}\left(1 - \exp\left(\frac{m(P-a)}{v}t\right)\right) + Y(0)\exp\left(\frac{m(P-a)}{v}t\right). \tag{9.7}$$

Solution (9.7) of equation (9.5) describes the dynamics of output within the model of natural growth without memory.

In the standard model of natural growth, which is described by equation (9.5), is supposed to perform the accelerator equation (9.2). Equations (9.2) and (9.5) contain only the first-order derivative with respect to time. It is known that the derivative of the first order is determined by the properties of differentiable functions of time only in infinitely small neighborhood of the time point. Because of this, the natural growth model (9.5) assumes an instantaneous change of output speed when changing the net investment. This means that the model of natural growth (9.5) neglects the effects of memory and lag. This is one of the basic assumptions used in the standard model of natural growth.

9.3 Memory effects by fractional derivatives and integrals

In articles [420, 440], the economic model of natural growth, which is described by equation (9.5), is generalized by taking into account the power-law memory. The memory is considered as a property that describes a dependence of the variables not only on the current state of the process, but also on the changes of these variables in the past. Memory can be considered as an average characteristic of process that describes the dependence of the process state at present time of the process state in the past.

We will not immediately postulate some generalization of equation (9.2), in the form of an accelerator with memory. First, we will use the interdependence (the duality [412, 397]) of the accelerator with memory and the multiplier with memory looks like. For this, we will describe the dependence of production volumes on investments in a form that takes into account the history of investment changes in the past.

To consider memory effects in natural growth model, we assume that the value of output $Y(t)$ at time t depends not only on the net investment $I(\tau)$ at the same time point $\tau = t$, but also depends on the changes $I(\tau)$ on a finite time interval in past $\tau \in (0, t)$. This is due to the fact that economic agents can remember the previous changes of investments $I(t)$ and the impact of these changes on the value of output $Y(t)$. To describe how memory affects the dependence of output from net investment, we can use the equation

$$Y(t) = \int_0^t M(t-\tau) I(\tau)\, d\tau + Y(0), \tag{9.8}$$

where $M(t)$ is the memory function that allows us to take into account the memory of the net investment. If $M(t) = 1/v$, then equation takes the form

$$Y(t) = \frac{1}{v} \int_0^t I(\tau)\, d\tau + Y(0), \tag{9.9}$$

which is equivalent to the equation of the standard accelerator (9.2). The equivalence of equations (9.2) and (9.9) follows from the fundamental theorem of calculus, which connects the differentiation and integration of the functions. According to this theorem, the differentiation of equation (9.9) gives equation (9.2):

$$\frac{dY(t)}{dt} = \frac{1}{v} \frac{d}{dt} \int_0^t I(\tau)\, d\tau + \frac{dY(0)}{dt} = \frac{1}{v} I(t). \tag{9.10}$$

Let us note some special cases of equation (9.8). If the function $M(t)$ is expressed by the Dirac delta-function ($M(t) = w\delta(t)$), then equation (9.8) becomes the standard equation of multiplier $Y(t) = wI(t)$. If the function $M(t)$ has the form $M(t) = w\delta(t - T)$, then equation (9.8) becomes the equation of multiplier with fixed-time delay $Y(t) = wI(t - T)$ [10, p. 25]. If the normalization condition holds for the function $M(t)$, then equation (9.8) can be interpreted as the equation of multiplier with the continuously distributed lag and the function $M(t)$ is called the weighting function [10]. In this case, the process passes through all states continuously without any loss.

Assuming a power-law form of memory fading, we can use the memory function in the form

$$M(t - \tau) = \frac{1}{\Gamma(\alpha)} \frac{w(\alpha)}{(t - \tau)^{1-\alpha}}, \tag{9.11}$$

where $\alpha > 0$ is the fading parameter, $t > \tau$, and $\Gamma(\alpha)$ is the gamma function. For function (9.11), equation (9.8) takes the form

$$Y(t) = w(\alpha)(I^{\alpha}_{RL;0+} I)(t), \tag{9.12}$$

where $I^{\alpha}_{RL;0+}$ is the left-sided Riemann–Liouville fractional integral of order $\alpha > 0$. In order to easily interpret the dimensions of the economic quantities, we can use the time t as a dimensionless variable by changing the variable $t \to tt_d$, where t_d is the unit of time (hour, day, month, year).

The left-sided Riemann–Liouville fractional integral is defined by the equation

$$(I^{\alpha}_{RL;0+} f)(t) = \frac{1}{\Gamma(\alpha)} \int_0^t \frac{f(\tau)\, d\tau}{(t - \tau)^{1-\alpha}}, \tag{9.13}$$

where $0 < \tau < t < T$. The function $f(\tau)$ is assumed to satisfy the condition $f(\tau) \in L_1(0, T)$, which implies that $f(\tau)$ is a Lebesgue integrable function and the inequality

$$\int_0^T |f(\tau)|\, d\tau < \infty \tag{9.14}$$

holds.

Equation (9.12) describes the multiplier with memory, and $w(\alpha)$ is a multiplier coefficient. This form of multiplier with memory allows us to use the fractional calculus and fractional differential equations to describe economic processes with fading memory.

In order to express the function of net investment $I(t)$ through the function $Y(t)$, we consider that action of the Caputo fractional derivative of order $\alpha > 0$ on equation (9.12). The Caputo fractional derivative is defined by the equation

$$(D^{\alpha}_{C;0+}Y)(t) = \frac{1}{\Gamma(n-\alpha)} \int_0^t \frac{Y^{(n)}(\tau)}{(t-\tau)^{\alpha-n+1}}\, d\tau, \tag{9.15}$$

where $n-1 < \alpha \leq n$, $Y^{(n)}(\tau)$ is the derivative of integer order $n \in \mathbb{N}$ of the function $Y(\tau)$ with respect to $\tau \in (0,t)$. Operator (9.15) exists if $Y(\tau) \in AC^n[0,T]$ (see Theorem 2.1 of [200, p. 92]), i. e., the function $Y(\tau)$ has derivatives of integer orders up to $(n-1)$-th order, which are continuous functions on the interval $[0,T]$, and the derivative $Y^{(n)}(\tau)$ is Lebesgue summable on the interval $[0,T]$.

The action of derivative (9.15) on equation (9.12) gives the expression

$$(D^{\alpha}_{C;0+}Y)(t) = w(\alpha)(D^{\alpha}_{C;0+}I^{\alpha}_{RL;0+}I)(t), \tag{9.16}$$

where we use that the action of the Caputo fractional derivative on a constant function ($Y(0) = \text{const}$) gives zero

$$(D^{\alpha}_{C;0+}Y(0))(t) = 0. \tag{9.17}$$

Note that the action of the Riemann–Liouville fractional derivative on a constant function yields nonzero.

It is known that the Caputo derivative is left inverse to the Riemann–Liouville integral (see Lemma 2.21 [200, p. 95]), and for any continuous function $f(t) \in C[0,T]$ the identity

$$(D^{\alpha}_{C;0+}I^{\alpha}_{RL;0+}f)(t) = f(t) \tag{9.18}$$

holds for any $\alpha > 0$, where $I^{\alpha}_{RL;0+}$ is the left-sided Riemann–Liouville fractional integral (9.13) and $D^{\alpha}_{C;0+}$ is the left-sided Caputo fractional derivative (9.15).

Using identity (9.18), equation (9.16) can be written as

$$(D^{\alpha}_{C;0+}Y)(t) = \frac{1}{v(\alpha)} I(t), \tag{9.19}$$

where $v(\alpha) = 1/w(\alpha)$.

As a result, the multiplier with memory (9.12) can be represented in the form of the accelerator with memory (9.19), where the coefficient of the accelerator is the inverse of the coefficient multiplier (9.12).

Accelerator equation (9.19) contains the standard equation of the accelerator and the multiplier, as special cases. To prove this, consider equation (9.19) for $\alpha = 0$ and $\alpha = 1$. Using property $(D^{1}_{C;0+}X)(t) = X^{(1)}(t)$ of the Caputo fractional derivative (see equation 2.4.14 in Theorem 2.1 of [200, p. 79]), equation (9.19) with $\alpha = 1$ gives equation (9.2) that describes the standard accelerator. Using the property $(D^{0}_{C;0+}Y)(t) = Y(t)$ (equation (2.4.19) of Theorem 2.1 of [200, p. 92]), equation (9.19) with $\alpha = 0$ can be written as $I(t) = wY(t)$, which is an equation of standard multiplier. As a result, the proposed concept of accelerator with memory generalizes the standard economic concepts of the accelerator and the multiplier [412, 397].

9.4 Equation of natural growth with memory and its solution

To take into account the fading memory in natural growth model, we use equation (9.19). Substituting expressions (9.3) and (9.4) into equation (9.19), we obtain

$$(D^{\alpha}_{C;0+}Y)(t) - \frac{m(P-a)}{v(\alpha)}Y(t) = -\frac{mb}{v(\alpha)}. \tag{9.20}$$

Equation (9.20) is the fractional differential equation with the Caputo fractional derivative of the order $\alpha > 0$. The existence of solutions of fractional differential equations with the Caputo derivative is described, for example, in [200, 90, 226]. Model of natural growth, which is based on equation (9.20), takes into account the power-law memory with fading parameter $\alpha \geq 0$. Equation (9.20) with $\alpha = 1$ gives equation (9.5), which describes the standard model of natural growth without memory.

Let us obtain the solution of equation (9.20) of the natural growth model that takes into account the effects of fading memory. It is known (see Theorem 5.15 in [200, p. 323]) that the fractional differential equation

$$(D^{\alpha}_{C;0+}Y)(t) - \lambda Y(t) = f(t), \tag{9.21}$$

where $f(t)$ is a real-valued function that is defined on the half-line $t > 0$ with initial conditions $Y^{(k)}(0) = c_k$, $(k = 0, \ldots, n-1)$, is solvable and has unique solution in the form

$$\begin{aligned} Y(t) = &\sum_{k=0}^{n-1} c_k t^k E_{\alpha,k+1}[\lambda t^{\alpha}] \\ &+ \int_0^t (t-\tau)^{\alpha-1} E_{\alpha,\alpha}[\lambda (t-\tau)^{\alpha}] f(\tau)\, d\tau, \end{aligned} \tag{9.22}$$

where $n-1 < \alpha \leq n$, $E_{\alpha,\beta}[z]$ is the two-parameter Mittag–Leffler function [200, p. 42], which is defined by the equation

$$E_{\alpha,\beta}[z] = \sum_{k=0}^{\infty} \frac{z^k}{\Gamma(\alpha k + \beta)}, \tag{9.23}$$

where $z, \beta \in \mathbb{C}$ and $\mathrm{Re}(\alpha) > 0$. The Mittag–Leffler function $E_{\alpha,\beta}[z]$ is a generalization of the exponential function e^z, since $E_{1,1}[z] = e^z$. It can be seen that equation (9.20) can be represented in form (9.21) with

$$\lambda = \frac{m(P-a)}{v(\alpha)}, \quad f(t) = -\frac{mb}{v(\alpha)}. \tag{9.24}$$

As a result, the solution of equation (9.20) has the form

$$\begin{aligned} Y(t) = {} & \sum_{k=0}^{n-1} Y^{(k)}(0) t^k E_{\alpha,k+1}\left[\frac{m(P-a)}{v(\alpha)} t^\alpha\right] \\ & - \frac{b}{(P-a)} \int_0^t (t-\tau)^{\alpha-1} E_{\alpha,\alpha}\left[\frac{m(P-a)}{v(\alpha)} (t-\tau)^\alpha\right] d\tau. \end{aligned} \tag{9.25}$$

The calculation of the integral in equation (9.25) can be realized by using the change of variable $\xi = t - \tau$, definition (9.23) of the Mittag–Leffler function and term by term integration. Alternatively, we can use equation 4.4.4 of [143, p. 61] to calculate the integral in equation (9.25). As a result, solution (9.25) takes the form

$$\begin{aligned} Y(t) = {} & \frac{b}{(P-a)}\left(1 - E_{\alpha,1}\left[\frac{m(P-a)}{v(\alpha)} t^\alpha\right]\right) \\ & + \sum_{k=0}^{n-1} Y^{(k)}(0) t^k E_{\alpha,k+1}\left[\frac{m(P-a)}{v(\alpha)} t^\alpha\right], \end{aligned} \tag{9.26}$$

where $n-1 < \alpha \leq n$, and $Y^{(k)}(0)$ are the values of the derivatives of the function $Y(t)$ at $t = 0$. Solution (9.26) describes the economic dynamics in the framework of natural growth model with power-law fading memory.

9.5 Some features of natural growth with memory

For $0 < \alpha \leq 1$, $(n = 1)$, solution (9.26) has the form

$$Y(t) = \frac{b}{(P-a)}\left(1 - E_{\alpha,1}\left[\frac{m(P-a)}{v(\alpha)} t^\alpha\right]\right) + Y(0) E_{\alpha,1}\left[\frac{m(P-a)}{v(\alpha)} t^\alpha\right]. \tag{9.27}$$

For $\alpha = 1$, expression (9.27) gives solution (9.7) since $E_{\alpha,1}[z] = e^z$, i. e., the solution of equation (9.26) with $\alpha = 1$ is exactly the same as the standard solution of equations of the natural growth model without memory. Equation (9.27) can be rewritten as

$$Y(t) = \frac{b\Gamma(\alpha)}{P-a} + \left(Y(0) - \frac{b\Gamma(\alpha)}{P-a}\right) E_{\alpha,1}\left[\frac{m(P-a)}{v(\alpha)} t^{\alpha}\right]. \tag{9.28}$$

If the condition

$$Y(0) - \frac{b}{(P-a)} > 0 \tag{9.29}$$

holds, then solution (9.28) describes the growth of output. If the inequality

$$Y(0) - \frac{b}{(P-a)} < 0 \tag{9.30}$$

is satisfied, then equation (9.28) describes a decline of output. There conditions can be used for the process with memory for $\alpha \in (0,1)$ and for the absence of memory, when $\alpha = 1$.

To illustrate the growth and decline of output $Y(t)$, which is described by equation (9.28) of the natural growth model with the fading memory at $0 < \alpha \leq 1$, we give Figures 9.1 and 9.2 for $\alpha = 0.8$ (see Plots 2) and $\alpha = 1$ (see Plots 1). Note that the other parameters of the models with and without memory are the same.

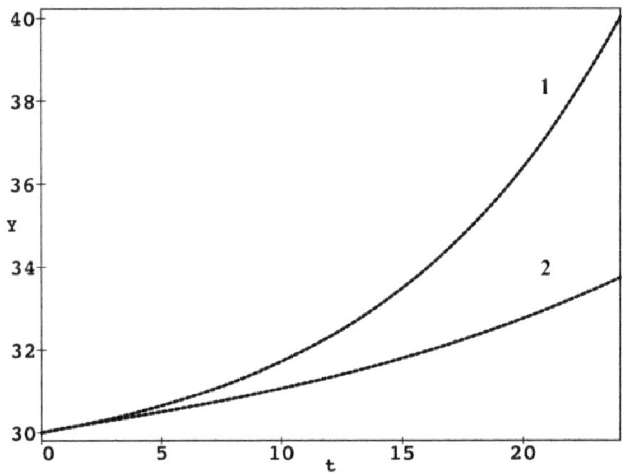

Figure 9.1: The effects of memory with fading parameter $\alpha \in (0,1)$ can lead to a slowing of the growth of economy compared to the standard model without memory: Solution (9.7) of equation (9.5), which describes the natural growth without memory ($\alpha = 1$), is present on Plot 1. Solution (9.28) of equation (9.20), which describes natural growth with memory of fading parameter $\alpha = 0.8$, is present on Plot 2. Plots built for the parameters $Y(0) = 30$, and $b = 290$, $P - a = 1$, $m = 0.2$, $v(\alpha) = v(1) = 2$.

The dynamics of output $Y(t)$, which is described by expression (9.28) of the natural growth model with the memory fading parameter $\alpha = 0.8$, is present on Figure 9.1 for

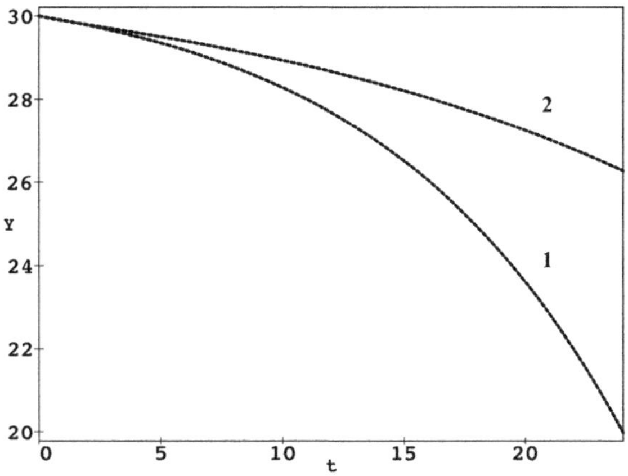

Figure 9.2: The effects of memory with fading parameter $\alpha \in (0, 1)$ can lead to a slowing of the decline of economy compared to the standard model without memory: Solution (9.7) of equation (9.5), which describes the natural growth without memory ($\alpha = 1$), is present on Plot 1. Solution (9.28) of equation (9.20), which describes natural growth model with memory of fading parameter $\alpha = 0.8$, is present on Plot 2, Plots built for the parameters $Y(0) = 30$, and $b = 310$, $P - a = 1$, $m = 0.2$, $v(\alpha) = v(1) = 2$.

$Y(0) = 30$, and $b = 290$, $P - a = 1$, $m = 0.2$, $v(\alpha) = 2$. We can see that the memory with fading parameter $\alpha = 0.8$ slow down the growth of output (see Plot 2 in Figure 9.1), in comparison with the process without memory ($\alpha = 1$) (see Plot 1 in Figure 9.1).

The output $Y(t)$, which is described by (9.28) of the natural growth model with the memory fading parameter $\alpha = 0.8$, is present on Figure 9.2 for $Y(0) = 30$, and $b = 310$, $P - a = 1$, $m = 0.2$, $v(\alpha) = 2$.

For $1 < \alpha \leq 2$, ($n = 2$), solution (9.26) has the form

$$\begin{aligned} Y(t) = & \frac{b\Gamma(\alpha)}{P-a} + \left(Y(0) - \frac{b\Gamma(\alpha)}{P-a}\right) E_{\alpha,1}\left[\frac{m(P-a)}{v(\alpha)} t^{\alpha}\right] \\ & + Y^{(1)}(0) t E_{\alpha,2}\left[\frac{m(P-a)}{v(\alpha)} t^{\alpha}\right], \end{aligned} \tag{9.31}$$

where $Y^{(1)}(0)$ is the value of the first-order derivative of the function $Y(t)$ at $t = 0$.

For the case $1 < \alpha < 2$ and $\lambda > 0$, the condition of growth with memory is represented by the inequality

$$Y(0) - \frac{b}{(P-a)} + \left(\frac{m(P-a)}{v(\alpha)}\right)^{-1/\alpha} Y^{(1)}(0) > 0. \tag{9.32}$$

The condition of decline with memory $1 < \alpha < 2$ is represented by the inequality

$$Y(0) - \frac{b}{(P-a)} + \left(\frac{m(P-a)}{v(\alpha)}\right)^{-1/\alpha} Y^{(1)}(0) < 0. \tag{9.33}$$

Note that conditions (9.32) and (9.33) can be applied to the case $b = 0$, when we have the homogenous fractional differential equation (see equation (9.27) with $b = 0$).

To illustrate the growth and decline of output $Y(t)$, which is described by equation (9.31) with memory fading parameter $\alpha \in (1, 2)$, we give Figures 9.3, 9.4 and 9.5 for $\alpha = 1.1$ (see Plots 2) and $\alpha = 1$ (see Plots 1).

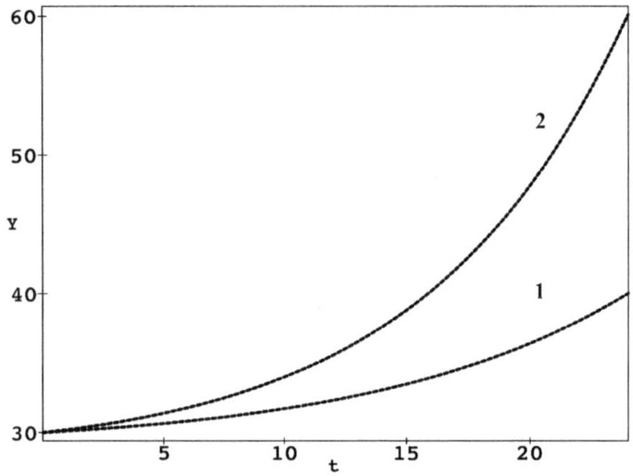

Figure 9.3: The effects of memory with fading parameter $\alpha \in (1, 2)$ can lead to faster growth of economy: Plot 1 presents solution (9.7) of equation (9.5) of the natural growth without memory. Plot 2 presents solution (9.31) of equation (9.20) of the natural growth with memory fading parameter $\alpha = 1.1$. Plots built for the parameters $Y(0) = 30$, $Y^{(1)}(0) = 0.1$ and $b = 290$, $P - a = 1$, $m = 0.2$, $v(\alpha) = v(1) = 2$.

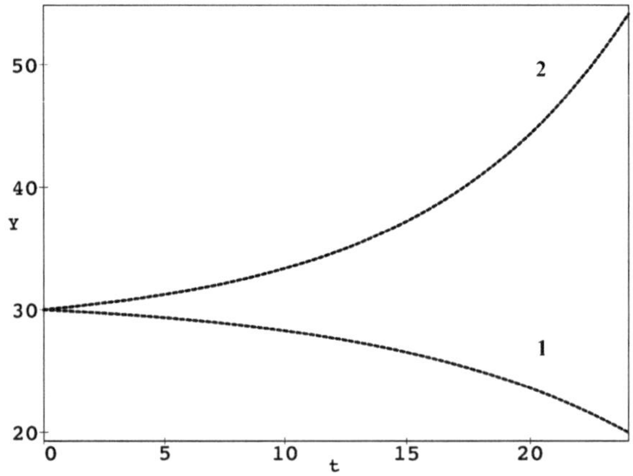

Figure 9.4: The effects of memory with fading parameter $\alpha \in (1, 2)$ can lead to a growth instead of decline compared to the standard model without memory: Plot 1 presents solution (9.7) of equation (9.5) of the natural growth without memory. Plot 2 presents solution (9.31) of equation (9.20) of the natural growth with memory fading parameter $\alpha = 1.1$. Plots built for the parameters $Y(0) = 30$, $Y^{(1)}(0) = 0.3$ and $b = 310$, $P - a = 1$, $m = 0.2$, $v(\alpha) = v(1) = 2$.

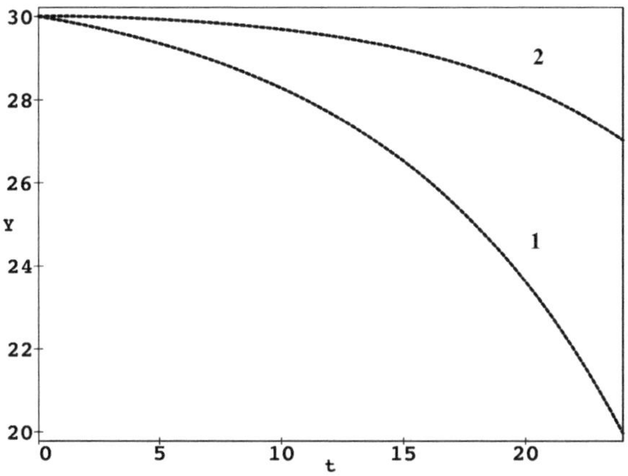

Figure 9.5: The effects of memory with fading parameter $\alpha \in (1, 2)$ can lead to a slowing of the decline of economy compared to the standard model without memory: Plot 1 presents solution (9.7) of equation (9.5) of the natural growth without memory. Plot 2 presents solution (9.31) of equation (9.20) of the natural growth with memory fading parameter $\alpha = 1.1$. Plots built for the parameters $Y(0) = 30$, $Y^{(1)}(0) = 0.1$ and $b = 310$, $P - a = 1$, $m = 0.2$, $v(\alpha) = v(1) = 2$.

The output $Y(t)$, which is described by (9.31) of the model with the memory fading parameter $\alpha = 1.1$, is present on Figure 9.3 for $Y(0) = 30$, $Y^{(1)}(0) = 0.1$ and $b = 290$, $P - a = 1$, $m = 0.2$, $v(\alpha) = 2$. We can see that the memory effect with parameter $\alpha = 1.1$ leads to the acceleration of output growth (see Plot 2 in Figure 9.3) compared to the case of absence of memory (see Plot 1 in Figure 9.3).

The output $Y(t)$, which is described by (9.31) of the natural growth model with the memory fading parameter $\alpha = 1.1$, is present on Figure 9.4 for $Y(0) = 30$, $Y^{(1)}(0) = 0.3$ and $b = 310$, $P - a = 1$, $m = 0.2$, $v(\alpha) = 2$. We can see that the memory effect with parameter $\alpha = 1.1$ can lead to the growth of output (see Plot 2 in Figure 9.4) instead of the decline that is described by the processes without memory (see Plot 1 in Figure 9.4).

The output $Y(t)$, which is described by (9.31) of the natural growth model with the memory fading parameter $\alpha = 1.1$, is present in Figure 9.5 for $Y(0) = 30$, $Y^{(1)}(0) = 0.1$ and $b = 310$, $P - a = 1$, $m = 0.2$, $v(\alpha) = 2$. We can see that the memory effect with parameter $\alpha = 1.1$ can lead to the slow down of decline of the output (see Plot 2 in Figure 9.5) compared to the decline without memory (see Plot 1 in Figure 9.5).

Solutions of the natural growth model with memory are given by equations (9.28) and (9.31) for the cases $\alpha \in (0, 1)$ and $\alpha \in (1, 2)$, respectively. These solutions with $\lambda \in (0, 1)$ are present in Figures 9.1–9.5 for different values of memory fading. It can be seen that the behavior of output is strongly depends on the presence or absence of the memory effect.

Figures illustrate that economic dynamics can strongly depend on the presence or absence of memory. Let describe some properties of this dynamics for the processes with power-law fading memory.

Statement 9.1 (Properties of processes with memory and $\lambda \in (0,1)$).
1. *The effects of memory with fading parameter* $\alpha \in (0,1)$ *can lead to a slowing of the growth of economy compared to the standard model without memory, when the other process parameters are unchanged. See plots of Figure 9.1.*
2. *The effects of memory with fading parameter* $\alpha \in (0,1)$ *can lead to a slowing of the decline of economy compared to the standard model without memory. See plots of Figure 9.2.*
3. *The effects of memory with fading parameter* $\alpha \in (1,2)$ *can lead to faster growth of economy, when the other process parameters are unchanged. See plots of Figure 9.3.*
4. *The effects of memory with fading parameter* $\alpha \in (1,2)$ *can lead to a growth instead of decline compared to the standard model without memory, when the other parameters of the models with and without memory are the same. The decline of economic processes can be replaced by the growth of economy, when the memory effects are taken into account. See plots of Figure 9.4.*
5. *The effects of memory with fading parameter* $\alpha \in (1,2)$ *can lead to a slowing of the decline of economy compared to the standard model without memory. See plots of Figure 9.5.*

Note that the other parameters of the models with and without memory are the same.

We should emphasize that the described changes in the behavior of economic processes are caused only by a change in the memory fading parameter, when the other parameters of the model remain unchanged.

Let us highlight the following positive effects due to the appearance of memory in the economic process. The memory can lead to the deceleration of recession (see Plot 2 of Figures 9.2 and 9.5) compared to the prediction of the standard model without memory (see Plot 1 of Figures 9.2 and 9.5). The memory can lead to the output grows is accelerating (see Plot 2 of Figure 9.3) compared to the prediction of the standard model without memory (see Plot 1 of Figure 9.3), or the output grows (see Plot 2 of Figure 9.4) instead of the output decline that is predicted by standard model without memory (see Plot 1 of Figure 9.4).

9.6 Conclusions

In general, in economic models we should take into account the memory effects that are based on the fact that economic agents can remember the history of changes of variables that characterize the economic process.

In the proposed model of natural growth with memory, we showed the presence of the following effects due to the appearance of memory in the economic process.

The memory with the fading parameter $\alpha \in (0,1)$ can lead to the following two effects: (1) a slowing of the growth of economy compared to the standard model without memory; (2) a slowing of the decline of economy compared to the standard model without memory.

The memory with the fading parameter $\alpha \in (1, 2)$ can lead to the following three effects: (1) a faster growth of economy to the standard model without memory; (2) a growth instead of decline compared to the standard model without memory; (3) a slowing of the decline of economy compared to the standard model without memory.

Note that when comparing models with memory ($\alpha \neq 1$) and without memory $\alpha = 1$, all other parameters were the same.

Let us emphasize the unusual memory effect: the decline of economic process can be replaced by the growth, when the memory effect is taken into account.

The proposed generalization of the model of natural growth, in which we take into account the effects of fading memory, show that the inclusion of memory effects can lead to significant changes in economic phenomena and processes.

10 Model of growth with constant pace and memory

In this chapter, we describe a generalization of model of economic growth with constant pace, which takes into account the effects of memory. This model is proposed in work [438]. We obtain fractional differential equations and their solutions, which describe the dynamics of the output caused by the changes of the net investments and effects of power-law fading memory. In addition to the constant pace model, we also consider the growth model with power-law price and memory, the model of price dynamics with memory, the model of dynamics of fixed assets with memory. This chapter is based on article [438].

10.1 Introduction

The standard growth model with constant pace can be considered as a special case of the standard model of natural growth if the price is assumed to be constant. A generalization of model of economic growth with constant pace, which takes into account the effects of memory, was proposed in work [438]. We obtain fractional differential equations and their solutions, which describe the dynamics of the output caused by the changes of the net investments and effects of power-law fading memory. In this chapter, we consider an example of the memory effects in the simplest economic model. A more detailed analysis, which is based on the asymptotic behavior of the solutions, will be given in the next chapter, where we describe the Harrod–Domar model with memory. We also describe the growth model with power-law price and memory, the growth model with two-parameter memory, the simple model of price dynamics with memory and the simple model of fixed assets dynamics with memory.

10.2 Standard model of growth with constant pace

Let us describe the standard model, which does not take into account the memory and lag. In the standard model of growth with constant pace, the following assumptions are used.

(1) The assumption of unsaturation of the consumer market, which means that all manufactured products are sold. In the simplest case, it is also assumed that the sales volume is not so high as to significantly affect the price P. This allows us to consider a fixed price ($P(t) = P = \text{const}$).
(2) Let $Y(t)$ be a function that describes the volume of production (the output), which was produced and sold at time t. It is known that an increase of the production volume $Y(t)$ is caused by the net investments $I(t)$, which is used to expand the production. The amount of net investment equal to the difference between the total investment and depreciation costs. To increase output, it is necessary that the net

https://doi.org/10.1515/9783110627459-010

investment $I(t)$ is greater than zero ($I(t) > 0$). In the case $I(t) = 0$, the investments only cover the cost of depreciation and the output level remains unchanged. In the case $I(t) < 0$, we have a reduction of fixed assets and, as a consequence, a decrease of output. Therefore, in the model of growth with constant pace is assumed that the speed of change of the production volume ($dY(t)/dt$) is directly proportional to the net investment $I(t)$. Mathematically, it is written by the differential equation

$$\frac{dY(t)}{dt} = \frac{1}{v} I(t), \tag{10.1}$$

where v is a positive constant, that is called the investment coefficient indicating the power of the accelerator [10, p. 62].

(3) Assuming that the amount of investment $I(t)$ is a fixed part of income $PY(t)$, we obtain

$$I(t) = mPY(t), \tag{10.2}$$

where m is the norm of the net investment ($0 < m < 1$), which describes a part of income that is spent on the net investment.

Substituting expression (10.2) into equation (10.1), we obtain

$$\frac{dY(t)}{dt} = \lambda Y(t), \tag{10.3}$$

where

$$\lambda = \frac{mP}{v}. \tag{10.4}$$

Differential equation (10.3) has the solution

$$Y(t) = Y(0) \exp(\lambda t). \tag{10.5}$$

Differential equation (10.3) describes the increase of output without restriction of growth [493, p. 81], [494]. Equation (10.3) is the equation of model of growth with constant pace.

Equations (10.1) and (10.3) contain only the first-order derivative of $Y(t)$ with respect to time. It is known that the derivatives of integer orders are determined by the properties of differentiable functions of time only in infinitely small neighborhood of the considered point of time. As a result, in this economic model, which is described by equation (10.3), the effects of memory and lag are neglected. The memory means a dependence of output at the present time on the investment changes in the past. Therefore, equation (10.3) cannot describe the memory effects.

10.3 Growth model with constant price and memory

In order to take into account a power-law memory, we propose to use the left-sided Caputo fractional derivative of order $\alpha > 0$ with respect to time. One of the important properties of the Caputo fractional derivatives is that the action of these derivatives on a constant function gives zero. Using only the left-sided fractional-order derivative, we take into account the history of changes of variable in the past, $\tau < t$. The right-sided Caputo derivatives are defined by integration over $\tau > t$ and, therefore, cannot take into account the past. The left-sided Caputo fractional derivative is defined by the equation

$$(D^{\alpha}_{C;0+}Y)(t) = \frac{1}{\Gamma(n-\alpha)} \int_0^t \frac{Y^{(n)}(\tau)\,d\tau}{(t-\tau)^{\alpha-n+1}}, \tag{10.6}$$

where $n-1 < \alpha \leq$, $Y^{(n)}(\tau)$ is the derivative of integer order $n \in \mathbb{N}$ of the function $Y(\tau)$ with respect to τ such that $0 < \tau < t < T$. In order for the operator to exist, it is necessary that the function $Y(\tau)$ should have integer-order derivatives up to $(n-1)$-th order, which are continuous functions on the interval $[0, T]$, and the derivative $Y^{(n)}(\tau)$ is Lebesgue summable on the interval $[0, T]$. In order to easily interpret the dimensions of the economic quantities, we can use the time t as a dimensionless variable by changing the variable $t \to tt_d$, where t_d is the unit of time (hour, day, month, year).

Using the equation of the accelerator with power-law memory,

$$(D^{\alpha}_{C;0+}Y)(t) = \frac{1}{v(\alpha)} I(t), \tag{10.7}$$

we get a generalization of equation (10.3) in the form

$$(D^{\alpha}_{C;0+}Y)(t) = \lambda Y(t), \tag{10.8}$$

where

$$\lambda = \frac{mP}{v(\alpha)}. \tag{10.9}$$

Fractional differential equation (10.8) takes into account one-parameter memory with power-law fading.

Let us consider the Cauchy problem for fractional differential equation (10.8) of order $\alpha > 0$ with the initial conditions

$$Y^{(k)}(0) = Y_k, \tag{10.10}$$

where $n-1 < \alpha < n$, λ is a real number, and $k = 1, \ldots, n-1$.

Remark 10.1. For this Cauchy problem, we can give the conditions for a unique solution $Y(t)$ in the space $C_{\gamma}^{\alpha,n-1}[0,T]$, where $0 \le t \le T$, $0 \le \gamma < 1$ and $\gamma \le \alpha$. This function space is defined by

$$C_{\gamma}^{\alpha,n-1}[0,T] = \{Y(t) \in C^n[0,T], (D_{C;0+}^{\alpha}Y)(t) \in C_{\gamma}[0,T]\}, \tag{10.11}$$

where $C_{\gamma}[0,T]$ is the weighted space of functions $Y(t)$ given on $[0,T]$, such that $t^{\gamma}Y(t) \in C[0,T]$. The space $C^n[0,T]$ is the space of functions $Y(t)$, which are continuously differentiable on $[0,T]$ up to order n. The space $C[0,T]$ is the space of functions $Y(t)$, which are continuous on $[0,T]$.

Using Theorem 4.3 of [200, p. 231], the Cauchy problem involving homogeneous fractional differential equation (10.8) and initial conditions (10.10) has a unique solution $Y(t) \in C_{\gamma}^{\alpha,n-1}[0,T]$ in the form

$$Y(t) = \sum_{k=0}^{n-1} Y_k t^k E_{\alpha,k+1}[\lambda t^{\alpha}], \tag{10.12}$$

where $E_{\alpha,\beta}[z]$ is the two-parameter Mittag–Leffler function [200, p. 42], which is defined by the equation

$$E_{\alpha,\beta}[z] = \sum_{k=0}^{\infty} \frac{z^k}{\Gamma(\alpha k + \beta)}. \tag{10.13}$$

The Mittag–Leffler function $E_{\alpha,\beta}[z]$ is a generalization of the exponential function e^z. The exponential function is a special case of this function $E_{1,1}[z] = e^z$. Therefore, for $\alpha = 1$, equation (10.12) gives solution (10.5), which describes the standard model without memory.

For $0 < \alpha < 1$, the solution of equation (10.8) has the form

$$Y(t) = Y(0)\, E_{\alpha,1}[\lambda t^{\alpha}]. \tag{10.14}$$

Solutions (10.12) and (10.14) describe the economic growth with constant pace and power-law fading memory.

10.4 Growth model with power-law price and memory

Let us consider the case [438, p. 42], when the price $P = P(t)$ is changed according to the power law

$$P(t) = pt^{\beta}, \tag{10.15}$$

where $\beta \ge 0$ and $p > 0$. In this case, we have the fractional differential equation

$$(D_{C;0+}^{\alpha}Y)(t) = \lambda t^{\beta} Y(t), \tag{10.16}$$

where the coefficient λ is defined by the equation $\lambda = mp/v(\alpha)$.

Using Theorem 4.4 of [200, p. 233], the Cauchy problem involving fractional differential equation (10.16) and initial conditions (10.10) has a unique solution $Y(t) \in C_{\gamma}^{\alpha,n-1}[0,T]$ in the form

$$Y(t) = \sum_{k=0}^{n-1} \frac{1}{k!} Y_k t^k E_{\alpha,1+\beta/\alpha,(\beta+k)/\alpha}[\lambda t^{\alpha+\beta}], \tag{10.17}$$

where $E_{ab,c}[z]$ is the generalized Mittag–Leffler function (see equation (1.9.19) in [200, p. 48]), which is proposed by Anatoly A. Kilbas and Megumi Saigo [197] (see also book [143]). This function is defined by the equation

$$E_{\alpha,b,c}[z] = \sum_{k=0}^{\infty} a_k(\alpha,b,c) z^k, \tag{10.18}$$

where $a_0(\alpha,b,c) = 1$ and

$$a_k(\alpha,b,c) = \prod_{j=0}^{k-1} \frac{\Gamma(\alpha(bk+c)+1)}{\Gamma(\alpha(bk+c+1)+1)} \tag{10.19}$$

for integer $k \geq 1$, where $\alpha > 0$, $m > 0$.

For $\beta = 0$, we have

$$E_{\alpha,1,k/\alpha}[\lambda t^{\alpha}] = k! E_{\alpha,k+1}[\lambda t^{\alpha}]. \tag{10.20}$$

Therefore, equation (10.17) with $\beta = 0$ gives (10.12).

For $0 < \alpha < 1$, the solution of equation (10.17) has the form

$$Y(t) = Y(0) E_{\alpha,1+\beta/\alpha,\beta/\alpha}[\lambda t^{\alpha+\beta}], \tag{10.21}$$

where we get (10.14) for the case $\beta = 0$.

10.5 Growth model with two-parameter memory

In economic models, we can take into account the presence of different types of memory fading, which characterize the different types of economic agents. For example, we can use the two-parameter description of the fading memory. Using the memory with two-parameter power-law fading, we can consider economic models to describe the memory effects in economy with two type of memory of economic agents.

Let us consider model with two-parameter power-law memory [438, pp. 42–43]. Using the concept of the accelerator with two-parameter power-law memory, we have

$$v_1(\alpha)(D_{C;0+}^{\alpha} Y)(t) + v_2(\beta)(D_{C;0+}^{\beta} Y)(t) = I(t), \tag{10.22}$$

where $\alpha > \beta > 0$, and μ, λ are real numbers. Equation (10.22) takes into account two-parameter memory with power-law fading. This equation can also be interpreted as the simultaneous (parallel) action of two accelerators with power-law memory, whose fading parameters are different. The coefficients $v_1(\alpha)$ and $v_2(\beta)$ are positive constants, that are the investment coefficients indicating the power of these accelerators.

Substituting expression (10.2) into equation (10.22), we obtain the fractional differential equation of the growth model with this type memory in the form

$$(D^{\alpha}_{C;0+}Y)(t) + \mu(D^{\beta}_{C;0+}Y)(t) = \lambda Y(t), \tag{10.23}$$

where μ, λ are real constants

$$\mu = \frac{v_2(\beta)}{v_1(\alpha)}, \quad \lambda = \frac{mP}{v_1(\alpha)}, \tag{10.24}$$

and $\alpha > \beta > 0$, $n-1 < \alpha \leq n$, $m-1 < \beta \leq m$, $m \leq n$, $0 \leq t \leq T$.

Remark 10.2. For $\alpha = \beta$, equation (10.23) gives equation (10.8) with $v(\alpha) = v_1(\alpha)+v_2(\alpha)$. For $\mu = 0$, equation (10.23) gives equation (10.8) with $v(\alpha) = v_1(\alpha)$. Therefore, the proposed model contains the growth model with constant price and one-parameter memory with power-law fading as a special case.

The solution of (10.23) is represented in terms of the generalized Wright function (the Fox–Wright function), $\Psi_{1,1}\left[{}^{(a,\alpha)}_{(b,\beta)} \mid z\right]$, which is defined [200, p. 56] by the equation

$$\Psi_{1,1}\left[{}^{(a,\alpha)}_{(b,\beta)} \mid z\right] = \sum_{k=0}^{\infty} \frac{\Gamma(\alpha k + a)}{\Gamma(\beta k + b)} \frac{z^k}{k!}. \tag{10.25}$$

Using Theorem 5.13 of [200, p. 314], where $\lambda \to -\mu$ and $\mu \to \lambda$, the solution of equation (10.23) has the form

$$Y(t) = \sum_{j=0}^{n-1} a_j Y_j(t), \tag{10.26}$$

where $Y_j(t)$, $j = 0, \ldots, n-1$ are defined by the following equations:

$$\begin{aligned} Y_j(t) = {} & \sum_{k=0}^{\infty} \frac{\lambda^k t^{k\alpha+j}}{\Gamma(k+1)} \Psi_{1,1}\left[{}^{(k+1,1)}_{(\alpha k+j+1,\alpha-\beta)} \mid -\mu t^{\alpha-\beta}\right] \\ & + \mu \sum_{k=0}^{\infty} \frac{\lambda^k t^{k\alpha+j+\alpha-\beta}}{\Gamma(k+1)} \Psi_{1,1}\left[{}^{(k+1,1)}_{(\alpha k+j+1+\alpha-\beta,\alpha-\beta)} \mid -\mu t^{\alpha-\beta}\right] \end{aligned} \tag{10.27}$$

for $j = 0, \dots, m-1$, and

$$Y_j(t) = \sum_{k=0}^{\infty} \frac{\lambda^k t^{k\alpha+j}}{\Gamma(k+1)} \Psi_{1,1}\left[{}^{(k+1,1)}_{(\alpha k+j+1,\alpha-\beta)} \mid -\mu t^{\alpha-\beta}\right] \tag{10.28}$$

for $j = m, \dots, n-1$.

For $0 < \beta < \alpha \leq 1$, the solution of equation (10.23) is written in the form

$$Y(t) = \sum_{k=0}^{\infty} \frac{\lambda^k t^{k\alpha}}{\Gamma(k+1)} \Psi_{1,1}\left[{}^{(k+1,1)}_{(\alpha k+1,\alpha-\beta)} \mid -\mu t^{\alpha-\beta}\right] + \mu \sum_{k=0}^{\infty} \frac{\lambda^k t^{k\alpha+\alpha-\beta}}{\Gamma(k+1)} \Psi_{1,1}\left[{}^{(k+1,1)}_{(\alpha k+1+\alpha-\beta,\alpha-\beta)} \mid -\mu t^{\alpha-\beta}\right]. \tag{10.29}$$

For $1 < \beta < \alpha \leq 2$, the solution of equation (10.23) has the form

$$Y(t) = a_0 Y_0(t) + a_1 Y_1(t), \tag{10.30}$$

where $Y_0(t)$ is defined by (10.29), and $Y_1(t)$ is defined by the equation

$$Y_1(t) = \sum_{k=0}^{\infty} \frac{\lambda^k t^{k\alpha+1}}{\Gamma(k+1)} \Psi_{1,1}\left[{}^{(k+1,1)}_{(\alpha k+2,\alpha-\beta)} \mid -\mu t^{\alpha-\beta}\right] + \mu \sum_{k=0}^{\infty} \frac{\lambda^k t^{k\alpha+1+\alpha-\beta}}{\Gamma(k+1)} \Psi_{1,1}\left[{}^{(k+1,1)}_{(\alpha k+2+\alpha-\beta,\alpha-\beta)} \mid -\mu t^{\alpha-\beta}\right]. \tag{10.31}$$

For $0 < \beta < 1 < \alpha \leq 2$, the solution of equation (10.23) is represented by equation (10.30) with $Y_0(t)$ in the form (10.29), and $Y_1(t)$ that is defined by the equation

$$Y_1(t) = \sum_{k=0}^{\infty} \frac{\lambda^k t^{k\alpha+1}}{\Gamma(k+1)} \Psi_{1,1}\left[{}^{(k+1,1)}_{(\alpha k+2,\alpha-\beta)} \mid -\mu t^{\alpha-\beta}\right]. \tag{10.32}$$

Note that these solutions can be represented through the general Mittag–Leffler (see equation (1.9.2) in [200, p. 45]) by using the equation

$$\Psi_{1,1}\left[{}^{(\gamma,1)}_{(\beta,\alpha)} \mid z\right] = \Gamma(\gamma) E^{\gamma}_{\alpha,\beta}[z]. \tag{10.33}$$

For the case of the multi-parameter power-law memory, we can use Theorem 5.14 of [200, pp. 319–320]. Two-parameter and multiparameter memory allows us to take into account the power-law fading of memory for different types of economic agents.

10.6 Simple model of price dynamics with memory

Let us consider the dynamics of price growth at a constant pace of inflation [438, p. 43]. We will assume that the price at time t is equal to $P(t)$. The inflation pace is assumed to be equal to the constant R. Then the price growth with power-law memory at constant pace of inflation can be described by the fractional differential equation

$$(D^{\alpha}_{C;0+}P)(t) = RP(t), \tag{10.34}$$

where $D^{\alpha}_{C;0+}$ is the Caputo derivative (10.6). For $\alpha = 1$, equation (10.34) takes the standard form

$$\frac{dP(t)}{dt} = RP(t). \tag{10.35}$$

Fractional differential equation (10.34) has the solution

$$P(t) = \sum_{k=0}^{n-1} P_k t^k E_{\alpha,k+1}[Rt^{\alpha}], \tag{10.36}$$

where $E_{\alpha,\beta}[z]$ is the two-parameter Mittag–Leffler function (10.13). Solution (10.36) describes the dynamics of price growth with power-law fading memory. For $\alpha = 1$, expression (10.36) takes the form

$$P(t) = P(0)\exp(Rt), \tag{10.37}$$

which is the solution of equation (10.35), which describes the price growth at a constant pace [493, p. 81] without memory effects.

10.7 Simple model of fixed assets dynamics with memory

In this section, we consider the dynamics of fixed assets, where we take into account the memory effects [438, pp. 43–44].

Let B be a coefficient of disposal of fixed assets. We assume that the investment is constant, which is equal to A monetary units. We can describe the dynamics of fixed assets, if the rate of change of the fixed assets is equal to the difference between investments and disposal of fixed assets. Let us denote the fixed assets at time $t \geq 0$ by $K(t)$. The dynamics of the fixed assets with power-law memory can be described by the fractional differential equation

$$(D^{\alpha}_{C;0+}K)(t) = A - BK(t), \tag{10.38}$$

where $D^{\alpha}_{C;0+}$ is the Caputo fractional derivative (10.6). For $\alpha = 1$, equation (10.38) takes the standard form

$$\frac{dK(t)}{dt} = A - BK(t). \tag{10.39}$$

Equation (10.39) describes the dynamics of fixed assets without memory [493, p. 82].

The solution of equation (10.38) has [200, p. 323] the form

$$K(t) = \sum_{k=0}^{n-1} K^{(k)}(0) t^k E_{\alpha,k+1}[-Bt^\alpha] + A \int_0^t (t-\tau)^{\alpha-1} E_{\alpha,\alpha}[-B(t-\tau)^\alpha]\, d\tau, \tag{10.40}$$

where $n-1 < \alpha \leq n$, $E_{\alpha,\beta}[z]$ is the two-parameter Mittag–Leffler function (10.13). Using identity 4.4.4 of [143, p. 61] to calculate the integral in solution (10.40), we get equation (10.40) in the form

$$K(t) = \frac{A}{B}(1 - E_{\alpha,1}[-Bt^\alpha]) + \sum_{k=0}^{n-1} K^{(k)}(0) t^k E_{\alpha,k+1}[-Bt^\alpha], \tag{10.41}$$

where $n-1 < \alpha \leq n$, and $K^{(k)}(0)$ are the values of the derivatives of the function $K(t)$ at $t = 0$. Solution (10.41) describes the dynamics of fixed assets with power-law fading memory.

For $0 < \alpha \leq 1$, $(n = 1)$, solution (10.41) has the form

$$K(t) = \frac{A}{B}(1 - E_{\alpha,1}[-Bt^\alpha]) + K(0) E_{\alpha,1}[-Bt^\alpha]. \tag{10.42}$$

Using $E_{1,1}[z] = e^z$, solution (10.42) with $\alpha = 1$ takes the form

$$K(t) = \frac{A}{B}(1 - \exp(-Bt)) + K(0)\exp(-Bt), \tag{10.43}$$

which describes the solution of equation (10.39) for the standard models of fixed assets without memory effects.

10.8 Conclusion

This chapter proposes economic models with power-law memory with one and two fading parameters. Note that the proposed models include standard models without memory as a special case, when the memory fading parameters take integer value equal to one. In economic models, we should take into account the memory effects that are based on the fact that economic agents remember the history of changes of variables that characterize the economic process. The proposed economic growth models with power-law memory show that the memory effects can play an important role in economic phenomena and processes.

11 Harrod–Domar growth model with memory

This chapter discusses the impact effects of power-law fading memory on economic growth in the framework of a generalization of the standard Harrod–Domar growth model. The Harrod–Domar growth model with power-law memory was first proposed in articles [421, 426] in 2016. For this model, the properties of the growth with memory is described in [450, 445, 453, 393]. In this chapter, we obtain solutions of the fractional differential equations of the proposed Harrod–Domar growth model with memory. Examples of dependence of macroeconomic dynamics on the memory effects are suggested. For the obtained solutions, the asymptotic behavior, which characterize the warranted rate of growth with memory, are described. Principles of economic dynamics with power-law memory are formulated. It has been shown that the effects of fading memory can change the economic growth rate and change dominant parameters, which determine growth rates. This chapter is based on results that are derived in works [421, 426, 450, 445, 453].

11.1 Introduction

Macroeconomic models are a powerful tool for theoretical studies of macroeconomic processes in global and national economies (e. g., see [10, 325, 327, 494, 115]). Among the main advantages of macroeconomic models, we note their availability for a detailed mathematical analysis, a possibility of studying macroeconomic processes with a small number of input data and small dimension, a possibility of rapid realization of multivariate calculations. These models also have important applications for developing the concept of economic growth, taking into account possible alternatives to economic policy and their long-term consequences.

In modern macroeconomics, a significant role is played by models that are described by differential equations with derivatives of integer orders. In the construction of macroeconomic models, different simplifying assumptions are assumed. Therefore, these models can have serious disadvantages, since they do not take into account some important aspects of economic phenomena and processes. Some of these disadvantages are partly caused by the restrictions of the mathematical tools. It is known that the derivatives of integer orders are determined by the properties of the differentiable function only in an infinitesimal neighborhood of the considered point. As a result, the differential equations with derivatives of integer orders with respect to time cannot describe processes with memory. In fact, these equations describe only such economic processes, in which agents actually have a total amnesia. In other words, economic models that use derivatives of integer orders can be applied only when economic agents forget the history of changes in economic indicators and factors during an infinitely small period of time.

https://doi.org/10.1515/9783110627459-011

Obviously, the assumption of lack of the memory in economic agents is a strong restriction for economic models.

The Harrod–Domar model with continuous time [170, 171, 172, 92, 93, 480] is one of the simplest classical models of economic growth ([10, pp. 64–69], [12, pp. 197–203], [146, pp. 46–59]). The standard Harrod–Domar model does not take into account memory effects. The generalization of this model, which was proposed in [421, 426] (see also [450, 445, 453, 393]), allows us to demonstrate the impact of memory effects on macroeconomic dynamics. The Harrod–Domar model can be considered as one-sectoral form of the dynamic Leontief model [219, 220] that is also called the input–output model. Note that a generalization of the dynamic Leontief model, in which memory is taken into account, was first proposed in works [437, 451, 433, 398].

In this chapter, we describe a generalization of the Harrod–Domar growth model, which takes into account effects of fading memory. The Harrod–Domar model with one-parameter power-law memory was first proposed in articles [421, 426] in 2016. The properties of the growth with memory is described in [450, 445, 453, 393]. For simplification, we consider the memory with power-law fading. This allows us to use the fractional derivatives of noninteger orders and fractional calculus. Equations of the proposed economic model are represented as fractional differential equations with derivatives of noninteger orders. We proposed solutions for the fractional differential equations of the Harrod–Domar growth model with memory are suggested. Examples of dependence of macroeconomic dynamics on memory are given. Asymptotic behaviors of the solutions, which characterize the warranted rate of growth with memory, are described. Using these solutions and its asymptotic behavior, we proposed the principles of economic dynamics with power-law memory [450, 445, 453, 393]. We prove that the effects of memory can change the economic growth rate, warranted rate and dominant parameters, which determine growth rates. The memory effect can lead to qualitative changes, including the emergence of economic growth instead of decrease and recession. This chapter is based on articles [421, 426] and [450, 445, 453, 393].

11.2 Harrod–Domar growth model without memory

Let us describe assumptions, which are used in the standard Harrod–Domar model without memory.

(1) The first assumption of the model is that exports and imports are equal to one another, and losses will be included in the used national income (input). Under these assumptions, the values of the produced national income (output) and the used national income (input) will be equal. As a result, the balance of production and distribution of gross product at any given time moment $t \geq 0$ is described by the linear equation

$$X(t) = AX(t) + Y(t), \tag{11.1}$$

where the function $X(t)$ describes the gross product, $Y(t)$ is the national income (output) and the coefficient A is the material consumption of the gross product ($0 < A < 1$). The coefficient A includes not only the current production costs, but also the replacement investment and the costs of capital repairs of basic production assets. Equation (11.1) can be written as the equation of linear multiplier

$$X(t) = mY(t), \tag{11.2}$$

where the coefficient $m = 1/(1 - A)$ is the multiplier of the gross product, which characterizes the ratio of gross product and national income.

(2) The second assumption of the macroeconomic model is that the used national income $Y(t)$ is divided into two parts: the investment $I(t)$ (accumulation of basic production assets); the nonproductive consumption $C(t)$ (nonproductive accumulation, growth of material current assets, state material reserves, losses). To be short, the functions $I(t)$ will be called "accumulation," and $C(t)$ will be called "consumption." As a result, we have the balance equation

$$Y(t) = I(t) + C(t), \tag{11.3}$$

where the independence of consumption behavior is assumed.

(3) The third assumption of the standard model without memory is the direct proportionality of production accumulation and the growth rate of the national income at the same time. This assumption gives a relationship between investment (accumulation) and the growth rate of gross product. This relationship is realized by the accelerator equation with coefficients of capital intensity. In the standard models, it is also assumed an instantaneous transformation of investment (capital investment) in the growth of gross product, that is, these models assume neglect of the delay and memory. The direct proportionality of the capital investment $I(t)$ and the growth rate of gross product $X(t)$ is described by the linear accelerator equation

$$I(t) = b\frac{dX(t)}{dt}, \tag{11.4}$$

where b is the capital intensity of gross product.

Using equation (11.2), we can write equality (11.4) in the form of direct proportionality between investment (productive accumulation) and the growth rate of the national income

$$I(t) = B\frac{dY(t)}{dt}, \tag{11.5}$$

where

$$B = bm = b(1 - A)^{-1} \tag{11.6}$$

is the accelerator coefficient that describes the capital intensity of national income growth [494]. It is also called the investment coefficient indicating the power of the accelerator [10, p. 65], which describes the capital intensity of the national income (incremental capital intensity, differential capital intensity).

Remark 11.1. In general, we can take into account memory effects for multiplier (11.2) by using the concept of multiplier with memory that is proposed in [412, 397, 441, 442], instead of the standard multiplier (11.2). For simplicity, we will not use this generalization in this chapter.

Remark 11.2. Basic assumption, which is usually used in the standard macroeconomic models, is the absence of memory, i. e., the full amnesia of all economic agents. In fact, equation (11.4) means that the capital investment $I(t)$ at time t is determined by the change of the gross product $X(t)$ in an infinitesimal neighborhood of the instant of time (i. e., it is determined by an infinitesimally close past). This assumes neglecting the memory effects. In other words, the instant amnesia for all economic agents is the main condition for the applicability of equations (11.4) and (11.5). This means that these equations assume the instant forgetting of the history of changes in the gross product and the made investments.

Let us give equation of the standard Harrod–Domar growth model for national income $Y(t)$. Substituting expression (11.5) into equation (11.3), we obtain

$$Y(t) = B\frac{dY(t)}{dt} + C(t). \tag{11.7}$$

Equation (11.7) describes the dynamics of the national income within framework of the standard Harrod–Domar growth model. Equation (11.7) shows that for a given parameter B, the dynamics of the national income $Y(t)$ is determined by the behavior of the function $C(t)$. Equation (11.2) allows us to describe the change of the gross product $X(t)$ by using the dynamics of national income $Y(t)$, which is described by solutions of equation (11.7). Mathematically, the standard model of the national income dynamics is described by linear nonhomogeneous differential equations of the first order.

The standard model assumes an instantaneous change in $X(t)$ when $Y(t)$ changes, and an instantaneous change in $I(t)$ when $Y^{(1)}(t)$ changes. This means that the standard model assumes the absence of memory. We can state that one of the strong restriction imposed on the standard model, which is described by equation (11.7), is the absence of memory. This assumption is a significant drawback of standard macroeconomic models. This drawback is due to the shortcomings of the mathematical apparatus used in standard macroeconomic models. Derivatives of the first order, which are used in the linear accelerator equations (11.4) and (11.5), imply an instantaneous change of the variable $I(t)$, when changing the growth rate of the gross product $X(t)$ and national income $Y(t)$. In other words, the investments at time t are determined by the properties of the national income in an infinitesimal neighborhood of this time point. Because of this, accelerator equation (11.5) does not account for the effects of memory about past changes in national income. As a result, differential equation (11.7) can be used only to describe an economy, in which all economic agents have an instant amnesia. This restriction substantially narrows the field of application of standard macroeconomic models to describe the real economic processes. In many cases, economic agents can remember the history of changes of the gross product, national income and investment. As a result, the history of changes can influence on decision-making by economic agents with memory.

11.3 Harrod–Domar growth model with power-law memory

To take into account the effects of power-law memory in macroeconomic models, we can apply the fractional calculus [335, 202, 308, 200, 90, 164, 165, 119]. The use of derivatives and integrals of noninteger orders allows us to generalize the standard equation (11.4), which describes the relationship between the investment $I(t)$ and the gross product $X(t)$. There are different types of derivatives of noninteger orders [335, 308, 200]. We will use the Caputo derivative [308, 200, 90] with respect to time, since the action of this derivative on a constant function gives zero. To take into account the history of changes of dynamic variables in the past, we use the left-sided fractional derivative. Note that the concept of economic accelerator with memory has been suggested in [412, 397, 441, 442]. The Harrod–Domar model with power-law memory was first proposed in works [421, 426, 450] in 2016.

Let us consider the Harrod–Domar model with memory. We will assume that the capital investment $I(t)$ at time t depend not only on the gross product $X(\tau)$ at the present time $\tau = t$, but also on the history of changes in the gross product $X(\tau)$ in the past at times $\tau \in (0, t)$. In this case, the equation of investment accelerator with memory can be written in the form

$$I(t) = b \int_0^t M(t-\tau) X^{(n)}(\tau)\, d\tau, \tag{11.8}$$

where $M(t-\tau)$ is the memory function. For $M(t-\tau) = \delta(t-\tau)$, equation (11.8) gives equation of the standard accelerator without memory and lag. If the function $M(\tau)$ describes memory with power-law fading in the form

$$M(t-\tau) = \frac{1}{\Gamma(n-\alpha)} (t-\tau)^{n-\alpha-1}, \tag{11.9}$$

then equation (11.9) of the accelerator with memory (11.8) can be represented in the form

$$I(t) = b(D^{\alpha}_{C;0+} X)(t), \tag{11.10}$$

where $(D^{\alpha}_{C;0+}X)(t)$ is the Caputo fractional derivative of the order $\alpha \geq 0$ that is defined by the equation

$$(D^{\alpha}_{C;0+}X)(t) = \frac{1}{\Gamma(n-\alpha)} \int_0^t (t-\tau)^{n-\alpha-1} X^{(n)}(\tau)\, d\tau, \tag{11.11}$$

where $n = [\alpha] + 1$ for noninteger values of α and $n = \alpha$ for integer values of α, $\Gamma(\alpha)$ is the gamma function and $t > 0$. The Caputo fractional derivative (21.10) of order $\alpha > 0$ exists almost everywhere on $[0, T]$, if $X(\tau) \in AC[0, T]$ (e. g., see Theorem 2.1 in [200, p. 92]), i. e., the function $X(\tau)$ has derivatives of integer orders up to $(n-1)$-th order, which are continuous functions on the interval $[0, T]$, and the derivative $X^{(n)}(\tau)$ is Lebesgue summable on the interval $[0, T]$.

Equation (11.10) assumes that the function $X(\tau)$ has integer-order derivatives up to $(n-1)$-th order, which are absolutely continuous functions on the interval $[0, T]$. Equation (11.2) allows us to write equation (11.10) for the national income in the form

$$I(t) = B(D^{\alpha}_{C;0+}Y)(t), \tag{11.12}$$

where $B = mb$ is the investment coefficient (capital intensity of national income growth) that takes into account the memory effects. In general, the investment coefficient depends on the parameter of memory fading, i. e., $B = B(\alpha)$. In order to have easily interpretable dimension of variables, we can use time t as a dimensionless variable.

For $\alpha = 1$, equation (11.12) gives equation (11.5), since the Caputo fractional derivative of the order $\alpha = 1$ coincides with the first derivative $(D^{1}_{C;0+}Y)(t) = dY(t)/dt$, [200, p. 92].

Substituting the expression for the investment $I(t)$, which is given by equation (11.12), into balance equation (11.3), we obtain the fractional differential equation

$$Y(t) = B(D^{\alpha}_{C;0+}Y)(t) + C(t). \tag{11.13}$$

For $\alpha = 1$, equation (11.13) gives equation (11.7).

Equation (11.13) describes the economic dynamics with one-parameter power-law memory. If the parameter B is given, then the dynamics of the national income $Y(t)$ is determined by the behavior of the function $C(t)$. As a result, this macroeconomic dynamics with memory is described by fractional differential equation of the non-integer order $\alpha > 0$.

11.4 General solution of model equation

To solve equation (11.13), we can use Theorem 5.15 of book [200, p. 323]. The fractional differential equation (11.13) can be written in the form

$$(D^{\alpha}_{C;0+}Y)(t) - B^{-1}Y(t) = -B^{-1}C(t), \tag{11.14}$$

where $n-1 < \alpha \leq n$. If $C(t)$ is a continuous function defined on the positive semi-axis ($t > 0$), then equation (11.14) has the solution

$$Y(t) = Y_C(t) + \sum_{k=0}^{n-1} Y^{(k)}(0) t^k E_{\alpha,k+1}[B^{-1} t^{\alpha}], \tag{11.15}$$

where $Y^{(k)}(0)$ are derivatives of the integer orders $k > 0$ of the function $Y(t)$ at $t = 0$, and

$$Y_C(t) = -B^{-1} \int_0^t (t-\tau)^{\alpha-1} E_{\alpha,\alpha}[B^{-1}(t-\tau)^{\alpha}] C(\tau)\, d\tau. \tag{11.16}$$

Here, $E_{\alpha,\beta}[z]$ is the two-parameter Mittag–Leffler function [200, 143] that is defined by the expression

$$E_{\alpha,\beta}[z] = \sum_{k=0}^{\infty} \frac{z^k}{\Gamma(\alpha k + \beta)}, \tag{11.17}$$

where $\Gamma(\alpha)$ is the gamma function, $\alpha > 0$, and β is an arbitrary real or complex number.

For $0 < \alpha \leq 1$, ($n = 1$), solution (11.15) of equation (11.14) has the form

$$\begin{aligned} Y(t) = &-B^{-1} \int_0^t (t-\tau)^{\alpha-1} E_{\alpha,\alpha}[B^{-1}(t-\tau)^{\alpha}] C(\tau)\, d\tau \\ &+ Y(0) E_{\alpha,1}[B^{-1} t^{\alpha}]. \end{aligned} \tag{11.18}$$

In the proposed growth model with memory, which is described by equation (11.14), the nonproductive consumption $C(t)$ is considered as a function, which is independent of national income and investment. We can consider that the function $C(t)$ is a constant in time, a power-law function, or it changes in a more complicated law.

If nonproductive consumption is represented as a fixed part of income $C(t) = cY(t)$, where c is the marginal propensity to consume, then equation (11.14) can be written in the form

$$(D^{\alpha}_{C;0+} Y)(t) = B^{-1}(1-c) Y(t). \tag{11.19}$$

In this case, the proposed model with memory coincides with the natural growth model with memory [420, 440].

Let us analyze the behavior of the solution for various functions $C(t)$ that describe the consumption dynamics. Models for these cases can be called closed and open economic models [146, p. 122–144]. These terms have been suggested for models without memory by Wassily W. Leontief, who won the Nobel Laureate in Economics in 1973. The closed model according to Leontief assumes the absence of external demand, when all inputs and outputs are spent on the production. In our case, the macroeconomic model will be called closed, if it assumes the absence of nonproductive consumption, that is, $C(t) = 0$. The consideration of this case allows us to estimate of the greatest possible growth rate in national income, which is restricted only by the value of the incremental capital intensity, i. e., by the material intensity and capital intensity of production.

11.5 Closed model with memory: rate of growth with memory

A macroeconomic model with memory is called closed if it assumes the absence of nonproductive consumption, $C(t) = 0$, [146, p. 122–144]. The assumption $C(t) = 0$ is unrealistic, but the consideration of this case allows us to estimate the greatest possible growth rate of national income, which is restricted only by the incremental capital intensity. For the case $C(t) = 0$, all resources are directed to investments, as a result of which the maximum possible rates of growth can be determined from the solution of the equation of economic model.

The equation of the closed model with memory has the form

$$(D^{\alpha}_{C;0+}Y)(t) - \lambda Y(t) = 0, \tag{11.20}$$

where $\lambda = B^{-1}$, and $n - 1 < \alpha \leq n$. The solution of equation (11.20) has the form

$$Y(t) = \sum_{k=0}^{n-1} Y^{(k)}(0) t^k E_{\alpha,k+1}[B^{-1}t^{\alpha}], \tag{11.21}$$

where $E_{\alpha,k+1}[z]$ is defined by equation (11.17).

For $0 < \alpha < 1$, solution (11.21) takes the form

$$Y(t) = Y(0)E_{\alpha,1}[B^{-1}t^{\alpha}]. \tag{11.22}$$

For $\alpha = 1$, solution (11.21) gives the equation

$$Y(t) = Y(0)\exp\{-t/B\}, \tag{11.23}$$

where the value $\lambda = B^{-1}$ is often called the "technological growth rate" [146, p. 49], and the Harrod's "warranted" rate of growth [10, p. 67], which is fixed by the structural constants B and m.

Let us now analyze the behavior of solution (11.21) at $t \to \infty$ to obtain the asymptotic behavior that determines the warranted rate of growth with memory. To describe the asymptotic behavior of the two-parameter Mittag–Leffler function $E_{\alpha,k+1}[\lambda t^{\alpha}]$ at $t \to \infty$, we can use equation (1.8.27) of the book [200, p. 43] for $0 < \alpha < 2$ in the form

$$\begin{aligned} E_{\alpha,\beta+1}[\lambda t^{\alpha}] = {} & \frac{\lambda^{-\beta/\alpha}}{\alpha} t^{-\beta} \exp(\lambda^{1/\alpha} t) \\ & - \sum_{j=1}^{m} \frac{\lambda^{-j}}{\Gamma(\beta + 1 - \alpha j)} \frac{1}{t^{\alpha j}} + O\left(\frac{1}{t^{\alpha(m+1)}}\right) \end{aligned} \tag{11.24}$$

for $t \to \infty$, where λ is a real number and we use the big O notation (asymptotic notation) that provides an upper bound on the growth rate of the following terms of the asymptotic series. Asymptotic equation (11.24) allows us to describe the behavior of national income $Y(t)$ at $t \to \infty$ in a macroeconomic model with power-law memory fading parameter $0 < \alpha < 2$. Substitution of expression (11.24) with $\lambda = B^{-1}$ and $\beta = k$ into (11.21) gives

$$\begin{aligned} Y(t) &= \sum_{k=0}^{n-1} Y^{(k)}(0) \frac{B^{k/\alpha}}{\alpha} \exp(B^{-1/\alpha} t) \\ &+ \sum_{k=0}^{n-1} \left(\sum_{j=1}^{m} \frac{Y^{(k)}(0) B^j}{\Gamma(k+1-\alpha j)} t^{k-\alpha j} + O\left(\frac{1}{t^{\alpha(m+1)-k}} \right) \right), \end{aligned} \tag{11.25}$$

where $0 < n-1 < \alpha < n$. Expression (11.25) describes the behavior of solution (11.21) at $t \to \infty$.

As a result, the warranted rate of growth for processes with power-law memory is equal to the value

$$\lambda_{\text{eff}}(\alpha) = \lambda^{1/\alpha} = B^{-1/\alpha}. \tag{11.26}$$

Rate (11.26) will be called the warranted rate of growth with memory (or the effective warranted rate of growth) for macroeconomic model with one-parameter power-law memory. This parameter characterizes the long-term memory effect of the economic process with power-law memory.

We see that the warranted rate of growth for economic model with one-parameter memory does not coincide with the growth rate $\lambda = B^{-1}$ of standard model without memory. Note that the warranted rate of growth with memory for parameter $\alpha = 1$ is equal to the warranted rate of growth without memory $\lambda_{\text{eff}}(1) = \lambda$.

Using equation (1.8.29) of the book [200, p. 43] (see also [100]) for $\alpha \geq 2$ and $t \to \infty$, we get the asymptotic expression

$$\begin{aligned} E_{\alpha,\beta+1}[\lambda t^{\alpha}] &= \frac{\lambda^{-\beta/\alpha}}{\alpha} t^{-\beta} \sum_{N} \exp(\lambda^{1/\alpha} t e^{(i\frac{2\pi N}{\alpha})}) e^{(-i\frac{2\pi \beta N}{\alpha})} \\ &- \sum_{k=1}^{m} \frac{\lambda^{-k}}{\Gamma(\beta+1-\alpha k)} \frac{1}{t^{\alpha k}} + O\left(\frac{1}{t^{\alpha(m+1)}} \right), \end{aligned} \tag{11.27}$$

where $\lambda = B^{-1}$ is a real number ($\arg(\lambda) = 0$) and the first sum is taken over all integer values of N that satisfy the condition $|N| \leq \alpha/4$. Using equation (11.27) and the inequality $|N| \leq \alpha/4$, we see that equation (11.24) can be used for $\alpha < 4$. As a result, the warranted rate of growth with memory for $\alpha < 4$ is described by equation (11.26). For the case $\alpha \geq 4$, the use of the Euler equation and equation (11.27) leads us to the warranted rate of growth with memory in the form

$$\lambda_{\text{eff}}(\alpha; N) = B^{-1/\alpha} \cos\left(\frac{2\pi N}{\alpha} \right), \tag{11.28}$$

where N runs through all integer values that satisfy the condition $|N| \leq \alpha/4$. Note that for $\alpha \geq 4$ there are several effective growth rates. However, the inequality $\cos(x) \leq 1$ leads us to the largest value

$$\max_N \{\lambda_{\text{eff}}(\alpha; N)\} = \lambda_{\text{eff}}(\alpha) = B^{-1/\alpha}, \tag{11.29}$$

which is the highest warranted rate of growth with memory for the fading parameter $\alpha \geq 4$.

Let us compare the warranted rate $\lambda_{\text{eff}}(\alpha)$ of growth for the proposed macroeconomic model with memory and the warranted rates $\lambda = \lambda_{\text{eff}}(1)$ of growth for the standard model without memory in Table 11.1.

Table 11.1: Comparison of warranted rates of growth for economic models with memory and without memory, where $\lambda_{\text{eff}}(\alpha)$ is the warranted rate of growth with memory, and λ is the warranted rate of growth without memory.

λ	B	$0 < \alpha < 1$	$\alpha > 1$
$0 < \lambda < 1$	$B > 1$	$\lambda_{\text{eff}}(\alpha) < \lambda$	$\lambda_{\text{eff}}(\alpha) > \lambda$
$\lambda = 1$	$B = 1$	$\lambda_{\text{eff}}(\alpha) = \lambda$	$\lambda_{\text{eff}}(\alpha) = \lambda$
$\lambda > 1$	$0 < B < 1$	$\lambda_{\text{eff}}(\alpha) > \lambda$	$\lambda_{\text{eff}}(\alpha) < \lambda$

Using Table 11.1, we can see that the account of memory effects can significantly change the warranted rates of growth in the macroeconomic models in comparison with the standard models, which do not take into account the memory effects. We also see that the warranted rates of growth may both increase and decrease. If the incremental capital intensity of the national income is less than one ($0 < B < 1$), then the warranted rate of growth of the economy can be much larger, when we consider the memory effects. For example, for $\alpha = 0.1$ and $B = 0.5$, the warranted rate of growth with memory ($\alpha = 0.1$) is equal to $\lambda_{\text{eff}}(0.1) = 1024$ instead of $\lambda_{\text{eff}}(1) = \lambda = 2$ for the standard model without memory ($\alpha = 1$), that is, 512 times more. This indicates that taking into account the memory in the macroeconomic model can change the warranted rate of growth by an order of magnitude or by several orders of magnitude.

As a result, we can formulate the following principle of macroeconomic dynamics with one-parameter power-law memory.

Principle 11.1 (Principle of changing of warranted rate by memory).

(a) *In macroeconomic growth models, the effect of one-parameter power-law memory with the fading parameter $\alpha > 0$ leads to the dependence of the warranted rate of growth on the parameter α, which is determined by the expression*

$$\lambda_{\text{eff}}(\alpha) = \lambda^{1/\alpha}, \tag{11.30}$$

where λ is the warranted rates of growth for the standard model without memory. The parameter $\lambda_{eff}(\alpha)$ is the warranted rates of growth with memory.

(b) *For small warranted rate of the standard growth model without memory ($0 < \lambda < 1$), the effect of one-parameter memory with fading parameter $\alpha > 0$ leads to decrease of the warranted rate of growth with memory for $0 < \alpha < 1$, and it leads to an increase of this warranted rate for $\alpha > 1$.*

(c) *For the large warranted rates ($\lambda > 1$) of the standard growth model without memory, the effect of power-law memory with fading parameter α leads to an increase of the warranted rates of growth with memory for $0 < \alpha < 1$, and it leads to a decrease of this warranted rates for $\alpha > 1$.*

We can state that neglecting the memory effects in macroeconomic models can greatly change the result. Accounting of memory can lead to new results for the same parameters of macroeconomic models.

11.6 Open model with memory and power-law consumption

The case of absence of nonproductive consumption $C(t) = 0$, which describes the closed macroeconomic model, is an unrealistic case. This case is considered to calculate the maximum possible (warranted, technological) rate of growth with memory. An important realistic particular case, which we consider for the macroeconomic growth model, is the power-law and constancy of the nonproductive consumption.

Let us describe some cases of the power-law behavior of the nonproductive consumption $C(t)$. The constancy of the consumption can be considered as a special case of the power-law function.

For the power-law dependence of consumption on time, an explicit analytical form of the solution can be obtained In addition, this solution can be used to describe the wide class of the consumption functions $C(t)$, which are described by power series or finite polynomials.

Let us consider the consumption function in the form

$$C(t) = Ct^{\mu-1}, \tag{11.31}$$

where $\mu > 0$. The constant consumption function ($C(t) = C$) is a special case of (11.31), when $\mu = 1$. Let us obtain the general solution of equation (11.14) with function (11.31). We note that expression (11.16) can be written in the form

$$Y_C(t) = -B^{-1}\int_0^t \tau^{\alpha-1}E_{\alpha,\alpha}[B^{-1}\tau^{\alpha}]C(t-\tau)\,d\tau. \tag{11.32}$$

To calculate integral (11.32) with the power-law function (11.31), we can use equation 4.10.8 from [143, p. 86], or equation (4.4.5) of [143, p. 61], where $\Gamma(\mu)$ should be used instead of $\Gamma(\alpha)$, in the form

$$\int_0^t \tau^{\beta-1} E_{\alpha,\beta}[\lambda\tau^{\alpha}](t-\tau)^{\mu-1}\, d\tau = \Gamma(\mu) t^{\beta+\mu-1} E_{\alpha,\beta+\mu}[\lambda t^{\alpha}], \tag{11.33}$$

where $\beta > 0$ and $\mu > 0$. Note that in equation (11.33) the exponent μ can be considered as a positive integer, or as a real positive number. For $\beta = \alpha$, equation (11.33) has the form

$$\int_0^t \tau^{\alpha-1} E_{\alpha,\alpha}[\lambda\tau^{\alpha}](t-\tau)^{\mu-1}\, d\tau = \Gamma(\mu) t^{\alpha+\mu-1} E_{\alpha,\alpha+\mu}[\lambda t^{\alpha}]. \tag{11.34}$$

Using expression (11.34), equation (11.32) with function (11.31) takes the form

$$Y_C(t) = -CB^{-1}\Gamma(\mu) t^{\alpha+\mu-1} E_{\alpha,\alpha+\mu}[B^{-1} t^{\alpha}], \tag{11.35}$$

where $\mu > 0$.

As a result, solution (11.15), (11.16) of equation (11.14) with consumption function (11.31) has the form

$$\begin{aligned} Y(t) = &\sum_{k=0}^{n-1} Y^{(k)}(0) t^k E_{\alpha,k+1}[B^{-1} t^{\alpha}] \\ &- CB^{-1}\Gamma(\mu) t^{\alpha+\mu-1} E_{\alpha,\alpha+\mu}[B^{-1} t^{\alpha}], \end{aligned} \tag{11.36}$$

where $0 < n-1 < \alpha \le n$. To describe the behavior of expression (11.35) at $t \to \infty$, we can use expression (11.24), (see also [200, p. 43], and [143, 100]). Substitution of (11.24) with $\lambda = B^{-1}$ and $\beta = \alpha + \mu - 1$ into (11.35) gives

$$\begin{aligned} Y_C(t) = &-\frac{C\Gamma(\mu) B^{(1-2\alpha-\mu)/\alpha}}{\alpha} \exp(B^{-1/\alpha} t) \\ &+ \sum_{j=1}^{m} \frac{C\Gamma(\mu) B^{j-1}}{\Gamma(\alpha(1-j)+\mu)} t^{\alpha(1-j)+\mu-1} + O\left(\frac{1}{t^{\alpha m+1-\mu}}\right), \end{aligned} \tag{11.37}$$

where $\mu > 0$ and we use the big O notation (asymptotic notation) that provides an upper bound on the growth rate of the following terms of the asymptotic series. Using expressions (11.37) and (11.25), we can see that the warranted rate of growth with memory, which is described by expression (11.36), is defined by equation (11.26), i. e., we have the equality

$$\lambda_{\text{eff}}(\alpha) = \lambda^{1/\alpha} = B^{-1/\alpha}. \tag{11.38}$$

This growth rate coincides with the warranted rate of growth with memory.

For expression (11.35), we can use equation (4.2.3) of [143, p. 57], which has the form

$$E_{\alpha,\beta}[z] = \frac{1}{\Gamma(\beta)} + zE_{\alpha,\alpha+\beta}[z]. \tag{11.39}$$

Using $z = \lambda t^{\alpha}$ and $\beta = \mu$, equation (11.39) takes the form

$$t^{\alpha}E_{\alpha,\alpha+\mu}[\lambda t^{\alpha}] = \frac{1}{\lambda}\left(E_{\alpha,\mu}[\lambda t^{\alpha}] - \frac{1}{\Gamma(\mu)}\right). \tag{11.40}$$

Then equation (11.33) can be written as

$$\int_0^t \tau^{\alpha-1}E_{\alpha,\alpha}[\lambda\tau^{\alpha}](t-\tau)^{\mu-1}\,d\tau = \frac{1}{\lambda}t^{\mu-1}(\Gamma(\mu)E_{\alpha,\mu}[\lambda t^{\alpha}] - 1). \tag{11.41}$$

Using equation (11.41) with $\lambda = B^{-1}$, expression (11.35) can be written in the form

$$Y_C(t) = Ct^{\mu-1}(1 - \Gamma(\mu)E_{\alpha,\mu}[B^{-1}t^{\alpha}]). \tag{11.42}$$

As a result, the equation of the macroeconomic model with power-law memory, whose fading parameter is α ($0 < n-1 < \alpha \leq n$) and the consumption function (11.31), has the solution

$$\begin{aligned} Y(t) = {} & Ct^{\mu-1}(1 - \Gamma(\mu)E_{\alpha,\mu}[B^{-1}t^{\alpha}]) \\ & + \sum_{k=0}^{n-1} Y^{(k)}(0)t^k E_{\alpha,k+1}[B^{-1}t^{\alpha}]. \end{aligned} \tag{11.43}$$

Solution (11.43) of equation (11.14) describes the dynamics of the national income in the framework of the model of economic growth with a power-law increase of consumption with one-parameter fading memory.

The obtained results can be generalized to the consumption functions $C(t)$, which are described by power series and finite polynomials. For example, using the function

$$C(t) = \sum_{m=1}^{N} C_m t^m, \tag{11.44}$$

the dynamics of national income is described by the equation

$$\begin{aligned} Y(t) = {} & \sum_{m=1}^{N} C_m t^m (1 - \Gamma(m+1)E_{\alpha,m+1}[B^{-1}t^{\alpha}]) \\ & + \sum_{k=0}^{n-1} Y^{(k)}(0)t^k E_{\alpha,k+1}[B^{-1}t^{\alpha}]. \end{aligned} \tag{11.45}$$

The interpolation by polynomials of finite degree allows us to apply the suggested methods for a wide class of consumption functions that describe real economic processes.

11.7 Model with memory and constant consumption

Let us consider the constant consumption function ($C(t) = C$) for solution of equation (11.14). This case allows us to clearly illustrate the dependence of economic growth on memory effects. The numerical simulations in the form of plots will be given in the next section.

Equation (11.14) with $C(t) = C = \text{const}$ has the form

$$(D^{\alpha}_{C;0+}Y)(t) - B^{-1}Y(t) = -B^{-1}C, \tag{11.46}$$

where $\lambda = B^{-1}$, and $n - 1 < \alpha \leq n$. Note that equation (11.46) can be represented in the form of the equation of the closed model for $Y_*(t) = Y(t) - C$ by using that $D^{\alpha}_{C;0+}C = 0$.

In the case $C(t) = C$, the solution of equation (11.46) has the form

$$Y(t) = C(1 - E_{\alpha,1}[B^{-1}t^{\alpha}]) + \sum_{k=0}^{n-1} Y^{(k)}(0)t^k E_{\alpha,k+1}[B^{-1}t^{\alpha}]. \tag{11.47}$$

This solution can be obtained from equation (11.43) by using $\mu = 0$. Solution (11.47) describes economic growth with one-parameter power-law memory and a constant nonproductive consumption.

For $0 < \alpha \leq 1$, ($n = 1$), solution (11.47) takes the form

$$Y(t) = C(1 - E_{\alpha,1}[B^{-1}t^{\alpha}]) + Y(0)E_{\alpha,1}[B^{-1}t^{\alpha}]. \tag{11.48}$$

For $1 < \alpha \leq 2$, ($n = 2$), solution (11.47) has the form

$$\begin{aligned} Y(t) &= C(1 - E_{\alpha,1}[B^{-1}t^{\alpha}]) \\ &\quad + Y(0)E_{\alpha,1}[B^{-1}t^{\alpha}] + Y^{(1)}(0)tE_{\alpha,2}[B^{-1}t^{\alpha}]. \end{aligned} \tag{11.49}$$

For $\alpha = 1$, the equality $E_{\alpha,1}[z] = \exp(z)$ leads solution (11.48) to the form

$$Y(t) = C(1 - \exp(t/B)) + Y(0)\exp(t/B), \tag{11.50}$$

which coincides with solution of equation for the standard model without memory.

Expression (11.50) describes economic growth in the framework of standard macroeconomic model with constant consumption ($C(t) = C$), that does not take into account memory effects ($\alpha = 1$). Expressions (11.47), (11.48), (11.49) represent the model of economic growth with constant consumption ($C(t) = C$), where we take into account the power-law memory with the fading parameter $\alpha > 0$.

Let us describe the asymptotic behavior of solution for the case of constant function $C(t) = C > 0$, where we will use $\lambda = B^{-1} > 0$. The behavior of solution (11.47) at $t \to \infty$ can be described by using the same asymptotic equations that is used for closed model ($C(t) = 0$), if we will consider $Y_*(t) = Y(t) - C$.

For $0 < \alpha < 1$, the asymptotic behavior of the solution is described by the equation

$$Y(t) = C + (Y(0) - C)\frac{1}{\alpha}\exp(\lambda^{1/\alpha}t) - (Y(0) - C)\left(\sum_{j=1}^{m}\frac{\lambda^{-j}t^{-\alpha j}}{\Gamma(1-\alpha j)} + O\left(\frac{1}{t^{\alpha(m+1)}}\right)\right). \quad (11.51)$$

For $1 < \alpha < 2$, the asymptotic behavior of the solution is described by the expression

$$Y(t) = C + (Y(0) - C + \lambda^{-1/\alpha}Y^{(1)}(0))\frac{1}{\alpha}\exp(\lambda^{1/\alpha}t) - \sum_{j=1}^{m}\left(Y(0) - C + \frac{Y^{(1)}(0)t}{1-\alpha j}\right)\frac{\lambda^{-j}t^{-\alpha j}}{\Gamma(1-\alpha j)} + O\left(\frac{1}{t^{\alpha(m+1)-1}}\right). \quad (11.52)$$

The conditions of growth and decline are important for macroeconomic models. Let us give these conditions for the proposed model with power-law memory, where we will use $\lambda = B^{-1} > 0$. In the case $1 < \alpha \le 2$, the condition of growth with memory is represented by the inequality

$$Y(0) - C + \lambda^{-1/\alpha}Y^{(1)}(0) > 0. \quad (11.53)$$

The condition of decline with memory is represented by the inequality

$$Y(0) - C + \lambda^{-1/\alpha}Y^{(1)}(0) < 0. \quad (11.54)$$

For the case $0 < \alpha \le 1$, we can use inequalities (11.53) and (11.54) with $Y^{(1)}(0) = 0$.

Note that conditions (11.53) and (11.54) can also be applied to the case $C = 0$, when we have the homogenous fractional differential equation that describes the closed model.

In the case of absence of memory ($\alpha = 1$), these conditions of growth and decline take the following form. The condition of growth without memory ($\alpha = 1$) and with memory at $0 < \alpha \le 1$ is represented by the inequality

$$Y(0) - C > 0. \quad (11.55)$$

The condition of decline without memory ($\alpha = 1$) and with memory $0 < \alpha \le 1$ is represented by the inequality

$$Y(0) - C < 0. \quad (11.56)$$

Let us note the case $Y^{(1)}(0) > 0$, where we take into account $C > 0$ and $\lambda = B^{-1} > 0$. If the inequalities

$$Y(0) < C < Y(0) + \lambda^{-1/\alpha}Y^{(1)}(0) \quad (11.57)$$

holds, the process without memory shows decline while the process with memory (at the same other parameters) demonstrates a growth. This condition means that the decline is replaced by the growth, when the memory effect is taken into account.

Let us formulate some properties of the growth processes with memory, when $C > 0$ and $0 < \lambda = B^{-1} < 1$.

1. The power-law memory with the fading parameter $\alpha \in (0, 1)$ can lead to decrease of the rates of growth for $\lambda \in (0, 1)$. The memory with $\alpha < 1$ can slow down the rate of economic growth.
2. The power-law memory with the fading parameter $\alpha \in (0, 1)$ can lead to decrease of the rates of decline for $\lambda \in (0, 1)$. For $0 < \alpha < 1$, the memory effects can slow down the pace of decline. The effects of memory can slow the rate of decline.
3. The power-law memory with the fading parameter $\alpha \in (1, 2)$ can lead to increase of growth rates.
4. The power-law memory with the fading parameter $\alpha \in (1, 2)$ can leads to a growth instead of decline if $\lambda \in (0, 1)$. The decline of processes can be replaced by the growth, when the memory effect is taken into account.
5. The power-law memory with the fading parameter $\alpha \in (1, 2)$ can lead to a slowdown in the rate of economic decline for $\lambda \in (0, 1)$. Memory effects may reduce the rate of decline. Memory with $\alpha > 1$ can decrease of the rate of slowdown of process.

As a result, we can formulate the following principles of economic growth for $\lambda \in (0, 1)$ and $0 < \alpha < 2$.

Principle 11.2 (Principle of growth with memory for fading $\alpha \in (0, 1)$). *The power-law memory with fading parameter $0 < \alpha < 1$ leads to a slowdown in the growth and decline of the economy for $\lambda \in (0, 1)$. The effect of memory with $0 < \alpha < 1$ leads to inhibition of growth and economic decline. The memory with small value of the parameter $0 < \alpha < 1$ leads to stagnation of the economy.*

Principle 11.3 (Principle of growth with memory for fading $\alpha \in (1, 2)$). *The power-law memory with fading parameter $1 < \alpha < 2$ leads to an improvement in economic dynamics for $\lambda \in (0, 1)$. The memory with $1 < \alpha < 2$ leads to positive results, such as a slowdown in the rate of decline, a replacement of the economic decline by its growth, an increase in the rate of economic growth.*

As a result, we conclude that accounting of the memory effects can give new type of behavior for the same parameters of macroeconomic models. The neglecting of the memory can lead to qualitatively different results, conclusions and predictions for economy. In studies of economy and the construction of macroeconomic models, we should take into account the effects of memory to obtain correct results and predictions.

For clarity, let us reformulate these principles by using such well-known stock symbols as a bull and a bear for describe the action of the memory effects on economy.

(1) *"A memory that has a fading parameter greater than one is a bull lifting both a bear with amnesia and a bull with amnesia."*
(2) *"A memory that has a fading parameter less than one is the bull lifting the bear with amnesia, and it is the bear pressing the bull with amnesia."*

Using the analogy "bull = economic growth" and "bear = economic decline," we can formulate the following rules that qualitatively describe the behavior of the economy with memory in terms of stock symbols.

Statement 11.1 (Rules of behavior of bulls and bears with memory).
(A) *Bears and bulls with memory, whose memory fading parameter is less than one, is weaker than bears and bulls with amnesia. Bears and bulls become weaker when a memory appears with fading parameter less than one.*
(B) *Bulls with memory, whose fading parameter is greater than one, are stronger than bulls with amnesia. Bulls become stronger when a memory appears with fading parameter greater than one.*
(C) *Bears with memory, the fading parameter of which is greater than one, is weaker than bears suffering from amnesia. Moreover, Bears with amnesia can become (turn into) bulls, when memory appears, whose fading parameters are larger than one. Bears become weaker or turn into bulls when a memory appears with fading parameter greater than one.*

Note that we have described only cases for $\lambda \in (0,1)$. If $\lambda > 1$, then we will also have nonstandard phenomena due to the consideration of memory in economic processes. For example, for $\alpha = 0.02$ and $\lambda = 1.2$, the warranted rate of growth with memory is equal to $\lambda_{\mathrm{eff}}(0.02) \approx 9100$ instead of $\lambda_{\mathrm{eff}}(1) = \lambda = 1.2$ for the standard model without memory ($\alpha = 1$), that is, it is almost seven and a half thousand times greater. This indicates that taking into account the memory in the macroeconomic model can change the warranted rate of growth by an order of magnitude or by several orders of magnitude.

The suggested principles of economics with memory can be used to give qualitative descriptions of economic processes with power-law memory. These rules and principles can be used for qualitative analysis of different ways of economic development of global and national economy in case of the presence of power-law memory effects.

11.8 Examples of memory effects for growth model

Let us give illustrative examples of behavior of the economy with memory. We will consider economy, which is described by equation (11.46), i. e., the model with one-parameter power-law memory and a constant nonproductive consumption. We give the plots of solution (11.47) for the cases $0 < \alpha < 1$ and $1 < \alpha < 2$, which are given by (11.48) and (11.49), in comparison with the plots of solution (11.50) of the standard model without memory ($\alpha = 1$).

Example 11.1. Let us consider solution (11.48) with $C = 19$, $Y(0) = 20$, and $\lambda = B^{-1} = 0.1$. Then expression (11.48) has the form

$$Y(t) = 19 + E_{\alpha,1}[0.1t^{\alpha}]. \tag{11.58}$$

For $\alpha = 0.1$, expression (11.58) describes the growth with memory. For $\alpha = 1$, expression (11.58) describes growth process without memory. The comparison of these processes is presented by Figure 11.1. It shows that the effects of memory can lead to decrease of the growth rates (the recession of growth rate). We see that the memory with $\alpha < 1$ can slow the growth rate. Memory effects can slowdown the rate of economic growth.

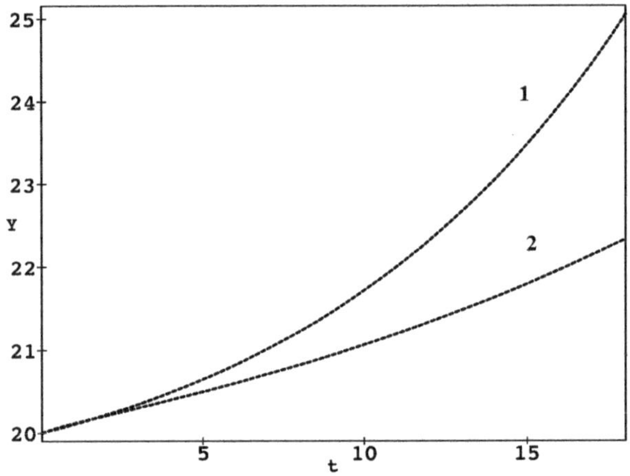

Figure 11.1: The national income $Y(t)$ for model without memory (Plot 1) and model with memory for $\alpha = 0.8$ (Plot 2) with $C = 19$, $Y(0) = 20$, $B^{-1} = 0.1$.

Example 11.2. Solution (11.48) with $C = 21$, $Y(0) = 20$, and $\lambda = B^{-1} = 0.1$, has the form

$$Y(t) = 21 - E_{\alpha,1}[0.1t^{\alpha}]. \tag{11.59}$$

For $\alpha = 0.1$, expression (11.59) describes the decline with memory. For $\alpha = 1$, expression (11.59) describes decline without memory. The comparison of these processes is presented by Figure 11.2. It shows that the effects of memory can lead to decrease of the rates of decline. We see that for $\alpha < 1$ the memory effects can slow down the pace of economic decline. The effects of memory can slow down a recession.

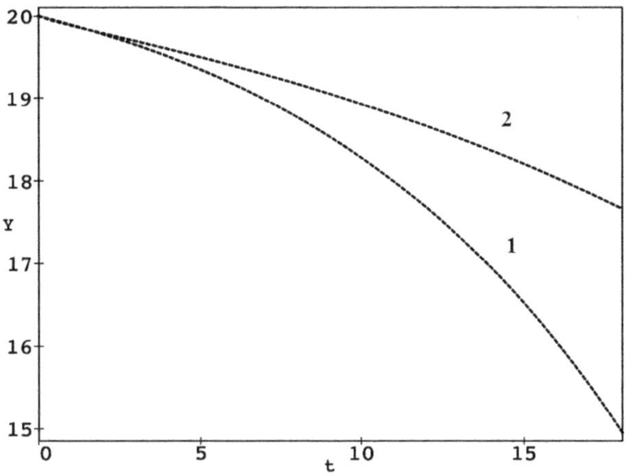

Figure 11.2: The national income $Y(t)$ for model without memory (Plot 1) and model with memory for $\alpha = 0.8$ (Plot 2) with $C = 21$, $Y(0) = 20$, $B^{-1} = 0.1$.

Example 11.3. Let us consider solution (11.49) with $C = 19$, $Y(0) = 20$, $Y^{(1)}(0) = 0.1$, and $\lambda = B^{-1} = 0.1$. Then expression (11.49) has the form

$$Y(t) = 19(1 - E_{\alpha,1}[0.1t^{\alpha}]) + 20E_{\alpha,1}[0.1t^{\alpha}] + 0.1tE_{\alpha,2}[0.1t^{\alpha}]. \tag{11.60}$$

For $\alpha = 1.1$, expression (11.60) describes the growth with memory. In this case, condition (11.53) holds. The comparison of this process and the growth without memory ($\alpha = 1$) is given in Figure 11.3. We see in Figure 11.3 that the effects of memory with $\alpha > 1$ can lead to increase of growth rates. Memory effects may increase the rate of economic growth.

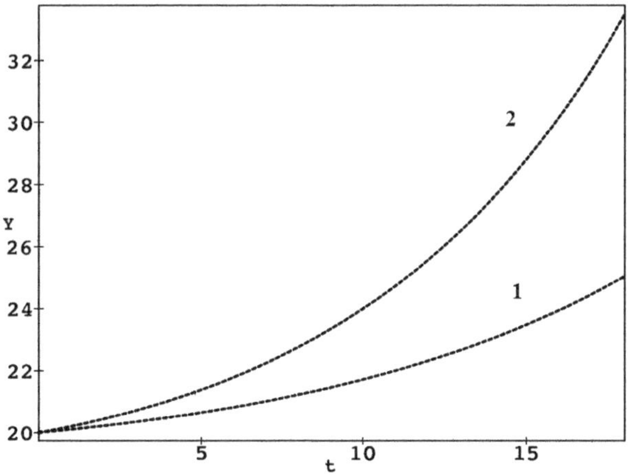

Figure 11.3: The national income $Y(t)$ for model without memory (Plot 1) and model with memory for $\alpha = 1.1$ (Plot 2) with $C = 19$, $Y(0) = 20$, $Y^{(1)}(0) = 0.1$, $\lambda = B^{-1} = 0.1$.

Example 11.4. Solution (11.49) with $C = 19$, $Y(0) = 20$, $Y^{(1)}(0) = 0.1$, and $\lambda = B^{-1} = 0.1$, has the form

$$Y(t) = 19(1 - E_{\alpha,1}[0.1t^{\alpha}]) + 20E_{\alpha,1}[0.1t^{\alpha}] + 0.3tE_{\alpha,2}[0.1t^{\alpha}]. \tag{11.61}$$

For $\alpha = 1.1$, expression (11.61) describes the growth with memory. In this case, condition (11.57) holds. The comparison of this process and the growth without memory ($\alpha = 1$) is given in Figure 11.4. We see that memory effect can give a growth instead of decreasing and decline. The effects of power-law memory with $\alpha > 1$ can lead to economic growth instead of a recession economy as it follows from the condition (11.57). The decline of the economy is replaced by the growth of the economy, when the memory effect is taken into account.

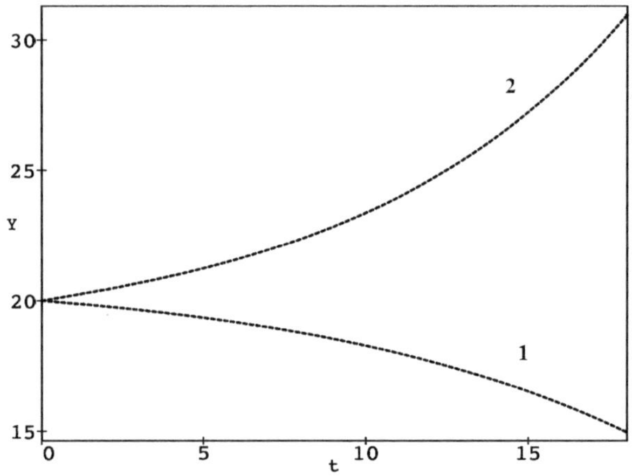

Figure 11.4: The national income $Y(t)$ for model without memory (Plot 1) and model with memory for $\alpha = 1.1$ (Plot 2) with $C = 21$, $Y(0) = 20$, $Y^{(1)}(0) = 0.3$, $\lambda = B^{-1} = 0.1$.

Example 11.5. Solution (11.49) with $C = 19$, $Y(0) = 20$, $Y^{(1)}(0) = 0.1$, and $\lambda = B^{-1} = 0.1$, has the form

$$Y(t) = 21(1 - E_{\alpha,1}[0.1t^{\alpha}]) + 20E_{\alpha,1}[0.1t^{\alpha}] + 0.1tE_{\alpha,2}[0.1t^{\alpha}]. \tag{11.62}$$

For $\alpha = 1.1$, expression (11.62) describes the growth with memory. In this case, condition (11.54) holds. The comparison of this process and the growth without memory ($\alpha = 1$) is given in Figure 11.5. It shows that the effects of power-law memory with $\alpha > 1$ can lead to a slowdown in the rate of economic decline. Memory effects may reduce the rate of decline. Memory with fading parameter $\alpha > 1$ can decrease of the rate of recession and increase the economic activity.

From Figures 11.1–11.5, it can be seen that the behavior of the income function is essentially dependent on the presence or absence of memory. Let us emphasize the properties of economic processes with power-law memory.

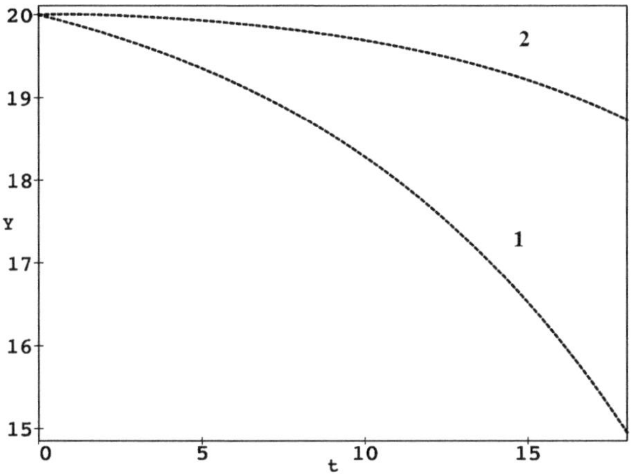

Figure 11.5: The national income $Y(t)$ for model without memory (Plot 1) and model with memory for $\alpha = 1.1$ (Plot 2) with $C = 21$, $Y(0) = 20$, $Y^{(1)}(0) = 0.1$, $\lambda = B^{-1} = 0.1$.

Statement 11.2 (Properties of processes with memory for $\lambda \in (0, 1)$).

1. *The effects of memory with $\alpha \in (0, 1)$ can lead to a slowing of the growth of economy, i. e., to decrease of the warranted growth rate ($\lambda_{\text{eff}}(\alpha) < \lambda$), when the other process parameters are unchanged. See plots of Figure* 11.1.
2. *The effects of memory with $\alpha \in (0, 1)$ can lead to a slowing of the decline of economy, i. e., to decrease of the warranted rate of growth ($\lambda_{\text{eff}}(\alpha) < \lambda$). See plots of Figure* 11.2.
3. *The effects of memory with $\alpha \in (1, 2)$ can lead to faster growth, i. e., to increase of the growth rate ($\lambda_{\text{eff}}(\alpha) > \lambda$). See plots of Figure* 11.3.
4. *The effects of memory with $\alpha \in (1, 2)$ can leads to a growth instead of decline, when the other process parameters are unchanged. The appearance of memory in the economic process can lead to the fact that recession is replaced by growth, even if other parameters of the process remain unchanged. See plots of Figure* 11.4.
5. *The effects of memory with $\alpha \in (1, 2)$ can lead to a slowing of the decline. Memory effects can slowdown a recession. See plots of Figure* 11.5.

As a result, the behavior of economy strongly depends on the absence or presence of memory effects and the value of memory fading parameter.

Figures 11.1–11.5 show that neglecting of the memory effects in macroeconomic models can change the result.

We see (Figures 11.1 and 11.2) that memory with fading parameters $\alpha < 1$ leads to a slowdown in the growth and decline of the economy. The effect of memory fading parameter with $\alpha < 1$ leads to inhibition of growth and decline. We can say that the memory with small value of the parameter $\alpha < 1$ leads to stagnation of the economy.

We also see (Figures 11.3–11.5) that taking into account the memory effect with $\alpha > 1$ leads to an improvement in economic dynamics. Memory with fading parameter

$\alpha > 1$ leads to positive results, such as a slowdown in the rate of decline, a replacement of the economic decline by its growth, an increase in the rate of economic growth.

11.9 Model with multiparameter memory

In previous sections, we considered macroeconomic model, in which the power-law fading of memory was characterized only by one fading parameter α. In a real economy, the memory fading parameters can be different for different types of economic agents and for different situations [450, 448, 412, 397, 399]. In addition, one can consider interpolations of memory functions in the form of linear combinations of power-law memory functions. In these cases, we can consider the simultaneous (parallel) action of several accelerators with power-law memory, the fading parameters of which are different. In these cases, to describe memory, we should use the linear combinations of the fractional derivatives with various noninteger orders.

11.10 Two-parameter power-law memory

Let us assume that the two-parameter power-law memory is characterized by the parameters $\alpha > 0$ and $\beta > 0$, and the accelerator with this memory (11.12) is described by equation

$$I(t) = B((D^{\alpha}_{C;0+}Y)(t) - \theta(D^{\beta}_{C;0+}Y)(t)), \tag{11.63}$$

where $\alpha > \beta$, $B = B(\alpha, \beta) > 0$ and $\theta \neq 1$. The parameter θ characterizes the influence of the memory, which fading parameter is $\beta > 0$, on the investment. The value of θ can be either positive or negative.

Substituting the expression of the investment $I(t)$, which is given by equation (11.63), into balance equation (11.3), we obtain the fractional differential equation

$$Y(t) = B(D^{\alpha}_{C;0+}Y)(t) - \theta B(D^{\beta}_{C;0+}Y)(t) + C(t). \tag{11.64}$$

Equation (11.64) can be rewritten as

$$(D^{\alpha}_{C;0+}Y)(t) - \theta(D^{\beta}_{C;0+}Y)(t) - B^{-1}Y(t) = -B^{-1}C(t), \tag{11.65}$$

where $\alpha > \beta$, $B^{-1} > 0$, and the parameter θ is a real number, which can be either positive or negative, in the general case. Fractional differential equation (11.65) determines the economic dynamics with two-parameter power-law memory.

To solve equation (11.65), we can use Theorem 5.16 of the book [200, pp. 323–324]. Equation (11.65) coincides with equation (5.3.73) of the book [200, p. 323], if we will use the notation $\lambda = \theta$, $\mu = B^{-1}$ and $f(t) = -B^{-1}C(t)$. As a result, for continuous function $C(t)$, which is defined on the positive semi-axis ($t > 0$), equation (11.63) with the parameters $0 < n-1 < \alpha \leq n$, and $m-1 < \beta \leq m$ (where $0 < \beta < \alpha$, $m \leq n$, $\alpha - n + 1 \leq \beta$) has the general solution

$$Y(t) = \sum_{j=0}^{n-1} C_j Y_j(t) + Y_C(t), \tag{11.66}$$

where C_j ($j = 0, \dots, n-1$) are the real constants that are determined by the initial conditions, and the function $Y_C(t)$ is defined as

$$Y_C(t) = -B^{-1} \int_0^t (t-\tau)^{\alpha-1} G_{\alpha,\beta,\theta,B}[t-\tau] C(\tau)\, d\tau. \tag{11.67}$$

The function $G_{\alpha,\beta,\theta,B}[\tau]$ is given by the equation

$$G_{\alpha,\beta,\theta,B}[\tau] = \sum_{k=0}^{\infty} \frac{\tau^{k\alpha}}{\Gamma(k+1)} B^{-k} \Psi_{1,1}\left[{}^{(k+1,1)}_{(\alpha k+\alpha,\alpha-\beta)} \mid \theta \tau^{\alpha-\beta}\right]. \tag{11.68}$$

The functions $Y_j(t)$ with $j = 0, \dots, m-1$ are represented by the expressions

$$\begin{aligned} Y_j(t) = {} & \sum_{k=0}^{\infty} \frac{t^{k\alpha+}}{\Gamma(k+1)} B^{-k} \Psi_{1,1}\left[{}^{(k+1,1)}_{(\alpha k+j+1,\alpha-\beta)} \mid \theta t^{\alpha-\beta}\right] \\ & - \theta \sum_{k=0}^{\infty} \frac{t^{k\alpha+j+\alpha-\beta}}{\Gamma(k+1)} B^{-k} \Psi_{1,1}\left[{}^{(k+1,1)}_{(\alpha k+j+1+\alpha-\beta,\alpha-\beta)} \mid \theta t^{\alpha-\beta}\right]. \end{aligned} \tag{11.69}$$

For $j = m, \dots, n-1$, the functions $Y_j(t)$ are defined by the equations

$$Y_j(t) = \sum_{k=0}^{\infty} \frac{t^{k\alpha+j}}{\Gamma(k+1)} B^{-k} \Psi_{1,1}\left[{}^{(n+1,1)}_{(\alpha k+j+1,\alpha-\beta)} \mid \theta t^{\alpha-\beta}\right]. \tag{11.70}$$

Here $\Psi_{1,1}$ is the generalized Wright functions (the Fox–Wright function) [200], which is defined by the equation

$$\Psi_{1,1}\left[{}^{(a,\alpha)}_{(b,\beta)} \mid z\right] = \sum_{k=0}^{\infty} \frac{\Gamma(\alpha k + a)}{\Gamma(\beta k + b)} \frac{z^k}{k!}. \tag{11.71}$$

Note that the three-parameter Mittag–Leffler function is a special case of the Fox–Wright function (see equation (1.9.1) in [200, p. 45]), such that

$$E^{\gamma}_{\rho,\beta}[z] = \frac{1}{\Gamma(\gamma)} \Psi_{1,1}\left[{}^{(\rho,1)}_{(\beta,\alpha)} \mid z\right]. \tag{11.72}$$

Equation (11.65) and its solution (11.66)–(11.70) describe the macroeconomic dynamics of the national income, where the memory is characterized by two-parameter power-law fading.

11.11 Open model with two-parameter memory

Let us consider the case when all national income is used to expand production, that is, the consumption is absent ($C(t) = 0$). Equation (11.65) with $C(t) = 0$ has the form

$$(D^{\alpha}_{C;0+}Y)(t) - \theta(D^{\beta}_{C;0+}Y)(t) - B^{-1}Y(t) = 0, \tag{11.73}$$

where $0 < \beta < \alpha \leq 2$. Solution of equation (11.73) can be used to estimate the greatest possible growth rate of the national income, when we take into account the two-parameter power-law fading memory.

Let us first consider equation (11.73) for the case $0 < \alpha \leq 1$ and $0 < \beta < \alpha$, by using Corollary 5.8 of [200, p. 317]. In this case, the solution of equation (11.73) with $0 < \beta < \alpha \leq 1$ has the form

$$Y(t) = C_1 Y_1(t), \tag{11.74}$$

where the constant C_1 is determined by the initial condition, and $Y_1(t)$ is given by the equation

$$Y_1(t) = \sum_{k=0}^{\infty} \frac{t^{k\alpha}}{\Gamma(k+1)} B^{-k} \Psi_{1,1}\Big[{}^{(k+1,1)}_{(\alpha k+1,\alpha-\beta)} \Big| \, \theta t^{\alpha-\beta}\Big] - \theta \sum_{k=0}^{\infty} \frac{t^{k\alpha+\alpha-\beta}}{\Gamma(k+1)} B^{-k} \Psi_{1,1}\Big[{}^{(k+1,1)}_{(\alpha k+1+\alpha-\beta,\alpha-\beta)} \Big| \, \theta t^{\alpha-\beta}\Big]. \tag{11.75}$$

Using $\Psi_{1,1}\Big[{}^{(a,\alpha)}_{(b,\beta)} \Big| \, 0\Big] = 1/\Gamma(\beta)$ and equations (11.74), we get

$$C_1 = \frac{\Gamma(\beta)Y(0)}{1-\theta}, \tag{11.76}$$

where $\theta \neq 1$.

Let us now consider equation (11.65) for the case $1 < \alpha \leq 2$ and $0 < \beta < \alpha$. Using Corollary 5.9 of [200, p. 317], the solution of equation (11.65) can be represented as a linear combination of $Y_1(t)$, which is given by (11.67), and $Y_2(t)$. If $0 < \beta \leq 1 < \alpha \leq 2$, then $Y_2(t)$ is defined by the expression

$$Y_2(t) = \sum_{k=0}^{\infty} \frac{t^{k\alpha+1}}{\Gamma(k+1)} B^{-k} \Psi_{1,1}\Big[{}^{(k+1,1)}_{(\alpha k+2,\alpha-\beta)} \Big| \, \theta t^{\alpha-\beta}\Big]. \tag{11.77}$$

If $1 < \beta < \alpha \leq 2$, then $Y_2(t)$ is defined by the equation

$$Y_2(t) = \sum_{k=0}^{\infty} \frac{t^{k\alpha+1}}{\Gamma(k+1)} B^{-k} \Psi_{1,1}\Big[{}^{(k+1,1)}_{(\alpha k+2,\alpha-\beta)} \Big| \, \theta t^{\alpha-\beta}\Big] - \theta \sum_{k=0}^{\infty} \frac{t^{k\alpha+1+\alpha-\beta}}{\Gamma(k+1)} B^{-k} \Psi_{1,1}\Big[{}^{(k+1,1)}_{(\alpha k+2+\alpha-\beta,\alpha-\beta)} \Big| \, \theta t^{\alpha-\beta}\Big]. \tag{11.78}$$

In the next section, we will consider some properties of the two-parameter model with $0 < \beta < \alpha \leq 2$.

To determine the effective warranted rates of growth (the warranted rate of growth with memory) for the two-parameter macroeconomic model (11.65), it is necessary to consider the asymptotic behavior of solutions (11.75), (11.77) and (11.78) at $t \to \infty$, which are represented as infinite sums of the Fox–Wright functions. Unfortunately, such asymptotic expressions are unknown.

Equations (11.75), (11.77) and (11.78) in the form of sums of the Fox–Wright functions also contain the argument $\theta t^{\alpha-\beta}$, which depends on the difference $\alpha - \beta$. This allows us to assume that the asymptotic behavior of solutions (11.75), (11.77) and (11.78) at $t \to \infty$ should be described by $\alpha - \beta$. Therefore, it can be assumed that the warranted rate of growth with memory for models with two-parameter power-law memory are determined by the difference $\alpha - \beta$ for two-parameter memory.

11.12 Model with multiparameter power-law memory

To take into account memory with $m \geq 2$ different power-law fading parameters, we can use the accelerator equation in the form

$$I(t) = B\left((D^{\alpha}_{C;0+}Y)(t) - \theta(D^{\beta}_{C;0+}Y)(t) - \sum_{k=1}^{m-2} \theta_k (D^{\alpha_k}_{C;0+}Y)(t) \right), \tag{11.79}$$

where $\alpha > \beta > \alpha_{m-2} > \cdots > \alpha_1 > 0$. Here, the parameters θ_k characterize how memory with fading parameter $\alpha_k > 0$ influences on the investment. We can consider θ_k and α_k with $k = 1, 2, \ldots, m$, by setting $\theta_m = 1$, $\theta_{m-1} = \theta$, $\alpha_m = \alpha$, $\alpha_{m-1} = \beta$. Macroeconomic models with multiparameter memory can more accurately describe the real economic processes with memory. In particular, a general memory function can be represented by interpolation method as a linear combination of power-law memory functions.

Expression (11.79) can be substituted into balance equation (11.3). As a result, we obtain the fractional differential equation

$$\begin{aligned} &(D^{\alpha}_{C;0+}Y)(t) - \theta(D^{\beta}_{C;0+}Y)(t) \\ &\quad - \sum_{k=1}^{m-2} \theta_k (D^{\alpha_k}_{C;0+}Y)(t) = -B^{-1}C(t), \end{aligned} \tag{11.80}$$

where we use the notation $\theta_0 = B^{-1}$, $\alpha_0 = 0$ and the property $(D^{0}_{C;0+}Y)(t) = Y(t)$.

The solutions of the equations of the macroeconomic model with m-parameter fading memory can be described by using Theorem 5.17 of the book [200, p. 324]. The solution of equation (11.80) with $C(t) = 0$ is described by equations (5.3.58–5.3.60) of Theorem 5.14 of the book [200, pp. 319–320]. Unfortunately, at the moment there is no exact analytical expression for the asymptotic behavior of the solution of the fractional differential equation (11.80).

However, despite the presence of $m \geq 2$ different memory fading parameters in equation (11.80), the solutions are determined only by the sums of the Fox–Wright functions with the argument $\theta t^{\alpha-\beta}$, which depends only on the difference $\alpha - \beta$ of the two leading fading parameters. In other words, the argument of the Fox–Wright functions does not depend on $\alpha_{m-2}, \dots, \alpha_1$. It depends on the same argument as the solution of the two-parameter model.

As a result, it can be assumed that the warranted rate of growth with m-parameter memory are determined by difference $\alpha - \beta$. This allows us to propose the following hypotheses for macroeconomic processes with multiparameter power-law memory.

Statement 11.3 (Hypothesis of seniority of memory fading parameters). *In macroeconomic models with multi-parameter power-law memory with fading parameters $\alpha > \beta > \alpha_{m-2} > \cdots > \alpha_1 > 0$, the warranted rate of growth with memory depends only on the difference of the two leading parameters $\alpha - \beta$, that is,*

$$\lambda_{\text{eff}}(\alpha, \beta, \alpha_{m-2}, \dots, \alpha_1) = \lambda_{\text{eff}}(\alpha - \beta). \tag{11.81}$$

Statement 11.4 (Hypothesis of changing of dependence in growth by memory). *The second largest parameter β of the power-law memory fading lead to a change of the dependence of the warranted rate of growth with memory, which expresses through the inverse value of incremental capital B^{-1}, on the influence factor θ, which corresponds to the second fading parameter β that characterizes the influence of memory on the value of investment. In symbolic form, the change of dependence can be represented as*

$$\lambda_{\text{eff}}(\alpha) = B^{-1/\alpha} \Longrightarrow \lambda_{\text{eff}}(\alpha - \beta) = \theta^{1/(\alpha-\beta)} \tag{11.82}$$

for $\alpha - \beta \neq 0$.

The proposed hypotheses can help us to describe the qualitative changes caused by taking into account the multiparameter power-law memory in macroeconomic processes. These hypotheses of multiparameter memory can be corrected when the exact expressions for the asymptotic behavior of the solutions (see Theorem 5.17 of [200, p. 324]) will be obtained for equation (11.80). However, the proposed hypotheses should be verified by numerical simulation of the initial fractional differential equations.

11.13 Conclusion

Economic models with memory can serve as a theoretical basis to develop the concept of economic growth, to study possible alternatives of economic policy and their long-term consequences, to predict the behavior of economic processes. The fractional calculus can be used us to build more adequate macroeconomic models of real economy. This mathematical tool allows us to take into account that economic agents can remember the history of changes in economic processes and take these changes into account, when they make decisions.

In this chapter, we proposed principles of economic dynamics with memory. These principles can be used to give qualitative descriptions of macroeconomic processes with power-law memory and to realize qualitative analysis of different ways of economic development. These principles can be used to study macroeconomic processes with memory by using this small number of initial data. Using these principles, we can make conclusions on the basis of the parameters of memory fading and the warranted rate of growth without memory. These principles can be used for qualitative analysis of different ways of economic development of global and national economy. The proposed principles allow us to draw conclusions about the rates of economic growth if fading parameters of memory were determined from economic data. We assume that the memory fading parameters can also be considered as controlling parameters for increasing economic growth in the construction of economic policies and measures of influence on the real economy.

12 Dynamic intersectoral Leontief models with memory

In this chapter, we describe the dynamic intersectoral model with power-law memory. A generalization of Leontief dynamic intersectoral model, to which memory effects are taken into account, was first proposed in articles [437, 451, 433, 398] in 2017–2018. The equations of open and closed intersectoral models, in which the memory effects are described by the Caputo derivatives of noninteger orders, are derived. We also derive solutions of these equations, which have the form of linear combinations of the Mittag–Leffler functions, which allow us to characterize warranted rates of growth with memory. Examples of intersectoral dynamics with power-law memory are suggested for two sectoral cases. We formulate two principles of intersectoral dynamics with memory: the principle of changing of warranted rates of growth and the principle of domination change. It has been shown that in the input–output economic dynamics the effects of fading memory can change the economic growth rate and dominant behavior of economic sectors. This chapter is based on articles [437, 451, 433, 398].

12.1 Introduction

Dynamic intersectoral macroeconomic models describe the behavior of gross and final products (the national income) in sectors of the economy. These models are based on the equations of input–output balance in monetary terms that describe the production and distribution of the gross and final products between sectors. Models take into account the intersectoral production links, the use of material resources, the production and distribution of the national income. In the balance equation, each sector is considered twice as a consumer and as a producer. This leads us to matrix form of the balance equations. A distinguishing feature of the intersectoral model is a description of the "input–output" balance equations in the matrix form. The matrix equation of intersectoral balance assumes that each product has only one production (one sector), and each production (industry) produces only one type of product. In the dynamic intersectoral models, the economic variables are described by matrices.

One of the most famous intersectoral model is the model developed by Wassily W. Leontief, who received the Nobel Memorial Prize in Economic Sciences in 1973 "for the development of the input–output method and for its application to important economic problems," [476]. Wassily W. Leontief describes the input–output model as early as in the 1930s. A comprehensive version was published in 1941 in the book "The Structure of American Economy, 1919–1929." An extended version appeared 10 years later [219, 220, 354, 355]. The Leontief dynamic model is an intersectoral model of growth of gross national product and national income [146, 309].

The standard dynamic intersectoral models use different assumptions. One of the assumptions is the neglect of the memory effects. In fact, the standard dynamic inter-

https://doi.org/10.1515/9783110627459-012

sectoral models assumed that economic agents cannot remember history of changes of the economic variables. As a result, we can say that these models describe only processes, in which all agents have full amnesia.

The dynamic intersectoral Leontief models with power-law memory was first proposed in articles [437, 451, 433, 398] in 2017–2018. We take into account the power-law memory in generalized dynamic intersectoral models with continuous time.

In this chapter, we describe dynamic intersectoral models with power-law memory based on works [437, 451, 433, 398]. In the equations of the suggested intersectoral model with memory, the memory is described by the Caputo derivatives of noninteger orders. This chapter presents and analyzes the solutions of these equations of the closed and open intersectoral dynamic models. These solutions are represented as linear combinations of the Mittag–Leffler functions, which allow us to characterize warranted rates of growth with memory. Examples of intersectoral dynamics with power-law memory are described for two sectoral cases. We formulate two principles of intersectoral dynamics with memory: the principle of changing of warranted rates and the principle of domination change. We prove that in the input–output economic dynamics the effects of fading memory can change the economic growth rate and dominant behavior of economic sectors.

12.2 Dynamic intersectoral model without memory

Let us consider standard intersectoral model without memory, where we assume that n kinds of products are produced and used. Each sector produces only one type of products, and each product is produced in a certain sector. Characteristics of production processes are assumed to be constant, that is, we will not take into account that progress can lead to changes in production technology. In addition, the import of goods and materials and the use of nonrenewable resources will not be considered in the standard model. Dynamic intersectoral model will be formulated in the framework of continuous time approach.

Let us describe a derivation of input–output balance of an equation of the dynamic Leontief model without memory (e. g., see Section 3.1 in [146, pp. 122–144]). In the Leontief model, the gross product (gross output) is described by the vector $X(t) = (X_k(t))$, and it is divided into two parts,

$$X(t) = Z(t) + Y(t), \tag{12.1}$$

where $Y(t) = (Y_k(t))$ is a vector of the final product; $Z(t) = (Z_k(t))$ is the vector of the intermediate product, where $k = 1, \ldots, n$ are production sectors. The final product is distributed to the investment and the nonproductive consumption

$$Y(t) = I(t) + C(t), \tag{12.2}$$

where $I(t) = (I_k(t))$ is a vector of investment; $C(t) = (C_k(t))$ is the vector of products of nonproductive consumption (including nonproductive accumulation), where $k =$

$1, \dots, n$ are production sectors. The dynamic Leontief model assumes that the balance equations (12.1) and (12.2) hold for any $t > 0$. These equations describe the dynamic equilibrium of the economy as a whole. For this reason, the dynamic Leontief model is a dynamic model of the "input–output" balance. Substituting the expression of the final product (12.2) into equation (12.1), we obtain the balance equation

$$X(t) = Z(t) + I(t) + C(t). \tag{12.3}$$

To get the Leontief model equation from the balance equation (12.3), it is necessary to eliminate the endogenous (internal) variables $Z(t)$ and $I(t)$. To do this, we should give dependence of $Z(t)$ and $I(t)$ on the exogenous variable $X(t)$.

The Leontief model assumes constancy of coefficients of direct material costs of production. The dependence of the intermediate product on the gross product is assumed in the form of direct proportionality. This allows us to express the vector of intermediate products $Z(t)$ through the multiplication of the matrix of direct material costs A and the vector of gross product $X(t)$ in the form of the matrix equation of the linear multiplier

$$Z(t) = AX(t), \tag{12.4}$$

where $A = (a_{ij})$ is the square matrix of n-th order with coefficients a_{ij}, which describe the direct material costs of i-th sector ($i = 1, \dots, n$) in the production of a unit of output j-th sector ($j = 1, \dots, n$). The matrix A is assumed to be constant, that is, it does not change with time. In the dynamic Leontief model, the coefficients a_{ij} include not only the direct material costs, but also the costs of disposal compensation and repair the basic production assets. Therefore, the elements of the main diagonal of the matrix A are nonzero.

The dynamic Leontief model is based on the assumption of the relationship between the accumulation and the growth of gross output. This relationship is represented by using matrix of capital intensity of production growth. In addition, it is assumed instantaneous transformation of investment in a gain of fixed assets and instantaneous of return of these funds in the production (in the outputs). Therefore, this model disregards lag and memory. Time is supposed continuous. This allows us to use the theory of differential equations. Dependence of the vector $I(t)$ of capital investments on the vector $X(t)$ of the gross product is described by the matrix equation of the linear accelerator

$$I(t) = B\frac{dX(t)}{dt}, \tag{12.5}$$

where $I(t) = (I_k(t))$ is the vector of investments, $B = (b_{ij})$ is square matrix of the n-th order of the coefficients b_{ij} of an incremental capital intensity of production. Coefficients b_{ij} describe the expenses of production of i-th sector for increase in production in j-th sector. Matrix B is assumed nondegenerate, i. e., the determinant of the matrix

B is different from zero. For the square matrix B, this condition is equivalent to the reversibility of the matrix, i. e., the existence of an inverse matrix B^{-1}. In general, the matrix B can have zero line, for example, for sectors, which produce only consumer goods. For the reversibility of the matrix B of intersectoral dynamic models, we can consider only the equation for sectors that form the fixed assets.

Substitution of expressions (12.4) and (12.5) into equation (12.3), we get the equation of the dynamic Leontief model for gross production (gross output) in the form of the matrix differential equation

$$B\frac{dX(t)}{dt} + (A - E)X(t) + C(t) = 0, \tag{12.6}$$

where E is the unit diagonal matrix of n-th order. The economic sense has only such solutions of equation (12.6), for which $X(t) \geq 0$.

The equation for the final product (national income), which is described by the vector $Y(t)$, can be obtained from equation (12.6). To do this, we substitute the expression of $Z(t)$, which is given by formula (12.4), into equation (12.1) and then we express $Y(t)$. As a result, we obtain the expression

$$Y(t) = (E - A)X(t). \tag{12.7}$$

Equation (12.7) allows us to find the vector $Y(t)$ of final product if the vector $X(t)$ of the gross product is given. Substituting (12.7), which is written in the form

$$X(t) = (E - A)^{-1}Y(t) \tag{12.8}$$

into equation (12.6) and taking into account the constancy of the matrix A, we get equation of the final product (national income) in the form

$$B(E - A)^{-1}\frac{dY(t)}{dt} - Y(t) + C(t) = 0, \tag{12.9}$$

where $B(E - A)^{-1}$ is the matrix of the full incremental capital intensity. Note that the economic models without the foreign trade have an economic meaning only if solutions of equation (12.9) satisfy the condition $Y(t) \geq 0$.

Let us note the some features of the dynamic intersectoral model.

(a) The coefficients of direct material costs a_{ij} and the coefficients incremental capital b_{ij} are assumed to be constant, i. e., the matrix A and B are constant in time. In the general case, this is not true, especially when we consider the long time intervals.
(b) Increase of production is instantaneous, when we change the investments. This means that we ignore the effects of the time delay between the final product and to the investment.
(c) Investments $I(t)$ at time t determined by the change in the gross product $X(t)$ only in an infinitesimal neighborhood of the point in time, i. e., the nearest infinitesimal past. This relationship between $I(t)$ and $X(t)$ implies neglect of memory ef-

fects. In other words, the instantaneous amnesia is actually assumed for all economic agents, i. e., the instant forgetting of the history of changes of the gross product and investments.

Intersectoral model is said to be closed if we assume the absence of non-productive consumption, i. e., $C(t) = 0$. The general solution of the system of differential equations (12.9) with $C(t) = 0$ can be written in the form

$$Y(t) = \sum_{k=1}^{n} c_k Y_k \exp(\lambda_k t), \tag{12.10}$$

where λ_k are eigenvalues of the matrix

$$\Lambda = (E - A)B^{-1}, \tag{12.11}$$

and the vectors $Y_k = (Y_{kj})$ are the eigenvectors, which correspond to the eigenvalues λ_k, i. e., we have the equality

$$\Lambda Y_k = \lambda_k Y_k, \tag{12.12}$$

where the coefficients c_k are determined from the initial condition

$$\sum_{k=1}^{n} c_k Y_k = Y(0). \tag{12.13}$$

The solution (12.10) is a linear combination of exponential functions $\exp(\lambda_k t)$ with different rates of growth λ_k. Therefore, the dynamics of the economic process along the path $Y(t) = Y(0)\exp(\lambda t)$ with the uniform growth rate λ for all sectors is impossible, in the general case. As a result, economic growth will be accompanied by constant structural changes. This fact significantly distinguishes the intersectoral model from the one-sector models such as the natural growth model, the Harrod–Domar model, the Keynes model and other. However, there is a correlation between the solution of the dynamic intersectoral models and the solutions of one-sector growth models. This relationship is based on the existence of an eigenvalue s_{max} with the highest absolute value for the matrix

$$S = B(E - A)^{-1} = \Lambda^{-1} \tag{12.14}$$

of the full incremental capital intensity. The existence of such eigenvalue follows from the Perron's theorem and its generalization by Frobenius. The Perron's theorem states [28] that for the positive matrix always exists a unique eigenvalue s_{max} with the highest absolute value. The number s_{max} is positive, and it is a simple root of the characteristic equation. The corresponding eigenvector can be chosen positive. The Frobenius theorem states [131, pp. 354–355] that an irreducible non-negative matrix always has a positive eigenvalue s_{max}, which is a simple root of the characteristic equation. The absolute value of all other eigenvalues do not exceed the number s_{max}, and this number corresponds to the eigenvector with positive components. The number s_{max} of the matrix S is often called the Frobenius–Perron number.

Definition 12.1. *In the intersectoral dynamic models the eigenvalue* $\lambda_s = 1/s_{max}$ *of the matrix* $\Lambda = S^{-1}$ *is called the warranted (technological) rate of growth.*

It is important to note that the eigenvectors Y_k, which corresponds to the eigenvalues λ_k that differ from λ_s ($\lambda_k \neq \lambda_s$), necessarily have the components with different signs.

Solutions (12.10) of the closed intersectoral model are combinations of exponential functions $\exp(\lambda_k t)$ with different rates of growth λ_k. In the solution, the term of (12.10), which has the maximum real part λ_k and corresponds to $c_k \neq 0$, begins to dominate at $t \to \infty$.

If the dominant is the term with $\lambda_s = 1/s_{max}$, then the rate of growth of final products tends to the warranted rate at $t \to \infty$ in all sectors. In this case, the sectoral structure of the national income in the limit ($t \to \infty$) will be determined by the proportions of the components of the corresponding eigenvector Y_k.

If the dominant term will have the growth rate $\lambda_k \neq \lambda_s = 1/s_{max}$, then the dynamics of $Y(t)$ at $t \to \infty$ will be determined by the corresponding eigenvector Y_k, the components of which have different signs. Therefore, this solution $Y(t)$ at $t \to \infty$ will necessarily have the negative components, which means that the solution loses the economic sense [146, p. 127]. As a result, solution (12.10) is economically unacceptable if the terms with the growth rate λ_k, which different from $\lambda_s = 1/s_{max}$, are dominated. This feature of the closed intersectoral models distinguishes this model from the one-sector model, in which the solutions are inadmissible because of the overstated (exorbitant) requirements to the growth of the consumption.

12.3 Dynamic intersectoral model with power-law memory

Equation (12.6), which describes the dynamic intersectoral model, assumes that the relationship between the vector $I(t)$ of capital and the vector $X(t)$ of the gross product is given by linear accelerator equation (12.5). Equation (12.5) contains derivative of the first order. The use of this derivative means an instantaneous change of the variable $I(t)$ at change of the variable $X^{(1)}(t)$. In other words, the investment at time t is determined by the properties of the vector of the gross product $X(t)$ in infinitely small neighborhood of this point in time. Therefore, accelerator equation (12.5) does not take into account the memory effects. As a result, the differential equations (12.6) and (12.9) can be used only to describe the economy, in which all economic agents have an instant amnesia. This restriction significantly reduces the scope of the intersectoral models to describe the real economic processes. In many cases, economic agents may remember the story of changes of the gross product and previous investments. Therefore, the neglect of the memory effects can lead to incorrect descriptions of economic processes.

To describe the power-law memory in dynamic intersectoral models, we should use a generalization of equation (12.5), which describes the relationship between the vector of investment $I(t)$ and the vector $X(t)$ of the gross product. To take into account the memory effects, the dependence of the vector $I(t)$ of capital investments on the vector $X(t)$ of the gross product can be described by the equations

$$I(t) = B \int_0^t M_X(t-\tau) X^{(n)}(\tau)\, d\tau, \qquad (12.15)$$

$$\int_0^t M_I(t-\tau) I(\tau)\, d\tau = BX^{(n)}(t), \qquad (12.16)$$

where $M_X(t-\tau)$ and $M_I(t-\tau)$ are functions, which are called the memory functions, and $X^{(n)}(\tau)$ is the derivative of the integer order $n \geq 0$. Equation (12.15) can be used to describe the time delay and memory effects with respect to the gross product. Equation (12.16) can describe the lag and memory effects with respect to the capital investments.

For processes without memory, the memory functions are expressed in terms of the Dirac delta-function in the form

$$M_X(t-\tau) = M_I(t-\tau) = \delta(t-\tau), \qquad (12.17)$$

where $\delta(t-\tau)$ is the Dirac delta-function.

Substitution of $M_X(t-\tau) = M_I(t-\tau) = \delta(t-\tau)$, into equations (12.15) and (12.16) with $n = 1$ gives equation (12.5) of economic accelerator $I(t) = BX^{(1)}(t)$ that describes process without memory and lag. In this case, the process connects the sequence of subsequent states of the economic process to the previous state only through the current state for each time t.

To describe the economic processes with power-law fading memory, we can use the memory functions in the form

$$M_X(t-\tau) = \frac{1}{\Gamma(\beta)} \frac{m(\beta)}{(t-\tau)^{1-\beta}}, \qquad (12.18)$$

$$M_I(t-\tau) = \frac{1}{\Gamma(\gamma)} \frac{m(\gamma)}{(t-\tau)^{1-\gamma}}, \qquad (12.19)$$

where $\Gamma(z)$ is the gamma function, β and γ are parameters that characterize the power-law fading, and $t > \tau$. For simplification, we assume that the parameters $m(\beta)$ and $m(\gamma)$ are equal to one.

Substitution of memory function (12.18) into equation (12.15) gives the fractional differential equations of the order $\alpha > 0$ in the form

$$I(t) = B(D^{\alpha}_{C;0+} X)(t), \qquad (12.20)$$

where $\alpha = n - \beta > 0$, $n - 1 < \alpha \leq n$, and $D^{\alpha}_{C;0+}$ is the left-sided Caputo fractional derivative of the order $\alpha \geq 0$ with respect to variable $t > 0$ [200, p. 92].

The Caputo fractional derivative of the order $\alpha \geq 0$ is defined by the equation

$$(D^{\alpha}_{C;0+}X)(t) = \frac{1}{\Gamma(n-\alpha)} \int_0^t \frac{X^{(n)}(\tau)\, d\tau}{(t-\tau)^{\alpha-n+1}}, \tag{12.21}$$

where $n-1 < \alpha \leq n$, $\Gamma(\alpha)$ is the gamma function, $X^{(n)}(\tau)$ is the derivative of the integer order $n \in \mathbb{N}$ of the function $X(\tau)$ with respect to time variable τ: $0 < \tau < t$. Expression (12.21) assumes that $Y(\tau) \in AC^n[0,T]$ (see Theorem 2.1 [200, p. 92]), i. e., the function $Y(\tau)$ has integer-order derivatives up to $(n-1)$-th order, which are continuous functions on the interval $[0,T]$, and the derivative $Y^{(n)}(\tau)$ is Lebesgue summable on the interval $[0,T]$. For $\alpha = 1$, equation (12.6) gives equation (12.5). To simplify the interpretation of the dimension of economic quantities, we can use the time t as a dimensionless variable by changing the variable $t \to tt_d$, where t_d is a time unit (hour, day, month year).

Substitution of memory function (12.19) into equation (12.16) gives the fractional integral equation in the form

$$(I^{\gamma}_{RL;0+}I)(t) = BX^{(n)}(\tau), \tag{12.22}$$

where $I^{\gamma}_{RL;0+}$ is the left-sided Riemann–Liouville fractional integral of the order $\gamma > 0$ with respect to time variable.

The left-sided Riemann–Liouville integral is defined [200, pp. 69–70] by the equation

$$(I^{\gamma}_{RL;0+}I)(t) = \frac{1}{\Gamma(\gamma)} \int_0^t (t-\tau)^{\gamma-1} I(\tau)\, d\tau, \tag{12.23}$$

where the function $I(t)$ is assumed to satisfy the condition $I(\tau) \in L_1[0,T]$ [335, 200].

It is known that the Caputo derivative is left-inverse to the Riemann–Liouville integral (see Lemma 2.21 in [200, p. 965]). In other words, for any continuous function $I(\tau) \in C[0,T]$, the equation

$$(D^{\gamma}_{C;0+} I^{\gamma}_{RL;0+} I)(t) = I(t), \tag{12.24}$$

is satisfied (see equation (2.4.32) in [200, p. 95]). As a result, the action of the Caputo fractional derivative $D^{\gamma}_{C;0+}$ on equation (12.22) gives

$$I(t) = B(D^{\alpha}_{C;0+}X)(t), \tag{12.25}$$

where $\alpha = n + \gamma$, and $D^{\alpha}_{C;0+}$ is the left-sided Caputo fractional derivative (12.21) of the order $\alpha \geq 0$ with respect to variable t. We can see that equations (12.20) and (12.25) are the same and differ only in the expressions for the parameter $\alpha > 0$, where $\alpha = n - \beta$ and $\alpha = n + \gamma$, respectively.

Equations (12.20) and (12.25) describe the same economic accelerator with the power-law memory, the fading parameter of which is $\alpha \geq 0$. As a result, we have an equation that allows us to describe the power-law memory with respect to the gross product and the capital investments. Therefore, we can use the linear equation of accelerator with power-law memory in the form

$$I(t) = B(D^{\alpha}_{C;0+}X)(t) \tag{12.26}$$

with $\alpha > 0$, and without further use of the parameters β, γ, and expressions $\alpha = n - \beta$, $\alpha = n + \gamma$.

For dynamic intersectoral model, equation (12.26) actually assumes a uniform parameter of memory fading for all sectors of economy. We first consider the dynamic models with a uniform fading parameter $\alpha \in (0, 2)$ for all sectors of economy. Then we will generalize the model to the case of sectoral memory, which is determined by the vector of fading parameters $\alpha = (\alpha_1, \dots, \alpha_n)$, where α_k is the parameter of the memory fading in the k-th sector of economy.

Substitution of equations (12.26) and (12.4) into balance equation (12.3), we get the fractional differential equation

$$B(D^{\alpha}_{C;0+}X)(t) + (A - E)X(t) + C(t) = 0. \tag{12.27}$$

Equation (12.27) describes dynamic intersectoral model, which takes into account power-law memory and generalizes the standard Leontief model that is described by equation (12.6). For $\alpha = 1$, equation (12.27) has the form (12.6).

The equation of the final product (national income) can be obtained from equation (12.27) by using expression (12.7) in the form

$$X(t) = (E - A)^{-1}Y(t). \tag{12.28}$$

If the matrix A is constant, then substituting expression (12.28) into equation (12.27), we get

$$B(E - A)^{-1}(D^{\alpha}_{C;0+}Y)(t) - Y(t) + C(t) = 0. \tag{12.29}$$

Equation (12.29) describes dynamics of final product $Y(t)$ in the dynamic intersectoral model with power-law memory. The matrix $B(E - A)^{-1}$ is called the matrix of the full incremental capital intensity.

12.4 Closed dynamic intersectoral model with memory

Let us consider closed dynamic model for the gross and final products. The closed models are described by equations (12.27) and (12.29) with zero nonproductive consumption ($C(t) = 0$). Using equation (12.27), the matrix equation for the gross product

$X(t)$ at $C(t) = 0$ has the form

$$B(D^{\alpha}_{C;0+}X)(t) + (A - E)X(t) = 0. \tag{12.30}$$

For $\alpha = 1$, equation (12.30) gives the equation of the standard closed dynamic intersectoral model without memory. Using equation (12.29) with $C(t) = 0$, we get the equation for the final product $Y(t)$ in the form

$$Y(t) = B(E - A)^{-1}(D^{\alpha}_{C;0+}Y)(t), \tag{12.31}$$

where $B(E-A)^{-1}$ is the matrix of the full incremental capital intensity. Equations (12.30) and (12.31) describe the dynamics of sectoral structure of the gross and final products in the closed dynamic intersectoral model with power-law memory. Solutions of closed model equations (12.30) and (12.31) characterize the extreme (limit) technological possibilities of production sectors, for given A and B, when the entire national income is directed to the expanded reproduction.

Equations (12.30) and (12.31) are the systems of linear fractional differential equations. If the matrix of the full incremental capital intensity is reversible, then equations (12.30) and (12.31) can be written as

$$(D^{\alpha}_{C;0+}X)(t) = B^{-1}(E - A)X(t), \tag{12.32}$$
$$(D^{\alpha}_{C;0+}Y)(t) = (E - A)B^{-1}Y(t). \tag{12.33}$$

Equations (12.32) and (12.33) describe a closed dynamic intersectoral model with the power-law memory, which has a uniform fading parameter for all sectors of economy. It is well known [200, p. 142] that the solution of the fractional differential equation

$$(D^{\alpha}_{C;0+}Y)(t) = \Lambda Y(t) \tag{12.34}$$

with the Caputo fractional derivative (12.21) of the order $\alpha \in (0, 1)$, and the matrix $\Lambda = (E - A)B^{-1}$, can be represented by the expression

$$Y(t) = E_{\alpha}[\Lambda t^{\alpha}]Y(0), \tag{12.35}$$

where $E_{\alpha}[\Lambda t^{\alpha}]$ is the matrix Mittag–Leffler function [200, p. 142] that is defined by the formula

$$E_{\alpha}[\Lambda t^{\alpha}] = \sum_{k=0}^{\infty} \frac{t^{\alpha k}}{\Gamma(\alpha k + 1)} \Lambda^k. \tag{12.36}$$

The general solution of fractional differential equations (12.34) with $n - 1 < \alpha \leq n$ can be written as

$$Y(t) = \sum_{k=1}^{n} c_k Y_k E_{\alpha}[\lambda_k t^{\alpha}], \tag{12.37}$$

where λ_k is the eigenvalues of the matrix $\Lambda = (E - A)B^{-1}$, and $Y_k = (Y_{kj})$ are the corresponding eigenvectors such that

$$\Lambda Y_k = \lambda_k Y_k. \tag{12.38}$$

In expression (12.36), the coefficients c_k are determined from the initial conditions by the equation

$$\sum_{k=1}^{n} c_k Y_k = Y(0). \tag{12.39}$$

Using $E_1[z] = \exp(z)$, we get that solution (12.37) with $\alpha = 1$ takes the form

$$Y(t) = \sum_{k=1}^{n} c_k Y_k \exp(\lambda_k t). \tag{12.40}$$

Therefore, solution (12.37) of the proposed model with memory contains solution (12.40) of the standard closed intersectoral model without memory as a special case that corresponds to $\alpha = 1$.

Similarly, the general solution of equation (12.32) can be written in the form

$$X(t) = \sum_{k=1}^{n} d_k X_k E_\alpha[\omega_k t^\alpha], \tag{12.41}$$

where ω_k are the eigenvalues of the matrix $\Omega = B^{-1}(E - A)$; the vectors $X_k = (X_{ki})$ are the eigenvectors of the matrix Ω, and the coefficients d_k are determined by the initial condition

$$\sum_{k=1}^{n} d_k X_k = X(0). \tag{12.42}$$

Using expression (12.8) and solution (12.37), the solution of equation (12.32) can be written as

$$X(t) = \sum_{k=1}^{n} c_k (E - A)^{-1} Y_k E_\alpha[\lambda_k t^\alpha], \tag{12.43}$$

which is equivalent to expression (12.41), where $\lambda_k = \omega_k$.

The solution is a linear combination of the Mittag–Leffler functions $E_\alpha[\lambda_k t^\alpha]$ with different eigenvalues λ_k. Therefore, in general, the dynamics of the economic process by the trajectory $Y(t) = Y(0)E_\alpha[\lambda t^\alpha]$, i. e., with the uniform parameter λ for all sectors, is impossible. It distinguishes the intersectoral model with memory and the one-sector models with memory. As a result, economic dynamics with memory will be realized with structural changes.

Note that there is a relationship between solutions (12.40), (12.41), (12.43) of the dynamic intersectoral model with memory and the solutions of the one-sector models with memory. This relationship is based, as well as for the standard model, on the existence of the eigenvalue s_{max} with the highest absolute value for the matrix $S = B(E - A)^{-1} = \Lambda^{-1}$ of the full incremental capital intensity. At the same time, absolute values of all other eigenvalues of the matrix S do not exceed the number s_{max}, which is the Frobenius–Perron number. The corresponding eigenvector may be selected so that all its components will be positive. In this case, the eigenvectors Y_k, which correspond to the eigenvalues $\lambda_k \neq \lambda_s$, have the components of different signs. The existence of such eigenvalue follows from the Perron's theorem [28, p. 319] and the Frobenius theorem [131, pp. 354–355]. Some roots λ_k of the characteristic equation of the matrix $\Lambda = (E-A)B^{-1}$ can be complex. Thus each complex root $\lambda_k = \mathrm{Re}(\lambda_k)+i\,\mathrm{Im}(\lambda_k)$ corresponds to the conjugate root $\lambda_k = \mathrm{Re}(\lambda_k)-i\,\mathrm{Im}(\lambda_k)$. Each pair of complex conjugate roots will be presented by a pair of terms in the solution. In this case, we have oscillations with the constant rate, which is equal to $\mathrm{Im}(\lambda_k)$, and the variable amplitude.

Let us now analyze the dominant behavior at $t \to \infty$ for dynamic intersectoral model with memory. For this, we use the asymptotic formula of the Mittag–Leffler functions $E_\alpha[\lambda_k t^\alpha]$ at $t \to \infty$. Using equations (3.4.14 and 3.4.15) of [143, pp. 25–26], for $0 < \alpha < 2$ and $t \to \infty$, we obtain the equation

$$E_\alpha[\lambda t^\alpha] = \frac{1}{\alpha}\exp(\lambda^{1/\alpha}t) - \sum_{k=1}^{m}\frac{\lambda^{-k}}{\Gamma(1-\alpha k)}\frac{1}{t^{\alpha k}} + O\left(\frac{1}{t^{\alpha(m+1)}}\right) \tag{12.44}$$

for real values of λ and for complex roots with $|\arg(\lambda)| \leq \theta$, where

$$\arg(\lambda) = \mathrm{arctg}\left(\frac{\mathrm{Im}(\lambda)}{\mathrm{Re}(\lambda)}\right), \quad \frac{\pi\alpha}{2} < \theta < \min\{\pi, \pi\alpha\}. \tag{12.45}$$

For complex values of λ, for which $\theta \leq |\arg(\lambda)| \leq \pi$, we have

$$E_\alpha[\lambda t^\alpha] = -\sum_{k=1}^{m}\frac{\lambda^{-k}}{\Gamma(1-\alpha k)}\frac{1}{t^{\alpha k}} + O\left(\frac{1}{t^{\alpha(m+1)}}\right), \tag{12.46}$$

where $\Gamma(1-\alpha) < 0$ for $0 < \alpha < 1$ and $\Gamma(1-\alpha) > 0$ for $1 < \alpha < 2$.

As a result, for the real values of λ, we find that the growth rate of the intersectoral model with memory does not coincide with the eigenvalues λ_k of the matrix Λ. The growth rate is equal to the values

$$\lambda_{k,\mathrm{eff}}(\alpha) = \lambda_k^{1/\alpha}, \tag{12.47}$$

which will be interpreted as the effective (warranted) rates of growth with memory. For complex values of λ, for which $\theta \leq |\arg(\lambda)| \leq \pi$, the dynamics at $t \to \infty$ is determined by the inverse proportionality $t^{-\alpha}$ instead of the exponential functions. In the presence of terms with the exponential behavior at $t \to \infty$, the contribution of the terms with $t^{-\alpha}$ can be neglected at $t \to \infty$.

Solution (12.37) of equation (12.33) is a combination of the Mittag–Leffler functions $E_\alpha[\lambda_k t^\alpha]$ with different parameters λ_k, which correspond to different effective growth rates $\lambda_{k,\text{eff}}(\alpha)$. In solution (12.37) at $t \to \infty$ will dominate the term with the maximum real part of $\lambda_{k,\text{eff}}(\alpha)$, for which $c_k \neq 0$ and $|\arg(\lambda)| \leq \theta$, where θ satisfies inequality (12.45). If the dominant term has the effective growth rate

$$\lambda_{s,\text{eff}}(\alpha) = (1/s_{\max})^{1/\alpha}, \tag{12.48}$$

where $s_{\max}$ is the Frobenius–Perron number of the matrix S, then the growth rate of final products in all sectors at $t \to \infty$ tends to the effective warranted rate of growth with memory (12.48). In this case, the sectoral structure of the national income at $t \to \infty$ is determined by the proportion between the components of the eigenvector Y_s, which corresponds to $\lambda_s = 1/s_{\max}$. If the dominant is a term with $\lambda_{k,\text{eff}}(\alpha) \neq \lambda_{s,\text{eff}}(\alpha)$, then the dynamics of $Y(t)$ at $t \to \infty$ is determined by the corresponding eigenvector Y_k, which components have different signs. Therefore, the solution $Y(t)$ at $t \to \infty$ will necessarily have negative components. This means that the solution loses an economic sense. As a result, solutions (12.37), (12.41), (12.43), where the terms with $\lambda_{k,\text{eff}}(\alpha) \neq \lambda_{s,\text{eff}}(\alpha)$ are dominated, are economically unacceptable and these solutions do not have an economic sense. Note that the solutions of closed one-sector models become unacceptable because of the large requirements for the growth of consumption.

A comparison of the growth rates of the intersectoral models with power-law memory and the standard model without memory, is given in Table 12.1 for the real values of λ_k and $0 < \alpha < 2$.

Table 12.1: Comparison of the growth rates of the intersectoral models with power-law memory and the standard model without memory.

λ_k	$0 < \alpha < 1$	$1 < \alpha < 2$
$0 < \lambda_k < 1$	$\lambda_{k,\text{eff}}(\alpha) < \lambda_k$	$\lambda_{k,\text{eff}}(\alpha) > \lambda_k$
$\lambda_k = 1$	$\lambda_{k,\text{eff}}(\alpha) = \lambda_k$	$\lambda_{k,\text{eff}}(\alpha) = \lambda_k$
$\lambda_k > 1$	$\lambda_{k,\text{eff}}(\alpha) > \lambda_k$	$\lambda_{k,\text{eff}}(\alpha) < \lambda_k$

Table 12.1 shows that the memory effects can significantly change the growth rates of the dynamic intersectoral models. At the same time, the growth rates can increase and decrease in comparison with the standard model. If the Frobenius–Perron number of the matrix of the full incremental capital intensity is less than one ($0 < s_{\max} < 1$), the corresponding growth rate can be much larger if we take into account the memory effects. For example, when $\alpha = 0.2$ and $s_{\max} = 0.5$, the growth rate in the model with memory is $\lambda_{s,\text{eff}}(0.5) = 32$ instead of $\lambda_s = 2$ for the standard model, i. e., it will be 16 times more.

These results allow us to formulate the following principle.

Principle 12.1 (Principle of changing of warranted rates of growth by memory). *For low warranted rate of growth without memory ($\lambda < 1$), the accounting for the memory* *with fading parameter $\alpha \in (0,1)$ leads to a decrease in the rate of economic growth. The accounting for the memory with fading parameter $\alpha \in (1,2)$ leads to an increase in the growth rate. For high warranted rates of growth without memory ($\lambda > 1$), the accounting for the memory with the fading parameter $\alpha \in (0,1)$ leads to an increase in economic growth, and it leads to a decrease in the growth rate for $\alpha \in (1,2)$.*

An example, which illustrates this principle, is presented in the next sections of this chapter.

For $1 < \alpha < 2$, the general solution of the fractional differential equation

$$(D^{\alpha}_{C;0+}Y)(t) = \Lambda Y(t) \tag{12.49}$$

with the Caputo fractional derivative of the order $1 < \alpha < 2$ can be written [200, p. 232] in the form

$$Y(t) = \sum_{k=1}^{n} c_{1k} Y_k E_{\alpha}[\lambda_k t^{\alpha}] + \sum_{k=1}^{n} c_{2k} Y_k t E_{\alpha,2}[\lambda_k t^{\alpha}], \tag{12.50}$$

where $E_{\alpha,2}[z]$ is the two-parameter Mittag–Leffler function (see equation (1.8.17) in [200, p. 42]), which is defined by the equation

$$E_{\alpha,\beta}[z] = \sum_{k=0}^{\infty} \frac{z^k}{\Gamma(\alpha k + \beta)}, \tag{12.51}$$

where $\mathrm{Re}(\alpha) > 0$, $\beta, z \in \mathbb{C}$. For $\beta = 1$, function (12.51) takes the form of Mittag–Leffler function (12.36), such that $E_{\alpha,1}[z] = E_{\alpha}[z]$. The numbers λ_k are eigenvalues of the matrix $\Lambda = (E - A)B^{-1}$, and $Y_k = (Y_{kj})$ are corresponding eigenvectors such that

$$\Lambda Y_k = \lambda_k Y_k. \tag{12.52}$$

The coefficients c_{1k} and c_{2k} are determined by the initial conditions in the form

$$\sum_{k=1}^{n} c_{1k} Y_k = Y(0), \tag{12.53}$$

$$\sum_{k=1}^{n} c_{2k} Y_k = Y^{(1)}(0), \tag{12.54}$$

where $Y(0)$ and $Y^{(1)}(0)$ are the national income and the speed of change of the national income at the initial time $t = 0$, respectively.

Let us give the asymptotic formulas for the two-parameter Mittag–Leffler function $E_{\alpha,2}[\lambda_k t^{\alpha}]$ at $t \to \infty$. Using equation (1.8.27 and 1.8.28) of [200, p. 43], we obtain for $0 < \alpha < 2$ and $t \to \infty$ the formula

$$E_{\alpha,2}[\lambda t^{\alpha}] = \frac{\lambda^{-1/\alpha}}{\alpha} t^{-1} \exp(\lambda^{1/\alpha} t) - \sum_{k=1}^{m} \frac{\lambda^{-k}}{\Gamma(2 - \alpha k)} \frac{1}{t^{\alpha k}} + O\left(\frac{1}{t^{\alpha(m+1)}}\right) \tag{12.55}$$

for real values of λ and the complex values of λ, for which $|\arg(\lambda)| \leq \theta$, where $\pi\alpha/2 < \theta < \min\{\pi, \pi\alpha\}$. For complex values of λ, for which $\theta \leq |\arg(\lambda)| \leq \pi$, we have

$$E_{\alpha,2}[\lambda t^{\alpha}] = -\sum_{k=1}^{m} \frac{\lambda^{-k}}{\Gamma(2-\alpha k)} \frac{1}{t^{\alpha k}} + O\left(\frac{1}{t^{\alpha(m+1)}}\right). \tag{12.56}$$

Asymptotic formulas (12.55) and (12.56) allow us to describe the dominant behavior at $t \to \infty$ in the intersectoral dynamic model with the power-law memory, which has the fading parameter $0 < \alpha < 2$.

12.5 Open dynamic intersectoral model with memory

The open dynamic intersectoral model with power-law memory is described by equations (12.27) and (12.29) with nonzero nonproductive consumption ($C(t) \neq 0$). Solutions of equations (12.27) and (12.29) were derived in articles [437, 451] (see also [433, 398]).

Let us consider an open dynamic model with memory for the vector of the final product (national income). This model is described by the matrix fractional differential equation (12.29). Using the reversibility of the matrix B, equation (12.29) can rewritten in the form

$$(D^{\alpha}_{C;0+}Y)(t) = (E - A)B^{-1}Y(t) - (E - A)B^{-1}C(t). \tag{12.57}$$

Equation (12.57) is inhomogeneous fractional differential equation, which describes the open dynamic intersectoral model with power-law memory.

Let us consider equation (12.57) with the vector-function of nonproductive consumption $C(t)$ in the form

$$C(t) = C(0)E_{\alpha}[rt^{\alpha}], \tag{12.58}$$

where we assumed that the components of nonproductive consumption grow with the uniform rate in all sectors. For $r = 0$, we have $C(t) = C(0) = \text{const}$. For $\alpha = 1$, expression (12.58) takes the form $C(t) = C(0)\exp(rt)$.

Let us consider the economic model that is described by equation (12.57), which takes into account the dynamics of nonproductive consumption and power-law memory. The general solution of fractional differential equations (12.57) can be represented as the sum of the general solution $Y_0(t)$ of the closed model, which is described by homogeneous equation (12.33), and a particular solution $Y_C(t)$ of the inhomogeneous equation (12.57), such that

$$Y(t) = Y_0(t) + Y_C(t). \tag{12.59}$$

A particular solution of equation (12.57) with $C(t)$ in the form (12.58) is given by the expression

$$Y_C(t) = (E - rS)^{-1} C(0) E_\alpha[rt^\alpha], \tag{12.60}$$

where $S = B(E - A)^{-1}$ is the matrix of the full incremental capital intensity.

For $0 < \alpha < 1$, the general solution of open dynamic intersectoral model with power-law memory can be represented in the form

$$Y(t) = \sum_{k=1}^{n} c_k Y_k E_\alpha[\lambda_k t^\alpha] + (E - rS)^{-1} C(0) E_\alpha[rt^\alpha], \tag{12.61}$$

where λ_k are eigenvalues of the matrix $\Lambda = (E-A)B^{-1}$, and $Y_k = (Y_{kj})$ are corresponding eigenvectors ($\Lambda Y_k = \lambda_k Y_k$). For the case $0 < \alpha < 1$, the coefficients c_k are determined from the initial condition by the equation

$$\sum_{k=1}^{n} c_k Y_k = Y(0) - (E - rS)^{-1} C(0). \tag{12.62}$$

If $C(0) = 0$, then solution (12.61) coincides with solution (12.37), and condition (12.62) becomes (12.39).

For $1 < \alpha < 2$, the solutions of open model (12.57) with the power-law memory are obtained similarly, i. e., by using solution (12.50) and conditions (12.53), (12.54) for the closed model (12.33) with the memory fading parameter $\alpha \in (1, 2)$.

To get a solution of the open growth model, we should find the eigenvalues λ_k of the matrix Λ, and the corresponding eigenvectors $Y_k = (Y_{kj})$. Then we should set the growth rate r of nonproductive consumption and find the values of the coefficients c_k, which correspond to the given growth rate r, from the system of algebraic equations (12.62).

Standard models without memory assume that r must not exceed the warranted rate of growth, due to the fact that the productivity condition of the matrix rS gives the inequality $r < \lambda_s$ [285] and [146, p. 130]. If $r > \lambda_s$, then $Y(t)$ grows faster than the term that contains λ_s. The matrix rS is unproductive and, therefore, the vector $Y_C(t)$ will have negative components. Since some components $Y_C(t)$ is negative, then we get negative components of the vector $Y(t)$, which means that this solution does not make economic sense. As a result, a necessary condition for the existence of economically interpretable solutions of national income is $r < \lambda_s$. In models with memory, which are characterized by uniform parameter of memory fading for all sectors, the condition $r < \lambda_s$ does not change.

12.6 Dynamic intersectoral model with sectoral memory

In the previous sections, we assumed a uniform parameter of memory fading for all sectors of economy. In general, the different sectors can be characterized by various fading parameters. In this case, the model can be called the intersectoral model with sectoral memory. Fading of the memory will be determined by the fading vector

$$\alpha = (\alpha_1, \ldots, \alpha_n). \tag{12.63}$$

To describe the dynamics of sectors with different fading parameters of power-law memory, we can use the matrix differentiation of noninteger order. For example, we can use the diagonal matrix operators

$$\widehat{D^{\alpha}_{C;0+}} = \operatorname{diag}(D^{\alpha_1}_{C;0+}, \ldots, D^{\alpha_n}_{C;0+}), \tag{12.64}$$

where the elements of this matrix are the operators $\widehat{D^{\alpha}_{kl}} = \delta_{kl} D^{\alpha_k}_{C;0+}$, where α_k is the fading parameter of the k-th sector of economy. For dynamic models with two sectors, operator (12.64) can be written in the form

$$\widehat{D^{\alpha}_{C;0+}} = \begin{pmatrix} D^{\alpha_1}_{C;0+} & 0 \\ 0 & D^{\alpha_2}_{C;0+} \end{pmatrix}, \tag{12.65}$$

and the action of operator (12.65) on the vector $Y(t)$ of the final product is given by the expression

$$(\widehat{D^{\alpha}_{C;0+}} Y)(t) = \begin{pmatrix} (D^{\alpha_1}_{C;0+} Y_1)(t) & 0 \\ 0 & (D^{\alpha_2}_{C;0+} Y_2)(t) \end{pmatrix}. \tag{12.66}$$

The closed dynamic intersectoral model with sectoral memory, which is characterized by the different fading parameters in the different sectors, can be described by the matrix fractional differential equation

$$(\widehat{D^{\alpha}_{C;0+}} Y)(t) = \Lambda Y(t), \tag{12.67}$$

where $\Lambda = (E - A)B^{-1}$. The solution of equation (12.67) cannot be represented by the equation $Y(t) = Y E_{\alpha}[\Lambda t^{\alpha}]$, where $Y = Y(0)$. The solution of equation (12.67) can be derived in the form

$$Y(t) = \widehat{E_{\alpha}}[\Lambda t^{\alpha}] Y, \tag{12.68}$$

where

$$\widehat{E_{\alpha}}[\Lambda t^{\alpha}] = \operatorname{diag}(E_{\alpha_1}[\Lambda t^{\alpha_1}], \ldots, E_{\alpha_n}[\Lambda t^{\alpha_n}]) \tag{12.69}$$

is the square diagonal matrix of the Mittag–Leffler functions (12.36). For the two-sector model, the matrix function $\widehat{E_{\alpha}}[\Lambda t^{\alpha}]$ can be written in the form

$$\widehat{E_{\alpha}}[\Lambda t^{\alpha}] = \begin{pmatrix} E_{\alpha_1}[\Lambda t^{\alpha_1}] & 0 \\ 0 & E_{\alpha_2}[\Lambda t^{\alpha_2}] \end{pmatrix}. \tag{12.70}$$

For the matrix function $\widehat{E_\alpha}[\Lambda t^\alpha]$, it is easy to check the property of the form

$$(\widehat{D^\alpha_{C;0+} E_\alpha}[\Lambda \tau^\alpha])(t) = \Lambda \widehat{E_\alpha}[\Lambda t^\alpha]. \tag{12.71}$$

Using equation (12.71), the vectors Y must satisfy the equation

$$(\Lambda - \lambda E) Y = 0. \tag{12.72}$$

In order to equation (12.72) holds for the nontrivial vector Y, it is necessary and sufficient that the matrix $(\Lambda - \lambda E)$ is singular, i. e., its determinant should be equal to zero: $|\Lambda - \lambda E| = 0$. For the roots λ_k of the characteristic equation $|\Lambda - \lambda E| = 0$, we can find the nonzero vectors $Y_k = (Y_{kj})$ by using equation (12.72). As a result, the general solution of fractional differential equations (12.67) with $0 < \alpha_k < 1$ for all $k = 1, \dots, n$ can be written in the form

$$Y(t) = \sum_{k=1}^{n} c_k \widehat{E_\alpha}[\lambda_k t^{\alpha_k}] Y_k. \tag{12.73}$$

In the component form, equation (12.73) can be written as

$$Y_j(t) = \sum_{k=1}^{n} c_k E_{\alpha_j}[\lambda_k t^{\alpha_j}] Y_{kj}, \tag{12.74}$$

where λ_k are eigenvalues of the matrix $\Lambda = (E - A)B^{-1}$, and $Y_k = (Y_{kj})$ are the corresponding eigenvectors ($\Lambda Y_k = \lambda_k Y_k$). The coefficients c_k are determined by the initial conditions

$$\sum_{k=1}^{n} c_k Y_k = Y(0). \tag{12.75}$$

Equation (12.73) at $t = 0$ gives (12.75) since $\widehat{E_\alpha}[0] = E$, where E is the unit diagonal matrix.

Using equation (12.44) for real values of λ, we find that the growth rates are given by the values

$$\lambda_{k,\text{eff}}(\alpha_k) = \lambda_k^{1/\alpha_k}. \tag{12.76}$$

Expression (12.76) is interpreted as the warranted rate of growth with memory of the k-th sector. In contrast to case (12.47), the warranted rates of growth with memory are not the same for different sectors of the economy in general. As a result, in the ordered series of the growth rates λ_k, the order can be changed, when we take into account the memory effects. For example, for a three sectoral model without memory, for which the growth rates satisfy the inequalities $\lambda_1 > \lambda_2 > \lambda_3$, consideration the sectoral memory can lead to values warranted rates of growth with memory in the form $\lambda_{3,\mathrm{eff}}(\alpha_3) > \lambda_{1,\mathrm{eff}}(\alpha_1) > \lambda_{2,\mathrm{eff}}(\alpha_2)$, for some values of the fading parameters α_1, α_2, α_3.

In solution (12.73) at $t \to \infty$, the term with k begins to dominate, if $\lambda_{k,\mathrm{eff}}(\alpha_k)$ has the maximum real part, $c_k \neq 0$, and the parameter θ satisfies inequality (12.45). If the dominant term has the effective growth rate

$$\lambda_{s,\mathrm{eff}}(\alpha_s) = (1/s_{\max})^{1/\alpha_s}, \tag{12.77}$$

where $s_{\max}$ is the Frobenius–Perron number of the matrix S, then the growth rate of the final product (national income) for all sectors will tend to the effective warranted rate (12.77) at $t \to \infty$. The sectoral structure of the national income in the limit will be determined by the proportions between the components of the eigenvector Y_s of matrix Λ, which corresponds to $\lambda_s = 1/s_{\max}$.

In general, the warranted rates of growth with memory for different sectors of the economy do not coincide. However, warranted rate of growth with memory may be the same in different sectors with different eigenvalues. The condition for the equality of effective growth with memory can be expressed by the equation

$$\frac{\alpha_k}{\alpha_l} = \frac{\ln(\lambda_k)}{\ln(\lambda_l)}. \tag{12.78}$$

If $\alpha_k \in (1,2)$ for all $k = 1, \dots, n$, then the general solution of the fractional differential equation

$$(\widehat{D^{\alpha}_{C;0+}} Y)(t) = \Lambda Y(t) \tag{12.79}$$

with the matrix Caputo fractional derivative (12.64) of order $1 < \alpha < 2$ can be written as

$$Y(t) = \sum_{k=1}^{n} c_{1k} \widehat{E_{\alpha}}[\lambda_k t^{\alpha}] Y_k + \sum_{k=1}^{n} c_{2k} t \widehat{E_{\alpha,2}}[\lambda_k t^{\alpha}] Y_k, \tag{12.80}$$

where

$$\widehat{E_{\alpha,2}}[\lambda t^{\alpha}] = \mathrm{diag}(E_{\alpha_1,2}[\lambda t^{\alpha_1}], \dots, E_{\alpha_n,2}[\lambda t^{\alpha_n}]) \tag{12.81}$$

is the square diagonal matrix of the two-parameter Mittag–Leffler functions (12.51).

If a part of the fading parameters belongs to the interval $\alpha_i \in (0,1)$, and the other part of the parameters is in the range $\alpha_j \in (1,2)$, then in solution (12.80) terms with $\widehat{E_{\alpha,2}}[\lambda_k t^{\alpha}]$ are absent for $\alpha_k \in (0,1)$.

The open dynamic intersectoral model with sectoral memory and nonzero nonproductive consumption ($C(t) \neq 0$) is described by the matrix fractional differential equation

$$(\widehat{D^{\alpha}_{C;0+}} Y)(t) = (E - A)B^{-1} Y(t) - (E - A)B^{-1} C(t). \tag{12.82}$$

Let us consider the vector function of nonproductive consumption $C(t)$ in the form

$$C(t) = \widehat{E_\alpha}[rt^\alpha]C(0), \tag{12.83}$$

where $\widehat{E_\alpha}[rt^\alpha]$ is the square diagonal matrix, and $C(0)$ is the vector, which defines nonproductive consumption in sectors at the initial time $t = 0$. Here, we do not impose the restriction that components of the vector $C(t)$ has the same constant growth rate. The components of the vector $C(t)$ has the form $C_k(t) = E_{\alpha_k}[r_k t^{\alpha_k}]C_k(0)$, where r_k is the growth rate of nonproductive consumption in the k-th sector.

The general solution of fractional differential equations (12.82) can be represented as the sum of the general solution $Y_0(t)$ of the closed model, which is described by equation (12.73), and a particular solution $Y_C(t)$ of the inhomogeneous equation (12.82), such that

$$Y(t) = Y_0(t) + Y_C(t). \tag{12.84}$$

As a result, the particular solution of equation (12.82) is given by the expression

$$Y_C(t) = (E - RS)^{-1}\widehat{E_\alpha}[rt^\alpha]C(0), \tag{12.85}$$

where $S = B(E - A)^{-1}$ is the matrix of the full incremental capital intensity, and $R = \mathrm{diag}(r_1, \ldots, r_n)$ is the matrix of the growth rate of nonproductive consumption. If $0 < \alpha_k < 1$ for all $k = 1, \ldots, n$, the general solution of the open dynamic intersectoral model with sectoral memory can be written in the form

$$Y(t) = \sum_{k=1}^{n} c_k \widehat{E_\alpha}[\lambda_k t^\alpha] Y_k + (E - RS)^{-1}\widehat{E_\alpha}[rt^\alpha]C(0), \tag{12.86}$$

where λ_k are eigenvalues of the matrix $\Lambda = (E - A)B^{-1}$, and $Y_k = (Y_{kj})$ are the corresponding eigenvectors ($\Lambda Y_k = \lambda_k Y_k$). The coefficients c_k are determined (for the case $\alpha_k \in (0, 1)$ for all $k = 1, \ldots, n$) by the initial conditions

$$\sum_{k=1}^{n} c_k Y_k = Y(0) - (E - RS)^{-1}C(0). \tag{12.87}$$

For $R = rE$, condition (12.87) takes standard form (12.62).

If $\alpha_k \in (1, 2)$ for all $k = 1, \ldots, n$, then the solutions of open model with sectoral memory are obtained analogously to the solutions from the closed model with $1 < \alpha_k < 2$ for all $k = 1, \ldots, n$.

A comparison of the growth rates of the intersectoral models with sectoral memory and the standard model without memory, is given in Table 12.2 for the real values of λ_k and $0 < \alpha_k < 2$.

Table 12.2 shows that the accounting of sectoral memory can change the growth rates of the economy and its sectors in the framework of the dynamic intersectoral

Table 12.2: Comparison of the growth rates of the intersectoral models with sectoral memory and the standard model without memory.

λ_k	$0 < \alpha_k < 1$	$1 < \alpha_k < 2$
$0 < \lambda_k < 1$	$\lambda_{k,\text{eff}}(\alpha_k) < \lambda_k$	$\lambda_{k,\text{eff}}(\alpha_k) > \lambda_k$
$\lambda_k = 1$	$\lambda_{k,\text{eff}}(\alpha_k) = \lambda_k$	$\lambda_{k,\text{eff}}(\alpha_k) = \lambda_k$
$\lambda_k > 1$	$\lambda_{k,\text{eff}}(\alpha_k) > \lambda_k$	$\lambda_{k,\text{eff}}(\alpha_k) < \lambda_k$

models. At the same time, the growth rates can increase and decrease in comparison with the standard intersectoral model without memory.

Using these results, we can formulate the following principle.

Principle 12.2 (Principle of domination change). *In intersectoral economic dynamics, the effects of fading sectoral memory can change the dominating behavior of economic sectors.*

For example, according to this principle, the inequality $\lambda_1 < \lambda_2$ of the standard model can leads to the inequality

$$\lambda_{1,\text{eff}}(\alpha_1) > \lambda_{2,\text{eff}}(\alpha_2) \tag{12.88}$$

of the model with power-law sectoral memory. More detailed example that illustrates this principle is presented in the next section about the third example.

12.7 First example of two-sectoral model with memory

For illustration, we present numerical examples of finding solutions of the closed dynamic intersectoral models with power-low memory. Solutions of the dynamic model of the final and gross products will be illustrated by the examples of two sectors of economy. For the calculations, we used the Maple computer algebra package.

In this section, we give the first illustrative example of dynamic two-sectoral Leontief model with power-law memory. Let us define the matrices A and B in the form

$$A = \begin{pmatrix} 0.1 & 0.2 \\ 0.2 & 0.3 \end{pmatrix}, \quad B = \begin{pmatrix} 0.4 & 0.4 \\ 1.0 & 0.5 \end{pmatrix}. \tag{12.89}$$

The inverse matrix B^{-1} and the matrix $(E - A)$ are represented by the expressions

$$B^{-1} = \begin{pmatrix} -2.5 & 2 \\ 5 & -2 \end{pmatrix}, \quad (E - A) = \begin{pmatrix} 0.9 & -0.2 \\ -0.2 & 0.7 \end{pmatrix}. \tag{12.90}$$

Then the products of these matrices have the form

$$\Lambda = (E - A)B^{-1} = \begin{pmatrix} -3.25 & 2.2 \\ 4.0 & -1.8 \end{pmatrix}, \tag{12.91}$$

$$\Omega = B^{-1}(E-A) = \begin{pmatrix} -2.65 & 1.9 \\ 4.9 & -2.4 \end{pmatrix}. \tag{12.92}$$

For matrices (12.91) and (12.92), equations (12.32) and (12.33) have the form

$$D^{\alpha}_{C;0+}\begin{pmatrix} Y_1(t) \\ Y_2(t) \end{pmatrix} = \begin{pmatrix} -3.25 & 2.2 \\ 4.0 & -1.8 \end{pmatrix}\begin{pmatrix} Y_1(t) \\ Y_2(t) \end{pmatrix}, \tag{12.93}$$

$$D^{\alpha}_{C;0+}\begin{pmatrix} X_1(t) \\ X_2(t) \end{pmatrix} = \begin{pmatrix} -2.65 & 1.9 \\ 4.9 & -2.4 \end{pmatrix}\begin{pmatrix} X_1(t) \\ X_2(t) \end{pmatrix}, \tag{12.94}$$

where $0 < \alpha < 1$. The characteristic equations for matrices (12.91) and (12.92) have the form

$$\begin{vmatrix} -3.25-\lambda & 2.2 \\ 4.0 & -1.8-\lambda \end{vmatrix} = 0, \tag{12.95}$$

$$\begin{vmatrix} -2.65-\omega & 1.9 \\ 4.9 & -2.4-\omega \end{vmatrix} = 0, \tag{12.96}$$

which lead to the quadratic equations

$$\lambda^2 + 5.05\lambda - 2.95 = 0, \tag{12.97}$$

$$\omega^2 + 5.05\omega - 2.95 = 0. \tag{12.98}$$

The roots of characteristic equations (12.97) and (12.98) are the numbers

$$\lambda_1 \approx 0.5287886305, \quad \lambda_2 \approx -5.578788631, \tag{12.99}$$

where $\omega_1 = \lambda_1$, $\omega_2 = \lambda_2$. As a result, we have equations for the eigenvectors

$$\begin{pmatrix} -3.25 & 2.2 \\ 4.0 & -1.8 \end{pmatrix} Y_k = \lambda_k Y_k, \tag{12.100}$$

$$\begin{pmatrix} -2.65 & 1.9 \\ 4.9 & -2.4 \end{pmatrix} X_k = \omega_k X_k, \tag{12.101}$$

where $k = 1, 2$, and $\lambda_k = \omega_k$ take values (12.99). Solutions of equations (12.100) and (12.101) give the eigenvectors

$$Y_1 = \begin{pmatrix} 0.5245339728 \\ 0.9009559150 \end{pmatrix}, \quad Y_2 = \begin{pmatrix} 0.6867206226 \\ -0.7269214446 \end{pmatrix}, \tag{12.102}$$

$$X_1 = \begin{pmatrix} 0.5716020012 \\ 0.9563168119 \end{pmatrix}, \quad X_2 = \begin{pmatrix} 0.5442405426 \\ -0.8389292175 \end{pmatrix}. \tag{12.103}$$

Components of eigenvectors (12.102) and (12.103) were obtained by using the Maple computer algebra package. Coefficients c_k can be derived from the initial conditions by using equation (12.39) in the form

$$c_1 Y_1 + c_2 Y_2 = Y(0). \tag{12.104}$$

Let us use the initial conditions for the final product in the form

$$Y(0) = \begin{pmatrix} Y_1(0) \\ Y_2(0) \end{pmatrix} = \begin{pmatrix} 40 \\ 40 \end{pmatrix}. \tag{12.105}$$

Then the system of linear equations (12.104) with vectors (12.102) has the form

$$\begin{cases} 0.5245339728c_1 + 0.6867206226c_2 = 40, \\ 0.9009559150c_1 - 0.7269214446c_2 = 40. \end{cases} \tag{12.106}$$

The system of algebraic equations (12.106) has the solution

$$c_1 \approx 15.05687769, \quad c_2 \approx 56.54568268. \tag{12.107}$$

As a result, we get the functions of the final product of the first and second sectors

$$\begin{cases} Y_1(t) = 29.660E_\alpha[0.52879t^\alpha] + 10.340E_\alpha[-5.5788t^\alpha], \\ Y_2(t) = 50.945E_\alpha[0.52879t^\alpha] - 10.945E_\alpha[-5.5788t^\alpha], \end{cases} \tag{12.108}$$

where we rounded the numerical parameters to five significant digits.

Using $E_\alpha[0] = 1$, we can see that solutions (12.108) at the initial time ($t = 0$) give the components of the final product $Y(0)$ for the first and second sectors in the form

$$Y_1(0) = 29.660 + 10.340 = 40, \tag{12.109}$$

$$Y_2(0) = 50.945 - 10.945 = 40, \tag{12.110}$$

which coincides with the initial conditions (12.105). For $\alpha = 1$, solution (12.108) takes the form

$$\begin{cases} Y_1(t) = 29.660\exp(0.52879t) + 10.340\exp(-5.5788t), \\ Y_2(t) = 50.945\exp(0.52879t) - 10.945\exp(-5.5788t), \end{cases} \tag{12.111}$$

where we use the property $E_1[z] = \exp(z)$. Equations (12.111) describe a two-sector model without memory.

For the vector $X(t)$ of gross product, the coefficients d_k are determined by the initial conditions

$$d_1X_1 + d_2X_2 = X(0), \tag{12.112}$$

where X_1 and X_2 are defined by expressions (12.103). Let us consider the initial conditions for the gross product in the form

$$X(0) = \begin{pmatrix} X_1(0) \\ X_2(0) \end{pmatrix} = \begin{pmatrix} 50 \\ 30 \end{pmatrix}. \tag{12.113}$$

Then the system of equations (12.112) with the vectors (12.103) and (12.113) has the form

$$\begin{cases} 0.571602001d_1 + 0.5442405426d_2 = 50, \\ 0.9563168119d_1 - 0.8389292175d_2 = 30. \end{cases} \tag{12.114}$$

Solving this system of linear equations, we get

$$d_1 \approx 58.27367714, \quad d_2 \approx 30.66778055. \tag{12.115}$$

Using the Maple to obtain the functions for gross product of the first and second sectors, we get

$$\begin{cases} X_1(t) = 33.309E_\alpha[0.52879t^\alpha] + 16.691E_\alpha[-5.5788t^\alpha], \\ X_2(t) = 55.728E_\alpha[0.52879t^\alpha] - 25.728E_\alpha[-5.5788t^\alpha], \end{cases} \tag{12.116}$$

where we rounded the numerical parameters to five significant digits. For $\alpha = 1$, solution (12.116) is written as the linear combination of exponents

$$\begin{cases} X_1(t) = 33.309\exp(0.52879t) + 16.691\exp(-5.5788t), \\ X_2(t) = 55.728\exp(0.52879t) - 25.728\exp(-5.5788t). \end{cases} \tag{12.117}$$

Equations (12.117) describe a two-sector model without memory.

Let us find the Frobenius–Perron number of the matrix $S = B(E-A)^{-1} = \Lambda^{-1}$ of the full incremental capital intensity. Using matrices (12.89), we get

$$S = B(E-A)^{-1} = \begin{pmatrix} 0.6101694915 & 0.7457627119 \\ 1.355932203 & 1.101694915 \end{pmatrix}. \tag{12.118}$$

We see that the matrix S is positive. Using the Maple package, we obtain the eigenvalues of the matrix (12.118) in the form

$$s_1 \approx 1.891114790, \quad s_2 \approx -0.179250383. \tag{12.119}$$

As a result, the Frobenius–Perron number is equal to $s_{max} = s_1 \approx 1.891114790$. The corresponding eigenvalue of the matrix $\Lambda = S^{-1}$, which describes the warranted rate of growth for the standard model without memory (with $\alpha = 1$), has the value

$$\lambda_s = \frac{1}{s_{max}} \approx 0.5287886305. \tag{12.120}$$

We can see that $\lambda_s = \lambda_1 \approx 0.5287886305$. In the closed intersectoral model (12.93), (12.94) with power-law memory, the growth rate is determined by the warranted rate of growth with memory

$$\lambda_{s,eff}(\alpha) = (1/s_{max})^{1/\alpha} = (0.5287886305)^{1/\alpha}. \tag{12.121}$$

Eigenvalue (12.120) corresponds to the eigenvector

$$Y_1 = \begin{pmatrix} 0.52453 \\ 0.90096 \end{pmatrix}, \tag{12.122}$$

which determines the structure of the vector $Y(t)$.

It should be noted that the second terms of expressions (12.108) and (12.116) contain the function $E_\alpha[-5.5788t^\alpha]$, the value of which decreases faster than the value of the first term $E_\alpha[0.52879t^\alpha]$ increases. Therefore, the second terms of (12.108) and (12.116) can be neglected for long intervals of time. In this case, the final product (national income) of the economic sectors will be described by the equations

$$\begin{cases} Y_1(t) \approx 29.660 E_\alpha[0.52879t^\alpha], \\ Y_2(t) \approx 50.945 E_\alpha[0.52879t^\alpha], \end{cases} \tag{12.123}$$

and the gross product in these sectors will be described by the equations

$$\begin{cases} X_1(t) \approx 33.309 E_\alpha[0.52879t^\alpha], \\ X_2(t) \approx 55.728 E_\alpha[0.52879t^\alpha]. \end{cases} \tag{12.124}$$

As a result, the growth rate of national income and sector structure quickly approaching to $\lambda_s = \lambda_1$ and Y_1, respectively. The growth rate of gross product and sector structure are fast approaching to $\omega_s = \omega_1 = \lambda_1 = \lambda_s$ and X_1, respectively.

As a result, for this example we can conclude the following.

(a) The process of the national income of each sector is described by two terms with the Mittag–Leffler functions;
(b) The second terms of functions of the gross and final products for both sectors are described by rapidly decreasing processes and therefore can be ignored on long time intervals;
(c) The unit value of fading parameter corresponds to the standard model with absence of memory.

12.8 Second example of two-sectoral model with memory

In this section, we give the second illustrative example of dynamic two-sectoral Leontief model with power-law memory.

Let us consider the matrices

$$A = \begin{pmatrix} 0.1 & 0.2 \\ 0.2 & 0.3 \end{pmatrix}, \quad B = \begin{pmatrix} 0.4 & 0.2 \\ 1.0 & 0.9 \end{pmatrix}. \tag{12.125}$$

The matrix A is taken with the same values of the coefficients, as in Example 1. The matrix $(E - A)$ and the inverse matrix B^{-1} are given by the expressions

$$(E - A) = \begin{pmatrix} 0.9 & -0.2 \\ -0.2 & 0.7 \end{pmatrix}, \quad B^{-1} = \begin{pmatrix} 5.625 & -1.25 \\ -6.25 & 2.5 \end{pmatrix}. \tag{12.126}$$

Then the product of these matrices has the form

$$\Lambda = (E - A)B^{-1} = \begin{pmatrix} 6.3125 & -1.625 \\ -5.5 & 2.0 \end{pmatrix}. \tag{12.127}$$

Equation (12.33) with matrix (12.127) has the form

$$D^{\alpha}_{C;0+} \begin{pmatrix} Y_1(t) \\ Y_2(t) \end{pmatrix} = \begin{pmatrix} 6.3125 & -1.625 \\ -5.5 & 2.0 \end{pmatrix} \begin{pmatrix} Y_1(t) \\ Y_2(t) \end{pmatrix}, \tag{12.128}$$

where we will assume $0 < \alpha < 1$. The characteristic equation for matrix (12.127) is written as

$$\begin{vmatrix} 6.3125 - \lambda & -1.625 \\ -5.5 & 2.0 - \lambda \end{vmatrix} = 0. \tag{12.129}$$

The roots of characteristic equation (12.129) are the numbers

$$\lambda_1 \approx 0.470206855, \quad \lambda_2 \approx 7.842293144. \tag{12.130}$$

The corresponding equations for the eigenvectors Y_k have the form

$$\begin{pmatrix} 6.3125 & -1.625 \\ -5.5 & 2.0 \end{pmatrix} Y_k = \lambda_k Y_k, \tag{12.131}$$

where $k = 1, 2$, and the values of λ_k are given by (12.130). Solutions of equation (12.131) are the eigenvectors

$$Y_1 = \begin{pmatrix} 0.3027353638 \\ 1.088411532 \end{pmatrix}, \tag{12.132}$$

$$Y_2 = \begin{pmatrix} 0.7281147014 \\ -0.6854553099 \end{pmatrix}. \tag{12.133}$$

Components of eigenvectors (12.133) were obtained by using the Maple package. Coefficients c_k can be derived from the initial conditions (12.105) in the form

$$\begin{cases} 0.3027353638 c_1 + 0.7281147014 c_2 = 40, \\ 1.088411532 c_1 - 0.6854553099 c_2 = 40. \end{cases} \tag{12.134}$$

System (12.134) has the solutions

$$c_1 \approx 56.54280044, \quad c_2 \approx 31.42704672. \tag{12.135}$$

As a result, the functions of the final product (national income) of the first and second sectors are given by the expressions

$$\begin{cases} Y_1(t) = 17.117 E_{\alpha}[0.47021 t^{\alpha}] + 22.882 E_{\alpha}[7.8423 t^{\alpha}], \\ Y_2(t) = 61.542 E_{\alpha}[0.47021 t^{\alpha}] - 21.542 E_{\alpha}[7.8423 t^{\alpha}], \end{cases} \tag{12.136}$$

where we rounded the numerical parameters to five significant digits. Comparing (12.136) with (12.108) of the first example, we can see that the first terms on the right-hand sides of the equations are changed a little. In particular,

$$\lambda_s = \frac{1}{s_{max}} \approx 0.470206855. \tag{12.137}$$

At the same time, the second term has become dominant. The warranted rate of growth with memory is

$$\lambda_{2,\text{eff}}(\alpha) = \lambda_2^{1/\alpha} = (7.842293144)^{1/\alpha}. \tag{12.138}$$

The second component of the vector $Y(t)$ quickly reaches zero and then rapidly decreases.

As a result, we can conclude that the solution of the closed model with matrices (12.125), gives unacceptable results, both for the standard model without memory and for the model with power-law memory.

12.9 Third example of two-sectoral model with memory

Let us consider intersectoral model with two sectors, in which the memory fading parameters of these sectors are different. We consider the matrices

$$A = \begin{pmatrix} 0.1 & 0.2 \\ 0.2 & 0.3 \end{pmatrix}, \quad B = \begin{pmatrix} 0.3 & 0.1 \\ 0.2 & 0.3 \end{pmatrix}, \tag{12.139}$$

where the matrix A is taken with the same coefficients, as in the First Example. The matrices $(E - A)$ and B^{-1} are given by the expressions

$$(E - A) = \begin{pmatrix} 0.9 & -0.2 \\ -0.2 & 0.7 \end{pmatrix}, \tag{12.140}$$

$$B^{-1} = \begin{pmatrix} 4.285714286 & -1.428571429 \\ -2.857142857 & 4.285714286 \end{pmatrix}. \tag{12.141}$$

Then the product of matrices (12.140) and (12.141) is given by the expression

$$\Lambda = (E - A)B^{-1} = \begin{pmatrix} 4.428571428 & -2.142857143 \\ -2.857142857 & 3.285714286 \end{pmatrix}. \tag{12.142}$$

Equation (12.33) with matrix (12.127) has the form

$$\widehat{D^{\alpha}_{C;0+}} \begin{pmatrix} Y_1(t) \\ Y_2(t) \end{pmatrix} = \begin{pmatrix} 4.428571428 & -2.142857143 \\ -2.857142857 & 3.285714286 \end{pmatrix} \begin{pmatrix} Y_1(t) \\ Y_2(t) \end{pmatrix}, \tag{12.143}$$

where $0 < \alpha_1 < 1$, $0 < \alpha_2 < 1$ and $\alpha_1 \neq \alpha_2$ in general. The characteristic equation for matrix (12.142) is written as

$$\left| \begin{array}{cc} 4.428571428 - \lambda & -2.142857143 \\ -2.857142857 & 3.285714286 - \lambda \end{array} \right| = 0. \tag{12.144}$$

The roots of characteristic equation (12.144) are the numbers

$$\lambda_1 \approx 1.317658738; \quad \lambda_2 \approx 6.396626976. \tag{12.145}$$

The corresponding equations for the eigenvectors Y_k have the form

$$\left(\begin{array}{cc} 4.428571428 & -2.142857143 \\ -2.857142857 & 3.285714286 \end{array} \right) Y_k = \lambda_k Y_k, \tag{12.146}$$

where λ_k is defined by values (12.145) and $k = 1, 2$. Solutions of equation (12.146) are the eigenvectors

$$Y_1 = \left(\begin{array}{c} 0.5728488169 \\ 0.8316385716 \end{array} \right), \quad Y_2 = \left(\begin{array}{c} 0.7365083952 \\ -0.6764284024 \end{array} \right). \tag{12.147}$$

Coefficients c_k are determined by the initial conditions

$$c_1 Y_1 + c_2 Y_2 = Y(0). \tag{12.148}$$

We consider the initial conditions for the vector of the final product in the form

$$Y(0) = \left(\begin{array}{c} Y_1(t) \\ Y_2(t) \end{array} \right) = \left(\begin{array}{c} 40 \\ 40 \end{array} \right). \tag{12.149}$$

Then the system of equations (12.148) with vectors (12.147) and (12.149) has the form

$$\left\{ \begin{array}{l} 0.5728488169 c_1 + 0.7365083952 c_2 = 40, \\ 0.8316385716 c_1 - 0.6764284024 c_2 = 40. \end{array} \right. \tag{12.150}$$

Solving the system of equations (12.150), we obtain the values

$$c_1 \approx 56.51747192, \quad c_2 \approx 10.35159019. \tag{12.151}$$

As a result, the final product (the national income) of the first and second sectors are given by the expressions

$$\left\{ \begin{array}{l} Y_1(t) = 32.376 E_{\alpha_1}[1.3177 t^{\alpha_1}] + 7.6240 E_{\alpha_1}[6.3966 t^{\alpha_1}], \\ Y_2(t) = 47.002 E_{\alpha_2}[1.3177 t^{\alpha_2}] - 7.0021 E_{\alpha_2}[6.3966 t^{\alpha_2}], \end{array} \right. \tag{12.152}$$

where we rounded the numerical parameters to five significant digits. For $\alpha_1 = \alpha_2 = 1$, equation (12.152) describes the standard model, since $E_1[z] = \exp(z)$.

Let us find the Frobenius–Perron number of the matrix $S = B(E-A)^{-1} = \Lambda^{-1}$ of the full incremental capital intensity. Using equation (12.139), we get

$$S = B(E-A)^{-1} = \begin{pmatrix} 0.3898305085 & 0.2542372881 \\ 0.3389830508 & 0.5254237289 \end{pmatrix}. \tag{12.153}$$

Using the Maple software package, we obtain the eigenvalues of this matrix (12.153) in the form

$$s_1 \approx 0.1563323927; \quad s_2 \approx 0.7589218447. \tag{12.154}$$

As a result, the Frobenius–Perron number is equal to

$$s_{\max} = \max\{s_1, s_2\} = s_2 \approx 0.7589218447. \tag{12.155}$$

The corresponding eigenvalue $\Lambda = S^{-1}$, which describes the warranted rate of growth for the standard model without memory (with $\alpha = 1$) has the form

$$\lambda_s = \frac{1}{s_{\max}} \approx 1.317658738. \tag{12.156}$$

Eigenvalue (12.156) corresponds to the eigenvector

$$Y_1 = \begin{pmatrix} 0.5728488169 \\ 0.8316385716 \end{pmatrix}, \tag{12.157}$$

which determines the structure of the vector $Y(t)$.

In the intersectoral model with power-law memory, the growth rate is determined by the warranted rate of growth with memory (12.77), which has the form

$$\lambda_{s,\mathrm{eff}}(\alpha_1) = \lambda_1^{1/\alpha_1} = (1.317658738)^{1/\alpha_1}. \tag{12.158}$$

In the same time, for the second sector the warranted rate of growth with memory (12.76) is equal to

$$\lambda_{2,\mathrm{eff}}(\alpha_2) = \lambda_2^{1/\alpha_2} = (6.396626976)^{1/\alpha_2}. \tag{12.159}$$

For $\alpha_1 = 0.1$ and $\alpha_2 = 0.9$, we get the values

$$\lambda_{s,\mathrm{eff}}(\alpha_1) = \lambda_1^{1/\alpha_1} = 15.777, \tag{12.160}$$

$$\lambda_{2,\mathrm{eff}}(\alpha_2) = \lambda_2^{1/\alpha_2} = 7.8614. \tag{12.161}$$

As a result, based on this example, we can formulate the following statements [437, 451]:

(I) For standard warranted rates of growth, we have the inequality $\lambda_s = \lambda_1 < \lambda_2$. As a result, the standard intersectoral model without memory gives unacceptable results. This statement is caused by the dominance of the trajectory $\exp(6.3966t)$ at $t \to \infty$, which leads to the fact that the second component of the vector $Y(t)$ quickly reaches zero and then rapidly decreases.

(II) For intersectoral model with sectoral memory, we have the reverse inequality $\lambda_{s,\mathrm{eff}}(\alpha_1) = \lambda_{1,\mathrm{eff}}(\alpha_1) > \lambda_{2,\mathrm{eff}}(\alpha_2)$. This intersectoral model with sectoral power-law memory, which based on the same matrices A and B as in the standard model, gives acceptable results with economic sense. This statement is based on the dominance of the trajectory $E_\alpha[1.3177t^\alpha]$ at $t \to \infty$, for which both components of the vector $Y(t)$ increase with the rate of (12.160).

(III) As a result, it is clear that the inclusion of sectoral memory into intersectoral model with matrices (12.139), may lead to changes of dominance. This means that memory effects can lead to qualitatively different results compared to the standard model. In accordance with the proposed principle of domination change, the effects of sectoral memory can change the dominating behavior of economic sectors.

12.10 Conclusion

Memory effects can play an important role in economics. Neglect of the memory effects in economic models, can lead to a distorted description of the real situation in the economy. Therefore, economic models should take into account the possibility of the impact of memory effects on the dynamics of economic processes. The proposed results can be considered as a theoretical and methodological basis for the application of dynamic intersectoral models with matrices of input–output balance to describe economic processes with power-law memory.

In this chapter, we proved that the inclusion of memory effects into the dynamic intersectoral models can lead to qualitatively new results at the same the matrices of the direct material costs, the incremental capital intensity of production and the full incremental capital intensity. The proposed approach to economic dynamics allows us to build more adequate dynamic intersectoral models of the economy. The economic models should takes into account that economic agents may remember the story of changes in economic processes and the agents can take into account these changes in making economic decisions. In addition, we assume that the parameters of the memory fading can be considered as control parameters to increase economic growth and the impacts on the real economy.

13 Market price dynamics with memory effects

In this chapter, we consider dynamics of market prices, in which power-law memory effects are taken into account. In the standard models of price dynamics, the memory effects are neglected. Memory be taken into account in dynamics of market prices, since amnesia and lack of memory about past price changes cannot be considered a characteristic property of merchants, buyers and suppliers. In this chapter, we propose new economic model of the dynamics of the market prices, which is a generalization of the Evans models. In the proposed model, we assume that economic entities (merchants, buyers, suppliers) can remember how stocks of goods and their prices have changed over time in the past. The properties of the solutions of the generalized Evans model with memory are described. Based on the properties of the solutions, we formulate the following principles: the principles of changing of characteristic time by memory; the principle of pricing with memory; the principle of changing of price relaxation time by memory; the principles of competition of deceleration by memory on short times. This chapter is based on the work [390].

13.1 Introduction

Prices change from time to time, and, in fact, rarely remain constant. To get an adequate picture of economies, we should have some assumptions, which allows us describe the price vary with time, and predict the price to enable us to determine it for various times, i. e., as a function of the time. In order for price dynamic model to be adequate and correct, we must take into account that economic actors can remember how stocks of goods and their prices have changed over time. Standard price models do not take into account the effect of memory on the demand for goods and their price.

In this chapter, we consider dynamics of market prices, in which power-law memory effects are taken into account. We propose a generalization of the Evans models of the price dynamics in the market for a single product, in which we assume that economic entities (merchants, buyers, suppliers) can remember how stocks of goods and their prices have changed over time.

13.2 Standard Evans model of price dynamics in market

The Evans model of price dynamics in market is one of the first dynamic models that describe the change in price. This model was proposed by Griffith C. Evans in works [116, 117] in 1924. The Evans model describes the price dynamics in the market for a single product (see also Chapter IV in [117, pp. 36–49]). This model is used for modeling the establishment of an equilibrium price in such a market. In the Evans model, there are three main groups of economic entities: merchants, buyers, suppliers [10, p. 15].

https://doi.org/10.1515/9783110627459-013

Buyers realize their intentions in purchases, and the volume of purchases is equal to sales. Suppliers realize their intentions in sales to merchants. Merchants have stocks of goods and realize their intentions in sales to buyers.

The actually realized supplies are added to the total mass of stocks, and the actual demand is met by stocks [10, p. 15]. It is assumed that the market is equal in buying and selling. In this case, the realized demand is their total value, that is, it is equal to the actual demand. For simplicity, it is assumed that traders always buy and sell at the same price.

The operation of the model begins with the fact that merchants make, setting prices for goods in accordance with stocks [10, p. 15]. There are two main cases [10, p. 15]. In the first case, prices are set depending on changes in stocks. In the second case, the prices are set in accordance with the level of stocks. In both cases, supply and demand functions are taken without delays and memory effects.

Let us describe the standard Evans model with continuous time. The brief description of this model is based on book [10] (see Sections 1.7 and 1.8 of [10, pp. 15–23] and book [493]).

We assume that the stocks $Q = Q(t)$ vary continuously over time like the other variables. Let $S(t)$ be supply (offer), $D(t)$ be demand for good, and $P(t)$ be price of goods. By definition, the change in stocks at time t is given by the differential equation

$$\frac{dQ(t)}{dt} = S(t) - D(t), \tag{13.1}$$

or by the equivalent integral equation

$$Q(t) = Q(0) + \int_0^t (S(\tau) - D(\tau))\, d\tau. \tag{13.2}$$

In the standard Evans model, the price $P(t)$ is set at each moment of time so that the rate of price increase is proportional to the rate of decrease in sticks (see Model III in [10, pp. 17–18]), i. e., we have the equation

$$\frac{dP(t)}{dt} = -\gamma \frac{dQ(t)}{dt}, \tag{13.3}$$

where the coefficient γ is given and positive. In other words, if we assume that the rate of price change is directly proportional to the excess of demand over supply, then we obtain the equation

$$\frac{dP(t)}{dt} = -\gamma (S(t) - D(t)). \tag{13.4}$$

The coefficient of proportionality γ is interpreted as the speed of response. The larger γ, the more rapid is the price reaction to a given deficit in supply [10, p. 21].

In the simplest case, supply and demand are considered as the linear functions of price

$$D(t) = d + aP(t), \tag{13.5}$$

$$S(t) = s + bP(t), \tag{13.6}$$

where a, b, c, d are constant parameters. The free terms d and s describe demand and supply, which do not depend on price. By virtue of this, d and s can be considered as positive constants. It is usually considered that at zero price, demand exceeds supply, i. e., $d > s$. Often, cases are considered when equation (13.4) describes a decrease in demand with rising price, and equation (13.5) describes that supply increases with price. In this case, we have $a < 0$, $b > 0$. However, other signs can be considered for these parameters (see Section 1.8 in the Allen's book [10, pp. 19–23]). Note that it is generally preferable to write $D(t) = d - aP(t)$ even a is negative ($a < 0$), rather than $D(t) = d - aP(t)$ with a positive ($a > 0$) [10, p. 2].

Substituting expressions (13.5) and (13.6) into equation (13.4), we obtain

$$\frac{dP(t)}{dt} + \gamma(b - a)P(t) = \gamma(d - s). \tag{13.7}$$

Equation (13.7) is a first-order differential equation that describes price dynamics within the framework of the Evans model that is described as Model III in Allen's book [10, pp. 17–18]. In the usual case, $a < 0$ and $b > 0$, so that $\gamma(b-a) > 0$ [10, p. 18]. In the physical sciences, equation (13.7) with $\gamma(b-a) > 0$ describes the universal law of one-decay processes and relaxation processes. The relaxation means that the perturbed process returns to equilibrium state. The relaxation process can be characterized by a relaxation time τ_{rel} and the rate of relaxation $\lambda_{\text{rel}} = 1/\tau_{\text{rel}}$. For equation (13.7), the relaxation time is equal to $\tau_{\text{rel}} = 1/(\gamma(b - a))$. The one-decay process is characterized by the half-life that is the time required for a quantity to reduce to half its initial value $T_{1/2} = \tau_{\text{rel}} \ln(2)$.

It should be emphasized that in general the price dynamics can also be characterized by other signs of parameters a and b, which may lead to $\gamma(b - a) < 0$ and to a deviation from the equilibrium state, that is, the instability of the process (see Section 1.8 in Allen's book [10, pp. 19–23]).

The equilibrium price $\overline{P}$ is determined by the stationarity condition $dP(t)/dt = 0$ for equation (13.4). Using equation (13.4), we obtain $D(t) - S(t) = 0$. For the linear functions (13.5) and (13.6), the equilibrium price is defined by equation (13.7) with $dP(t)/dt = 0$ in the form

$$\gamma(b - a)P(t) = \gamma(d - s). \tag{13.8}$$

As a result, we get the expression for the equilibrium price

$$\overline{P} = \frac{d - s}{b - a}. \tag{13.9}$$

The general solution of equation (13.7) has the form

$$P(t) = \overline{P} + Ce^{-\gamma(b-a)t}, \tag{13.10}$$

where C is the constant determined by the initial conditions.. Substituting of $t = 0$ into equation (13.10), we get $C = P(0) - \overline{P}$.

As a result, the solution of equation (13.7) has the form

$$P(t) = \overline{P} + (P(0) - \overline{P})e^{-\gamma(b-a)t}. \tag{13.11}$$

From equation (13.11), we can see that the price $P(t)$ monotonously tends to the equilibrium value $\overline{P}$, if $\gamma(b-a) > 0$ [10, p. 18]. This is true for all positive values of $\gamma(b-a) > 0$. The value of γ determines the rate of response or adaptation [10, p. 21]. The larger the value of $\gamma(b-a)$, the faster $P(t)$ approaches $\overline{P}$.

Allen considered a generalization of the Evans model (Model IV in [10, pp. 18–19]), which we will call the Evans–Allen model. In this model, the price is set at each moment in such a way that the rate of price increase $dP(t)/dt$ is proportional to the difference in inventory $Q(t)$ compared with a given level of stocks $\overline{Q}$. Therefore, we have the equation

$$\frac{dP(t)}{dt} = -\gamma(Q(t) - \overline{Q}). \tag{13.12}$$

Substituting (13.2) into equation (13.12), we get

$$\frac{dP(t)}{dt} = -\gamma(Q(0) - \overline{Q}) - \gamma \int_0^t (S(\tau) - D(\tau))\, d\tau. \tag{13.13}$$

Using that equation (13.12) for $t = 0$ gives $P^{(1)}(0) = -\gamma(Q(0) - \overline{Q})$, we can rewrite equation (13.13) as

$$\frac{dP(t)}{dt} = P^{(1)}(0) - \gamma \int_0^t (S(\tau) - D(\tau))\, d\tau. \tag{13.14}$$

Applying derivative of the first order to equation (13.14), we get

$$\frac{d^2P(t)}{dt^2} = -\gamma(S(t) - D(t)). \tag{13.15}$$

Equation (13.15) means that the acceleration of the price increase is proportional to the rate of decrease of stocks, i. e., $P^{(2)}(t) = -\gamma Q^{(1)}(t)$. Equation (13.15) differs from equation (13.4) in that instead of a first-order derivative, a second-order derivative is used. Substituting expressions (13.5) and (13.6) into equation (13.15), we obtain

$$\frac{d^2P(t)}{dt^2} + \gamma(b-a)P(t) = \gamma(d-s). \tag{13.16}$$

The second-order differential equation (13.16) describes price dynamics within the framework of the Evans–Allen model that is described as Model IV in Allen's book [10, pp. 18–19]. Equation (13.16) can be called the Evans–Allen equation [10]. In physics, equation (13.16) describes the undamped harmonic oscillator. The solution of equation (13.16) is described in Section 5.4 of [10, pp. 145–150]. In the usual case, ($a < 0$, $b > 0$, $\gamma > 0$), the term $\gamma(b - a)$ is positive and the general solution has the form

$$P(t) = \overline{P} + A\cos(\omega t - \varphi), \tag{13.17}$$

where $\omega = \sqrt{\gamma(b-a)}$, A and φ are constants to be given by initial conditions. The behavior of price $P(t)$ over time given by (13.17) is a regular oscillation with the frequency ω around the equilibrium level (13.9). In this case, the equilibrium level of price cannot be considered as unstable, since there is no progressive movement away from it. In addition, there is no tendency to return to equilibrium but only to oscillate regularly about it. To take into account in Model IV that the price $P(t)$ tends to the equilibrium price $\overline{P}$, i. e., for the relaxation process for price, the demand function is considered [10, p. 19] in the form

$$D(t) = d + aP(t) + cP^{(1)}(t), \tag{13.18}$$

where $c < 0$. In this case, price dynamics is described by the equation

$$\frac{d^2P(t)}{dt^2} - \gamma c P^{(1)}(t) + \gamma(b-a)P(t) = \gamma(d-s). \tag{13.19}$$

In physics, equation (13.19) describes the damped harmonic oscillator. The value of the damping ratio is described by the equation

$$\xi = \frac{\gamma c}{2\sqrt{\gamma(b-a)}}, \tag{13.20}$$

where $\gamma(b - a)$ is positive and $c < 0$, determines the behavior of the price. In the overdamped case ($\xi > 1$), the price returns (exponentially decays) to equilibrium price without oscillating. For larger values of the damping ratio (13.20), the price returns to equilibrium more slowly. In the critically damped ($\xi = 1$), the price returns to equilibrium price (13.9) as quickly as possible without oscillating (although overshoot can occur). In the underdamped ($\xi < 1$), the price oscillates with the amplitude gradually decreasing to zero and the frequency is a slightly different than the frequency in the undamped case.

13.3 Accounting for memory of excess of demand over supply

In the standard price dynamics, the memory effects and memory fading are neglected. At the same time, it is obvious that memory must be taken into account to describe the dynamics of market prices, because total amnesia about price changes in the past cannot be attributed to sellers, buyers, suppliers. From a mathematical point of view, the neglect of memory effects in price dynamics is based on the use only equations with derivatives of integer orders, which are determined by the properties of the function in an infinitely small neighborhood of the considered time. A powerful tool for describing memory effects is the fractional calculus of derivatives and integrals of noninteger orders. Using fractional calculus as a mathematical tool, a model of market price dynamics that takes into account power memory effects was proposed in article [390].

To take into account the memory effects in the Evans models, we will assume that the price $P(t)$ at time t is determined not only by the rate $Q^{(1)}(\tau) = dQ(\tau)/d\tau$ of decline in stocks of goods $Q(\tau)$ at time $\tau = t$, but also by the history of changes in $Q^{(1)}(\tau)$ at $\tau < t$, i. e., the differences in supply and demand $Q^{(1)}(\tau) = S(\tau) - D(\tau)$ on a finite time interval $\tau \in [0, t]$. This is due to the fact that the economic entities (merchants, buyers, suppliers) may be aware of changes in the stocks of the goods, the excess of demand over supply at previous points in time and they take this into account by changing the price of the goods at the present time.

Equations of the Evans models III and IV without memory [10] are described by the differential equation of the first order (13.7) and equation second order (13.16) for the linear case (13.5), (13.6), respectively. Based on the concept of fading memory [448, 450], we consider the generalization of these equations in the form

$$(D^{\alpha}_{C;0+}P)(t) + \gamma(b - a)P(t) = \gamma(d - s), \tag{13.21}$$

where $D^{\alpha}_{C;0+}$ is the Caputo fractional derivative of the order $\alpha > 0$.

Remark 13.1. Note that equations (13.7) and (13.16) are special cases of fractional differential equation (13.21) for $\alpha = 1$ and $\alpha = 2$, respectively.

Remark 13.2. In the general case, this approach to generalization of standard economic models cannot be considered as correct from economic point of view (for details, see [386] or [254, pp. 43–92]). It is not enough to generalize the differential equations that describe the dynamic model. It is necessary to generalize the whole scheme of obtaining (all steps of derivation) these equations from the basic principles, concepts and assumptions.

Let us consider the sequential derivation of the equations for the generalization of Models III and IV for the processes with memory.

First, we obtain a generalization of Evans model (Model III) described by equation (13.4). To take into account the memory effect, which manifests itself in the dependence of price on excess of demand over supply $S(\tau) - D(\tau)$ over a finite time interval, we should use the equation

$$\frac{dP(t)}{dt} = -\int_0^t \gamma(t - \tau)\frac{dQ(\tau)}{d\tau}\,d\tau = -\int_0^t \gamma(t - \tau)(S(\tau) - D(\tau))\,d\tau, \tag{13.22}$$

where $\gamma(t)$ can be interpreted as a memory function [448, 450]. This function takes into account that the behavior of the processes can depend not only at the present time, but also on the history of changes on finite time interval [448, 450].

If the function $\gamma(t)$ is expresses though the Dirac delta-function $\gamma(t-\tau) = \gamma\delta(t-\tau)$, then equation (13.22) takes the form (13.4). If the function $\gamma(t) = \gamma$, then equation (13.22) gives equation (13.13) with $Q(0) = \overline{Q}$.

If we assume the power-law fading of the memory, then we can use the memory function $\gamma(t)$ in the form

$$\gamma(t-\tau) = \frac{\gamma}{\Gamma(\alpha)}(t-\tau)^{\alpha-1}, \tag{13.23}$$

where $0 < \alpha < 1$ and $t > \tau$. In this case, equation (13.22) can be written as

$$\frac{dP(t)}{dt} = -\gamma(I^{\alpha}_{\mathrm{RL};0+}Q^{(1)})(t) = -\gamma(I^{\alpha}_{\mathrm{RL};0+}(S-D))(t), \tag{13.24}$$

where $I^{\alpha}_{\mathrm{RL};0+}$ is the Riemann–Liouville fractional integral of the order $\alpha > 0$. This integral is defined [200, pp. 69–70] by the equation

$$(I^{\alpha}_{\mathrm{RL};0+}f)(t) = \frac{1}{\Gamma(\alpha)}\int_0^t (t-\tau)^{\alpha-1}f(\tau)\,d\tau, \tag{13.25}$$

where $\Gamma(\alpha)$ is the gamma function, and the function $f(\tau)$ satisfies the condition $f(\tau) \in L_1[0,T]$ that implies the realization of the inequality $\int_0^T |f(\tau)|\,d\tau < \infty$.

In order to obtain a differential equation, which is an analogue of equation (13.4), we act on equation (13.24) by the left-sided Caputo derivative of the order $\alpha > 0$. This derivative is defined by the formula

$$(D^{\alpha}_{\mathrm{C};0+}f)(t) = \frac{1}{\Gamma(n-\alpha)}\int_0^t (t-\tau)^{n-\alpha-1}f^{(n)}(\tau)\,d\tau, \tag{13.26}$$

where $n = [\alpha]+1$ for noninteger values of $\alpha > 0$ and $n = \alpha$ for integer values of $\alpha > 0$. Expression (13.26) exists, if $f(\tau) \in AC^n[0,T]$ (see Theorem 2.1 in [200, p. 92]), i. e., the function $f(\tau)$ has integer-order derivatives up to $(n-1)$-th order, which are continuous functions on the interval $[0,T]$, and the derivative $f^{(n)}(\tau)$ is Lebesgue summable on the interval $[0,T]$.

The action of the fractional derivative (13.26) on equation (13.24) gives the expression

$$(D^{\alpha}_{\mathrm{C};0+}P^{(1)})(t) = -\gamma(D^{\alpha}_{\mathrm{C};0+}I^{\alpha}_{\mathrm{RL};0+}Q^{(1)})(t) = -\gamma Q^{(1)}(t) = -\gamma(S(t)-D(t)). \tag{13.27}$$

It is known that the Caputo fractional derivative is an operation that is the left inverse for the Riemann–Liouville integral [200, p. 95]. In other words, for any continuous function $f(t)$, the equality

$$(D^{\alpha}_{\mathrm{C};0+}I^{\alpha}_{\mathrm{RL};0+}f)(t) = f(t) \tag{13.28}$$

holds for any $\alpha > 0$, where $I^{\alpha}_{RL;0+}$ is the left-side Riemann–Liouville integral (13.25), and $D^{\alpha}_{C;0+}$ is the left-sided Caputo derivative (13.26).

Using the property of the Caputo fractional derivative

$$D^{\alpha}_{C;0+}D^n = D^{\alpha+n}_{C;0+}, \tag{13.29}$$

where $n \in \mathbb{N}$, the left side of equation (13.27) can be written as

$$(D^{\alpha}_{C;0+}P^{(1)})(t) = (D^{\alpha+1}_{C;0+}P)(t). \tag{13.30}$$

As a result, the equation of the generalized Evans model takes the form

$$(D^{\alpha+1}_{C;0+}P)(t) = -\gamma(S(t) - D(t)), \tag{13.31}$$

where $0 < \alpha \le 1$, and, therefore the order of the fractional derivative is $1 < \alpha + 1 \le 2$.

Similarly, we can derive a generalization of equation (13.13), which describes Model IV of the Allen's book [10], in the form

$$\frac{dP(t)}{dt} = P^{(1)}(0) - \int_0^t \gamma(t-\tau)(S(\tau) - D(\tau))\, d\tau. \tag{13.32}$$

If the kernel $\gamma(t - \tau)$ is defined by expression (13.23), then equation (13.32) is represented as

$$\frac{dP(t)}{dt} = P^{(1)}(0) - \gamma(I^{\alpha}_{RL;0+}(S - D))(t). \tag{13.33}$$

Acting on equation (13.33) by the left-sided Caputo derivative of order $\alpha > 0$ and using that the action of the Caputo derivative on a constant function is zero (see equation (2.4.31) in [200, p. 95]), we get the fractional differential equation

$$(D^{\alpha+1}_{C;0+}P)(t) = -\gamma(S(t) - D(t)) \tag{13.34}$$

that coincides with equation (13.31).

As a result, the equation of the proposed model, which is a generalization of the standard Evans models III and IV for processes with power-law memory, is described by the fractional differential equation

$$(D^{\alpha+1}_{C;0+}P)(t) = \gamma(D(t) - S(t)), \tag{13.35}$$

where $n-1 < \alpha+1 \le n$ and $n \in \mathbb{N}$. If $\alpha = 0$, then equation (13.35) takes the form (13.4) of the standard Evans equation. If $\alpha = 1$, then equation (13.35) takes the form of (13.15) of the Evans–Allen equation. These statements are based on the property of the Caputo fractional derivative [308, 200] in the form

$$(D^{n+1}_{C;0+}P)(t) = P^{(n+1)}(t) \tag{13.36}$$

for any integers $n \ge 0$, including $n = 0$ and $n = 1$.

As a result, equations (13.4) and (13.15), which describe two standard price dynamic models III and IV [10, pp. 17–19], are special cases of the fractional differential equation (13.35). The proposed generalized Evans model includes two standard models as special cases. The proposed model with memory also includes a wide range of intermediate models lying between the standard models III and IV, for $\alpha \in (0,1)$.

13.4 Evans model with memory: equation and solution

If the supply and demand are treated as linear functions of price (13.5) and (13.6), then the generalized Evans model with memory is described by the equation

$$(D^{\alpha+1}_{C;0+}P)(t) + \gamma(b-a)P(t) = \gamma(d-s). \tag{13.37}$$

Using the notation,

$$\lambda = -\gamma(b-a) \quad f = \gamma(d-s), \tag{13.38}$$

equation (13.37) can be rewritten as

$$(D^{\alpha+1}_{C;0+}P)(t) - \lambda P(t) = f(t), \tag{13.39}$$

where $f(t) = f$ and $n-1 < \alpha+1 \leq n$. If independent supply and demand (s and d) change over time ($s = s(t)$ and $d = d(t)$), then $f = \gamma(d-s)$ should be considered as a function of time.

In the general case, the parameter λ can be either negative (e. g., $\lambda < 0$ for $b > 0$ and $a < 0$), or positive. Similarly, the parameter $f = \gamma(d-s)$ can be both positive and negative, depending on the relationship between supply and demand, which are independent of price. If the independent demand d is less than the independent supply s, i. e., $d < s$, then the parameter $f < 0$.

We emphasize that equation (13.39) includes, as particular cases, equation (13.7) at $\alpha = 0$ and equation (13.16) at $\alpha = 1$.

The process that is described by equation (13.39) will be called relaxation if $\lambda < 0$. The process described by equation (13.39) will be called amplification if $\lambda > 0$. The parameter λ will be called the warranted rate of process without memory, i. e., $\lambda < 0$ is the rate of relaxation and $\lambda > 0$ is the rate of amplification. Types of processes and equilibrium states, depending on the relations between the parameters, are presented in Table 13.1.

To solve equation (13.39), we can use Theorem 5.15 of book [200, p. 323]. It is known [200, pp. 230–231] (see also [200, pp. 271,323]), that fractional differential equation (13.39) with $\lambda \in \mathbb{R}$ has the general solution

$$P(t) = \sum_{k=0}^{n-1} P^{(k)}(0)t^k E_{\alpha+1,k+1}[\lambda t^{\alpha+1}] + P_f(t), \tag{13.40}$$

Table 13.1: Here, $\lambda = -\gamma(b-a)$ and $\overline{P} > 0$ for all three cases (see Figures 2A, 2B, 2C in book [10, p. 22]).

	Parameters *a* and *b*	Parameters *d* and *s*	Rate λ	Type of process	Equilibrium price, Walrasian	Equilibrium price, Marshallian	Figures
1	$b > 0 > a$ $(b-a>0)$	$d > s > 0$ $(d-s>0)$	$\lambda < 0$	Relaxation	Walrasian stability	Marshallian stability	Fig. 2A
2	$a < b < 0$ $(b-a>0)$	$d > s > 0$ $(d-s>0)$	$\lambda < 0$	Relaxation	Walrasian stability	Marshallian instability	Fig. 2B
3	$b < a < 0$ $(b-a<0)$	$s > d > 0$ $(d-s<0)$	$\lambda < 0$	Amplification	Walrasian instability	Marshallian stability	Fig. 2C

where $P^{(k)}(0)$ is derivatives of the integer order $k > 0$ of the function $P(t)$ at $t = 0$, and

$$P_f(t) = \int_0^t (t-\tau)^{\alpha} E_{\alpha+1,\alpha+1}[\lambda(t-\tau)^{\alpha+1}] f(\tau)\, d\tau = \int_0^t \tau^{\alpha} E_{\alpha+1,\alpha+1}[\lambda\tau^{\alpha+1}] f(t-\tau)\, d\tau. \quad (13.41)$$

Here, $E_{\alpha,\beta}[z]$ is the two-parameter Mittag–Leffler function [143] that is defined by the expression

$$E_{\alpha,\beta}(z) = \sum_{k=0}^{\infty} \frac{z^k}{\Gamma(\alpha k + \beta)}, \quad (13.42)$$

where $\Gamma(\alpha)$ is the gamma function, $\alpha > 0$, and β is an arbitrary real or complex number. Note that the Mittag–Leffler function (13.42) is a generalization of the exponential function e^z, such that $E_{1,1}(z) = e^z$. For $\alpha \neq 1$, the Mittag–Leffler function $E_{\alpha,1}(z) = E_{\alpha}[z]$ satisfies the inequality

$$E_{\alpha}[\lambda t_1^{\alpha}] E_{\alpha}[\lambda t_2^{\alpha}] \neq E_{\alpha}[\lambda(t_1 + t_1)^{\alpha}] \quad (13.43)$$

that means the violation of the semigroup property that holds for exponential function [304, 111, 334].

For the constant function $f(t) = f$, expression (13.41) has the form

$$P_f(t) = f \int_0^t \tau^{\alpha} E_{\alpha+1,\alpha+1}[\lambda\tau^{\alpha+1}]\, d\tau. \quad (13.44)$$

To calculate integral (13.44), we can use equation 4.4.5 of [143, p. 61], where $\Gamma(\mu)$ should be used instead of $\Gamma(\alpha)$, in the form

$$\int_0^t \tau^{\beta-1} E_{\alpha,\beta}[\lambda\tau^{\alpha}]\, d\tau = t^{\beta} E_{\alpha,\beta+1}[\lambda t^{\alpha}]. \quad (13.45)$$

Then we can use equation 4.2.3 of [143, p. 57], which has the form

$$E_{\alpha,\beta}[z] = \frac{1}{\Gamma(\beta)} + zE_{\alpha,\alpha+\beta}[z]. \tag{13.46}$$

Equation (13.46) with $z = \lambda t^{\alpha}$ gives the equality

$$t^{\alpha} E_{\alpha,\alpha+\beta}[\lambda t^{\alpha}] = \frac{1}{\lambda}\left(E_{\alpha,\beta}[\lambda t^{\alpha}] - \frac{1}{\Gamma(\beta)} \right). \tag{13.47}$$

Using expression (13.47), equation (13.28) can be written as

$$P_f(t) = f\lambda^{-1}(E_{\alpha+1,1}[\lambda t^{\alpha+1}] - 1). \tag{13.48}$$

The parameters (13.38) allow us to express the multiplier through the equilibrium price (13.8) in the form

$$f\lambda^{-1} = -\frac{d-s}{b-a} = -\overline{P}, \tag{13.49}$$

and get equation (13.48) in the simple form

$$P_f(t) = \overline{P}(1 - E_{\alpha+1,1}[\lambda t^{\alpha+1}]). \tag{13.50}$$

As a result, solution (13.40) of equation (13.39) is written by the expression

$$P(t) = \overline{P}(1 - E_{\alpha+1,1}[\lambda t^{\alpha+1}]) + \sum_{k=0}^{n-1} P^{(k)}(0) t^k E_{\alpha,k+1}[\lambda t^{\alpha}], \tag{13.51}$$

where $0 < n-1 < \alpha+1 \leq n$, the parameters $\lambda \in \mathbb{R}$ and $\overline{P} \in \mathbb{R}_+$ are defined by expressions (13.38) and (13.49), respectively. Formula (13.51) is the solution of equation (13.39) for all $\alpha + 1 > 0$ and constant $f = \gamma(d-s)$.

For power-law memory with fading parameter $0 < \alpha \leq 1$, that is, for $1 < \alpha + 1 \leq 2$, $(n = 2)$, solution (13.51) of equation (13.39), which describes the price dynamics with power-law memory, has the form

$$P(t) = \overline{P} + (P(0) - \overline{P})E_{\alpha+1,1}[-\gamma(b-a)t^{\alpha+1}] + P^{(1)}(0) t E_{\alpha+1,2}[-\gamma(b-a)t^{\alpha+1}], \tag{13.52}$$

where $P^{(1)}(0)$ is the rate of price change at the initial time, described by the value of the first-order derivative of the price function $P(t)$ at $t = 0$. We should emphasize that solution (13.52) is true for both positive and negative values of $\lambda = -\gamma(b-a)$.

Note that equation (13.52) with $P^{(1)}(0) = 0$ coincides with the solution of equation (13.39) with $0 < \alpha + 1 \leq 1$ that has the form

$$P(t) = \overline{P} + (P(0) - \overline{P})E_{\alpha+1,1}[-\gamma(b-a)t^{\alpha+1}]. \tag{13.53}$$

For $\alpha = 0$, using $E_{1,1}[z] = e^z$, equation (13.53) gives (13.11), i. e., solution (13.53) with $\alpha = 0$ coincides with solution (13.11) of equation of the standard Evans model, which does not take into account memory effects. Therefore equation (13.39) and its solution (13.52) include equation (13.7) and solution (13.11) as a special case.

For $\alpha = 1$, solution (13.53) of fractional differential equation (13.39) with $1 < \alpha + 1 \leq 2$ gives the solution of equation (13.16).

13.5 Price dynamics with memory: amplification of market price

Let us consider the asymptotic behavior of the price dynamics with memory that is described by equation (13.52). We can use the asymptotic expressions the two-parameter Mittag–Leffler function $E_{\alpha,\beta}[z]$ with $0 < \alpha < 2$ at $|z| \to \infty$ [200, p. 43].

For $z > 0$, we have (see equation (1.8.27) of [200]) the asymptotic expression

$$E_{\alpha,\beta}[z] = \frac{1}{\alpha} z^{(1-\beta)/\alpha} \exp(z^{1/\alpha}) - \sum_{j=1}^{N} \frac{1}{\Gamma(\beta - \alpha j)} z^{-j} + O(z^{-(N+1)}). \tag{13.54}$$

For $z < 0$, which corresponds to $\lambda < 0$ in our model, we have (see equation (1.8.28) of [200]) the asymptotic expression

$$E_{\alpha,\beta}[z] = -\sum_{j=1}^{N} \frac{1}{\Gamma(\beta - \alpha j)} z^{-j} + O(z^{-(N+1)}), \tag{13.55}$$

where we use the big O notation (asymptotic notation) that provides an upper bound on the growth rate of the subsequent terms of the asymptotic series.

Using $z = \lambda t^{\alpha+1}$ with $\beta = 1$ (or $\beta = 2$) and $\alpha + 1$ instead of α, we get

$$\begin{aligned} E_{\alpha+1,1}[\lambda t^{\alpha+1}] = {} & \frac{1}{\alpha+1} \exp(\lambda^{1/(\alpha+1)} t) \\ & - \sum_{j=1}^{N} \frac{\lambda^{-j} t^{-j(\alpha+1)}}{\Gamma(1 - (\alpha+1)j)} + O(t^{-(\alpha+1)(N+1)}), \end{aligned} \tag{13.56}$$

and

$$\begin{aligned} E_{\alpha+1,2}[\lambda t^{\alpha+1}] = {} & \frac{1}{\alpha+1} \exp(\lambda^{1/(\alpha+1)} t) \\ & - \sum_{j=1}^{N} \frac{\lambda^{-j} t^{-j(\alpha+1)+1}}{\Gamma(2 - (\alpha+1)j)} + O(t^{-(\alpha+1)(N+1)+1}). \end{aligned} \tag{13.57}$$

These expressions allow us to obtain an asymptotic behavior of price $P(t)$ for the case $\lambda > 0$. For the case $\lambda < 0$, the first term with the exponent $\exp(\lambda^{1/(\alpha+1)} t)$ are absent in expressions (13.56) and (13.57).

As a result, for $\lambda > 0$ we get

$$P(t) = \overline{P} + \left(P(0) - \overline{P} + \frac{P^{(1)}(0)}{\lambda^{1/(\alpha+1)}} \right) \frac{\exp(\lambda^{1/(\alpha+1)} t)}{\alpha+1} + O(t^{-(\alpha+1)}). \tag{13.58}$$

Using expression (13.58) with $\lambda = -\gamma(b - a) > 0$, we can see that the rate of price increase with memory is defined by the equation

$$\lambda_{\text{eff}}(\alpha) = \lambda^{1/(\alpha+1)}, \tag{13.59}$$

where λ is the rate of price increase without memory.

The proposed rate of price increase with memory (13.59) is an analogue of the warranted rates of growth that are suggested in works [450, 445, 453, 393].

Instead of growth rates $\lambda > 0$, we can use the inverse value $\tau = \lambda^{-1}$, which can be interpreted as an characteristic time to achieve a business goal (sales price targets).

Using expression (13.59), the characteristic time of the amplification with memory in price dynamics is defined by the equation

$$\tau_{amp;\alpha+1} = \tau^{1/(\alpha+1)} = |\lambda|^{-1/(\alpha+1)} \tag{13.60}$$

for the memory fading parameter $1 < \alpha + 1 < 2$.

As a result, we find that characteristic time for amplification processes ($\lambda > 0$) with one-parameter memory does not coincide with the characteristic time of processes without memory. Note that for parameter $\alpha = 0$ the characteristic time (13.60) is equal to the characteristic time of amplification processes without memory $\tau_{amp;1} = \tau$. Equation (13.60) describes the characteristic time of amplification processes with memory fading parameter $\alpha > 0$.

Let us compare the characteristic time of processes with one-parameter power-law memory and the time of processes without memory ($\alpha = 0$). In Table 13.2, $\tau_{amp;\alpha+1}$ is the characteristic time of growth (or decline) with memory, and $\tau_{amp;1} = \tau$ is the characteristic time of processes without memory.

Table 13.2: Comparison of the characteristic time of amplification processes with memory and without memory.

Rate of amplification in price	**Characteristic time of amplification in price**	**Memory fading parameter $1 < \alpha + 1 < 2$**
$0 < \lambda < 1$	$\tau > 1$	$\tau_{amp;\alpha+1} > \tau$
$\lambda = 1$	$\tau = 1$	$\tau_{amp;\alpha+1} = \tau$
$\lambda > 1$	$0 < \tau < 1$	$\tau_{amp;\alpha+1} < \tau$

Using equation (13.60), we see that memory effects can change the characteristic time of amplification process ($\lambda > 0$). It can be both increased and decreased in comparison with the case of absent of memory.

Based on these results, we can formulate the following principle.

Principle 13.1 (Principles of changing of characteristic time by memory in amplification processes).

1. *The effects of one-parameter power-law memory with the noninteger fading parameter $1 < \alpha + 1 < 2$ lead to the characteristic time of amplification processes ($\lambda > 0$) in the form*

$$\tau_{amp;\alpha+1} = \tau^{1/(\alpha+1)}, \tag{13.61}$$

where $\tau = \lambda^{-1}$ is the characteristic time of processes without memory ($\alpha = 0$) for the same values of other parameters.

2. *For small characteristic time ($\tau < 1$) of amplification processes without memory ($\lambda > 0$, $\alpha = 0$), the memory effects with the parameter $1 < \alpha + 1 < 2$ lead to increase of the characteristic time.*
3. *For the large characteristic time ($\tau > 1$) of amplification processes without memory ($\lambda > 0$, $\alpha = 0$), the memory effects with the parameter $1 < \alpha + 1 < 2$ lead to decrease of the characteristic time.*

Let us give the conditions of price increase and price reduction for the proposed model with power-law memory and $\lambda > 0$. The condition of price increase in case of presence of memory with $\alpha + 1 \in (1, 2)$ has the form of the inequality

$$P(0) - \overline{P} + P^{(1)}(0)\lambda^{-1/(\alpha+1)} > 0. \tag{13.62}$$

The condition of price reduction in case of presence of memory with $\alpha + 1 \in (1, 2)$ is gives by the inequality

$$P(0) - \overline{P} + P^{(1)}(0)\lambda^{-1/(\alpha+1)} < 0, \tag{13.63}$$

where $\lambda = -\gamma(b - a)$ and $\overline{P} = (d - s)/(b - a)$.

The condition of price increase in case of absence of memory ($\alpha = 0$) is represented by the inequality $P(0) - \overline{P} > 0$. The condition of price reduction in case of absence of memory ($\alpha = 0$) is represented by the inequality $P(0) - \overline{P} < 0$.

Let us consider the case $P^{(1)}(0) > 0$, where $P(0) > 0$, $\overline{P} > 0$ and $\lambda > 0$. If the inequalities

$$P(0) < \overline{P} < P(0) + \lambda^{-1/\alpha}P^{(1)}(0) \tag{13.64}$$

are satisfied, the process without memory shows price reduction while the process with memory demonstrates a price increase at unchanged other parameters of the dynamic model. Condition (13.64) means that the price reduction is replaced by the price increase, when the memory effect is taken into account.

Let us formulate some properties of the price behavior with memory for the case $0 < \alpha < 1$ ($1 < \alpha + 1 < 2$) and $\lambda > 0$.

1. Pricing based on accounting for memory effects can lead to growth in rate of price increase.
2. Pricing with accounting of memory effects can leads to price increase instead of price reduction. The price reduction can be replaced by the price increase, when the memory effect is taken into account.
3. Pricing strategy, which takes into account the memory effects, can lead to a slowdown in the rate of price reduction. Memory effects may decline in the rate of price reduction.

As a result, we can formulate the following principle of price behavior in economy with memory.

Principle 13.2 (Principle of pricing with memory). *Pricing based on accounting of memory effects can lead to a revision of the price upward. The memory effects can lead to positive results, such as a decline in the rate of price reduction, a replacement of price reduction by its increase, a growth in rate of price increase.*

As a result, we conclude that accounting of the memory effects can give new type of behavior for the same parameters of price dynamic models. The neglecting of the memory can lead to qualitatively different results, conclusions and predictions. In pricing and the construction of models of price behavior, we should take into account the effects of memory to develop better pricing strategies and correct predictions.

13.6 Price dynamics with memory: oscillation of market price

In this section, the consideration the case of $\lambda < 0$ for the price dynamics with power-law memory is based on article [390], which is based on the mathematical results of works [244, 247, 140] and [387].

Let us consider price dynamics with memory that is described by the fractional differential equation

$$(D^{\alpha+1}_{C;0+}P)(t) = \lambda P(t) + f(t), \tag{13.65}$$

where $\lambda = -\gamma(b-a) < 0$, and $D^{\alpha+1}_{C;0+}$ is the Caputo fractional derivative. For $\alpha = 0$, equation (13.65) coincides with the relaxation equation (13.7) of the standard Model III with $\lambda < 0$. For $\alpha = 1$, equation (13.65) coincides with the oscillation equation (13.17) of the standard Model IV with $\lambda < 0$. For $0 < \alpha < 1$ $(1 < \alpha+1 < 2)$, the processes, which are described by equation (13.65), are called the fractional oscillation [244, 247, 140, 387].

The solution of equation (13.65) with $f(t) = f = \gamma(d-s)$, $1 < \alpha+1 \leq 2$ and $\lambda = -|\lambda| < 0$ has the form

$$\begin{aligned} P(t) = \overline{P} + (P(0) - \overline{P})E_{\alpha+1,1}[-|\lambda|t^{\alpha+1}] \\ + P^{(1)}(0)tE_{\alpha+1,2}[-|\lambda|t^{\alpha+1}]. \end{aligned} \tag{13.66}$$

The functions $t^k E_{\alpha+1,k+1}[-|\lambda|t^{\alpha+1}]$ with $k = 0$ and $k = 1$ can be represented as

$$E_{\alpha+1,1}[-|\lambda|t^{\alpha+1}] = f_{\alpha+1,0}(|\lambda|^{1/(\alpha+1)}t) + g_{\alpha+1,0}(|\lambda|^{1/(\alpha+1)}t), \tag{13.67}$$

and

$$tE_{\alpha+1,2}[-|\lambda|t^{\alpha+1}] = f_{\alpha+1,1}(|\lambda|^{1/(\alpha+1)}t) + g_{\alpha+1,1}(|\lambda|^{1/(\alpha+1)}t), \tag{13.68}$$

where the terms $f_{\alpha+1,k}(t)$ describe the relaxation, and $g_{\alpha+1,k}(t)$ describe the oscillations with exponentially decreasing amplitude. Expressions (13.67), (13.68) are called the Mainardi representation [387], which has been proposed in [244, 247, 140].

The functions $f_{\alpha+1,k}(t)$, which describe the relaxation, are defined by the equation

$$f_{\alpha+1,k}(|\lambda|^{1/(\alpha+1)}t) = \int_0^\infty K_{\alpha+1,k}(r,\lambda)\exp(-r|\lambda|^{1/(\alpha+1)}t)\,dr, \tag{13.69}$$

where

$$K_{\alpha,k}(r,\lambda) = \frac{(-1)^k}{\pi}\frac{|\lambda|^{-k/\alpha}r^{\alpha-k}\sin(\pi\alpha)}{r^{2\alpha}+2r^\alpha\cos(\pi\alpha)+1}. \tag{13.70}$$

In the case $1 < \alpha+1 < 2$, the function $f_{\alpha+1,0}(|\lambda|^{1/(\alpha+1)}t)$ tends to zero as t tends to infinity.

The functions $g_{\alpha+1,k}(t)$, which describe the oscillation, are defined by the equation

$$\begin{aligned} &g_{\alpha+1,k}(|\lambda|^{1/(\alpha+1)}t) \\ &= \frac{2}{\alpha+1}|\lambda|^{-k/(\alpha+1)}\exp\left(|\lambda|^{1/(\alpha+1)}t\cos\left(\frac{\pi}{\alpha+1}\right)\right) \\ &\quad\times\cos\left(|\lambda|^{1/(\alpha+1)}t\sin\left(\frac{\pi}{\alpha+1}\right)-\frac{k\pi}{\alpha+1}\right). \end{aligned} \tag{13.71}$$

The function (13.71) describes decline with oscillation, for which the circular frequency is represented by the expression

$$\omega_{\alpha+1} = |\lambda|^{1/(\alpha+1)}\sin\left(\frac{\pi}{\alpha+1}\right), \tag{13.72}$$

and the amplitude decays exponentially with the rate

$$\lambda_{\alpha+1} = |\lambda|^{1/(\alpha+1)}\cos\left(\frac{\pi}{\alpha+1}\right). \tag{13.73}$$

Using equations (13.67) and (13.68), solution (13.66) can be represented in the form

$$\begin{aligned} P(t) &= \overline{P} + (P(0)-\overline{P})f_{\alpha+1}(|\lambda|^{1/(\alpha+1)}t) \\ &\quad + (P(0)-\overline{P})g_{\alpha+1,0}(|\lambda|^{1/(\alpha+1)}t) \\ &\quad + P^{(1)}(0)f_{\alpha+1,1}(|\lambda|^{1/(\alpha+1)}t) + P^{(1)}(0)g_{\alpha+1,1}(|\lambda|^{1/(\alpha+1)}t). \end{aligned} \tag{13.74}$$

As a result, using expression (13.72), we get the characteristic time of the oscillation with memory

$$\tau_{\mathrm{osc},\alpha+1} = \tau^{1/(\alpha+1)}\left|\csc\left(\frac{\pi}{\alpha+1}\right)\right|. \tag{13.75}$$

Using equation (13.73), the characteristic time of exponentially damping amplitude of oscillations (damping of oscillation) is described by the equation

$$\tau_{\mathrm{dam},\alpha+1} = \tau^{1/(\alpha+1)} \left| \sec\left(\frac{\pi}{\alpha+1} \right) \right|, \tag{13.76}$$

where $1 < \alpha + 1 < 2$ and $\tau = |\lambda|^{-1}$ is a characteristic time of oscillation without memory ($\alpha = 1$). We note that the characteristic times $\tau_{\mathrm{osc},\alpha+1}$ and $\tau_{\mathrm{dam},\alpha+1}$ do not depend on the parameter $k \in \mathbb{N}$. Using that $|\sec(x)| \geq 1$ and $|\csc(x)| \geq 1$, we get the inequalities

$$\tau_{\mathrm{osc},\alpha+1} \geq \tau_{\mathrm{amp};\alpha+1}, \tag{13.77}$$

$$\tau_{\mathrm{dam},\alpha+1} \geq \tau_{\mathrm{amp};\alpha+1}, \tag{13.78}$$

where $\tau_{\mathrm{amp};\alpha+1} = \tau^{1/(\alpha+1)} = \lambda^{-1/(\alpha+1)}$ is the characteristic time of price increase in presence of memory with $1 < \alpha + 1 < 2$. In Table 13.3, we give the characteristic times for price dynamics with memory for $1 < \alpha + 1 < 2$.

Table 13.3: Characteristic times for price dynamics with memory in the case $1 < \alpha + 1 < 2$.

Type of processes	Sign λ	Characteristic times	Function
Amplification	$\lambda > 0$	$\tau_{\mathrm{amp};\alpha+1} = \tau^{1/(\alpha+1)}$	$E_{\alpha+1,1}[\lambda t^{\alpha+1}]$
Relaxation	$\lambda < 0$	$\tau_{\mathrm{rel};\alpha+1} = \tau^{1/\alpha}$	Relaxation by $f_{\alpha+1,k}(\lvert\lambda\rvert^{1/(\alpha+1)} t)$
Relaxation	$\lambda < 0$	$\tau_{\mathrm{osc},\alpha+1} = \tau^{1/(\alpha+1)} \lvert \csc(\frac{\pi}{\alpha+1}) \rvert$	Oscillation by $g_{\alpha+1,k}(\lvert\lambda\rvert^{1/(\alpha+1)} t)$
Relaxation	$\lambda < 0$	$\tau_{\mathrm{dam},\alpha+1} = \tau^{1/(\alpha+1)} \lvert \sec(\frac{\pi}{\alpha+1}) \rvert$	Oscillation damping by $g_{\alpha+1,k}(\lvert\lambda\rvert^{1/(\alpha+1)} t)$

13.7 Properties of relaxation with memory to equilibrium price

Let us consider some properties, including the characteristic time, for relaxation to equilibrium price, when power-law memory is taken into account. For this purpose, we will use the asymptotic behavior of the price function $P(t)$ at $t \to \infty$ for the case $\lambda < 0$ ($\lambda = -|\lambda|$). In the case $\lambda < 0$, we have

$$E_{\alpha,k+1}[z] = -\sum_{j=1}^{m} \frac{1}{\Gamma(k+1-\alpha j)} \frac{1}{z^j} + O\left(\frac{1}{z^{m+1}} \right) \tag{13.79}$$

for $|z| \to \infty$ and all $m \in \mathbb{N}$, where $z = \lambda t^{\alpha} = -|\lambda| t^{\alpha}$. Equation (13.79) describes the asymptotic behavior at infinity for the case $1 < \alpha + 1 < 2$. Note that this asymptotic behavior (13.79) is caused by the behavior of the function $f_{\alpha+1,k}(|\lambda|^{1/(\alpha+1)}t)$. The contribution of the function $g_{\alpha+1,k}(|\lambda|^{1/(\alpha+1)}t)$ is exponentially small at $t \to \infty$ for $0 < \alpha < 1$ $(1 < \alpha + 1 < 2)$.

The solution of the model equation with $\lambda = -|\lambda| < 0$ has the form

$$P(t) = \overline{P} + (P(0) - \overline{P})E_{\alpha+1,1}[-|\lambda|t^{\alpha+1}] \\ + P^{(1)}(0)tE_{\alpha+1,2}[-|\lambda|t^{\alpha+1}], \tag{13.80}$$

where $1 < \alpha + 1 < 2$.

Using equation (1.8.28) of [200, p. 43], in the form

$$E_{\alpha+1,1}[\lambda t^{\alpha+1}] = -\sum_{j=1}^{N} \frac{\lambda^{-j}}{\Gamma(1-(\alpha+1)j)} t^{-j(\alpha+1)} \\ + O(t^{-(\alpha+1)(N+1)}), \tag{13.81}$$

and

$$E_{\alpha+1,2}[\lambda t^{\alpha+1}] = -\sum_{j=1}^{N} \frac{\lambda^{-j}}{\Gamma(2-(\alpha+1)j)} t^{-j(\alpha+1)} \\ + O(t^{-(\alpha+1)(N+1)}), \tag{13.82}$$

we derive the behavior of solution (13.62) at $t \to \infty$ in the form

$$P(t) = \overline{P} - P^{(1)}(0)\frac{\lambda^{-1}t^{-\alpha}}{\Gamma(1-\alpha)} - (P(0) - \overline{P})\frac{\lambda^{-1}t^{-\alpha-1}}{\Gamma(-\alpha)} \\ - \sum_{j=2}^{N}\left(P(0) - \overline{P} + \frac{P^{(1)}(0)t}{1-(\alpha+1)j}\right)\frac{\lambda^{-j}t^{-(\alpha+1)j}}{\Gamma(1-(\alpha+1)j)} \\ + O\left(\frac{1}{t^{(\alpha+1)(N+1)-1}}\right), \tag{13.83}$$

where we use $\Gamma(z+1) = z\Gamma(z)$ and $0 < \alpha < 1$ $(1 < \alpha + 1 < 2)$.

If $P^{(1)}(0) \neq 0$, the slowest term is the second term of (13.83) tending to equilibrium when time tends to infinity. As a result, we get that relaxation processes with memory (for $1 < \alpha + 1 < 2$) exhibit an algebraic decay (the power-law decay) with the characteristic time

$$\tau_{\mathrm{rel},\alpha+1} = |\lambda|^{-1/\alpha} = \tau^{1/\alpha}, \tag{13.84}$$

where λ is the rate of growth (decline) for relaxation process without memory. We emphasize that the power of τ in (13.84) is $1/\alpha$ instead of $1/(\alpha+1)$. The relaxation ($\lambda < 0$) with memory $1 < \alpha + 1 < 2$ by decline with oscillation is realized, when the condition $P^{(1)}(0) > 0$ holds. The condition of the relaxation ($\lambda < 0$) by growth with oscillation has the form $P^{(1)}(0) < 0$.

As a result, if we take into account the power-law memory, the relaxation to equilibrium at $t \to \infty$ exhibit an algebraic (power-law) decay with characteristic time (13.84) for $1 < \alpha + 1 < 2$. Therefore, we can state that characteristic time of relaxation ($\lambda < 0$) to equilibrium price in the presence one-parameter memory does not coincide with the characteristic time of dynamics without memory ($\alpha = 1$).

13.8 Comparison of characteristic times for price dynamics

In this section, we compare the characteristic times of price dynamics for $\lambda < 0$ and $\lambda > 0$.

For $1 < \alpha + 1 < 2$, the characteristic time for price relaxation with memory to equilibrium price is equal to the value

$$\tau_{\mathrm{rel};\alpha+1} = \tau^{1/\alpha}, \tag{13.85}$$

and the characteristic time of oscillation damping is described by the equation

$$\tau_{\mathrm{dam};\alpha+1} = |\lambda|^{-1/(\alpha+1)} \left|\cos\left(\frac{\pi}{\alpha+1}\right)\right|^{-1} = \tau^{1/(\alpha+1)} \left|\sec\left(\frac{\pi}{\alpha+1}\right)\right|. \tag{13.86}$$

Equation (13.86) shows that characteristic time of relaxation with memory ($\lambda < 0$) to equilibrium price does not coincide the characteristic time of price amplification processes ($\lambda > 0$) with memory, in general, i. e., $\tau_{\mathrm{rel};\alpha+1} \neq \tau_{\mathrm{dam};\alpha+1}$ for $1 < \alpha + 1 < 2$.

As a result, we have the expression

$$\tau_{\mathrm{rel};\alpha+1} = \tau^{1/\alpha} \quad \text{for} \quad 1 < \alpha + 1 < 2, \tag{13.87}$$

where $\tau = |\lambda|^{-1}$ is the characteristic time of relaxation processes without memory. The parameter $\tau_{\mathrm{rel};\alpha+1}$ characterizes the rate of relaxation for processes with one-parameter power-law memory. As a result, we have the equality $\tau_{\mathrm{rel};\alpha+1} = \tau_{\mathrm{amp};\alpha}$ for $1 < \alpha + 1 < 2$.

Let us compare the characteristic times of amplification processes and the characteristic times of oscillation damping for relaxation processes (see Table 13.4). Note that the multiplier $|\sec(\pi/(\alpha+1))|$ with $1 < \alpha+1 < 2$ increases the characteristic time of oscillation damping $\tau_{\mathrm{dam};\alpha+1}$ in comparison of the characteristic time of amplification process with memory $\tau_{\mathrm{amp};\alpha}$ for $1 < \alpha + 1 < 2$.

Equation (13.86) shows that the closer the value of the fading parameter α to 2, the greater the characteristic time of oscillation damping. The characteristic time of oscillation damping with memory $1 < \alpha + 1 < 2$ is always equal to or longer than the characteristic time of amplification processes with the same memory fading $\tau_{\mathrm{ampl};\alpha+1}$, i. e., we have the inequality

$$\tau_{\mathrm{dam};\alpha+1} \geq \tau_{\mathrm{ampl};\alpha+1}. \tag{13.88}$$

Table 13.4: Comparison of the characteristic time $\tau_{dam;\alpha+1}$ for oscillation damping with memory and the characteristic time $\tau_{ampl;\alpha+1}$ of amplification process ($\lambda > 0$) with the same memory with $\alpha + 1 \in (1, 2)$.

Memory fading parameter $\alpha + 1 \in (1, 2)$	$\frac{\tau_{dam;\alpha+1}}{\tau_{ampl;\alpha+1}}$	**Memory fading parameter $\alpha + 1 \in (1, 2)$**	$\frac{\tau_{dam;\alpha+1}}{\tau_{ampl;\alpha+1}}$
1.1	1.04	1.7	3.65
1.2	1.15	1.8	5.76
1.3	1.34	1.9	12.1
1.4	1.60	1.95	24.8
1.5	2.00	1.99	126
1.6	2.61	1.995	254

Using the results of the previous subsections, we can formulate the following principles for relaxation processes of growth and decline with power-law memory with $\alpha \in (0, 1)$, i. e., $\alpha + 1 \in (1, 2)$.

Principle 13.3 (Principle of changing of price relaxation time by memory). *The effects of one-parameter power-law memory with the noninteger fading parameter $0 < \alpha+1 < 2$ lead to the dependence of the characteristic time of price relaxation on the parameter α in the form*

$$\tau_{rel;\alpha+1} = \tau^{1/\alpha} \quad \textit{for } 1 < \alpha + 1 < 2, \tag{13.89}$$

where $\tau = |\lambda|^{-1}$ is the characteristic (relaxation) time of processes without memory ($\alpha = 1$) for the same values of other parameters of the model.

Equation (13.89) means that the characteristic time for relaxation process ($\lambda < 0$) with memory fading parameter $\alpha + 1 \in (1, 2)$ is equal to the characteristic time of amplification process ($\lambda > 0$) with the memory fading α, i. e., we have the equality

$$\tau_{rel;\alpha+1} = \tau_{ampl;\alpha} \quad \text{for} \quad 1 < \alpha + 1 < 2. \tag{13.90}$$

For $\alpha + 1 > 2$, we have the amplification of the oscillations in relaxation processes ($\lambda < 0$) that can be interpreted as an instability.

13.9 Comparison of relaxation and oscillation damping of price

In this section, we compare the characteristic times of relaxation and oscillation damping of price dynamics with memory.

For $1 < \alpha + 1 < 2$, we have the following three characteristic times:

(1) The characteristic time of oscillation with memory

$$\tau_{osc;\alpha+1} = \tau^{1/(\alpha+1)} \left| \csc\left(\frac{\pi}{\alpha + 1} \right) \right|. \tag{13.91}$$

(2) The characteristic time of oscillation damping with memory

$$\tau_{\text{dam};\alpha+1} = \tau^{1/(\alpha+1)} \left| \sec\left(\frac{\pi}{\alpha+1} \right) \right|. \tag{13.92}$$

(3) The characteristic time of relaxation process with memory

$$\tau_{\text{rel};\alpha+1} = \tau^{1/\alpha}. \tag{13.93}$$

To describe price increase and price decrease, we should consider two characteristic times (13.92) and (13.93).

In Table 13.5, we describe the hierarchy of characteristic times $\tau_{\text{rel};\alpha+1}$ and $\tau_{\text{dam};\alpha+1}$ for relaxation processes with the memory fading parameters $\alpha+1$ close to the values 1 and 2 in the case $1 < \alpha + 1 < 2$, where $\tau = |\lambda|^{-1}$ is characteristic time of relaxation process without memory.

Table 13.5: Comparison of characteristic times $\tau_{\text{rel};\alpha+1}$ and $\tau_{\text{dam};\alpha+1}$ for price relaxation with memory.

Memory fading parameter $0 < \alpha < 1$ $(1 < \alpha + 1 < 2)$	Relaxation time without memory	Inequality for characteristic times
$\alpha \to 0+$	$\tau > 1$	$\tau_{\text{rel};\alpha+1} \gg \tau_{\text{dam};\alpha+1}$
$\alpha \to 0+$	$\tau < 1$	$\tau_{\text{rel};\alpha+1} \ll \tau_{\text{dam};\alpha+1}$
$\alpha \to 1-$	$\tau > 1$	$\tau_{\text{rel};\alpha+1} \ll \tau_{\text{dam};\alpha+1}$
$\alpha \to 1-$	$\tau < 1$	$\tau_{\text{rel};\alpha+1} \ll \tau_{\text{dam};\alpha+1}$

Using Table 13.5, we can see that there is an effect of competition of decelerations (slowing down) on short periods of time (short times). In this competition, the slower process wins for short times. For short times, the behavior of the entire process is controlled (determined) by a slower component of the process, that is, the component with larger characteristic time. Let us consider competition of oscillation damping and relaxation in short times. For example, at values of $(\alpha + 1)$ close to 2 for short time, the competition leads to deceleration by oscillation damping. Using Table 13.5, we see that the damping of the oscillation amplitude is slower than the relaxation for short times. At the same time, using asymptotic expressions, it can be seen that at long times (at $t \to \infty$) the behavior is determined by the relaxation time. As a result, we can formulate the following principles.

Principle 13.4 (Principles of competition of deceleration with memory on short times).

1. *For the memory fading with parameter α such that $\alpha_{\text{eq}} + 1 < \alpha + 1 < 2$, the characteristic time of relaxation processes is less than the characteristic time of oscillation damping, i. e., $\tau_{\text{rel};\alpha+1} < \tau_{\text{dam};\alpha+1}$ for $\alpha_{\text{eq}} + 1 < \alpha + 1 < 2$ and all $\tau > 0$.*

2. *At values of* $(\alpha+1)$ *close to* 2 *for short time, the competition leads to deceleration by oscillation damping for all* $\tau > 0$.
3. *For the memory fading with parameter* α *such that* $1 < \alpha+1 < \alpha_{eq}+1$ *and* $\tau > 1$, *the characteristic time of relaxation processes is greater than the characteristic time of oscillation damping, i. e.,* $\tau_{dam;\alpha+1} > \tau_{rel;\alpha+1}$ *if* $1 < \alpha+1 < \alpha_{eq}+1$ *and* $\tau > 1$.
4. *For* $1 < \alpha+1 < \alpha_{eq}+1$ *and* $\tau < 1$, *the characteristic time of relaxation processes is less than the characteristic time of oscillation damping, i. e.,* $\tau_{dam;\alpha+1} < \tau_{rel;\alpha+1}$ *if* $1 < \alpha+1 < \alpha_{eq}+1$ *and* $\tau < 1$.
5. *At values of* $(\alpha+1)$ *close to* 1 *for short time, the competition leads to deceleration by oscillation damping if* $\tau < 1$ *and to deceleration by relaxation if* $\tau > 1$.
6. *For* $\alpha+1 > 2$, *we have the amplification of the oscillations (increase of oscillation amplitude) in relaxation processes* ($\lambda < 0$) *that can be interpreted as instability.*

In these principles, we use the parameter α_{eq} that is defined by the equality

$$\tau_{dam;\alpha_{eq}+1} = \tau_{rel;\alpha_{eq}+1}, \tag{13.94}$$

where $\tau_{dam;\alpha_{eq}+1}$ and $\tau_{rel;\alpha_{eq}+1}$ are defined by equations (13.92) and (13.93), respectively. Equality (13.94) gives the transcendental equation

$$\left|\cos\left(\frac{\pi}{\alpha_{eq}+1}\right)\right|^{\alpha_{eq}(\alpha_{eq}+1)} = |\lambda|, \tag{13.95}$$

where $\tau = |\lambda|$. For example, $\alpha_{eq} \approx 1.78160$ for $|\lambda| = 0.1$, $\alpha_{eq} \approx 1.54850$ for $|\lambda| = 0.5$, and $\alpha_{eq} \approx 1.29296$ for $|\lambda| = 0.9$ [387].

As a result, we see that, we cannot neglect the characteristic time of the oscillation damping and take into account only the characteristic time of the relaxation in the general case. The characteristic time of oscillation damping should be taken into account for $\alpha > \alpha_{eq}$, where α_{eq} depends on the rate λ of relaxation processes without memory, i. e., it depends on the relaxation time of processes without memory τ. To illustrate these properties, the characteristic times for behavior of price $P(t)$ are given in Table 13.6.

Table 13.6: Comparison of characteristic times $\tau_{osc;\alpha+1}$, $\tau_{dam;\alpha+1}$ and $\tau_{rel;\alpha+1}$ for $\tau = 10$ and $\alpha_{eq}+1 \approx 1.78160$. Note that $\tau_{dam;\alpha+1} = \tau_{rel;\alpha+1}$ for $\alpha = \alpha_{eq}$.

$\alpha+1 \in (1,2)$	$\tau_{osc;\alpha+1}$	$\tau_{dam;\alpha+1}$	$\tau_{rel;\alpha+1}$
1.999	3.164	4026	10.023
1.9	3.371	40.69	2.915
1.7816	3.710	19.03	19.03
1.7	4.028	14.16	26.83
1.5	5.360	9.283	100
1.2	13.63	7.867	100000
1.1	28.79	8.454	10000000000

Table 13.6 and Figures 13.1–13.6 illustrate the fact that we cannot neglect the characteristic time of the oscillation damping and take into account only the characteristic time of the relaxation in the general case.

1. For $\alpha = 0.999 > \alpha_{eq}$, we have $\tau_{dam;\alpha+1} \approx 4026 \gg \tau_{rel;\alpha+1} \approx 10.023$ and $\tau_{rel;\alpha+1} \approx \tau$. Here, $\tau_{rel;\alpha+1}$ tends to $\tau = 10$, i. e., $\tau_{rel;\alpha+1} \to \tau$, when fading parameter $\alpha + 1$ tends to 2. See Figure 13.1.
2. For $\alpha = 0.9 > \alpha_{eq}$, we have $\tau_{dam;\alpha+1} \approx 40.69 > \tau_{rel;\alpha+1} \approx 2.915$ and $\tau_{dam;\alpha+1} > \tau > \tau_{rel;\alpha+1}$. See Figure 13.2.
3. For $\alpha = 0.7 < \alpha_{eq}$, we have $\tau_{dam;\alpha+1} \approx 14.16 < \tau_{rel;\alpha+1} \approx 26.83$. See Figure 13.3.
4. For $\alpha = 0.5 < \alpha_{eq}$, we have $\tau_{dam;\alpha+1} \approx 9.283 < \tau_{rel;\alpha+1} \approx 100$ and $\tau_{dam;\alpha+1} \approx \tau$. See Figure 13.4.
5. For $\alpha = 0.2 < \alpha_{eq}$, we have $\tau_{dam;\alpha+1} \approx 7.867 \ll \tau_{rel;\alpha+1} \approx 100000$. See Figure 13.5.
6. For $\alpha = 0.1 < \alpha_{eq}$, we have $\tau_{dam;\alpha+1} \approx 8.454 \ll \tau_{rel;\alpha+1} \approx 10000000000$ and $\tau_{dam;\alpha+1}$ tends to $\tau = 10$ from below, when the fading parameter $\alpha + 1$ tends to 1. See Figure 13.6.

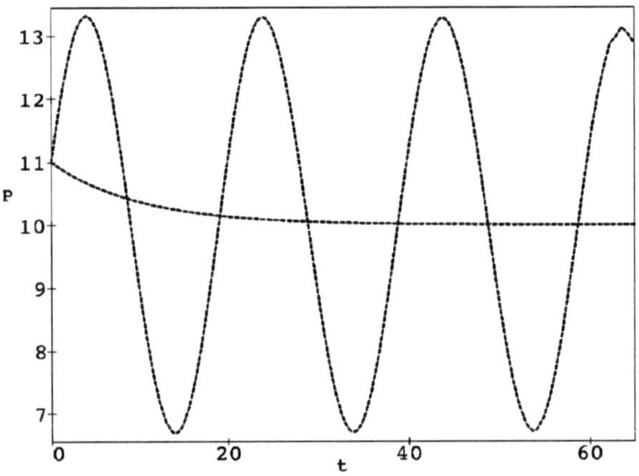

Figure 13.1: Comparison of relaxation with memory for $\alpha + 1 = 1.999$ and relaxation without memory ($\alpha + 1 = 1$), when $\overline{P} = 10$, $P(0) = 11$, $P^{(1)}(0) = 1$, $|\lambda| = 0.1$ ($\tau = 10$). For this case, $\tau_{dam;\alpha+1} \approx 4026 \gg \tau_{rel;\alpha+1} \approx 10.023$ and $\tau_{rel;\alpha+1} \approx \tau$.

We see that approach to the equilibrium state slows down with a decrease in the memory fading parameter. We can also see that damping of oscillations is the faster, the smaller the parameter of memory fading. For $1 < \alpha + 1 < 1.5$, the oscillations were very rapidly damped and $\tau_{dam;\alpha+1} \approx \tau$.

It is likely that the mechanisms of oscillation damping, which make the main contribution to deceleration, are first working. Then the mechanisms of deceleration due to relaxation are included (in the form of an algebraic decay that has the power-law form).

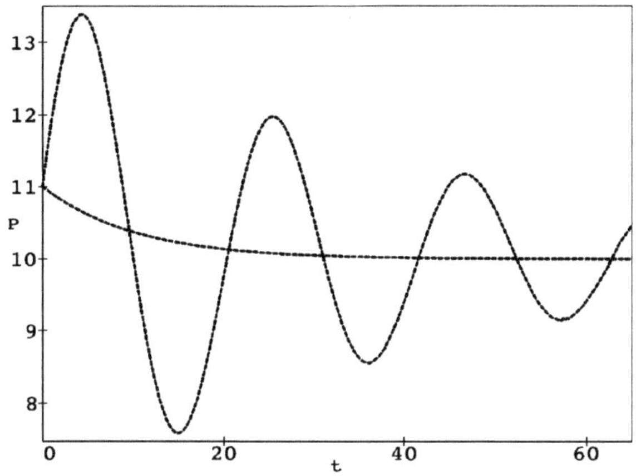

Figure 13.2: Comparison of relaxation with memory for $\alpha + 1 = 1.9$ and relaxation without memory ($\alpha + 1 = 1$), when $\overline{P} = 10$, $P(0) = 11$, $P^{(1)}(0) = 1$, $|\lambda| = 0.1$ ($\tau = 10$). For this case, $\tau_{\mathrm{dam};\alpha+1} \approx 40.69 > \tau_{\mathrm{rel};\alpha+1} \approx 2.915$ and $\tau_{\mathrm{dam};\alpha+1} > \tau > \tau_{\mathrm{rel};\alpha+1}$.

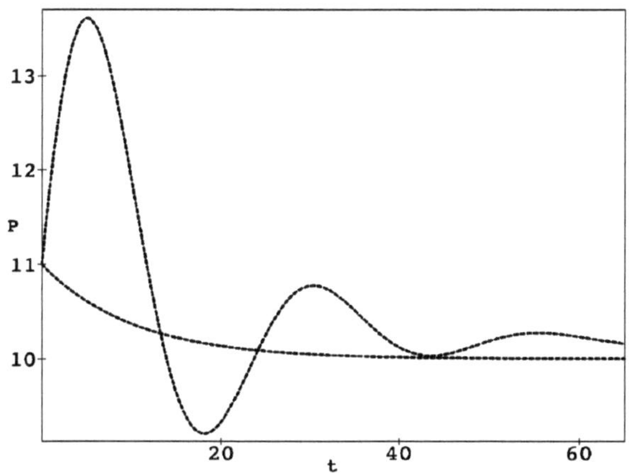

Figure 13.3: Comparison of relaxation with memory for $\alpha + 1 = 1.7$ and relaxation without memory ($\alpha + 1 = 1$), when $\overline{P} = 10$, $P(0) = 11$, $P^{(1)}(0) = 1$, $|\lambda| = 0.1$ ($\tau = 10$). For this case, $\tau_{\mathrm{dam};\alpha+1} \approx 14.16 < \tau_{\mathrm{rel};\alpha+1} \approx 26.83$.

The deceleration begins with the fast process. At the initial stage, the relaxation processes work ($\tau_{\mathrm{rel};\alpha+1} \gg \tau_{\mathrm{dam};\alpha+1}$) only under the condition that values of the memory fading parameter is close to one ($\alpha+1 \approx 1$, $\alpha > 0$) in processes with $|\lambda| < 1$, ($\tau > 1$). At the initial stage, the oscillation damping processes work ($\tau_{\mathrm{dam};\alpha+1} \gg \tau_{\mathrm{rel};\alpha+1}$) in processes with a memory fading parameter close to one ($\alpha \to 0$, $\alpha > 0$) and the relaxation rate is greater than one ($|\lambda| > 1$, $\tau < 1$), as well as in processes with fading parameter $\alpha \to 1$ ($\alpha < 1$) for all the relaxation rate $|\lambda|$. Therefore, we cannot neglect the characteristic time of the oscillation damping at the initial stage, in the general case.

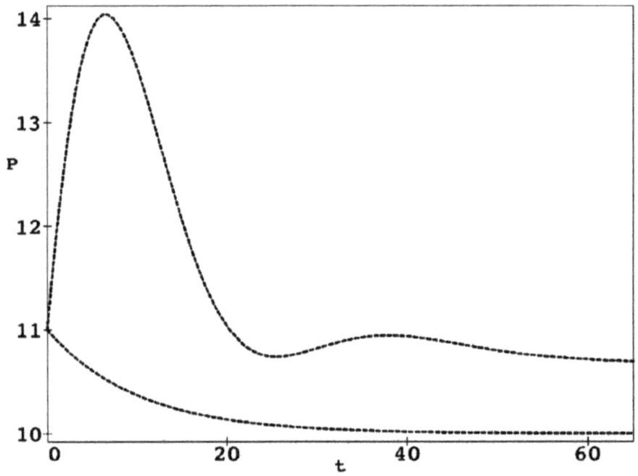

Figure 13.4: Comparison of relaxation with memory for $\alpha + 1 = 1.5$ and relaxation without memory ($\alpha + 1 = 1$), when $\overline{P} = 10$, $P(0) = 11$, $P^{(1)}(0) = 1$, $|\lambda| = 0.1$ ($\tau = 10$). For this case, $\tau_{\mathrm{dam};\alpha+1} \approx 9.283 < \tau_{\mathrm{rel};\alpha+1} \approx 100$ and $\tau_{\mathrm{dam};\alpha+1} \approx \tau$.

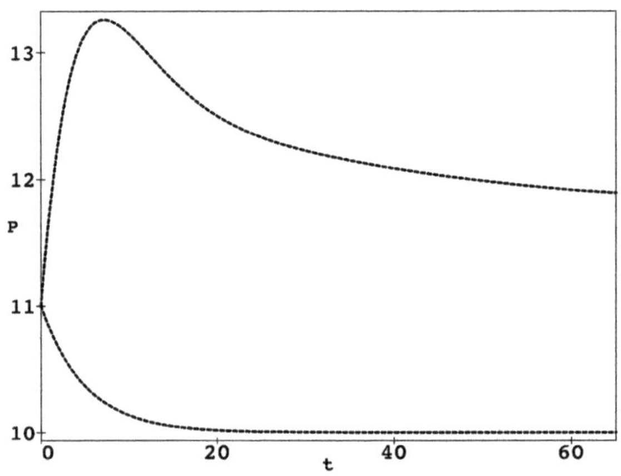

Figure 13.5: Comparison of relaxation with memory for $\alpha + 1 = 1.2$ and relaxation without memory ($\alpha + 1 = 1$), when $\overline{P} = 10$, $P(0) = 11$, $P^{(1)}(0) = 1$, $|\lambda| = 0.1$ ($\tau = 10$). For this case, $\tau_{\mathrm{dam};\alpha+1} \approx 7.867 \ll \tau_{\mathrm{rel};\alpha+1} \approx 100000$.

In description of price dynamics that take into account memory effects, competition between decelerations (oscillation damping and relaxation) arises over short periods of time (short times). At values of $\alpha + 1$ close to 2 for short times, competition leads to the winning of the damping of price oscillation since $\tau_{\mathrm{dam};\alpha+1} \gg \tau_{\mathrm{rel};\alpha+1}$. At values of $\alpha + 1$ close to 1 for short times, competition leads to the winning of the price relaxation if $|\lambda| < 1$ and the damping of price oscillation if $|\lambda| > 1$. At large (long) times (at $t \to \infty$) competition always leads to the winning of the price relaxation, if $\alpha + 1 \in (1, 2)$. For $\alpha + 1 > 2$, the price demonstrates an amplification of the oscillations that is interpreted as instability.

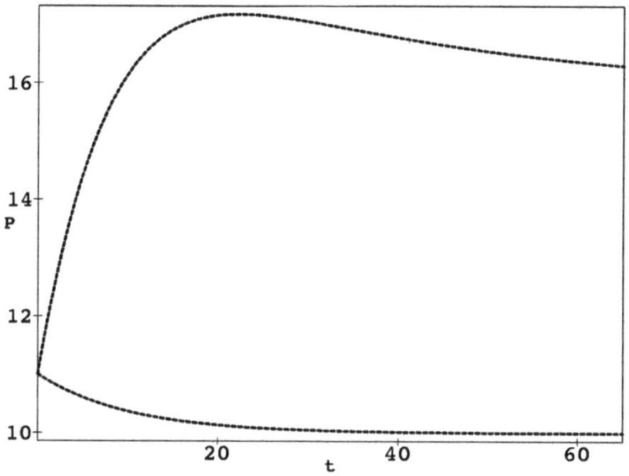

Figure 13.6: Comparison of relaxation with memory for $\alpha + 1 = 1.1$ and relaxation without memory ($\alpha + 1 = 1$), when $\overline{P} = 10$, $P(0) = 11$, $P^{(1)}(0) = 1$, $|\lambda| = 0.1$ ($\tau = 10$). For this case, $\tau_{\text{dam};\alpha+1} \approx 8.454 \ll \tau_{\text{rel};\alpha+1} \approx 10000000000$ and $\tau_{\text{dam};\alpha+1}$ tends to $\tau = 10$.

13.10 Conclusion

We generalized the Evans model of price dynamics in market, which is one of the first dynamic models describing the change in price. The standard Evans model does not take into account memory effects. We consider dynamics of market prices, in which power-law memory effects are taken into account. We propose new economic model of the price dynamics in the market for a single product. In this model, we assume that economic entities (merchants, buyers, suppliers) can remember how stocks of goods and their prices have changed over time. In the proposed model, we take into account that the behavior of the entities at time t can depend not only at the present time t, but also on the history of changes on finite time interval. The fractional differential equation, which describes the price dynamics with memory, and the expression of its exact solution are suggested. The asymptotic behavior of the proposed solutions is described. Based on the properties of the solutions, we formulate the following principles: the principles of changing of characteristic time by memory; principle of pricing with memory; principle of changing of price relaxation time by memory; principles of competition of deceleration with memory on short times. Using the results of the chapter, we can conclude that accounting of the memory effects can give new type of behavior for the remaining parameters of the models unchanged. The neglecting of the memory can lead to qualitatively different results, conclusions and predictions. In pricing and the construction of models of price behavior, we should take into account the effects of memory to develop better pricing strategies and correct predictions.

14 Cagan model of inflation with memory

In this chapter, we consider a generalization of the Cagan model. The standard model was proposed by Phillip D. Cagan to describe the dynamics of the actual inflation. We propose a generalization that takes into account the memory effects in dynamics of the actual inflation. In the standard Cagan model, the indicator of nervousness of economic agents, which characterizes the speed of revising the expectations, is represented as a constant parameter. We assume that the nervousness of economic agents can be caused not only by the current state of the process, but also by the history of its changes. In general, the speed of revising the expectations of inflation can depend on the history of changes in the difference between the real inflation rate and the rate expected by economic agents. The use of the memory function instead of the indicator of nervousness allows us to take into account the memory effects in the Cagan model. We consider dynamics of the actual inflation that takes into account memory with power-law fading. The fractional differential equation, which describes the proposed economic model with memory, and the expression of its exact solution are suggested. This chapter is based on work [388].

14.1 Introduction

Inflation and seigniorage are important effects that should be described in economics [326, 327, 487]. Sometimes inflation reaches high levels, and goes into hyperinflations, which are defined as inflation that exceeds 50 % per month. The basic cause of most cases of high inflation and hyperinflation is government's need to obtain seigniorage, i. e., revenue from printing money [43]. However even governments' need for seigniorage cannot account for hyperinflations [326, p. 543].

If the public does not immediately adjust its money holdings or its expectations of inflation to changes in the economic environment, then in the short run seigniorage is always increasing in money growth, and the government can obtain more seigniorage than the maximum sustainable amount, S^*. Thus hyperinflations arise when the government's seigniorage needs exceed S^* [43, 326]. Gradual adjustment of money holdings and gradual adjustment of expected inflation have similar implications for the dynamics of inflation.

One of the most famous models describing the actual inflation is the model that has been proposed by Phillip D. Cagan in work [43] in 1956 (see also books [487, pp. 157–159] and [326, pp. 543–547]). The Cagan model focuses on the case of gradual adjustment of money holdings. In this model, it is assumed that individuals desired money holdings are given by the Cagan money-demand function. The Cagan model was created to describe the processes of hyperinflation (see Section 10.8 in [326], or Section 11.9 in [327]). As the only factor in the demand for money, the Cagan model considers inflationary expectations. This assumption corresponds to a situation of

https://doi.org/10.1515/9783110627459-014

lack of economic growth, i. e., the Cagan model describes an economy with constant output.

In the standard Cagan model, the memory effects and memory fading are neglected. At the same time, it is obvious that to describe the expected rate of inflation, it is necessary to take into account memory and memory fading, since amnesia of economic agents is a strong restriction on the economic model.

In the standard Cagan model, the indicator of nervousness of economic agents, which characterizes the speed of revising the expectations, is represented as a constant parameter [487]. In general case, the nervousness of economic agents can be caused not only by the current state of the process, but also by the history of its changes. The speed of revising the expectations of inflation can depend on the history of changes in the difference between the real inflation rate and the rate expected by economic agents. The use of the memory function instead of the indicator of nervousness allows us to take into account the memory effects in the Cagan model.

In this chapter, we derive a generalization of the Cagan model, which is suggested in [43] (see also [487, pp. 157–159], [326, pp. 543–547]). In this generalization, the memory effects and memory fading are taken into account. Dynamics of the actual inflation, where we take into account memory with power-law fading, is described. The fractional nonlinear differential equation, which describes the proposed model, and the expression of its exact solution are suggested.

14.2 Standard Cagan model without memory

One of the well-known models, which describes the actual inflation, is the model proposed by Phillip D. Cagan in works [43]. In this section, we consider derivation of the standard Cagan model and the assumptions that are used in this model. In the Cagan model, which does not take into account memory effects, the following variables are used (e. g., see [487, pp. 157–158], [487, pp. 147–150], [326, pp. 543–547], and [327]):
- $M(t)$ is the nominal money supply (the money stock);
- $P(t)$ is the general price level;
- $m = M^{(1)}(t)/M(t)$ is the money supply growth rate;
- $z(t) = M(t)/P(t)$ is the real cash reserves (the stocks of money);
- $z^D(t)$ is the demand for real cash reserves;
- $z^S(t)$ is the supply of real cash reserves;
- $\pi(t)$ is the real rate of inflation (the rate of actual inflation);
- $\pi^e(t)$ is the expected rate of inflation.

Let us consider the assumptions of the standard Cagan model.
1. As an example of the relationship between inflation and steady-state seigniorage, we will consider the money-demand function proposed by Cagan in 1956 [43] (see also books [326, pp. 540–541] and [487, pp. 157]). The demand function for money

has the form

$$\left(\frac{M}{P}\right)^D = f(\pi^e) = \exp(-a\pi^e), \tag{14.1}$$

where $(M/P)^D$ is the demand for real cash reserves; π^e is the expected inflation rate; a is a parameter characterizing the elasticity of demand for money by inflation, $a > 0$. Note that the elasticity of money demand by inflation rate is described by expression $a\pi^e$.

2. The growth rate of the money supply is constant

$$\frac{M^{(1)}(t)}{M(t)} = m = \theta = \text{const}\,. \tag{14.2}$$

3. The key assumption of the model is that actual money holdings adjust gradually forward desired holdings [326, p. 543]. The rule for revising expectations in the standard Cagan model is given by the equation

$$\frac{d\pi^e(t)}{dt} = \beta(\pi(t) - \pi^e(t)), \tag{14.3}$$

where $\beta > 0$ is the constant. The parameter β is an indicator of nervousness of economic agents that characterizes the speed of revising the expectations [487]. The idea behind this assumption of gradual adjustment is that it is difficult for individuals to adjust their money holdings; for example, they may have made arrangements to make certain types of purchases using money. As a result, they adjust their money holdings toward the desired level only gradually [326, p. 543]. Equation (14.3) assumes that expectations are adaptive. If the real inflation rate (π) is higher than the rate expected by economic agents (π^e), then they will adjust their expectations for the future toward increasing inflation $d\pi^e(t)/dt > 0$, and vice versa, if $\pi - \pi^e < 0$, that is, $\pi < \pi^e$, then $d\pi^e(t)/dt < 0$.

4. Equilibrium condition in the money market has the form

$$\left(\frac{M}{P}\right)^D = \exp(-a\pi^e) = \left(\frac{M}{P}\right)^S = \frac{M}{P}. \tag{14.4}$$

5. The memory effects and memory fading are not taken into account in the standard model. This assumption is caused by the application of standard derivatives of first order in equations of the model.

Let us get the equation of the standard Cagan model and its solution.

Using logarithm identity (14.4), we get

$$-a\pi^e = \ln M - \ln P. \tag{14.5}$$

Taking the first derivative with respect to t, we obtain the equation

$$-a\frac{d\pi^e(t)}{dt} = \theta - \pi(t), \tag{14.6}$$

where we use the definition of θ by equation (14.2) and that $\pi(t) = P^{(1)}(t)/P(t)$. Multiplying equation (14.3) by the parameter $-a$, we get

$$-a\frac{d\pi^e(t)}{dt} = -a\beta(\pi(t) - \pi^e(t)). \tag{14.7}$$

Substitution of expression (14.5) into equation (14.7) gives

$$\theta - \pi(t) = -a\beta(\pi(t) - \pi^e(t)). \tag{14.8}$$

Solving algebraic equation (14.8), we get the expression

$$\pi(t) = \frac{\theta - a\beta\pi^e(t)}{1 - a\beta}. \tag{14.9}$$

Substitution of expression (14.9) into equation (14.7) (or using equations (14.3) and (14.6)) gives the equation for the expected rate of inflation in the form

$$\frac{d\pi^e(t)}{dt} = -\frac{\beta}{1 - a\beta}\pi^e(t) + \frac{\theta\beta}{1 - a\beta}. \tag{14.10}$$

Solution of linear differential equation (14.10) has the form

$$\pi^e(t) = \theta + (\pi^e(0) - \theta)\exp\left(\frac{-\beta t}{1 - a\beta}\right). \tag{14.11}$$

Expression (14.11) describes dynamics of the expected inflation without memory.

Taking the first derivative of equation (14.9), we have

$$\frac{d\pi(t)}{dt} = \frac{-a\beta}{1 - a\beta}\frac{d\pi^e(t)}{dt}. \tag{14.12}$$

Then using equation (14.6), we obtain

$$\frac{d\pi(t)}{dt} = \frac{\beta(\theta - \pi(t))}{1 - a\beta}. \tag{14.13}$$

Equation (14.13) can be written in the form

$$\frac{d\pi(t)}{dt} + \frac{\beta}{1 - a\beta}\pi(t) = \frac{\beta\theta}{1 - a\beta}. \tag{14.14}$$

Solution to linear differential equation (14.14) has the form

$$\pi(t) = \theta + (\pi(0) - \theta)\exp\left(\frac{-\beta t}{1 - a\beta}\right). \tag{14.15}$$

Comparing equations (14.10) and (14.14), we see that they have the same form and the same parameters. Because of this, we can conclude that the behavior of the actual and expected inflation coincide.

If the high inflationary economy is analyzed, we can assume that $\pi(0) > \theta$.

For the case $a\beta < 1$, using equation (14.11), we can see that $\pi(t) \to \theta$ for $t \to \infty$. This means that in a situation, where the coefficients characterizing the elasticity of money demand by inflation (a) and the rate of revision of inflation expectations (β) are not too high, the result of the standard Cagan model is consistent with the conclusion of the quantitative theory of money, i. e., in equilibrium $\pi = m = \theta$ [487].

If $a\beta > 1$, then $\pi(t) \to \infty$ for $t \to \infty$. In other words, if a or β is large, that is, agents greatly change the demand for money, when they revise their expectations or sharply change their expectations, then the economy may not come to an equilibrium state. In the first case, with rising inflation expectations, agents sharply reduce the demand for money, which leads to a further increase in inflation. In the second case, with rising inflation, agents sharply increase inflation expectations, which strengthen inflation processes. Inflation continues despite the stabilization of the money supply growth rate. To restore equilibrium in such an economy, it is necessary to carry out measures aimed at reducing the nervousness of economic agents. The Cagan model does not take into account the effect on the equilibrium of the dynamics of GDP. This drawback is overcome in the Bruno–Fisher model [42, 41], which can be used to detail the analysis of the equilibrium of the money market and the consequences of monetary policy. This model helps to assess the impact on inflationary equilibrium rate of inflation control measures such as reducing the budget deficit (e. g., see [487, pp. 159–169]). The Bruno–Fisher model [487, pp. 159–169] can also be generalized to the case of accounting for memory effects.

14.3 Generalization: Cagan model with memory

To generalize the standard model, it seems that a formal generalization of the model equation can be used by replacing the first-order derivative with derivative of noninteger order. The formal generalization of equation (14.14) can be suggested in the form

$$(D^{\alpha}_{C;0+}\pi)(t) + \frac{\beta}{1 - a\beta}\pi(t) = \frac{\beta\theta}{1 - a\beta}, \tag{14.16}$$

where $D^{\alpha}_{C;0+}$ is the Caputo fractional derivative. For $\alpha = 1$, equation (14.16) gives equation (14.14) of the standard Cagan model. However, if we carefully and consistently generalize the process of deriving the equation of the standard model from the basic assumption, then we will not get a fractional differential equation in form (14.16).

To obtain a generalized model that takes into account memory effects, we first consider a generalization of the standard equation of revising expectations (14.3) for the processes with memory.

At the same time, it is obvious that to describe expected rate of inflation, it is necessary to take into account memory and memory fading, since amnesia of economic agents is a strong restriction on the Cagan model. In the standard Cagan model, the indicator of nervousness of economic agents, which characterizes the speed of revising the expectations, is represented as a constant parameter [487]. In the general case, the nervousness of economic agents can be caused not only by the current state of the process, but also by the history of its changes. We assume that the speed of revising the expectations of inflation can depend on the history of changes in the difference between the real inflation rate and the rate expected by economic agents. Therefore, we assume that the revising expectations is described by the equation

$$\frac{d\pi^e(t)}{dt} = \int_0^t B(t-\tau)(\pi(\tau) - \pi^e(\tau))\, d\tau, \tag{14.17}$$

where the function $B(t-\tau)$ allows us to take into account a memory.

The nervousness of economic agents is caused not only by the current state of the process, but also by the history of its changes. The speed $d\pi^e(t)/dt$ of revising the expectations of inflation can depend on the history of changes in the difference between the real inflation rate (π) and the rate expected by economic agents (π^e). The use of the function $B(t-\tau)$ instead of the parameter β, allows us to take into account the memory effects and memory fading in the suggested generalization of the standard Cagan model.

If the function $B(t-\tau)$ is expressed through the Dirac delta-function $B(t-\tau) = \beta\delta(t-\tau)$, then equation (14.17) gives the standard equation (14.3) of Cagan model without memory.

If we assume the power-law form of the function

$$B(t-\tau) = \frac{\beta}{\Gamma(\alpha)}(t-\tau)^{\alpha-1}, \tag{14.18}$$

where α is the parameter of memory fading [448, 450]. The function $B(\tau)$ describes how the nervousness of economic agents (or the rate of revision of expectations) has changed in the past $\tau \in (0, t)$.

Remark 14.1. In equation (14.18), the parameter β characterizes the rate of revision of expectations or the nervousness of economic agents with memory. In the general case, the parameter β may depend on the parameter α of memory fading ($\beta = \beta(\alpha)$), which characterizes the economic agent. The functions $\beta(\alpha)$ can describe a distribution of the parameter of the memory fading on a set of economic agents. It is important for the economic models, since different types of economic agents can have various parameters of memory fading. In this case, we should consider the fractional operators, which depend on $\beta(\alpha)$, which is interpreted as a weight function (probability density function). Therefore, if the behavior of economic agents cannot be characterized by one parameter of memory fading, then we should use the fractional derivatives with distributed orders.

Using expression (14.18), equation (14.17) can be written in the form

$$\frac{d\pi^e(t)}{dt} = \beta(I^{\alpha}_{\mathrm{RL};0+}(\pi - \pi^e))(t), \tag{14.19}$$

where $I^{\alpha}_{\mathrm{RL};0+}$ is the Riemann–Liouville fractional integral of the order $\alpha > 0$.

The Riemann–Liouville fractional integral is defined [200, pp. 69–70] by the equation

$$(I^{\alpha}_{\mathrm{RL};0+}f)(t) = \frac{1}{\Gamma(\alpha)} \int_0^t (t-\tau)^{\alpha-1} f(\tau)\, d\tau, \tag{14.20}$$

where $\Gamma(\alpha)$ is the gamma function. In equation (14.20) the function $f(t)$ is assumed to satisfy the condition $f(\tau) \in L_1[0,T]$ [335].

Remark 14.2. We should note that equation (14.19) with Riemann–Liouville fractional integral can be considered as an approximation of the equations with generalized memory functions (14.17). In paper [380], we use the generalized Taylor series in the Trujillo–Rivero–Bonilla form for wide class of the memory functions. We proved that the equations with memory functions can be represented through the Riemann–Liouville fractional integrals (and the Caputo fractional derivatives) of noninteger orders.

The action of the Caputo derivative on equation (14.19) gives

$$\left(D^{\alpha}_{C;0+}\left(\frac{d\pi^e(\tau)}{d\tau}\right)\right)(t) = \beta(D^{\alpha}_{C;0+} I^{\alpha}_{\mathrm{RL};0+}(\pi - \pi^e))(t), \tag{14.21}$$

where $D^{\alpha}_{C;t_0+}$ is the Caputo fractional derivative of the order $\alpha > 0$. The Caputo fractional derivative is defined by the equation

$$(D^{\alpha}_{C;0+}f)(t) = \frac{1}{\Gamma(n-\alpha)} \int_0^t (t-\tau)^{n-\alpha-1} f^{(n)}(\tau)\, d\tau, \tag{14.22}$$

where $n = [\alpha] + 1$ for noninteger α, and $n = \alpha$ for integer α, $\Gamma(\alpha)$ is the gamma function, $f^{(n)}(\tau)$ is the derivative of the integer order $n = [\alpha] + 1$ with respect to τ. Operator (14.22) exists if $f(\tau) \in AC^n[0,T]$ (see Theorem 2.1 [200, p. 92]), i. e., the function $f(\tau)$ has integer-order derivatives up to $(n-1)$-th order, which are continuous functions on the interval $[0,T]$, and the derivative $f^{(n)}(\tau)$ is Lebesgue summable on the interval $[0,T]$.

Let us use the fact that the Caputo fractional derivative is the left inverse operator for the Riemann–Liouville fractional integral (see equation (2.4.32) of Lemma 2.21 of

[200, p. 95]), the has the form

$$(D^{\alpha}_{C;0+}(I^{\alpha}_{RL;0+}f))(t) = f(t), \tag{14.23}$$

if $\alpha > 0$, and $f(t) \in L_{\infty}(0,T)$ or $f(t) \in C[0,T]$.

As a result, equation (14.21) is represented in the form

$$(D^{\alpha+1}_{C;0+}\pi^e)(t) = \beta(\pi(t) - \pi^e(t)), \tag{14.24}$$

where we used the following property of the Caputo fractional derivative:

$$(D^{\alpha}_{C;0+}f^{(1)})(r) = (D^{\alpha+1}_{C;0+}f)(t), \tag{14.25}$$

which holds for $\alpha > 0$.

In the model with memory, we will assume the same equilibrium condition (14.4) in the money market as in the standard model, i. e., we use the expression

$$\exp(-a\pi^e) = \frac{M}{P}. \tag{14.26}$$

First, taking the logarithm and then differentiating, we obtain equation (14.6) that can be written in the form

$$\pi(t) = \theta + a\frac{d\pi^e(t)}{dt}. \tag{14.27}$$

Substitution of expression (14.27) into equation (14.24) gives the equation

$$(D^{\alpha+1}_{C;0+}\pi^e)(t) = \beta\left(\theta + a\frac{d\pi^e(t)}{dt} - \pi^e(t)\right). \tag{14.28}$$

As a result, we have the fractional differential equation for $\pi^e(t)$ in the form

$$(D^{\alpha+1}_{C;0+}\pi^e)(t) - a\beta\frac{d\pi^e(t)}{dt} + \beta\pi^e(t) = \beta\theta. \tag{14.29}$$

This equation describes the generalization of the Cagan model, in which we take into account the power-law memory. For $\alpha = 0$, equation (14.29) takes form (14.10) that describes the standard Cagan model.

Using equation (14.8) with $a \neq 0$ in the form

$$\pi^e(t) = \frac{\theta}{a\beta} - \frac{1 - a\beta}{a\beta}\pi(t), \tag{14.30}$$

and the property of the Caputo fractional derivative $(D^{\alpha}_{C;0+}\,\mathrm{const})(t) = 0$, we get the following equation for the actual inflation

$$(D^{\alpha+1}_{C;0+}\pi)(t) - a\beta\frac{d\pi(t)}{dt} + \beta\pi(t) = \beta\theta. \tag{14.31}$$

Comparing equations (14.31) and (14.29), we see that they have the same form and the same parameters if $a \neq 0$. Because of this, we can conclude that the dynamics of the actual and expected inflation with power-law memory coincide for the case $a \neq 0$.

14.4 Solution for equation of Cagan model with memory

Let us solve fractional differential equation (14.29). Using the fact that the Caputo derivative of a constant is equal to zero [200], and introducing new variable

$$y(t) = \pi^{e}(t) - \theta, \tag{14.32}$$

we can write equation (14.29) as a homogeneous equation in the form

$$(D^{\alpha+1}_{C;0+}y)(t) - a\beta y^{(1)}(t) + \beta y(t) = 0, \tag{14.33}$$

where $y^{(1)}(t) = dy(t)/dt$. The solution of equation (14.33) is given by Theorem 5.13 in [200, p. 314], and Corollary 5.9 in [200, p. 319], where $\alpha \to \alpha + 1$, $\beta \to 1$, $\lambda \to a\beta$, and $\mu \to -\beta$.

Let us consider the case

$$1 < n - 1 < \alpha + 1 \leq n. \tag{14.34}$$

The solution of equation (14.33) can be represented in terms of the generalized Wright function (the Fox–Wright function), $\Psi_{1,1}\big[{(a,\alpha) \atop (b,\beta)} \mid z\big]$, which is defined [200, p. 56] by the equation

$$\Psi_{1,1}\Big[{(a,\alpha) \atop (b,\beta)} \mid z\Big] = \sum_{k=0}^{\infty} \frac{\Gamma(\alpha k + a)}{\Gamma(\beta k + b)} \frac{z^k}{k!}. \tag{14.35}$$

Using Theorem 5.13 of [200, p. 314], the solution of equation (14.33) has the form

$$y(t) = \sum_{j=0}^{n-1} a_j y_j(t), \tag{14.36}$$

where $y_j(t)$, $j = 0, \dots, n-1$ are defined by the following equations:

$$\begin{aligned} y_0(t) = {} & \sum_{k=0}^{\infty} \frac{(-1)^k \beta^k t^{k(\alpha+1)}}{\Gamma(k+1)} \Psi_{1,1}\Big[{(k+1,1) \atop ((\alpha+1)k+1,\alpha)} \mid a\beta t^{\alpha}\Big] \\ & - a\beta \sum_{k=0}^{\infty} \frac{(-1)^k \beta^k t^{k(\alpha+1)+\alpha}}{\Gamma(k+1)} \Psi_{1,1}\Big[{(k+1,1) \atop ((\alpha+1)k+1+\alpha,\alpha)} \mid a\beta t^{\alpha}\Big], \end{aligned} \tag{14.37}$$

and

$$y_j(t) = \sum_{k=0}^{\infty} \frac{(-1)^k \beta^k t^{k(\alpha+1)+j}}{\Gamma(k+1)} \Psi_{1,1}\Big[{(k+1,1) \atop ((\alpha+1)k+j+1,\alpha)} \mid a\beta t^{\alpha}\Big] \tag{14.38}$$

for $j = 1, \dots, n-1$, where $n - 1 = [\alpha + 1]$ for noninteger values of α.

Note that the argument of the functions $\Psi_{1,1}$ is $z = a\beta t^{\alpha}$ instead of $z = a\beta t^{\alpha+1}$.

For $1 < \alpha + 1 \le 2$, i. e., $\alpha \in (0, 1)$, the solution is represented by the equation

$$y(t) = a_0 y_0(t) + a_1 y_1(t) \tag{14.39}$$

with $y_0(t)$ in the form

$$\begin{aligned} y_0(t) = &\sum_{k=0}^{\infty} \frac{(-1)^k \beta^k (\alpha) t^{k(\alpha+1)}}{\Gamma(k+1)} \Psi_{1,1}\left[{}^{(k+1,1)}_{((\alpha+1)k+1,\alpha)} \mid a\beta t^{\alpha}\right] \\ &- a\beta \sum_{k=0}^{\infty} \frac{(-1)^k \beta^k t^{k(\alpha+1)+\alpha}}{\Gamma(k+1)} \Psi_{1,1}\left[{}^{(k+1,1)}_{((\alpha+1)k+1+\alpha,\alpha)} \mid a\beta t^{\alpha}\right], \end{aligned} \tag{14.40}$$

and $y_1(t)$ is defined by the equation

$$y_1(t) = \sum_{k=0}^{\infty} \frac{(-1)^k \beta^k t^{k(\alpha+1)+1}}{\Gamma(k+1)} \Psi_{1,1}\left[{}^{(k+1,1)}_{((\alpha+1)k+2,\alpha)} \mid a\beta t^{\alpha}\right], \tag{14.41}$$

where $y(t)$ is defined by equation (14.32).

The Fox–Wright function that is used in equations (14.37) and (14.38) can be represented [200, p. 45] through the three-parameter Mittag–Leffler function [143], which is also called the Prabhakar function [132, 138], by the equation

$$\Psi_{1,1}\left[{}^{(\rho,1)}_{(\gamma,\alpha)} \mid z\right] = \Gamma(\rho) E^{\rho}_{\alpha,\gamma}[z], \tag{14.42}$$

where we will use $z = a\beta t^{\alpha}$ and $\rho = k+1$, and $\gamma = (\alpha+1)k+1$, $\gamma = (\alpha+1)k+1+\alpha$, $\gamma = (\alpha+1)k+1+j$, respectively.

For $1 < \alpha + 1 \le 2$ with $\alpha \in (0, 1)$, solution (14.39) is represented by the equation

$$y(t) = a_0 y_0(t) + a_1 y_1(t), \tag{14.43}$$

with $y_0(t)$ in the form

$$\begin{aligned} y_0(t) = &\sum_{k=0}^{\infty} (-1)^k \beta^k t^{k(\alpha+1)} E^{k+1}_{\alpha,(\alpha+1)k+1}[a\beta t^{\alpha}] \\ &- a\beta \sum_{k=0}^{\infty} (-1)^k \beta^k t^{k(\alpha+1)+\alpha} E^{k+1}_{\alpha,(\alpha+1)k+1+\alpha}[a\beta t^{\alpha}], \end{aligned} \tag{14.44}$$

and $y_1(t)$ is defined by the equation

$$y_1(t) = \sum_{k=0}^{\infty} (-1)^k \beta^k t^{k(\alpha+1)+1} E^{k+1}_{\alpha,(\alpha+1)k+2}[a\beta t^{\alpha}], \tag{14.45}$$

where $y(t)$ is defined by equation (14.32).

Solutions (14.43)–(14.45) describe the proposed generalization of the Cagan model, in which we take into account the memory effects and one-parameter memory fading. Expressions (14.36)–(14.38) (and equations (14.43)–(14.45) for $\alpha \in (0, 1)$) describe the dynamics of the expected inflation (and actual inflation if $a \neq 0$) that takes into account memory with power-law fading parameter $\alpha > 0$. We will describe the behavior of inflation and its properties in the next section for the special case $a = 0$.

14.5 Short-term behavior of expected inflation with memory

Using equations (14.31) and (14.29), we can see that the dynamics of the actual inflation $\pi(t)$ and the expected inflation $\pi^e(t)$ with memory coincide, if $a \neq 0$.

If $a = 0$, then using equation (14.27), we get that equation (14.31) takes the form

$$\pi(t) = \theta. \tag{14.46}$$

This means that actual inflation is constant and does not change over time.

If $a = 0$, then using equations (14.24) and (14.27), we get that equation (14.29) takes the form

$$(D^{\alpha+1}_{C;0+}\pi^e)(t) + \beta\pi^e(t) = \beta\theta. \tag{14.47}$$

Using equation (14.47), we see that the equilibrium expected rate of inflation is $\pi^e(t) = \theta$, i. e., the equilibrium state of expected inflation is actual inflation, if $a \neq 0$.

We see that the behavior of expected and actual inflation is different. The actual inflation does not change over time. The behavior of expected inflation is described by the fractional differential equation (14.47). Let us consider the inflation dynamics with power-law memory.

For $\alpha = 0$, equation (14.47) coincides with the standard relaxation equation. For $\alpha = 1$, equation (14.47) coincides with the standard oscillation equation. For $0 < \alpha < 1$, $(1 < \alpha + 1 < 2)$, the processes are described by equation (14.47) are called the fractional oscillation [244, 247, 140] and [387].

The solution of equation (14.47) with $1 < \alpha + 1 \leq 2$ has the form

$$\pi^e(t) = \theta + (\pi^e(0) - \theta)E_{\alpha+1,1}[-\beta t^{\alpha+1}] + \pi^{e(1)}(0)tE_{\alpha+1,2}[-\beta t^{\alpha+1}], \tag{14.48}$$

where $E_{\alpha,\beta}[z]$ is two-parameter Mittag–Leffler function [143].

Let us consider the short-term behavior of the expected inflation with power-law memory. In equation (14.48), the Mittag–Leffler functions can be represented [244, 247, 140] in the form

$$E_{\alpha+1,1}[-\beta t^{\alpha+1}] = f_{\alpha+1,0}(\beta^{1/(\alpha+1)}t) + g_{\alpha+1,0}(\beta^{1/(\alpha+1)}t), \tag{14.49}$$

and

$$tE_{\alpha+1,2}[-\beta t^{\alpha+1}] = f_{\alpha+1,1}(\beta^{1/(\alpha+1)}t) + g_{\alpha+1,1}(\beta^{1/(\alpha+1)}t), \tag{14.50}$$

where the terms $f_{\alpha+1,k}(t)$ describe the relaxation, and $g_{\alpha+1,k}(t)$ describe the oscillations with exponentially decreasing amplitude.

The functions $f_{\alpha+1,k}(t)$, which describes the relaxation, are defined by the equation

$$f_{\alpha+1,k}(\beta^{1/(\alpha+1)}t) = \int_0^\infty K_{\alpha+1,k}(r, -\beta)\exp(-r\beta^{1/(\alpha+1)}t)\,dr, \tag{14.51}$$

where

$$K_{\alpha,k}(r,-\beta) = \frac{(-1)^k}{\pi} \frac{\beta^{-k/\alpha} r^{\alpha-k} \sin(\pi\alpha)}{r^{2\alpha} + 2r^{\alpha}\cos(\pi\alpha) + 1}. \tag{14.52}$$

In the case $1 < \alpha + 1 < 2$, the function $f_{\alpha+1,0}(\beta^{1/(\alpha+1)}t)$ tends to zero as t tends to infinity.

The functions $g_{\alpha+1,k}(t)$, which describes the oscillation, are defined by the equation

$$g_{\alpha+1,k}(\beta^{1/(\alpha+1)}t) = \frac{2}{\alpha+1}\beta^{-k/(\alpha+1)} \exp\left(\beta^{1/(\alpha+1)} t \cos\left(\frac{\pi}{\alpha+1}\right)\right) \times \cos\left(\beta^{1/(\alpha+1)} t \sin\left(\frac{\pi}{\alpha+1}\right) - \frac{k\pi}{\alpha+1}\right). \tag{14.53}$$

Function (14.53) describes decline with oscillation, for which the circular frequency is given by the equation

$$\omega_{\alpha+1} = \beta^{1/(\alpha+1)} \sin\left(\frac{\pi}{\alpha+1}\right), \tag{14.54}$$

and the rate of exponentially decaying amplitude has the form

$$\lambda_{\alpha+1} = \beta^{1/(\alpha+1)} \cos\left(\frac{\pi}{\alpha+1}\right). \tag{14.55}$$

Using representation (14.49) and (14.50), solution (14.48) takes the form

$$\begin{aligned} \pi^e(t) = \theta &+ (\pi^e(0) - \theta) f_{\alpha+1}(\beta^{1/(\alpha+1)}t) \\ &+ (\pi^e(0) - \theta) g_{\alpha+1,0}(\beta^{1/(\alpha+1)}t) \\ &+ \pi^{e(1)}(0) f_{\alpha+1,1}(\beta^{1/(\alpha+1)}t) \\ &+ \pi^{e(1)}(0) g_{\alpha+1,1}(\beta^{1/(\alpha+1)}t). \end{aligned} \tag{14.56}$$

As a result, using expression (14.54), we get the characteristic time of the oscillation damping with memory

$$\tau_{\text{osc},\alpha+1} = \tau^{1/(\alpha+1)} \left|\csc\left(\frac{\pi}{\alpha+1}\right)\right|. \tag{14.57}$$

Using equation (14.55), the characteristic time of exponentially decaying amplitude of oscillations (damping of oscillation) is

$$\tau_{\text{dam},\alpha+1} = \tau^{1/(\alpha+1)} \left|\sec\left(\frac{\pi}{\alpha+1}\right)\right|, \tag{14.58}$$

where $1 < \alpha + 1 < 2$ and $\tau = \beta^{-1}$ is a characteristic time of oscillation without memory ($\alpha = 1$). Using that $|\sec(x)| \geq 1$ and $|\csc(x)| \geq 1$, we get the inequalities

$$\tau_{\text{osc},\alpha+1} \geq \beta^{-1/(\alpha+1)}, \quad \tau_{\text{dam},\alpha+1} \geq \beta^{-1/(\alpha+1)}, \tag{14.59}$$

where β is parameter that characterizes the nervousness of economic agents with memory.

14.6 Long-term behavior of expected inflation with memory

Let us consider the long-term behavior of the expected inflation with power-law memory. Using the asymptotic behavior of the solution, we describe the characteristic time, for relaxation to equilibrium state of expected inflation, when power-law memory is taken into account. For this purpose, we will use the asymptotic behavior of the inflation function $\pi^e(t)$ at $t \to \infty$. In the case $z \to -\infty$ with $z = -\beta t^{\alpha}$, we have the equation

$$E_{\alpha+1,k+1}[z] = -\sum_{j=1}^{m} \frac{1}{\Gamma(k+1-(\alpha+1)j)} \frac{1}{z^j} + O\left(\frac{1}{z^{m+1}}\right) \tag{14.60}$$

for $|z| \to \infty$ and all $m \in \mathbb{N}$. Equation (14.60) describes the asymptotic behavior at infinity for the case $1 < \alpha + 1 < 2$. Note that this asymptotic behavior (14.60) is caused by the behavior of the function $f_{\alpha+1,k}(\beta^{1/(\alpha+1)}t)$. The contribution of the function $g_{\alpha+1,k}(\beta^{1/(\alpha+1)}t)$ is exponentially small at $t \to \infty$ for $1 < \alpha + 1 < 2$.

Using equation (14.60) in the form

$$E_{\alpha+1,1}[-\beta t^{\alpha+1}] = -\sum_{j=1}^{N} \frac{(-\beta)^{-j} t^{-j(\alpha+1)}}{\Gamma(1-(\alpha+1)j)} + O(t^{-(\alpha+1)(N+1)}), \tag{14.61}$$

and

$$E_{\alpha+1,2}[-\beta t^{\alpha+1}] = -\sum_{j=1}^{N} \frac{(-\beta)^{-j} t^{-j(\alpha+1)}}{\Gamma(2-(\alpha+1)j)} + O(t^{-(\alpha+1)(N+1)}), \tag{14.62}$$

we obtain the behavior of expected inflation $\pi^e(t)$ at $t \to \infty$ in the form

$$\begin{aligned} \pi^e(t) &= \theta + \pi^{e(1)}(0) \frac{\beta^{-1} t^{-\alpha}}{\Gamma(1-\alpha)} + (\pi^e(0) - \theta) \frac{\beta^{-1} t^{-\alpha-1}}{\Gamma(-\alpha)} \\ &\quad - \sum_{j=2}^{N} \left(\pi^e(0) - \theta + \frac{\pi^{e(1)}(0) t}{1-(\alpha+1)j} \right) \frac{(-\beta)^{-j} t^{-(\alpha+1)j}}{\Gamma(1-(\alpha+1)j)} + O\left(\frac{1}{t^{(\alpha+1)(N+1)-1}}\right), \end{aligned} \tag{14.63}$$

where we use $\Gamma(z+1) = z\Gamma(z)$ and $0 < \alpha < 1$ $(1 < \alpha + 1 < 2)$. Note that the second term of expression (14.63) is $t^{-\alpha}$ has the form instead of $t^{-(\alpha+1)}$.

If $\pi^{e(1)}(0) \neq 0$, the slowest term is the second term of (14.63) tending to equilibrium when time tends to infinity. As a result, we get that relaxation processes with memory (for $1 < \alpha + 1 < 2$), exhibit the power-law decay with the characteristic time

$$\tau_{\text{rel},\alpha+1} = \tau^{1/\alpha} = \beta^{-1/\alpha}. \tag{14.64}$$

Note that expression (14.64), which characterizes the relaxation of expected inflation $\pi^e(t)$ to actual inflation $\pi(t) = \theta$, contains $\tau^{1/\alpha}$ instead of $\tau^{1/(\alpha+1)}$ as in expressions (14.57) and (14.58).

As a result, if we take into account the power-law memory, the relaxation of the expected inflation to actual inflation at $t \to \infty$ exhibits a power-law decay with characteristic time (14.64) for $1 < \alpha + 1 < 2$.

14.7 Properties of behavior of expected inflation with memory

Let us compare the characteristic times of relaxation and oscillation damping for the expected inflation with power-law memory for the case $a = 0$.

For behavior of the expected inflation with memory in the case $1 < \alpha + 1 < 2$ and $a = 0$, we have the following three characteristic times:

$$\tau_{\text{osc};\alpha+1} = \beta^{-1/(\alpha+1)} \left|\csc\left(\frac{\pi}{\alpha+1}\right)\right|, \tag{14.65}$$

$$\tau_{\text{dam};\alpha+1} = \beta^{-1/(\alpha+1)} \left|\sec\left(\frac{\pi}{\alpha+1}\right)\right|, \tag{14.66}$$

$$\tau_{\text{rel};\alpha+1} = \beta^{-1/\alpha}. \tag{14.67}$$

Let us define the parameter τ_{eq}, for which the characteristic time $\tau_{\text{dam}}(\alpha)$ of the oscillation damping with memory is equal to the characteristic time $\tau_{\text{rel}}(\alpha)$ of the relaxation with memory. This parameter τ_{eq} is defined by the equation

$$\tau_{\text{eq}} = \beta_{\text{eq}}^{-1}, \quad \text{where } \beta_{\text{eq}}^{-1} = \left|\cos\left(\frac{\pi}{\alpha+1}\right)\right|^{\alpha(\alpha+1)}. \tag{14.68}$$

Equation (14.68) defines a criterion parameter, in which some processes (relaxation and oscillation damping) with memory will be faster or slower relative to others. For the memory fading $\alpha_{\text{eq}} + 1 < \alpha + 1 < 2$, the characteristic time of relaxation processes is less than the characteristic time of oscillation damping, i. e., we have

$$\tau_{\text{rel}}(\alpha) < \tau_{\text{dam}}(\alpha) \quad \text{for } \alpha_{\text{eq}} + 1 < \alpha + 1 < 2. \tag{14.69}$$

For the memory fading $1 < \alpha < \alpha_{\text{eq}}$, the characteristic time of relaxation processes is greater than the characteristic time of oscillation damping

$$\tau_{\text{rel}}(\alpha) > \tau_{\text{dam}}(\alpha) \quad \text{for } 1 < \alpha + 1 < \alpha_{\text{eq}} + 1. \tag{14.70}$$

Let us give for comparison the characteristic times $\tau_{\text{osc}}(\alpha)$, $\tau_{\text{rel}}(\alpha)$ and $\tau_{\text{dam}}(\alpha)$ for $\beta_{\text{eq}} = 0.1$ in Table 14.1. Using the value $\beta_{\text{eq}} = 0.1$ for expression (14.68), we get $\alpha_{\text{eq}} + 1 = 1.7816$. In this case, $\tau_{\text{dam}}(\alpha_{\text{eq}}) = \tau_{\text{rel}}(\alpha_{\text{eq}}) = 19.03$ and $\tau_{\text{osc}}(\alpha_{\text{eq}}) = 3.710$.

Table 14.1 shows that we cannot neglect the characteristic time of the oscillation damping with memory and take into account only the characteristic time of the relaxation with memory, in the general case. It can be seen that approach to the equilibrium state slows down with a decrease in the memory fading parameter. It can be seen that damping of oscillations is the faster, the smaller the parameter of memory fading.

For expected inflation with memory fading $\alpha + 1 \in (1, 2)$, we should consider effect of slowing down competition of oscillation damping and relaxation in short time. There is an effect of competition of decelerations on short periods of time (short times).

Table 14.1: Comparison of the characteristic times $\tau_{osc}(\alpha)$, $\tau_{rel}(\alpha)$ and $\tau_{dam}(\alpha)$ for $\beta_{eq} = 0.1$.

$\alpha + 1 \in (1, 2)$	$\tau_{osc}(\alpha)$	$\tau_{dam}(\alpha)$	$\tau_{rel}(\alpha)$
1.999	3.164	4026	10.02
1.9	3.371	40.69	2.915
1.7	4.028	14.16	26.83
1.5	5.360	9.283	100
1.2	13.63	7.867	100 000
1.1	28.79	8.454	10 000 000 000

At values of $\alpha + 1$ close to 2 for short times, competition leads to the winning of by oscillation damping since $\tau_{dam}(\alpha) \gg \tau_{rel}(\alpha)$. At values of $\alpha + 1$ close to 1 for short times, competition leads to the winning of the relaxation for $0 < \beta < 1$ and the oscillation damping for $\beta > 1$. At large (long) times (at $t \to \infty$) competition always leads to the winning of the relaxation if $\alpha + 1 \in (1, 2)$. For $\alpha + 1 > 2$, the relaxation of inflation demonstrates an amplification of the oscillations that is interpreted as instability.

14.8 Conclusion

In the standard Cagan model, the memory effects and memory fading are not taken into account. The indicator of nervousness of economic agents, which characterizes the speed of revising the expectations, is represented as a constant parameter in standard model. We assume that the nervousness of economic agents can be caused not only by the current state of the process, but also by the history of its changes. The speed of revising the expectations of inflation can depend on the history of changes in the difference between the real inflation rate and the rate expected by economic agents. The use of the memory function instead of the indicator of nervousness allows us to take into account the memory effects in the Cagan model. We consider the behavior of the actual inflation that takes into account memory with power-law fading. The fractional differential equation, which describes the proposed economic model with memory, and the expression of its exact solution are suggested. Using the solutions, we describe the short-term and long-term behavior of the expected inflation with power-law memory. Three types of characteristic time have been formalized, allowing one to describe the behavior of fractional oscillation and relaxation of expected inflation for the case of constancy of actual inflation. We describe the effect of competition of decelerations for short-term and long-term behavior of the expected inflation with power-law memory. At values of $\alpha+1$ close to 2 for short times, competition leads to the winning of by oscillation damping. At values of $\alpha+1$ close to 1 for short times, competition leads to the winning of the relaxation for $0 < \beta < 1$ and the oscillation damping for $\beta > 1$. At long times competition always leads to the winning of the relaxation for $\alpha+1 \in (1, 2)$. For $\alpha+1 > 2$, the relaxation of inflation demonstrates an amplification of the oscillations that is interpreted as instability.

Part IV: **Nonlinear models of economics with memory**

15 Model of logistic growth with memory

In this chapter, we proposed the economic model of natural growth in a competitive environment with power-law memory, and its special case, the model of logistic growth with memory. This chapter is based on articles [386, 444].

15.1 Introduction

The logistic differential equation was initially proposed by Pierre F. Verhulst [492] in the form of the population growth model. In this model, the rate of reproduction is directly proportional to the product of the existing population and the amount of available resources. The logistic differential equation can be derived from economic model of natural growth in a competitive environment. The economic natural growth models are described by equations in which the rate of change of output is directly proportional to income. In the description of economic growth the competition effects are taken into account by considering the price as a function of the value of output. Model of natural growth in a competitive environment is often called a model of logistic growth [493, 494]. In this chapter, we proposed the economic model of natural growth in a competitive environment with power-law memory. This chapter is based on articles [386, 444].

15.2 Logistic growth model without memory

In this section, we describe the standard model of logistic growth, which does not take into account the memory effects.

Let $Y(t)$ be a function that describes the value of output at time t. Let $I(t)$ be a function that describes the investments made in the expansion of production. The value of $I(t)$ is the difference between the total investment and depreciation costs.

Let us describe the main assumptions that are used in the standard model of logistic growth without memory.

(1) It is assumed that all manufactured products are sold (the assumption of market unsaturation).

(2) It is assumed that the rate of change of output ($dY(t)/dt$) is directly proportional to the value of the net investment $I(t)$. This assumption is represented by the accelerator equation

$$I(t) = v \frac{dY(t)}{dt}, \tag{15.1}$$

where v is a positive constant that is the investment coefficient indicating the power of the accelerator (v is also called the accelerator coefficient), $1/v$ is the

https://doi.org/10.1515/9783110627459-015

marginal productivity of capital (rate of acceleration), and $dY(t)/dt$ is the first order derivative of the function $Y(t)$ with respect to time t. Equation (15.1) describes a linear accelerator that does not take into account the effects of lag and memory (e. g., see Section 3 in [10, pp. 62–63]).

(3) In the logistic growth model the price $P(t)$ is considered as a function of released product $Y(t)$, i. e., $P = P(Y(t))$. The function $P = P(Y)$ is usually considered as a decreasing function, that is, the increase of output leads to a decrease of price due to market saturation.

(4) Assuming that the amount of net investment is a fixed part of the income $PY(t)$, we get

$$I(t) = mPY(t), \tag{15.2}$$

where m is the norm of net investment ($0 < m < 1$), specifying the share of income, which is spent on the net investment. Equation (15.2) describes a linear multiplier that does not take into account the effects of lag and memory (e. g., see Chapter 2 of [10]).

(5) It is often assumed that the price as a function of output $Y(t)$ is linear, i. e., we have the expression

$$P(Y(t)) = b - aY(t), \tag{15.3}$$

where b is the price, which is independent of the output and a is the marginal price.

Substituting (15.2) into equation (15.1), we obtain

$$\frac{dY(t)}{dt} = \frac{m}{v} P(Y(t))Y(t). \tag{15.4}$$

Differential equation (15.4) describes the economic model of natural growth in a competitive environment.

In the linear case (15.3), equation (15.4) has the form

$$\frac{dY(t)}{dt} = \frac{m}{v}(b - aY(t))Y(t). \tag{15.5}$$

Equation (15.5) is the logistic differential equation, i. e., the ordinary differential equation of first order that describes the logistic growth. For $a = 0$, equation (15.5) describes the natural growth in the absence of competition.

The logistic growth model, which is described by equation (15.5), and the model of the natural growth in a competitive environment, which is described by equation (15.4), are based on the accelerator equation (15.1). Equations (15.1), (15.4) and (15.5) contain only the first-order derivative with respect to time. It is known that the derivative of the first order is determined by the properties of differentiable functions of time only in infinitely small neighborhood of the point of time. As a result, the models, which are described by equations (15.4) and (15.5), assume an instantaneous change of $Y^{(1)}(t) = dY(t)/dt$, when the net investment changes. This means not only neglecting the delay (lag) effects, but also the neglect of the memory effects, i. e., the neglect of dependence of output at the present time on the investment changes in the past. In other words, the model of logistic growth (15.5) does not take into account the effects of memory and delay.

15.3 Logistic growth model with memory

Let us generalize the standard accelerator equation (15.1) by taking into account the memory effects. The economic process at time $t > 0$ can depend on changes in the state of this process in the past, that is for $\tau \in (0, t)$. To describe the fading memory, we can use the generalized accelerator equation in the form

$$I(t) = \int_0^t v(t-\tau)Y^{(1)}(\tau)\,d\tau, \tag{15.6}$$

where $v(t-\tau)$ takes into account the impact of the history of changes in the dynamics of output $Y(\tau)$ on the net investment $I(t)$. The function $v(t-\tau)$ is interpreted as a memory function. In order to easily interpret the dimensions of the economic quantities, we can use the time t as a dimensionless variable by changing the variable $t \to tt_d$, where t_d is the unit of time (hour, day, month, year).

In paper [380], using the generalized Taylor series in the Trujillo–Rivero–Bonilla form for the memory function, we proved that for the wide class of memory functions can be can be approximately considered as a power function. Let us assume the power-law form of memory function

$$v(t-\tau) = \frac{v(\alpha)}{\Gamma(1-\alpha)}(t-\tau)^{-\alpha}, \tag{15.7}$$

where $\Gamma(\alpha)$ is the gamma function, and $v(\alpha)$ is a positive constant that is the investment coefficient indicating the power of the accelerator with memory.

As a result, the generalization (15.6) of the standard accelerator equation (15.1), which takes into account the memory effects of the order α, can be given in the form

$$I(t) = v(\alpha)(D^{\alpha}_{C;0+}Y)(t). \tag{15.8}$$

For $\alpha = 1$, equation (15.8) takes the form (15.1). We should emphasize that accelerator equation (15.8) includes the standard equation of the accelerator and the multiplier, as special cases for $\alpha = 0$ and $\alpha = 1$.

To consider a more general case, we can consider not only case $\alpha \in [0,1]$, but also take into account accelerators of a higher orders, that is, consider all kinds of positive values of the alpha parameter ($\alpha > 1$).

In equation (15.8), we use the left-sided Caputo derivative of order α with respect to time that is defined by the equation

$$(D^{\alpha}_{C;0+}Y)(t) = \frac{1}{\Gamma(n-\alpha)} \int_0^t \frac{Y^{(n)}(\tau)\,d\tau}{(t-\tau)^{\alpha-n+1}}, \tag{15.9}$$

where $n = [\alpha]+1$ for noninteger values of α, $Y^{(n)}(\tau)$ is the derivative of the integer order $n = [\alpha]+1$ of the function $Y(\tau)$ with respect to the variable τ: $0 < \tau < t$. For the existence of expression (15.9), the function $Y(\tau)$ must have the integer-order derivatives up to the $(n-1)$th-order, which are absolutely continuous functions on the interval $[0,t]$.

For integer orders $\alpha = n$, the Caputo derivatives coincide with the standard derivatives [308, p. 79], [200, pp. 92–93], i. e., $(D^{n}_{C;0+}Y)(t) = Y^{(n)}(t)$ and $(D^{0}_{C;0+}Y)(t) = Y(t)$.

In the natural growth model with a competitive environment, we take into account the power-law memory by using equation (15.8), which describes the accelerator with power-law memory. Substituting expression (15.2), where $P(Y(t))$, into equation (15.8), we obtain

$$(D^{\alpha}_{C;0+}Y)(t) = \frac{m}{v(\alpha)} P(Y(t))Y(t), \tag{15.10}$$

where $(D^{\alpha}_{C;0+}Y)(t)$ is the Caputo derivative (15.9) of the order $\alpha \geq 0$ of the function $Y(t)$ with respect to time. Equation (15.10) is the fractional differential equation with the Caputo derivative of the order $\alpha > 0$. The proposed model of natural growth in a competitive environment with memory is based on equation (15.10), which takes into account the memory with power-law fading. For $\alpha = 1$, equation (15.10) takes the form of equation (15.4), which describes a model of natural growth in a competitive environment without memory.

In the case of linearity of the price,

$$P(Y(t)) = b - aY(t), \tag{15.11}$$

equation (15.10) has the form

$$(D^{\alpha}_{C;0+}Y)(t) = \frac{m}{v(\alpha)} (b - aY(t))Y(t). \tag{15.12}$$

Equation (15.12) is the nonlinear fractional differential equation that describes the economic model of the logistic growth with memory. For $\alpha = 1$, equation (15.12) takes the form of equation (15.5), which describes the logistic growth without memory effects.

If $a = 0$, then equation (15.12) takes the form of the equation of natural growth with memory

$$(D^{\alpha}_{C;0+}Y)(t) = \frac{mb}{v(\alpha)} Y(t), \tag{15.13}$$

where b is the price, which does not depend on the value of output. Using Theorem 4.3 of [200, p. 231], we obtain the solution of equation (15.13) in the form

$$Y(t) = \sum_{k=0}^{n-1} Y^{(k)}(0) t^k E_{\alpha,k+1}\left[\frac{mb}{v(\alpha)} t^{\alpha}\right], \tag{15.14}$$

where $n-1 < \alpha \leq n$, $Y^{(k)}(0)$ is the derivative of integer order k of the function $Y(t)$ at $t = 0$, and $E_{\alpha,\beta}[z]$ is the two-parameter Mittag–Leffler function that is defined by the equation

$$E_{\alpha,\beta}[z] = \sum_{k=0}^{\infty} \frac{z^k}{\Gamma(\alpha k + \beta)}. \tag{15.15}$$

The Mittag–Leffler function $E_{\alpha,\beta}(z)$ is a generalization of the exponential function e^z, such that $E_{1,1}(z) = e^z$.

If $a \neq 0$ and $b \neq 0$, we can use the variable $z(t)$ and the parameter μ, which are defined by the equations

$$Z(t) = \frac{a}{b} Y(t), \quad \mu(\alpha) = \frac{m}{v(\alpha)}. \tag{15.16}$$

Then equation of logistic growth (15.12) is represented in the form

$$(D^{\alpha}_{C;0+} Z)(t) = \mu(\alpha)(1 - Z(t))Z(t). \tag{15.17}$$

Equation (15.17) is the fractional logistic differential equation that is a generalization of the standard logistic differential equation of the first order.

In paper [503], Bruce J. West proposed an analytical expression of the solution for the fractional logistic equation with $\alpha \in (0, 1)$ in the form

$$Z(t) = \sum_{k=0}^{\infty} \left(\frac{Z(0) - 1}{Z(0)}\right) E_{\alpha}[-k\mu(\alpha)t^{\alpha}], \tag{15.18}$$

where $E_{\alpha}[z]$ is the Mittag–Leffler function. As it has been proved by I. Area, J. Losada, J. Nieto in [17], the function (15.18) cannot be the solution to (15.15). This fact is caused by the violation the semigroup property by the Mittag–Leffler function, i. e., we have (e. g., see [365, 304], and [111, 334]) the inequality

$$E_{\alpha}[\lambda(t+s)^{\alpha}] \neq E_{\alpha}[\lambda t^{\alpha}] E_{\alpha}[\lambda s^{\alpha}] \tag{15.19}$$

for $\alpha \in (0, 1)$, and real constant λ. In [17] it has been proved that equation (15.18), which is proposed in [503], is not an exact solution of the fractional logistic equation (15.17). At present time, an exact analytical solution to the fractional logistic differential equation has not been obtained.

We should note that the violation of the standard semigroup property is an important characteristic of processes with memory that should be taken into account in economic models. Neglect of this nonstandard property of the dynamics with memory can lead to errors.

15.4 Logistic growth with memory

The logistic differential equation can be derived from economic model of natural growth in a competitive environment. Differential equation that describes logistic growth in competitive environment without memory has the form

$$\frac{dY(t)}{dt} = \frac{m}{v}(b - aY(t))Y(t), \tag{15.20}$$

where $v = v(1)$.

Equation (15.20) is the logistic differential equation, i. e., the ordinary differential equation of first order that describes the logistic growth. For $a = 0$, equation (15.20) describes the natural growth in the absence of competition. If $a \neq 0$ and $b \neq 0$, we can use the variable $Z(t)$ and the parameter μ that are defined by the expressions

$$Z(t) = \frac{a}{b}Y(t), \quad \mu = \frac{m}{v}. \tag{15.21}$$

For $\alpha = 1$, equation (15.20) gives the standard equation of the logistic growth model in the form

$$\frac{dZ(t)}{dt} = \mu Z(t)(1 - Z(t)), \tag{15.22}$$

where $\mu = \mu(1)$. This is the standard logistics differential equation. The solution of equation (15.22) has the form

$$Z(t) = \frac{Z(0)}{Z(0) + (1 - Z(0))\exp(-\mu t)} = \frac{Z(0)\exp(\mu t)}{1 + Z(0)(\exp(\mu t) - 1)}. \tag{15.23}$$

Using the variable,

$$u(t) = \frac{1}{Z(t)}, \tag{15.24}$$

equation (15.22) can be represented as

$$\frac{du(t)}{dt} = \mu(1 - u(t)). \tag{15.25}$$

Equations (15.22) and (15.25) are integer-order differential equations that are equivalent (s-equivalent, solution equivalent) for wide class of functions $Z(t)$.

Let us consider the generalization of equations (15.22) and (15.25), which are represented in the form

$$(D^{\alpha}_{C;0+}Z)(t) = \mu(\alpha)Z(t)(1 - Z(t)), \tag{15.26}$$

$$(D^{\alpha}_{C;0+}u)(x) = \mu(\alpha)(1 - u(t)). \tag{15.27}$$

Equations (15.26) and (15.27) cannot be considered as equivalent equations [386] for non-integer values of $\alpha > 0$.

The analytical expression for solution of nonlinear fractional differential equation (15.26) for the function $Z(t)$ is still unknown at the moment.

Let us obtain the solution of equation (15.27). Using Theorem 5.15 of [200, p. 323], we can state that the fractional differential equation

$$(D^{\alpha}_{C;0+}u)(t) - \lambda u(t) = f(t), \tag{15.28}$$

with $\lambda \in \mathbb{R}$, and initial conditions $u^{(k)}(0) = c_k$, $(k = 0, \dots, n-1)$, where $n - 1 < \alpha \leq n$ has the general solution in the form

$$u(t) = \sum_{k=0}^{n-1} c_k t^k E_{\alpha,k+1}[\lambda t^{\alpha}] + u_C(t), \tag{15.29}$$

where

$$u_C(t) = \int_0^t (t-\tau)^{\alpha-1} E_{\alpha,\alpha}[\lambda (t-\tau)^{\alpha}] f(\tau)\, d\tau, \tag{15.30}$$

where $n - 1 < \alpha \leq n$, $E_{\alpha,\beta}[z]$ is the two-parameter Mittag–Leffler function [200, p. 42], [143, p. 56] which is defined by the equation

$$E_{\alpha,\beta}(z) = \sum_{k=0}^{\infty} \frac{z^k}{\Gamma(\alpha k + \beta)}. \tag{15.31}$$

The Mittag–Leffler function $E_{\alpha,\beta}(z)$ is a generalization of the exponential function e^z, since $E_{1,1}(z) = e^z$.

For equation (15.27), we have the solution that is described by expressions (15.29) and (15.30) with the parameters

$$\lambda = -\mu(\alpha), \quad f(t) = \mu(\alpha). \tag{15.32}$$

Let us consider the constant function $f(t) = \mu(\alpha) =$ const for solution (15.29) and (15.30).

Statement 15.1. *For the case* (15.32), *expression* (15.30) *of* $u_C(t)$ *can be written in the form*

$$u_C(t) = 1 - E_{\alpha,1}[-\mu(\alpha) t^{\alpha}]. \tag{15.33}$$

Proof. In this case (15.32), expression (15.30) of $u_C(t)$ takes the form

$$u_C(t) = \mu(\alpha) \int_0^t (t-\tau)^{\alpha-1} E_{\alpha,\alpha}[-\mu(\alpha)(t-\tau)^{\alpha}]\, d\tau. \tag{15.34}$$

Using the change of variable $\xi = t - \tau$, equation (15.34) is represented in the form

$$u_C(t) = \mu(\alpha) \int_0^t \xi^{\alpha-1} E_{\alpha,\alpha}[-\mu(\alpha)\xi^{\alpha}]\, d\xi. \tag{15.35}$$

Let us use equation 4.4.4 of [143, p. 61], which has the form

$$\int_0^t \tau^{\beta-1} E_{\alpha,\beta}[\lambda\tau^{\alpha}]\, d\tau = t^{\beta} E_{\alpha,\beta+1}[\lambda t^{\alpha}], \tag{15.36}$$

where $\beta > 0$. For $\beta = \alpha$, equation (15.36) takes the form

$$\int_0^t \tau^{\alpha-1} E_{\alpha,\alpha}[-\mu(\alpha)\tau^{\alpha}]\, d\tau = t^{\alpha} E_{\alpha,\alpha+1}[-\mu(\alpha)t^{\alpha}]. \tag{15.37}$$

Then, we can use equation 4.2.3 of [143, p. 57], which has the form

$$E_{\alpha,\beta}[z] = \frac{1}{\Gamma(\beta)} + zE_{\alpha,\alpha+\beta}[z]. \tag{15.38}$$

For $\beta = 1$, equation (15.38) gives $zE_{\alpha,\alpha+1}[z] = E_{\alpha,1}[z] - 1$, and, consequently, we obtain the equality

$$t^{\alpha} E_{\alpha,\alpha+1}[\lambda t^{\alpha}] = \frac{1}{\lambda}(E_{\alpha,1}[\lambda t^{\alpha}] - 1). \tag{15.39}$$

Equation (15.39) allows us to write integral (15.37) in the form

$$\int_0^t \tau^{\alpha-1} E_{\alpha,\alpha}[\lambda\tau^{\alpha}]\, d\tau = \frac{1}{\lambda}(E_{\alpha,1}[\lambda t^{\alpha}] - 1), \tag{15.40}$$

which leads to expression (15.41) for $\lambda = -\mu(\alpha)$.

Therefore, we have

$$\int_0^t \xi^{\alpha-1} E_{\alpha,\alpha}[-\mu(\alpha)\xi^{\alpha}]\, d\xi = -\mu^{-1}(\alpha)(E_{\alpha,1}[-\mu(\alpha)t^{\alpha}] - 1). \tag{15.41}$$

Using equation (15.41), expression (15.35) can be written as

$$u_C(t) = 1 - E_{\alpha,1}[-\mu(\alpha)t^{\alpha}]. \tag{15.42}$$

□

As a result, solution of equation (15.29) has the form

$$u(t) = -(1 - E_{\alpha,1}[-\mu(\alpha)t^{\alpha}]) + \sum_{k=0}^{n-1} u^{(k)}(0)t^{k} E_{\alpha,k+1}[-\mu(\alpha)t^{\alpha}]. \tag{15.43}$$

Let us give some special cases of solution (15.43).

For $0 < \alpha \leq 1$, ($n = 1$), solution (15.43) takes the form

$$u(t) = (1 - E_{\alpha,1}[-\mu(\alpha)t^{\alpha}]) + u(0)E_{\alpha,1}[-\mu(\alpha)t^{\alpha}]. \tag{15.44}$$

For $1 < \alpha \leq 2$, ($n = 2$) solution (15.43) has the form

$$\begin{aligned} u(t) = &(1 - E_{\alpha,1}[-\mu(\alpha)t^{\alpha}]) + u(0)E_{\alpha,1}[-\mu(\alpha)t^{\alpha}] \\ &+ u^{(1)}(0)tE_{\alpha,2}[-\mu(\alpha)t^{\alpha}]. \end{aligned} \tag{15.45}$$

For $\alpha = 1$, the equality $E_{\alpha,1}[z] = e^{z}$ leads solution (15.44) to the form

$$u(t) = (1 - \exp(-\mu t)) + u(0)\exp(-\mu t), \tag{15.46}$$

where $\mu = \mu(1)$.

For $\alpha \in (0,1)$, the solution of linear equation (15.27) for the function $u(t)$ is given in the form

$$u(t) = 1 + (u(0) - 1)E_{\alpha}[-\mu(\alpha)t^{\alpha}], \tag{15.47}$$

where $\mu(\alpha) > 0$. Therefore, we obtain

$$Z(t) = \frac{1}{u(t)} = \frac{Z(0)}{Z(0) + (1 - Z(0))E_{\alpha}[-\mu(\alpha)t^{\alpha}]}. \tag{15.48}$$

For $\alpha = 1$, using that $E_1[-\mu(1)t] = \exp(-\mu t)$, equation (15.48) is the standard solution (15.23).

As a result, we can see that equations (15.26) and (15.27) have different nonequivalent solutions. The generalizations of standard models without memory violate the equivalence of fractional differential equations of appropriate economic models. The generalizations of equivalent models without memory can give nonequivalent dynamic models with memory. Accounting for memory effects violates the solution equivalence of economic models.

15.5 Conclusion

In formulation of economic models with memory, we should take into account that generalizations of equivalent representations of standard economic models, which are described by equivalent differential equations, as a rule, lead to different models with memory that have equations with nonequivalent solutions. This, in a sense, is analogous to the situation in quantum theory when quantization of equivalent classical models leads to nonequivalent quantum theories.

This property of generalizations is caused by the violation of the standard rule (the chain rule, product rule, and other rules and properties) for fractional derivatives of noninteger order.

This nonequivalence of equations in economics generates difficulties in the description of the processes with memory. Note that an additional difficulty of generalizations arises due to the presence of a large number of different types of fractional derivatives and integrals [386].

Due to this fact, the correct and self-consistent derivation of equations of economic models with memory and the economic justification of existence of memory for one or another variable, are of fundamental importance for economics with memory.

16 Kaldor-type model of business cycles with memory

In this chapter, we consider a generalization of the Kaldor-type model of business cycles and its special cases. The standard models of business cycles do not take into account the effects of memory. We derive the equations of the generalized Cagan model that, takes into account the power-law memory. In this derivation, we take into account the nonstandard properties of the fractional derivative, which actually restrict the form of the equations of this model with memory. This chapter is based on work [386].

16.1 Introduction

In this chapter, we consider a generalization of the Kaldor-type model of business cycles and its special cases based on the Van der Pol equation [62, 68, 67]. We demonstrate that the violation of the standard chain rule gives a restriction in the generalization of dynamic models. Economic models, which are based on the Van der Pol equation, are considered as prototypes of model for complex economic dynamics [62, 68, 67]. Nonlinear dynamic models are used to explain irregular and chaotic behavior of complex economic and financial processes [67, 127, 128, 232, 233]. This Chapter is based on work [386] (see also book [254, pp. 43–92]).

16.2 Kaldor-type model of business cycles without memory

In the framework of Keynesian approach to theory of national income, Nicholas Kaldor formulated [187, 188, 62] the first nonlinear model of endogenous business cycles. Kaldor consider the interactions between the investment $I(Y)$ and the savings $S(Y)$, where $Y = Y(t)$ denotes national income. Using the fact that the linear functions $I(Y)$ and $S(Y)$ cannot describe processes of business cycle, Kaldor proposed nonlinear form for $I(Y)$ and $S(Y)$, which leads to oscillatory processes of business cycles [68, 67].

Let us derive the equation of the Kaldor model of business cycles without memory by using approach proposed by Chang and Smyth [62] (see also [68, 67, 127]). In the Kaldor model, instead of the standard accelerator equation $I(t) = vY^{(1)}(t)$, the dependence of investments on the rate of change of national income is considered in the form

$$I(Y,K) - S(Y,K) = vY^{(1)}(t), \tag{16.1}$$

which takes into account the savings, where $K = K(t)$ denotes the capital stock, $Y = Y(t)$ is the national income, v is the accelerator coefficient, and $Y^{(1)}(t)$ denotes time

https://doi.org/10.1515/9783110627459-016

derivative of the first order. The parameter $a = 1/v$ is an adjustment coefficient. In this model, it is assumed that

$$I_K(Y,K) = \frac{\partial I(Y,K)}{\partial K} < 0, \tag{16.2}$$

and

$$S_K(Y,K) = \frac{\partial S(Y,K)}{\partial K} > 0. \tag{16.3}$$

Differentiation of equation (16.1) with respect to time and using the standard chain rule in the form

$$\begin{aligned} \frac{d}{dt} I(Y(t),K(t)) &= \frac{\partial I(Y,K)}{\partial Y} \frac{dY(t)}{dt} + \frac{\partial I(Y,K)}{\partial K} \frac{dK(t)}{dt} \\ &= I_Y(Y,K) Y^{(1)}(t) + I_K(Y,K) K^{(1)}(t), \end{aligned} \tag{16.4}$$

we obtain the equation

$$\begin{aligned} vY^{(2)}(t) = (I_Y(Y,K) - S_Y(Y,K)) Y^{(1)}(t) \\ + (I_K(Y,K) - S_K(Y,K)) K^{(1)}(t). \end{aligned} \tag{16.5}$$

In works [68, 67], it is assumed that the actual change in the capital stock is determined by savings decisions, i. e., we have the equation

$$S(Y,K) = K^{(1)}(t). \tag{16.6}$$

Substituting equation (16.6) into equation (16.5) gives

$$\begin{aligned} vY^{(2)}(t) =& (I_Y(Y,K) - S_Y(Y,K)) Y^{(1)}(t) \\ &+ (I_K(Y,K) - S_K(Y,K)) S(Y,K). \end{aligned} \tag{16.7}$$

It is also assumed [68, 67] that the function $I(Y,K)$ is linear in $K(t)$ and savings is independent of the capital stock, i. e., the function $S(Y,K) = S(Y)$. Therefore, $S_K(Y,K) = 0$, and the expression $(I_K(Y,K) - S_K(Y,K))$ is independent of the capital stock $K(t)$ and equation (16.7) takes the form

$$vY^{(2)}(t) = (I_Y - S_Y)(Y) Y^{(1)}(t) + I_K(Y) S(Y). \tag{16.8}$$

Using the variables,

$$y(t) = Y(t) - \overline{Y}, \tag{16.9}$$

where $\overline{Y}$ is the equilibrium value, equation (16.8) can be rewritten [68, 67] in the form of the Lienard equation

$$y^{(2)}(t) + g(y(t)) y^{(1)}(t) + f(y(t)) = 0 \tag{16.10}$$

that is used in physics to describe the dynamics of a spring-mass system.

Using the assumption [68, 67] that the shapes of the investment and savings functions are symmetric, the following form of the function of their difference:

$$g(y) = \mu(y^2 - 1), \tag{16.11}$$

and the linear form

$$f(y) = y, \tag{16.12}$$

we get the Van der Pol equation

$$y^{(2)}(t) + \mu(y^2(t) - 1)y^{(1)}(t) + y(t) = 0. \tag{16.13}$$

Equation (16.13) is used in economic modeling of the business cycles in the framework of standard nonlinear economic models with continuous time. The Van der Pol equation (16.13) can be written in the two-dimensional form

$$\begin{cases} y^{(1)} = x, \\ x^{(1)} = \mu(1 - y^2)x - y(t). \end{cases} \tag{16.14}$$

This form of the Van der Pol equation is used in computer simulation on the phase space.

16.3 Kaldor-type model of business cycles with memory

To generalize equation (16.13) for the case of processes with memory, we cannot simply replace the derivatives of integer order by fractional derivatives to get the fractional Van der Pol equation

$$(D^{\alpha}_{C;0+}X)(t) + \mu(y^2(t) - 1)(D^{\beta}_{C;0+}X)(t) + y(t) = 0, \tag{16.15}$$

where $\alpha > \beta > 0$.

To correctly generalize the standard model, it is necessary to take into account the process of obtaining equation (16.13) and (16.14) from equation (16.1). Note that the replacement of the derivatives of the integer order in equations (16.1) and (16.5) by fractional derivatives also does not allow us to obtain the fractional differential equation (16.15). This is because the derivation of equation (16.13) from equations (16.1) and (16.5), we must use the standard chain rules in the form

$$D^1_t F(Y(t), K(t)) = F_Y(Y, K)Y^{(1)}(t) + F_K(Y, K)K^{(1)}(t), \tag{16.16}$$

where $D^1_t = d/dt$.

The chain rule for fractional derivative has more complicated form (see equation (2.209) in Section 2.7.3 of [308] and [377]). In order to get around this problem, we restrict ourselves to the assumption of the presence of a memory only for equation (16.1).

We will assume that the excess of investment over saving, i. e., the difference $I(Y,K) - S(Y,K)$ is determined by changes in the growth rate of the national income in the past

$$I(Y(t),K(t)) - S(Y(t),K(t)) = \int_0^t v(t-\tau)Y^{(1)}(\tau)\,d\tau, \tag{16.17}$$

where the time variable is considered as dimensionless variable. For the case $v(t-\tau) = v\delta(t-\tau)$, equation (16.17) gives equation (16.1) of the standard model.

Let us consider power-law memory fading in the form

$$v(t-\tau) = \frac{v(\alpha)}{\Gamma(1-\alpha)}(t-\tau)^{-\alpha}, \tag{16.18}$$

where $0 < \alpha \leq 1$, and $\Gamma(\alpha)$ is the gamma function, $(D^{\alpha}_{C;0+}Y)(t)$ is the Caputo fractional derivative

$$\begin{aligned}(D^{\alpha}_{C;0+}Y)(t) &= (I^{n-\alpha}_{RL;0+}Y^{(n)})(t)\\ &= \frac{1}{\Gamma(n-\alpha)}\int_0^t (t-\tau)^{n-\alpha-1}Y^{(n)}(\tau)\,d\tau,\end{aligned} \tag{16.19}$$

where $n = [\alpha]+1$ for $\alpha \notin \mathbb{N}$ and $n = \alpha$ for $\alpha \in \mathbb{N}$, and the function $Y(\tau)$ has integer-order derivatives $Y^{(j)}(\tau)$, $j = 1,\ldots,(n-1)$, that are absolutely continuous.

Equation (16.17) with kernel (16.18) can be rewritten through the Caputo fractional derivative

$$I(Y(t),K(t)) - S(Y(t),K(t)) = v(\alpha)(D^{\alpha}_{C;0+}Y)(t). \tag{16.20}$$

Action of the first-order derivative $D^1_t = d/dt$ with respect to time on equation (16.20) and using the standard chain rule, we obtain

$$\begin{aligned}v(\alpha)D^1_t(D^{\alpha}_{C;0+}Y)(t) =&(I_Y(Y,K) - S_Y(Y,K))Y^{(1)}(t)\\ &+ (I_K(Y,K) - S_K(Y,K))K^{(1)}(t).\end{aligned} \tag{16.21}$$

Substituting equation (16.8) into equation (16.21) gives

$$\begin{aligned}v(\alpha)D^1_t(D^{\alpha}_{C;0+}Y)(t) = &(I_Y(Y,K) - S_Y(Y,K))Y^{(1)}(t)\\ &+ (I_K(Y,K) - S_K(Y,K))S(Y,K).\end{aligned} \tag{16.22}$$

Using the assumption that are proposed in paper [68, 67], equation (16.22) takes the form

$$v(\alpha)D^1_t(D^{\alpha}_{C;0+}Y)(t) = (I_Y - S_Y)(Y)Y^{(1)}(t) + I_K(Y)S(Y). \tag{16.23}$$

Using the variable $y(t)$, which is defined by expression (16.9), equation (16.23) takes the form

$$D_t^1(D_{C;0+}^{\alpha}y)(t) + g(y)y^{(1)}(t) + f(y) = 0. \tag{16.24}$$

Note that the inequality

$$D_t^1(D_{C;0+}^{\alpha}y)(t) \neq (D_{C;0+}^{\alpha+1}y)(t) \tag{16.25}$$

is satisfied for noninteger values of α.

To obtain two-dimensional form of fractional differential equation (16.24), we can use the Riemann–Liouville fractional derivative that is defined by the equation

$$\begin{aligned}(D_{RL;0+}^{\alpha}Y)(t) &= D_t^n(I_{RL;0+}^{n-\alpha}Y)(t)\\ &= \frac{1}{\Gamma(n-\alpha)}\frac{d^n}{dt^n}\int\limits_0^t (t-\tau)^{n-\alpha-1}Y(\tau)\,d\tau.\end{aligned} \tag{16.26}$$

Using expression (16.26), we can get the equalities

$$\begin{aligned}D_t^1(D_{C;0+}^{\alpha}Y)(t) &= D_t^1(I_{RL;0+}^{1-\alpha}Y^{(1)})(t)\\ &= ((D_t^1 I_{RL;0+}^{1-\alpha})Y^{(1)})(t) = (D_{RL;0+}^{\alpha}Y^{(1)})(t).\end{aligned} \tag{16.27}$$

This allows us to rewrite (16.24) as the fractional differential equation with the Riemann–Liouville fractional derivative. As a result, the Kaldor-type model of business cycles with power-law memory can be described by the fractional Van der Pol equation in the form

$$(D_{RL;0+}^{\alpha}y^{(1)})(t) + g(y)y^{(1)}(t) + f(y) = 0. \tag{16.28}$$

Equation (16.28) can be written in the two-dimensional form

$$\begin{cases} D_t^1 y = x,\\ D_{RL;0+}^{\alpha}x = \mu(1-y^2)x - y.\end{cases} \tag{16.29}$$

This form of the fractional Van der Pol equation can be used for computer simulation of the Kaldor-type model of business cycles with power-law memory.

We should note that initial conditions for the equation with the Riemann–Liouville fractional derivative are formulated for the values $(D_{RL;0+}^{\alpha-k}x)(0)$, where $k = 1,\dots,n-1$ and $n = [\alpha]+1$ (see expression 3.1.3 and 3.1.3 in the book [200, pp. 136–137]).

The second equation of system (16.29) can be represented through the Caputo fractional derivative by using equations (2.4.6) and (2.4.8) of book [200, p. 91], that has the form

$$(D_{RL;0+}^{\alpha}x)(t) = (D_{C;0+}^{\alpha}x)(t) + \sum_{k=0}^{n-1}\frac{x^{(k)}(0)}{\Gamma(k+1-\alpha)}t^{k-\alpha}, \tag{16.30}$$

where $n = [\alpha] + 1$. For $0 < \alpha < 1$, equation takes the form

$$(D^{\alpha}_{\mathrm{RL};0+}x)(t) = (D^{\alpha}_{C;0+}x)(t) + \frac{x(0)}{\Gamma(1-\alpha)}t^{-\alpha}. \tag{16.31}$$

Using equation (16.31), we can represent the two-dimensional form (16.29) as

$$\begin{cases} D^1_t y = x, \\ D^{\alpha}_{C;0+}x = \mu(1-y^2)x - y - \sum_{k=0}^{n-1} \frac{x^{(k)}(0)}{\Gamma(k+1-\alpha)} t^{k-\alpha}. \end{cases} \tag{16.32}$$

In this case, we can use initial conditions of the differential equation with the Caputo fractional derivative for the values $x^{(k)}(0)$, where $k = 0, \dots, n-1$ and $n-1 < \alpha \leq n$ (see expression 3.1.33 in book [200, p. 140]).

For $0 < \alpha < 1$, system of equations (16.32) has the form

$$\begin{cases} D^1_t y = x, \\ D^{\alpha}_{C;0+}x = \mu(1-y^2)x - y - \frac{x(0)}{\Gamma(1-\alpha)} t^{-\alpha}. \end{cases} \tag{16.33}$$

This form of the fractional Van der Pol equation can be also used in computer simulation of the Kaldor-type model of business cycles with power-law memory.

16.4 Conclusion

The Kaldor-type model of business cycles, which are based on the Kaldor nonlinear investment-savings functions [68, 67, 127, 128], and the Goodwin nonlinear accelerator-multiplier [179, 139, 256, 10], can be reduced to the Van der Pol equation, which describes damped oscillations [68, 67, 127, 128]. The equations of the standard Kaldor-type model of business cycles and the standard Van der Pol oscillator equation are differential equations of integer order. As a result, these models cannot take into account memory effects. Formal generalizations of this equation can be obtained by replacing the integer derivatives by the fractional derivatives. The resulting equation can take into account memory effects. However, this way is not correct since it do not take into account how this (standard) equation is derived. This type of errors is described in paper [386, pp. 17–19]. The consistent derivation of the equation leads us to the fact that we can get only one fractional derivative, and the another derivative remains integer. Moreover, the fractional derivative is the Riemann–Liouville derivative, and not Caputo. Note that for the Riemann–Liouville fractional derivatives the initial conditions have nonstandard form.

The proposed fractional differential equation of Van der Pol can be used in computer simulation of the Kaldor-type models of business cycles with power-law memory. We note that in computer modeling it is important to study what changes in the behavior of the economy occur when memory is taken into account.

17 Solow models with power-law memory

In this chapter, we consider generalizations of various Solow growth models. In the first section after the Introduction, we discuss a possibility of generalizing the standard Solow–Swan model by taking into account the fading memory. We demonstrate that the violation of the standard product (Leibniz) rule for fractional derivatives gives a restriction in the construction of such generalizations. In the second section, we proposed a generalization of the model of long-run growth, in which we take into account the power-law memory, and the power-law form of the dynamics of the labor and knowledge. In the third section, we consider a generalization of the standard growth model for closed economy without capital depreciation. For the suggested model with memory, we derive nonlinear fractional differential equation and its explicit analytical solution. In this chapter, we prove that the rate of growth with memory can be greater than the rate of growth without memory. In other words, we prove for nonlinear growth models with memory that growth rates can be changed by memory effects. This chapter is based on work [386, 383] (see also book [254, pp. 43–92]).

17.1 Introduction

One of the most famous models among the models proposed by Solow is the Solow–Swan growth model [24, 325, 327]. This model is a dynamic single-sector model of economic growth (see, Solow and Swan articles [350, 351, 357], and books [325, 327, 493, 494]). In this model, the economy is considered without structural subdivisions, i. e., as one-sector model. It is assumed that the economy produces only universal products, which can be consumed both in the non-production and production consumptions, i. e., can be consumed, or invested. Exports and imports are not taken into account. This model describes the capital accumulation, labor or population growth, and increases in productivity, which is commonly called the technological progress. The Solow–Swan model can be used to estimate the separate effects on economic growth of capital, labor and technological change.

The Solow–Swan model is a generalization of the Harrod–Domar model, which includes a productivity growth as new effect. This relatively simple growth model was independently proposed by Robert M. Solow and Trevor W. Swan in 1956 [350, 351, 357]. In 1987, Solow was awarded the Nobel Memorial Prize in Economic Sciences for his contributions to the theory of economic growth [477]. Mathematically, the Solow–Swan model is actually represented by nonlinear ordinary differential equation, which describes the evolution of the per capita stock of capital.

The standard Solow–Swan model with continuous time can be represented in the form of the nonlinear ordinary differential equation

$$k^{(1)}(t) = -(\rho + \delta)k(t) + sf(k(t)), \tag{17.1}$$

https://doi.org/10.1515/9783110627459-017

where $k(t) = K(t)/L(t)$ is the per capita capital; $K(t)$ is capital expenditure; δ is the capital retirement ratio; s is the rate of accumulation; ρ is the rate of increase in labor resources. The supply of labor changes is described as $L(t) = L_0 \exp(\rho t)$ at a constant rate $\rho \in (-1, +1)$. The function $f(k(t))$ describes the labor productivity, which is usually considered in the form $f(k(t)) = Ak^b(t)$ with $b \in (0, 1)$.

In the standard Solow–Swan model, the memory effects and memory fading are neglected. From a mathematical point of view, the neglect of memory effects in standard models is due to the fact that to describe the economic process we use only equations with derivatives of integer orders. These derivatives are determined by the properties of the function in an infinitely small neighborhood of the considered time. A very effective and powerful tool for describing memory effects is derivatives and integrals of noninteger orders.

The formal generalization of equation (17.1) can be realized by replacing the first-order derivative by the fractional derivative of the order $\alpha > 0$ in this equation. In this case, we get the equation

$$(D^{\alpha}_{C;0+}k)(t) = -(\rho + \delta)k(t) + sf(k(t)), \tag{17.2}$$

where $D^{\alpha}_{C;0+}$ is the Caputo fractional derivative, for example.

Unfortunately, the consistent construction of the generalization of the standard model equation (17.1) cannot give a fractional differential equation in the form (17.2). One of the obstacles to such generalizations is the nonstandard properties of fractional derivatives [367, 287, 377, 376, 382, 338, 81], including violation of the standard product (Leibniz) rule. These nonstandard property complicate the generalization of the classical Solow models, which do not take into account memory effects.

The need to take into account the nonstandard properties of fractional derivatives and integrals leads us [386] to the following: *It is not enough to generalize the differential equations that describe standard models. It is necessary to generalize the all steps of derivation of these equations from the basic principles, concepts and assumptions* (see the derivability principle in [386]).

Therefore, we will consider the derivation of the standard Solow models. To do this, we first briefly describe the sequential derivation of the equation for the standard Solow models.

17.2 Solow–Swan model with memory

In this section, we discuss a possibility of generalizing the standard Solow–Swan model, which was independently proposed by Robert M. Solow and Trevor W. Swan in 1956 [350, 351, 357]. In generalization of the standard model, we take into account fading memory. We demonstrate the fact that the violation of the standard product (Leibniz) rule for fractional derivatives, which is main characteristic property of these operators, gives a restriction in the construction of such generalizations.

17.2.1 Standard Solow–Swan model with continuous time

The standard Solow–Swan model is a classical nonlinear economic model that is actively used in modern economics [493, 494, 325, 327, 115]. In the Solow–Swan model, the state of the economy is given by the following five endogenous state variables (defined within the model):
- $Y(t)$ is the final product (production capacity);
- $L(t)$ is the labor input (available labor resources);
- $K(t)$ describes the capital reserves (capital expenditure, production assets);
- $I(t)$ is the investment (investment rates);
- $C(t)$ is the amount of nonproductive consumption (instant consumption).

All these variables are functions of time t, which is assumed to be continuous. In addition, the Solow–Swan model uses exogenous indicators (defined outside the model):
- $\rho \in (-1, +1)$ is the rate of increase in labor resources;
- $\delta \in [0, 1]$ is the capital retirement ratio;
- $s \in [0, 1]$ is the rate of accumulation (the share of the final product used for investment).

These exogenous indicators are considered constant in time. The rate of accumulation is considered as a controlling parameter. It is assumed that the production and labor resources are fully used in the production of the final product. The final product at each moment in time is a single-valued function of the capital and labor

$$Y = F(K(t), L(t)). \tag{17.3}$$

In the standard Solow–Swan model, it is assumed that $Y = F(K, L)$ is a linearly homogeneous function satisfying the constant scale, for which the following property holds:

$$F(zK, zL) = zF(K, L). \tag{17.4}$$

The production function $F(K, L)$ of the national economy is often specified to be a function of the Cobb–Douglas type.

The final product $Y(t)$ is used for nonproductive consumption $C(t)$ and investment $I(t)$. Therefore, we have the equation

$$Y(t) = C(t) + I(t). \tag{17.5}$$

The accumulation rate $s \in [0, 1]$ is the fraction of the final product used for investment [325, 327, 24], which leads to the equation

$$I(t) = sY(t). \tag{17.6}$$

Equation (17.6) is the equation of the multiplier without memory.

If we assume that the speed of increase in labor resources is proportional to the available labor resources, then taking into account the growth rate of employed $\rho \in (-1, +1)$, we can write the differential equation

$$L^{(1)}(t) = \rho L(t), \tag{17.7}$$

where $L^{(1)}(t) = dL(t)/dt$ is the derivative of first order. Equation (17.7) with the initial condition $L(0) = L_0$, has the solution $L(t) = L_0 \exp(\rho t)$, where L_0 is the labor resources at the beginning of observation at $t = 0$. The equation of labor resources can also be considered in the form of the logistic equation (e. g., see [120]) instead of (17.7).

Capital stock may change for two reasons: investment causes an increase in capital stock; depreciation or disposal of capital causes a decrease in its reserves. If we assume that the retirement of capital occurs with a constant retirement rate of $\delta \in [0, 1]$, then the capital dynamics is described by the equation

$$K^{(1)}(t) = I(t) - \delta K(t). \tag{17.8}$$

Substitution of expressions from (17.3) and (17.6) into equation (17.8) gives

$$K^{(1)}(t) = sF(K(t), L(t)) - \delta K(t), \tag{17.9}$$

where $s \in [0, 1]$ is the coefficient of the economic multiplier without memory and lag.

To obtain the equation of the standard Solow–Swan model, the following relative variables are introduced. The first such variable is the labor productivity that is defined by the expression

$$y(t) = \frac{Y(t)}{L(t)} = \frac{F(K(t), L(t))}{L(t)} = F(K(t)/L(t), 1) = f(k), \tag{17.10}$$

where we use the property of the linear homogeneity of the production function that is described by equation (17.4).

The dynamics of the output of the final product depends on the amount of the capital per employed person, the per capita capital (capital endowment) that is defined as

$$k(t) = \frac{K(t)}{L(t)}. \tag{17.11}$$

Substitution of $K(t) = k(t)L(t)$ into equation (17.9) gives

$$(k(t)L(t))^{(1)} = sF(k(t)L(t), L(t)) - \delta k(t)L(t). \tag{17.12}$$

Using the standard product (Leibniz) rule for derivative of the first order

$$(k(t)L(t))^{(1)} = k^{(1)}(t)L(t) + k(t)L^{(1)}(t), \tag{17.13}$$

and the property of the linearly homogeneity (17.4), equation (17.12) is rewritten in the form

$$k^{(1)}(t)L(t) + k(t)L^{(1)}(t) = sf(k(t))L(t) - \delta k(t)L(t). \tag{17.14}$$

Using equation (17.7) for the labor resources, we obtain

$$k^{(1)}(t) = -(\rho + \delta)k(t) + sf(k(t)). \tag{17.15}$$

Equation (17.15) is the differential equation of the standard Solow–Swan model.

The behavior of the indicators of the standard Solow–Swan model is determined by the ordinary differential equation (17.15) of the first order and the dynamics of labor resources (17.7).

17.2.2 Generalization of Solow–Swan model

To take into account effects of fading memory in the Solow–Swan model, we can use fractional derivatives of noninteger orders. The consistent derivation of equation (17.15) of the standard Solow–Swan model, which was described in the previous subsection, allows us to see the difficulties in the generalization of this model. It is well known that we cannot use the standard product (Leibniz) rule for fractional derivative. Unfortunately, the standard form of the product (Leibniz) rule for fractional derivatives of noninteger orders is violated

$$(D^{\beta}_{C;0+}k(t)L(t)) \neq (D^{\beta}_{C;0+}k)(t)L(t) + k(t)(D^{\beta}_{C;0+}L)(t) \tag{17.16}$$

for $\alpha \neq 1$. For $\alpha = 1$, we standard product rule (17.13) holds. As a result, we cannot obtain fractional generalization of the differential equation (17.15) for the per capita capital $k(t) = K(t)/L(t)$.

We emphasize that the violation of the standard product rule is a characteristic property of all type of the fractional derivatives of noninteger orders [367, 81]. Note that the implementation of the standard product rule for an operator means that this operator is a differential operator of integer order [367, 81], and such operators cannot describe the effects of memory and nonlocality [382].

As a result, the generalization of the standard Solow–Swan model, which will take into account the power-law memory effects, should be represented as the system of the fractional differential equations

$$\begin{cases} (D^{\alpha}_{C;0+}L)(t) = \rho L(t), \\ (D^{\beta}_{C;0+}K)(t) = sF(K(t), L(t)) - \delta K(t), \end{cases} \tag{17.17}$$

where $n - 1 < \alpha \leq n$ and $m - 1 < \beta \leq m$. The dynamics with memory for the per capita capital $k(t)$ can be described as the ratio $K(t)/L(t)$ of solutions of these two fractional differential equations.

For production function of the national economy in the form the Cobb–Douglas function

$$F(K,L) = AK^{a}(t)L^{1-a}(t), \tag{17.18}$$

we get system (17.17) in the form

$$\begin{cases} (D^{\alpha}_{C;0+}L)(t) = \rho L(t), \\ (D^{\beta}_{C;0+}K)(t) = sAK^{a}(t)L^{1-a}(t) - \delta K(t). \end{cases} \tag{17.19}$$

The first fractional differential equation of system (17.19), which describes the labor resources, has the solution (see Theorem 5.15 of [200, p. 323]) in the form

$$L(t) = \sum_{k=0}^{n-1} L^{(k)}(0)t^{k}E_{\alpha,k+1}[\rho t^{\alpha}], \tag{17.20}$$

where $n-1 < \alpha \leq n$, $L^{(k)}(0)$ is integer-order derivatives of orders $k \geq 0$ at $t = 0$, and $E_{\alpha,k+1}[\rho t^{\alpha}]$ is the two-parameter Mittag–Leffler function [143]. In the case, $0 < \alpha \leq 1$ ($n = 1$) equation (17.20) takes the form

$$L(t) = L(0)E_{\alpha,1}[\rho t^{\alpha}]. \tag{17.21}$$

For $\alpha = 1$, equation (17.21) gives the standard solution $L(t) = L_0 \exp(\rho t)$, where $L_0 = L(0)$.

Using equation (17.21) that describes dynamics of the labor resources, we can obtain the nonlinear fractional differential equation for the capital expenditure $K(t)$ in the form

$$(D^{\beta}_{C;0+}K)(t) = sAK^{a}(t)L_0^{1-a}(E_{\alpha,1}[\rho t^{\alpha}])^{1-a} - \delta K(t). \tag{17.22}$$

In the case $\alpha = 1$, this equation takes the form

$$(D^{\beta}_{C;0+}K)(t) = sAK^{a}(t)L_0^{1-a}e^{\rho(1-a)t} - \delta K(t). \tag{17.23}$$

To describe the capital expenditure dynamics, it should be used the computer simulations of nonlinear fractional differential equation (17.22) or (17.23).

Note that the nonlinear fractional differential equations (17.22) and (17.23) can be represented as Volterra integral equations by using the results of the articles of Anatoly A. Kilbas and Sergei A. Marzan [196, 195]. In the space $C^{r}[0,T]$ of continuously differentiable function, the Cauchy problem for fractional differential equation

$$(D^{\beta}_{C;0+}K)(t) = G(t,K(t)), \tag{17.24}$$

where $n-1 < \beta \leq n$, and $G(t, K(t))$ is the expression of the right-hand side of equations (17.22) and (17.23), is equivalent (see Theorem 3.24 of [200, pp. 199–202]) to the Volterra integral equation

$$K(t) = \sum_{m=0}^{n-1} \frac{K^{(m)}(0)}{m!} t^m + \frac{1}{\Gamma(\alpha)} \int_0^t (t-\tau)^{\beta-1} G(\tau, K(\tau))\, d\tau. \tag{17.25}$$

This equivalence exists if the following conditions hold: (1) $G(t, K(t)) \in C_\gamma[0, T]$ with $0 \leq \gamma < 1$ and $\gamma \leq \beta$; (2) $K(t) \in C^r[0, T]$, where $r = n$ for integer values of β, ($\beta \in \mathbb{N}$) and $r = n-1$ for noninteger values of β, ($\beta \notin \mathbb{N}$).

Using expression of the right-hand side of equations (17.22), equation (17.25) is written in the form

$$K(t) = \sum_{m=0}^{n-1} \frac{K^{(m)}(0)}{m!} t^m + \frac{sAL_0^{1-a}}{\Gamma(\alpha)} \int_0^t (t-\tau)^{\beta-1} K^a(\tau) e^{\rho(1-a)\tau}\, d\tau \\ - \frac{\delta}{\Gamma(\alpha)} \int_0^t (t-\tau)^{\beta-1} K(\tau)\, d\tau. \tag{17.26}$$

As a result, we can conclude that the generalizations of the Solow–Swan model, which take into account the effects of power-law memory, are described by the system of equations (17.17) (or system (17.19), or equation (17.22)), and not equation (17.2).

At the same time, equation (17.2), which is a formal generalization of the equation of the standard model, does not have economic significance due to the violation of the principle of derivability [386] (see also book [254, pp. 43–92]).

17.3 Solow model of long-run growth with memory

In this section, we consider a generalization of the model of long-run growth, which is considered in Solow's paper "A contribution to the theory of economic growth," [350]. In this model, the capital depreciation is neglected. In the proposed generalization, we take into account the power-law memory. We also assume the power-law form of the dynamics of the labor and knowledge. Exact analytical solution of the nonlinear fractional differential equation, which describes the proposed model, is suggested.

17.3.1 Long-run growth without memory and capital depreciation

Let us describe standard Solow model, which is considered in Solow's paper [350, pp. 66-67]. The Solow model uses four variables: output $Y(t)$, capital $K(t)$, labor $L(t)$, and knowledge $A(t)$. At any time, the economy has some amounts of capital, labor, and knowledge, which are combined to produce output. The production function is considered in the form

$$Y(t) = F(K(t), A(t)L(t)). \tag{17.27}$$

The Solow model uses assumptions about how the stocks of labor $L(t)$, knowledge $A(t)$, and capital $K(t)$ change over time. The initial levels of capital, labor, and knowledge are taken as given, and are assumed to be strictly positive. In the standard model, the labor and knowledge grow are described [326, p. 13], by the equations

$$L^{(1)}(t) = \rho L(t), \tag{17.28}$$

$$A^{(1)}(t) = gA(t), \tag{17.29}$$

where ρ and g are exogenous parameters and $f^{(1)}(t)$ denotes a first derivative with respect to time. The solutions of equations (17.28) and (17.29) can be represented in the form

$$L(t) = L(t_0)\exp\{\rho(t - t_0)\}, \quad A(t) = A(t_0)\exp\{g(t - t_0)\}. \tag{17.30}$$

Let us consider a power-law generalization of equations (17.28) and (17.29) in the form

$$L^{(1)}(t) = \rho L^{q}(t), \tag{17.31}$$

$$A^{(1)}(t) = gA^{p}(t), \tag{17.32}$$

where q and p are exogenous parameters. For $q = 1$ and $p = 1$, equations (17.31) and (17.32) give standard equations (17.28) and (17.29).

Let us obtain solutions of equations (17.31) and (17.32) for the case $q \neq 1$ and $p \neq 1$. Equation (17.31) can be written as

$$L^{-q}(t)L^{(1)}(t) = \rho. \tag{17.33}$$

Using the standard chain rule, we get

$$\frac{d}{dt}\big(L^{1-q}(t)\big) = \rho(1 - q). \tag{17.34}$$

The solution of equation (17.34) has the form

$$L^{1-q}(t) = \rho(1 - q)t + c, \tag{17.35}$$

where $c = L^{1-q}(0)$. For simplicity, we will assume that $L(0) = 0$. Then $c = 0$, and equation (17.35) is given as

$$L(t) = \big(\rho(1 - q)t\big)^{1/(1-q)}, \tag{17.36}$$

where $t > 0$ and $\rho(1-q) > 0$. For $t = t_0 > 0$, equation (17.36) has the form

$$L(t_0) = (\rho(1-q)t_0)^{1/(1-q)}. \tag{17.37}$$

Using expression (17.37), we can represent solution (17.36) by the expression

$$L(t) = L(t_0)\left(\frac{t}{t_0}\right)^{1/(1-q)}. \tag{17.38}$$

Analogously, we get

$$A(t) = (g(1-p)t)^{1/(1-p)}, \tag{17.39}$$

where we assume that $A(0) = 0$. Then we represent solution (17.39) in the form

$$A(t) = A(t_0)\left(\frac{t}{t_0}\right)^{1/(1-p)}. \tag{17.40}$$

Output $Y(t)$ is divided between consumption $C(t)$ and investment $I(t)$. In standard model, the fraction $s \in [0,1]$ of output devoted to investment is exogenous and constant such that

$$I(t) = sY(t). \tag{17.41}$$

Equation (17.41) describes economic multiplier without memory and lag.

In addition, existing capital depreciates at rate $\delta \in [0,1]$. Therefore, in the standard model, we have

$$K^{(1)}(t) = sY(t) - \delta K(t). \tag{17.42}$$

Let us consider the model without capital depreciation, i. e., $\delta = 0$. Then equation (17.42) takes the form

$$K^{(1)}(t) = sY(t). \tag{17.43}$$

Substitution of (17.27) into equation (17.43) gives

$$K^{(1)}(t) = sF(K(t), A(t)L(t)), \tag{17.44}$$

where $L(t)$ and $A(t)$ are described by equations (17.38) and (17.40). Substitution of (17.38) and (17.40) into equation (17.44) gives

$$K^{(1)}(t) = sF(K(t), A(t_0)L(t_0)(t/t_0)^{1/(1-p)+1/(1-q)}), \tag{17.45}$$

where $q \neq 1$ and $p \neq 1$.

The solution of equation (17.45), which describes a model without memory, will be obtained as a special case of the generalized model, when the memory fading parameter is equal to one $\alpha = 1$ that means the absence of memory.

17.3.2 Long-run growth with power-law memory

We should note that equation (17.43) cannot take into account the memory, since the derivative $K^{(1)}(t)$ is determined by behavior of $Y(\tau)$ only at the same time instant $\tau = t$, and does not take into account the history of changes of $Y(\tau)$ in the past, when $\tau \in (0, t)$.

To take into account of the changes of $Y(\tau)$ in the past, instead of equation (17.41), we can consider the equation of the multiplier with memory in the form

$$I(t) = \int_0^t s(t,\tau) Y(\tau)\, d\tau, \tag{17.46}$$

where the function $s(t,\tau)$ allows us to take into account the changes of output $Y(\tau)$ in the past $\tau \in (0,t)$. The function $s(t,\tau)$ is interpreted a memory function. In order to have standard dimensions of economic quantities, the time variable t is considered as dimensionless parameter, i. e., we change the variable $t_{\text{old}} \to t_{\text{new}} = t_{\text{old}}/t_c$, where t_c is a characteristic time of processes.

Substitution of expression (17.46) into equation (17.8) with $\delta = 0$ gives

$$K^{(1)}(t) = \int_0^t s(t,\tau) Y(\tau)\, d\tau. \tag{17.47}$$

Let us consider the power-law memory. In this case, the memory function can be described by the equation

$$s(t,\tau) = \frac{s}{\Gamma(\mu)} (t-\tau)^{\mu-1}, \tag{17.48}$$

where s is the savings parameter, and $\Gamma(\alpha)$ is the gamma function. For function (17.48), equation (17.47) can be written in the form

$$K^{(1)}(t) = s(I^{\mu}_{\text{RL};0+} Y)(t), \tag{17.49}$$

where $I^{\mu}_{\text{RL};0+}$ is the Riemann–Liouville fractional integral [200] of the order $\mu > 0$. Note that equation (17.49) with the Riemann–Liouville fractional integral can be considered [380] as an approximation of equations (17.47) with generalized memory functions $s(t,\tau)$.

Using the property that the Caputo fractional derivative is the left inverse operator for the Riemann–Liouville fractional integral (see Lemma 2.21 of [200, p. 95]), the action of the Caputo derivative on equation (17.49) gives

$$(D^{\mu+1}_{C;0+} K)(t) = sY(t), \tag{17.50}$$

where $D^{\mu}_{C;0+}$ is the Caputo fractional derivative [200] of the order $\mu > 0$.

As a result, to take into account the power-law memory, we can use the equation

$$(D^{\alpha}_{C;0+}K)(t) = sY(t), \tag{17.51}$$

where $D^{\alpha}_{C;0+}$ is the Caputo fractional derivative of the order $\alpha = \mu + 1 > 1$ with $\mu > 0$, i. e., $\alpha > 1$. In the general case, we can consider equation (17.51) for the order $\alpha > 0$ including $\alpha \in (0, 1)$.

Let us consider the specific example of a production function. We will use the Cobb–Douglas function. In this case, equation (17.27) takes the form

$$Y(t) = K^{a}(t)(A(t)L(t))^{b}, \tag{17.52}$$

where we can consider $b = 1 - a$. This production function is easy to analyze, and it appears to be a good first approximation to actual production functions. Therefore, the Cobb–Douglas function is useful for modeling.

Substitution of expression (17.52) into equation (17.51) gives

$$(D^{\alpha}_{C;0+}K)(t) = s(A(t)L(t))^{b}K^{a}(t). \tag{17.53}$$

Then substituting expressions (17.38) and (17.40) into equation (17.53), we obtain the equation

$$(D^{\alpha}_{C;0+}K)(t) = \lambda(t_0)t^{\beta}K^{a}(t), \tag{17.54}$$

where

$$\lambda(t_0) = sA^{b}(t_0)L^{b}(t_0)t_0^{-\beta}, \tag{17.55}$$

$$\beta = \frac{b}{1-p} + \frac{b}{1-q}. \tag{17.56}$$

Equation (17.54) is the nonlinear fractional differential equation that describes the generalization of the Solow model without capital depreciation, where we take into account the power-law fading memory.

Using equations 3.5.47 and 3.5.48 of Section 3.5.3 in [200, p. 209], equation (17.54) with $\lambda(t_0) \neq 0$, $a, q, p \neq 1$ and $\alpha > 0$ has the solution

$$K(t) = \left(\frac{\Gamma(\alpha - \gamma(\alpha) + 1)}{\lambda(t_0)\Gamma(1 - \gamma(\alpha))}\right)^{1/(a-1)} t^{\alpha - \gamma(\alpha)}, \tag{17.57}$$

where

$$\gamma(\alpha) = \frac{\beta + a\alpha}{a - 1}, \quad \alpha - \gamma(\alpha) = \frac{\alpha + \beta}{1 - a}, \tag{17.58}$$

and $\lambda(t_0)$ is defined by (17.55), β is defined by equation (17.56).

For $t = t_0$, equation (17.57) has the form

$$K(t_0) = \left(\frac{\Gamma(\alpha - \gamma(\alpha) + 1)}{\lambda(t_0)\Gamma(1 - \gamma(\alpha))} \right)^{1/(a-1)} t_0^{\alpha-\gamma(\alpha)}. \tag{17.59}$$

Using equation (17.59), solution (17.57) can be written as

$$K(t) = K(t_0)\left(\frac{t}{t_0}\right)^{\alpha-\gamma(\alpha)}. \tag{17.60}$$

The Caputo fractional derivative of the power-law function is given (see Example 3.1 of book [90, p. 49]) by the equation

$$(D^{\alpha}_{C;0+}(\tau)^{\delta})(t) = \frac{\Gamma(\delta + 1)}{\Gamma(\delta - \alpha + 1)}(t)^{\delta-\alpha}, \tag{17.61}$$

if $\delta > n - 1$, and

$$(D^{\alpha}_{C;0+}(\tau)^{\delta})(t) = 0, \tag{17.62}$$

if $\delta = 0, 1, \dots n - 1$. In all remaining cases ($\delta < n - 1$ such that $\delta \neq 0, 1, \dots, n - 1$), the integral in the definition of the Caputo fractional derivative is improper and divergent.

Remark 17.1. Using equation (17.61), it is directly verified that expression (17.57) is the explicit solution of equation (17.54) if $\delta = \alpha - \gamma(\alpha) > n-1$, where $n-1 = [\alpha]$ for noninteger values of $\alpha > 0$. As a result, the condition $\alpha - \gamma(\alpha) > n - 1$ (or $(\alpha + \beta)/(a - 1) > n - 1$) with $t > 0$ should be satisfied for parameters of equation (17.57) with the Caputo derivative instead of $\alpha - \gamma(\alpha) \geq 0$ that is used in [200].

Remark 17.2. Note that in equations 3.5.48 and 3.5.49 of [200, p. 209] there are typos: the signs of the parameters $\gamma(\alpha)$ and α in the gamma functions must be opposite. In equation (17.57), these signs are correct.

Taking into account these Remarks and Propositions 3.8 and 3.9 of [200, pp. 209–210], we can formulate the following conditions for the existence of solution (17.57) for equation (17.54).

Statement 17.1. *Nonlinear fractional differential equation* (17.54) *with* $\alpha \in (n - 1, n)$, $n \in \mathbb{N}$, *has the solution, which is given by equations* (17.57), (17.58), *if the following conditions are satisfied:*

$$\begin{cases} \alpha \in (0, 1), \\ \beta > -\alpha - (n - 1)(a - 1), \end{cases} \quad or \quad \begin{cases} a > 1, \\ \beta < -\alpha - (n - 1)(a - 1). \end{cases} \tag{17.63}$$

Here, we take into account that $\delta = \alpha - \gamma(\alpha) > n - 1$ instead of $\delta = \alpha - \gamma(\alpha) > 0$ that is used in Propositions 3.8 and 3.9 of [200].

Conditions (17.63) for the parameters, under which solution (17.57) exists, are important for applications in economic models.

17.3.3 Rate of growth with power-law memory

For solution (17.57), the rate of growth with power-law memory at $t > t_0$ is approximately equal to

$$R(\alpha) = \frac{\alpha - \gamma(\alpha)}{t} = \frac{\alpha + \beta}{(1-a)t}, \tag{17.64}$$

where $a \neq 1$. For case $\alpha = 1$, which corresponds to the absence of memory, equation (17.64) can be written in the form

$$R(1) = \frac{1 - \gamma(1)}{t} = \frac{1 + \beta}{(1-a)t}. \tag{17.65}$$

Let us consider the case when we have growth with memory, i. e., the increasing function $K(t)$ for noninteger values of α. In this case, the growth rate with memory, $R(\alpha)$ should be positive for noninteger values of $\alpha \in (n-1, \alpha)$, $n \in \mathbb{N}$, i. e., we have the inequality

$$R(\alpha) > 0. \tag{17.66}$$

Solutions (17.57), (17.58) exist only if the following conditions are satisfied

$$\alpha - \gamma(\alpha) = \frac{\alpha + \beta}{1 - a} > n - 1, \quad \gamma(\alpha) < 1. \tag{17.67}$$

Obviously, using equations (17.67) and (17.64), we get that condition (17.66) has the form

$$R(\alpha) = \frac{\alpha - \gamma(\alpha)}{t} = \frac{\alpha + \beta}{(1-a)t} > 0, \tag{17.68}$$

which will be satisfied for any solutions for $t > 0$. Mathematically, this statement follows from the fact that if $\alpha - \gamma(\alpha) > n - 1$ with $n \in \mathbb{N}$, then $\alpha - \gamma(\alpha) > 0$.

As a result, we can formulate the following principle.

Principle 17.1 (Principle of inevitability of growth with memory). *For economic models with memory, which are described by equations* (17.53), (17.31), (17.32), *with $a \neq 1$, $q \neq 1$ and $p \neq 1$, the rate of growth with power-law memory* (17.64) *is positive*

$$R(\alpha) = \frac{\alpha + \beta}{(1-a)t} = \frac{\alpha(1-p)(1-q) + \beta(2-p-q)}{(1-a)(1-p)(1-q)t} > 0 \tag{17.69}$$

for noninteger values of α. As a result, the inclusion of power-law memory effects leads to the inevitability of capital growth $K(t)$.

In addition, we have the lower boundary of the growth rates for processes with memory, which is described by the following statement.

Statement 17.2. *The rate of growth with power-law memory for processes, which are described by equation* (17.54), *has the lower boundary that is defined by the inequality*

$$R(\alpha) > \frac{n-1}{t}, \tag{17.70}$$

where $n = [\alpha] + 1$ *for noninteger values of* α.

This statement is based on the fact that solution exist only if the condition

$$\alpha - \gamma(\alpha) = \frac{\alpha + \beta}{(1-a)} > n - 1 \tag{17.71}$$

holds. Using definition (17.64) of the rate of growth with memory, we get (17.70).

Remark 17.3. Solutions, in which the decline (recession) is realized, do not exist within the framework of the suggested model with memory. In this model, the processes with memory cannot have negative values of the rate of growth with power-law memory. Note that for processes without memory, which are described by equation (17.64) with $\alpha = 1$, the negative growth rates ($R(1) < 0$) can be realized.

Let us consider the condition under which the growth rate with memory, $R(\alpha)$, is greater than the growth rate without memory, $R(1)$. This condition is represented by the inequality

$$R(\alpha) > R(1). \tag{17.72}$$

Using expressions (17.64) and (17.65), inequality (17.72) gives

$$R(\alpha) - R(1) = \frac{\alpha - 1}{(1-a)t} > 0, \tag{17.73}$$

where we assume that $a \in (0, 1)$.

Taking into account conditions (17.63) for the existence of solution (17.57) and that $n = 1$ for $\alpha \in (0, 1)$, we can formulate the following statement.

Statement 17.3. *The rate* $R(\alpha)$ *of growth with memory for the capital* $K(t)$ *is greater than the rate* $R(1)$ *of growth without memory* $R(\alpha) > R(1)$, *if the following conditions are satisfied*

$$\begin{cases} \alpha \in (0,1), \\ a > 1, \\ \beta < -\alpha, \end{cases} \quad \text{or} \quad \begin{cases} \alpha \in (n-1, n), \quad n \in \mathbb{N},\ n \neq 1, \\ a \in (0,1), \\ \beta > -\alpha + (n-1)(1-a). \end{cases} \tag{17.74}$$

Let us consider the question of how much growth with memory can be greater than growth without memory. To describe this, we give the ratio of growth rates $R(\alpha)$ and $R(1)$. This ratio has the form

$$\frac{R(\alpha)}{R(1)} = \frac{\alpha + \beta}{1 + \beta}, \tag{17.75}$$

where we assume that $R(1) > 0$, conditions (17.74) are satisfied, and β is given by (17.56). Note that $\alpha \in (0,1)$, in inequality $R(\alpha) > R(1)$ is also satisfied because $a > 1$ and $\alpha + \beta < 1 + \beta < 0$.

17.3.4 Dynamics of capital per unit of effective labor

The amount of capital per unit of effective labor [326, p. 10], is defined by the equation

$$k(t) = \frac{K(t)}{A(t)L(t)}. \tag{17.76}$$

The solution of capital in form (17.60) and expressions (17.38) and (17.40) allow us to describe the amount of capital per unit of effective labor (17.76) by the equation

$$k(t) = \frac{K(t_0)}{A(t_0)L(t_0)}\left(\frac{t}{t_0}\right)^{\alpha-\gamma(\alpha)-1/(1-q)-1/(1-p)}. \tag{17.77}$$

For the capital per unit of effective labor $k(t)$, the rate of growth with power-law memory at $t > t_0$ is approximately equal to

$$r(\alpha) = \frac{\alpha - \gamma(\alpha)}{t} - \frac{1}{t}\left(\frac{1}{1-q} + \frac{1}{1-p}\right), \tag{17.78}$$

where $q \neq 1$, $p \neq 1$, $a \neq 1$, and β is defined by equation (17.55). Expression (17.78) can be represented in the form

$$r(\alpha) = \frac{\alpha+\beta}{(1-a)t} - \frac{\beta}{bt} = \frac{b(\alpha+\beta) - \beta(1-a)}{(1-a)bt}. \tag{17.79}$$

For $\alpha = 1$, which describes the case of absence memory, expression (17.79) takes the form

$$r(1) = \frac{b(1+\beta) - \beta(1-a)}{(1-a)bt}. \tag{17.80}$$

Expression (17.80) describes the rate of growth without memory for $a \neq 1$.

Let us consider the condition under which the growth rate with memory, $r(\alpha)$, is greater than the growth rate without memory, $r(1)$. This condition is represented by the inequality

$$r(\alpha) > r(1). \tag{17.81}$$

Using equations (17.79) and (17.80), inequality (17.81) takes the form

$$r(\alpha) - r(1) = \frac{\alpha - 1}{(1-a)t} > 0. \tag{17.82}$$

Inequality (17.82) coincides with inequality (17.73). Therefore, the Statement 17.3 and the system of inequalities (17.74) also should be satisfied for dynamics of the capital per unit of effective labor.

For the capital per unit of effective labor $k(t)$, the rate $r(\alpha)$ of growth with memory is greater than the rate $r(1)$ of growth without memory $r(\alpha) > r(1)$ if the conditions (17.74) are satisfied.

Let us consider the question of how much growth with memory can be greater than growth without memory. Using equations (17.79) and (17.80), the ratio of $r(\alpha)$ and $r(1)$ is expressed by the equation

$$\frac{r(\alpha)}{r(1)} = \frac{\alpha b + \beta(a + b - 1)}{b + \beta(a + b - 1)}, \tag{17.83}$$

where we assume that $r(1) > 0$, and the parameter β is defined by equation (17.55).

For $b = 1 - a$ with $a \in (0, 1)$, we have the linearly homogeneous production functions Cobb–Douglas. In this case, the rate of growth with power-law memory (17.79) takes the form

$$r(\alpha) = \frac{\alpha}{(1 - a)t}. \tag{17.84}$$

Using that $\alpha \in (n - 1, n)$, we get

$$r(\alpha) > \frac{n - 1}{(1 - a)t} \tag{17.85}$$

for noninteger values of α. In this case, the ratio of $r(\alpha)$ and $r(1)$ is expressed by the equation

$$\frac{r(\alpha)}{r(1)} = \alpha. \tag{17.86}$$

Using that $\alpha \in (n - 1, n)$, we obtain the inequality

$$\frac{r(\alpha)}{r(1)} > n - 1 \tag{17.87}$$

for noninteger values of α.

As a result, we find that the growth rate with memory is greater than the growth rate without memory by α times. This allows us to formulate the following principle, where we assume that conditions (17.74) are satisfied.

Principle 17.2 (Principle of changing growth rates by memory). *In the case $b = 1 - a$ with $a \in (0, 1)$ and conditions (17.74) are satisfied, the rate of growth with power-law memory for capital per unit effective labor can be greater than the rate of growth without memory ($r(\alpha) > r(1)$) in α times, if the parameter of memory fading α is more than one ($\alpha > 1$). The growth rate with memory is less than the growth rate without memory, i. e., $r(\alpha) < r(1)$, if the parameter α is less than one ($0 < \alpha < 1$).*

As a result, we can conclude that the fading memory can significantly change growth rates. Therefore, we should not neglect the memory in economic models.

17.4 Solow–Lucas model of closed economy with memory

In this section, we consider a generalization of the standard growth model for closed economy without capital depreciation that is considered by Robert E. Lucas in paper "Making a Miracle" [238]. In the generalized model, we take into account the memory with power-law fading. The suggested model with memory is described by nonlinear fractional differential equation. The explicit expression of analytical solution of this nonlinear equation is obtained.

17.4.1 Solow–Lucas model for closed economy without memory

Let us consider the Solow model for closed economy that is described by Robert E. Lucas in [238, pp. 253–254] (see also Section 3.2 in [239, pp. 73–75]). We will call this model the Solow–Lucas model. Note that the Sveriges Riksbank Prize in Economic Sciences in Memory of Alfred Nobel 1995 was awarded to Robert E. Lucas Jr. "for having developed and applied the hypothesis of rational expectations, and thereby having transformed macroeconomic analysis and deepened our understanding of economic policy" [478, 475].

Let us consider an economy that uses physical capital $k(t)$ and the human capital $h(t)$ to produce a single good $y(t)$, dynamics of which is described by the equation

$$y(t) = Ak^{a}(t)\big[uh(t)\big]^{1-a}. \tag{17.88}$$

Here, the human capital input is multiplied by u, where u is the fraction of time that people spend on producing goods.

The growth of human capital $h(t)$ depends on the amount of time spent on production, adjusted for quality (see equation (13) in [237]) in the form

$$\frac{dh(t)}{dt} = \delta(1-u)h(t). \tag{17.89}$$

The solution of equation (17.89) can be written as

$$h(t) = h(0)\exp\{\delta(1-u)t\}, \tag{17.90}$$

where $h(0)$ is the human capital at time $t = 0$.

The growth of physical capital depends on the savings rate s, such that we have the equation

$$\frac{dk(t)}{dt} = sy(t) \tag{17.91}$$

that is interpreted as an economic accelerator without memory and lag. Substitution of expressions (17.88) and (17.90) into equation (17.91) gives

$$\frac{dk(t)}{dt} = sAu^{1-a}h^{1-a}(0)\exp\{\delta(1-a)(1-u)t\}k^{a}(t). \tag{17.92}$$

In this model, it is assumed that the variable s and u are considered as given constants. The model, which is described by equations (17.88)–(17.92), will turn out to be a Solow model rewritten for a closed economy [239, 238].

In this model, the rate of technological changes (the average Solow balance) is described by the expression

$$\lambda_0 = \delta(1-a)(1-u), \tag{17.93}$$

and the initial technology level equal to $Ah^{1-a}(0)$.

The long run growth rate of both capital and production per worker is $\delta(1-u)$. The growth rate of human capital, and the ratio of physical and human capital converge to constant values. In the long run, the level of income is proportional to the economy's initial stock of human capital.

Remark 17.4. Note that in the Solow–Lucas model, the average values of $k(t)$, $h(t)$ and $y(t)$ are taken as original variables to build a model, in contrast to the Solow–Swan model that use $K(t)$, $L(t)$, $Y(t)$ as original variables. In standard models, the use of these two types of variables is equivalent. Unlike standard models, in models with memory the use of different types of variables leads to different models that are not equivalent in the general case [386]. This nonequivalence is associated with the violation of the standard form of the product rule for the fractional derivative of noninteger orders. As a consequence, we have a violation of the standard relationship of the rates of change of absolute (volume) and relative (specific) indicators.

17.4.2 Solow–Lucas model for closed economy with memory

In the proposed generalization of the standard Solow-Lucas model, we will use the generalizations of equations (17.88), (17.89) and (17.91).

(1) We assume that economy, which uses physical capital $k(t)$ and the human capital $h(t)$ to produce a single good $y(t)$, is described by the equation

$$y(t) = Ak^{a}(t)[uh(t)]^{b}, \tag{17.94}$$

where u is the fraction of time people spend producing goods. If $b = 1-a$, then equation (17.94) gives equation (17.88).

(2) The growth of human capital is assumed to be the power law in the form

$$\frac{dh(t)}{dt} = \delta(1-u)h^{\theta}(t). \tag{17.95}$$

If $\theta = 1$, then equation (17.95) gives equation (17.89) of the standard model.

(3) Equation (17.91) does not take into account the memory effects, since the derivative $dk(t)/dt$ is determined by behavior of $y(\tau)$ only at the same time instant $\tau = t$,

and does not take into account the history of changes of $y(\tau)$ in the past, when $\tau \in (0, t)$. To take into account of the changes of $y(\tau)$ in the past, we can use the equation

$$\frac{dk(t)}{dt} = \int_0^t s(t,\tau) y(\tau)\, d\tau, \tag{17.96}$$

where the function $s(t,\tau)$ allows us to take into account that the growth of physical capital $k(t)$ depends on the changes of $y(\tau)$ in the past at $\tau \in (0, t)$. The function $s(t,\tau)$ can be interpreted as a memory function. In order to have standard dimensions of economic quantities, the time variable t is considered as dimensionless parameter, i. e., we changed the variable $t_{\text{old}} \to t_{\text{new}} = t_{\text{old}}/t_c$, where t_c is a characteristic time of economic process.

An important property of memory is described by the principle of memory fading, which states that the increasing of the time interval leads to a decrease in the corresponding contribution [450]. Let us consider the power-law form of the memory fading. In this case, the memory function can be described by the equation

$$s(t,\tau) = \frac{s}{\Gamma(\mu)}(t-\tau)^{\mu-1}, \tag{17.97}$$

where s is the savings rate, $\Gamma(\alpha)$ is the gamma function. For function (17.97), equation (17.96) can be written in the form

$$\frac{dk(t)}{dt} = s(I^{\mu}_{RL;0+} y)(t), \tag{17.98}$$

where $I^{\mu}_{RL;0+}$ is the Riemann–Liouville fractional integral of the order $\mu > 0$ [200, pp. 69–70].

We should note that (17.98) with the Riemann–Liouville fractional integral can be considered as an approximation of the equations with generalized memory functions (17.96). In paper [380], using the generalized Taylor series in the Trujillo–Rivero–Bonilla form for the memory function, we proved that equation (17.96) for a wide class of memory functions can be represented through the Riemann–Liouville fractional integrals (and the Caputo fractional derivatives) of noninteger orders.

The action of the Caputo derivative on equation (17.98) gives

$$(D^{\mu}_{C;0+} k^{(1)})(t) = s(D^{\mu}_{C;0+} I^{\mu}_{RL;0+} y)(t), \tag{17.99}$$

where $D^{\mu}_{C;0+}$ is the Caputo fractional derivative of the order $\mu > 0$ [200, pp. 90–99].

In equation (17.99), we can use the property that the Caputo fractional derivative is the left inverse operator for the Riemann–Liouville fractional integral (see the equation of Lemma 2.21 in [200, p. 95]), in the form

$$(D^{\mu}_{C;0+} I^{\mu}_{RL;0+} y)(t) = y(t) \tag{17.100}$$

for $y(t) \in L_{\infty}(0,t)$ or $y(t) \in C[0,t]$, and the property of the Caputo fractional derivative

$$(D^{\mu}_{C;0+}k^{(1)})(t) = (D^{\mu+1}_{C;0+}k)(t). \tag{17.101}$$

Using expressions (17.100) and (17.101), equation (17.99) can be written as

$$(D^{\alpha}_{C;0+}k)(t) = sy(t), \tag{17.102}$$

with $\alpha = \mu + 1$. If $\alpha = 1$, then equation (17.102) take the form (17.91).

Substitution (17.94) into (17.102) gives equation that describes the dynamics with memory for physical capital $k(t)$ in the form

$$(D^{\alpha}_{C;0+}k)(t) = sAk^{a}(t)[uh(t)]^{b}, \tag{17.103}$$

where behavior of $h(t)$ is given by equation (17.95). Equation (17.103) describes the proposed generalization of the Solow–Lucas model for closed economy with power-law memory.

For $\alpha = 1$, equation (17.103) describes the Solow–Lucas model without memory, where $b \neq 1 - a$ and $\theta \neq 1$.

17.4.3 Growth rates of closed economy with memory

Let us obtain solution of the nonlinear fractional differential equation (17.95) that describe the Solow–Lucas model of closed economy with power-law memory.

We first solve equation (17.95) with $\theta \neq 1$ for human capital. Equation (17.95) can be written as

$$\frac{d}{dt}(h^{1-\theta}(t)) = \delta(1-u)(1-\theta). \tag{17.104}$$

Therefore, the solution of equation (17.104) with $\theta \neq 1$ has the form

$$h^{1-\theta}(t) = \delta(1-u)(1-\theta)t + c, \tag{17.105}$$

where we can use $c = h^{1-\theta}(0)$.

For simplicity, we will assume that $h(0) = 0$ and $h(t_0) = 1$. Then we get $c = 0$ and solution (17.105) can be written in the form

$$h(t) = (\delta(1-u)(1-\theta))^{1/(1-\theta)}t^{1/(1-\theta)}. \tag{17.106}$$

Equation (17.106) with $t = t_0 > 0$ has the form

$$h(t_0) = (\delta(1-u)(1-\theta))^{1/(1-\theta)}t_0^{1/(1-\theta)}, \tag{17.107}$$

where $\delta(1-u)(1-\theta) > 0$. Assuming that $h(t_0) = 1$, we get that

$$t_0 = \frac{1}{\delta(1-u)(1-\theta)}, \tag{17.108}$$

and solution (17.106) can represent by the equation

$$h(t) = h(t_0)\left(\frac{t}{t_0}\right)^{1/(1-\theta)}. \tag{17.109}$$

Substitution (17.109) into (17.103) gives

$$(D^{\alpha}_{C;0+}k)(t) = sA\big[uh(t_0)\big]^b \left(\frac{t}{t_0}\right)^{b/(1-\theta)} k^a(t), \tag{17.110}$$

where $\alpha = \mu + 1$ with $\mu > 0$. Equation (17.110) can be written in the form

$$(D^{\alpha}_{C;0+}k)(t) = \lambda t^{b/(1-\theta)} k^a(t), \tag{17.111}$$

where $\theta \neq 1$ and the constant λ is defined as

$$\lambda = sA\big[uh(t_0)\big]^b t_0^{-b/(1-\theta)}. \tag{17.112}$$

Using equations (3.5.47) and (3.5.48) of Section 3.5.3 in [200, p. 209], the equation

$$(D^{\alpha}_{C;0+}k)(t) = \lambda t^{\beta} k^a(t), \tag{17.113}$$

with $\lambda \neq 0$, $a \neq 1$, $\alpha = \mu + 1 > 1$, and

$$\beta = \frac{b}{1-\theta}, \tag{17.114}$$

has the solution

$$k(t) = \left(\frac{\Gamma(\alpha - \gamma(\alpha) + 1)}{\lambda \Gamma(1 - \gamma(\alpha))}\right)^{1/(a-1)} t^{\alpha - \gamma(\alpha)}, \tag{17.115}$$

where λ is defined by equation (17.112), and

$$\gamma(\alpha) = \frac{\beta + a\alpha}{a-1}, \quad \alpha - \gamma(\alpha) = \frac{\beta + \alpha}{1-a}. \tag{17.116}$$

The condition for the existence of a solution can be written as

$$\alpha - \gamma(\alpha) > n - 1, \quad \gamma(\alpha) < 1. \tag{17.117}$$

As a result, the rate of growth with power-law memory ($\alpha \neq 1$) at time t is approximately equal to

$$R(\alpha) = \frac{\alpha - \gamma(\alpha)}{t} = \frac{\beta + \alpha}{(1-a)t} = \frac{\alpha + b - \alpha\theta}{(1-a)(1-\theta)t} \tag{17.118}$$

for $t > t_0 > 0$ and $\theta \neq 1$, $a \neq 1$, where we use (17.114).

For $\alpha = 1$, expression (17.118) takes the form

$$R(1) = \frac{\beta + 1}{1 - a} = \frac{1 + b - \theta}{(1 - a)(1 - \theta)t}. \tag{17.119}$$

Expression (17.119) describes the rate of growth without memory for the case $\theta \neq 1$, $a \neq 1$.

Using that solution (17.115) exists if conditions (17.117) are satisfied, i. e., $\alpha - \gamma(\alpha) > n - 1$, we get

$$R(\alpha) > \frac{n - 1}{t} \tag{17.120}$$

for noninteger values of $\alpha \in (n - 1, n)$, and $t > 0$.

As a result, we get that $R(\alpha) > 0$ for noninteger values of α. Therefore, the rate of growth with power-law memory is positive for $t > 0$, and noninteger values of α.

We should note that the generalized model, which is described by equation (17.96) assumes that $\alpha = \mu + 1$, where $\mu > 0$. Therefore, we may not consider the case with $0 < \alpha < 1$.

Let us consider the condition, under which the rate of growth with memory, $R(\alpha)$, is greater than the rate of growth without memory, $R(1)$, that is,

$$R(\alpha) > R(1). \tag{17.121}$$

Using expressions (17.118) and (17.119), inequality (17.121) gives

$$R(\alpha) - R(1) = \frac{\alpha - 1}{(1 - a)t} = \frac{\mu}{(1 - a)t} > 0, \tag{17.122}$$

where we assume that $a \in (0, 1)$.

Using equation (17.122), we can state that for $a \in (0, 1)$, we have $R(\alpha) > R(1)$ for all $t > 0$. As a result, we can formulate the following principle.

Principle 17.3 (Principle of changing growth rates by memory). *If a solution of equation* (17.113) *for the physical capital $k(t)$ exists and $a < 1$, then the rate of growth with power-law memory, $R(\alpha)$, is greater than the rate of growth without memory, $R(1)$, i. e., $R(\alpha) > R(1)$.*

Using conditions (17.74) for the existence of solution (17.115), and that $\alpha = \mu + 1$ with $\mu > 0$, $m = n - 1$, we can formulate the following statement.

Statement 17.4. *The rate $R(\alpha)$ of growth with memory for the physical capital $k(t)$ is greater than the rate $R(1)$ of growth without memory ($R(\alpha) > R(1)$), if the following conditions are satisfied*

$$\begin{cases} \mu \in (m - 1, m), \quad m \in \mathbb{N}, \\ a \in (0, 1), \\ \beta > -\mu - 1 + m(1 - a). \end{cases} \tag{17.123}$$

The ratio of $R(\alpha)$ and $R(1)$, which are given by expressions (17.118) and (17.119), is expressed by the equation

$$\frac{R(\alpha)}{R(1)} = \frac{\alpha(1-\theta)+b}{1-\theta+b} = \frac{1-\theta+b+\mu(1-\theta)}{1-\theta+b}. \tag{17.124}$$

For the special case $0 < 1-\theta = b < 1$, equation (17.124) gives

$$\frac{R(\alpha)}{R(1)} = \frac{\alpha+1}{2} = 1+\frac{\mu}{2}, \tag{17.125}$$

where we use $\alpha = \mu + 1$. In particular, if $1-\theta = b \in (0,1)$, then $R(3.6)$ will be 2.3 times larger than $R(1)$.

In this case, we see that the effects of memory can increase the growth rate by more than two times in comparison with the standard Solow–Lucas model without memory.

As a result, we can conclude that the memory effects can significantly change growth rates and we should not neglect the memory in economic models.

17.5 Conclusion

In this chapter, we consider generalizations of various Solow growth models. In the proposed generalizations of the standard models we take into account the memory with power-law fading. We demonstrate that the violation of the standard product (Leibniz) rule for fractional derivatives gives a restriction in the construction of such generalizations. For the suggested model with memory, we derive nonlinear fractional differential equations and its explicit analytical solutions. We prove that the rate of growth with memory can be greater than the rate of growth without memory. For nonlinear growth models with power-law memory, the growth rates can be changed by memory effects. We prove that the rate of growth with power-law memory $R(\alpha)$ is greater than the rate of growth without memory $R(1)$, i. e., $R(\alpha) > R(1)$, if the parameter of memory fading α is more than one ($\alpha > 1$). Therefore, we conclude that the memory effects can significantly change growth rates and we should not neglect the memory in economic models.

18 Lucas model of learning with memory

In this chapter, we consider generalization of the Lucas model of learning (learning-by-doing) that is described in the paper "Making a Miracle" by Robert E. Lucas, who was awarded the Nobel Prize in Economic Sciences in 1995. The equation of the standard Lucas model is nonlinear differential equation of the first order. In the standard learning model, the memory effects and memory fading are not taken into account. For the first time, a learning model, which take into account memory, was proposed in article [391, 389, 381]. We propose [391, 389, 381], a generalization of the Lucas model of learning that takes into account memory with power-law fading. The fractional nonlinear differential equation, which describes the learning-by-doing with memory, is proposed. The expression of exact solution for this equation is obtained. Based on the exact solution of the model equation, we show that growth rates can be changed by memory. This chapter is based on the work [391, 389, 381].

18.1 Introduction

The possible connections between learning by doing and productivity growth in an economy are important in economics [326, 239, 89]. Economic models of learning (and learning-by-doing) with continuous time are usually described by differential equations with derivatives of integer orders. Example of such models is well-known classical model of learning (learning-by-doing) that is described by Robert E. Lucas in the paper "Making a Miracle" [238, pp. 262–263] (see also book [239, pp. 84–90]). Note that the Sveriges Riksbank Prize in Economic Sciences in Memory of Alfred Nobel 1995 was awarded to Robert E. Lucas Jr. "for having developed and applied the hypothesis of rational expectations, and thereby having transformed macroeconomic analysis and deepened our understanding of economic policy" [478, 475].

In the standard learning model [239, 238], the memory effects and memory fading are neglected. At the same time, it is obvious that to describe learning, it is necessary to take into account memory and memory fading, since amnesia and learning are incompatible. From a mathematical point of view, the neglect of memory effects in standard learning models is due to the fact that only equations with derivatives of integer orders are used to describe the learning processes. These derivatives are determined by the properties of the function in an infinitesimal neighborhood of the time under consideration and cannot describe such a nonlocal in time phenomenon as memory. An effective and powerful tool for describing memory effects is fractional calculus of derivatives and integrals of noninteger orders [335, 202, 308, 200, 90, 164, 165].

We proposed a generalization of the Lucas model of learning-by-doing that takes into account memory with power-law fading. The fractional nonlinear differential equation, which describes the learning with memory, and the expression of its exact

https://doi.org/10.1515/9783110627459-018

solution are suggested. We proved that growth rate can be substantially changed by memory.

18.2 Standard Lucas model of learning without memory

In order to describe the possible connection between evidence of learning on individual product lines and productivity growth in an economy as a whole, Robert E. Lucas considered [239, 238] the labor-only technology

$$X(t) = kn(t)Z^{a}(t), \tag{18.1}$$

where $X(t)$ is the rate of production of a good, k is the productivity parameter that depends on the units in which labor input and output are measured, and $n(t)$ is employment. The variable $Z(t)$ represents cumulative experience in the production of this good. Dynamics of cumulative experience is described [239, 238] by the differential equation of the first order

$$\frac{dZ(t)}{dt} = n(t)Z^{a}(t). \tag{18.2}$$

The initial value $Z(t_0)$ describes the experience on the date t_0, when production was begun. Usually it is assumed that the initial value $Z(t_0)$ is greater than or equal to one.

The general solution of (18.2) has the form

$$Z(t) = \left(Z^{1-a}(t_0) + (1-a) \int_{t_0}^{t} n(\tau)\, d\tau \right)^{1/(1-a)}, \tag{18.3}$$

where a characterizes the learning rate.

Let us describe the dynamics of production of a single good in the framework of this model.

In the case, when the employment $n(t)$ is constant over time and $n(t) = n_c$, equations (18.1) and (18.3) gives the dynamics of production in the form

$$X(t) = kn_c(Z^{1-a}(t_0) + (1-a)n_c(t-t_0))^{a/(1-a)}. \tag{18.4}$$

Solution (18.4) means that production grows without bound, and the rate of productivity growth declines monotonically from $an_cZ^{a-1}(t_0)$ to zero. For any initial productivity level $Z(t_0) > 1$ and any employment level, the productivity at time $t > t_0$ is an increasing function of the learning rate a.

Assuming that $n_c(t-t_0)$ is large relative to initial experience, the rate of productivity growth over time equal to $t-t_0$ after the start of production is approximately equal to the value

$$R_Z = \frac{a}{(1-a)(t-t_0)}, \tag{18.5}$$

which is given in book [239, p. 86], and paper [238, p. 263].

Let us consider the case, when the dynamics of the employment $n(t)$ has the power-law form

$$n(t) = n_0(t - t_0)^{\beta}. \tag{18.6}$$

Then equation (18.3) gives the expression

$$Z(t) = \left(Z^{1-a}(t_0) + \frac{1-a}{\beta+1} n_0 (t - t_0)^{\beta+1} \right)^{1/(1-a)}. \tag{18.7}$$

Equations (18.1) and (18.7) lead to the fact that production is described by the expression

$$X(t) = k n_0 (t - t_0)^{\beta} \left(Z^{1-a}(t_0) + \frac{1-a}{\beta+1} n_0 (t - t_0)^{\beta+1} \right)^{a/(1-a)}. \tag{18.8}$$

If $n_0(t - t_0)^{\beta+1}$ is large relative to initial experience, the rate of productivity growth is approximately equal to

$$R_X = \frac{a+\beta}{(1-a)(t-t_0)}, \tag{18.9}$$

where $t - t_0$ is the time after the start of production at t_0. If $\beta = 0$, then equation (18.9) gives expression (18.5).

18.3 Generalized Lucas model of learning with memory

The presence of memory in an economic process means that the behavior of the process depends not only on the characteristics at the present time, but also on the history of changes in these characteristics over a finite time interval. The concept of memory for economic processes is described in [450, 448, 399].

Equation (18.2) does not take into account the memory effects and memory fading, since the derivative $dZ(t)/dt$ is determined by behavior of $n(\tau)Z^a(\tau)$ only at the same time instants and does not take into account the history of changes of $n(\tau)Z^a(\tau)$ in the past. To take into account the changes of $n(\tau)Z^a(\tau)$ for $\tau \in (t_0, t)$, in article [391, 389], we proposed to use the equation

$$\frac{dZ(t)}{dt} = \int_{t_0}^{t} M(t-\tau) n(\tau) Z^a(\tau)\, d\tau, \tag{18.10}$$

where $M(t)$ is the memory function. The time variable t is considered as dimensionless parameter, i. e., we change the variable $t_{old} \to t_{new} = t_{old}/t_c$, where t_c is a characteristic time (e. g., 1 year, month or day).

An important property of memory is described by the principle of memory fading, which states that an increase in the time interval leads to a decrease in the corresponding contribution [450]. Let us assume the power-law form of the memory fading. In this case, the memory function can be described by the expression

$$M(t-\tau) = \frac{1}{\Gamma(\mu)}(t-\tau)^{\mu-1}, \tag{18.11}$$

where $\Gamma(\mu)$ is the gamma function. In this case, equation (18.10) can be written in the form

$$\frac{dZ(t)}{dt} = (I^{\mu}_{\mathrm{RL};t_0+} nZ^{a})(t), \tag{18.12}$$

where $I^{\mu}_{\mathrm{RL};t_0+}$ is the Riemann–Liouville fractional integral of the order $\mu > 0$.

The Riemann–Liouville fractional integral is defined [200, pp. 69–70] by the equation

$$(I^{\mu}_{\mathrm{RL};t_0+} Z)(t) = \frac{1}{\Gamma(\mu)} \int_{t_0}^{t} (t-\tau)^{\mu-1} Z(\tau)\, d\tau, \tag{18.13}$$

where $t_0 < t < t_f$. In equation (18.13) the function $Z(t)$ is assumed to satisfy the condition $Z(t) \in L_1(t_0, t_f)$.

We should note that (18.12) with the Riemann–Liouville fractional integral can be considered as an approximation of the equations with generalized memory functions (18.10). In paper [380], using the generalized Taylor series in the Trujillo–Rivero–Bonilla form for the memory function, we proved that the equations with memory functions can be represented through the Riemann–Liouville fractional integrals (and the Caputo fractional derivatives) of noninteger orders.

Let us use the fact that the Caputo fractional derivative is the left inverse operator for the Riemann–Liouville fractional integral (see equation (2.4.32) of Lemma 2.21 in [200, p. 95]), which is represented in the form

$$(D^{\mu}_{C;t_0+}(I^{\mu}_{\mathrm{RL};t_0+} Z))(t) = Z(t), \tag{18.14}$$

if $\mu > 0$, and $Z(t) \in L_{\infty}(t_0, t_f)$ or $Z(t) \in C[t_0, t_f]$. The Caputo fractional derivative is defined by the equation

$$(D^{\mu}_{C;t_0+} Z)(t) = (I^{n-\mu}_{\mathrm{RL};t_0+} D^{n}_{\tau} Z)(t) = \frac{1}{\Gamma(n-\mu)} \int_{t_0}^{t} (t-\tau)^{n-\mu-1} Z^{(n)}(\tau)\, d\tau, \tag{18.15}$$

where $n = [\mu] + 1$ for noninteger valued of μ, and $n = \mu$ for $\mu \in \mathbb{N}$ (see equation (2.4.3) in [200]), $t_0 < t < t_f$, and $Z^{(n)}(\tau) = D^{n}_{\tau} Z(\tau) = d^{n} Z(\tau)/d\tau^{n}$ is the derivative of the integer order n with respect to τ. Operator (18.15) exists if $Z(\tau) \in AC^{n}[t_0, t_f]$ (see Theorem 2.1 in

[200, p. 92]), i. e., the function $Z(\tau)$ has integer-order derivatives up to $(n-1)$-th order, which are continuous functions on the interval $[t_0, t_f]$, and the derivative $Z^{(n)}(\tau)$ is Lebesgue summable on the interval $[t_0, t_f]$.

The action of the Caputo derivative (18.15) on equation (18.12) gives

$$(D^{\mu+1}_{C;t_0+}Z)(t) = n(t)Z^a(t). \tag{18.16}$$

Here, we use that

$$(D^{\mu}_{C;t_0+}Z^{(1)})(t) = (D^{\mu+1}_{C;t_0+}Z)(t), \tag{18.17}$$

where $\mu > 0$. In general, we should take into account the violation of semigroup properties for the repeated action of fractional and integer derivatives. Note that equation (18.17) is based on the property of the Caputo fractional derivative, in which the semigroup property for left action of fractional derivative on the integer derivative holds. Using definition (18.15) of the Caputo fractional derivative, we get

$$\begin{aligned}(D^{\mu}_{C;t_0+}f^{(k)})(t) &= (I^{n-\mu}_{RL;t_0+}D^n_\tau f^{(k)})(t) = (I^{n-\mu}_{RL;t_0+}D^{n+k}_\tau f)(t) = (I^{n+k-(\mu+k)}_{RL;t_0+}D^{n+k}_\tau f)(t)\\ &= (D^{\mu+k}_{C;t_0+}f)(t)\end{aligned} \tag{18.18}$$

for all $n-1 < \mu < n$ and $n, k \in \mathbb{N}$ if $f \in AC^{n+k}[t_0, t_f]$. Note that this property is violated for the Riemann–Liouville fractional derivatives. For the Riemann–Liouville operators, we have the semigroup properties for the left action of integer derivative on the fractional derivative (see Property 2.3 in [200, p. 74]).

Equation (18.16) is fractional differential equation of the order $\mu + 1 > 1$. In order to study a more general case, we can consider the equation of the model in the form

$$(D^{\alpha}_{C;t_0+}Z)(t) = n(t)Z^a(t), \tag{18.19}$$

where $D^{\alpha}_{C;t_0+}$ is the Caputo fractional derivative of the order $\alpha > 0$. For $\alpha = 1$, equation (18.19) gives the standard equation (18.2). For $\alpha = \mu + 1$, where $\mu > 0$, equation (18.19) gives equation (18.16).

Let us consider the power-law form of employment changes

$$n(t) = n_0(t - t_0)^\beta, \tag{18.20}$$

where $\beta \in \mathbb{R}$. In this case, equation (18.19) takes the form

$$(D^{\alpha}_{C;t_0+}Z)(t) = n_0(t - t_0)^\beta Z^a(t). \tag{18.21}$$

Using equations (3.5.47) and (3.5.48) of Section 3.5.3 in [200, p. 209], and Propositions 3.8 and 3.9 in [200, pp. 209–210], we can state that equation (18.21) with $n_0 \neq 0$, $a \neq 1$ and $\alpha > 0$ for wide class of functions has the solution

$$Z(t) = \left(\frac{n_0\Gamma(1-\gamma(\alpha))}{\Gamma(\alpha-\gamma(\alpha)+1)}\right)^{1/(1-a)}(t-t_0)^{\alpha-\gamma(\alpha)}, \tag{18.22}$$

where

$$\gamma(\alpha) = \frac{\beta + a\alpha}{a - 1}, \quad \alpha - \gamma(\alpha) = \frac{\alpha + \beta}{1 - a}. \tag{18.23}$$

The Caputo fractional derivative of the power-law function is given (see Example 3.1 of book [90, p. 49]) by the equation

$$(D^{\alpha}_{C;t_0+}(\tau - t_0)^{\delta})(t) = \frac{\Gamma(\delta + 1)}{\Gamma(\delta - \alpha + 1)}(t - t_0)^{\delta-\alpha}, \tag{18.24}$$

if $\delta > n - 1$, and this derivative is equal to zero, if $\delta = 0, 1, \dots, n - 1$. In all remaining cases ($\delta < n-1$ such that $\delta \neq 0, 1, \dots, n-1$), the integral in the definition of the Caputo fractional derivative is improper and divergent.

Remark 18.1. Using equations (18.24), it is directly verified that expression (18.22) is the explicit solution of equation (18.21) if $\delta = \alpha - \gamma(\alpha) > n - 1$, where $n = [\alpha] + 1$ for noninteger values of $\alpha > 0$. As a result, the conditions $\delta = \alpha - \gamma(\alpha) > n - 1$ for $t > t_0$ should be satisfied for parameters of equation (18.22) with the Caputo derivative instead of the inequality $\alpha - \gamma(\alpha) \geq 0$ that is used in [200].

Remark 18.2. Note that in equations (3.5.48) and (3.5.49) of [200, p. 209], the signs of the parameters $\gamma(\alpha)$ and α in the gamma functions must be opposite. In equation (18.22), these signs are correct.

Taking into account these Remarks and Propositions 3.8 and 3.9 of [200, pp. 209–210], we can formulate the following conditions for the existence of solution (18.22) for equation (18.23).

Statement 18.1. *Nonlinear fractional differential equation* (18.21) *with* $\alpha \in (n-1, n)$, $n \in \mathbb{N}$, *has the solution, which is given by equations* (18.22), (18.23), *if the following conditions are satisfied:*

$$\begin{cases} a \in (0,1), \\ \beta > -\alpha - (n-1)(a-1), \end{cases} \quad or \quad \begin{cases} a > 1, \\ \beta < -\alpha - (n-1)(a-1) \end{cases} \tag{18.25}$$

Here, we take into account that $\delta = \alpha - \gamma(\alpha) > n - 1$ instead of $\delta = \alpha - \gamma(\alpha) > 0$ that is used in Propositions 3.8 and 3.9 of [200].

18.4 Cumulative experience: growth with power-law memory

For solution (18.22), the rate of growth with power-law memory at $t > t_0$ is approximately equal to

$$R_Z(\alpha) = \frac{\alpha - \gamma(\alpha)}{t - t_0} = \frac{\alpha + \beta}{(1 - a)(t - t_0)}, \tag{18.26}$$

where $a \neq 1$. For case $\alpha = 1$, which corresponds to the absence of memory, equation (18.26) can be written in the form

$$R_Z(1) = \frac{\alpha - \gamma(1)}{t - t_0} = \frac{1 + \beta}{(1 - a)(t - t_0)}. \tag{18.27}$$

Let us consider the case when we have growth with memory, i. e., the function $Z(t)$ is increasing for noninteger values of α. In this case, the growth rate with memory, $R_Z(\alpha)$, should be positive for noninteger values of $\alpha \in (n - 1, \alpha)$, $n \in \mathbb{N}$. Then we have the inequality

$$R_Z(\alpha) > 0. \tag{18.28}$$

Solution (18.22), where the parameters are defined by (18.23), exists only if the following conditions are satisfied:

$$\alpha - \gamma(\alpha) = \frac{\alpha + \beta}{1 - a} > n - 1, \quad \gamma(\alpha) < 1. \tag{18.29}$$

Obviously, using equations (18.29) and (18.26), we get that condition (18.28), which has the form

$$R_Z(\alpha) = \frac{\alpha - \gamma(\alpha)}{t - t_0} = \frac{\alpha + \beta}{(1 - a)(t - t_0)} > 0, \tag{18.30}$$

will be satisfied for any solutions if $t > t_0$. Mathematically, this statement follows from the fact that if the inequality $\alpha - \gamma(\alpha) > n - 1$ holds for $n \in \mathbb{N}$, then $\alpha - \gamma(\alpha) > 0$.

As a result, we can formulate the following principle.

Principle 18.1 (Principle of inevitability of growth with memory). *For the model of learning with memory, which is described by equation* (18.21) *with $a \neq 1$, the rate* (18.26) *of growth with power-law memory is positive*

$$R_Z(\alpha) = \frac{\alpha + \beta}{(1 - a)(t - t_0)} > 0 \tag{18.31}$$

for noninteger values of α. As a result, the power-law memory effects leads to the inevitability of growth of the cumulative experience $Z(t)$ in the production of the good.

In addition, there is a lower boundary of the growth rates for processes with memory. The rate of growth with power-law memory for processes, which are described by equation (18.21), has the lower boundary that is defined by the inequality

$$R_Z(\alpha) > \frac{n - 1}{t - t_0}, \tag{18.32}$$

where $n = [\alpha] + 1$ for noninteger values of α.

This statement is based on the fact that solution (18.22) exists only if the condition

$$\alpha - \gamma(\alpha) > n - 1 \tag{18.33}$$

holds. Using definition (18.26) of the rate of growth with memory, we get (18.32).

Remark 18.3. We emphasize that solutions, in which the decline (recession) of $Z(t)$ is realized, does not exist within the framework of the suggested model with memory that is described by equation (18.21). In this model, the processes with memory cannot have negative values of the rate of growth with power-law memory. Note that for processes without memory, which are described by equation (18.26) with $\alpha = 1$, the negative growth rates ($R_Z(1) < 0$) can be realized.

Let us consider the condition under which the growth rate with memory, $R_Z(\alpha)$, is greater than the growth rate without memory, $R_Z(1)$, i. e., we have the following inequality:

$$R_Z(\alpha) > R_Z(1). \tag{18.34}$$

Using equations (18.26) and (18.27), we get the inequality

$$R_Z(\alpha) - R_Z(1) = \frac{\alpha - 1}{(1-a)(t-t_0)} > 0, \tag{18.35}$$

where we assume that $a \in (0,1)$.

Taking into account conditions (18.25) for the existence of solution (18.22) and that $n = 1$ for $\alpha \in (0,1)$, we can formulate the following statement.

Statement 18.2. *The rate $R_Z(\alpha)$ of growth with memory for the cumulative experience $Z(t)$ in the production of the good is greater than the rate $R_Z(1)$ of growth without memory if the conditions*

$$\begin{cases} \alpha \in (0,1), \\ a > 1, \\ \beta < -\alpha, \end{cases} \quad \textit{or} \quad \begin{cases} \alpha \in (n-1,n), \quad n \in \mathbb{N},\ n \neq 1, \\ a \in (0,1), \\ \beta > -\alpha + (n-1)(1-a) \end{cases} \tag{18.36}$$

are satisfied for $t > t_0$.

In the next subsection, we will consider the rate of productivity growth with memory.

18.5 Productivity growth with memory

Substituting (18.20) and (18.22) into (18.1), we get the rate of productivity growth in the form

$$X(t) = kn_0(t-t_0)^{\beta}\left(\frac{n_0\Gamma(1-\gamma(\alpha))}{\Gamma(\alpha-\gamma(\alpha)+1)}\right)^{a/(1-a)}(t-t_0)^{a(\alpha-\gamma(\alpha))}. \tag{18.37}$$

Expression (18.37) means that we have production growth without upper bound, if the following condition is satisfied

$$a(\alpha - \gamma(\alpha)) + \beta > 0. \tag{18.38}$$

Using equation (18.23), condition (18.38) gives the inequality

$$\frac{a(\beta+\alpha)}{1-a}+\beta=\frac{a\alpha+\beta}{1-a}>0. \tag{18.39}$$

As a result, in case (18.39) the rate of productivity growth at $t > t_0$ is approximately equal to

$$R_X(\alpha)=\frac{a\alpha+\beta}{(1-a)(t-t_0)}. \tag{18.40}$$

For $\alpha = 1$, expression (18.40) takes the form

$$R_X(1)=\frac{a+\beta}{(1-a)(t-t_0)}. \tag{18.41}$$

Equation (18.41) describes the rate of growth without memory. Therefore, equation (18.40) with $\alpha = 1$ coincides with expression (18.9).

Using condition $\alpha - \gamma(\alpha) > n - 1$, where $n = [\alpha] + 1$ for noninteger values of α, we can formulate the following principle of inevitability of growth.

Principle 18.2 (Principle of inevitability of growth for process with memory). *For model of learning with memory, which are described by equations* (18.21) *with $a \in (0, 1)$, rate* (18.40) *of growth with power-law memory is positive $R_X(\alpha) > 0$ for noninteger values of α if $\beta \geq -a(n-1)$ and $t > t_0$.*

As a result, the inclusion of power-law memory effects into the Lucas learning model can lead to the inevitability of growth for $Z(t)$ and $X(t)$.

We also can formulate the following principle.

Principle 18.3 (Principle of changing growth rates by memory). *In the case $a \in (0, 1)$, the rate of growth with power-law memory can be greater than the rate of growth without memory ($R_X(\alpha) > R_X(1)$) for $t > t_0$, if the parameter of memory fading α is more than one ($\alpha > 1$) and $\beta \geq -a(n-1)$, where $n = [\alpha] + 1$ for noninteger values of α.*

Considering the suggested model of learning with power-law memory that is described by equation (18.16) of the order $\alpha = \mu + 1$ with $\mu \in (0, 1)$, we get that the rate of growth with power-law memory can be greater than the rate of growth without memory if $\beta \geq 0$.

Let us give a generalization of the numerical examples, which is considered in the book [238, p. 86], and in the paper [239, p. 263]. Using the value $a = 0.2$ and $\beta = 0$, the estimated productivity growth one year ($t - t_0 = 1$) after the start of production is equal to $R_X = 0.25$ for the standard learning model without memory ($\alpha = 1$ or $\mu = 0$). In the proposed model of learning with memory, the estimated productivity growth with memory $R_X(\alpha)$ can take the following values $R_X(0.01) = 0.0025$, $R_X(0.4) = 0.1$, $R_X(1.2) = 0.3$, $R_X(1.99) \approx 0.5$, $R_X(3.6) = 0.9$, $R_X(8.8) = 2.2$. We see that memory effects can almost double the growth rate of the standard model, i. e., it can be equal $1.99R$, if $\beta \approx 0$.

As a result, we can conclude that in learning models, memory effects can significantly change growth rates and we should not neglect the memory in economic models.

18.6 Conclusion

In general, in dynamic models of learning we should take into account the memory effects since economic agents remember the history of changes of parameters and variables that characterize the economic process. We proposed a generalization of the Lucas model of learning-by-doing that takes into account the fading memory. We derive the nonlinear fractional differential equations that describe the learning with memory, and then we obtain exact solutions of these equations. We demonstrate that the estimated productivity growth rate can be significantly changed by memory effects. The proposed model of learning with memory shows that the memory effects can play an important role in economic phenomena and processes including the productivity growth in an economy as a whole.

19 Self-organization of processes with memory

In this chapter, we propose a generalization of the slaving principle and the method of adiabatic elimination of fast relaxing variables for processes with power-law memory. Simple models to describe the distinctive features of the self-organized processes with power-law memory are considered. We have proved that when the memory fading parameters change, the processes with memory can significantly change its rates (by many orders of magnitude) and change the characteristic times both upward and downward. Therefore, the parameters of memory fading can be considered as control parameters that change of characteristic time of processes. Principles, which describe specific features of self-organization of processes with memory, are proposed. This chapter is based on work [387].

19.1 Introduction

Self-organization is a process of formation of ordered spatial or temporal structures that does not require external organizing influences [284], [159, 161, 160, 163], [270, 271, 272]. We can say that a deterministic process usually evolves toward an equilibrium state, which can be described in terms of an attractor in a basin of surrounding states [18]. Self-organization is realized in many social and economic processes [513, 207, 18, 353]. A self-organizing system is a dynamic adaptive system in which the memorization of information (the accumulation of experience) is expressed in a change in the structure of the system. Therefore, we can assume that memory is an important part of self-organization processes. The term "memory" means property that characterizes a dependence of the process state at a given time on the history of state changes in the past. Therefore, to describe the processes with memory we should take into account not only the current state and the previous states that is infinitely close to it, but all the previous states of the considered process on a finite (or infinite) time interval. Mathematically, this means that to describe such a process it is not enough to use the integer-order derivatives of the state variables with respect to time, since these derivatives are determined by the properties of the function in an infinitesimal neighborhood of instant of time. Because of this, equations with integer-order derivatives cannot describe processes with memory, i. e., these derivatives can only be applied to processes in which the history of changes of states is forgotten almost instantly. It can be said that these equations describe only processes with total amnesia or infinitesimally fast amnesia. We should use new mathematical tools to describe the processes with memory that depend on the previous history of their changes. For example, we can use different types of derivatives and integrals of noninteger orders. The theory of equations with integrals and derivatives of noninteger order [335, 202, 308, 200, 90, 164, 165] is a powerful tool to describe processes with fading memory.

https://doi.org/10.1515/9783110627459-019

Possible important generalizations of economic models with memory are connected with consideration economic agents as self-organizing sets of objects with memory.

In works [453, 450, 393, 451], we proved that the rates of growth with memory do not coincide with the growth rates λ without memory, in general. The rate of growth with power-law memory is given by the expression $\lambda_{\text{eff}}(\alpha) = \lambda^{1/\alpha}$, where $\alpha > 0$ is a parameter of memory fading that is described as the order of fractional derivative or integral. The standard rate of growth without memory corresponds to $\alpha = 1$, i. e., $\lambda_{\text{eff}}(1) = \lambda$. Accounting for memory effects can lead to significantly change the growth rate [445, 453, 450, 393, 437, 451]. Note that taking into account the memory can change the rate by several orders of magnitude. Using the principle of domination change, which is proposed in [445, 453, 450, 393, 437, 451], we can state that the inequality $\lambda_1 < \lambda_2$ for models without memory can leads to the inequality

$$\lambda_{1,\text{eff}}(\alpha_1) > \lambda_{2,\text{eff}}(\alpha_2) \tag{19.1}$$

for the models with memory. This allows us to assume that method of adiabatic elimination of fast relaxing variables, which is a special form of the slaving principle, can be generalized for the processes with memory.

In article [387], a generalization of the slaving principle [159, 161, 160, 163] and [162, 505] in the form of the adiabatic method of elimination of fast relaxing variables is suggested for processes with power-law memory. The distinctive features of the self-organized processes with power-law memory are described. We prove that change of memory fading parameters can lead to significant change of the characteristic times by many orders of magnitude. Some principles that describe distinctive features of self-organization with memory are proposed in paper [387].

In this chapter, we formulate principles of self-organization with power-law memory. We propose the following principles: the principle of self-organization arising by memory, the principle of changing the characteristic time by memory, the principle of changing the order parameter by memory, the principle of generation of hierarchical structure by memory. We consider a generalization of the Haken example, where we take into account power-law memory. We prove an adiabatic exclusion of the variable of rapidly damped processes (boss replacement principle) and significant changes of characteristic times by power-law memory (principle of birth of boss and slave by memory). An example of self-organization by memory toward fractional logistic growth is suggested.

19.2 Nonlinear equations of processes with power-law memory

To simplify, we will consider two processes with one-parameter power-law memory, which are described by the fractional differential equations:

$$(D^{\alpha_1}_{C;0+}X)(t) = \lambda_1 X(t) + N_1(X(t), Y(t), t), \tag{19.2}$$

$$(D^{\alpha_2}_{C;0+}Y)(t) = \lambda_2 Y(t) + N_2(X(t), Y(t), t), \tag{19.3}$$

where $D^{\alpha_1}_{C;0+}$ and $D^{\alpha_2}_{C;0+}$ are the left-sided Caputo fractional derivatives of the orders $n-1 < \alpha_1 \leq n$ and $m-1 < \alpha_2 \leq m$, respectively, λ_1 and λ_2 are coefficients (parameters) that are described by real numbers ($\lambda_1, \lambda_2 \in \mathbb{R}$), the functions N_1 and N_2 describe nonlinear part of equations.

The left-sided Caputo fractional derivative $(D^{\alpha}_{C;0+}Y)(t)$ of the order $\alpha \geq 0$ [308, 200, 90] is defined by equation

$$(D^{\alpha}_{C;0+}Y)(t) = \frac{1}{\Gamma(n-\alpha)} \int_0^t \frac{Y^{(n)}(\tau)\, d\tau}{(t-\tau)^{\alpha-n+1}}, \tag{19.4}$$

where $n = [\alpha] + 1$ if α is noninteger and $n = \alpha$ if α is integer, $\Gamma(\alpha)$ is the Euler gamma function, $Y^{(n)}(\tau)$ is the integer-order derivative of $Y(\tau)$ with respect to $\tau \in (0, T)$. Here, it is assumed that $Y(\tau)$ has derivatives up to $(n-1)$-th order that are absolutely continuous for $\tau \in [0, T]$. The Caputo fractional derivative (19.4) with $\alpha = n \in \mathbb{N}$ is the standard integer-order derivative $(D^{n}_{C;0+}Y)(t) = d^n Y(t)/dt^n$.

Note that we use the left-sided Caputo derivative with respect to time for the following reason. One of the main distinguishing features of the Caputo fractional derivatives is that action of these derivatives on a constant function gives zero. Using only the left-sided fractional derivative of noninteger order, we take into account the history of changes of processes in the past. The process at time $\tau = t$ can depend on changes in the state of this process in the past, that is for $\tau < t$. The right-sided Caputo derivatives are defined by integration over $\tau > t$. In order to easily interpret the dimensions of the economic quantities, we can use the time t as a dimensionless variable by changing the variable $t \to t t_d$, where t_d is the unit of time (hour, day, month, year).

We will also consider the nonlinear fractional differential equation

$$(D^{\alpha}_{C;0+}Y)(t) = F(Y(t), t), \tag{19.5}$$

where $n-1 < \alpha \leq n$. Equation (19.5) with $\alpha = n$ takes the form of the differential equation of the integer order

$$\frac{d^n Y(t)}{dt^n} = F(Y(t), t). \tag{19.6}$$

Note that equation (19.6) cannot describe processes with memory since equations are determined by the properties of the differentiable function $Y(t)$ only in an infinitesimal neighborhood of the time moment t.

The processes described by equation (19.5) will be called processes with memory if the alpha takes on noninteger value ($\alpha \notin \mathbb{N}$). The processes, which are described by equation (19.6) will be called processes without memory if the alpha is equal to one or other positive integer value ($\alpha = 1$, $\alpha \in \mathbb{N}$).

Kilbas and Marzan [196, 195] have proven that fractional differential equation (19.5) is equivalent to the nonlinear Volterra integral equation of the second kind. Using Theorem 3.24 of [200, pp. 199–202], we can state that in the space $C^r[0,T]$ of continuously differentiable functions the Cauchy problem for the ordinary fractional differential equation (19.5) is equivalent to the Volterra integral equation

$$Y(t) = \sum_{k=0}^{n-1} \frac{Y^{(k)}(0)}{k!} t^k + \frac{1}{\Gamma(\alpha)} \int_0^t (t-\tau)^{\alpha-1} F(Y(\tau),\tau)\, d\tau, \tag{19.7}$$

if the function $F(Y(\tau),\tau) \in C_\gamma[0,T]$ with $0 \leq \gamma < 1$ and $\gamma \leq \alpha$, the variable $Y(\tau) \in C^r[0,T]$, where $r = n$ for integer values of α, ($\alpha \notin \mathbb{N}$) and $r = n-1$ for noninteger values of α ($\alpha \notin \mathbb{N}$). This statement is a generalization of the equivalence of equation (19.6) and the Volterra integral equation

$$Y(t) = \sum_{k=0}^{n-1} \frac{Y^{(k)}(0)}{k!} t^k + \frac{1}{(n-1)!} \int_0^t (t-\tau)^{n-1} F(Y(\tau),\tau)\, d\tau. \tag{19.8}$$

Note that equation (19.8) cannot describe the processes with memory since this integral equation is equivalent to the differential equation that contains derivatives of integer orders only [382]. To solve the nonlinear Volterra integral equations of the second kind (19.7), which describes processes with power-law memory, it is possible to use the method of successive approximations (e. g., see Section 4 of [482, 483]).

19.3 Slaving principle of self-organization with memory

The slaving principle states a possibility to express the rapidly damped variables, which are called enslaved variables (fast process), through slowly changing variables, which are called order parameters (slow process). The terms fast and slow processes will be used by us in the following sense: the characteristic time of slow process much more than characteristic time of fast process, $\tau_{\text{slow}} \gg \tau_{\text{fast}}$.

The slaving principle plays an important role in self-organization. This principle was first proposed by Hermann Haken [159, 161, 160, 163]. The most formalized formulation of the slaving principle has been given in [160, 163]. The elimination of the rapidly damped variables is an important advantage of the slaving principle, since it allows realize reduction of the degrees of freedom. The slaving principle allows us to remove the stable variables and we thus obtain equations for the order parameter alone.

The principle of adiabatic elimination of the rapidly damped variables is a special form of the slaving principle. The adiabatic approximation allows us to express the rapidly damped variable $Y(t)$ explicitly by the slowly changing variable $X(t)$. This is interpreted as $Y(t)$ is slaved by $X(t)$. The variable $X(t)$ play the role of the order parameter, which slaves all the other variables $Y(t)$ that can be called enslaved variables.

The slaving principle means that processes, which occur slowly over a long period of time, determine an environment (background, or “climate” in economy) for the changes of other (fast) processes. Therefore, we can assume that processes with long-term memory can control other processes with short memory and without memory. In reality, the influence of memory effects on the subordination of some variables to other variables in processes with memory is more complex.

In the article [387], we proposed a generalization of the slaving principle in the form of adiabatic elimination of fast relaxing variables for processes with power-law memory. We consider simple models to describe the distinctive features of the self-organization in processes with power-law memory. We have proved that when the memory fading parameters change, the processes with memory can significantly (by many orders of magnitude) change its rates and the relaxation times both upward and downward. Therefore, the parameters of memory fading can be considered as control parameters, because its change can make fast variables slow, and slow variables can make fast. Moreover, the hierarchy of memory fading parameters can change the hierarchy of characteristic times and relaxation times of processes.

For processes with memory, the slaving principle, according to which the variable of the order parameters change slowly and the enslaved variables react quickly, get a new interpretation. For processes with memory, the possibility to express one variable through another is due to the fact that a rapidly damped process quickly reaches a steady (equilibrium) state. In equilibrium (steady) states of processes with memory, there is no memory, because one variable at a time t is associated with another variable at the same time, $Y(t) = F(X(t), t)$. In this case, changing one variable instantly affects the other variable, without any delay, lag or memory. The equation that expresses the variable $Y(t)$ explicitly by the other variable $X(t)$, which is interpreted as an order parameter, means that the processes $\{Y(t), t \geq 0\}$ follow immediately the processes $\{X(t), t \geq 0\}$. As a result, we can formulate the following principle.

Principle 19.1 (Principle of absence of memory in equilibrium states). *Memory is absent in the steady (equilibrium) states of processes with memory.*

This principle states that equilibrium (steady) states of processes with memory do not have memory.

Let us consider two interrelated processes with one-parameter power-law memory that are described by the fractional differential equations

$$(D^{\alpha_1}_{C;0+}X)(t) = \lambda_1 X(t) + N_1(X(t), Y(t), t), \tag{19.9}$$

$$(D^{\alpha_2}_{C;0+}Y)(t) = \lambda_2 Y(t) + N_2(X(t), Y(t), t), \tag{19.10}$$

where $D^{\alpha_1}_{C;0+}$ and $D^{\alpha_2}_{C;0+}$ are the Caputo fractional derivative of orders $n-1 < \alpha_1 \leq n$ and $m-1 < \alpha_2 \leq m$, λ_1 and λ_2 are coefficients (control parameter) that are described by real numbers, $\lambda_1, \lambda_2 \in \mathbb{R}$.

In the linear approximation, processes that are described by equations (19.9) and (19.10) have two different characteristic times

$$\tau_1(\alpha_1) = |\lambda_1|^{-1/\alpha_1} = \tau_1^{1/\alpha_1}, \tag{19.11}$$

$$\tau_2(\alpha_2) = |\lambda_2|^{-1/\alpha_y} = \tau_2^{1/\alpha_2}. \tag{19.12}$$

Let us assume that process, which is described by equation (19.10), is damped in the absence of the process (19.9). This means that $\lambda_2 < 0$. Let us also assume that $\tau_2(\alpha_2) \ll \tau_1(\alpha_1)$, i. e., equation (19.10) describes the fast process. The evolution begins with the fast process that has the relaxation time $\tau_2(\alpha_2)$. A fast damped process (19.10) with the relaxation time $\tau_2(\alpha_2)$ very quickly leads to a state that is a dynamic equilibrium. The equation of this equilibrium has the form

$$\lambda_2 Y(t) + N_2(X(t), Y(t), t) = 0. \tag{19.13}$$

Using equation (19.13) of dynamic equilibrium, we can express the variable $Y(t)$ of the second process through the variable $X(t)$ of the first process. Let us denote this dependence in the form $Y(t) = Z(X(t), t)$. The substitution of $Y(t) = Z(X(t), t)$ into first equation (19.9) makes it possible to substantially simplify the problem of describing the processes. This substitution gives the equation

$$(D^{\alpha_1}_{C;0+} X)(t) = \lambda_1 X(t) + N_1(X(t), Z(X(t), t), t). \tag{19.14}$$

As a result, the condition $\tau_2(\alpha_2) \ll \tau_1(\alpha_1)$ allows us to use the method of the adiabatic exclusion of the variable, which is based on the allocation of characteristic time scales. As a result, we can state that behavior of the interrelated processes is determined by the evolution of the slow process. Slow processes manage fast processes. That is why the variable $X(t)$ is usually called the order parameter and the variable $Y(t)$ is called the enslaved variable.

Let us note an most important features of the process with memory. In articles [445, 453, 450, 393, 437, 451], it has been proposed the principle of domination change, according to which in intersectoral economic dynamics the effects of fading sectoral memory can change the dominating behavior of economic sectors. For example, according to this principle, the inequality $\lambda_1 > \lambda_2$ of the standard model can leads to the inequality $\lambda_1(\alpha_1) < \lambda_2(\alpha_2)$ of the model with power-law sectoral memory. Therefore, we can state that the inequality $\tau_1 < \tau_2$ of the standard two sectoral model without memory ($\alpha_1 = \alpha_2 = 1$) can leads to the inequality $\tau_1(\alpha_1) > \tau_2(\alpha_2)$ of the model with power-law sectoral memory. For example, $\tau_1 = 2$ and $\tau_2 = 243$ (i. e., $\tau_1 < \tau_2$) can lead to $\tau_1(\alpha_1) = 1024$ and $\tau_2(\alpha_2) = 9$ (that is $\tau_1(\alpha_1) > \tau_2(\alpha_2)$) for the memory fading parameters $\alpha_1 = 0.1$, $\alpha_2 = 2.5$. This fact can be used in the adiabatic approximation for processes similar to those considered in articles [437, 451]. Moreover, as it will be demonstrate below, the inequality $\tau_1 \ll \tau_2$ for processes without memory can lead to the inequality $\tau_1(\alpha_1) \gg \tau_2(\alpha_2)$ for the processes with memory, when other parameters are not changed.

As a result, we can state that memory fading parameters can be used as control parameters, which allows us to turn on and off the self-organization of processes. This feature is an important characteristic property of processes of self-organization with memory. We can say that the parameter of memory fading can determine which variable (process) will be the order parameter.

We can formulate the following principles of self-organization with memory.

Principle 19.2 (Principle of self-organization arising by memory). *The appearance of memory in a process can generate a self-organization. The appearance (or change) of memory in the process can lead to the self-organization of this process, even if all other parameters of the process remain unchanged.*

Principle 19.3 (Principle of changing the characteristic time by memory). *The appearance of memory or memory changes can lead to a significant acceleration or slowdown of the process, that is, the characteristic time of the process under consideration can be substantially increased or decreased by memory.*

Principle 19.4 (Principle of changing the order parameter by memory). *Changing the memory fading parameter, when other parameters of the process are unchanged, can lead to the fact that other variable becomes order parameter. In processes with memory the variables can change roles of order parameter and enslaved parameter: The order parameter $X(t)$ and enslave variable $Y(t)$, which describe processes without memory, can become enslave variable and order parameter respectively, when memory appears in the process.*

Principle 19.4 states that the appearance of memory or the change of the memory fading parameter can change the role of processes (variables), i. e., the order parameters and enslaves variables can change their roles. For Principle 19.4, we can use the words “Who was nothing will become everything.” These are the closing words of the first strophe of the National Anthem of the Soviet Union (USSR) from 1918 to 1944, and then the official anthem of the Communist Party of the Soviet Union from 1944 to 1991.

In practice, a hierarchical structure can arise from an unstructured group of processes due to the appearance of memory with different fading parameters. This property allows us to formulate the principle of generation of hierarchical structure by memory. Then the proposed principle will be illustrated by numerical example.

Principle 19.5 (Principle of generation of hierarchical structure by memory). *For memoryless processes with the same characteristic times (τ_k, where $k = 1, 2, \ldots, n$), the hierarchy of memory fading parameters can lead to the appearance of a hierarchy of characteristic time for relaxation and amplification processes. The hierarchy of memory fading* *parameters α_k, where $k = 1, 2, \dots, n$, that can be ordered so that*

$$\alpha_1 > \alpha_2 > \alpha_3 > \cdots > \alpha_n, \tag{19.15}$$

can lead to a hierarchy of characteristic times $\tau_k(\alpha_k)$, where $k = 1, 2, \dots, n$ that are ordered so that

$$\tau_1(\alpha_1) \ll \tau_2(\alpha_2) \ll \tau_3(\alpha_3) \ll \cdots \ll \tau_n(\alpha_n) \tag{19.16}$$

for amplification and relaxation processes.

In economics, a hierarchical structure of processes may arise, in which characteristic times $\tau_k(\alpha_k)$, where $k = 1, 2, \dots, n$ can be ordered so that

$$\tau_1(\alpha_1) \ll \tau_2(\alpha_2) \ll \tau_3(\alpha_3) \ll \cdots \ll \tau_n(\alpha_n), \tag{19.17}$$

when these processes without memory have the equal characteristic times

$$\tau_1(1) = \tau_2(1) = \tau_3(1) = \cdots = \tau_n(1). \tag{19.18}$$

This means that this hierarchy is caused by memory only.

For the hierarchical structure, one can apply the adiabatic elimination procedure first to the variables $Y_1(t)$, connected with $\tau_1(\alpha_1)$, leaving us with the other variables. Then we can apply this methods to the variables $Y_2(t)$, connected with $\tau_2(\alpha_2)$ and so on.

Example 19.1. Let us consider the processes without memory, for which the rates were the same and equal to $|\lambda| = 0.1$ for all processes ($|\lambda_k| = |\lambda| = 0.1$ for all $k = 1, 2, 3, 4$). Suppose that these processes have such power-law memory, for which the fading parameters of all processes are different. Let us consider the hierarchy of memory fading parameters α_k, where $k = 1, 2, 3, 4$, that can be ordered so that

$$1 > \alpha_1 = 0.5 > \alpha_2 = 0.2 > \alpha_3 = 0.1 > \alpha_4 = 0.05. \tag{19.19}$$

The characteristic times of amplification and relaxation processes ($\lambda > 0$ and $\lambda < 0$ respectively) with power-law memory are given in the form

$$\tau_{\mathrm{rel}}(\alpha) = \tau_{\mathrm{amp}}(\alpha) = |\lambda|^{1/\alpha} \quad \text{for } 0 < \alpha < 1. \tag{19.20}$$

Therefore, for power-law memory we get a hierarchy of the characteristic times $\tau_k(\alpha_k)$ in the form

$$\tau_1(\alpha_1) = 100 \ll \tau_2(\alpha_2) = 10^5 \ll \tau_3(\alpha_3) = 10^{10} \ll \tau_4(\alpha_4) = 10^{20}. \tag{19.21}$$

It should be emphasized that characteristic times of these processes in the absence of memory are equal to $\tau = 10$ for all processes (for $k = 1, 2, 3, 4$).

19.4 Variable exception without using the adiabatic method

In Haken's books [159, 161, 160, 163], an example of system that is described by two nonlinear differential equations with derivatives of first order is considered. We can consider a generalization of this example for the case of power-law memory by using the fractional derivatives of the orders $\alpha_1 > 0$ and $\alpha_2 > 0$ instead of the derivatives of first order. Let us consider two interrelated processes with memory that are described by the equations

$$(D^{\alpha_1}_{C;0+}X)(t) = \lambda_1 X(t) + aX(t)Y(t), \tag{19.22}$$

$$(D^{\alpha_2}_{C;0+}Y)(t) = \lambda_2 Y(t) + bX^2(t), \tag{19.23}$$

where the parameters of the first process (λ_1, a) and the second process (λ_2, b) are real numbers. The parameters α_1 and α_2, which describe memory fading in these processes, take positive values.

First, we consider the possibility of excluding one of the variables without using the adiabatic method. In system of two nonlinear fractional differential equations (19.22) and (19.23), we can exclude the second variable without any approximations. Using Theorem 5.15 of [200, p. 323], where $x = t$, $y = Y$, $f(t) = bX^2(t)$ we can get the solution of equation (19.23) in the form

$$Y(t) = \sum_{k=0}^{n-1} Y^{(k)}(0) t^k E_{\alpha_2,k+1}[\lambda_2 t^{\alpha_2}] + b\int_0^t (t-\tau)^{\alpha_2-1} E_{\alpha_2,\alpha_2}[\lambda_2 (t-\tau)^{\alpha_2}] X^2(\tau)\, d\tau, \tag{19.24}$$

where $n-1 < \alpha \leq n$, $Y^{(k)}(0)$ is derivative of the integer order $k \geq 0$ of the function $Y(t)$ at $t = 0$, and $E_{\alpha,\beta}[z]$ is the two-parameter Mittag–Leffler function.

Let us give some special cases of solution (19.24) of equation (19.23) for different values of the memory fading parameter α.

For $0 < \alpha_2 \leq 1$, $(n = 1)$, solution (19.24) of equation (19.23) has the form

$$Y(t) = Y(0) E_{\alpha_2,1}[\lambda_2 t^{\alpha_2}] + b\int_0^t (t-\tau)^{\alpha_2-1} E_{\alpha_2,\alpha_2}[\lambda_2 (t-\tau)^{\alpha_2}] X^2(\tau)\, d\tau. \tag{19.25}$$

For $\alpha_2 = 1$, solution (19.24) takes the form

$$Y(t) = Y(0)\exp(\lambda_2 t) + b\int_0^t \exp(\lambda_2 (t-\tau)) X^2(\tau)\, d\tau. \tag{19.26}$$

For $1 < \alpha_2 \leq 2$ $(n = 2)$, expression (19.24) takes the form

$$Y(t) = Y(0)E_{\alpha_2,1}[\lambda_2 t^{\alpha_2}] + Y^{(1)}(0)tE_{\alpha_2,1}[\lambda_2 t^{\alpha_2}] + b\int_0^t (t-\tau)^{\alpha_2-1}E_{\alpha_2,\alpha_2}[\lambda_2(t-\tau)^{\alpha_2}]X^2(\tau)\,d\tau. \quad (19.27)$$

As a result, in equation (19.22) the variable $Y(t)$ is expressed in terms of the variable $X(t)$. Equation (19.24) can be interpreted as a fact that behavior of the variable $Y(t)$ is determined by the changes of the variable $X(\tau)$ on the interval $(0, t)$. For noninteger values of parameter α_2, equation (19.27) describes processes with memory. Note that equation (19.24) with $\alpha_2 = n \in \mathbb{N}$ cannot describe processes with memory [382].

Then eliminating the variable $Y(t)$ from equation (19.22), we get

$$(D^{\alpha_1}_{C;0+}X)(t) = \lambda_1 X(t) + aX(t)\sum_{k=0}^{n-1} Y^{(k)}(0)t^k E_{\alpha_2,k+1}[\lambda_2 t^{\alpha_2}] + abX(t)\int_0^t (t-\tau)^{\alpha_2-1}E_{\alpha_2,\alpha_2}[\lambda_2(t-\tau)^{\alpha_2}]X^2(\tau)\,d\tau. \quad (19.28)$$

However, the exact analytical solution of nonlinear fractional differential equation (19.28) is currently unknown, and, as a consequence, we cannot describe dynamics of the process that is given by equations (19.22) and (19.23).

19.5 Adiabatic exclusion of variable for rapid damping

Let us describe the method of the adiabatic exclusion of the variable of rapidly damped processes with memory. This method that allows us to simplify the considered problem, which is described by equations (19.22) and (19.23). We will assume that the process $Y(t)$ is rapidly damped process, i. e., $\lambda_2 < 0$ and $\tau_2(\alpha_2) \ll \tau_1(\alpha_1)$. The dynamic equilibrium of the process $Y(t)$ is described by the equation $\lambda_2 Y(t) + bX^2(t) = 0$ that allows us express the variable $Y(t)$ in the form

$$Y(t) = \frac{b}{|\lambda_2|}X^2(t). \quad (19.29)$$

Equation (19.29) means that changing one variable $X(t)$ instantly affects the other variable $Y(t)$, without any delay or memory. This dependence, which is described by equation (19.29), shows the absence of memory.

The adiabatic approximation allows us to express $Y(t)$ explicitly through the variable $X(t)$. This fact is interpreted that $Y(t)$ is slaved by $X(t)$. The adiabatic exclusion of the variable $Y(t)$ of rapidly damped process allows us to simplify our problem. The substitution expression (19.29) into equation (19.22) gives the fractional differential equation for $X(t)$ in the form

$$(D^{\alpha_1}_{C;0+}X)(t) = \lambda_1 X(t) + \frac{ab}{|\lambda_2|} X^3(t). \tag{19.30}$$

As a result, instead of two equations for $X(t)$ and $Y(t)$, we can consider only a single equation for $X(t)$, which can contain derivative of integer order if α_1 is integer and α_2 is noninteger. Then we may express the enslaved variables $Y(t)$ according to the slaving principle. In this case, the variable $X(t)$ is called the order parameter. Equation (19.30) describes the behavior of the order parameter $X(t)$.

The stationary solutions of equation (19.30) for order parameter $X(t)$, which describe the steady states of process, are defined by the equation

$$\lambda_1 X(t) + \frac{ab}{|\lambda_2|} X^3(t) = 0. \tag{19.31}$$

Equation (19.31) shows that two completely different kinds of solutions (two kind of steady states) occur depending on the sign of the product $(ab\lambda_1)$ of control parameters. If the product of these parameters satisfies the condition $(\lambda_1 ab) > 0$, we have only trivial stationary solution $X(t) = 0$. For the case $(ab\lambda_1) < 0$, the stationary solutions have the form

$$X(t) = 0, \quad X(t) = \pm \left| \frac{\lambda_1 \lambda_2}{ab} \right|^{1/2}. \tag{19.32}$$

The question arises as to which new states the process will pass into. If the process is initially at the steady state $X(0) = 0$, then it remains in this state for an infinitely long time and self-organization cannot be realized. For this reason, we should have a certain initial pushes, which will transfer the process to another steady state. This can be achieved by fluctuations and random external influence (random external force).

There is a relationship between the loss of linear stability, the appearance of order parameters and the validity of the slaving principle. We can consider the case, when the control parameter λ_1 is changed. In this case, the process can lose linear stability. For example, changing the sign of a parameter λ_1 means that it becomes very small. Then corresponding characteristic time

$$\tau_1(\alpha_1) = |\lambda_1|^{-1/\alpha_1} \tag{19.33}$$

becomes very large, and the slaving principle can be applied. In this case, the behavior of a process can be governed only by the order parameters at points, where structural changes (bifurcation) occur. However, equation (19.33) shows that the parameter $\alpha_1 > 0$, which characterizes the memory fading, can change this situation.

We can consider the case, when the fading parameter α_1 or α_2 is changed. In this case, the inequalities for characteristic times $\tau_2(\alpha_2)$, $\tau_1(\alpha_1)$ can change the inequality sign on the opposite. For example, the inequality $\tau_2(\alpha_2) \ll \tau_1(\alpha_1)$ can take the form $\tau_2(\alpha'_2) \gg \tau_1(\alpha'_1)$ for new parameters of memory fading. In this case, another variable can become an order parameter. We can state that enslaved processes, in which memory arose, may become the order parameter. It is possible say in words from the National Anthem of the Soviet Union "Who was nothing will become everything." In other words, changing the value of the parameter of memory fading can lead to the replacement of the order parameter. We can say "A slave with memory can become a boss, and a boss becomes a slave."

Principle 19.6 (Boss replacement principle). *A slave (enslaved) process with memory can become a boss (order parameter), and a boss becomes a slave. The inequality $\tau_2(\alpha_2) \ll \tau_1(\alpha_1)$ for characteristic times can take the form $\tau_2(\alpha'_2) \gg \tau_1(\alpha'_1)$ when the memory fading parameters are changed.*

This principle is a new formulation of the principle of changing the order parameter by memory.

19.6 Significant changes of characteristic times by memory

Let us consider examples of two interrelated processes that are described by equations (19.22) and (19.23) in the form

$$(D^{\alpha_1}_{C;0+}X)(t) = \lambda_1 X(t) + aX(t)Y(t), \tag{19.34}$$

$$(D^{\alpha_2}_{C;0+}Y)(t) = \lambda_2 Y(t) + bX^2(t), \tag{19.35}$$

where $\lambda_1 > 0$, $\lambda_2 < 0$ and we assume that values of the parameters λ_1 and λ_2 are equal without sign, i. e., $|\lambda_1| = |\lambda_2|$. In this case, the characteristic times of these processes without memory ($\alpha_1 = \alpha_2 = 1$) are equal $\tau_1 = \tau_2$, where $\tau_k = \tau_k(1)$ with $k = 1, 2$.

Example 19.2. Let us consider equations (19.34) and (19.35) for the case $|\lambda_1| = |\lambda_1| = 10$. If the memory is absent ($\alpha_1 = \alpha_2 = 1$), then the characteristic times of these processes without memory are the same $\tau_1 = \tau_2 = 0.1$. Let us assume that process, which is described by the variable $X(t)$, has no memory ($\alpha_1 = 1$), and the variable $Y(t)$ describes a process with memory, such that the fading parameter is equal to $\alpha_2 = 0.1$. In this case, equations (19.34) and (19.35) have form

$$X^{(1)}(t) = 10X(t) - aX(t)Y(t), \tag{19.36}$$

$$(D^{\alpha_2}_{C;0+}Y)(t) = -10Y(t) + bX^2(t), \tag{19.37}$$

where $\alpha_2 = 0.1$. In this case, the characteristic times of these processes are the following:

$$\tau_1(\alpha_1) = \tau_1(1) = \tau_1 = 0.1 \gg \tau_2(\alpha_2) = \tau_2(0.1) = 10^{-10}. \tag{19.38}$$

We can have the same inequalities even if both processes have a memory with different fading parameters. For example, $\tau_1(0.9) \gg \tau_2(0.1)$ or $\tau_1(1.2) \gg \tau_2(0.1)$ for $\tau_1 = \tau_2$.

As a result, we see that in this example, the memory effects with the fading parameter $0 < \alpha \ll 1$ significantly decrease the characteristic time of relaxation to the equilibrium state if the parameters satisfies the inequality $|\lambda_1| = |\lambda_2| > 1$.

Example 19.3. Let us consider equations (19.34) and (19.35) for the case $|\lambda_1| = |\lambda_1| = 0.1$. If the memory is absent ($\alpha_1 = \alpha_2 = 1$), then the characteristic times of these processes without memory are the same $\tau_1 = \tau_2 = 10$. Let us assume that the variable $X(t)$ describes a process with memory such that fading parameter is equal to $\alpha_1 = 0.1$, and the variable $Y(t)$ describes processes without memory ($\alpha_2 = 1$). In this case, equations (19.34) and (19.35) have form

$$(D^{\alpha_1}_{C;0+}X)(t) = 0.1X(t) - aX(t)Y(t), \tag{19.39}$$

$$Y^{(1)}(t) = -0.1Y(t) + bX^2(t), \tag{19.40}$$

where $\alpha_1 = 0.1$. In this case, we have the characteristic times

$$\tau_1(\alpha_1) = \tau_1(1) = \tau_1 = 10 \ll \tau_2(\alpha_2) = \tau_2(0.1) = 10^{10}. \tag{19.41}$$

Note that we can have the same type of inequalities ($\tau_1(\alpha_1) \ll \tau_2(\alpha_2)$ for $\tau_1 = \tau_2$) even if both processes have a memory with different fading parameters. For example, $\tau_1(0.9) \ll \tau_2(0.1)$ or $\tau_1(1.2) \ll \tau_2(0.1)$.

As a result, in this example, we see that the memory effects with the fading parameter $0 < \alpha \ll 1$ lead to a significant increase in the characteristic time of process relaxation to the equilibrium state if $|\lambda_1| = |\lambda_2| < 1$.

In the suggested examples, we started with memoryless processes, in which there were no fast and slow variables, since their relaxation times were equal to the same values. Then we showed that taking into account memory effects can lead to the appearance of fast and slow variables (processes). We proved that memory effects with the fading parameters $0 < \alpha \ll 1$ can lead to a significant increase and decrease of the characteristic time, during which the process comes to an equilibrium state. We showed that if the relaxation times of two processes are equal in the absence of memory, then the presence of memory in one of the processes can lead to the inequality $\tau_1(\alpha_1) \ll \tau_2(\alpha_2)$ or $\tau_1(\alpha_1) \gg \tau_2(\alpha_2)$.

As a result, we can formulate the following principle.

Principle 19.7 (Principle of birth of boss and slave by memory). *Two identical processes with amnesia (without memory) at the appearance of memory can be divided into slow and fast processes (the order parameter and enslaved variable).*

This principle states that relaxation times can disappear when memory appears. The appearance of memory can lead to the birth of bosses and enslaved subordinates.

As a result, we can state that memory fading parameter can be considered as a control parameter, the change of which can enable and disable the self-organization processes.

19.7 Self-organization by memory toward logistic growth

Let us consider economic model of two interacting sectors (or two economic processes) with memory that is described by two fractional differential equations:

$$\begin{cases} (D^{\alpha_1}_{C;0+}X)(t) = X(t)F_1(X(t), Y(t)) \\ (D^{\alpha_2}_{C;0+}Y)(t) = Y(t)F_2(X(t), Y(t)). \end{cases} \quad (19.42)$$

For $\alpha_1 = \alpha_2 = 1$, the functions F_1 and F_2 are interpreted as the respective growth rates of these two sectors. In the case $\alpha_1 = \alpha_2 = 1$, the general model (19.42) is often called Kolmogorov's predator–prey model. Let us use the expansion of the functions $F_k(X(t), Y(t))$ of two variables in the Taylor series. For simplicity, we will consider only the linear approximation

$$F_k(X, Y) = \lambda_k + a_k X + b_k Y, \quad (19.43)$$

where

$$a_k = \left(\frac{F_k(X, Y)}{\partial X}\right)_{X=Y=0}, \quad b_k = \left(\frac{F_k(X, Y)}{\partial Y}\right)_{X=Y=0}. \quad (19.44)$$

For convenience of further consideration, we relabel constants a_k and b_k such that

$$a_1 = d, \quad b_1 = -a, \quad a_2 = -b, \quad b_2 = c. \quad (19.45)$$

In this case, the fractional differential equations of the system (19.42) take the form

$$(D^{\alpha_1}_{C;0+}X)(t) = \lambda_1 X(t) - aX(t)Y(t) + dX^2(t), \quad (19.46)$$

$$(D^{\alpha_2}_{C;0+}Y)(t) = \lambda_2 Y(t) - bX(t)Y(t) + cY^2(t). \quad (19.47)$$

Remark 19.1. If $d = 0$, $c = 0$ and $\lambda_1 > 0$, $\lambda_2 < 0$, $a > 0$, $b < 0$ are real parameters, then equations (19.46), (19.47) take the form of the fractional Lotka–Volterra equations [85] that can also be called as the fractional predator–prey equations.

We will assume that process in one of the sectors is fast and linearly stable, and the other is slow and linearly unstable.

Let us give examples of the dynamics of such processes.

Example 19.4. Let us assume that the first process, which is described by the variable $X(t)$, is a slow process and it is linearly unstable, i. e., $\lambda_1 > 0$ and $\tau_1(\alpha_1) \gg \tau_2(\alpha_2)$. The second process, which is described by the variable $Y(t)$, will be assumed to be a fast damped process, i. e., $\lambda_2 < 0$ and $\tau_2(\alpha_2) \ll \tau_1(\alpha_1)$. Let us use the method of the adiabatic exclusion of the variable of rapidly damped process. The dynamic equilibrium of the process $Y(t)$ is described by the equation

$$\lambda_2 Y(t) - bX(t)Y(t) + cY^2(t) = 0, \tag{19.48}$$

which allows us to obtain the following expression:

$$Y(t) = \frac{b}{c}X(t) - \frac{\lambda_2}{c}. \tag{19.49}$$

Substitution of expression (19.49) into equation (19.46) gives

$$(D^{\alpha_1}_{C;0+}X)(t) = \left(\lambda_1 + \frac{a\lambda_2}{c}\right)X(t) - \left(\frac{ab}{c} - d\right)X^2(t). \tag{19.50}$$

Equation (19.50) can be represented as the fractional logistic equation (e. g., see [444] and references therein) in the form

$$(D^{\alpha_1}_{C;0+}X)(t) = r_x X(t)\left(1 - \frac{1}{K_x}X(t)\right), \tag{19.51}$$

where

$$r_x = \lambda_1 + \frac{a\lambda_2}{c} = \frac{c\lambda_1 + a\lambda_2}{c}, \tag{19.52}$$

$$K_x = \left(\lambda_1 + \frac{a\lambda_2}{c}\right)\left(\frac{ab}{c} - d\right)^{-1} = \frac{a\lambda_2 + c\lambda_1}{ab - cd}. \tag{19.53}$$

The relaxation time of the process, which is described by equation (19.51), is defined by the equation

$$\tau_r(\alpha_1) = r_x^{-1/\alpha_1} = \left(\lambda_1 + \frac{a\lambda_2}{c}\right)^{-1/\alpha_1}. \tag{19.54}$$

Note that equation (19.51) can be represented as

$$(D^{\alpha_1}_{C;0+}Z_x)(t) = r_x Z_x(t)(1 - Z_x(t)), \tag{19.55}$$

where

$$Z_x(t) = \frac{1}{K_x}X(t) = \left(\frac{ab}{c} - d\right)\left(\lambda_1 + \frac{a\lambda_2}{c}\right)^{-1} X(t) = \frac{ab - cd}{a\lambda_2 + c\lambda_1}X(t). \tag{19.56}$$

Fractional logistic equation (19.51) has two steady states $X(t) = 0$ and $X(t) = K_x$. This equation describes the evolution of the process with memory from one stationary state to another.

Example 19.5. Let we consider the case, when the first process $X(t)$ is fast damped process, i. e., $\lambda_2 < 0$ and $\tau_1(\alpha_1) \ll \tau_2(\alpha_2)$ the second process $Y(t)$ is a slow process that is linearly unstable, i. e., $\lambda_2 > 0$ and $\tau_2(\alpha_2) \gg \tau_1(\alpha_1)$.

Example 19.5 differs from Example 19.4 only in the values of the memory fading parameters (α_1, α_2) and in the opposite sign of the parameters λ_1, λ_2. Using the method of the adiabatic exclusion of the variable of rapidly damped processes $Y(t)$, we get the fractional logistic equation

$$(D^{\alpha_2}_{C;0+}Y)(t) = r_y Y(t)\left(1 - \frac{1}{K_y} Y(t)\right), \tag{19.57}$$

where

$$r_y = \lambda_2 + \frac{b\lambda_1}{d} = \frac{b\lambda_1 + d\lambda_2}{d}, \tag{19.58}$$

$$K_y(t) = \left(\lambda_1 + \frac{b\lambda_2}{d}\right)\left(\frac{ab}{d} - c\right)^{-1} = \frac{d\lambda_1 + b\lambda_2}{ab - cd}. \tag{19.59}$$

As a result, we get that answer to the question about which of the processes (variables) will be the order parameter: The result is determined by values of memory fading parameters of considered processes.

If the parameter of memory fading of the fractional logistic equation is equal to one ($\alpha_1 = 1$ for equation of $X(t)$ or $\alpha_2 = 1$ for equation of $Y(t)$), then memory is absent in this process. Then we can write the solution of equation (19.51) with $\alpha_1 = 1$ in the form

$$X(t) = \frac{K_x X(0)\exp(r_x t)}{K_x + X(0)(\exp(r_x t) - 1)}. \tag{19.60}$$

Function (19.60) is called the logistic function.

It is clear from equation (19.60) that if the process $X(t)$ is initially in steady state $X(0) = 0$, then it remains in this state for an infinitely long time ($X(t) = 0$ for all $t > 0$) and self-organization cannot be realized. For this reason, to get a self-organization, we should have a certain initial perturbation ($X(t) \neq 0$). This perturbation can arise due to fluctuations and random external influence (random force) that will allow to start changes leading to another steady state. The final state is given by the equation

$$\lim_{t\to\infty} X(t) = \lim_{t\to\infty} \frac{K_x X(0)\exp(r_x t)}{K_x + X(0)(\exp(r_x t) - 1)} = K_x. \tag{19.61}$$

As a result, if the process is not in steady state ($X(t) \neq 0$), then the fractional logistic equation (19.51) describes dynamics of the order parameter $X(t)$ to new equilibrium state ($X(t) \to K_x$).

Example 19.6. Let us consider numerical example of this dynamics (transfer to new steady state) by using the differential equations (19.46) and (19.47) in the form

$$X^{(1)}(t) = 2X(t) - 15X(t)Y(t) + 15X^2(t), \tag{19.62}$$

$$(D^{\alpha_2}_{C;0+}Y)(t) = -10Y(t) - 20X(t)Y(t) + 10Y^2(t), \tag{19.63}$$

where $\alpha_1 = 1$, $\alpha_2 = 0.2$. Then characteristic times take the following values:

$$\tau_1 = 0.5, \quad \tau_1(1) = 0.5, \tag{19.64}$$

$$\tau_2 = 0.2, \quad \tau_2(\alpha_2) = 0.00001. \tag{19.65}$$

The parameters of the logistic equation is equal to $r_x = 0.5$, $K_x = 5$. Note that relaxation time (19.54) of the order parameter is equal to $\tau_r(\alpha_1) = 2$ and $\tau_r(\alpha_1) \gg \tau_2(\alpha_2) = 0.00001$ for $\tau_r \approx \tau_2(1)$. The function $X(t)$, which is described by expression (19.60), has the form

$$X(t) = \frac{5X(0)\exp(0.5t)}{5 + X(0)(\exp(0.5t) - 1)}. \tag{19.66}$$

Fractional logistic equation were investigated by using numerical simulation in different papers. However, still not found an exact expression for the solution of the fractional logistic equation. Recently, West has published paper [503], in which he proposed an expression of the solution for the fractional logistic equation with $0 < \alpha < 1$ in the form

$$Z(t) = \sum_{k=0}^{\infty} \left(\frac{Z(0) - 1}{Z(0)} \right) E_\alpha[-r_x(\alpha)t^\alpha], \tag{19.67}$$

where $E_\alpha[z]$ is the Mittag–Leffler function [200, p. 42], [143], which is defined by the equation

$$E_\alpha[z] = \sum_{k=0}^{\infty} \frac{z^k}{\Gamma(\alpha k + 1)}. \tag{19.68}$$

The Mittag–Leffler function $E_\alpha[z]$ is a generalization of the exponential function $\exp(z)$, since $E_1[z] = \exp(z)$. As it has been proved by Area, Losada and Nieto in [17], that function (19.67) is not the solution to (19.55). The main reason is the violation the semi-group property by the Mittag–Leffler function, i. e., we have the inequality

$$E_\alpha[\lambda(t + s)^\alpha] \neq E_\alpha[\lambda t^\alpha] E_\alpha[\lambda s^\alpha] \tag{19.69}$$

for $0 < \alpha < 1$ and real constant λ. In [17], it has been proved that function (19.67), which is recently proposed by West, is not an exact solution of the fractional logistic equation. Note that violation of the semigroup property (19.69) is an important characteristic property of the processes with memory [386].

Since an explicit solution to a fractional logistic equation is not known at present time, approximate solutions and computer simulation should be used to solve it. A review of various methods for approximate solving the fractional logistic equation is given in paper [333]. To solve fractional logistic equation as a nonlinear fractional differential equation, we can also use the background field and mean field methods that are proposed in paper [374].

As a result, we can state that it is the values of the memory fading parameters for the considered processes that determine which of the variables $X(t)$ or $Y(t)$ will be the order parameter.

19.8 Conclusion

In this chapter, we describe generalizations of the slaving principle and the method of adiabatic elimination of fast relaxing variables in the processes with power-law memory, which are proposed in [387]. We described main distinctive features of the self-organized processes with memory. It is proved that the change of the memory fading parameters can lead to significant change on many orders of magnitude of the characteristic times of processes with memory. A slave (enslaved) process with memory can become a boss (order parameter), and a boss becomes a slave. The inequality $\tau_2(\alpha_2) \ll \tau_1(\alpha_1)$ for characteristic times can take the form $\tau_2(\alpha_2') \gg \tau_1(\alpha_1')$ when the memory fading parameters are changed ($\alpha_1 \to \alpha_1'$, $\alpha_2 \to \alpha_2'$). Two identical processes without memory can be divided into slow and fast processes (the order parameter and enslaved variable) when memory appears. The appearance (or change) of memory in the process (or system) can lead to the self-organization, even if all other parameters of the process and system remain unchanged. The appearance of memory or memory changes can lead to a significant acceleration or slowdown of the process, that is, the characteristic time of the process can be substantially increased or decreased by memory. Change the memory fading parameter, when other parameters do not change, can lead to the fact that other variable becomes order parameter. In processes with memory, the variables can change roles of order parameter and enslaved parameter. The appearance of memory or the change of the memory fading parameter can change the role of processes (variables), i. e., the order parameters and enslaves variables can change their roles. In memoryless processes with the same relaxation times, the emergence of a hierarchy of memory fading parameters can lead to a hierarchy of characteristic times for relaxation and amplification processes. The hierarchy of memory fading parameters α_k, where $k = 1, 2, \ldots, n$, which can be ordered so that $\alpha_1 > \alpha_2 > \alpha_3 > \cdots > \alpha_n$, can lead to a hierarchy of characteristic times $\tau_k(\alpha_k)$, where $k = 1, 2, \ldots, n$ in the form

$$\tau_1(\alpha_1) \ll \tau_2(\alpha_2) \ll \tau_3(\alpha_3) \ll \cdots \ll \tau_n(\alpha_n) \tag{19.70}$$

for amplification and relaxation processes at the same values of $\tau_k = \tau_k(1)$.

An important direction in the development of theory of self-organization for economic processes with memory is the construction of models, in which the economic agents are represented as a self-organizing set of interacting objects with memory. The proposed approach an principles can be used to generalize standard models of self-organization in economy, which are described in [513, 207, 353], to take into account the memory effects.

Part V: **Advanced models: distributed lag and memory**

20 Multipliers and accelerators with lag and memory

In this chapter, we proposed multipliers and accelerators that take into account simultaneous action of the continuously distributed lag and fading memory. These concepts and their application were proposed in articles [400] and [404, 402, 407, 405]. This chapter is based on these articles.

20.1 Introduction

The continuously distributed lag has been considered in economics starting with the works of Michal A. Kalecki [189] and Alban W. H. Phillips [305, 306]. The continuous uniform distribution of delay time was considered in dynamic models of business cycles by Michal A. Kalecki in 1935 [189] (see also Section 8.4 of [10, pp. 251–254]). The concepts of the accelerator and multiplier with exponentially distributed lag were proposed by Alban W. H. Phillips [305, 306] (see also Sections 3.4, 3.5 and 8.7 in [10]). The exponential distribution of delay time was proposed in the economic growth models by Alban W. H. Phillips [305, 306] in 1954. Then operators with continuously distributed lag were considered by Roy G. D. Allen in the book [8] in 1956 (see also [9, 10]). In the general case, the distribution of delay time can be described by various distributions of delay time [400], not just exponential distributions.

The concepts of the accelerator and multiplier with exponentially distributed lag were proposed in articles [412, 397] (see also [441, 442, 431]). Then a generalization of these concepts for economic processes with simultaneous action of the distributed lag and fading memory effects were proposed in articles [400] and [404, 402, 407, 405], in which economic models with memory and lag are suggested.

The time delay (lag) is caused by finite speeds of processes, i. e., the change of one variable does not lead to instant changes of another variable. Therefore, the distributed lag (time delay) cannot be interpreted as a memory in processes. For example, the change in electromagnetic field at the point A of observation is delayed with respect to the change in the sources of the field located at the point B at the time $t = |AB|/c$, where c is the speed of propagation of disturbances and $|AB|$ is the distance between points A and B. It is known that the processes of propagation of the electromagnetic field in a vacuum do not mean the presence of memory in this process. Similarly, with a good organization of work, delays in the delivery of goods from the warehouse cannot be explained by forgetting, or the fading of memory. This lag can be explained by the finiteness of the speed of some processes, for example, the speed of transport.

In this chapter, we consider economic concepts and mathematical tools that can describe the simultaneous action of two effects: memory with power-law fading and lag with continuous distribution of delay time. The mathematical tools are fractional differential and integral operators with distributed lag and memory proposed in [400].

https://doi.org/10.1515/9783110627459-020

These operators are used to formulate of economic models that are proposed in [403, 405, 404, 402, 407]. The models with fading memory and distributed lag will be described in the following chapters. In this chapter, we consider only concepts of the accelerator and multiplier with fading memory and distributed lag.

20.2 General multipliers and accelerators with distributed lag

The operators with continuously distributed lag were described by R. G. D. Allen [8] in 1956. The continuously (exponentially) distributed lag are described in Section 1.9 of [10, pp. 23–29], Section 5.8 of [10, pp. 166–170], and Section 5.5 of [12, pp. 88–93]. The exponential distribution is the continuous analogue of the geometric distribution, and its basic property is memoryless. These operators are often used to describe economic processes with time delay [12].

Let us define the general form of economic concepts of multipliers and accelerators with distributed lag, which do not take into account the effects of memory.

The standard multiplier with fixed time delay is defined by the equation

$$Z(t) = m(\tau)(T_\tau Y)(t) = m(\tau)Y(t-\tau), \tag{20.1}$$

where $\tau > 0$ is the delay time. Here, we use the translation (shift) operator T_τ that is defined [335, pp. 95–96], by the expression

$$(T_\tau Y)(t) = Y(t-\tau). \tag{20.2}$$

The standard accelerator with fixed time delay is described by the equation

$$Z(t) = a(\tau)(T_\tau Y^{(1)})(t) = a(\tau)Y^{(1)}(t-\tau), \tag{20.3}$$

where $\tau > 0$ is the delay time.

In the general case, the delay time $\tau > 0$ can be considered as a random variable, which is distributed by probability law (distribution) on positive semi-axis [400]. In this case, we can define the general multiplier with continuously distributed lag in the following form.

Definition 20.1. The general multiplier with continuously distributed delay time (the general multiplier with lag) is described by the equation

$$Z(t) = \int_0^\infty M_T(\tau)Y(t-\tau)\,d\tau, \tag{20.4}$$

where $Y(t)$ and $M_T(t)$ are piecewise continuous functions on $\mathbb{R}$ and the integral converges. The kernel $M_T(\tau)$ is the probability density function (the weighting function) that satisfies the conditions

$$M_T(\tau) \geq 0, \quad \int_0^\infty M_T(\tau)\, d\tau = 1 \tag{20.5}$$

for $\tau \in (0, \infty)$.

To describe the general multiplier (20.4) with distributed lag in compact form, we define [400] the translation operator $\mathbf{T}_M$ with the continuously distributed lag by the equation

$$(\mathbf{T}_M Y)(t) = \int_0^\infty M_T(\tau)(T_\tau Y)(t)\, d\tau = \int_{-\infty}^t M_T(t-\tau) Y(\tau)\, d\tau, \tag{20.6}$$

where $M_T(\tau)$ is the probability density function [400, 403, 405, 404, 402, 407]. Using operator (20.6), general multiplier (20.4) is represented in the form

$$Z = (\mathbf{T}_M Y)(t). \tag{20.7}$$

Using operator (20.6), we can also define the general multiplier with distributed lag.

Definition 20.2. The general accelerator with continuously distributed delay time (the general accelerator with lag) is described by the equation

$$Z(t) = (\mathbf{T}_M Y^{(n)})(t) = \int_0^\infty M_T(\tau) Y^{(n)}(t-\tau)\, d\tau, \tag{20.8}$$

where $Y(t)$ and $M_T(t)$ are piecewise continuous functions on $\mathbb{R}$ and the integral converges. The kernel $M_T(\tau)$ is the probability density function (the weighting function) that satisfies the conditions (20.5).

Note that equation (20.8) with $n = 0$ gives equation (20.4) of the generalized multiplier with distributed lag. Equation (20.8) with $n = 1$ and $M_T(t-\tau) = a\delta(t-\tau)$ gives equation of the standard accelerator without lag.

As a probability density function $M_T(\tau)$, we can use functions of the continuous uniform, exponential, gamma, and other distributions [411, 124, 214, 488].

20.3 Accelerator with uniformly distributed lag

Let us consider the continuous uniform distribution of delay time. For the first time in economics, the uniform distributed lag was proposed in the work of Michal A. Kalecki in 1935 [189] (see also Sections 8.4 of [10, pp. 251–254]). The continuous uniform distribution of delay time was used by M. A. Kalecki to propose the economic models of business cycles.

Let us consider the continuous uniform distribution of delay time.

The probability density function $M_T(\tau)$ of the continuous uniform distribution of delay time is defined by the equation

$$M_T(\tau) = \begin{cases} 1/\theta & \tau \in [0,\theta], \\ 0 & \tau \notin [0,\theta], \end{cases} \tag{20.9}$$

where $\theta > 0$.

The multiplier with uniformly distributed lag is defined by the equation

$$Z(t) = m(\theta) \int_0^{\infty} M_T(\tau) Y(t-\tau)\, d\tau, \tag{20.10}$$

where $M_T(t)$ is expressed by (20.9). Substitution of expression (20.9) into equation (20.10) gives

$$Z(t) = m(\theta)\frac{1}{\theta} \int_0^{\theta} Y(t-\tau)\, d\tau = m(\theta)\frac{1}{\theta} \int_{t-\theta}^{t} Y(\tau)\, d\tau. \tag{20.11}$$

For example, the Kalecki multiplier with uniformly distributed lag is defined (see equation (2) in Section 8.4 of [10, p. 252]) by the equation

$$I(t) = \frac{1}{\theta} \int_{t-\theta}^{t} B(\tau)\, d\tau. \tag{20.12}$$

This multiplier describes relationship between the planned investments $B(t)$ and the actually implemented investments $I(t)$.

Let us define the accelerator with uniform distribution of delay time.

Definition 20.3. The accelerator with uniformly distributed lag is described by the equation

$$Z(t) = a(\theta) \int_0^{\infty} M_T(\tau) Y^{(n)}(t-\tau)\, d\tau, \tag{20.13}$$

where the probability density function $M_T(t)$ is defined by expression (20.9).

Substitution of expression (20.9) into equation (20.13) gives

$$Z(t) = a(\theta)\frac{1}{\theta} \int_0^{\theta} Y^{(n)}(t-\tau)\, d\tau = a(\theta)\frac{1}{\theta} \int_{t-\theta}^{t} Y^{(n)}(\tau)\, d\tau. \tag{20.14}$$

Let us consider the case $n = 1$. Then accelerator (20.13) with $n = 1$ can be described by the equation

$$Z(t) = a(\theta)(\mathbf{D}_T^1 X)(t), \tag{20.15}$$

where $\mathbf{D}_T^1$ is the derivative of first order with uniformly distributed lag that is defined by the equation

$$(\mathbf{D}_T^1 X)(t) = \int_0^{\infty} M_T(\tau) X^{(1)}(t-\tau)\, d\tau, \tag{20.16}$$

and the weighting function $M_T(\tau)$ is defined by (20.9). Substitution of (20.9) into (20.16) gives

$$(\mathbf{D}_T^1 X)(t) = \frac{1}{\theta} \int_0^{\theta} X^{(1)}(t-\tau)\, d\tau. \tag{20.17}$$

Using the change of variable and the Newton–Leibniz formula, we obtain

$$(\mathbf{D}_T^1 X)(t) = \frac{1}{\theta} \int_{t-\theta}^{t} X^{(1)}(z)\, dz = \frac{1}{\theta}(X(t) - X(t-\theta)). \tag{20.18}$$

For $t = n\theta$, using $X_n = X(n\theta)$ and $\theta = 1$, equation (20.18) can be written in the form

$$(\mathbf{D}_T^1 X)(n) = X_n - X_{n-1}. \tag{20.19}$$

As a result, the first-order derivative with uniformly distributed lag is the backward finite difference (see also Section 8.4 in [10, p. 253]). We can formulate the following statement.

Statement 20.1. *Discrete economic models, which are described by equations with finite differences, can be interpreted as models with continuous time and continuous uniform distribution of delay time (lag).*

This representation can be used in different applications. For example, discrete economic models, which are considered, for example, in the book of Allen [10], could be considered as models with continuous uniformly distributed lag.

Remark 20.1. The proposed interpretation of the finite differences can be used as one of the steps in generalizing discrete models. In the first step, a discrete economic model, which is described by equation with finite difference, is represented as a continuous-time model with continuously distributed lag, in which the weighting function $M_T(\tau)$ is defined by expression (20.9). In the second step, we can consider the weighting functions $M_T(\tau)$ that describe other types of lag distributions, for example, such as the exponential, gamma, Weibull and other distributions of delay time.

Remark 20.2. We should note that this interpretation is possible due to the fact that the finite differences are not exact analogues of derivatives [378] and [441, 442, 431].

20.4 Multiplier and accelerator with exponential lag

Exponentially distributed lag in the economics was first considered by Alban W. H. Phillips [305, 306] in 1954. Economic models with continuously distributed lags in the exponential form and corresponding differential equations were proposed in works [305, 306]. The operators with exponentially distributed lag were described by Roy G. D. Allen [8] in 1956 (see also [10]). The exponentially distributed lag are considered in Section 1.9 of [10, pp. 23–29], Section 5.8 of [pp. 166–170]Allen63, and Section 5.5 of [pp. 88–93]Allen68.

Recently, operators with exponentially distributed lag were defined in works of Caputo and Fabrizio [56, 57], where it was misinterpreted as a derivative of a fractional order [400, 384, 6]. We can state that the differential operator (20.116) of integer order with exponentially distributed lag coincides with the Caputo–Fabrizio operator of the order $\alpha = n - 1 + \lambda/(\lambda + 1)$ [400, 384]. The Caputo–Fabrizio operator cannot describe the memory effects [400, 384].

The multiplier with exponentially distributed lag is defined by the equation

$$Z(t) = m\lambda \int_0^{\infty} \exp\{-\lambda\tau\} Y(t-\tau)\, d\tau, \tag{20.20}$$

where $m = m(\lambda) > 0$ is the multiplier coefficient. Equation (20.20) can be written as

$$Z(t) = m(\mathbf{D}_T^{\lambda,0} Y)(t), \tag{20.21}$$

where $\mathbf{D}_T^{\lambda,0}$ is the translation operator with exponentially distributed lag

$$(\mathbf{D}_T^{\lambda,0} Y)(t) = (\mathbf{T}_M Y)(t) = \lambda \int_0^{\infty} \exp\{-\lambda\tau\} Y(t-\tau)\, d\tau. \tag{20.22}$$

This equation can be rewritten in the equivalent form

$$(\mathbf{D}_T^{\lambda,0} Y)(t) = (\mathbf{T}_M Y)(t) = \lambda \int_{-\infty}^{t} \exp\{-\lambda(t-\tau)\} Y(\tau)\, d\tau. \tag{20.23}$$

The accelerator with exponentially distributed lag can be described by the equation

$$Z(t) = a\lambda \int_0^{\infty} \exp\{-\lambda\tau\} Y^{(n)}(t-\tau)\, d\tau, \tag{20.24}$$

where $a = a(\lambda) > 0$ is the accelerator coefficient. Equation (20.24) can be written as

$$Z(t) = a(\mathbf{D}_{T;C}^{\lambda,n} Y)(t), \tag{20.25}$$

where $\mathbf{D}^{\lambda,n}_{T;C}$ is the translation operator with exponentially distributed lag

$$(\mathbf{D}^{\lambda,n}_{T;C}Y)(t) = \lambda \int_{t_0}^{t} \exp\{-\lambda(t-\tau)\} Y^{(n)}(\tau)\, d\tau, \tag{20.26}$$

where $n \in \mathbb{N}$. We can supplement definition (20.26) by including the case $n = 0$. Then equation (20.22) will be a special case of equation (20.26).

Let us consider an example of the accelerator with exponentially distributed lag. The investment accelerator with exponentially distributed lag is defined by the equation

$$I(t) = v\lambda \int_{-\infty}^{t} \exp\{-\lambda(t-\tau)\} Y^{(1)}(\tau)\, d\tau, \tag{20.27}$$

where $v > 0$ is the accelerator coefficient. Using operator (20.26), equation (20.27) can be written as

$$I(t) = v(\mathbf{D}^{\lambda,1}_{T,C}Y)(t). \tag{20.28}$$

Equation (20.28) describes investment accelerator with the exponentially distributed lag, for which the speed response is equal to $\lambda = 1/T$.

The exponential weighting function is actively used in macroeconomic models with distributed lag and the continuous time [10, 12]. It is known that, under certain conditions, equations with continuously distributed lag are equivalent to differential equations with standard derivatives of integer orders. These differential equations are usually more convenient and easier to apply in comparison with the integro-differential equations that describe the distributed lag. For example, differential equation (20.29) is used in the Phillips model of multiplier-accelerator that takes into account two exponentially distributed lags (for details, see [10, pp. 72–74]).

In standard macroeconomic models, the differential equations of exponentially distributed lag are used instead of equations with integro-differential operators. For example, the economic accelerator (20.27), (20.28) with the exponential lag is usually considered [10, p. 63] in the form

$$I^{(1)}(t) = -\lambda(I(t) - vY^{(1)}(t)). \tag{20.29}$$

Equation (20.29) is called the differential equations of the exponential lag [10, p. 27]. The use of differential equations of integer orders instead of the integro-differential operators (20.27), (20.28) is caused by difficulties in using operators (20.27) and (20.28).

Note that for other weight functions (probability density functions), equations for an accelerator with distributed delay time cannot be represented as differential equations in the general case [400, 384].

20.5 General multipliers with distributed lag and memory

To describe the simultaneous manifestation of effects of the distributed lag and fading memory, we can use a composition of the translation operator (20.4) and integration of noninteger order. This mathematical tool allows you to define new concepts, namely the concept of multiplier with lag and memory.

Definition 20.4. The general multiplier with continuously distributed lag and power-law memory (the general multiplier with lag and memory) is described by the equation

$$Z(t) = m \int_0^\infty M_T(\tau)(I^\alpha_{L;+}Y)(t-\tau)\,d\tau, \tag{20.30}$$

where $I^\alpha_{L;+}$ is the Liouville fractional integral of the order $\alpha \in \mathbb{R}_+$, and $M_T(\tau)$ is the probability density function (the weighting function) that describe distribution of the delay time τ and satisfies condition (20.5). Here we can assume that $(I^\alpha_{L;+}Y)(t)$ and $M_T(t)$ are piecewise continuous functions on $\mathbb{R}$, for which integral (20.30) converges.

In equation (20.30), we use the Liouville fractional integral $I^\alpha_{L;+}$ that is defined [200, pp. 87–90] by the expression

$$(I^\alpha_{L;+}Y)(t) = \frac{1}{\Gamma(\alpha)} \int_{-\infty}^{t} (t-\tau)^{\alpha-1} Y(\tau)\,d\tau, \tag{20.31}$$

where $\alpha \geq 0$ is the order of the fractional integral, $\Gamma(\alpha)$ is the gamma function.

To simplify the representation of multipliers with memory and lag, we define the following fractional integral operator with lag. The Liouville fractional integral with distributed time delay can be defined [400] by the equation

$$(\mathbf{I}^\alpha_{T;L;+}Y)(t) = \int_0^\infty M_T(\tau)(I^\alpha_{L;+}Y)(t-\tau)\,d\tau, \tag{20.32}$$

where $(I^\alpha_{L;+}Y)(t)$ is the Liouville fractional integral (20.31) and $M_T(\tau)$ is the weighting function that satisfies the conditions (20.5).

In this case, the general multiplier with distributed lag and power-law memory can be represented by the equation

$$Z(t) = m(\mathbf{I}^\alpha_{T;L;+}Y)(t), \tag{20.33}$$

where the operator $\mathbf{I}^\alpha_{T;L;+}$ is defined by expression (20.32).

For economic processes, in which the effects of power-law memory exist on a finite time interval, we can use general multiplier that is based on the Riemann–Liouville fractional integral.

Definition 20.5. The general multiplier with continuously distributed lag and power-law memory (the general multiplier with lag and memory) on an finite time interval is described by the equation

$$Z(t) = m \int_0^t M_T(\tau)(I^{\alpha}_{\mathrm{RL};0+} Y)(t-\tau)\, d\tau, \tag{20.34}$$

where $I^{\alpha}_{\mathrm{RL};0+}$ is the Riemann–Liouville fractional integral of the order $\alpha \in \mathbb{R}_+$, and $M_T(\tau)$ is the probability density function (the weighting function) that describe distribution of the delay time τ and satisfies condition (20.5). Here, we can assume that $(I^{\alpha}_{\mathrm{RL};0+} Y)(t)$ and $M_T(t)$ are piecewise continuous functions on $\mathbb{R}$, for which integral (20.34) converges.

The Riemann–Liouville fractional integral with distributed time delay can be defined [400] by the equation

$$(\mathbf{I}^{\alpha}_{T;\mathrm{RL};0} Y)(t) = \int_0^{\infty} M_T(\tau)(I^{\alpha}_{\mathrm{RL};0+} Y)(t-\tau)\, d\tau, \tag{20.35}$$

where $(I^{\alpha}_{\mathrm{RL};0+} Y)(t)$ is the Riemann–Liouville fractional integral of the order $\alpha \in \mathbb{R}_+$, and $M_T(\tau)$ is the weighting function that satisfies the conditions (20.5). Here, we can assume that $(I^{\alpha}_{\mathrm{RL};0+} Y)(t)$ and $M_T(t)$ are piecewise continuous functions on $\mathbb{R}$, for which the integral converges. The Riemann–Liouville fractional integral $I^{\alpha}_{\mathrm{RL};0+}$ is defined [200, pp. 69–70] by the expression

$$(I^{\alpha}_{\mathrm{RL};0} Y)(t) = \frac{1}{\Gamma(\alpha)} \int_0^t (t-\tau)^{\alpha-1} Y(\tau)\, d\tau, \tag{20.36}$$

where $\alpha \geq 0$ is the order of the fractional integral, $\Gamma(\alpha)$ is the gamma function, and $\tau \in [0, t]$. Note that for the correctness of the definition (20.35), we should take into account that the fractional integral (20.36) is defined for the case $t > 0$ [200, p. 69].

As a result, using operator (20.35), the general multiplier with distributed lag and power-law memory on finite time interval (20.34) can be written in the form

$$Z(t) = m(\mathbf{I}^{\alpha}_{T;\mathrm{RL};0+} Y)(t). \tag{20.37}$$

Let us prove that the general multiplier with distributed lag and power-law memory can be interpreted as a multiplier with some more general type of fading memory.

Statement 20.2. *The general multiplier with distributed lag and power-law memory on finite time interval* (20.37) *can be represented by the equation*

$$Z(t) = m \int_0^t M^{\alpha}_{\mathrm{TRL}}(t-\tau) Y(\tau)\, d\tau, \tag{20.38}$$

where $M^{\alpha}_{\mathrm{TRL}}(t-\tau)$ is the new memory function (the memory-and-lag function) that is defined by the expression

$$M^{\alpha}_{\mathrm{TRL}}(t) = \int_0^t M_T(t-\tau) M^{\alpha}_{\mathrm{RL}}(\tau)\, d\tau, \tag{20.39}$$

where $M^{\alpha}_{\mathrm{RL}}(\tau)$ is the kernels of the Riemann–Liouville fractional integral and $M_T(\tau)$ is the weighting function.

Proof. Using the Laplace convolution $*$, equation (20.37) can be represented in the form

$$Z(t) = m(M^{\lambda,a}_T * (I^{\alpha}_{\mathrm{RL};0+} Y))(t). \tag{20.40}$$

It is known that the Riemann–Liouville fractional integral (20.36) can be also represented as the Laplace convolution

$$(I^{\alpha}_{\mathrm{RL};0+} Y)(t) = \frac{1}{\Gamma(\alpha)} \int_0^t M^{\alpha}_{\mathrm{RL}}(t-\tau) Y(\tau)\, d\tau = (M^{\alpha}_{\mathrm{RL}} * Y)(t), \tag{20.41}$$

where

$$M^{\alpha}_{\mathrm{RL}}(t-\tau) = \frac{(t-\tau)^{\alpha-1}}{\Gamma(\alpha)} \tag{20.42}$$

is the kernel of the Riemann–Liouville fractional integral. This allows us to represent equation (20.40) in the form

$$Z(t) = m(M_T * (M^{\alpha}_{\mathrm{RL}} * Y))(t). \tag{20.43}$$

The associativity of the Laplace convolution

$$(M_T * (M^{\alpha}_{\mathrm{RL}} * Y))(t) = ((M_T * M^{\alpha}_{\mathrm{RL}}) * Y)(t) \tag{20.44}$$

leads to the fact that equation (20.43) takes the form

$$Z(t) = m(M^{\alpha}_{\mathrm{TRL}} * Y)(t), \tag{20.45}$$

where $M^{\alpha}_{\mathrm{TRL}}(t)$ is the memory-and-lag function

$$M^{\alpha}_{\mathrm{TRL}}(t) = (M_T * M^{\alpha}_{\mathrm{RL}})(t) = \int_0^t M_T(t-\tau) M^{\alpha}_{\mathrm{RL}}(\tau)\, d\tau. \tag{20.46}$$

This allows us to represent (20.45) in the form (20.38). □

Remark 20.3. In the two proposed definitions, we use the Liouville and Riemann–Liouville fractional integrals only. In the general case, we can consider other types of fractional integrals for the description of the general multiplier with lag and memory.

20.6 General accelerator with distributed lag and memory

To describe the simultaneous manifestation of effects of the distributed lag and power-law fading memory, also should use a composition of the translation operator (20.4) and fractional derivatives of noninteger order. This mathematical tool allows you to define new concepts, namely the concept of an accelerator with lag and memory.

Let us use the fact that the Caputo fractional derivative can be represented as a sequential action of the integer-order derivative and the Riemann–Liouville fractional integral

$$(\mathbf{D}^{\alpha}_{C;0+}Y)(t) = (\mathbf{I}^{\alpha}_{\mathrm{RL};0+}Y^{(n)})(t), \tag{20.47}$$

where $Y^{(n)}(\tau) = d^nY(\tau)/d\tau^n$ and $n-1 < \alpha < n$. Using representation (20.47) and the definition of the general multiplier with lag and memory (20.34), we can define the general accelerator with lag and memory in the form

$$Z(t) = a(\mathbf{I}^{n-\alpha}_{T;C;0+}Y^{(n)})(t). \tag{20.48}$$

As a result, we get the following definition.

Definition 20.6. The general accelerator with distributed lag and power-law memory can be described by the equation

$$Z(t) = a(\mathbf{D}^{\alpha}_{T;C;0+}Y)(t), \tag{20.49}$$

where $\mathbf{D}^{\alpha}_{T;C;0+}$ is the Caputo fractional derivative with distributed lag

$$(\mathbf{D}^{\alpha}_{T;C;0+}Y)(t) = \int_0^t M_T(\tau)(D^{\alpha}_{C;0+}Y)(t-\tau)\,d\tau. \tag{20.50}$$

Using the Laplace convolution $*$, equation (20.48) can be represented in the form

$$Z(t) = a(M^{\lambda,a}_T * (D^{\alpha}_{C;0+}Y))(t). \tag{20.51}$$

Similar to the general multiplier with lag and memory, the general accelerators with lag and memory can be defined by the equation

$$Z(t) = a\int_0^t M^{n-\alpha}_{\mathrm{TRL}}(t-\tau)Y^{(n)}(\tau)\,d\tau, \tag{20.52}$$

where $n = [\alpha] + 1$ for noninteger values of α, and $n = \alpha$ for integer values.

Remark 20.4. The nonnegative memory function $M_T(\tau)$, which satisfies the normalization condition, can be interpreted as the probability density function defined on the positive semi-axis ($\tau \in \mathbb{R}_+$). In general, we can use different probability distributions on the positive semi-axis $\mathbb{R}_+$ to describe continuous distributions of delay time. For example, we can use the uniform distribution, the exponential distribution, the Erlang distribution, the gamma distribution, the Weibull–Gnedenko distribution and other distributions on the positive semi-axis [411, 124, 214, 488].

20.7 Multiplier with gamma distributed lag and memory

Let us consider the concept of the multiplier with distributed lag and power-law memory, in which the delay time is described by the gamma distribution.

The probability density function of the gamma distribution is defined by the expression

$$M_T^{\lambda,a}(\tau) = \begin{cases} \frac{\lambda^a \tau^{a-1}}{\Gamma(a)} \exp(-\lambda\tau) & \text{if } \tau > 0, \\ 0 & \text{if } \tau \le 0 \end{cases} \tag{20.53}$$

with the shape parameter $a > 0$ and the rate parameter $\lambda > 0$. If $a = 1$, function (20.53) describes the exponential distribution.

In economics, the gamma distribution (20.53) is used to take into account waiting times, when there is a sharp increase in the average delay time. For example, this distribution is applied to describe the delays in payments and delays orders in queues. The gamma distribution allows us to take into account the likelihood of risk events. The distribution also describes the time of receipt of the order for the enterprise, the service life of device components and time between store visits.

Definition 20.7. The multiplier with gamma distributed lag and power-law memory is represented by the equation

$$Z(t) = m(\alpha, a, \lambda)(\mathbf{I}^{\alpha,a,\lambda}_{T;RL;0+} Y)(t), \tag{20.54}$$

where $\mathbf{I}^{\alpha,a,\lambda}_{T;RL;0+}$ is the Riemann–Liouville fractional integral with gamma distribution of delay time that is defined as

$$(\mathbf{I}^{\alpha,a,\lambda}_{T;RL;0+} Y)(t) = \int_0^t M_T^{\lambda,a}(\tau)(I^{\alpha}_{RL;0+} Y)(t-\tau)\, d\tau, \tag{20.55}$$

and the kernel $M_T^{\lambda,a}(\tau)$ is defined by equation (20.53).

Multiplier (20.54) can be interpreted as a multiplier with memory function, which is described by the confluent hypergeometric Kummer function $F_{1,1}(a;b;z)$. This function is defined (see [200, pp. 29–30]) by the equation

$$F_{1,1}(a;c;z) = \sum_{k=0}^{\infty} \frac{\Gamma(a+k)\Gamma(c)}{\Gamma(a)\Gamma(c+k)} \frac{z^k}{k!}, \tag{20.56}$$

where $a, z \in \mathbb{C}$, $\mathrm{Re}(c) > \mathrm{Re}(a) > 0$ such that $c \neq 0, -1, -2, \ldots$.

Statement 20.3. *The multiplier* (20.54) *with gamma distributed lag and power-law memory can be represented in the form*

$$Z(t) = m(\alpha, a, \lambda) \int_0^t M^{\lambda,a;\alpha}_{TRL}(t-\tau) Y(\tau)\, d\tau, \tag{20.57}$$

where $M^{\alpha,a,\lambda}_{TRL}(t-\tau)$ *is defined by equation*

$$M^{\lambda,a;\alpha}_{TRL}(t) = (M^{\lambda,a}_{T} * M^{\alpha}_{RL})(t) = \frac{\lambda^a \Gamma(a)}{\Gamma(a+\alpha)} t^{a+\alpha-1} F_{1,1}(a; a+\alpha; -\lambda t), \tag{20.58}$$

where $\alpha > 0$ *is the order of integration and the parameters* $a > 0$, $\lambda > 0$ *describe the shape and rate of the gamma distribution, respectively. Here* $M^{\alpha}_{RL}(t)$ *is kernel* (20.42) *of the Riemann–Liouville integral, and* $M^{\lambda,a}_{T}(t)$ *is the probability density function* (20.53)*. The parameter* $m = m(\lambda, a; \alpha)$ *is the multiplier coefficient.*

Proof. The multiplier (20.54) with gamma distributed lag and power-law memory can be expressed thought the Laplace convolution of memory function and the weighting function

$$(\mathbf{I}^{\alpha,a,\lambda}_{T;RL;0+} Y)(t) = (M^{\lambda,a}_{T} * (M^{\alpha}_{RL} * Y))(t), \tag{20.59}$$

where the memory function

$$M^{\alpha}_{RL}(t) = \frac{(t-\tau)^{\alpha-1}}{\Gamma(\alpha)} \tag{20.60}$$

is the kernel of the Riemann–Liouville fractional integral (20.36). The associativity of the Laplace convolution

$$(M^{\lambda,a}_{T} * (M^{\alpha}_{RL} * Y))(t) = ((M^{\lambda,a}_{T} * M^{\alpha}_{RL}) * Y)(t), \tag{20.61}$$

gives

$$(M^{\lambda,a}_{T} * (M^{\alpha}_{RL} * Y))(t) = (M^{\lambda,a;\alpha}_{TRL} * Y)(t), \tag{20.62}$$

where $M^{\lambda,a;\alpha}_{TRL}(t)$ is the memory-and-lag function

$$M^{\lambda,a;\alpha}_{TRL}(t) = (M^{\lambda,a}_{T} * M^{\alpha}_{RL})(t). \tag{20.63}$$

This allows us to represent operator (20.59) in the form

$$(\mathbf{I}^{\alpha,a,\lambda}_{T;RL;0+} Y)(t) = \int_0^t M^{\lambda,a;\alpha}_{TRL}(t-\tau) Y(\tau)\, d\tau, \tag{20.64}$$

where $M^{\alpha,a,\lambda}_{TRL}(t-\tau)$ is defined by equation (20.63). Note that operator (20.64) can be represented as

$$(\mathbf{I}^{\alpha,a,\lambda}_{T;RL;0+} Y)(t) = (\mathbf{T}^{\lambda,a}_{M}(I^{\alpha}_{RL;0+} Y))(t). \tag{20.65}$$

Let us obtain an explicit form of the memory-and-lag function $M_{\mathrm{TRL}}^{\lambda,a;\alpha}(t)$. For this purpose, we can use equation (2.3.6.1) of [316, p. 324], that has the form

$$\int_0^t (t-\tau)^{\alpha-1}\tau^{\beta-1}\exp(-\lambda\tau)\,d\tau = \frac{\Gamma(\alpha)\Gamma(\beta)}{\Gamma(\alpha+\beta)} t^{\alpha+\beta-1} F_{1,1}(\beta;\alpha+\beta;-\lambda t), \tag{20.66}$$

where $\mathrm{Re}(\alpha) > 0$, $\mathrm{Re}(\beta) > 0$ and $F_{1,1}(a;b;z)$ is the confluent hypergeometric Kummer function (20.56).

Using equation (20.66), the memory-and-lag function (20.63) can be written as

$$M_{\mathrm{TRL}}^{\alpha,a,\lambda}(t) = \frac{\lambda^a\Gamma(a)}{\Gamma(a+\alpha)} t^{a+\alpha-1} F_{1,1}(a;a+\alpha;-\lambda t) \tag{20.67}$$

that defines the kernel of operator (20.64). In general, function (20.67) can be interpreted as a new memory function.

As a result, multiplier (20.54) with power-law memory and gamma distributed lag can represented in the form

$$Z(t) = m(\alpha,a,\lambda)\frac{\lambda^a\Gamma(a)}{\Gamma(a+\alpha)}\int_0^t \tau^{\alpha+a-1}F_{1,1}(a;a+\alpha;-\lambda\tau)Y(t-\tau)\,d\tau, \tag{20.68}$$

where $\alpha > 0$ is the order of integration and the parameters $a > 0$ and $\lambda > 0$ describe the shape and rate of the gamma distribution, respectively. □

Remark 20.5. Note that equality

$$F_{1,1}(a;c;z) = \Gamma(c)E_{1,c}^{a}(z) \tag{20.69}$$

allows us to represent the memory function (20.67) through the three parameter Mittag–Leffler functions (see equation (5.1.18) of [143, p. 99]).

20.8 Multiplier with memory through Abel-type operator

In mathematics, the fractional integral operators with the Kummer function in the kernel is known (see equation (37.1) in [335, p. 731]). This integral operator (the Abel-type fractional integral) is defined by the equation

$$(\mathbf{I}_{AT;0+}^{\alpha,\beta,\lambda}Y)(t) = \frac{1}{\Gamma(\alpha)}\int_0^t (t-\tau)^{\alpha-1}F_{1,1}(\beta;\alpha;\lambda(t-\tau))Y(\tau)\,d\tau, \tag{20.70}$$

where $\mathrm{Re}(\alpha) > 0$, β, λ are complex parameters ($\alpha,\beta,\lambda \in \mathbb{C}$) and $F_{1,1}(a;b;z)$ is the confluent hypergeometric Kummer function (20.56).

The confluent hypergeometric Kummer function $F_{1,1}(\beta;\alpha;\lambda(t-\tau))$ can be considered as a memory function in the multipliers and accelerators with memory.

Let us consider the multiplier with memory function that is expressed through the confluent hypergeometric Kummer function.

Definition 20.8. The multiplier with the memory function

$$M_{AT}(t-\tau) = \frac{1}{\Gamma(\alpha)}(t-\tau)^{\alpha-1}F_{1,1}(\beta;\alpha;\lambda(t-\tau)) \tag{20.71}$$

is represented by the equation

$$Z(t) = m(\alpha,\beta,\lambda)(\mathbf{I}^{\alpha,\beta,\lambda}_{AT;0+}Y)(t), \tag{20.72}$$

where the operator $\mathbf{I}^{\alpha,\beta,\lambda}_{AT;0+}$ is defined by equation (20.70).

Remark 20.6. Comparing the multiplier with gamma distributed lag and power-law memory (20.54) with the multiplier (20.72), we can see that multiplier (20.54) is a special case of (20.72). This statement follows form the fact that the Riemann–Liouville fractional integral with gamma distributed lag (20.68) can be expressed through the Abel-type fractional integral (20.70) by the equation

$$(\mathbf{I}^{\alpha,a,\lambda}_{T;\mathrm{RL};0+}Y)(t) = \lambda^{a}\Gamma(a)(\mathbf{I}^{a+\alpha,a,-\lambda}_{AT;0+}Y)(t), \tag{20.73}$$

where $\lambda > 0$, $a > 0$, $\alpha > 0$.

As a result, we can formulate the following statement.

Statement 20.4. *Not all types of memory functions describe pure memory effects only. For example, memory function* (20.71) *can describe the combined action of two effects, namely, distributed lag and power-law memory.*

Remark 20.7. The Abel-type fractional integral (20.70) can be represented as an infinite series of the Riemann–Liouville fractional integrals

$$I^{\alpha,\beta,\lambda}_{AT;0+} = \sum_{k=0}^{\infty}\frac{(\beta)_k}{k!}\lambda^k I^{\alpha+k}_{\mathrm{RL};0+}, \tag{20.74}$$

where $(\beta)_k = \Gamma(\beta+k)/\Gamma(\beta)$ is the Pochhammer symbol. Expression (20.74) is called the Neumann generalized series (see equation 37.10 of [335, p. 732]), which characterizes the structure of the AT fractional integral operator (20.70). Using equations (20.73) and (20.74), the Riemann–Liouville fractional integral with gamma distributed lag (20.68) can be represented as the series

$$(\mathbf{I}^{\alpha,a,\lambda}_{T;\mathrm{RL};0+}Y)(t) = \sum_{k=0}^{\infty}\frac{\Gamma(a+k)}{\Gamma(k+1)}(-1)^k\lambda^{k+a}(I^{\alpha+a+k}_{\mathrm{RL};0+}Y)(t). \tag{20.75}$$

As a result, we can formulate the following statements.

Statement 20.5. *The action of the multiplier with memory function* (20.71) *is equivalent to the parallel action of infinite numbers of multipliers with simple power-law memory. The action of multiplier* (20.72) *is equivalent to the parallel action of the multipliers with simple power-law memory:*

$$Z(t) = \sum_{k=0}^{\infty} m_{AT;k}(\alpha, \beta, \lambda)(I^{\alpha+k}_{\mathrm{RL};0+} Y)(t), \tag{20.76}$$

where

$$m_{AT;k}(\alpha, \beta, \lambda) = m(\alpha, \beta, \lambda)\frac{(\beta)_k}{k!}\lambda^k. \tag{20.77}$$

Using the fact that the Riemann–Liouville fractional integral is a special case of the Abel-type fractional integral (20.70), we can formulate the following statement as a consequence.

Statement 20.6. *The action of the multiplier with gamma distributed lag and power-law memory* (20.54) *is equivalent to the parallel action of multipliers with simple power-law memory*

$$Z(t) = \sum_{k=0}^{\infty} (-1)^k m_{\mathrm{TRL};k}(\alpha, a, \lambda)(I^{\alpha+a+k}_{\mathrm{RL};0+} Y)(t), \tag{20.78}$$

where

$$m_{\mathrm{TRL};k}(\alpha, a, \lambda) = m(\alpha, a, \lambda)\frac{\Gamma(a+k)}{\Gamma(k+1)}\lambda^{k+a}, \tag{20.79}$$

and $\alpha > 0$, $a > 0$, $\lambda > 0$.

Remark 20.8. Using equation (20.75) and the semigroup property of the Riemann–Liouville fractional integrals, we can get the semigroup property for the Abel-type fractional integral (20.70) in the form

$$\mathbf{I}^{\alpha_1,\beta_1,\lambda}_{AT;0+}\mathbf{I}^{\alpha_1\beta_2,\lambda}_{AT;0+} = \mathbf{I}^{\alpha_1+\alpha_2,\beta_1+\beta_2,\lambda}_{AT;0+}. \tag{20.80}$$

Equality (20.80) directly follows from equation 37.14 of [p. 733]FC1.

This remark allows us to proposed an important principle of superposition for the multipliers (20.72), where we use equality (20.80).

Principle 20.1 (Principle of superposition of multipliers with AT memory function). *The multiplier effect, which is created at a given moment by sequential action of two linear multipliers with memory function* (20.71)*, is equivalent to the action of linear multiplier, for which the first two parameters is equal to the sum of first two parameters of these multipliers.*

Symbolically, this principle can be written in the following form. The sequential action of two linear multipliers with memory function (20.71), which are described by the equations

$$Y(t) = m_1(\alpha_1, \beta_1, \lambda)(\mathbf{I}^{\alpha_1,\beta_1,\lambda}_{AT;0+}X)(t), \quad (20.81)$$

$$Z(t) = m_2(\alpha_2, \beta_2, \lambda)(\mathbf{I}^{\alpha_2,\beta_2,\lambda}_{AT;0+}Y)(t) \quad (20.82)$$

is equivalent to the action of the multiplier with memory function (20.71) in the form

$$Z(t) = m(\alpha, \beta, \lambda)(\mathbf{I}^{\alpha,\beta,\lambda}_{AT;0+}X)(t), \quad (20.83)$$

where

$$\alpha = \alpha_1 + \alpha_2, \quad \beta = \beta_1 + \beta_2, \quad (20.84)$$

and

$$m(\alpha, \beta, \lambda) = m_1(\alpha_1, \beta_1, \lambda)m_2(\alpha_2, \beta_2, \lambda) \quad (20.85)$$

for all $\alpha_1, \alpha_2 > 0$ and $\beta_1, \beta_2 > 0$.

Remark 20.9. Using equation (20.73), we can get the semigroup property for the Riemann–Liouville fractional integral with gamma distributed lag (20.68) in the form

$$\mathbf{I}^{\alpha,a,\lambda}_{T;RL;0+}\mathbf{I}^{\beta,b,\lambda}_{T;RL;0+} = B(\alpha, \beta)\mathbf{I}^{\alpha+\beta,a+b,\lambda}_{T;RL;0+}, \quad (20.86)$$

where

$$B(\alpha, \beta) = \frac{\Gamma(\alpha)\Gamma(\beta)}{\Gamma(\alpha + \beta)} \quad (20.87)$$

is the beta function. Equality (20.86) directly follows from equation (37.14) of [335, p. 733].

Using the fact that the Riemann–Liouville fractional integral with gamma distributed lag is a special case of the Abel-type fractional integral (20.70), we can formulate the following principle as a consequence.

Principle 20.2 (Principle of superposition for multipliers with memory and lag). *The sequential action of two linear multipliers with power-law memory and gamma distributed lag, which are described by the equations*

$$Y(t) = m_1(\alpha, a, \lambda)(\mathbf{I}^{\alpha,a,\lambda}_{T;RL;0+}X)(t), \quad (20.88)$$

$$Z(t) = m_2(\beta, b, \lambda)(\mathbf{I}^{\beta,b.\lambda}_{T;RL;0+}Y)(t) \quad (20.89)$$

is equivalent to the action of the multiplier with memory and lag in the form

$$Z(t) = m(\alpha, \beta, a, b, \lambda)(\mathbf{I}^{\alpha+\beta, a+b, \lambda}_{T;\mathrm{RL};0+} X)(t), \tag{20.90}$$

where

$$m(\alpha, \beta, a, b, \lambda) = B(\alpha, \beta) m_1(\alpha, a, \lambda) m_2(\beta, b, \lambda) \tag{20.91}$$

for all $\alpha, \beta > 0$ and $a, b > 0$.

Remark 20.10. Using Theorem 37.1 of [335, p. 733], and equation (20.73), we can state that the suggested fractional integral with lag (20.68) has the same range in $L_p(t_0, t_1)$ as the Riemann–Liouville fractional integrals and it is bounded from $L_p(t_0, t_1)$ onto $I^{\alpha}_{\mathrm{RL};0+}[L_p(t_0, t_1)] \subset L_p(t_0, t_1)$.

Remark 20.11. Using the condition of the invertibility of the AT operators (20.70), which is described by Theorem 37.2 of [335, p. 736], and equation (37.32) of [335, p. 734], we get that the equation

$$Z(t) = m(\alpha, \beta, \lambda)\, (\mathbf{I}^{\alpha,\beta,\lambda}_{AT;0+} Y)(t) \tag{20.92}$$

can be represented in the form

$$Y(t) = \frac{a(\alpha, \beta, \lambda)}{\lambda^{a}\Gamma(a)} e^{-\lambda t} D^{a}_{\mathrm{RL};0+}(e^{\lambda t} D^{\alpha}_{\mathrm{RL};0+} Z)(t), \tag{20.93}$$

where $a(\alpha, \beta, \lambda) = 1/m(\alpha, \beta, \lambda)$ $D^{\alpha}_{\mathrm{RL};0+}$ and $D^{a}_{\mathrm{RL};0+}$ are the Riemann–Liouville fractional derivatives of orders $\alpha > 0$ and $a > 0$, respectively. These derivatives can be defined by the Laplace convolution as

$$(D^{\alpha}_{\mathrm{RL};0+} X)(t) = D^{n}_{t}(M^{n-\alpha}_{\mathrm{RL}} * X)(t), \tag{20.94}$$

where $D^{n}_{t} = d^{n}/dt^{n}$, $n \in \mathbb{N}$. Note that equation (20.92) can be interpreted as an equation of economic multiplier with power-law memory and distributed lag [412, 397]. In this case, equation (20.93) can be interpreted as an equation of accelerator with memory [412, 397]. The equation of multiplier with memory is a reversible such that the dual (inverse) equation describes an accelerator with memory (see Section 4 of [412, 397]).

20.9 Accelerators with gamma distributed lag and memory

Let us define the accelerator with power-law memory and continuously distributed lag, in which delay time is described by the gamma distribution (20.53).

Definition 20.9. The accelerator with power-law memory and gamma distributed lag is described by the equation

$$Z(t) = a(\alpha, a, \lambda)(\mathbf{D}^{\alpha,a,\lambda}_{T;C;0+} Y)(t), \tag{20.95}$$

where $\mathbf{D}^{\alpha,a,\lambda}_{T;C;0+}$ is the Caputo fractional derivative with gamma distributed lag that is defined as

$$(\mathbf{D}^{\alpha,a,\lambda}_{T;C;0+} Y)(t) = \int_0^t M_T^{\lambda,a}(\tau)(D^{\alpha}_{C;0+} Y)(t-\tau)\, d\tau. \tag{20.96}$$

Remark 20.12. Fractional differential operator (20.96) can be expressed through the Riemann–Liouville fractional integral (20.68) with gamma distributed lag in the form

$$(\mathbf{D}^{\lambda,a,\alpha}_{T;C;0+} Y)(t) = (\mathbf{I}^{\lambda,a;n-\alpha}_{T;RL;0+} Y^{(n)})(t), \tag{20.97}$$

where $n-1 < \alpha \leq n$.

The proposed accelerator with memory and lag can be represented as the accelerator with the memory function that is expressed through the confluent hypergeometric Kummer function.

Statement 20.7. *The accelerators with gamma distributed lag and power-law memory can be represented in the form*

$$Z(t) = a(\alpha, a, \lambda) \int_0^t M^{\lambda,a;n-\alpha}_{\mathrm{TRL}}(t-\tau) Y^{(n)}(\tau)\, d\tau, \tag{20.98}$$

where $n-1 < \alpha \leq n$, and the kernel $M^{\lambda,a;n-\alpha}_{\mathrm{TRL}}(t)$ is defined in the form

$$M^{\lambda,a;n-\alpha}_{\mathrm{TRL}}(t) = \frac{\lambda^a \Gamma(a)}{\Gamma(a+n-\alpha)} t^{a+n-\alpha-1} F_{1,1}(a; a+n-\alpha; -\lambda t). \tag{20.99}$$

Proof. The accelerator equation (20.98) can be rewritten in the form of the Laplace convolution

$$Z(t) = a(\alpha, a, \lambda)(M_T^{\lambda,a} * (M^{n-\alpha}_{\mathrm{RL}} * Y^{(n)}))(t), \tag{20.100}$$

where $M_T^{\lambda,a}$ is the probability density function of the gamma distribution, and $M^{n-\alpha}_{\mathrm{RL}}$ is the kernel of the Caputo fractional derivative. Using that the convolution is an associative operation, we get

$$(M_T^{\lambda,a} * (M^{n-\alpha}_{\mathrm{RL}} * Y^{(n)}))(t) = ((M_T^{\lambda,a} * M^{n-\alpha}_{\mathrm{RL}}) * Y^{(n)})(t). \tag{20.101}$$

Therefore, equation can be rewritten in the form

$$Z(t) = a(\alpha, a, \lambda)(M^{\lambda,a;n-\alpha}_{\mathrm{TRL}} * Y^{(n)})(t), \tag{20.102}$$

where $M^{\lambda,a;n-\alpha}_{\mathrm{TRL}}(t)$ is defined by the equation

$$M^{\lambda,a;n-\alpha}_{\mathrm{TRL}}(t) = (M_T^{\lambda,a} * M^{n-\alpha}_{\mathrm{RL}})(t). \tag{20.103}$$

To get the explicit form of function (20.103), we can use equation (20.66). Then we obtain that the memory-and-lag function (20.103) is represented in the form (20.99). □

Remark 20.13. Using the Laplace transform of the Caputo fractional derivative and the gamma distribution function, we get [400] the Laplace transform of the Caputo fractional derivative with gamma distributed lag in the form

$$(\mathcal{L}(\mathbf{D}^{\alpha,a,\lambda}_{T;C;0+}Y)(t))(s) = \frac{\lambda^a}{(s+\lambda)^a}\left(s^\alpha(\mathcal{L}Y)(s) - \sum_{j=0}^{n-1} s^{\alpha j-1}Y^{(j)}(0)\right), \qquad (20.104)$$

where $n-1 < \alpha \leq n$.

Remark 20.14. Let us note that operator (20.96) with $a = 1$ describes the Caputo fractional derivative with exponentially distributed lag [400]. Using the fact that the Caputo fractional derivatives with integer values $\alpha = n \in \mathbb{N}$ are the integer-order derivatives $(D^\alpha_{C;0+}Y)(t) = Y^{(n)}(t)$, operator (20.95) with $a = 1$ and $\alpha = n$ describes the integer-order derivatives with the exponential distribution [400].

Operator (20.95) can be interpreted as a new generalized operator with memory function, which contains the confluent hypergeometric function. Proposed accelerator, which are defined by equations (20.95) and (20.98), can be used to describe the simultaneously action of the fading memory with power-law fading and distributed lag with the gamma distribution of delay time.

20.10 Exponentially distributed lag and power-law memory

In economics, the exponentially distributed lag is actively used (e. g., see [10]). The exponential distribution can be considered as a special case of the gamma distribution, when $a = 1$. For the exponential distribution, the probability density function is

$$M^\lambda_T(\tau) = M^{\lambda,1}_T(\tau) = \begin{cases} \lambda\exp(-\lambda\tau) & \tau > 0, \\ 0 & \tau \leq 0, \end{cases} \qquad (20.105)$$

where $\lambda > 0$ is rate parameter. The parameter $\lambda > 0$ is also called the speed of response [10, p. 27]. As an alternative parameter to the speed of response for the exponential lag, we can consider the time-constant $T = 1/\lambda$. For the exponentially distributed lag, the parameter T is the average time (length) of the delay [10, p. 27], since the expected value of t for (20.105) is equal to T.

The exponential distribution describes many phenomena such as waiting time for an insurance event; time of receipt of the order for the enterprise; time between visits by shop or supermarkets; the service life of components of devices.

Exponential distribution describes the time between events in a Poisson point process, i. e., a process in which events occur continuously and independently at a constant average rate. Exponential distribution can be considered as a continuous analogue of the geometric distribution [10], and it has the key property of being memoryless.

Definition 20.10. The multiplier with exponentially distributed lag and power-law memory is represented by the equation

$$Z(t) = m(\alpha,\lambda)(\mathbf{I}^{\alpha,\lambda}_{T;\mathrm{RL};0+}Y)(t), \tag{20.106}$$

where $\mathbf{I}^{\alpha,\lambda}_{T;\mathrm{RL};0+}$ is the Riemann–Liouville fractional integral with exponential distribution of delay time that is defined as

$$(\mathbf{I}^{\alpha,\lambda}_{T;\mathrm{RL};0+}Y)(t) = \int_0^t M^{\lambda}_T(\tau)(I^{\alpha}_{\mathrm{RL};0+}Y)(t-\tau)\,d\tau, \tag{20.107}$$

and the kernel $M^{\lambda}_T(\tau)$ is defined by equation (20.105). The parameter $\alpha > 0$ describes the memory fading and the parameter $\lambda > 0$ is the rate of the exponential distribution.

The multiplier (20.106) can be represented as the multiplier with memory function, which is expressed through the confluent hypergeometric Kummer function $F_{1,1}(a;b;z)$, that is defined by equation (20.56).

Statement 20.8. *The multiplier with exponentially distributed lag and power-law memory can be represented in the form*

$$Z(t) = m(\alpha,\lambda)\int_0^t M^{\lambda,\alpha}_{\mathrm{TRL}}(t-\tau)Y(\tau)\,d\tau, \tag{20.108}$$

where $M^{\alpha;\lambda}_{\mathrm{TRL}}(t-\tau)$ is defined by the equation

$$M^{\lambda,\alpha}_{\mathrm{TRL}}(t) = (M^{\lambda}_T * M^{\alpha}_{\mathrm{RL}})(t) = \frac{\lambda}{\Gamma(\alpha+1)}t^{a+\alpha-1}F_{1,1}(1;\alpha+1;-\lambda t). \tag{20.109}$$

Here, $M^{\alpha}_{\mathrm{RL}}(t)$ is the kernel (20.42) of the Riemann–Liouville integral, and $M^{\lambda}_T(t)$ probability density function (20.105). The parameter $m = m(\lambda,\alpha)$ is the multiplier coefficient.

Let us define the accelerator with power-law memory and continuously distributed lag, in which delay time is described by the exponential distribution (20.105).

Definition 20.11. The accelerator with power-law memory and exponentially distributed lag is described by the equation

$$Z(t) = a(\alpha,\lambda)(\mathbf{D}^{\alpha,\lambda}_{T;C;0+}Y)(t), \tag{20.110}$$

where $\mathbf{D}^{\alpha,\lambda}_{T;C;0+}$ is the Caputo fractional derivative with exponentially distributed lag that is defined as

$$(\mathbf{D}^{\alpha,\lambda}_{T;C;0+}Y)(t) = \int_0^t M^{\lambda}_T(\tau)(D^{\alpha}_{C;0+}Y)(t-\tau)\,d\tau, \tag{20.111}$$

and $M^{\lambda}_T(\tau)$ is defined by equation (20.105).

Remark 20.15. Fractional differential operator (20.111) can be expressed through the Riemann–Liouville fractional integral (20.68) with exponentially distributed lag in the form

$$(\mathbf{D}^{\lambda,\alpha}_{T;C;0+}Y)(t) = (\mathbf{I}^{\lambda,n-\alpha}_{T;RL;0+}Y^{(n)})(t), \tag{20.112}$$

where $n-1 < \alpha \leq n$.

The proposed accelerator with memory and lag can be represented as the accelerator with the memory function that is expressed through the confluent hypergeometric Kummer function.

Statement 20.9. *The accelerators with exponentially distributed lag and power-law memory* (20.110) *can be represented in the form*

$$Z(t) = a(\alpha, a, \lambda) \int_0^t M^{\lambda,n-\alpha}_{\mathrm{TRL}}(t-\tau) Y^{(n)}(\tau)\, d\tau, \tag{20.113}$$

where $n-1 < \alpha \leq n$, and the kernel $M^{\lambda,n-\alpha}_{\mathrm{TRL}}(t)$ is defined by the expression

$$M^{\lambda,a;n-\alpha}_{\mathrm{TRL}}(t) = \frac{\lambda}{\Gamma(n-\alpha+1)} t^{n-\alpha} F_{1,1}(1; n-\alpha+1; -\lambda t). \tag{20.114}$$

The fractional derivative with exponentially distributed lag (20.111) has the form

$$(\mathbf{D}^{\lambda,\alpha}_{T,C,0+}Y)(t) = \lambda \int_0^t \exp\{-\lambda\tau\}(D^{\alpha}_{C;0+}Y)(t-\tau)\, d\tau, \tag{20.115}$$

where $\lambda > 0$ is the rate parameter of exponential distribution and $(D^{\alpha}_{C;0+}Y)(t)$ is the Caputo fractional derivative. For integer values $\alpha = n \in \mathbb{N}$, equation takes the form

$$(\mathbf{D}^{\lambda,n}_{T,C,0+}Y)(t) = \lambda \int_0^t \exp\{-\lambda(t-\tau)\} Y^{(n)}(\tau)\, d\tau. \tag{20.116}$$

Operator (20.116) can be interpreted as the integer-order derivative with exponentially distributed delay time. We can expand the definition by including the case $\alpha = 0$.

Remark 20.16. We can consider generalized multipliers and accelerators, in which the memory function is represented through the three parameter Mittag–Leffler function (also called the Prabhakar function). Note that the confluent hypergeometric Kummer function can be represented through the three parameter Mittag–Leffler (Prabhakar) function by the equation

$$F_{1,1}(a, x : z) = E^{a}_{1,c}[z], \tag{20.117}$$

where $E^{a}_{b,c}[z]$ is the three parameter Mittag–Leffler function [143].

Remark 20.17. The integral operators with the generalized (three parameter) Mittag–Leffler function $E^{a}_{b,c}[z]$ in the kernels are proposed by Tilak R. Prabhakar [314] in 1971. These integral operators are called the Prabhakar fractional integral. For the first time, the fractional derivative of Riemann–Liouville type, which has the Prabhakar function in the kernel, was proposed by Anatoly A. Kilbas, Megumi Saigo and Ram K. Saxena in the work [199] in 2004. This operator is called the Prabhakar fractional derivative of Riemann–Liouville type. However, this operator was not proposed in the works of Prabhakar. This operator can be called as the Kilbas–Saigo–Saxena (KSS) fractional derivative. Note that the KSS operators are left-inverse for the Prabhakar fractional integrals [199]. The fractional integrals and derivatives of the Caputo type (the regularized Prabhakar fractional derivative) with the Prabhakar function in the kernels was proposed by Mirko D'Ovidio, Federico Polito in [96] in 2013 (see also [94, 95]). Then fractional derivatives and integrals began to be studied in various papers (e. g., see [198, 132, 137, 138]). An application of these operators in economics with memory is proposed in papers [406, 391]. We should emphasize that the main property of any generalized (fractional) derivative is to be a left-inverse operator to the corresponding generalized (fractional) integral operator. This requirement is important for a self-consistent mathematical theory of the fractional operators to have a general fractional calculus of these operators. In economics, this property leads to the duality of the concepts of multiplier with memory and the accelerator with memory.

20.11 Conclusion

In economics, we can have a simultaneous action of two different types of phenomena. For example, in economic processes we can have the simultaneous action of two phenomena, such as the continuously distributed lag and fading memory. To describe these processes, we can use the combination of two or more different type of operators. The proposed fractional operators, which are combinations of fractional derivatives (or integrals) and translation operator with distributed delay time. These operators can be important for applications in economics.

In the next chapters, we will consider an application of the proposed multipliers and accelerators with memory and lag to describe generalizations of the Harrod–Domar model [404], the Keynes model [402, 407] and the Phillips models [405].

21 Harrod–Domar model with memory and distributed lag

In this chapter, we consider a macroeconomic growth model, which takes into account memory with power-law fading and gamma distributed lag. This model, which is first proposed in [404], is a generalization of the standard Harrod–Domar growth model. Fractional differential equations of this generalized model that takes into account memory and lag are suggested. For these equations, we obtain solutions, which describe the macroeconomic growth of national income with fading memory and distributed delay time. This chapter is based on article [404].

21.1 Introduction

The standard Harrod–Domar growth model has been proposed by Roy Harrod [170, 171, 172] and Evsey Domar [92, 93] in 1946–1947. A generalization of this model by taking into account the exponentially distributed lag without memory was proposed by William Phillips [305, 306] in 1954 (see also [8, 9]). The Harrod–Domar growth model with power-law memory was suggested by authors [421, 426] in 2016 (see also [450]). The simultaneous consideration of the effects of memory and time delay is important for the description of economic processes. In the paper [404], we propose a generalization of the Harrod–Domar growth model, which describes the dynamics of national income with fading memory and continuously distributed lag.

As a starting point of this Chapter, we take the Harrod–Domar model without memory and lag, which is described in the Allen's book [10, pp. 64–69]. For simplification, we consider one-parameter power-law memory and gamma distributed time delay. We use operators that are the composition of fractional differentiation and continuously distributed translation (shift). We consider the fractional differential equations for this generalization of the Harrod–Domar growth model. We obtain solutions of these equations that describe the macroeconomic growth of national income with power-law fading memory and gamma distributed time-delay.

21.2 Harrod–Domar growth model without memory and lag

The Harrod–Domar model with continuous time describes the dynamics of national income $Y(t)$, which is determined by the sum of the nonproductive consumption $C(t)$, the induced investment $I(t)$ and the autonomous investment $A(t)$. The balance equation of this model has the form

$$Y(t) = C(t) + I(t) + A(t). \tag{21.1}$$

In the standard Harrod–Domar model [170, 171, 172, 92, 93] (see also [8, 10]) of the growth without memory, the following assumptions are used:

https://doi.org/10.1515/9783110627459-021

(a) In the Harrod–Domar model, the autonomous investment $A(t)$ is considered as an exogenous variable, which is independent of national income $Y(t)$.
(b) In the model without memory, the consumption $C(t)$ is a linear function of national income that is described by the linear equation

$$C(t) = cY(t), \tag{21.2}$$

where c is the marginal propensity to consume ($0 < c < 1$). Equation (21.2) described the linear multiplier without memory and lag.
(c) In the standard Harrod–Domar model of the growth without memory, it is assumed that induced investment $I(t)$ is determined by the rate of the national income. This assumption is described by the linear differential equation

$$I(t) = vY^{(1)}(t), \tag{21.3}$$

where v is the positive constant, which is called the investment coefficient indicating the power of accelerator or the capital intensity of the national income. Here, $Y^{(1)}(t) = dY(t)/dt$ is the first-order derivative of the function $Y(t)$. Equation (21.3) means that the induced investment is a constant proportion of the current rate of change of income. Equation (21.3) described the linear accelerator without memory and lag.

Substitution of equations (21.2) and (21.3) into (21.1) gives

$$Y(t) = cY(t) + vY^{(1)}(t) + A(t). \tag{21.4}$$

Equation (21.4) can be written as

$$vY^{(1)}(t) = sY(t) - A(t), \tag{21.5}$$

where $s = 1 - c$ is the marginal propensity to save ($0 < s < 1$).

The economic dynamics, which is represented by equation (21.5), can be qualitatively described in the following form. If independent investments $A(t)$ grow, for example, due to the sudden appearance of large inventions, the multiplier $A(t)/(1-c)$ gives rise to a corresponding increase in output [10, p. 65], where c is the marginal propensity to consume ($0 < c < 1$). The expansion of output drives the accelerator and is accompanied by the appearance of other (induced) investments. In turn, these additional investments increase (“multiply”) the products due to the economic multiplier, and a new cycle begins. In general, the result is a progressive increase of national income.

An important characteristic of macroeconomic growth models is the Harrod’s warranted rate of growth output [10], which is also called the technological growth rate [146, p. 49]. The warranted (technological) rate describes the growth rate in the case of the constant structure of the economy and the absence of external influences. The constant structure means that the parameters of the model are constants (e. g., s, v are constants). The absence of external influences means the absence of exogenous variables ($A(t) = 0$). Mathematically, the warranted rate of growth in the long term can be described by the asymptotic behavior of the solution of the homogeneous differential equation for the macroeconomic model.

In the standard Harrod–Domar model, the solution of equation (21.5) with $A(t) = 0$ has the form $Y(t) = Y(0)\exp(\omega t)$. Therefore, the warranted rate of growth for this standard model is described by the value $\omega = s/v$.

Equation (21.5) defines the Harrod–Domar model without memory and lag, where the behavior of the national income $Y(t)$ is determined by the dynamics of the autonomous investment $A(t)$. The solution of equation (21.5) depends on what is assumed about the change of autonomous expenditure over time. The solution of equation (21.5) and its analysis are given in [8, 9, 10].

21.3 Harrod–Domar growth model with memory

The standard Harrod–Domar model is described by first-order differential equation (21.5), which is based on the multiplier equation (21.2) and the accelerator equation (21.3). Equation (21.2) assumes that the consumption $C(t)$ changes instantly when income changes. The derivative of the first order, which is used in equation (21.3), implies an instantaneous change of the investment $I(t)$ when changing the growth rate of the national income $Y(t)$. Because of this, accelerator equation (21.3) does not take into account memory and lag. As a result, equation (21.5) can be used only to describe an economy in which all economic agents have an instantaneous amnesia. This restriction substantially narrows the field of application of macroeconomic models to describe the real economic processes. In many cases, economic agents can remember the history of changes of the national income and investment and this fact influences the decision-making by economic agents. The Harrod–Domar model with one-parameter power-law memory was first proposed in the articles [421, 426, 450] in 2016.

Let us consider the Harrod–Domar model with memory. We can use the equation of investment accelerator with memory in the form

$$I(t) = v \int_0^t M(t-\tau) Y^{(n)}(\tau)\, d\tau, \tag{21.6}$$

where $M(t-\tau)$ is the memory function or the weighting function (the probability density function) that describes the lag. For $M(t-\tau) = \delta(t-\tau)$, equation (21.6) gives equation (21.3) of the standard accelerator without memory and lag. Substituting expression (21.6) into balance equation (21.1), and using expression (21.2), we obtain the integro-differential equation

$$v \int_0^t M(t-\tau) Y^{(n)}(\tau)\, d\tau = sY(t) - A(t). \tag{21.7}$$

For $M(t-\tau) = \delta(t-\tau)$, equation (21.7) gives equation (21.5) that describes the standard Harrod–Domar model without memory and lag.

Equation (21.7) determines the dynamics of the national income within the framework of the Harrod–Domar macroeconomic model of growth with memory (and the time delay). If the parameter s, v is given, then the dynamics of national income $Y(t)$ is determined by the behavior of the autonomous investment $A(t)$.

If the function $M(\tau)$ describes memory with power-law fading in the form,

$$M(t-\tau) = M_{\mathrm{RL}}^{n-\alpha}(t-\tau) = \frac{1}{\Gamma(n-\alpha)}(t-\tau)^{n-\alpha-1}, \tag{21.8}$$

then equation (21.8), which describes the accelerator with memory, has the form

$$I(t) = v(D_{C;0+}^{\alpha} Y)(t), \tag{21.9}$$

where $(D_{C;0+}^{\alpha} Y)(t)$ is the Caputo fractional derivative of the order $\alpha \geq 0$ that is defined by the equation

$$(D_{C;0+}^{\alpha} Y)(t) = (M_{\mathrm{RL}}^{n-\alpha} * Y^{(n)})(t) = \frac{1}{\Gamma(n-\alpha)} \int_0^t (t-\tau)^{n-\alpha-1} Y^{(n)}(\tau)\, d\tau, \tag{21.10}$$

where $*$ is the Laplace convolution, $n = [\alpha] + 1$ for noninteger values of α and $n = \alpha$ for integer values of α, $\Gamma(\alpha)$ is the gamma function and $t > 0$. The Caputo fractional derivative (21.10) of order $\alpha > 0$ exists almost everywhere on $[0, T]$, if $Y(\tau) \in AC[0, T]$ (e. g., see Theorem 2.1 [200, p. 92]). The condition $Y(\tau) \in AC[0, T]$ means that the function $Y(\tau)$ has integer-order derivatives up to $(n-1)$-th order, which are continuous functions on the interval $[0, T]$, and the derivative $Y^{(n)}(\tau)$ is a Lebesgue summable on $[0, T]$.

In general, the capital intensity depends on the parameter of memory fading, i. e., $v = v(\alpha)$. For equation (21.8), the Harrod–Domar model with power-law memory is described by the fractional differential equation

$$v(D_{C;0+}^{\alpha} Y)(t) = sY(t) - A(t). \tag{21.11}$$

We can say that the fading parameter $0 < \alpha < 1$ corresponds to subgrowth and the parameter $1 < \alpha < 2$ corresponds to supergrowth. This interpretation is based on the results of our works [453, 450, 445]. In macroeconomic models, the parameter $0 < \alpha < 1$ leads to a slowdown (inhibition) of economic growth [453, 450, 393, 387, 445]. The fading parameter $1 < \alpha < 2$ can lead to an increase in the economic growth and to growth instead of decline [453, 450, 393, 387, 445].

The parameters of memory fading are interpreted as the orders of fractional derivatives and integrals. The parameter of memory fading can be determine by analyzing the long-range time dependence in time series based on economic data. Note that we can use the criteria of the existence of power-law memory which are proposed in [399] (see also [413]).

Note that the Harrod–Domar model with one-parameter power-law memory was proposed in [421, 426, 450]. The solutions of the fractional differential equation (21.11) and its properties are also described in [453, 450, 393, 387, 445]. We proved [453, 450, 393, 387, 445] that the warranted rates of growth for macroeconomic models with one-parameter memory do not coincide with the growth rates $\omega = s/v$ of standard Harrod–Domar model. The warranted rate of growth with memory is equal to the value $\omega_{\mathrm{eff}}(\alpha) = \omega^{1/\alpha}$, where $\alpha > 0$ characterizes power-law fading of memory. In the works [453, 450, 393, 387, 445], we demonstrate that the account of memory effects can significantly change the warranted rates of growth. The principles of changing of warranted rates of growth by power-law memory have been suggested in [453, 393, 387, 445]. The warranted rates may both increase and decrease in comparison with the standard Harrod–Domar model without memory. The accounting of the memory can give a new type of behavior for the same parameters of the macroeconomic model. The memory with the fading parameter $\alpha \in (0, 1)$ leads to a slowdown in the growth and decline of the economy. In other words, the effect of fading memory with $\alpha \in (0, 1)$ leads to inhibition of economic growth and decline. The memory with the fading parameter $\alpha > 1$ leads to an improvement in economic dynamics, such as a slowdown in the rate of decline, a replacement of the economic decline by its growth, and an increase in the rate of economic growth.

21.4 Equation for growth with memory and lag

Assuming that induced investment $I(t)$ depends on the power-law memory and continuously distributed delay time, which is described by the gamma distribution, we can use the equation of accelerator with memory and lag in the form

$$I(t) = v(\mathbf{D}^{\alpha,a;\lambda}_{T;C;0+} Y)(t), \tag{21.12}$$

where $\mathbf{D}^{\alpha,a;\lambda}_{T;C;0+}$ is the Caputo fractional derivative with gamma distributed lag that is defined by the equation

$$(\mathbf{D}^{\alpha,a;\lambda}_{T;C;0+} Y)(t) = \frac{\lambda^a \Gamma(a)}{\Gamma(a+n-\alpha)} \int_0^t (t-\tau)^{n-\alpha+a-1} F_{1,1}(a; a+n-\alpha; -\lambda(t-\tau)) Y^{(n)}(\tau)\, d\tau, \tag{21.13}$$

where $n - 1 < \alpha \leq n$.

Substitution of equations (21.2) and (21.12) into (21.1) gives the macroeconomic growth model with memory and lag. The Harrod–Domar model with power-law memory and gamma distributed lag is described by the fractional differential equation

$$v(\mathbf{D}_{T;C;0+}^{\alpha,a;\lambda} Y)(t) = sY(t) - A(t), \tag{21.14}$$

where $\mathbf{D}_{T;C;0+}^{\alpha,a;\lambda}$ is the fractional derivative of order $0 < \alpha < 2$ ($n = [\alpha]+1$), with the shape parameter $a > 0$ and the rate $\lambda > 0$. For the Erlang distribution, the shape parameter is integer number ($a = m \in \mathbb{N}$).

For simplification, we rewrite equation (21.14) in the form

$$(\mathbf{D}_{T;C;0+}^{\alpha,a;\lambda} Y)(t) = \omega Y(t) + F(t), \tag{21.15}$$

where

$$\omega = \frac{s}{v}, \quad F(t) = -\frac{1}{v} A(t). \tag{21.16}$$

The general solution of the nonhomogeneous equation (21.15) has the form

$$Y(t) = Y_0(t) + Y_F(t). \tag{21.17}$$

The function $Y_0(t)$ is the solution of the homogeneous equation

$$(\mathbf{D}_{T;C;0+}^{\alpha,a;\lambda} Y)(t) = \omega Y(t), \tag{21.18}$$

where ω is defined by equation (21.16). The function $Y_F(t)$ is a particular solution of the nonhomogeneous equation. This particular solution has the form

$$Y_F(t) = \int_0^t G_\alpha[t-\tau] F(\tau)\, d\tau = -\frac{1}{v} \int_0^t G_\alpha[t-\tau] A(\tau)\, d\tau, \tag{21.19}$$

where $G_\alpha[t-\tau]$ is the fractional analog of the Green function [200, pp. 281, 295].

Let us obtain the solution for the homogeneous fractional differential equation (21.18). The Laplace transform of equation (21.18) has the form

$$\frac{\lambda^a}{(s+\lambda)^a} \left(s^\alpha (\mathcal{L}Y)(s) - \sum_{j=0}^{n-1} s^{\alpha-j-1} Y^{(j)}(0) \right) = \omega (\mathcal{L}Y)(s), \tag{21.20}$$

where $\omega = s/v$. Then we get

$$(\mathcal{L}Y)(s) = \sum_{j=0}^{n-1} \frac{\lambda^a s^{\alpha-j-1}}{\lambda^a s^\alpha - \omega (s+\lambda)^a} Y^{(j)}(0) = \sum_{j=0}^{n-1} \frac{s^{\alpha-j-1}}{s^\alpha - \mu (s+\lambda)^a} Y^{(j)}(0), \tag{21.21}$$

where $\mu = \omega\lambda^{-a}$. Let us define the special function $S^{\gamma}_{\alpha,\delta}[\mu,\lambda|t]$ in the form

$$S^{\gamma}_{\alpha,\delta}[\mu,\lambda|t] = -\sum_{k=0}^{\infty} \mu^{-(k+1)} t^{\delta(k+1)-\alpha k-\gamma-1} E^{\delta(k+1)}_{1,\delta(k+1)-\alpha k-\gamma}[-\lambda t], \tag{21.22}$$

where $E^{\gamma}_{\alpha,\beta}[z]$ is the three parameter Mittag–Leffler functions [143]. Using the equation $F_{1,1}(a;c;z) = \Gamma(c)E^{a}_{1,c}[z]$ (see equation (5.1.18) of [143, p. 99]) in the form

$$E^{\delta(k+1)}_{1,\delta(k+1)-\alpha k-\gamma}[-\lambda t] = \frac{1}{\Gamma(\delta(k+1)-\alpha k-\gamma)} F_{1,1}(\delta(k+1); \delta(k+1)-\alpha k-\gamma, -\lambda t), \tag{21.23}$$

function (21.22) can be represented as an infinite series with the confluent hypergeometric Kummer functions $F_{1,1}(a;b;z)$.

The S-function $S^{\gamma}_{\alpha,\delta}[\mu,\lambda|t]$ allows us to represent solutions of the fractional differential equations with derivatives of noninteger order with gamma distributed lag. In the S-function (21.22), the parameter $\delta > 0$ is interpreted as the shape of the gamma distribution, the parameter $\lambda > 0$ is interpreted as the rate of the gamma distribution, and the parameter $\alpha > 0$ is interpreted as a parameter of memory fading.

Using the inverse Laplace transform (see equation (5.4.9) of [26]) in the form

$$\left(\mathcal{L}^{-1}\left(\frac{s^{a}}{(s+b)^{c}}\right)\right)(s) = \frac{1}{\Gamma(c-a)} t^{c-a-1} F_{1,1}(c; c-a, -bt), \tag{21.24}$$

where $\mathrm{Re}(c-a) > 0$, we can proof [400] that the Laplace transform of the S-function (21.22) has the form

$$\mathcal{L}(S^{\gamma}_{\alpha,\delta}[\mu,\lambda|t])(s) = \frac{s^{\gamma}}{s^{\alpha} - \mu(s+\lambda)^{\delta}}. \tag{21.25}$$

Note that we can use equation (5.1.6) of [143, p. 98] (or equation (3) of [132, p. 316], and equation (2.5) in [314, p. 8]), in the form

$$(\mathcal{L}(t^{\beta-1}E^{\gamma}_{\alpha,\beta}[\lambda t^{\alpha}]))(s) = \frac{s^{\alpha\gamma-\beta}}{(s^{\alpha}-\lambda)^{\gamma}}, \tag{21.26}$$

where $\mathrm{Re}(s) > 0$, $\mathrm{Re}(\beta) > 0$ and $|s|^{\alpha} > |\lambda|$. Equation (21.26) can be used instead of equation (21.24) to represent the S-function though generalized Mittag–Leffler functions instead of the confluent hypergeometric functions.

As a result, the solution of homogenous fractional differential equation (21.18) is represented in the form

$$Y_0(t) = \sum_{j=0}^{n-1} S^{\alpha-j-1}_{\alpha,a}[\omega\lambda^{-a}, \lambda|t] Y^{(j)}(0), \tag{21.27}$$

where $S_{\alpha,a}^{\alpha-j-1}[\omega\lambda^{-a},\lambda|t]$ is defined by equation (21.22). Here, $a > 0$ and $\lambda > 0$ are the shape and rate parameters of the gamma distribution, respectively. For $0 < \alpha < 1$, solution (21.27) can be written in the form

$$Y_0(t) = -\sum_{k=1}^{\infty} \omega^{-k}\lambda^{ak} t^{(\alpha+a)k} E^{ak}_{1,(\alpha+a)k+1}[-\lambda t]Y(0), \tag{21.28}$$

where $E^{\gamma}_{\alpha,\beta}[\lambda t^{\alpha}]$ is the three parameter Mittag–Leffler function [143].

Note that for special case of the absence of time delay (lag) the solution of equation (21.15) is expressed thought the two parameter Mittag–Leffler function [143], in which the argument depends on ωt^{α}, instead of the rate parameter $\lambda > 0$ of the gamma distribution.

Let us obtain the particular solution of equation (21.15). This particular solution is represented in form (21.19), where $G_{\alpha}[t-\tau]$ is the fractional analog of the Green function [200, pp. 281, 295]. Expression (21.19) yields a solution $Y_F(t)$ for nonhomogeneous equation (21.15) with zero initial conditions, $Y^{(j)}(0) = 0$ for $j = 0, \dots (n-1)$. The Laplace transform of equation (21.15) with conditions $Y^{(j)}(0) = 0$ for all $j = 0, \dots (n-1)$ has the form

$$(\mathcal{L}Y)(s) = \frac{(s+\lambda)^a}{\lambda^a s^{\alpha} - \omega(s+\lambda)^a}(\mathcal{L}F)(s). \tag{21.29}$$

Using the transformations

$$\begin{aligned}\frac{(s+\lambda)^a}{\lambda^a s^{\alpha} - \omega(s+\lambda)^a} &= \frac{1}{\lambda^a s^{\alpha}/(s+\lambda)^a - \omega} = -\frac{1}{\omega}\frac{1}{1-\frac{\lambda^a s^{\alpha}}{\omega(s+\lambda)^a}} = -\frac{1}{\omega}\sum_{k=0}^{\infty}\left(\frac{\lambda^a s^{\alpha}}{\omega(s+\lambda)^a}\right)^k \\ &= -\frac{1}{\omega}\sum_{k=0}^{\infty}\frac{\lambda^{ak}}{\omega^k}\frac{s^{\alpha k}}{(s+\lambda)^{ak}},\end{aligned} \tag{21.30}$$

where

$$\left|\frac{\lambda^a s^{\alpha}}{\omega(s+\lambda)^a}\right| < 1 \tag{21.31}$$

we get

$$\begin{aligned}(\mathcal{L}Y)(s) &= -\sum_{k=0}^{\infty}\frac{\lambda^{ak}}{\omega^{k+1}}\frac{s^{\alpha k}}{(s+\lambda)^{ak}}(\mathcal{L}F)(s) = -\frac{1}{\omega}(\mathcal{L}F)(s) - \sum_{k=1}^{\infty}\frac{\lambda^{ak}}{\omega^{k+1}}\frac{s^{\alpha k}}{(s+\lambda)^{ak}}(\mathcal{L}F)(s) \\ &= -\frac{1}{\omega}(\mathcal{L}F)(s) - \sum_{k=0}^{\infty}\frac{\lambda^{a(k+1)}}{\omega^{k+2}}\frac{s^{\alpha(k+1)}}{(s+\lambda)^{a(k+1)}}(\mathcal{L}F)(s).\end{aligned} \tag{21.32}$$

Using equation (5.4.8) of [26, p. 238] of the inverse Laplace transform in the form

$$\left(\mathcal{L}^{-1}\left(\frac{s^a}{(s+b)^c}\right)\right)(s) = \frac{1}{\Gamma(c-a)}t^{c-a-1}F_{1,1}(c; c-a, -bt), \tag{21.33}$$

where $a = \alpha(k+1)$, $c = a(k+1)$, $b = \lambda$, we obtain the expression

$$G_\alpha[t-\tau] = -\frac{1}{\omega}\delta(t-\tau) - \frac{1}{\omega}\sum_{k=0}^{\infty}\frac{1}{\mu^{k+1}}\frac{(t-\tau)^{(k+1)(a-\alpha)-1}}{\Gamma((k+1)(a-\alpha))} \times F_{1,1}(a(k+1);(k+1)(a-\alpha),-\lambda(t-\tau)), \tag{21.34}$$

where $-(k+1)(a-\alpha) \notin \mathbb{N}$. Using the special function $S^{\gamma}_{\alpha,\delta}[\mu,\lambda|t]$ with $\delta = a$, $\gamma = \alpha$, we get the representation

$$G_\alpha[t-\tau] = -\frac{1}{\omega}\delta(t-\tau) + \frac{1}{\omega}S^{\alpha}_{\alpha,a}[\mu,\lambda|t-\tau]. \tag{21.35}$$

As a result, we get the expression

$$Y_F(t) = \frac{1}{v\omega}A(t) - \frac{1}{v\omega}\int_0^t S^{\alpha}_{\alpha,a}[\mu,\lambda,t-\tau]A(\tau)\,d\tau, \tag{21.36}$$

which describes the particular solution $Y_F(t)$ of the nonhomogeneous equation (21.15) for gamma distribution of delay time.

Remark 21.1. For the Erlang distribution, the shape parameter is integer number ($a = m \in \mathbb{N}$) and we have

$$(\mathcal{L}Y)(s) = \sum_{k=0}^{m}\binom{m}{k}\frac{s^k\lambda^{m-k}}{\lambda^m s^\alpha - \omega(s+\lambda)^m}(\mathcal{L}F)(s) = \sum_{k=0}^{m}\binom{m}{k}\frac{\lambda^{-k}s^k}{s^\alpha - \mu(s+\lambda)^m}(\mathcal{L}F)(s), \tag{21.37}$$

where $\mu = \omega\lambda^{-a} = \omega\lambda^{-m}$, and we used the binomial expansion

$$(s+\lambda)^m = \sum_{k=0}^{m}\binom{m}{k}s^k\lambda^{m-k}. \tag{21.38}$$

Using the special function $S^{\gamma}_{\alpha,\delta}[\mu,\lambda|t]$, we can write

$$(\mathcal{L}Y)(s) = \sum_{k=0}^{m}\binom{m}{k}\lambda^{-k}\mathcal{L}(S^{k}_{\alpha,m}[\mu,\lambda|t])(s)(\mathcal{L}F)(s). \tag{21.39}$$

Therefore, the fractional Green function for the Erlang distribution of lag has the form

$$G_\alpha[t-\tau] = \sum_{k=0}^{m}\binom{m}{k}\lambda^{-k}S^{k}_{\alpha,m}[\mu,\lambda|t-\tau] = -\sum_{k=0}^{m}\sum_{j=0}^{\infty}\binom{m}{j}\frac{\lambda^{-j}(t-\tau)^{m(k+1)-\alpha k-k-1}}{\mu^{k+1}\Gamma(m(k+1)-\alpha k-k)} \times F_{1,1}(m(k+1);m(k+1)-\alpha k-k,-\lambda(t-\tau)), \tag{21.40}$$

where $\mu = \omega\lambda^{-a} = \lambda^{-a}s/v$. Using the three parameter Mittag–Leffler function, we obtain the particular solution $Y_F(t)$ of nonhomogeneous equation (21.15) for the Erlang distribution of delay time in the form

$$Y_F(t) = \frac{1}{v}\int_0^t \sum_{k=0}^{m}\sum_{j=0}^{\infty}\binom{m}{j}\frac{\lambda^{-j}}{\mu^{k+1}}(t-\tau)^{m(k+1)-\alpha k-k-1}E^{m(k+1)}_{1,m(k+1)-\alpha k-k}(-\lambda(t-\tau))A(\tau)\,d\tau, \quad (21.41)$$

where $A(t)$ is the exogenous variable that describes autonomous investment.

Mathematically, the warranted rate in the long term can be described by the asymptotic behavior of the solution of the homogeneous differential equation for the macroeconomic model. Let us consider the warranted rate of growth of national income by using the asymptotic behavior of the solution of the homogeneous fractional differential equation. This solution is represented through the special function $S^{\gamma}_{\alpha,\delta}[\mu,\lambda|t]$ which is defined as an infinite sum of the confluent hypergeometric Kummer function or the generalized Mittag–Leffler function. For these functions, asymptotic expressions exist. The asymptotic behavior of the confluent hypergeometric Kummer function $\Phi(a;c;z) = F_{1,1}(a;c;z)$ at infinity $z \to -\infty$ is known (e. g., see equations (1.6.17–1.6.9) in [200, p. 29]). However, at present time, there are no asymptotic expressions of the function $S^{\gamma}_{\alpha,\delta}[\mu,\lambda|t]$, which is an infinite series of the confluent hypergeometric Kummer function. In the paper [404], on the basis of asymptotic expressions for the confluent hypergeometric Kummer function, we proposed an assumption about the behavior of the solutions. Equations of solution allows us to assume that the warranted rate of growth has the power-law form with the power $-\alpha + j$, where $j \in 0,\dots,n-1$ is a smallest values at which $Y^{(j)}(0) \neq 0$. As a result, the growth of the national income with memory and distributed lag has the power-law type instead of the exponential type of growth with memory in absence of lag [453, 450, 393, 387, 445], where the warranted rate of growth is $\omega_{\text{eff}}(\alpha) = \omega^{1/\alpha}$. Therefore, we can assume that the distributed lag (time delay) suppresses the effects of fading memory. We assume that some information about the behavior of the national income for simultaneous effects of memory and lag can be obtained by computer simulation of the proposed model.

21.5 Conclusion

The standard Harrod–Domar growth model, which is proposed in papers [170, 171, 172, 92, 93], describes the behavior of national income in the absence of distributed lag and memory. A generalization of this model by taking into account the exponentially distributed lag without memory was proposed by Phillips in [305, 306]. The Harrod–Domar growth model with power-law memory was suggested by authors in

[421, 426, 450]. In this chapter, based on work [421, 426], we have described a generalization of Harrod–Domar growth model by taking into account one-parameter power-law memory and gamma distributed delay time. We obtain the fractional differential equation of this generalized model, where we use fractional derivatives with distributed lag to take into account the memory and lag in the economic accelerator. The solution, which describes the behavior of national income, has been proposed. We assume that the asymptotic behavior of national income with memory and distributed lag demonstrates the power-law type of growth instead of the exponential growth for processes with memory in the absence of time delay. The warranted rate of growth with memory and without lag is equal to the value $\omega_{\mathrm{eff}}(\alpha) = \omega^{1/\alpha}$, where $\alpha > 0$ is a memory fading parameter. The memory effects can significantly accelerate the growth rate of the economy. We can assume that the continuously distributed lag suppresses the influence of memory effects. However, the question of whether taking into account the influence of power-law memory in processes with distributed lag leads to economic growth or decline, and under what conditions this occurs, remains an open question at present. We assume that some information about this question for simultaneous effects of memory and lag can be obtained by computer simulations of the proposed model.

22 Dynamic Keynesian model with memory and lag

In this chapter, we consider a generalization of the standard Keynesian macroeconomic model is first proposed in [402, 407]. In this model of economic growth, we take into account the fading memory and continuous distribution of delay time. To take into account the simultaneous action of the fading memory and gamma distributed lag, we use the integral and integro-differential operators with the confluent hypergeometric Kummer function in the kernel. These operators allow us to formulate the concepts of economic accelerator and multiplier, in which the memory and lag are taken into account. Fractional differential equation, which describes dynamics of national income in this generalized model, is suggested. The solution of this fractional differential equation is obtained. This chapter is based on works [402, 407].

22.1 Introduction

In works [402, 407], we suggest a generalization of one of the most famous models of economic growth, which is associated with the founder of modern macroeconomic theory, John M. Keynes [191, 193, 194, 192]. In suggested generalization, we take into account two types of phenomena: (I) memory with power-law fading and (II) continuously distributed lag with gamma distribution of delay time.

The continuously distributed lag has been considered in economics starting with the works of Michal A. Kalecki [189] and Alban W. H. Phillips [305, 306]. The dynamic model of business cycles with continuous uniform distribution of delay time was considered by Michal A. Kalecki in 1935 [189] (see also Section 8.4 of [10, pp. 251–254]). The economic growth models with exponential distributed lag were proposed by Alban W. H. Phillips [305, 306] in 1954 (see also Sections 3.4 and 3.5 of [10, pp. 69–74]). The operators with continuously distributed lag were considered by Roy G. D. Allen in 1956 [8] (see also [10, pp. 23–29]).

The time delay (lag) is caused by finite speeds of processes, i. e., the change of one variable does not lead to instant changes of another variable. Therefore, the distributed lag (time delay) cannot be considered as a memory in processes. For example, the change of electromagnetic field in vacuum at the point of observation is delayed with respect to the change in the sources of the field located at another point. It is known that the propagation of the electromagnetic field in a vacuum does not mean the presence of memory in this process.

An application of advanced mathematical methods of fractional calculus in Keynesian economic models with continuous time was first proposed in articles [423, 427, 422] in 2016. The fractional differential equations of the dynamic Keynesian model with power-law memory and their solutions have been considered. The continuously distributed lag was not discussed in these works. To take into account the simultaneous action of the fading memory and distributed lag, we should use the integral

https://doi.org/10.1515/9783110627459-022

and integro-differential operators that are proposed in [400]. Using these mathematical tools, a generalization of the standard Keynesian macroeconomic model was proposed in [402, 407].

For macroeconomics, it is important to simultaneously take into account lagging and memory phenomena. In this chapter, we consider memory with power-law fading and lag with gamma distribution of delay time. The memory is described by the Riemann–Liouville fractional integrals and the Caputo fractional derivatives. The distributed lag is described by the generalization of the translation operator, in which the delay time is considered as a random variable that is distributed by probability law (distribution) on positive semi-axis. The composition of these operators can be represented as the Abel-type integral and integro-differential operators with the confluent hypergeometric Kummer function in the kernel. Using these operators, we propose the fractional differential equation for generalized dynamic Keynesian model, which describes the fractional dynamics of national income. We obtain solution of this equation that describes the macroeconomic growth with power-law fading memory and gamma distribution of delay time.

22.2 Standard dynamic Keynesian model

Let us consider the standard dynamical Keynesian model with continuous time. In the Keynesian model, the following variables are used to describe the dynamics of the revenue and expenditure parts of the economy: $Y(t)$ is the national income; $G(t)$ is the government expenditure; $C(t)$ is the consumption expenditure; $I(t)$ is the investment expenditure. The total expenditure $E(t)$ is defined as the sum

$$E(t) = C(t) + I(t) + G(t). \tag{22.1}$$

In dynamic equilibrium, we have

$$Y(t) = E(t). \tag{22.2}$$

In this case, the balance equation establishes the equality of the national income to the sum of all expenditures

$$Y(t) = C(t) + I(t) + G(t). \tag{22.3}$$

In the Keynesian model, it is assumed that the consumption expenditure at time t depends on the income level at the same time. This means that the model neglects the effects of time delay and memory. The consumption expenditure $C(t)$ is regarded as an endogenous variable equal to the amount of domestic consumption of some part of the national income and final consumption independent of income. As a result, the consumption expenditure $C(t)$ is described by the linear equation of the economic multiplier

$$C(t) = m(t)Y(t) + b(t), \tag{22.4}$$

where $m(t)$ is the multiplier factor that describes the marginal propensity to consume $(0 < m(t) < 1)$, and the function $b(t) > 0$ describes the autonomous consumption that does not depend on income. The expression $m(t)Y(t)$ describes the part of consumption that depends on income.

In the static model, the investment expenditure and government expenditure are considered as exogenous variables. In the dynamic Keynesian model, the investment expenditure $I(t)$ is treated as endogenous and it is assumed to depend on the level of income [494, pp. 95–97]. In the standard model, the investment expenditure $I(t)$ is determined by the rate of change of the national income. This assumption is described by the equation of the economic accelerator

$$I(t) = v(t)Y^{(1)}(t), \tag{22.5}$$

where $v(t)$ is the rate of acceleration, which characterizes the level of technology and state infrastructure. Here, $Y^{(1)}(t) = dY(t)/dt$ is the first-order derivative of the income function $Y(t)$ with respect to time variable t.

In the Keynesian model, government expenditure $G(t)$, the propensity to consume $m(t)$, the rate of acceleration $a(t)$, and the autonomous consumption $b(t)$ are exogenous variables that are specified as external to the model. These variables, as functions of time, are assumed to be given.

The purpose of the dynamic Keynesian model is to describe the behavior of the national income. For this, it is necessary to find the national income $Y(t)$, as a function of time t. Substituting the multiplier equation (22.4) and the accelerator equation (22.5) into the balance equation (22.3), we obtain

$$Y(t) = m(t)Y(t) + b(t) + v(t)Y^{(1)}(t) + G(t). \tag{22.6}$$

Equation (22.6) can be written in the form

$$\frac{dY(t)}{dt} - \frac{1 - m(t)}{v(t)} Y(t) = -\frac{G(t) + b(t)}{v(t)}. \tag{22.7}$$

Equation (22.7) of the dynamic Keynesian model is a nonhomogeneous linear differential equation with a first-order derivative.

We see that the functions $G(t)$ and $b(t)$ are included in equation (22.7) as a sum. This could be expected, since $G(t)$ is an independent expenditure on investment, that is independent of income, and $b(t)$ is an independent expenditure on consumption, which is also not dependent on national income. These two types of expenditure complement each other [494, pp. 95–98]. Therefore, it is convenient to use the sum

$$G_b(t) = G(t) + b(t), \tag{22.8}$$

which describes the independent expenditure. In this case, the consumption function $C(t) = m(t)Y(t)$ is that part of consumption that depends on income. Therefore, we can rewrite the equation of the standard dynamic Keynesian model in the form

$$\frac{dY(t)}{dt} = \frac{1 - m(t)}{v(t)} Y(t) - \frac{G_b(t)}{v(t)}, \qquad (22.9)$$

where $0 < 1 - m(t) < 1$, and $v(t) > 0$. Equation (22.9) is the nonhomogeneous first-order differential equation that describes the standard dynamic Keynesian model, which does not take into account the effects of memory and time delay.

22.3 Dynamic Keynesian model with memory

In the standard Keynesian model, equation (22.9) implies an instantaneous change in the investment expenditure, when the rate of growth of national income changes. This means that the equation of the standard model does not take into account the effects of memory and delay. Mathematically, this is due to the fact that the standard model equation is the differential equation of the first order. The derivative of the first order, which is used in accelerator equations (22.5), implies an instantaneous change of the investment expenditure $I(t)$, when changing the rate of the national income $Y(t)$. Because of this, accelerator equation (22.5) does not take into account memory and lag. Multiplier equation (22.4) also assumes that the consumption expenditure $C(t)$ changes instantly, when national income changes. As a result, equation (22.9) can describe an economy, in which agents have no memory. This fact greatly limits the applicability of the standard model to describe the real processes in economy. To expand the scope of the model, we should take into account that economic agents can remember the history of changes of the national income and the investment expenditure, and it affects the behavior of these agents.

Generalization of the standard Keynesian model, in which a memory is taken into account, was proposed by the authors [423, 427, 422]. Let us briefly describe this generalized model with memory. The equation of investment accelerator with memory can be written as

$$I(t) = \int_0^t V(t, \tau) Y^{(n)}(\tau)\, d\tau, \qquad (22.10)$$

where $V(t, \tau)$ can be interpreted as the function that characterizes time lag between appearance and implementation of new technologies, obsolescence of technology and infrastructure. The function $V(t, \tau)$ is considered as a generalization of the function $v(t)$, which is used in the standard Keynesian model.

To simplify our consideration, we can assume that the function $V(t, \tau)$ is representable in the form

$$V(t, \tau) = v(t) M(t - \tau), \qquad (22.11)$$

where $M(t-\tau)$ is the memory function, or the weighting function, which is interpreted as the probability density function. Using equations (22.11) and (22.10), we get

$$I(t) = v(t) \int_0^t M(t-\tau) Y^{(n)}(\tau)\, d\tau. \tag{22.12}$$

Equation (22.5) is as a special case of equation (22.12) for the case $M(t-\tau) = \delta(t-\tau)$ and $n = 1$. Substitution of the investment $I(t)$ in the form of expression (22.12), and the consumption expenditure $C(t)$ in form (22.4) into balance equation (22.3), we get

$$v(t) \int_0^t M(t-\tau) Y^{(n)}(\tau)\, d\tau = (1-m(t))Y(t) - G_b(t), \tag{22.13}$$

where $G_b(t)$ is defined by equation (22.8). Equation (22.9) of the standard Keynesian model without memory and lag can be obtained from (22.13) by using $M(t-\tau) = \delta(t-\tau)$ and $n = 1$.

Equation (22.13) can describe the dynamics of the national income within the framework of the Keynesian model of growth with memory and lag. If the parameters $m(t)$ and $v(t)$ are given, the growth of national income $Y(t)$ is conditioned by the behavior of the independent expenditure (22.8).

Let us consider the memory with one-parameter power-law fading that is described by the function

$$M(t-\tau) = M_{\mathrm{RL}}^{n-\alpha}(t-\tau) = \frac{1}{\Gamma(n-\alpha)}(t-\tau)^{n-\alpha-1}, \tag{22.14}$$

where $\Gamma(\alpha)$ is the gamma function and $n-1 < \alpha \leq n$. Using expressions (22.14), equation (22.14) describes the accelerator with memory in the form

$$I(t) = v(t)(D_{C;0+}^{\alpha} Y)(t), \tag{22.15}$$

where $(D_{C;0+}^{\alpha} Y)(t)$ is the Caputo fractional derivative [200]. In general, the rate of acceleration $v(t)$ depends on the parameter of memory fading, i. e., $v(t) = v(t,\alpha)$. The parameter $\alpha > 0$ is interpreted as a fading parameter of power-law memory [450]. The concept of accelerator with memory [412, 397] allows us to get the equation of the Keynesian model with power-law memory in the form of the fractional differential equation

$$(D_{C;0+}^{\alpha} Y)(t) = \frac{1-m(t)}{v(t)} Y(t) - \frac{G_b(t)}{v(t)}, \tag{22.16}$$

where $D_{C;0+}^{\alpha}$ is the Caputo fractional derivative that can be represented by the Laplace convolution

$$(D_{C;0+}^{\alpha} Y)(t) = (M_{\mathrm{RL}}^{n-\alpha} * Y^{(n)})(t) = \frac{1}{\Gamma(n-\alpha)} \int_0^t (t-\tau)^{n-\alpha-1} Y^{(n)}(\tau)\, d\tau, \tag{22.17}$$

where $n = [\alpha]+1$ for $\alpha \notin \mathbb{N}$ and $n = \alpha$ for $\alpha \in \mathbb{N}$, and the function $Y(\tau)$ has integer-order derivatives $Y^{(j)}(\tau)$, $j = 1, \dots, (n-1)$, that are absolutely continuous.

The solution of equation (22.16) with constant values of $m(t) = m$ and $v(t) = v$ has the form

$$Y(t) = \sum_{j=0}^{n-1} Y^{(j)}(0) t^j E_{\alpha,j+1}\left[\frac{1-m}{v} t^{\alpha}\right] - \frac{1}{v}\int_0^t (t-\tau)^{\alpha-1} E_{\alpha,\alpha}\left[\frac{1-m}{v}(t-\tau)^{\alpha}\right] G_b(t)\, d\tau, \quad (22.18)$$

where $n-1 < \alpha \le n$. Solution (22.18) of the fractional differential equation (22.16) and it properties are described in [423, 427, 422] (see also [453, 393, 387]).

22.4 Equation of Keynesian model with memory and lag

Let us assume that the relationship between the investment expenditure $I(t)$ and the national income $Y(t)$ takes into account memory and lag effects. For the power-law fading memory and gamma distribution of the delay time, we can use the equation

$$I(t) = v(t)(\mathbf{D}^{\alpha,a,\lambda}_{T;C;0+} Y)(t), \quad (22.19)$$

where $\mathbf{D}^{\alpha,a,\lambda}_{T;C;0+}$ is the Caputo fractional derivative with distributed lag. Equation (22.19) describes the economic accelerator with the power-law fading memory and the gamma distributed lag.

Substituting expressions (22.19) and (22.4) into balance equation (22.3), we obtain the fractional differential equation of the generalized Keynesian model with power-law memory and gamma distribution of delay time in the form

$$(\mathbf{D}^{\alpha,a,\lambda}_{T;C;0+} Y)(t) = \frac{1-m(t)}{v(t)} Y(t) - \frac{G_b(t)}{v(t)}, \quad (22.20)$$

where α is the parameter of memory fading, $a > 0$ is the shape parameter and $\lambda > 0$ is the rate parameter of the gamma distribution of the delay time.

Consider the case of constant parameters $v(t) = v$ and $m(t) =$. Then equation (22.20) of the suggested Keynesian model with memory and lag is described by the fractional differential equation

$$(\mathbf{D}^{\alpha,a,\lambda}_{T;C;0+} Y)(t) = \omega Y(t) + F(t), \quad (22.21)$$

where

$$\omega = \frac{1-m}{v}, \quad F(t) = -\frac{1}{v} G_b(t). \quad (22.22)$$

The general solution of equation (22.21) can be written in the form

$$Y(t) = Y_0(t) + Y_F(t), \quad (22.23)$$

where $Y_0(t)$ is the solution of equation (22.21) with $F(t) = 0$, i. e., the homogeneous equation

$$(\mathbf{D}^{\alpha,a,\lambda}_{T;C;0+}Y)(t) = \omega Y(t), \tag{22.24}$$

and $Y_F(t)$ is particular solution of (22.21) that can be represented in the form

$$Y_F(t) = \int_0^t G_\alpha[t-\tau]F(\tau)\,d\tau, \tag{22.25}$$

where $G_\alpha[t-\tau]$ is the generalized Green function [200, pp. 281, 295]. Equation (22.25) yields the solution $Y_F(t)$ for equation (22.21) with initial conditions, $Y^{(j)}(0) = 0$ for all $j = 0, \dots, (n-1)$.

Theorem 22.1. *The fractional differential equation*

$$(\mathbf{D}^{\alpha,a,\lambda}_{T;C;0+}Y)(t) = \omega Y(t) + F(t), \tag{22.26}$$

where $\mathbf{D}^{\alpha,a,\lambda}_{T;C;0+}$ is the Caputo fractional derivative with gamma distributed lag, in which $\alpha > 0$, $a > 0$ and $\lambda > 0$ are the fading, shape and rate parameters respectively, has the solution

$$\begin{aligned} Y(t) = &\sum_{j=0}^{n-1} S^{\alpha j-1}_{\alpha,a}[\omega\lambda^{-a},\lambda|t]Y^{(j)}(0) + \frac{1}{\omega}F(t) \\ &- \frac{1}{\omega}\int_0^t S^{\alpha}_{\alpha,a}[\omega\lambda^{-a},\lambda|t-\tau]F(\tau)\,d\tau, \end{aligned} \tag{22.27}$$

where $n = [\alpha]+1$, and $S^{\gamma}_{\alpha,\delta}[\mu,\lambda|t]$ is the special function that is defined by the expression

$$\begin{aligned} S^{\gamma}_{\alpha,\delta}[\mu,\lambda t] = &-\sum_{k=0}^{\infty} \frac{t^{\delta(k+1)-\alpha k-\gamma-1}}{\mu^{k+1}\Gamma(\delta(k+1)-\alpha k-\gamma)} \\ &\times F_{1,1}(\delta(k+1);\delta(k+1)-\alpha k-\gamma,-\lambda t), \end{aligned} \tag{22.28}$$

where $F_{1,1}(a;b;z)$ is the confluent hypergeometric Kummer function (20.56).

Proof. The first step is to find a solution of the homogeneous equation (22.24). Using the Laplace transform of equation (22.24), we get

$$\frac{\lambda^a}{(s+\lambda)^a}\left(s^\alpha(\mathcal{L}Y)(s) - \sum_{j=0}^{n-1} s^{\alpha j-1}Y^{(j)}(0)\right) = \omega(\mathcal{L}Y)(s). \tag{22.29}$$

Then we can write

$$(\mathcal{L}Y)(s) = \sum_{j=0}^{n-1} \frac{s^{\alpha j-1}}{s^\alpha - \mu(s+\lambda)^a}Y^{(j)}(0), \tag{22.30}$$

where $\mu = \omega\lambda^{-a}$. Using equation 5.4.9 of [26] in the form

$$\left(\mathcal{L}^{-1}\left(\frac{s^a}{(s+b)^c}\right)\right)(s) = \frac{1}{\Gamma(c-a)} t^{c-a-1} F_{1,1}(c; c-a, -bt), \tag{22.31}$$

where $\mathrm{Re}(c-a) > 0$, we get [400] the Laplace transform of the function (22.28) as

$$\mathcal{L}(S^{\gamma}_{\alpha,\delta}[\mu, \lambda|t])(s) = \frac{s^{\gamma}}{s^{\alpha} - \mu(s+\lambda)^{\delta}}. \tag{22.32}$$

Using equation (22.32), solution $Y(t) = Y_0(t)$ with (22.30) for the homogenous fractional differential equation (22.24) has the form

$$Y_0(t) = \sum_{j=0}^{n-1} S^{\alpha j-1}_{\alpha,a}[\omega\lambda^{-a}, \lambda|t] Y^{(j)}(0), \tag{22.33}$$

where $S^{\alpha j-1}_{\alpha,a}[\omega\lambda^{-a}, \lambda|t]$ is defined by equation (22.28).

The second step is to find a particular solution (22.25) of equation (22.26). The Laplace transform of equation (22.26) with conditions $Y^{(j)}(0) = 0$ for all $j = 0, \ldots (n-1)$ gives the expression

$$\frac{\lambda^a}{(s+\lambda)^a} s^{\alpha} (\mathcal{L}Y)(s) = \omega(\mathcal{L}Y)(s) + (\mathcal{L}F)(s) \tag{22.34}$$

that can be rewritten in the form

$$(\mathcal{L}Y)(s) = \frac{(s+\lambda)^a}{\lambda^a s^{\alpha} - \omega(s+\lambda)^a} (\mathcal{L}F)(s). \tag{22.35}$$

The equality

$$\frac{(s+\lambda)^a}{\lambda^a s^{\alpha} - \omega(s+\lambda)^a} = -\frac{1}{\omega} + \frac{1}{\omega} \frac{s^{\alpha}}{s^{\alpha} - \mu(s+\lambda)^a}, \tag{22.36}$$

where $\mu = \omega\lambda^{-a}$, gives

$$(\mathcal{L}Y)(s) = -\frac{1}{\omega}(\mathcal{L}F)(s) + \frac{1}{\omega} \frac{s^{\alpha}}{s^{\alpha} - \mu(s+\lambda)^a} (\mathcal{L}F)(s). \tag{22.37}$$

Using equation (22.32) with $\delta = a$, and $\gamma = \alpha$, we have

$$G_{\alpha}[t-\tau] = -\frac{1}{\omega}\delta(t-\tau) + \frac{1}{\omega} S^{\alpha}_{\alpha,a}[\mu, \lambda|t-\tau]. \tag{22.38}$$

As a result, we obtain

$$Y_F(t) = \frac{1}{\omega}F(t) - \frac{1}{\omega} \int_0^t S^{\alpha}_{\alpha,a}[\mu, \lambda|t-\tau] F(\tau)\, d\tau \tag{22.39}$$

that describes the particular solution of equation (22.26).

Substitution of (22.33) and (22.39) into (22.23) gives (22.27). □

As a result, the solution of equation (22.26) of the Keynesian model with one-parameter power-law memory and gamma distributed lag for constant $v(t) = v$ and $m(t) = m$ is described by the expression

$$Y(t) = \sum_{j=0}^{n-1} S_{\alpha,a}^{\alpha j-1}[\omega\lambda^{-a}, \lambda|t] Y^{(j)}(0) - \frac{1}{1-m} G_b(t) + \frac{1}{1-m} \int_0^t S_{\alpha,a}^{\alpha}[\omega\lambda^{-a}, \lambda|t-\tau] G_b(\tau)\, d\tau, \tag{22.40}$$

where $G_b(\tau) = G(\tau) + g(\tau)$ and $\omega = (1-m)/v$.

22.5 Asymptotic behavior of growth with memory and lag

In economic theory, the important concept of the Harrod's warranted rate of growth [10] is used. The warranted rate characterizes growth for the case when the following two conditions are satisfied. The first condition is the constancy of the structure of the economy. This condition means that the parameters of the model do not change over time. In the Keynesian model, we should consider the parameters $m(t)$, $b(t)$ and $m(t)$ as constants. The second condition is the absence of external influences. This condition means the absence of exogenous variables. We also should consider the case of the absence of the independent expenditure ($G_b(t) = 0$). For a long-term case, the warranted rate of growth can be obtained from the asymptotic expression of the solution of the homogeneous differential equation of the macroeconomic model. In the standard Keynesian model without memory and lag, the solution of equation (22.9) with $G_b(t) = 0$ has the form $Y(t) = Y(0)\exp(\omega t)$. Therefore, the warranted rate of growth for this model is described by the value

$$R = \omega = \frac{1-m}{v}, \tag{22.41}$$

when we use $G_b(t) = 0$.

The Keynesian model with memory was suggested in articles [423, 427, 422] in 2016 (see also [453, 450, 393, 387]). For this growth model, the fractional differential equation, its solution and properties have been described. We proved [453, 450, 393, 387, 445] that the warranted rate of growth with power-law memory is equal to the value

$$R(\alpha) = \omega_{\text{eff}}(\alpha) = \omega^{1/\alpha} = \left(\frac{1-m}{v}\right)^{1/\alpha}, \tag{22.42}$$

where $\alpha > 0$ is parameter of power-law memory fading and ω is the rate of growth without memory ($\alpha = 1$). The warranted rates of growth for the models with memory do not coincide with the growth rates of standard Keynesian model, in the general case. The memory effects can significantly change the growth rates of economy

[453, 450, 393, 387, 445], and lead to new type of behavior for the same parameters of economic model. The principles of changing of growth rates by memory were proposed in [453, 450, 393, 387]. The memory effects can both increase and decrease the warranted rates in comparison with the standard Keynesian model. For the memory fading parameter $\alpha < 1$, we get a slowdown in the growth and decline of the economy. We can state that memory with $\alpha < 1$ leads to inhibition of economic growth or decline, i. e., we have stagnation of the economy if $\alpha < 1$. For the memory fading with $\alpha > 1$, we have an improvement in economy. In this case, the memory effect leads either to the slowdown in the decline rate or to the replacement of the decline by a growth, or to the increase in the growth rate. Using equation (5.1.18) of [143, p. 99], in the form $F_{1,1}(a; c; z) = \Gamma(c)E^{a}_{1,c}(z)$, we can also get an asymptotic expressions by using asymptotic expression of the three parameter Mittag–Leffler functions $E^{\delta(k+1)}_{1,\delta(k+1)-\alpha k-\gamma}[-\lambda t]$.

The solution of the Keynesian model with memory and lag is represented as an infinite sum of the confluent hypergeometric Kummer function or the generalized Mittag–Leffler function. For these functions, asymptotic expressions exist. However, at present, there are no asymptotic expressions of an infinite series of these functions that describe the obtained solutions. In the works [402, 407], on the basis of asymptotic expressions for the confluent hypergeometric Kummer function, assumptions were made about the behavior of the solutions. Equations of solution allow us to assume that the warranted rate of growth has the power-law form with the power $-\alpha + j$, where $j \in 0, \ldots, n-1$ is a smallest values at which $Y^{(j)}(0) \neq 0$. As a result, the growth of the national income with memory and distributed lag has the power-law type instead of the exponential type of growth with memory in absence of lag [453, 450, 393, 387], where warranted rate of growth is $\omega_{\text{eff}}(\alpha) = \omega^{1/\alpha}$. Therefore, we can assume that the distributed lag (time delay) suppresses the effects of fading memory. We assume that new information on the behavior of the economy for the case of simultaneous effects of memory and lag can be obtained in computer modeling of the proposed models.

22.6 Conclusion

The standard Keynesian model describes the dynamics of national income in absence of memory and distributed lag. The Keynesian model with power-law memory has been suggested by authors [423, 427, 422]. The effects of continuously distributed lag is not considered in [423, 427, 422]. In this chapter, which is based on the our works [402, 407], we generalize the Keynesian model with memory by taking into account gamma distribution of delay time. To take into account the distributed lag, we use the operators that are composition of translation operator with distributed delay time and the fractional derivatives. These operators allow us to take into account the memory and lag in the economic accelerator. These operators are the integro-differential operators with the confluent hypergeometric Kummer function in the kernel. The solution of the fractional differential equation, which describes the fractional dynamics of national income, was suggested. We assume that the asymptotic behavior of economic processes with memory and distributed lag demonstrates the power-law behavior. In the absence of delays, the processes with fading memory demonstrate the exponential behavior. The warranted rate of growth with memory is equal to the value $\omega_{\text{eff}}(\alpha) = \omega^{1/\alpha}$, where $\alpha > 0$ is a memory fading parameter. The memory effects can significantly accelerate the growth rate of the economy by several orders of magnitude. The fading memory can lead to an increase in the growth rate in processes without lag. The appearance of distributed lag does not accelerate growth due to the memory effect. Moreover, the lag can suppress the effect of fading memory. The distributed lag can lead to slower growth. We assume that new information about the simultaneous action of memory and lag effects can be obtained by the computer simulations based on the suggested models.

23 Phillips model with distributed lag and memory

In this chapter, we consider two generalizations of the Phillips model of multiplier-accelerator, which are first proposed in article [405]. These generalized models take into account the effects of fading memory and distributed lag. In the first generalization, we consider the model, where we replace the exponential weighting function by the power-law memory function. In this model, we consider two power-law fading memory, one on the side of the accelerator (induced investment responding to changes in output with memory fading parameter α) and the other on the supply side (output responding to demand with memory fading parameter β). To describe power-law memory, we use the fractional derivatives in the accelerator equation and the fractional integral in multiplier equation. For the proposed model, a solution of the fractional differential equation is suggested. In the second generalization of the Phillips model of multiplier-accelerator, we consider the power-law memory in addition to the continuously (exponentially) distributed lag. The solution of equation, which describes generalized Phillips model of multiplier-accelerator with distributed lag and power-law memory, is proposed. This chapter is based on article [405].

23.1 Introduction

Macroeconomic models with continuous time are usually described by differential equations with derivatives of integer orders [8, 9, 10] [11, 12, 325, 327]. Examples of such models are well-known classical models, such as the natural growth model, the logistic model, the Harrod–Domar model, the Keynes model and the Leontief model. Generalizations of such models are models, in which the time lags are taken into account. For example, a generalization of the Harrod–Domar model with continuous time, which takes into account a continuously (exponentially) distributed lag, was proposed by Alban William Housego Phillips in 1954 [305] (see also [306, pp. 134–168], [10, pp. 69–74] and [12, pp. 328–333]). In the Phillips models, a dynamic multiplier and accelerator are considered with a continuous (exponential) lag. The operators with exponentially distributed lag have been suggested by R. G. D. Allen in 1956 [8] (see also [10, 12]). The continuously (exponentially) distributed lag is described in Section 1.9 of [10, pp. 23–29], Section 5.8 of [10, pp. 166–170] and Section 5.5 of [12, pp. 88–93]. These operators are often used to describe economic processes with lag.

The existence of the time delay (lag) is connected with the fact that the processes take place with a finite speed, and the change of one variable does not lead to instant changes of another variable that depends on it. The time lag takes into account the final rate of economic processes in time. Therefore, these processes cannot be interpreted as memory effects. Memory means that there is a variable at time t that depends on the history of changes of another variable at $\tau \in (t_0, t)$. In the general case,

https://doi.org/10.1515/9783110627459-023

it is important to take into account the simultaneous action of distributed lag and the memory effects in economic models.

This chapter is based on paper [405], where we proposed two generalization of the Phillips model of multiplier-accelerator by taking into account memory with power-law fading. In the first generalization, we consider the model, where we replace the exponential weighting function by the power-law memory function, to describe memory by using the fractional derivatives in the accelerator equation and the fractional integral in multiplier equation. In the second generalization, we consider the power-law memory in addition to the continuously (exponentially) distributed lag. The solution of the fractional differential equations that describe proposed models with memory are obtained.

23.2 Phillips model with power-law memory and without lag

Let us consider the Phillips model with power-law memory that is proposed in [405]. For the first time, the concepts of an economic accelerator with memory and an economic multiplier with memory were proposed in 2016 [412, 397] (see also [441, 442, 431, 432]). Let us consider the investment accelerator with power-law memory. If $I(t)$ is actual induced investment at time $t > t_0$ in response to changes in output $Y(\tau)$ during the time interval (t_0, t), then $I(t)$ can be given by the equation

$$I(t) = v(D^{\alpha}_{C;t_0} Y)(t), \tag{23.1}$$

where $v > 0$ is the investment coefficient and $D^{\alpha}_{C;t_0}$ is the Caputo fractional derivative. Equation (23.1) described the economic accelerator with one-parametric power-law memory. The total demand without memory and lags is

$$Z(t) = C(t) + I(t) + A(t), \tag{23.2}$$

where $C(t) = cY(t)$ is the planned consumption, $A(t)$ is the autonomous expenditure (investment and consumption). In this model, we can use $\omega = 1 - c$, the marginal propensity to save instead of the marginal propensity to consume c. Then we have the equation

$$Z(t) = cY(t) + I(t) + A(t). \tag{23.3}$$

On the supply side, the response of output $Y(t)$ to demand $Z(t)$ is not assumed as instantaneous, and we take into account the power-law fading memory

$$Y(t) = (I^{\beta}_{RL;t_0} Z)(t), \tag{23.4}$$

where $I^{\beta}_{RL;t_0}$ is the Riemann–Liouville fractional integral of the order $\beta > 0$. Equation (23.4) described the economic multiplier with power-law fading memory [412, 397].

Let us use the property (see equation (2.4.32) of Lemma 2.21 in [200, p. 95]), which states that the Caputo fractional derivative is an operation inverse to the Riemann–Liouville fractional integral, such that

$$D^{\beta}_{C;t_0}(I^{\beta}_{RL;t_0}Y)(t) = Y(t) \tag{23.5}$$

for $\beta > 0$ and $Y(t) \in L_{\infty}(t_0, t_1)$ or $Y(t) \in C(t_0, t_1)$. Acting by the Caputo fractional derivative of order $\beta > 0$ on equation (23.4), we obtain

$$(D^{\beta}_{C;t_0}Y)(t) = Z(t). \tag{23.6}$$

Equations (23.1), (23.3) and (23.6) define the Phillips model with two power-law fading of memory, one on the side of the accelerator (induced investment responding to changes in output with memory fading parameter α) and the other on the supply side (output responding to demand with memory fading parameter β). As a result, the generalized Phillips model with memory and without lag is described by the system

$$\begin{cases} I(t) = v(D^{\alpha}_{C;t_0}Y)(t), \\ Z(t) = cY(t) + I(t) + A(t), \\ Z(t) = (D^{\beta}_{C;t_0}Y)(t). \end{cases} \tag{23.7}$$

An equation for the output $Y(t)$ is obtained by eliminating $Z(t)$ and $I(t)$ from system of equations (23.7) in the form

$$(D^{\beta}_{C;t_0}Y)(t) = cY(t) + +v(D^{\alpha}_{C;t_0}Y)(t) + A(t). \tag{23.8}$$

Fractional differential equation (23.8) determines the dynamics of the output within the framework of the proposed generalization of the Phillips model with two fading memory effects. To solve equation (23.8), we can use Theorem 5.16 of the book [200, pp. 323–324].

For simplification, we will consider the case $t_0 = 0$.

For $\alpha > \beta$, equation (23.8) with $t_0 = 0$ can be rewritten in the form

$$(D^{\alpha}_{C;0}Y)(t) - v^{-1}(D^{\beta}_{C;0}Y)(t) + cv^{-1}Y(t) = -v^{-1}A(t). \tag{23.9}$$

Then equation (23.9) coincides with equation (5.3.73) of book [200, pp. 323–324] if we will use the notation $\lambda = v^{-1}$, $\mu = -cv^{-1}$ and $f(t) = -v^{-1}A(t)$.

For $\beta > \alpha$, equation (23.8) with $t_0 = 0$ can be rewritten in the form

$$(D^{\beta}_{C;0}Y)(t) - v(D^{\alpha}_{C;0}Y)(t) - cY(t) = A(t). \tag{23.10}$$

Then equation (23.10) coincides with equation (5.3.73) of book [200, p. 323], if we will use the notation $\lambda = v$, $\mu = c$ and $f(t) = A(t)$, parameter β instead of parameter α and vice versa.

Let us consider solutions of equation (23.9). The solutions of equation (23.10) can be obtained similarly to the solution of equation (23.9).

For continuous function $A(t)$, which is defined on the positive semi-axis ($t > 0$), equation (23.9) with the parameters $0 < n-1 < \alpha \leq n$, $m-1 < \beta \leq m$ (where $0 < \beta < \alpha$, $m \leq n$, $\alpha - n + 1 \geq \beta$), has the general solution (Theorem 5.16 of [200, pp. 323–324]) in the form

$$Y(t) = \sum_{j=0}^{n-1} c_j Y_j(t) + Y_A(t), \tag{23.11}$$

where c_j ($j = 0, \dots, n-1$) are the real constants that are determined by the initial conditions. The function $Y_A(t)$ is defined as

$$Y_A(t) = -v^{-1} \int_0^t (t-\tau)^{\alpha-1} G_{\alpha,\beta,1/v,-c/v}[t-\tau] A(\tau)\, d\tau, \tag{23.12}$$

where the function $G_{\alpha,\beta,1/v,-c/v}[\tau]$ is given by the equation

$$G_{\alpha,\beta,1/v,-c/v}[\tau] = \sum_{k=0}^{\infty} \frac{\tau^{k\alpha}}{\Gamma(k+1)} \left(-\frac{c}{v}\right)^k \Psi_{1,1}\left[{}^{(k+1,1)}_{(\alpha k+\alpha,\alpha-\beta)} \mid v^{-1}\tau^{\alpha-\beta}\right]. \tag{23.13}$$

The functions $Y_j(t)$ with $j = 0, \dots, m-1$, where $m = [\beta] + 1$ for noninteger values of β and $m = \beta$ for integer values of α, are represented by the expressions

$$\begin{aligned} Y_j(t) = {} & \sum_{k=0}^{\infty} \frac{t^{k\alpha+j}}{\Gamma(k+1)} \left(-\frac{c}{v}\right)^k \Psi_{1,1}\left[{}^{(k+1,1)}_{(\alpha k+j+1,\alpha-\beta)} \mid v^{-1} t^{\alpha-\beta}\right] \\ & - v^{-1} \sum_{k=0}^{\infty} \frac{t^{k\alpha+j+\alpha-\beta}}{\Gamma(k+1)} \left(-\frac{c}{v}\right)^k \Psi_{1,1}\left[{}^{(k+1,1)}_{(\alpha k+j+1+\alpha-\beta,\alpha-\beta)} \mid v^{-1} t^{\alpha-\beta}\right]. \end{aligned} \tag{23.14}$$

For $j = m, \dots, n-1$, the functions $Y_j(t)$ are defined by the equations

$$Y_j(t) = \sum_{k=0}^{\infty} \frac{t^{k\alpha+j}}{\Gamma(k+1)} \left(-\frac{c}{v}\right)^k \Psi_{1,1}\left[{}^{(k+1,1)}_{(\alpha k+j+1,\alpha-\beta)} \mid v^{-1} t^{\alpha-\beta}\right]. \tag{23.15}$$

Here, $\Psi_{1,1}$ is the generalized Wright functions (the Fox–Wright function) [200, pp. 54–58] that is defined by the equation

$$\Psi_{1,1}\left[{}^{(a,\alpha)}_{(b,\beta)} \mid z\right] = \sum_{k=0}^{\infty} \frac{\Gamma(\alpha k + a)}{\Gamma(\beta k + b)} \frac{z^k}{k!}. \tag{23.16}$$

Note that the Fox–Wright functions, which are used in equations (23.13), (23.14), (23.15), (23.16), can be represented (e. g., see equation (1.9.1) of [200, p. 45] and equation (5.1.37) of [143, p. 105]) through the three parameter Mittag–Leffler functions $E^{\gamma}_{\alpha,\beta}$

by the equation

$$\Psi_{\rho,1}\left[\begin{smallmatrix}(1,1)\\(\beta,\alpha)\end{smallmatrix} \mid z\right] = \Gamma(\rho)E^{\rho}_{\alpha,\beta}[z] = \sum_{k=0}^{\infty} \frac{(\rho)_k}{\Gamma(\alpha k + \beta)} \frac{z^k}{k!}, \tag{23.17}$$

where $\alpha > 0$, $\alpha, \beta, \rho, z \in \mathbb{C}$ in general. The generalized Mittag–Leffler function $E^{\rho}_{\alpha,\beta}[z]$, which is also called the Prabhakar function, is defined (e. g., see equation (5.1.1) of [143, p. 97] and equation (1.9.1) of [200, p. 45]) by the equation

$$E^{\rho}_{\alpha,\beta}[z] = \sum_{k=0}^{\infty} \frac{(\rho)_k}{\Gamma(\alpha k + \beta)} \frac{z^k}{k!}, \tag{23.18}$$

where $(\rho)_k$ is the Pochhammer symbol.

Equation (23.9) and its solution (23.11)–(23.15) describe the generalized Phillips model of multiplier-accelerator with power-law memory that is characterized by two fading parameters.

Let us consider the special case when the autonomous expenditure (investment and consumption) is absent ($A(t) = 0$). Equation (23.9) with $A(t) = 0$ has the form

$$(D^{\alpha}_{C;0}Y)(t) - v^{-1}(D^{\beta}_{C;0}Y)(t) + cv^{-1}Y(t) = 0, \tag{23.19}$$

where $0 < \beta < \alpha \leq 2$. Equation (23.19) can be used to estimate the greatest possible growth rate, when we take into account the two power-law fading memory.

For $0 < \beta < \alpha \leq 1$, equation (23.19) has (see Corollary 5.8 in [200, p. 317]) the solution

$$Y(t) = c_1 Y_1(t), \tag{23.20}$$

where the constant c_1 is determined by the initial condition, and $Y_1(t)$ is given by the equation

$$\begin{aligned} Y_1(t) = {} & \sum_{k=0}^{\infty} \frac{t^{k\alpha}}{\Gamma(k+1)} \left(-\frac{c}{v}\right)^k \Psi_{1,1}\left[\begin{smallmatrix}(k+1,1)\\(\alpha k+1,\alpha-\beta)\end{smallmatrix} \mid v^{-1}t^{\alpha-\beta}\right] \\ & - v^{-1} \sum_{k=0}^{\infty} \frac{t^{k\alpha+\alpha-\beta}}{\Gamma(k+1)} \left(-\frac{c}{v}\right)^k \Psi_{1,1}\left[\begin{smallmatrix}(k+1,1)\\(\alpha k+1+\alpha-\beta,\alpha-\beta)\end{smallmatrix} \mid v^{-1}t^{\alpha-\beta}\right]. \end{aligned} \tag{23.21}$$

Using

$$\Psi_{1,1}\left[\begin{smallmatrix}(a,\alpha)\\(b,\beta)\end{smallmatrix} \mid 0\right] = \frac{1}{\Gamma(\beta)}, \tag{23.22}$$

we get

$$Y(0) = \frac{c_1}{\Gamma(\beta)}(1 - v^{-1}). \tag{23.23}$$

Then equation (23.23) gives the constant

$$C_1 = \frac{\Gamma(\beta)Y(0)}{1 - v^{-1}}, \tag{23.24}$$

where $v \neq 1$.

For $0 < \beta \leq 1 < \alpha \leq 2$, equation (23.19) has the solution (see Corollary 5.9 of [200, p. 317]). This solution can be represented as a linear combination of $Y_1(t)$, which is given by (23.21), and $Y_2(t)$, which is defined by the expression

$$Y_2(t) = \sum_{k=0}^{\infty} \frac{t^{k\alpha+1}}{\Gamma(k+1)} \left(-\frac{c}{v}\right)^k \Psi_{1,1}\left[{}^{(k+1,1)}_{(\alpha k+2,\alpha-\beta)} \mid v^{-1} t^{\alpha-\beta}\right]. \tag{23.25}$$

For $1 < \beta < \alpha \leq 2$, equation (23.19) has the solution that is a linear combination of $Y_1(t)$, which is given by equation (23.21), and

$$\begin{aligned} Y_2(t) = {} & \sum_{k=0}^{\infty} \frac{t^{k\alpha+1}}{\Gamma(k+1)} \left(-\frac{c}{v}\right)^k \Psi_{1,1}\left[{}^{(k+1,1)}_{(\alpha k+2,\alpha-\beta)} \mid v^{-1} t^{\alpha-\beta}\right] \\ & - v^{-1} \sum_{k=0}^{\infty} \frac{t^{k\alpha+1+\alpha-\beta}}{\Gamma(k+1)} \left(-\frac{c}{v}\right)^k \Psi_{1,1}\left[{}^{(k+1,1)}_{(\alpha k+2+\alpha-\beta,\alpha-\beta)} \mid v^{-1} t^{\alpha-\beta}\right]. \end{aligned} \tag{23.26}$$

Note that to determine the effective warranted rates (the warranted rates of growth with memory) for the generalized Phillips model of multiplier-accelerator with power-law memory, it is necessary to consider the asymptotic behavior of solutions at $t \to \infty$, which are represented as infinite sums of the Fox–Wright functions. The asymptotic formulas for the Fox–Wright functions were described in paper [297] (see also [298, 300, 301, 302, 299]). However, there is currently no exact analytical expression describing the asymptotic behavior of the solution, which is described by infinite sums of the Fox–Wright functions.

23.3 Phillips model with distributed lag and memory

The standard Phillips growth model of multiplier-accelerator takes into account two continuously distributed lags, one on the side of the accelerator (induced investment responding to changes in output with speed λ_1) and the other on the supply side (output responding to demand with speed λ_2). In the Phillips growth model, the lags are distributed exponentially and the model does not take into account the memory effects. We consider generalization of this model by taking into account the power-law memory. We will use the concept of generalized accelerators with memory [412, 397] and the fractional differential operators with continuously distributed lag that are proposed in [400, 405].

Let us give a lagged version of the investment accelerator with memory. If $I(t)$ is actual induced investment at time t in response to changes in output $Y(t)$, then $I(t)$ is given by the equation

$$I(t) = v(\mathbf{D}^{\lambda_1,\alpha}_{T,C} Y)(t), \tag{23.27}$$

where $v > 0$ is the investment coefficient and $\mathbf{D}^{\lambda_1,\alpha}_{T,C}$ is the fractional derivative with exponentially distributed lag. The total demand without lags is

$$Z(t) = C(t) + I(t) + A(t), \tag{23.28}$$

where $C(t) = cY(t)$ is the planned consumption, $A(t)$ is the autonomous expenditure (investment and consumption). We can use $\omega = 1 - c$, the marginal propensity to save instead of the marginal propensity to consume $c > 0$. Then we have the equation

$$Z(t) = cY(t) + I(t) + A(t). \tag{23.29}$$

On the supply side, we assume that the response of output $Y(t)$ to demand $Z(t)$ is described by the equation with continuously distributed lag

$$Y(t) = (\mathbf{D}^{\lambda_2,0}_{T} Z)(t), \tag{23.30}$$

where $\mathbf{D}^{\lambda_2,0}_{T}$ is the translation operator of the exponentially distributed lag.

Equations (23.27), (23.29) and (23.30) define the Phillips model with power-law memory and two continuously distributed lags, one on the supply side (output responding to demand with speed λ_2) and the other on the side of the accelerator (induced investment responding to changes in output with speed λ_1). As a result, the Phillips model with distributed lag and memory is described by the system

$$\begin{cases} I(t) = v(\mathbf{D}^{\lambda_1,\alpha}_{T,C} Y)(t), \\ Z(t) = cY(t) + I(t) + A(t), \\ Y(t) = (\mathbf{D}^{\lambda_2,0}_{T} Z)(t). \end{cases} \tag{23.31}$$

An equation in the output $Y(t)$ is obtained by eliminating $Z(t)$ and $I(t)$ from system of equations (23.31). As a result, we get the equation

$$c(\mathbf{D}^{\lambda_2,0}_{T} Y)(t) + v(\mathbf{D}^{\lambda_2,0}_{T}(\mathbf{D}^{\lambda_1,\alpha}_{T,C} Y))(t) - Y(t) = -(\mathbf{D}^{\lambda_2,0}_{T} A)(t). \tag{23.32}$$

Equation (23.32) describes generalized Phillips model of multiplier-accelerator with distributed lag and power-law memory.

For $\alpha = 1$, equation (23.32) describes the standard Phillips model of multiplier-accelerator with exponentially distributed lag [10]. In the standard model, $A(t)$ is often takes as a given constant ($A(t) = A$) [10].

Equation (23.32) is nonhomogeneous equation. The general solution of this nonhomogeneous equation is given by the following statement.

The solution of equation (23.32) of the generalized Phillips model with lag and memory has the form

$$Y(t) = Y_0(t) + Y_A(t), \tag{23.33}$$

where $Y_0(t)$ is the solution of the corresponding homogeneous equation such that

$$Y_0(t) = -\sum_{j=0}^{n-1}\sum_{k=1}^{\infty} \frac{(v\lambda_1\lambda_2)^k t^{(2-\alpha)k+j}}{\Gamma((2-\alpha)k+j+1)} \cdot \Phi_2[k,k;(2-\alpha)k+j+1;-\lambda_1 t,-\omega\lambda_2 t]Y^{(j)}(0). \tag{23.34}$$

The function $Y_A(t)$ is a particular solution of the nonhomogeneous equation that is represented in the form

$$Y_A(t) = \int_0^t G_\alpha[t-\tau]A(\tau)\,d\tau, \tag{23.35}$$

where $G_\alpha[t-\tau]$ is the fractional analog of the Green function [200, pp. 281, 295]. For equation (23.32), the fractional Green function has the form

$$\begin{aligned} G_\alpha[t] = {} & \sum_{k=0}^{\infty} \frac{\lambda_2(v\lambda_1\lambda_2)^k t^{(2-\alpha)k}}{\Gamma((2-\alpha)k+1)}\Phi_2[k+1,k+1;(2-\alpha)k+1;-\lambda_1 t,-\omega\lambda_2 t] \\ & + \sum_{k=0}^{\infty} \frac{v^k(\lambda_1\lambda_2)^{k+1} t^{(2-\alpha)k+1}}{\Gamma((2-\alpha)k+2)}\Phi_2[k+1,k+1;(2-\alpha)k+2;-\lambda_1 t,-\omega\lambda_2 t]. \end{aligned} \tag{23.36}$$

Here, $\Phi_2[a,b;y;x,y]$ is the two-variable confluent hypergeometric function (e. g., see Section 5.7 in [113, pp. 222–229] and [114]) that is defined by the series

$$\Phi_2[a,b;c;x,y] = \sum_{n,m=0}^{\infty} \frac{\Gamma(a+n)\Gamma(b+m)\Gamma(c)}{\Gamma(a)\Gamma(b)\Gamma(c+n+m)} \frac{x^n}{n!}\frac{y^m}{m!}. \tag{23.37}$$

The proof of this statement is given in [405].

The expressions of $Y_0(t)$, $Y_A(t)$ and $G_\alpha[t]$ give the solution of the generalized the Phillips model of multiplier-accelerator that takes into account the exponentially distributed lag and power-law fading memory.

The computer simulation of the suggested equations of the Phillips models with fading memory and distributed lag can be used for modeling of economy.

23.4 Conclusion

Generalizations of the Phillips model of multiplier-accelerator with exponentially distributed lag are suggested, where we take into account power-law fading memory. In the proposed models, we use memory on the side of the accelerator, where induced investment responding to changes in output with memory fading parameter α. On the supply side, we consider the output responding to demand with memory fading parameter β. The fractional integro-differential equations of these generalized models are solved. We used the fractional differential and integral operators with continuously distributed lag [400]. The suggested approach, which is based on these operators, allows separately and simultaneously to take into account the effects of continuously distributed lag and power-law fading memory. The various and economic models with memory such as the model of natural growth, the growth model with constant pace, the Harrod–Domar model, the dynamic intersectoral models, the logistic growth model, the time-dependent growth model, can be generalized by taking into account the continuously distributed lags. The computer simulation of the proposed economic models with fading memory and distributed lag can be used for modeling of real processes in the economy.

Part VI: **Advanced models: discrete time approach**

24 Discrete accelerator with memory

In this chapter, we propose discrete accelerators with memory to describe the economic processes with the power-law memory and the periodic sharp splashes (periodic kicks). In approach of continuous time, the memory is described by fractional-order differential equations. In the discrete time approach, the accelerators with memory can be described by fractional differences of noninteger-orders and discrete maps with memory, which are derived from the fractional-order differential equation without approximations. In order to derive the discrete maps with memory, we use the equivalence of fractional-order differential equations and the Volterra integral equations. To describe discrete accelerators, we use the capital stock adjustment principle, which has been suggested by Matthews. This chapter is based on articles [417, 436, 444].

24.1 Introduction

One of the basic concepts of macroeconomics is the accelerator [10, 12, 212, 494]. Accelerator equations can be represented in the frameworks of the discrete time and continuous time approaches. We consider an exact correspondence between these two approaches for economic processes with power-law memory. Initially, we prove that the discrete accelerator equations, which contain standard finite differences, can be derived from the differential equation with the periodic sharp splashes (periodic kicks). Using the generalization of these equations, which take into account power-law memory, we derive the discrete accelerator equations with memory in the form of discrete maps with memory.

24.2 Standard accelerator in continuous and discrete terms

Let us consider an example of the linear accelerator without memory in the continuous-time and discrete-time terms. In continuous-time terms, the simplest equation of the linear accelerator [10, p. 62] is the continuous linear form without memory is

$$\frac{dY(t)}{dt} = \frac{1}{v} I(t), \tag{24.1}$$

where $dY(t)/dt$ is the rate of output (income), $I(t)$ is the rate of induced investment, and v is a positive constant, which is called the investment coefficient [10, p. 62] that indicates the power of the accelerator. The coefficient v is also called the accelerator coefficient or the capital intensity of the income growth rate [494, p. 91], where $1/v$ is the capital productivity incremental or marginal productivity of capital [494, p. 91]. Equation (24.1) means that the induced investment is proportional to the current rate of change of output.

https://doi.org/10.1515/9783110627459-024

In discrete-time terms, the linear accelerator without memory can be written [10, p. 63] in the form of the linear difference equation

$$Y_n - Y_{n-1} = \frac{T}{v} I_n, \tag{24.2}$$

where $Y_n = Y(nT)$ and $I_n = I(nT)$, with the positive constant T indicating the time scale. If $T = 1$, then $t = n$ and $Y_n = Y_t$. In this case, equation (24.2) has the form

$$I_t = v(Y_t - Y_{t-1}). \tag{24.3}$$

Equation (24.2) means that induced investment depends on the current change in output [10, p. 63].

24.3 Capital stock adjustment principle

There is an alternative approach to the discrete accelerator equation, which is proposed by Matthews in the form of the capital stock adjustment principle [257, 11].

Let us consider the capital stock $K(t)$, which can be taken as varying, and so the level of net investment $I(t)$ depends on $K(t)$ [12, p. 68]. The investment $I(t)$ must depend on profits, both as an indicator of profitability of production (or the level of demand), and as one of the sources of funds available to finance investment. In the general case, we can take income $Y(t)$, instead of profits, as an indicator of the level of demand and of the availability of finance. On this approach, we can write the investment function $I = I(Y, K)$, where we ignored the influence of the interest rate. In the linear case, the investment function is linear in $Y(t)$ and $K(t)$, and we have

$$I(Y(t), K(t)) = aY(t) - bK(t), \tag{24.4}$$

where a and b are positive coefficient of the investment function. This case corresponds to the capital stock adjustment principle proposed by Matthews [257, 11].

We can approach to the accelerator equation by starting from the linear investment function (24.4) on the basis of the Matthews' capital stock adjustment principle [257, 11]. Considering the particular case of this principle in which $b = 1$ and $T = 1$, we have [11, p. 73] the equation

$$aY_t = K_t + I_t = K_{t+1}. \tag{24.5}$$

Then, we have $K_{t+1} = aI_t$ and $K_t = aI_{t-1}$ that give

$$I_t = K_{t+1} - K_t = a(Y_t - Y_{t-1}). \tag{24.6}$$

As a result, the accelerator equation is obtained as a particular case ($a = v$, $b = 1$, $T = 1$) of the capital stock adjustment principle.

Let us consider the Harrod–Domar model (see Section 11 in book [11, pp. 197–218]), which is translatable into discrete terms. The variables for a sequence of periods $t = 0, T, 2T, \dots$ are output (income) Y_n as the flow at $t = nT$, the capital stock K_n timed at the beginning of the period, and the investment in period $t = nT$ is given by I_n [11, p. 204]. In the fixed-coefficients version of the Harrod–Domar model, the "investment = saving" equation (see equation (1) of Section 11.4 in [11, p. 204]) has the form

$$K_{n+1} - K_n = sTY_n, \tag{24.7}$$

where the parameter s is the constant propensity to save. For $T = 1$, equation (24.7) has the form

$$K_{t+1} - K_t = sY_t. \tag{24.8}$$

In the continuous time approach, equation (24.7) is considered (see equation (1) of Section 11.2 in [11, p. 199]) in the form

$$\frac{dK(t)}{dt} = sY(t), \tag{24.9}$$

where $dK(t)/dt$ is the derivative of first order of the capital stock function $K(t)$.

This formulation can be weakened [11, pp. 204–205] by dropping explicit reference to the capital stock and the production function. Then the equation of full capacity (see equation (4) of Section 11.4 in [11, p. 205]) has the form

$$Y_{n+1} - Y_n = \frac{T}{v} I_n, \tag{24.10}$$

where v is a fixed coefficient of the production function. For $T = 1$, equation (24.10) has the form

$$I_t = v(Y_{t+1} - Y_t). \tag{24.11}$$

In the continuous time terms, equation (24.10) is considered (see equation (1) of Section 11.2 in [11, p. 199]) in the form (24.1), where $dY(t)/dt$ is the derivative of first order of the income function $Y(t)$.

The multiplier-accelerator version is a variant of (24.10), in which the full-capacity condition is reversed [11, p. 205] in its lag, lead interpretation to give the investment function by equation (24.4).

24.4 Connection of discrete and continuous time terms

Equations (24.2), (24.10) and (24.7) cannot be considered as exact discrete analogs of equation (24.1) and (24.9). This is caused by the fact that the standard finite differences, such as the forward difference

$$\Delta_f^1 Y(t) = Y(t+1) - Y(t), \tag{24.12}$$

and the backward difference

$$\Delta_b^1 Y(t) = Y(t) - Y(t-1) \tag{24.13}$$

do not have the same basic characteristic properties as the derivatives of first order [378, 371]. For example, the standard Leibniz rule (the product rule) is violated for these finite differences [378, 371]. For example, the equality

$$\Delta_f^1(X(t)Y(t)) \neq (\Delta_f^1 X(t))Y(t) + X(t)(\Delta_f^1 Y(t)) \tag{24.14}$$

does not hold, in the general case.

Using the approach, which is suggested in [361, 362] and [363, 364, pp. 409–453], we can propose the differential equation of the accelerator that gives discrete time analogs of equations (24.2), (24.10) and (24.7), which corresponds to the capital stock adjustment principle. Discrete equations (24.2), (24.10) and (24.7) can be derived from the suggested differential equation without the use of any approximations.

In the continuous-time terms, the standard accelerators can be described by the equations

$$\frac{dK(t)}{dt} = sY(t), \tag{24.15}$$

$$\frac{dY(t)}{dt} = \frac{1}{v} I(t), \tag{24.16}$$

which coincides with equations (24.9) and (24.1).

Let us consider the linear accelerators with the periodic sharp splashes (kicks) that is described by the differential equations

$$\frac{dK(t)}{dt} = sY(t) \sum_{k=1}^{\infty} \delta\left(\frac{t}{T} - k\right), \tag{24.17}$$

$$\frac{dY(t)}{dt} = \frac{1}{v} I(t) \sum_{k=1}^{\infty} \delta\left(\frac{t}{T} - k\right), \tag{24.18}$$

where $\delta(z)$ is the Dirac delta-function, which is a distribution ("generalized function") [229, 134]. The delta functions describe the periodic sharp splashes (kicks). It should be noted that distributions ("generalized functions") are treated as continuous functionals on a space of test functions. These functionals are continuous in a suitable topology on the space of test functions. Therefore, equations (24.17) and (24.18) should be understood in a generalized sense, i. e., on the space of test functions.

To derive a discrete equation from equations (24.17) (similarly, for equation (24.18)), we can use the fundamental theorem of standard calculus and the Newton–Leibniz formula in the form

$$\int_0^t f^{(1)}(\tau)\, d\tau = f(t) - f(0), \tag{24.19}$$

where $f^{(1)}(\tau) = df(\tau)/d\tau$ is the derivative of first order.

Using integration of equation (24.17) from 0 to t, where $nT < t < (n+1)T$, we get the discrete equation

$$K_{n+1} = K_0 + sT \sum_{k=1}^{n} Y_k, \tag{24.20}$$

where $K(0) = K_0$, and

$$Y_k = Y(kT - 0) = \lim_{\varepsilon \to 0+} Y(kT - \varepsilon), \tag{24.21}$$

$$K_{n+1} = K\big((n+1)T - 0\big) = \lim_{\varepsilon \to 0+} K\big((n+1)T - \varepsilon\big). \tag{24.22}$$

In equation (24.20), we can replace $n+1$ by n to get the form

$$K_n = K_0 + sT \sum_{k=1}^{n-1} Y_k. \tag{24.23}$$

Subtracting equation (24.23) from equation (24.20), we obtain

$$K_{n+1} - K_n = sTY_n, \tag{24.24}$$

which coincides with equation (24.7).

As a result, we can conclude that discrete time equations (24.7) and (24.10) correspond to differential equations (24.17) and (24.18), respectively. This allows us to formulate the following statement.

Statement 24.1. *In the continuous-time terms, the standard discrete accelerators (24.23) and (24.24) describe economic processes with the periodic sharp splashes (kicks), which are represented by the Dirac delta-functions. The standard discrete accelerators without memory correspond to the continuous-time dynamics with the periodic sharp splashes.*

Therefore, we can state that the discrete accelerator (24.7) actually describes the economic dynamics with the periodic sharp splashes of the output or the periodic sharp splashes of propensity to save s. The standard discrete accelerator (24.10) actually describes the economic dynamics with the periodic sharp splashes of the net investment or periodic sharp splashes of capital productivity.

24.5 Continuous-time accelerator with power-law memory

To take into account the power-law memory effect in acceleration principle and the Matthews capital stock adjustment principle, we can use concept of the accelerators of with memory [412, 397]. Equations (24.15) and (24.16) can be generalized by using accelerators with memory [412, 397], which describes the relationship between the net investment (the output) and the margin output (the capital stock) of noninteger order. In order to easily interpret the dimensions of the economic quantities, we can use the time t as a dimensionless variable by changing the variable $t \to tt_d$, where t_d is the unit of time (hour, day, month, year). These generalizations of the standard accelerator equations (24.15) and (24.16), which take into account the power-law memory with fading parameter α, can be given [412, 397] in the form

$$(D^{\alpha}_{C;0+}Y)(t) = \frac{1}{v}I(t), \tag{24.25}$$

$$(D^{\alpha}_{C;0+}K)(t) = sY(t), \tag{24.26}$$

where $D^{\alpha}_{C;0+}$ is the left-sided Caputo derivative of the order $\alpha > 0$,

Definition 24.1. The left-sided Caputo derivative of the order $\alpha > 0$ is defined by the expression

$$(D^{\alpha}_{C;0+}K)(t) = \frac{1}{\Gamma(n-\alpha)} \int\limits_0^t \frac{K^{(n)}(\tau)\,d\tau}{(t-\tau)^{\alpha-n+1}}, \tag{24.27}$$

where $n-1 < \alpha \leq n$, $\Gamma(\alpha)$ is the gamma function, $K^{(n)}(\tau)$ is the derivative of the integer order n of the function $K(\tau)$ with respect to τ. For existence of expression (24.27), the function $K(\tau)$ must has integer-order derivatives up to $(n-1)$-th order, which are continuous functions on the interval $[0, T]$, and $K^{(n)}(\tau)$ is Lebesgue summable on the interval $[0, T]$.

For integer $\alpha = n \in \mathbb{N}$, the Caputo derivatives coincide with standard derivatives of the integer order n, i. e., $(D^{n}_{C;0+}K)(t) = K^{(n)}(t)$. It should be noted that equations (24.25) and (24.26) with $\alpha = 1$ take the form (24.15) and (24.16), respectively.

Remark 24.1. The accelerator equations (24.25) and (24.26) include the standard equations of the accelerators as special cases for $\alpha = 1$, respectively. Using the property $(D^{1}_{C;0+}Y)(t) = Y^{(1)}(t)$ of the Caputo fractional derivative [200, p. 79], formula (24.25) with $\alpha = 1$ gives equation (24.1) that describes the standard accelerator.

Let us consider equations of the linear accelerators with the periodic sharp splashes (kicks) and power-law memory that have the form

$$(D^{\alpha}_{C;0+}Y)(t) = \frac{1}{v}I(t)\sum_{k=1}^{\infty}\delta\left(\frac{t}{T}-k\right), \tag{24.28}$$

$$(D^{\alpha}_{C;0+}K)(t) = sY(t)\sum_{k=1}^{\infty}\delta\left(\frac{t}{T}-k\right). \tag{24.29}$$

These equations should be considered in a generalized sense, i. e., on the space of test functions. In the framework of continuous-time terms, equations (24.28) and (24.29) are interpreted as accelerator equations for economic processes with power-law memory and periodic sharp splashes.

The action of the left-sided Riemann–Liouville fractional integral of the order α on equations (24.28) and (24.29) is defined on the space of test functions on the half-axis by using the adjoint operator approach [335, pp. 154–157]. The left-sided Riemann–Liouville fractional integration provides operation inverse [200, pp. 96–97] to the left-sided Caputo fractional differentiation, which is used in equations (24.28) and (24.29). Lemma 2.22 of [200, pp. 96–97] is a basis of the equivalence of fractional differential equations and the Volterra integral equations [200, pp. 199–208]. For fractional differential equations (24.28) and (24.29), this equivalence should be considered on the space of test functions [335, pp. 154–157], since these equations contain the Dirac delta functions.

24.6 Discrete-time accelerator with power-law memory

Let us obtain an equation of discrete accelerator with memory, which corresponds to fractional differential equation (24.29). For this purpose, we use Theorem 18.19 of [363, p. 444], which is valid for any positive order $\alpha > 0$ and which was initially suggested in [361, 362]. This theorem is based on the equivalence of fractional differential equations and the Volterra integral equations in the generalized sense, i. e., on a space of test functions. Using Theorem 18.19 of [363, p. 444], we can formulate the following statement, which is a consequence of Theorem 18.19.

Theorem 24.1. *The Cauchy problem with the differential equation*

$$(D^{\alpha}_{C;0+}K)(t) = sY(t)\sum_{k=1}^{\infty}\delta\left(\frac{t}{T}-k\right), \tag{24.30}$$

and the initial conditions $K^{(k)}(0) = K_0^{(k)}$ $(k = 0, 1, \dots, N-1)$, where $N-1 < \alpha < N$, is equivalent to the discrete equation

$$\begin{aligned} K_{n+1}^{(m)} = &\sum_{k=0}^{N-m-1}\frac{T^k}{k!}K_0^{(k+m)}(n+1)^k \\ &+ \frac{s^{\alpha-m}}{\Gamma(\alpha-m)}\sum_{k=1}^{n}(n+1-k)^{\alpha-1-m}Y_k, \end{aligned} \tag{24.31}$$

where $Y^{(m)}(t) = d^mY(t)/dt^m$,

$$Y_k^{(m)} = \lim_{\varepsilon\to 0+} Y^{(m)}(kT-\varepsilon), \tag{24.32}$$

and $m = 0, 1, \dots, N-1$. Equation (24.31) is the discrete map with memory. Equation (24.31) defines the accelerator with memory in the framework of discrete time terms.

Remark 24.2. We should emphasize that discrete equation (24.31) is derived from the fractional differential equation (24.29) without the use of any approximations, i. e., it is an exact discrete analog of the fractional differential equation (24.29). Equations (24.31) define a discrete map with power-law memory fading parameter $\alpha > 0$.

For $0 < \alpha < 1$, $(N = 1)$, discrete map (24.31) is described by the equation

$$K_{n+1} = K_0 + \frac{sT^\alpha}{\Gamma(\alpha)} \sum_{k=1}^{n} (n+1-k)^{\alpha-1} Y_k. \tag{24.33}$$

We can formulate the following statement for map (24.33).

Statement 24.2. *Equation* (24.33) *of the discrete map* (24.31) *with* $0 < \alpha < 1$ *can be represented in the form*

$$K_{n+1} - K_n = \frac{sT^\alpha}{\Gamma(\alpha)} Y_n + \frac{sT^\alpha}{\Gamma(\alpha)} \sum_{k=1}^{n-1} V_\alpha(n-k) Y_k, \tag{24.34}$$

where $V_\alpha(z)$ *is defined by*

$$V_\alpha(z) = (z+1)^{\alpha-1} - (z)^{\alpha-1}. \tag{24.35}$$

Proof. In equation (24.33), the replacement $n+1$ by n gives

$$K_n = K_0 + \frac{sT^\alpha}{\Gamma(\alpha)} \sum_{k=1}^{n-1} (n-k)^{\alpha-1} Y_k. \tag{24.36}$$

Subtracting equation (24.36) from equation (24.33), we obtain discrete map (24.34). □

Equation (24.34) is a generalization of equation (24.7) for the case of power-law memory with $0 < \alpha < 1$.

Similarly, fractional differential equation (24.28) gives the discrete equation

$$Y_{n+1} - Y_n = \frac{T^\alpha}{v\Gamma(\alpha)} I_n + \frac{T^\alpha}{v\Gamma(\alpha)} \sum_{k=1}^{n-1} V_\alpha(n-k) I_k, \tag{24.37}$$

which is a generalization of equation (24.10) for the case of power-law memory with $0 < \alpha < 1$.

Equations (24.34) and (24.37) describe the discrete analogs of accelerators with memory of the order $0 < \alpha < 1$. For $\alpha = 1$, we can use $V_1(z) = 0$, and equations (24.34) and (24.37) give the discrete maps, which coincide with equation (24.7) and (24.10) respectively.

Applications of these type discrete accelerators with memory is suggested in article [444], and it is described in next chapter.

24.7 Conclusion

As a result, we can conclude that fractional differential equations (24.28) and (24.29) exactly correspond to fractional difference equations (24.34) and (24.37), respectively.

In the continuous time approach, the discrete accelerators with memory describe economic dynamics with memory and the periodic sharp splashes (kicks). Discrete accelerator (24.34) actually describes the economic processes with memory and the periodic sharp splashes of the output (or the periodic sharp splashes of the propensity to save). Discrete accelerator (24.37) actually describes the economic dynamics with memory and the periodic sharp splashes of the net investment (or the capital productivity).

We emphasize that there is no exact correspondence between the standard discrete-time and continuous-time accelerators. These accelerators are connected only asymptotically. In continuous time approach, the standard discrete accelerators correspond exactly to the differential equations with the periodic sharp splashes (kicks), which are represented by the delta functions. Therefore, the standard discrete accelerators without memory correspond to the continuous-time dynamics with the periodic sharp splashes. The proposed discrete accelerators with memory describe the economic processes with power-law memory and periodic sharp splashes.

The suggested discrete accelerators with memory can be used to describe economic processes with power-law memory in discrete time terms.

25 Comparison of discrete and continuous accelerators

In this chapter, we describe discrete-time accelerator that is based on the exact finite differences. It is known that equations of the continuous-time and standard discrete-time models have different solutions and can predict the different behavior of the economy. For proposed discrete-time approach to economic models, equations with exact finite differences have the same solutions as the corresponding continuous-time models and these discrete and continuous models describe the same behavior of the economy. For example, it is known that the standard Harrod–Domar growth models with continuous and discrete time are not equivalent [10, 11]. Using the Harrod–Domar growth model as an example, we show that equations of the continuous-time model and the suggested exact discrete model have the same solutions, and these models predict the same behavior of the economy. This Chapter is based on articles [431, 432] (see also [441, 442]).

25.1 Introduction

Economic accelerator is a fundamental concept of macroeconomic theory [10, 11, 494]. Accelerators can be considered in the models with continuous and discrete time. The continuous-time accelerators are described by using equations with derivative of the first order. The discrete-time accelerators are described by using the equations with finite differences.

One of the simplest macroeconomic models, in which the concept of the accelerator is used, is the Harrod–Domar growth model proposed in works [170, 171, 172, 92, 93]. It is known that the Harrod–Domar growth model with continuous time [10, pp. 64–69] and the Harrod–Domar growth model with discrete time [10, pp. 74–79] are not equivalent (see also [11, pp. 198–207]). A similar situation occurs with other economic models. The discrete-time economic models cannot be considered as exact discrete analogs of continuous-time models. The equations of these models have different solutions and can predict the different behavior of the economy. In this regard, it is important to understand the reasons for the lack of equivalence of discrete and continuous models.

It is well known that the standard finite differences of integer orders cannot be considered as an exact discretization of the integer derivatives [40] (see also articles [378, 371] and references therein). Therefore, the discrete-time accelerator equation with the standard finite differences cannot be considered as an exact discrete analog of the accelerator equation, which contains the derivative of first order. To define discrete-time accelerators that are exact discrete analogs of continuous-time accelerators, we should consider an exact correspondence between the continuous and discrete time approaches.

https://doi.org/10.1515/9783110627459-025

The problem of exact discretization of the differential equations of integer orders was formulated by Renfrey B. Potts [312, 313] and Ronald E. Mickens [266, 268, 269], [267, 16, 2]. It has been proved that for differential equations there are finite-difference discretizations such that the local truncation errors are zero. A main disadvantage of this approach to discretization is that the suggested differences depend on the parameters and type of the considered differential equation. In addition, these differences do not have the same algebraic properties as the integer derivatives. Recently, a new approach to the exact discretization was suggested in article [378, 371], [441, 442, 431, 432]. This approach is based on the principle of universality and the algebraic correspondence principle [378, 371]. The exact finite differences have a property of universality if they do not depend on the form and parameters of the considered differential equations. An algebraic correspondence means that the exact finite differences should satisfy the same algebraic relations as the derivatives. In this chapter, we propose the discrete-time economic accelerators without memory that are based on the exact finite differences. This chapter is based on articles [431, 432] (see also [441, 442]).

25.2 Comparison of accelerator without memory

In macroeconomics, the accelerator describes how much the change in the value of the economic variable (e. g., the induced investment $I(t)$) in response of a single relative increase of another variable (e. g., the income $Y(t)$). The formulation of the accelerator depends on whether continuous or discrete analysis is used. The simplest expression of the linear accelerator in the continuous form without lags [10, pp. 62–63] has the form

$$I(t) = v\frac{dY(t)}{dt}, \tag{25.1}$$

where $dY(t)/dt$ is the rate of output (income), and $I(t)$ is the rate of induced investment, and v is a positive constant, the investment coefficient indicating the power of the accelerator. Equation (25.1) means that induced investment is a constant proportion of the current rate of change of output.

In discrete form, the linear accelerator without lags can be written [10, p. 63] in the form

$$I_t = v(Y_t - Y_{t-1}), \tag{25.2}$$

in which the unit step ($T = 1$) is supposed and $Y_t = Y(t)$ for integer values of t. In the discrete approach with an arbitrary step $T > 0$, the linear accelerator can be written in the form

$$I_n = \frac{v}{T}(Y_n - Y_{n-1}), \tag{25.3}$$

where $Y_n = Y(nT)$, $I_n = I(nT)$ and T is a positive constant indicating the time scale. If $T = 1$, then $t = n$ and $Y_n = Y_t$. In this case, equation (25.3) takes the form (25.2). Equations (25.2) and (25.3) mean that induced investment depends on the current change in output [10, p. 63].

Using the standard the backward finite difference, equation (25.2) can be written as

$$I(t) = v\Delta_b^1 Y(t), \tag{25.4}$$

where the backward difference of the first order is defined by the equation

$$\Delta_b^1 Y(t) = Y(t) - Y(t-1). \tag{25.5}$$

Equations (25.2), (25.3) and (25.4) cannot be considered as exact discrete analogs of equation (25.1). This is caused by that the standard finite differences, such as backward difference (25.5), and the forward difference

$$\Delta_f^1 Y(t) = Y(t+1) - Y(t), \tag{25.6}$$

do not have the same basic characteristic properties as the derivatives of first order [378, 371]. For example, the standard Leibniz rule (the product rule), which is a characteristic property of first-order derivative, is violated for the standard finite differences [378, 371], that is, we have the inequality

$$\Delta_b^1(X_1(t)X_2(t)) \neq (\Delta_b^1 X_1(t))X_2(t) + X_1(t)(\Delta_b^1 X_2(t)). \tag{25.7}$$

For the backward difference, the product rule has the nonstandard form

$$\begin{aligned}\Delta_b^1(X_1(t)X_2(t)) = (\Delta_b^1 X_1(t))X_2(t) \\ + X_1(t)(\Delta_b^1 X_2(t)) - (\Delta_b^1 X_1(t))(\Delta_b^1 X_2(t)).\end{aligned} \tag{25.8}$$

For comparison, we give the action of the derivative and the standard finite difference on some elementary functions in the form of Table 25.1.

Table 25.1: Actions of derivative and standard finite difference of first orders.

$Y(t)$	$dY(t)/dt$	$\Delta_b^1 Y(t)$
$\exp(\lambda t)$	$\lambda\exp(\lambda t)$	$\frac{\exp(\lambda)-1}{\exp(\lambda)}\exp(\lambda t)$
$\sin(\lambda t)$	$\lambda\cos(\lambda t)$	$2\sin(\lambda t - \frac{\lambda}{2})\cos(\frac{\lambda}{2})$
$\cos(\lambda t)$	$-\lambda\sin(\lambda t)$	$-2\sin(\lambda t - \frac{\lambda}{2})\sin(\frac{\lambda}{2})$
t^2	$2t$	$2t - 1$
t^3	$3t^2$	$3t^2 - 3t + 1$

In Table 25.1, we can see that the action of standard difference Δ_b^1 does not coincide with the action of first derivative, in general. As a result, in the general case, the solutions of the equations with standard finite differences do not coincide with solutions of the differential equations, which are derived by the replacement of the standard finite differences by the derivatives of the same orders [378, 371].

The nonequivalence of the action of derivatives and standard finite differences leads to the fact that macroeconomic models with discrete time are not equivalent to the corresponding models with continuous time. In the next section, we demonstrate the nonequivalence of the standard continuous and discrete macroeconomic models by using the Harrod–Domar growth models [431, 432].

25.3 Harrod–Domar growth models: continuous and discrete

In this section, we briefly describe the model, the Harrod–Domar growth model with continuous time [10, pp. 64–69] and the Harrod–Domar growth model with discrete time [10, pp. 74–79], and then we will compare the solutions to prove nonequivalence of these models (see also [11, pp. 198–207]).

25.3.1 Continuous time approach

Let us describe the Harrod–Domar growth model with continuous time [10, pp. 64–69]. If autonomous investment $A(t)$ grow, for example, as a result of the sudden appearance of large inventions, the multiplier gives a corresponding increase $A(t)/(1-c)$ in output, where c is the marginal value of propensity to consume $(0 < c < 1)$. The expansion of output activates the accelerator and leads to further (induced) investment. These additional investments increase output due to the multiplier effect and another cycle begins.

The standard Harrod–Domar model describes the interaction of the multiplier and the accelerator in the absence of delays (lags) and it is used the simplest form of an accelerator. In a continuous time approach, all variables are taken as continuous functions of time and relations are assumed linear. If we select independent (autonomous) expenditures for both consumption and capital investment, the basic condition (balance equation) can be written in the form

$$Y(t) = C(t) + I(t) + A(t), \tag{25.9}$$

where $Y(t)$ is the output (income), $C(t)$ is the consumption, $I(t)$ is the induced investment, and $A(t)$ is the autonomous investment. Here, we can use the consumption function $C(t) = cY(t)$ and accelerator equation (25.1) with $0 < c < 1$ and $v > 0$. As a result, we get the equation

$$Y(t) = cY(t) + v\frac{dY(t)}{dt} + A(t). \tag{25.10}$$

Equation (25.10) can be rewritten in the form

$$\frac{dY(t)}{dt} = \lambda Y(t) - \frac{1}{v}A(t), \tag{25.11}$$

where $\lambda = s/v$ and $s = 1 - c$ is the marginal propensity to save. Equation (25.11) is the differential equation, whose solution describes the dynamics of output $Y(t)$ over time. The solution of (25.11) depends on the dynamics of autonomous expenditure $A(t)$ over time. Let us consider the case of the fixed autonomous expenditure ($A(t) = A = \text{const}$). Let $y(t)$ be the deviation of income from the fixed level A/s, i. e., $y(t) = Y(t) - A/s$. Then, using $dy(t)/dt = dY(t)/dt$, equation (25.11) can be rewritten in the form

$$\frac{dy(t)}{dt} = \lambda y(t), \tag{25.12}$$

where $\lambda = s/v$. The solution of equation (25.12) has the form

$$y(t) = y(0)\exp(\lambda t), \tag{25.13}$$

where $y(0)$ is a constant that describes the initial income level. Using $y(t) = Y(t) - A/s$, we get the solution of equation (25.11) with $A(t) = A$ in the form

$$Y(t) = \frac{A}{s} + \left(Y(0) - \frac{A}{s}\right)\exp(\lambda t). \tag{25.14}$$

Solution (25.14) expresses continuous growth of output or income with a constant growth rate $\lambda = s/v > 0$. Usually the marginal propensity to save $s = 1 - c$ is quite small in comparison with the investment coefficient v. In this case, the growth rate $\lambda = s/v$ is a positive fraction that may be quite small.

25.3.2 Discrete time approach

Let us describe the Harrod–Domar growth model with discrete time [10, pp. 74–76]. The main Harrod assumption is that saving plans, rather than consumption plans, are realized. This is one possible assumption that leads to the introduction of delays. In the linear case, when we exclude any autonomous expenditure $A_t = 0$, the saving function has the form $S_t = sY_{t-1}$, where s is the constant marginal propensity to save. Generally speaking, this is the expected ratio. But, as savings plans are implemented, S_t is also the actual value of savings. The expected consumption will be equal to $(1 - s)Y_{t-1}$, the actual consumption is determined by the formula

$$C_t = Y_t - S_t = Y_t - sY_{t-1}. \tag{25.15}$$

The balance equation, which connects the actual values of the model, is analogous to equation (25.9) of continuous model and it has the form

$$Y_t = C_t + I_t + A_t, \tag{25.16}$$

where I_t is induced investment, and A_t is independent investment. Therefore, we have equation $S_t = I_t + A_t$, which expresses the actual equality of savings and investment. Let us consider the important case, when there are no autonomous investments. For $A_t = 0$, the actual investment, which all are induced, is given by expression

$$I_t = S_t = sY_{t-1}. \tag{25.17}$$

Expected induced investments express the action of the accelerator without lag in the form

$$J_t = v(Y_t - Y_{t-1}). \tag{25.18}$$

The further specification of the model depends on the relationship between the expected investment J_t and actual investments I_t. The growth rate of output Y_t is given by the equilibrium condition that investment plans are always realized ($J_t = I_t$) for all t. Since saving plans are assumed realized in the first place, this is the special type of situation in which saving and investment are always the same, expected and actual. This condition is expressed by the equation

$$v(Y_t - Y_{t-1}) = sY_{t-1}. \tag{25.19}$$

Using the backward difference $\Delta_b^1 Y(t) = Y(t) - Y(t-1)$, equation (25.19) can be written in the form

$$\Delta_b^1 Y(t) = \lambda Y_{t-1}, \tag{25.20}$$

where $\lambda = s/v$. Equation (25.19) also can be written in the form $Y_t = (1+\lambda)Y_{t-1}$. The solution of difference equation [10, p. 76] has the form

$$Y_t = Y_0(1+\lambda)^t = Y_0 \exp(t \ln(1+\lambda)). \tag{25.21}$$

Solution (25.21) expresses continuous growth of output or income with the constant relative speed $\ln(1+\lambda)$.

Let us consider the case of the fixed autonomous expenditure ($A(t) = A = \text{const}$). The equation has the form

$$\Delta_b^1 Y(t) = \lambda Y_{t-1} - \frac{A}{v}. \tag{25.22}$$

The solution of equation (25.22) can be given by the expression

$$Y_t = \frac{A}{s} + \left(Y_0 - \frac{A}{s}\right) \exp(t \ln(1+\lambda)), \tag{25.23}$$

which described the growth of income with the constant growth rate $\ln(1+\lambda)$.

25.3.3 Comparison of discrete and continuous models

As a result, we obtain the following expressions of the warranted growth rates:

$$\lambda_{CM} = \lambda, \quad \lambda_{SDM} = \ln(1+\lambda) \tag{25.24}$$

for the continuous model (CM) and the standard discrete model (SDM). Therefore, in the general case, we have the inequality

$$\lambda_{CM} \neq \lambda_{SDM}. \tag{25.25}$$

If we take into account the step $T \neq 1$, solution (25.21) takes the form $Y_t = Y_0(1 + \lambda T)^{t/T}$. Only in the limit $T \to 0$, we get $Y(t) = Y(0)\exp(\lambda t)$, where we use $\lim_{x\to 0}(1 + x)^{1/x} = e$.

It is easy to see by direct substitution that expression (25.14) is not a solution of the difference equation (25.22), since

$$\Delta_b^1 \exp(\lambda t) \neq \lambda \exp(\lambda t). \tag{25.26}$$

As a result, we can see that the growth rate $\ln(1 + \lambda)$ of the discrete models does not coincide with growth rate $\lambda = s/v$ of the continuous model.

The similar situation occurs with other macroeconomic growth models, including the natural growth model, the Keynes model, the dynamic intersectoral model of Leontief, and others.

Using the Harrod–Domar growth model as an example, we show that the discrete-time macroeconomic models, which are based on standard differences, cannot be considered as exact discrete analogs of continuous-time models. The equations of these models can have different solutions and can predict the different behavior of the economy. In the next section, we propose the discrete-time economic accelerators that allow us to propose discrete macroeconomic models. These proposed discrete-time models can be considered as exact discretization of the corresponding continuous-time models. In addition, these discrete models predict the same behavior of the economy as the corresponding continuous-time macroeconomic models.

25.4 Concept of exact discretization

In order to have difference equations of the accelerator, which can be considered as exact discrete analogs of equation (25.1), we propose the requirement on difference operators in the form of the correspondence principle [378, 371].

Principle 25.1 (Correspondence principle for differences). *The finite differences, which are exact discretization of derivatives of integer orders, should satisfy the same algebraic characteristic relations as these derivatives. This means that the correspondence* *between the differences and derivatives lies not so much in the limiting condition, when the step tends to zero ($T \to 0$) as in the fact that the mathematical operations should obey in many cases the same mathematical laws.*

The exact discrete analogs of the derivatives should have the same basic characteristic properties as these derivatives [378, 371], including primarily the following:

(1) The product rule (the Leibniz rule) is a characteristic property of the derivatives of integer orders. Therefore, the exact discretization of the derivatives should satisfy this rule. The Leibniz rule should be the main characteristic property of exact discrete analogs of the derivatives.

(2) The exact discretization should satisfy the semigroup property. For example, the exact finite difference of second-order should be equal to the repeated action of the exact differences of the first order.

(3) The exact differences of power-law functions should give the same expression as an action of derivatives. This allows us to consider the exact correspondence of derivatives and differences on the space of entire functions.

Remark 25.1. We should note that these properties are not satisfied for the standard finite differences and nonstandard differences, which are proposed by Mickens in [266, 268, 269].

In articles [378, 371] (see also [431, 432] and [441, 442]), new approach to exact discretization and new difference operators were proposed. This approach is based on new difference operators, which can be considered as an exact discretization of derivatives of integer and noninteger orders. These differences do not depend on the form and parameters of considered differential equations in contrast from non-standard differences of Mickens. Using suggested exact differences, we can get an exact discretization of differential equation of integer and noninteger orders. The suggested approach to exact discretization allows us to obtain difference equations that exactly correspond to the differential equations. We consider not only an exact correspondence between the equations, but also exact correspondence between solutions. The suggested exact differences are used in articles [431, 432] (see also [441, 442]) to propose the exact discrete-time analogs of the continuous-time equations of the accelerators.

As a result, we can formulate the correspondence principle for discrete accelerator.

Principle 25.2 (Correspondence principle for discrete accelerator). *The discrete-time accelerator, which is exact discretization of continuous-time accelerator, and the continuous-time accelerator must have the same algebraic characteristic relations. This means that the correspondence between the discrete-time and continuous-time economic models of the same type should lie not so much in the limiting condition, when the step tends to zero ($T \to 0$) as in the fact that concepts of these two models should obey in many cases the same properties.*

25.5 Exact discrete accelerator and multiplier without memory

In this section, we describe exact discrete analogs of standard accelerators and multiplier without memory.

Let us consider a space of entire function $E(\mathbb{R})$ on the real axis $\mathbb{R}$. We will assume that $X(t) \in E(\mathbb{R})$, and we will use the notation $X(n) \in E(\mathbb{Z})$, where $E(\mathbb{Z})$ is the space of entire function over the field of integer scalars $\mathbb{Z}$. It is known that any function $X(t) \in E(\mathbb{R})$ can be represented in the form of the power series

$$X(t) = \sum_{k=0}^{\infty} X_k t^k, \tag{25.27}$$

where the coefficients X_k satisfy the condition $\lim_{k\to\infty} \sqrt[k]{X_k} = 0$ and $t \in \mathbb{R}$. It is obvious that $X(n) \in E(\mathbb{Z})$ if $X(t) \in E(\mathbb{R})$.

Let us define the difference operator Δ_t^k of the positive integer order k on the function space $E(\mathbb{Z})$.

Definition 25.1. The linear operator Δ_t^k will be called the exact finite difference of integer order, if the following condition is satisfied: If $X(t), Y(t) \in E(\mathbb{R})$ and the differential equation

$$\frac{d^k Y(t)}{dt^k} = \lambda X(t) \tag{25.28}$$

holds for all $t \in \mathbb{R}$, then the difference equation

$$\Delta_t^k Y(n) = \lambda X(n) \tag{25.29}$$

holds for all $n \in \mathbb{Z}$.

In papers [378, 371], the exact finite-differences of integer order were obtained in explicit form. The exact finite difference of the first order is defined by the formula

$$\Delta_t^1 X(t) = \sum_{m=1}^{\infty} \frac{(-1)^m}{m} (X(t - Tm) - X(t + Tm)), \tag{25.30}$$

where the sum implies the Cesaro or Poisson–Abel summation [121, 123, 122, 169] and [378, pp. 55–56].

Equation (25.28) with $k = 1$ represents the standard equation of the continuous-time accelerator. Equation (25.29) with the exact difference (25.30) represents the exact discrete analog of the standard accelerator, which is represented by equation (25.28) with $k = 1$.

Exact finite difference of second and next integer orders can be defined by the recurrence formulas

$$\Delta_t^{k+1} X(t) = \Delta_t^1 (\Delta_t^k X(t)). \tag{25.31}$$

As a result, we get

$$\Delta_1^2 X(t) = -\sum_{m=1}^{\infty} \frac{2(-1)^m}{m^2}(X(t-Tm)+X(t+Tm)) - \frac{\pi^2}{3}X(t). \tag{25.32}$$

For the arbitrary positive integer order n, the exact difference is written by the equation

$$\Delta_t^n X(t) = \sum_{m=1}^{\infty} M_n(m)(X(t-Tm)+(-1)^n X(t+Tm)) - M_n(0)X(t), \tag{25.33}$$

where the kernel $M_n(m)$ is given by the equation

$$M_n(m) = \sum_{k=0}^{[\frac{n+1}{2}]+1} \frac{(-1)^{m+k}\Gamma(n+1)\pi^{n-2k-2}}{\Gamma(n-2k+1)m^{2k+2}}\left((n-2k)\cos\left(\frac{\pi n}{2}\right) + \pi m \sin\left(\frac{\pi n}{2}\right)\right) \tag{25.34}$$

for $m \neq 0$, and by the expression

$$M_n(0) = \frac{\pi^n}{n+1}\cos\left(\frac{\pi n}{2}\right). \tag{25.35}$$

Here, we take into account that $1/\Gamma(-z) = 0$ for positive integer z.

The Fourier transform of the exact difference operator (25.33) has the form

$$F\{\Delta_t^n X(t)\}(\omega) = (i\omega t)^n F\{X(t)\}(\omega) \tag{25.36}$$

for all positive integer order n.

An important characteristic property of the exact finite difference of the first order is the Leibniz rule on the space of entire functions [378, 371], i. e.,

$$\Delta_t^1(X(t)Y(t)) = (\Delta_t^1 X(t))Y(t) + X(t)(\Delta_t^1 Y(t)) \tag{25.37}$$

for all $X(t), Y(t) \in E(\mathbb{Z})$. Note that the rule (25.37) rule is not satisfied for standard finite differences of first order [378, 371]. For exact finite difference of integer order k, the Leibniz rule has the form

$$\Delta_t^k(X(t)Y(t)) = \sum_{j=0}^{k} \binom{k}{j}(\Delta_t^{k-j}X(t))(\Delta_t^j Y(t)), \tag{25.38}$$

which is an exact analog of the rule for derivative of the integer order k.

To compare finite differences and derivatives, Table 25.2 shows the action of the derivatives $dX(t)/dt$, the standard finite differences $\Delta_b^1 X(t) = X(t) - X(t-T)$ and the exact finite difference $\Delta_t^1 X(t)$ on some elementary functions $X(t)$, where we use $T = 1$ for simplification.

Note that the elementary functions that are considered in Table 25.2 are examples of entire functions. In [378, 371], it is proved that the action of the exact finite

Table 25.2: The action of derivative, standard and exact finite differences on some elementary functions.

$X(t)$	$dX(t)/dt$	$\Delta_b^1 X(t)$	$\Delta_T^1 X(t)$
$\exp(\lambda t)$	$\lambda \exp(\lambda t)$	$\frac{\exp(\lambda)-1}{\exp(\lambda)} \exp(\lambda t)$	$\lambda \exp(\lambda t)$
$\sin(\lambda t)$	$\lambda \cos(\lambda t)$	$2\sin(\lambda t - \frac{\lambda}{2})\cos(\frac{\lambda}{2})$	$\lambda \cos(\lambda t)$
$\cos(\lambda t)$	$-\lambda \sin(\lambda t)$	$2\sin(\lambda t - \frac{\lambda}{2})\sin(\frac{\lambda}{2})$	$-\lambda \sin(\lambda t)$
t^2	$2t$	$2t-1$	$2t$
t^3	$3t^2$	$3t^2-3t+1$	$3t^2$

differences Δ_t^1 on the space of entire functions coincides with the action of the first derivative. As a result, solutions of equations with exact differences coincide with solutions of a wide class of differential equations [378, 371]. Equivalence of the actions of derivatives and exact finite differences leads to the equivalence of a wide class of macroeconomic models with discrete and continuous time, if exact finite differences are used in discrete models.

The exact difference analog of the differential equation

$$X(t) = v\frac{dY(t)}{dt}, \tag{25.39}$$

which describes the standard accelerator without memory, has the form

$$X(t) = v(\Delta_t^1 Y)(t). \tag{25.40}$$

Using expression (25.30), equation (25.40) can be written as

$$X(t) = v\sum_{k=1}^{\infty} \frac{(-1)^k}{k}(Y(t-Tk) - Y(t+Tk)). \tag{25.41}$$

Using the Newton–Leibniz theorem, equation (25.39) gives

$$Y(t) = Y(0) + \frac{1}{v}\int_0^t X(\tau)\,d\tau. \tag{25.42}$$

The exact difference analog of the integral equation (25.42), which corresponds to (25.39), has the form

$$Y(t) = \frac{1}{v}\sum_{k=1}^{\infty} \frac{Si(\pi k)}{\pi}(X(t-Tk) - X(t+Tk)), \tag{25.43}$$

where $Si(\pi k)$ is the sine integral. In equation (25.43), we use the exact difference Δ_t^{-1} of the first negative order that can be considered as an exact discrete analog of the antiderivative [378, 371], such that the relations

$$(\Delta_t^1 \Delta_t^{-1} X)(t) = X(t), \quad (\Delta_t^{k+1} \Delta_t^{-1} X)(t) = (\Delta_t^k X)(t) \tag{25.44}$$

hold for all $X(t) \in E(\mathbb{Z})$.

25.6 Comparison of standard and exact discrete models

Let us demonstrate the equivalence of the continuous and discrete Harrod–Domar growth models. For this purpose, we shall use the concept of an exact discrete accelerator [441, 442, 431, 432].

Discrete equation, which is exact discrete analog of the Harrod–Domar model with continuous time, can be rewritten in the form

$$(\Delta_T^1 Y)(t) = \lambda Y(t) - \frac{1}{v} A(t), \tag{25.45}$$

where $\lambda = s/v$ and $s = 1 - c$ is the marginal propensity to save. The solution of equation (25.45) with $A(t) = A = \text{const}$ has the form

$$Y(t) = \frac{A}{s} + \left(Y(0) - \frac{A}{s} \right) \exp(\lambda t). \tag{25.46}$$

The fact that function (25.46) is a solution of the exact-difference equation (25.45) can be verified by direct substitution of this function into equation (25.45) and using the following equalities:

$$\Delta_T^1 \exp(\lambda t) = \lambda \exp(\lambda t), \quad \Delta_T^1 (A/s) = 0. \tag{25.47}$$

Solution (25.46) coincides with solution (25.14) of equation (25.11) of the Harrod–Domar model with continuous time. As a result, we can state the discrete Harrod–Domar growth model with the exact differences, is equivalent to the continuous Harrod–Domar growth model, which is based on the differential equation.

As a result, using the Harrod–Domar growth model as an example, we proved that equations of the continuous-time models and the corresponding discrete-time models, which are based on the suggested exact differences, have the same solutions. These discrete and continuous macroeconomic models describe the same behavior of the economy.

Let us give an illustration of the difference between the proposed approach and the standard approach by simple computer simulation of output (income) growth. We will compare the Harrod–Domar growth model with continuous time [10, pp. 64–69], the standard Harrod–Domar growth model with discrete time [10, pp. 74–79], and the suggested exact discretization of the Harrod–Domar model with continuous time [431, 432].

The comparison of the growth in the continuous model, the exact discrete and the standard discrete models will be illustrated by simple numerical examples of the output growth, which is described by equations (25.14), (25.23), (25.46) with $A = 0$.

The comparison of the output growth of the continuous model (CM), the standard discrete model (SDM), and the exact discrete model (EDM) is given by Table 25.3. The first column specifies the warranted growth rate of CM; the second column gives the

Table 25.3: Comparison of the growth of the continuous model (CM), the standard discrete model (SDM) and the exact discrete model (EDM). The values of (25.50) and (25.51) characterize the differences between standard and exact discrete models.

CM	EDM	SDM	D (%)	G (times)
0.1	0.1	0.095	4.69	1.048
0.3	0.3	0.262	12.54	1.457
0.5	0.5	0.405	18.90	2.574
0.7	0.7	0.531	24.19	5.440
0.9	0.9	0.642	28.68	13.22
1.1	1.1	0.742	32.55	35.90
1.3	1.3	0.833	35.93	106.8
1.5	1.5	0.916	38.91	342.7
1.7	1.7	0.993	41.57	1173
1.9	1.9	1.065	43.96	4242

warranted growth rate of EDM. Note that the warranted growth rates of CM and EDM coincide. The third column gives the warranted growth rate of SDM. The fourth column specifies the difference between the warranted growth rates of CM and EDM on the one hand, and the warranted growth rate of SDM on the other hand in percentages. The fifth column describes how many times the growth in output at $t = 10$ with $T = 1$ is greater for the ED model in comparison with the SD model.

The warranted growth rates of these models are defined by the equations

$$\begin{aligned} \lambda_{\mathrm{CM}} &= \lambda, \\ \lambda_{\mathrm{EDM}} &= \lambda, \\ \lambda_{\mathrm{SDM}} &= \ln(1+\lambda) \end{aligned} \tag{25.48}$$

for the continuous model (CM) the exact discrete model (EDM), and the standard discrete model (SDM). We can see the relations

$$\lambda_{\mathrm{EDM}} = \lambda_{\mathrm{CM}}, \quad \lambda_{\mathrm{EDM}} \neq \lambda_{\mathrm{SDM}}. \tag{25.49}$$

For a quantitative description of the differences between the rates λ_{EDM} and λ_{SDM}, we use the following two characteristics:

$$D = \frac{|\lambda_{\mathrm{EDM}} - \lambda_{\mathrm{SDM}}|}{\lambda_{\mathrm{EDM}}} 100\,\% = \frac{|\lambda - \ln(1+\lambda)|}{\lambda} 100\,\%. \tag{25.50}$$

The value of D shows how many percent the warranted growth rate of the standard discrete model is less than the warranted growth rate of the exact discrete model.

$$G = \frac{Y_{\mathrm{CM}}(t)}{Y_{\mathrm{SDM}}(t)} = \frac{Y_{\mathrm{EDM}}(t)}{Y_{\mathrm{SDM}}(t)} = \frac{\exp(\lambda t)}{(1+\lambda)^t}. \tag{25.51}$$

The value of G shows how many times the output of the exact discrete model is greater than the output of the standard discrete model at $t = 10$, $T = 1$, with $A = 0$.

Note that Table 25.3 shows rounded values.

Let us give some explanations to the numerical results that are given in Table 25.3.

(1) For example, if the warranted growth rate of EDM and CM is equal to $\lambda = 0.3$, then the growth rate of SDM is $\ln(1 + \lambda) \approx 0.262$, i. e., the growth rate of the standard discrete model is less than the growth rate of the continuous model by more than 12 %. The growth rates of CM and EDM coincide. As a result, for example the output at $t = 10$, $T = 1$ differs almost one and a half times in standard and exact discrete models with $A = 0$.
(2) If the growth rate of EDM and CM is $\lambda = 0.9$, then the warranted growth rate of SDM is equal to $\ln(1+\lambda) \approx 0.642$, i. e., the growth rate of the discrete model is less than the growth rate of the continuous model by more than 28 %. In this case, for $t = 10$ the output growth differs by more than 13 times for standard discrete (SDM) and exact discrete models (EDM) with $A = 0$. The output growth rates of CM and EDM coincide.
(3) If the warranted growth rate of EDM and CM is equal to $\lambda = 1.7$, then the growth rate of SDM is $\ln(1+\lambda) \approx 0.993$, i. e., the growth rate of the standard discrete model is less than the growth rate of the continuous model by more than 41 percent. As a result, the output at $t = 10$, $T = 1$ differs more than 1000 times in standard and exact discrete models with $A = 0$. The warranted growth rates of CM and EDM coincide.

As a result, we have that the differences between the standard discrete model and the exact discrete model (the standard continuous models) can be significant for the magnitude of output growth. Moreover, the growth of the output may differ not only in several times, but also by an order of magnitude (see the fifth column (G) of 25.3).

25.7 Conclusion

We describe new approach to the exact discretization of the continuous-time macroeconomic models, which is proposed in [431, 432]. This approach is based on the exact finite differences that are suggested in [378, 371] (see also [441, 442, 431, 432]). The proposed exact differences satisfy the principle of universality and the algebraic correspondence principle [378, 371]. The finite differences have a property of universality if they do not depend on the form and parameters of the considered differential equations. An algebraic correspondence means that the exact finite differences should satisfy the same algebraic relations as the derivatives. We proved that equations of the continuous-time macroeconomic models and the corresponding discrete-time models, which are based on the suggested exact differences, can have the same solutions.

These discrete and continuous economic models can describe the same behavior of the economy.

The exact fractional differences allow us to propose the exact discrete-time analogs of the continuous-time equations of the accelerator and multiplier with power-law memory that are described by the Liouville fractional integrals and derivatives. The discrete macroeconomic models, which are used exact fractional differences, can be equivalent to the continuous models of processes with memory, which is described by the Liouville fractional derivatives.

It is known that the equations of the continuous and standard discrete models have different solutions and can predict the different behavior of the economy. Using the well-known Harrod–Domar growth models, we proved the advantage of the proposed approach. A numerical comparison of the output growth for solutions of the model equations showed a significant difference between the standard and exact discrete models. Note that the output of the exact discrete model can be greater than the output of the standard discrete model more than 1000 times at $t = 10$, $T = 1$. In this case, the warranted growth rates of continuous and exact discrete models coincide.

26 Exact discrete accelerator and multiplier with memory

In this chapter, we describe economic accelerator and multiplier with fading memory in the framework of the discrete-time approach. A relationship of the discrete-time and continuous-time equations of accelerator with memory is considered. Exact discrete analogs of the fractional differential equations of accelerator are suggested by using the exact fractional differences of noninteger orders. This chapter is based on articles [441, 442] (see also [431, 432]).

26.1 Introduction

To describe economic processes with memory Granger, Joyeux and Hosking [155, 180] proposed the so-called fractional differencing and integrating for discrete time models (e. g., see books [29, 295, 30, 481, 231, 485], and reviews [21, 303, 23, 135]). The fractional operators of Granger, Joyeux, Hosking were proposed and then began to be used in economics up to the present time without any connection with the fractional calculus and the well-known fractional differences. The approach, which is based on the discrete operators of Granger, Joyeux, and Hosking, is the most common among economists [21, 303, 23, 135]. In mathematics, the Granger–Joyeux fractional differencing and integrating are the well-known Grunwald–Letnikov fractional differences, which were suggested in 1867 and 1868 in the works [157, 222]. The approach based on discrete operators, which are proposed by Granger, Joyeux and Hosking, is restricted by one type of fractional finite differences. In addition, this approach is used without a real connection with the development of fractional calculus in the last 200 years.

We should note that the Fourier transforms of the Grunwald–Letnikov fractional differences do not have the power-law form (see equation (20.5) of [335, p. 373]). As a result, the Grunwald–Letnikov fractional differences (and therefore the fractional differencing and integrating of Granger, Joyeux and Hosking) cannot be considered as an exact tool to describe the power-law memory (for details, see [378]). The Grunwald–Letnikov fractional differences lead us to insensitivity of the mathematical tool with respect to different short term shocks, since the Fourier transform of these fractional differences satisfy the power law in the neighborhood of zero only.

The basic concepts in macroeconomics are accelerator and multiplier [10, 11]. The accelerator with memory and the multiplier with memory were proposed in articles [412, 397] within the continuous-time approach. The discrete-time accelerator for economic processes with the power-law memory was suggested in [417, 436, 444] for the case of the periodic sharp splashes (periodic kicks). The discrete-time accelerators with memory, which is based on exact fractional differences, are proposed in articles [441, 442].

https://doi.org/10.1515/9783110627459-026

It is well known that the standard finite differences of integer orders cannot be considered as an exact discretization of the derivatives of integer orders. Therefore, equations with the standard finite difference cannot be considered as exact discrete analogs of equations, which contain the derivatives of the integer orders. The problem of exact discretization of the differential equations of integer orders was formulated by Renfrey B. Potts [312, 313] and Ronald E. Mickens [266, 268, 269], [267, 16, 2]. These works prove that for differential equations there are finite-difference discretizations such that the local truncation errors are zero. A main disadvantage of this approach to discretization is that the suggested nonstandard differences depend on the parameters and type of the considered differential equation. In addition, these differences do not have the same algebraic properties as the integer derivatives.

In articles [378, 371] (see also [441, 442, 431, 432]), the exact fractional differences were suggested. The exact finite differences have a property of universality, which means that these differences do not depend on the form and parameters of the considered differential equations. For integer orders, the exact finite differences satisfy the same algebraic relations as the derivatives of integer orders. In articles [441, 442] (see also [431, 432]), we propose the discrete-time economic accelerators that described by equations with the exact finite differences.

In this chapter, we consider a relationship of the discrete-time and continuous-time descriptions of accelerators with power-law memory. The continuous-time accelerator and multiplier with power-law memory can be described by the equations with the fractional derivatives and integrals [412, 397]. We use the Liouville fractional derivatives and integrals of noninteger orders. The Fourier transform of the Liouville fractional derivatives and integrals has the power-law form. This allows us to consider the exact fractional differences as an exact discretization of the Liouville fractional derivatives and integrals [441, 442]. Then these exact fractional differences can be used to derive exact discrete analogs of equations of the accelerator and multiplier with power-law memory. This chapter is based on articles [441, 442].

26.2 Continuous-time accelerator and multiplier with memory

To consider memory effects in economic model, we assume that the value of the economic variable $Y(t)$ at time t depends not only on the another variable $X(\tau)$ at the same time point $\tau = t$, but it also depends on the changes $X(\tau)$ in the past $\tau \in (-\infty, t]$. This dependence may appear because economic agents can remember the previous changes of the variable $X(\tau)$ and the impact of these changes on another variable $Y(t)$.

The general formulation of the memory for economics is the following. In the economic process with memory, there is a variable $Y(t)$, which depends on the history of the change of the variable $X(\tau)$ at $\tau \in (-\infty, t)$. This formulation can be represented by the symbolic expression

$$Y(t) = F^t_{-\infty}(X(\tau)). \tag{26.1}$$

In equation (26.1), the symbol $F^t_{-\infty}$ denotes a method that allows to find the value of $Y(t)$ for any time t, if it is known $X(\tau)$ for $\tau \in (-\infty, t]$. We can say that $F^t_{-\infty}$ is an operator, which is a mapping from one space of functions to another. This operator transforms each history of changes of $X(\tau)$ for $\tau \in (-\infty, t]$ into the appropriate history of changes of $Y(\tau)$ with $\tau \in (-\infty, t]$.

We will consider linear operators such that expression (26.1), the dependence of $Y(t)$ from $X(\tau)$, which takes into account the memory, will be described by the integral equation

$$Y(t) = \int_{-\infty}^{t} M(t,\tau)X(\tau)\,d\tau, \tag{26.2}$$

where $M(t,\tau)$ is the memory function that allows us to take into account the memory in economic processes. Equation (26.2) can be interpreted as an equation of economic multiplier with memory of general type.

However, not all kinds of functions $M(t,\tau)$ can be used to describe processes with memory [450, 399]. To describe the memory with power-law fading, we can use the memory function in the form

$$M(t,\tau) = \frac{1}{\Gamma(\alpha)} \frac{m}{(t-\tau)^{1-\alpha}}, \tag{26.3}$$

where $\Gamma(\alpha)$ is the gamma function, $\alpha > 0$ is a parameter that characterize the power-law of fading, m is a positive real number, and $t > \tau$. In order to easily interpret the dimensions of the economic quantities, we can use the time t as a dimensionless variable by changing the variable $t \to tt_d$, where t_d is the unit of time (hour, day, month, year).

Substitution of expression (26.3) into equation (26.2) gives the fractional integral equation of the order $\alpha > 0$ in the form

$$Y(t) = m(I^\alpha_{L;+}X)(t), \tag{26.4}$$

where $I^\alpha_{L;+}$ is the left-sided Liouville fractional integral of the order $\alpha > 0$ with respect to time variable. This integral is defined [335, pp. 93–119], [200, p. 87] by the equation

$$(I^\alpha_{L;+}X)(t) = \frac{1}{\Gamma(\alpha)} \int_{-\infty}^{t} \frac{X(\tau)\,d\tau}{(t-\tau)^{1-\alpha}}, \tag{26.5}$$

where $\Gamma(\alpha)$ is the gamma function. The Liouville integration (26.5) of the order $\alpha = 1$ gives the standard integration of first order

$$(I^1_{L;+}X)(t) = \int_{-\infty}^{t} X(\tau)\,d\tau. \tag{26.6}$$

Equation (26.4) describes the economic multiplier with power-law memory, and the parameter m is the coefficient of the multiplier [412, 397].

In order to express the function $X(t)$ through the function $Y(t)$, we act on equation (26.4) by the Liouville fractional derivative of the order $\alpha > 0$, which is defined [335, pp. 93–109], [200, p. 87] by the equation

$$(D^{\alpha}_{L;+}Y)(t) = \frac{1}{\Gamma(n-\alpha)} \frac{d^n}{dt^n} \int_{-\infty}^{t} \frac{Y(\tau)\,d\tau}{(t-\tau)^{\alpha-n+1}}, \tag{26.7}$$

where $n = [\alpha] + 1$ and $\tau < t$. Here, function $Y(\tau)$ must have the derivatives of integer orders up to the $(n-1)$ order, which are absolutely continuous functions on the interval $(-\infty, t]$.

The action of the Liouville derivative (26.7) on equation (26.4) gives the expression

$$(D^{\alpha}_{L;+}Y)(t) = m(D^{\alpha}_{L;+}I^{\alpha}_{L;+}X)(t). \tag{26.8}$$

It is known that the Liouville fractional derivative is left inverse to the Liouville fractional integral [200, p. 89], and for any function $X(t) \in L_1(-\infty, +\infty)$ the identity

$$(D^{\alpha}_{L;+}I^{\alpha}_{L;+}X)(t) = X(t) \tag{26.9}$$

holds for any $\alpha > 0$, where $I^{\alpha}_{L;+}$ is the left-sided Liouville fractional integral (26.5) and $D^{\alpha}_{L;+}$ is the left-sided Liouville fractional derivative (26.7).

Remark 26.1. Note that we can use the Marchaud fractional derivative instead of the Liouville fractional derivative [335, pp. 93–109]. The Marchaud fractional derivative [335, pp. 109–119] is more convenient than the Liouville fractional derivative, since they allow more freedom for the function $X(t)$ at infinity [335, p. 110].

Using identity (26.9), equation (26.8) can be written as

$$X(t) = v(D^{\alpha}_{L;+}Y)(t), \tag{26.10}$$

where $v = 1/m$. For $\alpha = 1$, equation (26.10) takes the form $X(t) = v dY(t)/dt$, which is the equation of the standard economic accelerator. As a result, the multiplier (26.4) with power-law memory can be represented in the form of the accelerator with memory (26.10), where the coefficient of the accelerator is inversed to the coefficient of the multiplier [412, 397].

Accelerator equation (26.10) contains the standard equation of the accelerator and the multiplier, as special cases. For example, using the property $(D^{1}_{L;+}X)(t) = X^{(1)}(t)$ of the Liouville fractional derivative [200, p. 87], equation (26.10) with $\alpha = 1$ gives equation $X(t) = vY^{(1)}(t)$ that describes the standard accelerator. Using the property $(D^{0}_{L;+}Y)(t) = Y(t)$ [200, p. 87], equation (26.10) with $\alpha = 0$ can be written as $X(t) =$

$vY(t)$, that is a standard equation of multiplier. As a result, the accelerator with memory (26.10) generalizes the standard economic concepts of the accelerator and the multiplier [412, 397].

It should be emphasized that the Fourier transform of the Liouville fractional integral and derivative has the power law form (see [335, p. 137] and [200, p. 90]). The Fourier transform F of the Liouville fractional integral (26.5) is described by the equation

$$F\{(I^{\alpha}_{L;+}X)(t)\}(\omega) = (i\omega)^{-\alpha}F\{X(t)\}(\omega). \tag{26.11}$$

The Fourier transform of the Liouville fractional derivative (26.7) is represented by the expression

$$F\{(D^{\alpha}_{L;+}Y)(t)\}(\omega) = (i\omega)^{\alpha}F\{Y(t)\}(\omega). \tag{26.12}$$

In equations (26.11) and (26.12), we use the Fourier transform in the form

$$F\{X(t)\}(\omega) = \int_{-\infty}^{+\infty} X(t)e^{-i\omega t}\, dt, \tag{26.13}$$

where a negative sign in front $i\omega t$ is used in equation (25.29). Note that book [200] uses a positive sign in front $i\omega t$ [335, p. 10]. Therefore, in equations (26.11) and (26.12), we use $(i\omega)^{\mp\alpha}$ instead of $(-i\omega)^{\mp\alpha}$, which is used in [200, p. 90]. In equations (26.11) and (26.12), the power-law form $(i\omega)^{\mp\alpha}$ means

$$(i\omega)^{\alpha} = |\omega|^{\alpha}\exp\left(i\alpha\pi\frac{\operatorname{sgn}(\omega)}{2}\right). \tag{26.14}$$

For $\omega > 0$, equation (26.14) has the form

$$(i\omega)^{\alpha} = \omega^{\alpha}\left(\cos\left(\frac{\pi\alpha}{2}\right) + i\sin\left(\frac{\pi\alpha}{2}\right)\right). \tag{26.15}$$

Using equation (26.11), we get the Fourier transform of equation (26.4) of the multiplier with memory in the form

$$F\{Y(t)\}(\omega) = m(i\omega)^{-\alpha}F\{X(t)\}(\omega). \tag{26.16}$$

Using equation (26.12), we get the Fourier transform of the accelerator with memory (26.10) in the form

$$F\{X(t)\}(\omega) = v(i\omega)^{\alpha}F\{Y(t)\}(\omega). \tag{26.17}$$

One can see that equations (26.16) and (26.17) coincide if $v = 1/m$. As a result, we see that the Fourier transforms of the accelerator and multiplier with memory have the power-law form exactly.

26.3 Standard discrete time approach to memory in economics

To describe economic processes with memory Granger, Joyeux and Hosking [155, 180] proposed the so-called fractional differencing and integrating for discrete time models without any connection with the fractional calculus and the well-known fractional differences of noninteger orders. In fact, these fractional differencing and integrating are the well-known Grunwald–Letnikov fractional differences, which were suggested in 1867 and 1868 in works [157, 222]. Unfortunately, this approach is used without an explicit connection with the development of fractional calculus in the last 200 years. The Grunwald–Letnikov fractional differences are actively used in the fractional calculus [335, pp. 371–388], [308, pp. 43–62], [200, pp. 121–123].

We can state that the main tool, which used to describe memory in discrete-time economic models, is the Grunwald–Letnikov fractional differences. The Grunwald–Letnikov fractional difference $\Delta^{\alpha}_{\mathrm{GL};T}$ of order α with the step T is defined by the equation

$$\Delta^{\alpha}_{\mathrm{GL};T}X(t) = (1 - L_T)^{\alpha}X(t) = \sum_{m=0}^{\infty} (-1)^m \binom{\alpha}{m} X(t - mT), \tag{26.18}$$

where $L_T X(t) = X(t - T)$ is fixed-time delay, T is the positive time-constant, and

$$\binom{\alpha}{m} = \frac{\Gamma(\alpha + 1)}{\Gamma(\alpha - m + 1)\Gamma(m + 1)} \tag{26.19}$$

are the generalized binomial coefficients [200, pp. 26–27] that can be written (see equation (1.48) of [335, p. 14]) in the form

$$\binom{\alpha}{m} = \frac{(-1)^{m-1}\alpha\Gamma(m - \alpha)}{\Gamma(1 - \alpha)\Gamma(m + 1)}. \tag{26.20}$$

Using expression (26.20), equation (26.18) can be represented in the form

$$\Delta^{\alpha}_{\mathrm{GL};T} = (1 - L_T)^{\alpha} = \sum_{m=0}^{\infty} \frac{\Gamma(m - \alpha)}{\Gamma(-\alpha)\Gamma(m + 1)} L^m_T, \tag{26.21}$$

which is usually used in articles [21, 303, 23, 135] and books [29, 295, 30, 481, 231, 485] of economists.

Remark 26.2. The Grunwald–Letnikov fractional difference (26.18) converges [335, p. 372] for $\alpha < 0$, if the function $X(t)$ satisfies the inequality $|X(t)| \le c(1 + |t|)^{-\mu}$, where $\mu > \alpha$. In this case, we can use (26.18) as a discrete fractional integration in the nonperiodic case.

It is known that in the nonperiodic case the Fourier transform F of the Grunwald–Letnikov fractional difference (26.18) is given [335, p. 373] by the formula

$$F\{\Delta^{\alpha}_{\mathrm{GL};T}X(t)\}(\omega) = (1 - \exp(i\omega t))^{\alpha} F\{X(t)\}(\omega). \tag{26.22}$$

For $\alpha = 1$, equation (26.18) gives the standard difference of the first order such that $\Delta^1_{GL;1}X(n) = X(n) - X(n-1)$. For this standard difference, the Fourier transform is also given by equation (26.22) with $\alpha = 1$ and $T = 1$. It is well known that the standard finite differences of integer orders cannot be considered as an exact discretization of the integer derivatives [378, 371].

Therefore, the discrete-time accelerator equations with standard finite differences cannot be considered as exact discrete analogs of the continuous-time accelerator equation, which contains the first derivatives.

The fractional differences (26.18) cannot be considered as an exact discrete (difference) analog of the accelerator and multiplier with power-law memory, which are described by equations (26.10) and (26.4), since the Fourier transform for the Grunwald–Letnikov fractional differences are not a power law, i. e.,

$$F\{\Delta^{\alpha}_{GL;T}X(t)\}(\omega) \neq (i\omega t)^{\alpha}F\{X(t)\}(\omega). \tag{26.23}$$

We emphasize that difference (26.18) satisfies a power law only asymptotically at $\omega > 0$.

As a result, the Grunwald–Letnikov fractional differences $\Delta^{\alpha}_{GL;T}$ of order α cannot correspond exactly to the power-law memory, which is described in the continuous-time approach. The Grunwald–Letnikov fractional differences lead us to insensitivity of the mathematical tools with respect to different short term shocks, since the Fourier transform of these differences satisfy power law in the neighborhood of zero only.

26.4 Exact discrete accelerator and multiplier with memory

In this section, we describe exact discrete analogs of accelerators and multiplier with memory.

In order to have the power law for the Fourier transform of the fractional difference of order α, we can use the exact fractional differences, which are proposed in [378, 371].

The exact fractional difference is defined [378, 371] by the equation

$$(\Delta^{\alpha}_{T}X)(t) = \sum_{m=-\infty}^{\infty} M_{\alpha}(m)X(t - mT), \tag{26.24}$$

where $\alpha \geq -1$. The memory function $M_{\alpha}(m)$ of the exact fractional differences (26.24) is expressed by the generalized hypergeometric functions $F_{1,2}(a; b, c; z)$ instead the gamma functions, which are used in the Grunwald–Letnikov fractional differences (26.20). The memory function $M_{\alpha}(m)$ of exact fractional differences (26.24) can be represented by the equation

$$M_{\alpha}(m) = \cos\left(\frac{\pi\alpha}{2}\right)M^{+}_{\alpha}(m) + \sin\left(\frac{\pi\alpha}{2}\right)M^{-}_{\alpha}(m), \tag{26.25}$$

where the odd and even memory functions are given in the form

$$M_\alpha^+(m) = \frac{\pi^\alpha}{\alpha+1} F_{1,2}\left(\frac{\alpha+1}{2}; \frac{1}{2}, \frac{\alpha+3}{2}; -\frac{\pi^2 m^2}{4}\right), \quad (\alpha > -1), \tag{26.26}$$

$$M_\alpha^-(m) = -\frac{\pi^{\alpha+1} m}{\alpha+2} F_{1,2}\left(\frac{\alpha+2}{2}; \frac{3}{2}, \frac{\alpha+4}{2}; -\frac{\pi^2 m^2}{4}\right), \quad (\alpha > -2). \tag{26.27}$$

Here, $F_{1,2}(a; b, c; z)$ is the generalized hypergeometric function that is defined as

$$F_{1,2}(a; b, c; z) = \sum_{k=0}^{\infty} \frac{\Gamma(a+k)\Gamma(b)\Gamma(c)}{\Gamma(a)\Gamma(b+k)\Gamma(c+k)} \frac{z^k}{k!}. \tag{26.28}$$

For $\alpha = -1$, the exact fractional difference is defined by equation (25.43).

Using equation (26.28), the memory function (26.25) can be represented in the form

$$M_\alpha(m) = \sum_{k=0}^{\infty} \frac{(-1)^k \pi^{2k+\alpha+\frac{1}{2}} m^{2k}}{2^{2k} k! \Gamma(k+\frac{1}{2})} \left(\frac{\cos(\frac{\pi\alpha}{2})}{\alpha+2k+1} - \frac{m\pi \sin(\frac{\pi\alpha}{2})}{(\alpha+2k+2)(2k+1)} \right) \tag{26.29}$$

for all $m \in \mathbb{Z}$. For $\alpha = n$, the function (26.29) gives expressions (25.34) and (25.35).

For $\alpha < 0$, equation (26.24) with memory function (26.29) defines the discrete fractional integration.

Using the exact fractional differences, we can get the equations of the accelerators and multipliers with memory for the discrete time approach. The discrete equation of the accelerator with memory (26.10) has the form

$$X(n) = v(\Delta_T^\alpha Y)(n), \tag{26.30}$$

where $\alpha > 0$. The discrete equation of the multiplier with memory (26.4) has the form

$$Y(n) = m(\Delta_T^{-\alpha} X)(n), \tag{26.31}$$

where $\alpha > 0$.

The exact fractional differences of the order $\alpha \geq -1$ are defined by equation (26.24) The exact finite differences (26.24) can be considered as exact discretization of the left-sided Liouville fractional derivatives and integrals, which are defined by equations (26.5) and (26.7).

The Fourier transform of the exact fractional differences of order α in the power-law form

$$F\{\Delta_T^\alpha X(t)\}(\omega) = (i\omega t)^\alpha F\{X(t)\}(\omega). \tag{26.32}$$

For $\alpha = n \in \mathbb{N}$, the Fourier transform for the exact fractional differences of the integer order $\alpha = n$ gives the power-law form

$$F\{\Delta_T^n X(t)\}(\omega) = (i\omega t)^n F\{X(t)\}(\omega). \tag{26.33}$$

For $\alpha = 1$, equation (26.30) gives the exact discrete analog of the equation of the standard accelerator

$$X(n) = v(\Delta_T^1 Y)(n), \tag{26.34}$$

which can be written in form (25.41) with $t = n$.

26.5 Conclusion

The discrete fractional calculus, which is based on the exact fractional differences, allows us to consider an exact correspondence between the discrete time and continuous time economic models with power-law memory. The Fourier transform of the proposed exact fractional differences have the power-law form, unlike the Grunwald–Letnikov fractional differences (and therefore the fractional differencing and integrating of Granger, Joyeux and Hosking). Therefore, the Grunwald–Letnikov fractional differences lead us to insensitivity of the mathematical tool with respect to different short term shocks, since the Fourier transform of these difference operators satisfy the power law in the neighborhood of zero only. The suggested approach, which is based on the exact fractional differences, can be used to describe processes with power-law memory in finance and economics within the discrete-time models.

27 Logistic map with memory from economic model

In this chapter, generalizations of the economic model of logistic growth, which take into account the effects of memory and periodic sharp splashes, is suggested. Memory means that an economic process at any given time depends not only on their states at that time, but also on their states at previous times. To describe the fading memory, we use derivatives and integrals of noninteger orders. The periodic sharp splashes of the price are mathematically described by the sum of delta-functions. Using the equivalence of fractional differential equations and the Volterra integral equations, we obtain discrete maps with memory that are exact discrete analogs of fractional differential equations. We derive logistic map with memory and its generalizations from the fractional differential equations, which describe the economic natural growth with competition, power-law fading memory and periodic sharp splashes. We proposed generalized logistic models and corresponding discrete maps with memory that take into account nonzero values of price between sharp splashes. This chapter is based on article [444] (see also [417, 436]).

27.1 Introduction

The logistic differential equation was initially proposed in the population growth model by Pierre F. Verhulst [492]. In this model, the rate of reproduction is directly proportional to the product of the existing population and the amount of available resources. This differential equation is actively used in economic growth models (e. g., see [216, 136]). The logistic map is considered as a discrete analog of this differential equation. The logistic map, which is a simple quadratic map, demonstrates complicated dynamics, which can be characterized as universal and chaotic [343, 260, 27, 511].

The logistic differential equation can be derived from economic model of natural growth in a competitive environment [493, pp. 84–90]. The economic natural growth models are described by equations, in which the rate of change of output is directly proportional to income. In the description of economic growth, the competition effects are taken into account by considering the price as a function of the value of output. Model of natural growth in a competitive environment is often called a model of logistic growth. Logistic equation with continuously distributed lag and its application in economics is given in [403]. Natural growth in a competitive environment with memory was first proposed in [444].

27.2 Logistic growth with memory and periodic kicks

To take into account the power-law memory effects in the natural growth model with a competitive environment, we propose [444] the equation

https://doi.org/10.1515/9783110627459-027

$$(D^{\alpha}_{C;0+}Y)(t) = \frac{m}{v}P(Y(t))Y(t), \tag{27.1}$$

where $(D^{\alpha}_{C;0+}Y)(t)$ is the Caputo derivative (27.2) of the order $\alpha \geq 0$ of the function $Y(t)$ with respect to time.

The left-sided Caputo derivative of order $\alpha > 0$ is defined by the equation

$$(D^{\alpha}_{C;0+}Y)(t) = \frac{1}{\Gamma(n-\alpha)} \int_0^t \frac{Y^{(n)}(\tau)\, d\tau}{(t-\tau)^{\alpha-n+1}}, \tag{27.2}$$

where $n-1 < \alpha \leq n$, $\Gamma(\alpha)$ is the gamma function, $Y^{(n)}(\tau)$ is the derivative of the integer order $n = [\alpha]+1$ of the function $Y(\tau)$ with respect to the variable τ: $0 < \tau < t$. For integer orders $\alpha = n$, the Caputo derivatives coincide with the standard derivatives of integer orders, i. e., $(D^{n}_{C;0+}Y)(t) = Y^{(n)}(t)$ and $(D^{0}_{C;0+}Y)(t) = Y(t)$ [308, p. 79], [200, pp. 92–93].

The model of natural growth in a competitive environment, which is based on equation (27.1), takes into account the effects of memory with power-law fading. For $\alpha = 1$, equation (27.1) describes a model of natural growth in a competitive environment without memory.

In the case of linearity of the price $P(Y(t)) = b - aY(t)$, equation (27.1) has the form

$$(D^{\alpha}_{C;0+}Y)(t) = \frac{m}{v(\alpha)}(b - aY(t))Y(t). \tag{27.3}$$

Equation (27.3) is the nonlinear fractional differential equation that describes the economic model of the logistic growth with memory. For $\alpha = 1$, equation (27.3) takes the form of equation that describes the standard logistic growth without memory.

Let us consider sudden changes of price in the form of price splashes that can be represented by Gaussian functions with zero mean and small variance. It is known that the delta-function can be considered as a limit of a family of Gaussian functions with zero mean, when the variance becomes smaller. For simplicity, we assume that the price splashes are periodic with period $T > 0$ and we will describe them by the Dirac delta-function, which is a generalized function [229, 134]. In general, it is possible to consider different values of the intervals between the bursts of the price. The suggested assumption allows us to consider the price function, which takes into account the periodic sharp splashes of the price, in the form

$$P(Y(t)) = -F(Y(t)) \sum_{k=1}^{\infty} \delta\left(\frac{t}{T} - k\right), \tag{27.4}$$

where $F(Y(t))$ is a continuous function of the output $Y(t)$ and $\delta(t)$ is the Dirac delta-function, which is a generalized function [229, 134]. The right-hand side of equation (27.4) makes sense if the function $F(Y(t))$ is continuous at the points $t = kT$.

Substituting expression (27.4) into equation (27.1), we get

$$(D^{\alpha}_{C;0+}Y)(t) = -\frac{m}{v}F(Y(t))Y(t) \sum_{k=1}^{\infty} \delta\left(\frac{t}{T} - k\right). \tag{27.5}$$

Equation (27.5) describes economic processes of natural growth in a competitive environment with memory and sharp splashes.

Fractional differential equation (27.5) contains the Dirac delta-functions, which are the generalized functions [229, 134]. The generalized functions are treated as functionals on a space of test functions. These functionals are continuous in a suitable topology on the space of test functions. Therefore, equation (27.5) for any positive order $\alpha > 0$ should be considered in a generalized sense, i. e., on the space of test functions, which are continuous. In equation (27.5), the product of the delta-function and the function $F(Y(t))Y(t)$ is meaningful, if the function $F(Y(t))Y(t)$ is continuous at the points $t = kT$. We can use $F(Y(t-\varepsilon))Y(t-\varepsilon)$ with $0 < \varepsilon < T$ ($\varepsilon \to 0+$) instead of $F(Y(t))Y(t)$ to make a sense of the right side of equation (27.5) for the case $0 < \alpha < 1$, when $Y(kT - -0) \neq Y(kT + 0)$ [106, 107, 108].

27.3 Economic and logistic laps with memory

Let us derive discrete maps with memory from fractional differential equations of the proposed nonlinear economic models with power-law memory [444]. To derive discrete maps with memory from fractional differential equation (27.5), we can use Theorem 18.19 of book [363, p. 444], which is valid for any positive order $\alpha > 0$ and which was initially suggested in [361, 362, 364]. The applicability of this theorem for $0 < \alpha < 1$ was noted in [106, 107, 108]. Theorem 18.19 is based on the equivalence of fractional differential equations and the Volterra integral equations. Note that Lemma 2.22 of [200, pp. 96–97] is the basis of this equivalence of fractional differential equations and the Volterra integral equations [200, pp. 199–208]. This lemma states that the left-sided Riemann–Liouville fractional integration provides operation, which is inverse [200, pp. 96–97] to the left-sided Caputo fractional differentiation that is used in equation (27.5). The action of the left-sided Riemann-Liouville fractional integral of the order α on equation (27.5) is defined on the space of test functions on the half-axis by using the adjoint operator approach [335, pp. 154–157]. For fractional differential equation (27.5), the equivalence of fractional differential equation (27.5) and the Volterra integral equations should be considered in the generalized sense, i. e., for the fractional differential equation with the generalized function on the space of test functions.

Theorem 18.19 of [363, p. 444], which is valid for any positive order $\alpha > 0$, states that the Cauchy problem for fractional differential equation (27.5) can be represented by the discrete map with memory.

Theorem 27.1. *The Cauchy problem with fractional differential equation* (27.5) *with* $\alpha > 0$ *and the initial conditions* $Y^{(k)}(0) = Y_0^{(k)}$, *where* $k = 0, 1, \dots, N-1$, *and* $N-1 < \alpha < N$

($N \in \mathbb{N}$), is equivalent to the following discrete map with memory:

$$Y_{n+1}^{(s)} = \sum_{k=0}^{N-s-1} \frac{T^k}{k!} Y_0^{(k+s)} (n+1)^k - \frac{mT^{\alpha-s}}{\nu\Gamma(\alpha-s)} \sum_{k=1}^{n} (n+1-k)^{\alpha-1-s} F(Y_k) Y_k, \tag{27.6}$$

where $s = 0, 1, \ldots, N-1$, $Y^{(s)}(t) = d^s Y(t)/dt^s$, *and*

$$Y_k^{(s)} = Y^{(s)}(kT-0) = \lim_{\varepsilon \to 0+} Y^{(s)}(kT-\varepsilon). \tag{27.7}$$

Proof. This theorem is proved in [363, pp. 444–445]. □

Example 27.1. For $0 < \alpha < 1$, ($N = 1$), the discrete map (27.6) is described by the equation

$$Y_{n+1} = Y_0 - \frac{mT^\alpha}{\nu\Gamma(\alpha)} \sum_{k=1}^{n} (n+1-k)^{\alpha-1} F(Y_k) Y_k, \tag{27.8}$$

where n takes the positive integer values.

Example 27.2. For $1 < \alpha < 2$, ($N = 2$), discrete map (27.6) is defined by the equation

$$Y_{n+1} = Y_0 + Y_0^{(1)}(n+1)T - \frac{mT^\alpha}{\nu\Gamma(\alpha)} \sum_{k=1}^{n} (n+1-k)^{\alpha-1} F(Y_k) Y_k, \tag{27.9}$$

$$Y_{n+1}^{(1)} = Y_0^{(1)} - \frac{mT^{\alpha-1}}{\nu\Gamma(\alpha-1)} \sum_{k=1}^{n} (n+1-k)^{\alpha-2} F(Y_k) Y_k. \tag{27.10}$$

Equations (27.6) and its special cases (27.8), (27.9) define a discrete map with power-law memory of the order $\alpha > 0$. This map describes the natural growth in a competitive environment with memory and sharp splashes. We emphasize that discrete equation (27.6) is derived from the fractional differential equation (27.5) without the use of any approximations, i. e., it is an exact discrete analog of the fractional differential equation (27.5). If we will use $F(Y_k) = aY_k - b$, then equation (27.6) defines the logistic map with power-law memory of the order $\alpha > 0$.

Statement 27.1. *Discrete map* (27.6) *with memory can be represented in the form*

$$Y_{n+1} = Y_n - \frac{mT^\alpha}{\nu\Gamma(\alpha)} F(Y_n) Y_n - \frac{mT^\alpha}{\nu\Gamma(\alpha)} \sum_{k=1}^{n-1} V_\alpha(n-k) F(Y_k) Y_k, \tag{27.11}$$

where $V_\alpha(z)$ *is defined by*

$$V_\alpha(z) = (z+1)^{\alpha-1} - (z)^{\alpha-1}. \tag{27.12}$$

Proof. For $0 < \alpha < 1$, $(N = 1)$, discrete map (27.6) is described by the equation

$$Y_{n+1} = Y_0 - \frac{mT^{\alpha}}{v\Gamma(\alpha)} \sum_{k=1}^{n} (n+1-k)^{\alpha-1} F(Y_k) Y_k, \tag{27.13}$$

where n takes the positive integer values. We can write equation (27.13) in the form

$$Y_{n+1} = Y_0 - \frac{mT^{\alpha}}{v\Gamma(\alpha)} F(Y_n) Y_n - \frac{mT^{\alpha}}{v\Gamma(\alpha)} \sum_{k=1}^{n-1} (n+1-k)^{\alpha-1} F(Y_k) Y_k. \tag{27.14}$$

Replacing $n+1$ by n in equation (27.13), we get

$$Y_n = Y_0 - \frac{mT^{\alpha}}{v\Gamma(\alpha)} \sum_{k=1}^{n-1} (n-k)^{\alpha-1} F(Y_k) Y_k. \tag{27.15}$$

Subtracting equation (27.15) from equation (27.14), we obtain discrete map (27.11). □

For $F(Y_k) = aY_k - b$, discrete maps (27.11) and (27.9)–(27.10) describe the logistic growth with power-law memory for $0 < \alpha < 1$ and $1 < \alpha < 2$, respectively. Equations (27.11) and (27.9)–(27.10) describe a generalized logistic maps with power-law memory for the fading parameter $\alpha \in (0, 1)$ and $\alpha \in (1, 2)$.

Remark 27.1. For $\alpha = 1$, we can use $V_1(z) = 0$, and equation (27.11) gives the discrete map

$$Y_{n+1} = Y_n - \frac{m}{v} TF(Y_n) Y_n, \tag{27.16}$$

which describes the natural growth in a competitive environment with sharp splashes without taking into account a memory. Using $F(Y_k) = aY_k - b$, equation (27.16) gives the logistic map

$$Y_{n+1} = \left(1 + \frac{mbT}{v}\right) Y_n - \frac{maT}{v} Y_n^2, \tag{27.17}$$

which describes the logistic economic growth without the memory effects, but with the sharp splashes of price. Equation (27.17) can be written as

$$Y_{n+1} = \left(1 + \frac{mbT}{v}\right) Y_n \left(1 - \frac{maT}{v + mbT} Y_n\right). \tag{27.18}$$

If $a \neq 0$, we can use the variable Z_n and the parameter λ, which are defined by the equations

$$Z_n = \frac{maT}{v + mbT} Y_n, \quad \lambda = 1 + \frac{mbT}{v}. \tag{27.19}$$

Then equation (27.18) is represented in the form

$$Z_{n+1} = \lambda Z_n (1 - Z_n). \tag{27.20}$$

Equation (27.20) is the standard logistic map [343, 260, 27]. This map is used to describe different economic processes [493, 275, 474, 63, 32].

As a result, we can state that standard logistic map (27.20) can be derived from the logistic differential equation without approximation, if the price function is given in form (27.4), i. e., when the price behavior is described by the periodic sharp splashes in the form of the delta-function.

Let us consider logistic map with memory (27.11), where $0 < \alpha < 1$. For $F(Y_k) = aY_k - b$, equation (27.11) takes the form

$$Y_{n+1} = \left(1 + \frac{mbT^{\alpha}}{v\Gamma(\alpha)}\right)Y_n - \frac{maT^{\alpha}}{v\Gamma(\alpha)}Y_n^2 - \frac{mT^{\alpha}}{v\Gamma(\alpha)}\sum_{k=1}^{n-1} V_{\alpha}(n-k)(aY_k - b)Y_k. \tag{27.21}$$

Equation (27.21) can be written as

$$Y_{n+1} = \left(1 + \frac{mbT^{\alpha}}{v\Gamma(\alpha)}\right)Y_n\left(1 - \frac{maT^{\alpha}}{v\Gamma(\alpha) + mbT^{\alpha}}Y_n\right) - \frac{mbT^{\alpha}}{v\Gamma(\alpha)}\sum_{k=1}^{n-1} V_{\alpha}(n-k)Y_k\left(1 - \frac{a}{b}Y_k\right). \tag{27.22}$$

Using the variable $Z_n(\alpha)$ and the parameters $\lambda(\alpha)$, $\mu(\alpha)$, $\rho(\alpha)$, which are defined by the expressions

$$Z_n(\alpha) = \frac{maT^{\alpha}}{v\Gamma(\alpha) + mbT^{\alpha}}Y_n, \quad \lambda(\alpha) = 1 + \frac{mbT^{\alpha}}{v\Gamma(\alpha)}, \tag{27.23}$$

$$\mu(\alpha) = \frac{mbT^{\alpha}}{v\Gamma(\alpha)}, \quad \eta(\alpha) = \frac{v\Gamma(\alpha) + mbT^{\alpha}}{m}bT^{\alpha}, \tag{27.24}$$

we can represent equation (27.22) in the form

$$Z_{n+1}(\alpha) = \lambda(\alpha)Z_n(\alpha)(1 - Z_n(\alpha)) - \mu(\alpha)\sum_{k=1}^{n-1} V_{\alpha}(n-k)Z_k(\alpha)(1 - \eta(\alpha)Z_k(\alpha)). \tag{27.25}$$

Equation (27.25) describes the logistic map with power-law memory for fading parameter $\alpha \in (0, 1)$.

The similar form can be derived for equations (27.6) and (27.9)–(27.10) with $\alpha \in (1, 2)$.

The logistic map with memory (27.25) as well as equations (27.6) is an exact discrete analog of fractional differential equations (27.5). It should be emphasized that equations of discrete maps (27.6) and (27.25) are obtained from equation (27.5) without any approximations (for details, see [361, 362] and Chapter 18 of [363]).

Using equation (27.4), we can see that the logistic map with memory (27.25) describes a special case of economic dynamics, when price is close to zero between bursts. This behavior of price is very unusual for real economic processes. Therefore, the discrete map (27.25) is a toy model, but it can be used to study some properties caused by bursts of price and nonlinearity.

27.4 Generalized economic and logistic maps with memory

The proposed continuous-time model, which is based on equation (27.4), describes a very special case of economic dynamics, when price is close to zero between bursts. This behavior of price can rarely correspond to real economic processes. Therefore, this model cannot be applied to describe real economy, but it can be used to describe some of their general properties. In article [444], we proposed a family of new economic models that allows us to describe the real behavior of price. These suggested models and corresponding discrete maps with memory take into account nonzero values of price between bursts of price.

Let us consider the price function, which takes into account nonzero behavior of price and the periodic sharp splashes of price, in the form

$$P(Y(t)) = pG(Y(t)) - qF(Y(t)) \sum_{k=1}^{\infty} \delta\left(\frac{t}{T} - k\right), \tag{27.26}$$

where $G(Y(t))$ is a continuous function of output $Y(t)$ such that the antiderivative of the expression $(G(y)y)^{-1}$ is differentiable with respect to the variable y, and the function $F(Y(t))$ is continuous at the points $t = kT$. The parameter $q = 1 - p$ can be considered as a measure of the impact of sharp surges on price.

For example, the function $G(Y(t))$ can be considered in the following form:

(a) the constant function $G(Y(t)) = P_0$;
(b) the direct proportionality $G(Y(r)) = \rho Y(t)$;
(c) the power-law case $G(Y(t)) = \rho Y^j(t)$.

In general, the coefficients P_0 and ρ, which do not depend on $Y(t)$, are functions of time ($P_0 = P_0(t), \rho = \rho(t)$).

Equation (27.26) generalizes price equation (27.4) and the standard case without periodic sharp splashes of price. For $p = 1$, $q = 0$, equation (27.26) corresponds to the standard case that is described by the equation $P(Y(t)) = G(Y(t))$. For $p = 0$, $q = 1$, equation (27.26) takes form (27.4) that corresponds to equation (27.5) with $\alpha = 1$.

Substituting (27.26) into equation (27.1) with $\alpha = 1$, we obtain

$$\frac{dY(t)}{dt} = p\frac{m}{v}G(Y(t))Y(t) - q\frac{m}{v}F(Y(t))Y(t) \sum_{k=1}^{\infty} \delta\left(\frac{t}{T} - k\right). \tag{27.27}$$

We can consider the functions $G(Y(t))$ such that equation (27.27) can be represented in the form

$$\frac{dR(Y(t))}{dt} = p\frac{m}{v}C(t) - q\frac{m}{v}F_G(Y(t))\sum_{k=1}^{\infty}\delta\Big(\frac{t}{T} - k\Big), \tag{27.28}$$

where $C(t)$ is the function, which is independent of the output $Y(t)$, the function F_G is the fraction

$$F_G(Y(t)) = F\Big(\frac{Y(t)}{G(Y(t))}\Big), \tag{27.29}$$

and the function $R(Y(t))$ is defined by the equation

$$R(Y) = \int_0^Y (G(Y)Y)^{-1}\, dy. \tag{27.30}$$

Let us give simple examples of the function $R(Y(t))$:

(a) If $G(Y(t)) = P_0$, then $R(Y(t)) = \ln(Y(t))$ and $C = P_0$;

(b) If $G(Y(r)) = \rho Y(t)$, then $R(Y(t)) = -\rho Y^{-1}(t)$ and $C = \rho$;

(c) If $G(Y(t)) = \rho Y^j(t)$ with $j \neq 0$, then $R(Y(t)) = -(\rho/j)Y^{-j}(t)$ and $C = \rho$.

For the economic processes with power-law memory, the generalization of equation (27.28) has the form

$$(D^{\alpha}_{C;0+}R(Y))(t) = p\frac{m}{v}C(t) - q\frac{m}{v}F_G(Y(t))\sum_{k=1}^{\infty}\delta\Big(\frac{t}{T} - k\Big), \tag{27.31}$$

where $N - 1 < \alpha < N$. For $0 < \alpha < 1$, we can use $F_G(Y(t - \varepsilon))$ with $0 < \varepsilon < T$ ($\varepsilon \to 0+$) instead of $F_G(Y(t))$.

Statement 27.2. *Fractional differential equation* (27.31) *can be represented in the form of the discrete maps with memory*

$$\begin{aligned} R^{(s)}_{n+1} &= \sum_{k=0}^{N-s-1}\frac{T^k}{k!}R^{(k+s)}_0 (n+1)^k + p\frac{m}{v}C^{(\alpha-s)}_{n+1} \\ &\quad - q\frac{mT^{\alpha-s}}{v\Gamma(\alpha-s)}\sum_{k=1}^{n}(n+1-k)^{\alpha-1-s}F_G(Y_k), \end{aligned} \tag{27.32}$$

where $R^{(s)}(t) = d^sR(Y(t))/dt^s$, $R^{(s)}_0 = R^{(s)}(0)$ *and*

$$R^{(s)}_k = R^{(s)}(kT - 0) = \lim_{\varepsilon\to 0+} R^{(s)}(kT - \varepsilon), \tag{27.33}$$

$$Y^{(s)}_k = Y^{(s)}(kT - 0), \tag{27.34}$$

$$C^{(\alpha-s)}_{n+1} = (I^{\alpha-s}_{RL;0+}C)((n+1)T), \tag{27.35}$$

and $I^{\alpha-s}_{RL;0+}$ is the Riemann–Liouville integration of the order $\alpha - s > 0$, $s = 0, 1, \dots, N-1$, with $N-1 < \alpha < N$.

Proof. Let us integrate equation (27.31) by the Riemann–Liouville fractional integral $I^{\alpha}_{RL;0+}$ of the order $\alpha > 0$ with respect to $nT < t < (n+1)T$. Then we get

$$(I^{\alpha}_{RL;0+}D^{\alpha}_{0+}R(Y))(t) = p\frac{m}{v}(I^{\alpha}_{RL;0+}C)(t) - q\frac{m}{v}I^{\alpha}_{RL;0+}F_G(Y(t))\sum_{k=1}^{\infty}\delta\left(\frac{t}{T}-k\right). \quad (27.36)$$

Using equation 2.4.42 of Lemma 2.22 of [200, p. 96], equation (27.36) takes the form

$$R(Y(t)) - \sum_{k=0}^{N-1}\frac{t^k}{k!}R^k(Y(0)) = p\frac{m}{v}(I^{\alpha}_{RL;0+}C)(t) - q\frac{mT}{v\Gamma(\alpha)}\sum_{k=1}^{n}F_G(Y(kT))(t-kT)^{\alpha-1}. \quad (27.37)$$

Then using transformations from the proof of Theorem 18.19 of [363, p. 444] and formula 2.2.28 of [200, p. 83] in the form

$$(D^s I^{\alpha}_{RL;0+}C)(t) = (I^{\alpha-s}_{RL;0+}C)(t) \quad (27.38)$$

with $s < \alpha$, equation (27.37) gives the economic discrete map with memory in form (27.32). □

Example 27.3 (The case $C(t) = C$). For example, using Theorem 18.19 of [363, p. 444] and the formula $I^{\alpha}_{RL;0+}1 = t^{\alpha}/\Gamma(\alpha+1)$, equation (27.37) with constant $C(t) = C$ gives the economic discrete map with memory in the form

$$R^{(s)}_{n+1} = \sum_{k=0}^{N-s-1}\frac{T^k}{k!}R^{(k+s)}_0(n+1)^k + p\frac{CmT^{\alpha-s}}{v\Gamma(\alpha+1-s)}(n+1)^{\alpha-s} - q\frac{mT^{\alpha-s}}{v\Gamma(\alpha-s)}\sum_{k=1}^{n}(n+1-k)^{\alpha-1-s}F_G(Y_k), \quad (27.39)$$

where $s = 0, 1, \dots, N-1$, with $N-1 < \alpha < N$, $R^{(s)}(t) = d^sR(Y(t))/dt^s$, $R^{(s)}_0 = R^{(s)}(0)$ and

$$R^{(s)}_k = R^{(s)}(kT-0) = \lim_{\varepsilon\to 0+}R^{(s)}(kT-\varepsilon), \quad Y^{(s)}_k = Y^{(s)}(kT-0). \quad (27.40)$$

Remark 27.2. For $p = 0$ and $q = 1$, equations (27.32) and (27.39) with $R(Y(t)) = Y(t)$ give the discrete map (27.6). The "economic" discrete maps with memory, which are defined by equations (27.32) and (27.39), describe the economic model of natural growth in a competitive environment with memory and sharp splashes. These discrete maps are exact discrete analogs of the fractional differential equation (27.31).

Remark 27.3. Discrete map (27.39) can be represented in another form. Using equation (27.39) with replacement $n+1$ by n, and subtraction the result from equation (27.39), we get the discrete map with memory $\alpha > 0$ in the form

$$R_{n+1}^{(s)} = R_n^{(s)} + \sum_{k=0}^{N-s-1} \frac{T^k}{k!} R_0^{(k+s)} V_{k-1}(n) + p\frac{CmT^{\alpha-s}}{v\Gamma(\alpha+1-s)} V_{\alpha+1-s}(n) - q\frac{mT^{\alpha-s}}{v\Gamma(\alpha-s)} F_G(Y_n)$$
$$- q\frac{mT^{\alpha-s}}{v\Gamma(\alpha-s)} \sum_{k=1}^{n-1} V_{\alpha-s}(n-k) F_G(Y_k), \tag{27.41}$$

where $V_\alpha(z)$ is defined by

$$V_\alpha(z) = (z+1)^{\alpha-1} - (z)^{\alpha-1}, \tag{27.42}$$

where $s = 0, 1, \ldots, N-1$.

Remark 27.4. For $0 < \alpha < 1$, $(N = 1)$, discrete map (27.39) has the form

$$R_{n+1} = R_n + p\frac{CmT^\alpha}{v\Gamma(\alpha+1)} V_{\alpha+1}(n)$$
$$- q\frac{mT^\alpha}{v\Gamma(\alpha)} F_G(Y_n) - q\frac{mT^\alpha}{v\Gamma(\alpha)} \sum_{k=1}^{n-1} V_\alpha(n-k) F_G(Y_k). \tag{27.43}$$

Let us give some simple examples of economic discrete maps with memory (27.39) for $0 < \alpha < 1$, i. e., discrete map (27.43) with various functions $G(Y(t))$.

Example 27.4 (The case $G(Y(t)) = P_0$). For $0 < \alpha < 1$ and $G(Y(t)) = P_0$, discrete map (27.39) is described by the equation

$$\ln(Y_{n+1}) = \ln(Y_0) + p\frac{P_0 mT^\alpha}{v\Gamma(\alpha+1)} (n+1)^\alpha$$
$$- q\frac{mT^\alpha}{v\Gamma(\alpha)} \sum_{k=1}^{n} (n+1-k)^{\alpha-1} F(Y_k). \tag{27.44}$$

Example 27.5 (The case $G(Y(t)) = \rho Y^j(t)$). For $0 < \alpha < 1$ and $G(Y(t)) = \rho Y^j(t)$ with $j \neq 0$, economic discrete map (27.39) is described by the equation

$$Y_{n+1}^{-j} = Y_0^{-j} - p\frac{\rho mT^\alpha}{jv\Gamma(\alpha+1)} (n+1)^\alpha$$
$$- q\frac{mT^\alpha}{v\Gamma(\alpha)} \sum_{k=1}^{n} (n+1-k)^{\alpha-1} F(Y_k) Y_k^{-j}. \tag{27.45}$$

Example 27.6 (The case $G(Y(t)) = \rho Y^{-1}(t)$). For $j = -1$, equation (27.45) has the form

$$Y_{n+1} = Y_0 - p\frac{\rho m T^{\alpha}}{jv\Gamma(\alpha+1)}(n+1)^{\alpha} - q\frac{mT^{\alpha}}{v\Gamma(\alpha)}\sum_{k=1}^{n}(n+1-k)^{\alpha-1}F(Y_k)Y_k. \tag{27.46}$$

For $p = 0$ and $q = 1$, equation (27.46) gives discrete map (27.8).

Remark 27.5. Discrete maps with memory (27.44)–(27.46) can be rewritten in the form similar to (27.43). For example, using equation (27.45) with replacement $n+1$ by n, and subtraction the result from equation (27.45), we get the discrete map with memory for $\alpha \in (0,1)$ in the form

$$Y_{n+1}^{-j} = Y_n^{-j} - p\frac{\rho m T^{\alpha}}{jv\Gamma(\alpha+1)}V_{\alpha+1}(n) - q\frac{mT^{\alpha}}{v\Gamma(\alpha)}F(Y_n)Y_n^{-j} - q\frac{mT^{\alpha}}{v\Gamma(\alpha)}\sum_{k=1}^{n-1}V_{\alpha}(n-k)F(Y_k)Y_k^{-j}, \tag{27.47}$$

where $V_\alpha(z)$ is defined by

$$V_{\alpha}(z) = (z+1)^{\alpha-1} - (z)^{\alpha-1}. \tag{27.48}$$

For $j = -1$, equation (27.47) has the form

$$Y_{n+1} = Y_n - p\frac{\rho m T^{\alpha}}{jv\Gamma(\alpha+1)}V_{\alpha+1}(n) - q\frac{mT^{\alpha}}{v\Gamma(\alpha)}F(Y_n)Y_n - q\frac{mT^{\alpha}}{v\Gamma(\alpha)}\sum_{k=1}^{n-1}V_{\alpha}(n-k)F(Y_k)Y_k. \tag{27.49}$$

For $p = 0$ and $q = 1$, equation (27.49) gives discrete map (27.11).

Let us consider some examples of the discrete map with memory (27.32) with non-constant function $C(t)$.

Example 27.7 (The case $C(t) = Ct^{\beta}$). For the power function $C(t) = Ct^{\beta}$, where $\beta > -1$ and $0 < \alpha < 1$, $(N = 1)$, we can use equation 2.1.16 of [200, p. 71] for the Riemann–Liouville integration

$$I^{\alpha}_{\mathrm{RL};0+}t^{\beta} = \frac{\Gamma(\beta+1)}{\Gamma(\alpha+\beta+1)}t^{\beta+\alpha}, \tag{27.50}$$

where $\beta > -1$. For $\beta = 0$, we have the equation $I^{\alpha}_{\mathrm{RL};0+}1 = t^{\alpha}/\Gamma(\alpha+1)$. Then the discrete map (27.32) with $0 < \alpha < 1$ is described by the equation

$$R_{n+1} = R_0 + p\frac{CmT^{\alpha+\beta}\Gamma(\beta+1)}{v\Gamma(\alpha+\beta+1)}(n+1)^{\alpha+\beta} - q\frac{mT^{\alpha}}{v\Gamma(\alpha)}\sum_{k=1}^{n}(n+1-k)^{\alpha-1}F(Y_k). \quad (27.51)$$

For $\beta = 0$, map (27.51) takes form (27.39).

Example 27.8 (The case $Ct^{\beta-1}E_{\mu,\beta}(\gamma t^{\mu})$). Let us consider the function

$$C(t) = Ct^{\beta-1}E_{\mu,\beta}(\gamma t^{\mu}), \quad (27.52)$$

where

$$E_{\mu,\beta}(z) = \sum_{k=0}^{\infty}\frac{z^k}{\Gamma(\mu k+\beta)} \quad (27.53)$$

is the two-parameter Mittag–Leffler function [200, p. 42], which is a generalization of the exponential function. In this case, we can use equation (2.2.51) [200, p. 86] to get the discrete map with memory. For example, if $0 < \alpha < 1$, then the discrete map with memory, which corresponds to fractional differential equation (27.37) with $C(t) = Ct^{\beta-1}E_{\mu,\beta}(\gamma t^{\mu})$, is defined by the equation

$$R_{n+1} = R_0 + p\frac{CmT^{\alpha+\beta-1}}{v\Gamma(\alpha+1)}(n+1)^{\alpha+\beta-1}E_{\mu,\alpha+\beta}(\gamma T^{\mu}(n+1)^{\mu}) - q\frac{mT^{\alpha-s}}{v\Gamma(\alpha-s)}\sum_{k=1}^{n}(n+1-k)^{\alpha-1-s}F_G(Y_k). \quad (27.54)$$

Using $F(Y_k) = aY_k - b$, equations (27.32)–(27.54) give the generalized logistic maps with memory, which describe exact discrete analogs of the economic model of generalized logistic growth in a competitive environment with memory and sharp splashes.

27.5 Conclusion

First, we briefly describe what is proposed in this chapter.

(1) We proposed new economic models of the logistic growth (the natural growth in competitive environment), which take into account the power-law memory and sharp splashes. These continuous-time economic models are described by fractional differential equations (27.5) and (27.31) with delta-functions.

(2) Using approach, which has been proposed in works [410, 361, 362, 363], we derived exact discrete analogs of fractional differential equations (27.5) and (27.31). As a result, we got the discrete time representation of these economic models in the form of the discrete maps with memory (27.6) and (27.32).

(3) We can state that the discrete maps (27.6), (27.11)–(27.10), and the logistic maps with memory (27.21), (27.25) are special types of the universal map with memory suggested in [361, 362, 363, 444]. We proved that logistic map with memory (27.21), (27.25) and economics maps (27.6), (27.11)–(27.10) describe a very special case of economic dynamics, when price is close to zero between bursts. This behavior of price is unusual for real economic processes. Therefore, this map can be considered as a toy model of real economic processes.
(4) In order to have a more realistic description of the behavior of price, we proposed an economic model, which is described by equation (27.31), and corresponding discrete maps that are closer to real economic dynamics of price. The discrete maps with memory, which are described by equations (27.32)–(27.54), take into account nonzero values of price between bursts of price. The suggested discrete maps (27.32)–(27.54) contain the discrete maps (27.6), (27.11)–(27.10), and the logistic map with memory (27.21), (27.25) as special cases.
(5) The suggested discrete maps with memory (27.32)–(27.54) are exact discrete analogs of the corresponding fractional differential equations (27.31). These maps and equations are the discrete-time and continuous-time representations of the model of the economic growth in competitive environment with memory and sharp splashes.

We now make some remarks and comments related to these results.

It should be noted that some generalizations of logistic map, which are proposed by other scientists to take into account memory effects, are not exact discrete analogs of differential equations that describe the logistic growth with memory.

We proposed [361, 362, 363, 364, 444] the discrete maps with memory as the exact discrete analogs of fractional differential equations, which describe the economic growth with competition, memory and sharp splashes. This relationship of the fractional differential equations and the discrete maps distinguishes the suggested maps with memory from all other discrete maps with memory. The suggested discrete maps with memory are derived from the economic models, which also highlight these discrete maps.

It is known that the standard logistic map (27.20), which does not take into account the memory effects, can give the chaotic behavior [343, pp. 33–67] and [260, 27]. Using logistic map (27.20), which is derived from economic model (27.5) with $\alpha = 1$, we can state that the sudden changes of price in the form of price splashes could lead to deterministic chaotic phenomena. The suggested logistic map with memory and its generalizations, can demonstrate a new chaotic behavior.

The discrete maps with memory, which are exact discrete analogues of the fractional differential equations, were first proposed in works [410, 361, 362, 363]. Then, this approach, which is based on the equivalence of the fractional differential equations and the discrete maps with memory, has been applied in works [364, 395, 110, 444] [101, 102, 103, 105, 104] [106, 107, 108, 109] to describe properties of the discrete maps with memory. Computer simulations of some discrete maps with memory were realized in [364, 395, 110] [101, 102, 103, 105, 104] [106, 107, 108, 109]. New types of chaotic behavior and new kinds of attractors have been found in these works. Therefore, these types of deterministic chaotic behavior can describe some properties of the price behavior in the simple economic models, which are described by discrete maps (27.6), (27.11)–(27.10), and logistic map with memory (27.21), (27.25).

Some properties of the fractional logistic maps, which can be represented in the form (27.21), are investigated by computer simulation in [102, 103, 105, 104, 106, 107, 108, 109]. In paper [444], we proved that the logistic map with memory (27.21) and (27.25), and the economics maps (27.6) describe a very special case of economic dynamics, when price is close to zero between bursts. This behavior of price is very unusual for real economic processes. Therefore, the map with memory, which is described by (27.6), (27.11)–(27.10), and logistic map with memory (27.21), (27.25), can be considered only as toy models of real economic processes, but it can be used to study some properties caused by bursts of price and nonlinearity.

In order to have a more realistic description of the behavior of price, we propose economic models and corresponding discrete maps (27.32)–(27.54) that are closer to real economic dynamics of price. These suggested discrete maps with memory, which are described by equations (27.32)–(27.54), take into account nonzero values of price between bursts of price. In paper [444], we derive the generalized logistic map with memory and the economic discrete maps (27.32)–(27.54) from the economic model of natural growth in a competitive environment with memory and sharp splashes. The suggested discrete maps with memory are exact discrete analogs of the fractional differential equation (27.31) of economic dynamics. To study the properties of generalized logistic and economic discrete maps with memory (27.32)– (27.54), computer simulation is required. The computer simulations of the suggested discrete maps with memory, which describe the natural growth in competitive environment with memory and sharp splashes, can allow us to describe new types of economic phenomena.

Part VII: **Advanced models: generalized memory**

28 Economics model with generalized memory

Concept of memory can be described by the memory function that is a kernel of the integro-differential operator. In this chapter, we consider a possibility to use the fractional calculus, for the case of memory functions that do not have a power-law form. Using the generalized Taylor series in the Trujillo–Rivero–Bonilla (TRB) form for the memory function, we represent the integro-differential equations with memory functions by fractional integral and differential equations with derivatives and integrals of noninteger orders. This allows us to describe general economic dynamics with memory by the methods of fractional calculus. We prove that equation of the generalized accelerator with the TRB memory function can be represented by as a composition of actions of the accelerator with simplest power-law memory and the multiparametric power-law multiplier. As an example of application of the suggested approach, we consider new generalizations of the Harrod–Domar growth model and the Lucas model of learning. This chapter is based on the articles [380, 391, 389].

28.1 Introduction

A powerful tool for describing the effects of power-law memory is the fractional calculus. The fractional derivatives and integrals of noninteger orders can be used to describe processes with memory.

In economic models with memory, we consider mainly the power-law fading memory. To describe processes with wider types of memory, it is important to use a wider class of memory function. For this purposes, we proposed to use the fractional Taylor series. In work [399] (see also [408, 366, 368]), the expansion into fractional Taylor series was realized for the Fourier transform of the memory functions, i. e., in the frequency domain. In the article [380], we use the fractional Taylor series in the time domain.

In this chapter, we describe an approach [380] that allows us to consider a wide class of memory functions by using the methods of fractional calculus. For this purpose, we use the generalized Taylor series in the Trujillo–Rivero–Bonilla (TRB) form [484]. An application of this series allows us to consider wide class of memory functions as kernels of fractional integral and differential operators of noninteger orders. We proposed concepts of multipliers and accelerators with generalized memory. We prove that equation of the generalized accelerator with the memory of TRB type can be represented by as a composition of actions of the accelerator with power-law memory and the multiplier with multi-parametric power-law memory. As an application of the proposed approach, we consider generalizations of the Harrod–Domar growth model and the Lucas model of learning with memory of TRB type [380, 391, 389].

https://doi.org/10.1515/9783110627459-028

28.2 Generalized multiplier and accelerator with memory

For economic processes with memory, there is an indicator $Y(t)$ that depends on the changes of a factor $X(\tau)$ on a finite time interval $0 \leq \tau \leq t$. The reason to take into account the memory effects is the fact that the economic agents remember the previous changes of the factor $X(t)$ and their influence on changes of the indicator $Y(t)$. The economic process is a process with memory if there exists at least one variable $Y(t)$, which depends on the history of the change of $X(\tau)$ at $\tau \in [0, t]$. We will consider the time interval starting from zero and not with minus infinity, since any economic process exists only during a finite time interval. Therefore, the start time of the process can be selected as $t = 0$. It is obvious that the models of processes with memory will be determined by time interval that is considered in the model. In a sense, this fact is analogous to the fact that the function is determined by the domain of definition.

Let us give the definition of the generalized multiplier with memory, which has been suggested in [412, 450].

Definition 28.1. The generalized multiplier with memory is the dependence of a variable $Y(t)$ at the time $t \geq 0$ on the history of the change of a variable $X(\tau)$ on a finite time interval $0 \leq \tau \leq t$ such that

$$Y(t) = F_0^t(X(\tau)), \tag{28.1}$$

where F_0^t denotes an operator that specifies the value of $Y(t)$ for any time $t \geq 0$, if $X(\tau)$ is known for $\tau \in [0, t]$. The operator F_0^t transforms each history of changes of $X(\tau)$ for $\tau \in [0, t]$ into the corresponding history of changes of $Y(\tau)$.

For simplification, we will consider the linear generalized multiplier with memory that is described by the equation

$$Y(t) = \int_0^t M(t-\tau)X(\tau)\,d\tau, \tag{28.2}$$

where the function $M(t-\tau)$ characterizes the memory. The function $M(t)$ is called the memory function. Note that memory cannot be described by an arbitrary function [450], for which integral (28.2) exists. For the memory function $M(t-\tau) = m\delta(t-\tau)$, where $\delta(t-\tau)$ is the Dirac delta-function, equation (28.2) has the form of the standard multiplier equation $Y(t) = mX(t)$. For the function $M(t-\tau) = m\delta(t-\tau-T)$, where T is the time constant, equation (28.2) describes the linear multiplier with fixed-time delay.

The generalized concept of the economic accelerator for processes with memory takes into account that the indicator $Y(t)$ depends on the changes of the integer derivatives of the factor $X^{(k)}(\tau)$, $(k = 1, 2, \ldots, n)$ on a finite time interval $\tau \in [0, t]$.

The generalized accelerator with memory can be considered as a dependence of a variable $Y(t)$ at the time $t \geq 0$ on the histories of the changes of another variable

$X^{(n)}(\tau) = d^n X(\tau)/d\tau^n$ with $n \in \mathbb{N}$ on a finite time interval $0 \leq \tau \leq t$. For simplification, we will consider the linear accelerator with the memory function $M(t-\tau)$. The definition of the generalized accelerator with memory has been suggested in [412, 450].

Definition 28.2. The generalized linear accelerator with memory function $M(t-\tau)$ is the dependence of a variable $Y(t)$ at the time $t \geq 0$ on the history of the change of another variable $X^{(n)}(\tau) = d^n X(\tau)/d\tau^n$ with $n \in \mathbb{N}$ on finite time interval $0 \leq \tau \leq t$ such that

$$Y(t) = F_0^t(X^{(n)}(\tau)) = \int_0^t M(t-\tau) X^{(n)}(\tau)\, d\tau. \tag{28.3}$$

The second type of the generalized linear accelerators with memory function $M(t-\tau)$, can be defined as a derivative of the positive integer order n of the generalized linear multiplier (28.2) in the form

$$Y(t) = \frac{d^n}{dt^n} F_0^t(X(\tau)) = \frac{d^n}{dt^n} \int_0^t M(t-\tau) X(\tau)\, d\tau, \tag{28.4}$$

where $\Gamma(\alpha)$ is the gamma function, n is positive integer number and the function $M(t-\tau)$ describes a memory. Here, F_0^t denotes a linear operator, which specifies values of $Y(t)$ for any time $t > 0$, if $X^{(n)}(\tau)$ is known for $\tau \in [0, t]$, respectively.

For the memory function $M(t-\tau)$, which is represented by the Dirac delta-function ($M(t-\tau) = a\delta(t-\tau)$), equation (28.3) with $n = 1$ has the form of the standard accelerator equation without memory, $Y(t) = aX^{(1)}(t)$.

Note that equations (28.3) and (28.4) of the generalized linear accelerator with memory can be interpreted as a composition of actions of the standard accelerators without memory and the generalized linear multiplier with memory.

For the simplest power-law memory fading, the memory function can be considered in the form

$$M(t-\tau) = c(\beta)(t-\tau)^{\beta}, \tag{28.5}$$

where $c(\beta) > 0$ and $\beta > -1$ are constants. Using the parameters

$$\alpha = \beta + 1, \quad m(\alpha) = c(\alpha - 1)\Gamma(\alpha), \tag{28.6}$$

memory function (28.5) takes the form

$$M(t-\tau) = M_{\alpha}(t-\tau) = \frac{m(\alpha)}{\Gamma(\alpha)}(t-\tau)^{\alpha-1}, \tag{28.7}$$

where $\alpha > 0$, and $m(\alpha)$ is the numerical coefficient, $t > \tau$. Substituting expression (28.7) into equation (28.2), we get multiplier equation (28.2) in the form

$$Y(t) = m(\alpha)(I^{\alpha}_{\mathrm{RL};0+}X)(t), \tag{28.8}$$

where $I^{\alpha}_{RL;0+}$ is the left-sided Riemann–Liouville fractional integral of the order $\alpha > 0$ with respect to the time variable t. This fractional integral is defined [335, 200] by the equation

$$(I^{\alpha}_{RL;0+}X)(t) = \frac{1}{\Gamma(\alpha)} \int_0^t (t-\tau)^{\alpha-1} X(\tau)\, d\tau, \tag{28.9}$$

where $0 < \tau < t < T$. The function $X(t)$ is assumed to satisfy the condition $X(\tau) \in L_1[0,T]$ [335, 200]. For the order $\alpha = 1$, the Riemann–Liouville fractional integral (28.9) is the standard integral of the first order.

Equation (28.8) describes the equation of economic multiplier with simplest power-law memory (SPL memory), for which fading is described by the parameter $\alpha \geq 0$, and $m(\alpha)$ is a positive constant indicating the multiplier coefficient.

To describe the accelerators with SPL memory, we can use the memory function in the form (28.5). Using the parameters $\alpha = n - \beta - 1$ and

$$c(\beta) = c(n-\alpha-1) = \frac{a(\alpha)}{\Gamma(n-\alpha)}, \tag{28.10}$$

i. e., $a(\alpha) = c(n-\alpha-1)\Gamma(n-\alpha)$, the memory function takes the form

$$M(t-\tau) = M_{n-\alpha}(t-\tau) = \frac{a(\alpha)}{\Gamma(n-\alpha)}(t-\tau)^{n-\alpha-1}, \tag{28.11}$$

where $n = [\alpha] + 1$, $\alpha > 0$, and $a(\alpha)$ is the numerical coefficient. Substitution of expression (28.11) into equation (28.3) gives the linear accelerator equation in the form

$$Y(t) = a(\alpha)(D^{\alpha}_{C;0+}X)(t), \tag{28.12}$$

where $D^{\alpha}_{C;0+}$ is the left-sided Caputo fractional derivative of the order $\alpha \geq 0$ that is defined [200, p. 92] by the equation

$$(D^{\alpha}_{C;0+}X)(t) = (I^{n-\alpha}_{RL;0+}X^{(n)})(t) = \frac{1}{\Gamma(n-\alpha)} \int_0^t (t-\tau)^{n-\alpha-1} X^{(n)}(\tau)\, d\tau, \tag{28.13}$$

where $\Gamma(\alpha)$ is the gamma function, and $X^{(n)}(\tau)$ is the derivative of the integer order $n = [\alpha] + 1$ (and $n = \alpha$ for integer values of α) of the function $X(\tau)$ with respect to the time variable τ, and $I^{n-\alpha}_{RL;0+}$ is the left-sided Riemann–Liouville fractional integral (28.9) of the order $n - \alpha > 0$. In equation (28.13), it is assumed that the function $X(t)$ has derivatives up to $(n-1)$th order, which are absolutely continuous functions on the interval $[0, T]$. Equation (28.12) describes the economic accelerator with memory with the power-law fading of the order $\alpha \geq 0$, where $a = a(\alpha)$ is a positive constant indicating the power of this accelerator.

As a second example of the simplest linear accelerator with memory that is represented in the form (28.4), we can consider the equation

$$Y(t) = a(\alpha)(D^{\alpha}_{\mathrm{RL};0+}X)(t), \tag{28.14}$$

where $D^{\alpha}_{\mathrm{RL};0+}$ is the left-sided Riemann–Liouville fractional derivative of order $\alpha \geq 0$ of the function $X(t)$, which is defined [200, p. 70] by the equation

$$(D^{\alpha}_{\mathrm{RL};0+}X)(t) = \frac{d^n}{dt^n}(I^{n-\alpha}_{\mathrm{RL};0+}X)(t) = \frac{1}{\Gamma(n-\alpha)}\frac{d^n}{dt^n}\int\limits_0^t (t-\tau)^{n-\alpha-1}X(\tau)\,d\tau, \tag{28.15}$$

where $\Gamma(\alpha)$ is the gamma function, $0 < \tau < t < T$ and $n = [\alpha] + 1$. A sufficient condition of the existence of fractional derivatives (28.15) is $X(t) \in AC^{n-1}[0,T]$. The space $AC^n[0,T]$ consists of functions $X(t)$, which have continuous derivatives up to order $n-1$ on $[0,T]$ with absolutely continuous functions $X^{(n-1)}(t)$ on the interval $[0,T]$ (see Definition 1.3 in [335, p. 3]).

Note that the Riemann–Liouville derivatives of orders $\alpha = 1$ and $\alpha = 0$ give the expressions $(D^{1}_{\mathrm{RL};0+}X)(t) = X^{(1)}(t)$ and $(D^{0}_{\mathrm{RL};0+}X)(t) = X(t)$, respectively (see equation (2.1.7) in [200, p. 70]).

28.3 Generalized Taylor series for memory function

In this book, we consider the memory functions of power-law type, which allows us to apply fractional calculus in the description of economic processes with memory. In the general case, to describe different types of processes with memory we must consider a wider class of memory functions.

In the paper [380], we proposed a wide class of memory functions that allows us to use fractional calculus for economic processes with memory. These memory functions are characterized by the fact that they can be represented as a generalized Taylor series. For this purpose, it is most convenient to use the Taylor formula in the Trujillo–Rivero–Bonilla (TRB) form, which was proposed in [484], with the Riemann–Liouville fractional derivatives. Convenience of this form of the generalized Taylor formula is based on the fact that the first term of this series is a power-law function.

Let us define a type of the memory functions that allows us to describe the generalized multipliers and accelerators with memory by using the fractional integrals and derivatives. This type of memory functions will be called the Trujillo–Rivero–Bonilla type (TRB type).

Definition 28.3. The memory function $M(t)$ is called the function of the Trujillo–Rivero–Bonilla type (TRB type) if it is a continuous function on the interval $(0,T]$ satisfying the following conditions:

1. $(D^{\alpha}_{RL;0+})^k M(t) \in C((0,T])$ and $(D^{\alpha}_{RL;0+})^k M(t) \in I^{\alpha}_0([0,T])$ for all $k = 1, \dots N$.
2. $(D^{\alpha}_{RL;0+})^{N+1} M(t)$ is continuous on $(0,T]$.
3. If $\alpha < 1/2$, then for each positive integer k, such that $1 \leq k \leq N$ and $(k+1)\alpha < 1$, the derivatives $(D^{\alpha}_{RL;0+})^{k+1} M(t)$ is γ-continuous in $t = 0$ for some γ, which satisfies the inequality $1 - (k+1)\alpha \leq \gamma \leq 1$ or 0-singular of order α.

This definition is based on the conditions that are used in Theorem 4.1 of [484, p. 261].

Let us define the set $I^{\alpha}_0([0,T])$, the concepts of the γ-continuous and the 0-singularity of order α, which are used in the definition of the memory function of the TRB type.

– The function $M(t)$ is called 0-singular of order α if there exists real numbers α an $k \neq 0$, such that

$$\lim_{t \to 0+} \frac{M(t)}{t^{\alpha-1}} = k < \infty. \tag{28.16}$$

– The symbol $I^{\alpha}_0([0,T])$ with $\alpha \in (0,1)$ denotes the set of functions $M(t) \in F([0,T])$, for which the Riemann–Liouville fractional integral $(I^{\alpha}_{RL:0+} M)(t)$ exists and it is finite for all $t \in [0,T]$. Here $F([0,T])$ is the set of real functions of a single real variable with domain in $[0,T]$.
– The function $M(t)$, which is a Lebesgue measurable function in $[0,T]$, is called γ-continuous in $t = \tau \in [0,T]$, if there exists $\lambda \in [0, 1-\gamma)$, for which $g(t) = |t - \tau|^{\lambda} M(t)$ is continuous function in τ. As usually it is said that $M(t)$ is a γ-continuous function on $(0,T]$, if $M(t)$ is γ-continuous for every $t \in (0,T]$, and it is denoted by the symbol $C_{\gamma}((0,T])$.

The TRB memory functions can be represented as series of power-law memory functions. Using Theorem 4.1 of [484, p. 261], with $a = 0$ and $x = t - \tau$, the memory function $M(t)$ of TRB type with $t \in [0,T]$ can be represented as the generalized Taylor series with the left-sided Riemann–Liouville fractional derivatives of order $0 \leq \alpha \leq 1$ in the Trujillo–Rivero–Bonilla form.

Theorem 28.1. *For all $x = (t - \tau) \in (0,T)$, the memory function $M(t-\tau)$ of the TRB type can be represented by the series*

$$M(t-\tau) = \sum_{k=0}^{N} \frac{c_k}{\Gamma((k+1)\alpha)} (t-\tau)^{(k+1)\alpha-1} + \frac{((D^{\alpha}_{RL;0+})^{N+1} M)(\xi_N)}{\Gamma((N+1)\alpha+1)} (t-\tau)^{(N+1)\alpha}, \tag{28.17}$$

where $0 \leq \xi_N \leq t$ and

$$c_k = \Gamma(\alpha)(t^{1-\alpha}(D^{\alpha}_{RL;0+})^k M(t))(0+) = (I^{1-\alpha}_{RL;0+}(D^{\alpha}_{RL;0+})^k M)(0+) \tag{28.18}$$

for each positive integer $k \in [0,N]$.

Proof. The proof of this theorem is given in [484]. □

The last term of (28.17) is called the remainder term $R_{N\alpha}(t, 0+)$. The remainder term tends to zero at $N \to \infty$.

Let us give two examples for $N = 0$ and $N = 1$.

Example 28.1. For $N = 0$, the memory function is represented in the form

$$M(t-\tau) = \frac{c_0}{\Gamma(\alpha)}(t-\tau)^{\alpha-1} + \frac{(D^{\alpha}_{RL;0+}M)(\xi_0)}{\Gamma(\alpha+1)}(t-\tau)^{\alpha}, \tag{28.19}$$

where $0 \le \xi_0 \le T$, and

$$c_0 = \Gamma(\alpha) \lim_{t\to 0+} \frac{M(t)}{t^{\alpha-1}}. \tag{28.20}$$

Example 28.2. For $N = 1$, we have the expression

$$M(t-\tau) = \frac{c_0}{\Gamma(\alpha)}(t-\tau)^{\alpha-1} + \frac{c_1}{\Gamma(2\alpha)}(t-\tau)^{2\alpha-1} + \frac{((D^{\alpha}_{RL;0+})^2M)(\xi_1)}{\Gamma(2\alpha+1)}(t-\tau)^{2\alpha}, \tag{28.21}$$

where $0 \le \xi_1 \le t$.

Using Remark 4.1 of [484, p. 262], all memory functions $M(t)$ of the TRB type, can be represented by the generalized Taylor's series

$$M(t-\tau) = (t-\tau)^{\alpha-1} \sum_{k=0}^{\infty} \frac{c_k}{\Gamma((k+1)\alpha)}(t-\tau)^{k\alpha} \tag{28.22}$$

that holds for all $t \in (0, T]$, and this series converges. The functions, which can be represented by equation (28.22), are called α-analytic in $t = 0$ [484].

28.4 Multiplier with memory of TRB type

Substituting expression (28.17) into equation (28.2) gives the multiplier equation

$$\begin{aligned} Y(t) = {} & \sum_{k=0}^{N} \frac{c_k}{\Gamma((k+1)\alpha)} \int_0^t (t-\tau)^{(k+1)\alpha-1} X(\tau)\, d\tau \\ & + \frac{((D^{\alpha}_{RL;0+})^{N+1}M)(\xi_N)}{\Gamma((N+1)\alpha+1)} \int_0^t (t-\tau)^{(N+1)\alpha} X(\tau)\, d\tau. \end{aligned} \tag{28.23}$$

Using equation (28.9) of the Riemann–Liouville fractional integral in the form

$$\int_0^t (t-\tau)^{\alpha-1} X(\tau)\, d\tau = \Gamma(\alpha)(I^{\alpha}_{RL;0+}X)(t), \tag{28.24}$$

we can rewrite equation (28.23) as

$$Y(t) = \sum_{k=0}^{N} m_k(\alpha)(I^{(k+1)\alpha}_{RL;0+}X)(t) + m_{R;N}(\alpha)(I^{(N+1)\alpha+1}_{RL;0+}X)(t), \tag{28.25}$$

where the coefficients $m_k(\alpha)$ and $m_{R;k}(\alpha)$ are multiplier coefficients that are defined by the equations

$$m_k(\alpha) = (I^{1-\alpha}_{\mathrm{RL};0+}(D^{\alpha}_{\mathrm{RL};0+})^k M)(0+) = \Gamma(\alpha)(t^{1-\alpha}(D^{\alpha}_{\mathrm{RL};0+})^k M(t))(0+), \tag{28.26}$$

$$m_{R;k}(\alpha) = ((D^{\alpha}_{\mathrm{RL};0+})^k M)(\xi_{k-1}), \tag{28.27}$$

where $0 \le \xi_{k-1} \le t$. There multiplier coefficients are defined by the derivatives of the memory function $M(t)$.

Let us define the integral operators J^{α}_N by the equation

$$(J^{\alpha}_N X)(t) = \sum_{k=0}^{N} m_k(\alpha)(I^{(k+1)\alpha}_{\mathrm{RL};0+} X)(t) + m_{R;N}(\alpha)(I^{(N+1)\alpha+1}_{\mathrm{RL};0+} X)(t), \tag{28.28}$$

where $m_k(\alpha)$ and $m_{R;k}(\alpha)$ are defined by equations (28.26) and (28.27). Using the integral operators J^{α}_N, the generalized multiplier equation (28.25) for the TRB memory can be written in the compact form

$$Y(t) = (J^{\alpha}_N X)(t), \tag{28.29}$$

where $N \in \mathbb{N}$.

Example 28.3. For $N = 0$, using the series for the memory function of the Trujillo–Rivero–Bonilla type, the linear generalized multiplier (28.2) with memory (28.19) takes the form

$$Y(t) = \frac{c_0}{\Gamma(\alpha)} \int_0^t (t-\tau)^{\alpha-1} X(\tau)\, d\tau + \frac{(D^{\alpha}_{\mathrm{RL};0+} M)(\xi_0)}{\Gamma(\alpha+1)} \int_0^t (t-\tau)^{\alpha} X(\tau)\, d\tau. \tag{28.30}$$

Using equation (28.24), we can rewrite equation (28.30) in the form

$$Y(t) = m_0(\alpha)(I^{\alpha}_{\mathrm{RL};0+} X)(t) + m_{R;1}(\alpha)(I^{\alpha+1}_{\mathrm{RL};0+} X)(t), \tag{28.31}$$

where $m_k(\alpha)$ and $m_{R;k}(\alpha)$ are defined by equations (28.26) and (28.27).

Example 28.4. For $N = 1$, we use the memory (28.21). Then the generalized multiplier (28.2) is described by the equation

$$Y(t) = \frac{c_0}{\Gamma(\alpha)} \int_0^t (t-\tau)^{\alpha-1} X(\tau)\, d\tau + \frac{c_1}{\Gamma(2\alpha)} \int_0^t (t-\tau)^{2\alpha-1} X(\tau)\, d\tau + \frac{((D^{\alpha}_{\mathrm{RL};0+})^2 M)(\xi_1)}{\Gamma(\alpha+1)} \int_0^t (t-\tau)^{2\alpha} X(\tau)\, d\tau. \tag{28.32}$$

Using equation (28.24), we rewrite expression (28.32) in the form

$$Y(t) = m_0(\alpha)(I^{\alpha}_{\mathrm{RL};0+}X)(t) + m_1(\alpha)(I^{2\alpha}_{\mathrm{RL};0+}X)(t) + m_{R;2}(\alpha)(I^{2\alpha+1}_{\mathrm{RL};0+}X)(t), \tag{28.33}$$

where $m_k(\alpha)$ and $m_{R;k}(\alpha)$ are defined by equations (28.26) and (28.27).

As a result, we can formulate the following principle for the multiplier with generalized memory.

Principle 28.1 (Principle of decomposability of multiplier with generalized memory). *Equation of the generalized multiplier with memory of the TRB type can be represented as a multiplier with multi-parametric power-law memory, i. e., as a parallel action (a sum of action) of the multipliers with simplest power-law memory.*

Equations (28.29) and (28.31), (28.33) describe multipliers with memory, for which the memory functions have the form different from power-law form. These equations can be applied for wide class of memory functions of the Trujillo–Rivero–Bonilla type. This representation of multiplier allows us to describe wide class of processes with memory by using the methods of fractional calculus.

Taking into account the contribution of the remainder terms of Taylor's series ($m_{R;k}(\alpha) \neq 0$) in the multiplier and accelerator equations, we get more general models of considered processes. Neglecting these terms can be used only in a narrower class of models. Models, which do not take into account the contribution of terms with $m_{R;k}(\alpha) \neq 0$, can easily be obtained from the proposed equations of multipliers and accelerators with memory. If for considered processes it is possible to neglect the remainder term, then we can eliminate this term by setting the corresponding coefficient $m_{R;N}(\alpha)$ equal to zero.

In multiplier equations (28.29) and (28.31), (28.33), we saved the remainder terms of the Taylor's series ($m_{R;N}(\alpha) \neq 0$) to consider a more general class of processes with memory in economics.

28.5 Accelerator with memory of TRB type

Let us consider the linear generalized accelerators with memory function $M(t-\tau)$ that are represented by equations (28.3) and (28.4), where n is positive integer number and the function $M(t-\tau)$ describes the memory of the TRB type. Using the generalized Taylor series with the left-sided Riemann–Liouville fractional derivatives of order $0 \leq \beta \leq 1$ in the Trujillo–Rivero–Bonilla form, equations (28.3) and (28.4) takes the form

$$Y(t) = \sum_{k=0}^{N} m_k(\beta)(I^{(k+1)\beta}_{\mathrm{RL};0+}X^{(n)})(t) + m_{R;N}(\beta)(I^{(N+1)\beta+1}_{\mathrm{RL};0+}X^{(n)})(t), \tag{28.34}$$

$$Y(t) = \sum_{k=0}^{N} m_k(\beta)(I^{(k+1)\beta}_{\mathrm{RL};0+}X)^{(n)}(t) + m_{R;N}(\beta)(I^{(N+1)\beta+1}_{\mathrm{RL};0+}X)^{(n)}(t), \tag{28.35}$$

where $m_k(\beta)$ and $m_{R;k}(\beta)$ are coefficients that are defined by the equations

$$m_k(\beta) = (I^{1-\beta}_{\mathrm{RL};0+}(D^{\beta}_{\mathrm{RL};0+})^k M)(0+) = \Gamma(\beta)(t^{1-\beta}(D^{\beta}_{\mathrm{RL};0+})^k M(t))(0+), \tag{28.36}$$

$$m_{R;k}(\beta) = ((D^{\beta}_{\mathrm{RL};0+})^k M)(\xi_{k-1}). \tag{28.37}$$

Let us consider the case $\beta = n-\alpha$, where $\beta \in (0,1)$, i. e., $n-1 < \alpha < n$. In this case, we will take into account that the Riemann–Liouville and Caputo fractional derivatives are given by the equations

$$\frac{d^n}{dt^n}(I^{n-\alpha}_{\mathrm{RL};0+}X)(t) = (D^{\alpha}_{\mathrm{RL};0+}X)(t), \tag{28.38}$$

$$(I^{n-\alpha}_{\mathrm{RL};0+}X^{(n)})(t) = (D^{\alpha}_{C;0+}X)(t). \tag{28.39}$$

These expressions allow us to write the generalized accelerator equation as a series of the Riemann–Liouville and Caputo fractional derivatives. To realize this representation, we can use the semigroup property of the Riemann–Liouville fractional integration. Using equation 2.21 of [335, p. 34] (see also Lemma 2.3 of [200, p. 73]), we have the equality

$$(I^{\alpha+\beta}_{\mathrm{RL};0+}X)(t) = (I^{\alpha}_{\mathrm{RL};0+}I^{\beta}_{\mathrm{RL};0+}X)(t), \tag{28.40}$$

that is satisfied in any point for $X(t) \in C[0,T]$, where $\alpha > 0$ and $\beta > 0$. Equality (28.40) holds in almost every point for $X(t) \in L_1[0,T]$ and $X(t) \in L_p[0,T]$ with $1 \le p \le \infty$. If $\alpha + \beta \le 1$ equality (28.40) holds for $X(t) \in L_1[0,T]$ at any point of $[0,T]$. Using equality (28.40), we can obtain the equations

$$I^{(k+1)(n-\alpha)}_{\mathrm{RL};0+} = I^{n-\alpha}_{\mathrm{RL};0+}I^{k(n-\alpha)}_{\mathrm{RL};0+} = I^{k(n-\alpha)}_{\mathrm{RL};0+}I^{n-\alpha}_{\mathrm{RL};0+}, \tag{28.41}$$

$$I^{(N+1)(n-\alpha)+1}_{\mathrm{RL};0+} = I^{n-\alpha}_{\mathrm{RL};0+}I^{N(n-\alpha)+1}_{\mathrm{RL};0+} = I^{N(n-\alpha)+1}_{\mathrm{RL};0+}I^{n-\alpha}_{\mathrm{RL};0+}. \tag{28.42}$$

As a result, we get equations (28.34) and (28.35) of the generalized accelerators with memory of TRB type in the form

$$\begin{aligned} Y(t) = {} & \sum_{k=0}^{N} m_k(n-\alpha)(D^{\alpha}_{\mathrm{RL};0+}I^{k(n-\alpha)}_{\mathrm{RL};0+}X)(t) \\ & + m_{R;N}(n-\alpha)(D^{\alpha}_{\mathrm{RL};0+}I^{N(n-\alpha)+1}_{\mathrm{RL};0+}X)(t), \end{aligned} \tag{28.43}$$

$$\begin{aligned} Y(t) = {} & \sum_{k=0}^{N} m_k(n-\alpha)(I^{k(n-\alpha)}_{\mathrm{RL};0+}D^{\alpha}_{C;0+}X)(t) \\ & + m_{R;N}(n-\alpha)(I^{N(n-\alpha)+1}_{\mathrm{RL};0+}D^{\alpha}_{C;0+}X)(t), \end{aligned} \tag{28.44}$$

where $n-1<\alpha<n$ and $m_{R;N}(n-\alpha)$, $m_{R;N}(n-\alpha)$ are defined by equations (28.36) and (28.37), respectively.

These equations allow us to describe processes with memory by methods of fractional calculus for wide class of memory functions.

Using Property 2.2 of [200, p. 74], we have the equality

$$(D^{\alpha}_{\mathrm{RL};0+}I^{\beta}_{\mathrm{RL};0+}X)(t)=(I^{\beta-\alpha}_{\mathrm{RL};0+}X)(t), \tag{28.45}$$

if $\beta>\alpha>0$ and $X(t)\in L_p[0,T]$ with $1\le p\le\infty$. For example, if $\beta=n-\alpha$, then the inequality $\beta>\alpha>0$ means $\alpha<\frac{1}{2}$ and $n=1$. The equality leads to the fact that, starting with a certain value of k, the terms of the generalized accelerator with memory of TRB type will contain only the fractional integrals, which describe the multiplicative effect with memory.

Using the integral operators J^{α}_{N}, which is defined by equation (28.28), the generalized accelerator equations (28.43) and (28.44) can be written in the compact form

$$Y(t)=(D^{\alpha}_{\mathrm{RL};0+}J^{n-\alpha}_{N}X)(t), \tag{28.46}$$

$$Y(t)=(J^{n-\alpha}_{N}D^{\alpha}_{C;0+}X)(t). \tag{28.47}$$

Equations (28.46) and (28.47) can be interpreted as combinations of sequential actions of the accelerator with simplest power-law (SPL) memory and the multiplier with memory of the TRB type. Equation (28.46) describes the situation, when the multiplier with TRB memory acts first, and then the accelerator with SPL memory acts. Equation (28.47) describes a situation, when the accelerator with SPL memory acts first, and then the multiplier with TRB memory acts. Therefore, we can call the generalized accelerator, which is described by equation (28.46), as the generalized AM-accelerator with TRB memory. The generalized accelerator, which is described by equation (28.47), is the generalized MA-accelerator with TRB memory.

Example 28.5. For the cases $N=0$ and $N=1$, the equations of the generalized AM-accelerator are

$$Y(t)=m_0(n-\alpha)(D^{\alpha}_{\mathrm{RL};0+}X)(t)+m_{R;1}(n-\alpha)(D^{\alpha}_{\mathrm{RL};0+}I^{1}_{\mathrm{RL};0+}X)(t), \tag{28.48}$$

$$\begin{aligned}Y(t)=&\,m_0(n-\alpha)(D^{\alpha}_{\mathrm{RL};0+}X)(t)+m_1(n-\alpha)(D^{\alpha}_{\mathrm{RL};0+}I^{n-\alpha}_{\mathrm{RL};0+}X)(t)\\&+m_{R;2}(n-\alpha)(D^{\alpha}_{\mathrm{RL};0+}I^{n-\alpha+1}_{\mathrm{RL};0+}X)(t),\end{aligned} \tag{28.49}$$

where $n-1<\alpha<n$.

Example 28.6. For $N=0$ and $N=1$, the equations of the generalized MA-accelerator have the form

$$Y(t)=m_0(n-\alpha)(D^{\alpha}_{C;0+}X)(t)+m_{R;1}(n-\alpha)(I^{1}_{\mathrm{RL};0+}D^{\alpha}_{C;0+}X)(t), \tag{28.50}$$

$$
\begin{aligned}
Y(t) = {} & m_0(n-\alpha)(D^{\alpha}_{C;0+}X)(t) + m_1(n-\alpha)(I^{n-\alpha}_{\mathrm{RL};0+}D^{\alpha}_{C;0+}X)(t) \\
& + m_{R;2}(n-\alpha)(I^{n-\alpha+1}_{\mathrm{RL};0+}D^{\alpha}_{C;0+}X)(t),
\end{aligned} \tag{28.51}
$$

where $n-1 < \alpha < n$.

In equations (28.46), (28.47) and (28.50), (28.51), we take into account the contribution of the remainder terms of Taylor's series ($m_{R;N}(n-\alpha) \neq 0$) in the accelerator equations, to have a possibility consider more general class of processes with memory. If in the considered problem it is possible to neglect the remainder term, then we can use the coefficient $m_{R;N}(\alpha)$ equal to zero in the accelerator equations.

Using the representation of the memory function that is given by equation (28.22), we can write the accelerator with TRB memory in the form

$$
Y(t) = (D^{\alpha}_{\mathrm{RL};0+}J^{n-\alpha}_{\infty}X)(t), \tag{28.52}
$$

$$
Y(t) = (J^{n-\alpha}_{\infty}D^{\alpha}_{C;0+}X)(t), \tag{28.53}
$$

where the operator $J^{n-\alpha}_{\infty}$ is defined as

$$
(J^{n-\alpha}_{\infty}X)(t) = \sum_{k=0}^{\infty} m_k(n-\alpha)(I^{k(n-\alpha)}_{\mathrm{RL};0+}X)(t). \tag{28.54}
$$

Note that

$$
\lim_{N\to\infty}(J^{\alpha}_{N}X)(t) = (I^{\alpha}_{\mathrm{RL};0+}J^{\alpha}_{\infty}X)(t), \tag{28.55}
$$

where J^{α}_{N} is defined by equation (28.28).

The coefficients $m_k(n-\alpha)$ are defined by equation (28.36). However, to simulate real processes it is more convenient to use accelerators of the form given by equations (28.46) and (28.47).

As a result, we can formulate the following principle for the accelerator with generalized memory.

Principle 28.2 (Principle of decomposability of accelerator with generalized memory). *Equation of the generalized accelerator with memory of the TRB type can be represented as sequential actions of the accelerator with simplest power-law memory and the multiplier with multiparametric power-law memory.*

This principle allows us to formulate macroeconomic models for wide class of memory functions by using the methods of fractional calculus.

28.6 Harrod–Domar growth model with memory of TRB type

Let us generalize the standard Harrod–Domar growth model with continuous time [170, 171, 172, 92, 93] by taking into account memory of TRB type. The balance equation of this model has the form

$$Y(t) = I(t) + C(t), \tag{28.56}$$

where $Y(t)$ is the national income, $I(t)$ is the investment, $C(t)$ is the non-productive consumption. In the standard Harrod–Domar model of the growth without memory, it is assumed that investment is determined by the growth rate of the national income. This assumption is described by the accelerator equations

$$I(t) = B\frac{dY(t)}{dt}, \tag{28.57}$$

where B is the accelerator coefficient, which describes the capital intensity of the national income. Substitution of equation (28.57) into (28.56) gives

$$Y(t) = B\frac{dY(t)}{dt} + C(t). \tag{28.58}$$

Equation (28.58) defines the Harrod–Domar model without memory, where the behavior of the national income $Y(t)$ is determined by the dynamics of non-productive consumption $C(t)$.

Equation (28.57) is differential equation of the first order. This means instantaneous change of the investment $I(t)$ when changing the growth rate of the national income $Y(t)$. Therefore, equations (28.57) and (28.58) do not take into account the memory effects. The Harrod–Domar model with simplest power-law (SPL) memory has been considered in [421, 426, 450]. Let us consider the Harrod–Domar model with memory of the TRB type, which is proposed in [380].

The equation of investment accelerator with TRB memory can be written in the form

$$I(t) = \int_0^t B(t,\tau)Y^{(n)}(\tau)\,d\tau, \tag{28.59}$$

where the constant parameter B of the capital intensity is replaced by the function $B(t,\tau)$ to take into account the dependence of the investment $I(t)$ on the history of changes of the national income $Y(\tau)$ in $\tau < t$.

For simplicity, we consider the case

$$B(t,\tau) = BM(t-\tau), \tag{28.60}$$

where $M(t-\tau)$ is the memory function of the TRB type. Using expression (28.60), equation (28.59) is rewritten in the form

$$I(t) = B\int_0^t M(t-\tau)Y^{(n)}(\tau)\,d\tau. \tag{28.61}$$

For $n = 1$ and $M(t-\tau) = \delta(t-\tau)$, equation (28.61) gives equation (28.57) of the standard accelerator without memory.

Substituting the investment $I(t)$, which is given by formula (28.61), into balance equation (28.56), we obtain the equation

$$Y(t) = B \int_0^t M(t-\tau) Y^{(n)}(\tau)\, d\tau + C(t). \tag{28.62}$$

For $M(t-\tau) = \delta(t-\tau)$ and $n = 1$, equation (28.62) gives equation (28.58).

Equation (28.62) determines the dynamics of the national income within the framework of the Harrod–Domar macroeconomic model with memory. If the parameter B is given, then the dynamics of national income $Y(t)$ is determined by the behavior of the function $C(t)$.

Let us use the representation of the generalized MA-accelerator with $N = 0$ in the form

$$\begin{aligned}\int_0^t M(t-\tau) Y^{(n)}(\tau)\, d\tau &= m_0 (D^{\alpha}_{C;0+} Y)(t) \\ &\quad + m_{R;1} (I^1_{RL;0+} D^{\alpha}_{C;0+} Y)(t),\end{aligned} \tag{28.63}$$

where m_0 and $m_{R;1}$ are defined by equations (28.36) and (28.37).

As a result, equation (28.62) takes the form

$$Y(t) = B m_0 (D^{\alpha}_{C;0+} Y)(t) + B m_{R;1} (I^1_{RL;0+} D^{\alpha}_{C;0+} Y)(t) + C(t). \tag{28.64}$$

To get solutions of this equation, we should consider two cases: $\alpha > 1$ and $0 < \alpha < 1$, $(n = 1)$.

First case: For $\alpha > 1$, we can use the definition of the Caputo fractional derivative in the form

$$(D^{\alpha}_{RL;0+} Y)(t) = (I^{n-\alpha}_{RL;0+} D^n Y)(t), \tag{28.65}$$

where $n = [\alpha] + 1$ for noninteger values of α. Using the semigroup property of the Riemann–Liouville fractional integration (see equation (2.21) of the book [335, p. 34] and Lemma 2.3 of the book [200, p. 73]), we can write

$$\begin{aligned}(I^1_{RL;0+} D^{\alpha}_{RL;0+} Y)(t) &= (I^1_{RL;0+} I^{n-\alpha}_{RL;0+} D^n Y)(t) \\ &= (I^{n-\alpha+1}_{RL;0+} D^n Y)(t) = (I^{n-\alpha}_{RL;0+} I^1_{RL;0+} D^n Y)(t) \\ &= (I^{n-\alpha}_{RL;0+} I^1_{RL;0+} D^1 Y^{(n-1)})(t) = (I^{n-\alpha}_{RL;0+} (Y^{(n-1)} - Y^{(n-1)}(0)))(t) \\ &= (I^{n-\alpha}_{RL;0+} Y^{(n-1)})(t) - Y^{(n-1)}(0)(I^{n-\alpha}_{RL;0+} 1)(t)\end{aligned} \tag{28.66}$$

$$= (I^{(n-1)-(\alpha-1)}_{\mathrm{RL};0+} Y^{(n-1)})(t) - Y^{(n-1)}(0)(I^{n-\alpha}_{\mathrm{RL};0+} 1)(t)$$

$$= (D^{\alpha-1}_{\mathrm{RL};0+} Y)(t) - Y^{(n-1)}(0) \frac{1}{\Gamma(n-\alpha+1)} t^{n-\alpha},$$

where we use equation (2.1.16) of the book [200, p. 71] and the standard Newton–Leibniz equation $I^1_{\mathrm{RL};0+} D^1 f(t) = f(t) - f(0)$.

As a result, we get the equation

$$(I^1_{\mathrm{RL};0+} D^{\alpha}_{\mathrm{RL};0+} Y)(t) = (D^{\alpha-1}_{\mathrm{RL};0+} Y)(t) - \frac{Y^{(n-1)}(0)}{\Gamma(n-\alpha+1)} t^{n-\alpha}, \tag{28.67}$$

Equation (28.67) allows us to represent equation (28.64) in the form

$$Y(t) = Bm_0 (D^{\alpha}_{C;0+} Y)(t) + Bm_{R;1} (D^{\alpha-1}_{C;0+} Y)(t) + F(t), \tag{28.68}$$

where $\alpha > 1$ and

$$F(t) = C(t) - \frac{Y^{(n-1)}(0)}{\Gamma(n-\alpha+1)} t^{n-\alpha}. \tag{28.69}$$

Equation (28.68) can be rewritten in the form

$$(D^{\alpha}_{C;0+} Y)(t) + \frac{m_{R;1}}{m_0} (D^{\alpha-1}_{C;0+} Y)(t) - \frac{1}{Bm_0} Y(t) = -\frac{1}{Bm_0} F(t). \tag{28.70}$$

Fractional differential equation (28.70) determines the dynamics of the national income within the framework of the proposed macroeconomic model with memory of TRB type with $N = 0$.

To solve equation (28.70), we can use Theorem 5.16 of book [200, pp. 323–324]. Equation (28.70) coincides with equation (5.3.73) of book [200, p. 323], when we use the notation

$$\lambda = -\frac{m_{R;1}}{m_0}; \quad \mu = \frac{1}{Bm_0}; \quad f(t) = -\frac{1}{Bm_0} F(t); \quad \beta = \alpha - 1 > 0. \tag{28.71}$$

As a result, for continuous function $f(t)$, which is defined on the positive semiaxis ($t > 0$), equation (28.70) with the parameters $n-1 < \alpha \leq n$ and $0 \leq n-2 < \beta = \alpha-1 \leq n-1$ is solvable [200, pp. 323–324], and it has the general solution

$$Y(t) = \sum_{j=0}^{n-1} c_j Y_j(t) + Y_C(t), \tag{28.72}$$

where c_j, ($j = 0, \dots, n-1$) are the real constants that are determined by the initial conditions, the function $Y_C(t)$ is defined as

$$Y_C(t) = -\frac{1}{Bm_0} \int_0^t (t-\tau)^{\alpha-1} G_{\alpha,\alpha-1,B,m_0,m_{R;1}}[t-\tau] F(\tau)\, d\tau, \tag{28.73}$$

the function $G_{\alpha,\alpha-1,B,m_0,m_{R;1}}[\tau]$ is given by the equation

$$G_{\alpha,\alpha-1,B,m_0,m_{R;1}}[\tau] = \sum_{k=0}^{\infty} \frac{\tau^{k\alpha}}{\Gamma(k+1)} \frac{1}{(Bm_0)^k} \Psi_{1,1}\left[\begin{smallmatrix}(k+1,1)\\(\alpha k+\alpha,1)\end{smallmatrix} \mid -\frac{m_{R;1}}{m_0}\tau\right]. \tag{28.74}$$

The functions $Y_j(t)$ with $j = 0, \ldots, n-2$ are represented by the expression

$$Y_j(t) = \sum_{k=0}^{\infty} \frac{t^{k\alpha+j}}{\Gamma(k+1)} \frac{1}{(Bm_0)^k} \Psi_{1,1}\left[\begin{smallmatrix}(k+1,1)\\(\alpha k+j+1,1)\end{smallmatrix} \mid -\frac{m_{R;1}}{m_0}t\right] + \frac{m_{R;1}}{m_0} \sum_{k=0}^{\infty} \frac{t^{k\alpha+j+\alpha-\beta}}{\Gamma(k+1)} \frac{1}{(Bm_0)^k} \Psi_{1,1}\left[\begin{smallmatrix}(k+1,1)\\(\alpha k+j+2,1)\end{smallmatrix} \mid -\frac{m_{R;1}}{m_0}t\right]. \tag{28.75}$$

For $j = n-2$ and $j = n-1$, the functions $Y_j(t)$ are defined by the equations

$$Y_j(t) = \sum_{k=0}^{\infty} \frac{t^{k\alpha+j}}{\Gamma(k+1)} \frac{1}{(Bm_0)^k} \Psi_{1,1}\left[\begin{smallmatrix}(k+1,1)\\(\alpha k+j+1,1)\end{smallmatrix} \mid -\frac{m_{R;1}}{m_0}t\right]. \tag{28.76}$$

Here, $\Psi_{1,1}$ is the generalized Wright functions (the Fox–Wright function) [200, p. 56], which is defined by the equation

$$\Psi_{1,1}\left[\begin{smallmatrix}(a,\alpha)\\(b,\beta)\end{smallmatrix} \mid z\right] = \sum_{k=0}^{\infty} \frac{\Gamma(\alpha k+a)}{\Gamma(\beta k+b)} \frac{z^k}{k!}. \tag{28.77}$$

Note that the generalized Mittag–Leffler function (the Prabhakar function) [143] is a special case of the Fox–Wright function (see equation (1.9.1) of [200, p. 45]), that is,

$$\Psi_{1,1}\left[\begin{smallmatrix}(\rho,1)\\(\beta,\alpha)\end{smallmatrix} \mid z\right] = \Gamma(\rho) E^{\rho}_{\alpha,\beta}[z]. \tag{28.78}$$

Equation (28.70) and its solution (28.72)–(28.76) describe the macroeconomic dynamics of the national income, where the memory is TRB-type memory and the fading parameter $\alpha > 1$.

Second case: For $0 < \alpha < 1$ ($n = 1$), we can take the first derivative of the left and right sides of this equation (28.64), to obtain

$$Y^{(1)}(t) = Bm_0 D^1 (D^{\alpha}_{C;0+} Y)(t) + Bm_{R;1} (D^{\alpha}_{C;0+} Y)(t) + C^{(1)}(t), \tag{28.79}$$

where we use

$$D^1 I^1_{RL;0+} Y(t) = Y(t); \quad C^{(1)}(t) = \frac{dC(t)}{dt}; \quad D^1 = \frac{d}{dt}. \tag{28.80}$$

Using equation (2.4.6) of book [200, p. 91], we have

$$(D^{\alpha}_{C;0+} Y)(t) = (D^{\alpha}_{RL;0+} Y)(t) - \sum_{k=0}^{n-1} \frac{Y^{(k)}(0)}{\Gamma(k-\alpha+1)} t^{k-\alpha}, \tag{28.81}$$

where $n = [\alpha] + 1$ for non-integer values of $\alpha > 0$. Then using equation (2.1.34) of Property 2.3 in book [200, p. 74] in the form

$$D^1(D^{\alpha}_{\mathrm{RL};0+}Y)(t) = (D^{\alpha+1}_{\mathrm{RL};0+}Y)(t), \tag{28.82}$$

and

$$\frac{d}{dt}\sum_{k=0}^{n-1}\frac{Y^{(k)}(0)}{\Gamma(k-\alpha+1)}t^{k-\alpha} = \sum_{k=0}^{n-1}\frac{Y^{(k)}(0)(k-\alpha)}{\Gamma(k-\alpha+1)}t^{k-\alpha-1} = \sum_{k=0}^{n-1}\frac{Y^{(k)}(0)}{\Gamma(k-\alpha)}t^{k-\alpha-1}, \tag{28.83}$$

we can use the definition of the Riemann–Liouville fractional derivative in the form

$$(D^{\alpha}_{\mathrm{RL};0+}Y)(t) = D^n(I^{n-\alpha}_{\mathrm{RL};0+}Y)(t). \tag{28.84}$$

Then we use equation 2.1.60 of book [200, p. 91] in the form

$$(D^{\alpha+1}_{RL;0+}Y)(t) = (D^{\alpha+1}_{C;0+}Y)(t) + \sum_{k=0}^{n}\frac{Y^{(k)}(0)}{\Gamma(k-\alpha)}t^{k-\alpha-1}. \tag{28.85}$$

This leads to the expression

$$\begin{aligned} D^1(D^{\alpha}_{C;0+}Y)(t) = (D^{\alpha+1}_{C;0+}Y)(t) - \sum_{k=0}^{n-1}\frac{Y^{(k)}(0)}{\Gamma(k-\alpha)}t^{k-\alpha-1} \\ + \sum_{k=0}^{n}\frac{Y^{(k)}(0)}{\Gamma(k-\alpha)}t^{k-\alpha-1}. \end{aligned} \tag{28.86}$$

Equation (28.86) can be written in the form

$$D^1(D^{\alpha}_{C;0+}Y)(t) = (D^{\alpha+1}_{C;0+}Y)(t) + \frac{Y^{(n)}(0)}{\Gamma(n-\alpha)}t^{n-\alpha-1}. \tag{28.87}$$

As a result, we have the fractional differential equation

$$Bm_0(D^{\alpha+1}_{C;0+}Y)(t) + Bm_{R;1}(D^{\alpha}_{C;0+}Y)(t) - Y^{(1)}(t) = -C^{(1)}(t) - \frac{Y^{(n)}(0)}{\Gamma(n-\alpha)}t^{n-\alpha-1}, \tag{28.88}$$

where $0 < \alpha < 1$, $(n = 1)$. This equation can be rewritten in the form

$$(D^{\alpha+1}_{C;0+}Y)(t) - \frac{1}{Bm_0}Y^{(1)}(t) + \frac{m_{R;1}}{m_0}(D^{\alpha}_{C;0+}Y)(t) = -\frac{1}{Bm_0}F(t). \tag{28.89}$$

Using the notation

$$\lambda = \frac{1}{Bm_0}; \quad \delta = A_0 = -\frac{m_{R;1}}{m_0}; \tag{28.90}$$

$$\alpha > \alpha + 1; \quad \beta = 1; \quad \alpha_0 = \alpha; \quad \mu = 0, \tag{28.91}$$

$$F(t) = C^{(1)}(t) + \frac{Y^{(n)}(0)}{\Gamma(n-\alpha)} t^{n-\alpha-1}, \tag{28.92}$$

the solution of equation (28.89), which corresponds to the case $0 < \alpha < 1$ ($n = 1$) and $N = 0$, is described by Example 5.23 of [200, p. 326], and Theorem 5.17 of the book [200, p. 324].

As a result, for continuous function $F(t)$, which is defined on the positive semi-axis ($t > 0$), equation (28.89) has the general solution (see Example 5.23 of [200, p. 326]) in the form

$$Y(t) = c_0 Y_0(t) + c_1 Y_1(t) + Y_C(t), \tag{28.93}$$

where c_0 and c_1 are the real constants that are determined by the initial conditions, the function $Y_C(t)$ is defined as

$$Y_C(t) = -\frac{1}{Bm_0} \int_0^t (t-\tau)^{\alpha-1} G_{\alpha+1,1,\alpha,B,m_0,m_{R;1}}[t-\tau] F(\tau)\, d\tau, \tag{28.94}$$

the function $G_{\alpha+1,1,\alpha,B,m_0,m_{R;1}}[\tau]$ is given by the equation

$$G_{\alpha+1,1,\alpha,B,m_0,m_{R;1}}[\tau] = \sum_{k=0}^{\infty} \frac{\tau^k}{\Gamma(k+1)} \left(-\frac{m_{R;1}}{m_0}\right)^k \Psi_{1,1}\left[\begin{smallmatrix}(n+1,1)\\(\alpha+k+1,\alpha)\end{smallmatrix} \,\middle|\, \frac{1}{Bm_0}\tau^\alpha\right]. \tag{28.95}$$

The function $Y_0(t)$ is represented by the expression

$$\begin{aligned} Y_0(t) = {} & \sum_{k=0}^{\infty} \frac{t^k}{\Gamma(k+1)} \left(-\frac{m_{R;1}}{m_0}\right)^k \Psi_{1,1}\left[\begin{smallmatrix}(k+1,1)\\(k+1,\alpha)\end{smallmatrix} \,\middle|\, \frac{1}{Bm_0}t^\alpha\right] \\ & - \frac{1}{Bm_0} t^\alpha \sum_{k=0}^{\infty} \frac{t^k}{\Gamma(k+1)} \left(-\frac{m_{R;1}}{m_0}\right)^k \Psi_{1,1}\left[\begin{smallmatrix}(k+1,1)\\(\alpha+k+1,\alpha)\end{smallmatrix} \,\middle|\, \frac{1}{Bm_0}t^\alpha\right] \\ & + \frac{m_{R;1}}{m_0} t^{1-\alpha} \sum_{k=0}^{\infty} \frac{t^k}{\Gamma(k+1)} \left(-\frac{m_{R;1}}{m_0}\right)^k \Psi_{1,1}\left[\begin{smallmatrix}(k+1,1)\\(k+2,\alpha)\end{smallmatrix} \,\middle|\, \frac{1}{Bm_0}t^\alpha\right], \end{aligned} \tag{28.96}$$

and $Y_1(t)$ is given by the equation

$$Y_1(t) = \sum_{k=0}^{\infty} \frac{t^{k+1}}{\Gamma(k+1)} \left(-\frac{m_{R;1}}{m_0}\right)^{k+1} \Psi_{1,1}\left[\begin{smallmatrix}(k+1,1)\\(k+2,\alpha)\end{smallmatrix} \,\middle|\, \frac{1}{Bm_0}t^\alpha\right]. \tag{28.97}$$

Equation (28.89) and its solution (28.93)–(28.97) describe the macroeconomic dynamics of the national income with memory of the TRB type and the fading parameter $0 < \alpha < 1$.

The solutions of the equations of the macroeconomic model with TRB memory with $N \geq 1$ can be described by using Theorem 5.17 of [200, p. 324].

28.7 Lucas model of learning with memory of TRB type

In order to describe the possible connection between evidence of learning on individual product lines and productivity growth in an economy, Robert E. Lucas considers [239, 238] the labor-only technology

$$X(t) = kn(t)Z^{a}(t), \tag{28.98}$$

where $X(t)$ is the rate of production of a good, k is the productivity parameter that depends on the units in which labor input and output are measured, and $n(t)$ is employment. The variable $Z(t)$ represents cumulative experience in the production of this good. Cumulative experience is defined [239, 238] by the differential equation of the first order

$$\frac{dZ(t)}{dt} = n(t)Z^{a}(t), \tag{28.99}$$

and the initial value $Z(t_0)$ of the experience variable on the date to when production was begun. Usually it is assumed that the initial value $Z(t_0)$ is greater than or equal to one.

The general solution of (28.99) has the form

$$Z(t) = \left(Z^{1-a}(t_0) + (1-a) \int_{t_0}^{t} n(\tau)\, d\tau \right)^{1/(1-a)} . \tag{28.100}$$

Let us describe the implications of this model for the dynamics of production of a single good.

In the case, when the employment $n(t)$ is constant at n_c over time. Then (28.98) and (28.100) imply that the production follows:

$$X(t) = kn_c(Z^{1-a}(t_0) + (1-a)n_c(t-t_0))^{a/(1-a)}. \tag{28.101}$$

Solution (28.101) means that production growth without bound, and the rate of productivity growth declines monotonically from $an_cZ^{a-1}(t_0)$ to zero. For any initial productivity level $Z(t_0) > 1$ and any employment level (or path), the productivity at date t is an increasing function of the learning rate a.

Assuming that $n_c(t-t_0)$ is large relative to initial experience, the rate of productivity growth is approximately equal to the value

$$R_X = \frac{a}{(1-a)(t-t_0)}, \tag{28.102}$$

where $t - t_0$ is the time after the start of production at t_0. Expression (28.102) is given in the book [239, p. 86], and the paper [238, p. 263].

Let us consider the case, when the employment $n(t)$ is described by the power-law expression

$$n(t) = n_0(t - t_0)^{\delta}. \tag{28.103}$$

Then equations (28.98) and (28.100) give the production

$$X(t) = kn_0(t - t_0)^{\delta}\left(Z^{1-a}(t_0) + \frac{1-a}{\delta+1}n_0(t - t_0)^{\delta+1}\right)^{a/(1-a)}. \tag{28.104}$$

If $n_0(t - t_0)^{\delta+1}$ is large relative to initial experience, then the rate of productivity growth is approximately equal to

$$R_X = \frac{a+\delta}{(1-a)(t-t_0)}, \tag{28.105}$$

where $t-t_0$ is the time after the start of production at t_0. If $\delta = 0$, then equation (28.105) gives expression (28.102).

The presence of memory in an economic process means that the behavior of the process depends not only on the present time, but also on the history of changes on finite time interval [450].

Equation (18.2) does not take into account the memory effects and memory fading, since the derivative $dZ(t)/dt$ is determined by behavior of $n(\tau)Z^a(\tau)$ only at $\tau = t$, and does not take into account the history of changes of $n(\tau)Z^a(\tau)$ in the past ($\tau < t$). To take into account of the changes of $n(\tau)Z^a(\tau)$ in the past, we can consider the equation

$$\frac{dZ(t)}{dt} = \int_{t_0}^{t} M(t-\tau)n(\tau)Z^a(\tau)\,d\tau, \tag{28.106}$$

where $M(t)$ is the memory function.

Let us consider the TRB type of the memory functions $M(t - \tau)$ that allows us to describe economic processes with memory for wide class of memory by using the fractional calculus.

The TRB-type memory functions can be represented as series of power-law memory functions. Using Theorem 4.1 of [484, p. 261], with $a = 0$ and $x = t - \tau$, the memory function $M(t)$, which belongs to TRB type, can be represented by the generalized Taylor series [484].

Using representation (28.17), we can consider the memory function $M(t-\tau)$ of the TRB type in the form

$$M(t-\tau) = \sum_{k=0}^{N} \frac{c_k}{\Gamma((k+1)\alpha)}(t-\tau)^{(k+1)\alpha-1} + \frac{((D^{\alpha}_{RL;0+})^{N+1}M)(\xi_N)}{\Gamma((N+1)\alpha+1)}(t-\tau)^{(N+1)\alpha}, \tag{28.107}$$

where last term in equation (28.107) is the remainder term with ξ_N such that $0 \leq \xi_N \leq t$, the operator $D^{\alpha}_{\mathrm{RL};0+}$ is the left-sided Riemann–Liouville fractional derivative of the order $0 \leq \alpha \leq 1$, and the coefficients c_k are defined by the equations

$$c_k = \Gamma(\alpha)(t^{1-\alpha}(D^{\alpha}_{\mathrm{RL};0+})^k M(t))(0+). \tag{28.108}$$

The proof of this statement is given in [484].

Example 28.7. For $N = 0$, memory function (28.107) is represented in the form

$$M(t-\tau) = \frac{c_0}{\Gamma(\alpha)}(t-\tau)^{\alpha-1} + \frac{(D^{\alpha}_{\mathrm{RL};0+}M)(\xi_0)}{\Gamma(\alpha+1)}(t-\tau)^{\alpha}, \tag{28.109}$$

where $0 \leq \xi_0 \leq T$, and

$$c_0 = \Gamma(\alpha) \lim_{t\to 0+} \frac{M(t)}{t^{\alpha-1}}. \tag{28.110}$$

In cases, when the remainder term can be neglected, the memory function (28.109) takes the form

$$M(t-\tau) = \frac{c_0}{\Gamma(\alpha)}(t-\tau)^{\alpha-1}. \tag{28.111}$$

Example 28.8. In the general case, all memory functions of TRB type can be represented (see Remark 4.1 of [484, p. 262]) by the generalized Taylor's series

$$M(t-\tau) = (t-\tau)^{\alpha-1} \sum_{k=0}^{\infty} \frac{c_k}{\Gamma((k+1)\alpha)}(t-\tau)^{k\alpha} \tag{28.112}$$

that holds for all $t \in (0, T]$, such that the series converges. The functions, which can be represented by equation (28.112), are called α-analytic in $t = 0$ [484].

As example of the memory functions, which are α-analytic, we can consider the kernel of the Abel-type fractional integral (see equation 37.1 in [335, p. 731]) and the kernel of the Prabhakar fractional integral [314, 199].

It should be noted that criteria for the existence of power-law type (PLT) memory in economic processes are suggested in paper [399]. We give the criterion of existence of power-law long-range dependence in time, and the criterion of existence of PLT memory for frequency domain.

28.8 Memory function of Prabhakar type

Let us consider the memory functions that contains the Prabhakar function [314, 199], which is also called the three-parameter Mittag–Leffler function [143].

For example, if the coefficients c_k of expression (28.112) have the form

$$c_k = \frac{\Gamma(\alpha k + \alpha)(\gamma)_k}{\Gamma(\rho k + \mu) k!}, \tag{28.113}$$

then equation (28.112) of the memory function defines the kernel of the Prabhakar fractional integral [314, 199] that has the form

$$M_{PI}(t - \tau) = (t - \tau)^{\mu - 1} E^{\gamma}_{\rho,\mu}[\omega(t - \tau)^{\rho}], \tag{28.114}$$

where $E^{\gamma}_{\rho,\mu}[z]$ is the Prabhakar function (the generalized Mittag–Leffler function [143]). This function was introduced by Tilak R. Prabhakar [314], and it can be represented in the form

$$E^{\gamma}_{\rho,\mu}[z] = \sum_{k=0}^{\infty} \frac{(\gamma)_k}{\Gamma(\rho k + \mu)} \frac{z^k}{k!}, \tag{28.115}$$

where $\rho, \mu, \gamma \in \mathbb{C}$ and $\mathrm{Re}(\rho) > 0$, and $(\gamma)_k$ is the Pochhammer symbol

$$(\gamma)_k = \gamma(\gamma + 1)(\gamma + 2) \cdots (\gamma + n - 1) = \frac{\Gamma(\gamma + k)}{\Gamma(\gamma)}. \tag{28.116}$$

The Prabhakar fractional integral is defined by the equation

$$(\mathcal{E}^{\gamma}_{\rho,\mu,\omega,a+} f)(t) = \int_a^t (t - \tau)^{\mu - 1} E^{\gamma}_{\rho,\mu}[\omega(t - \tau)^{\rho}] f(\tau)\, d\tau, \tag{28.117}$$

where $\rho, \mu, \gamma, \omega \in \mathbb{C}$ such that $\mathrm{Re}(\rho), \mathrm{Re}(\mu) > 0$.

We should note that operator (28.117) is bounded [199] in the space $L(a, b)$ of Lebesgue measurable functions on a finite interval $[a, b]$ of the real line and in the space $C[a, b]$ of continuous functions on $[a, b]$.

Let us describe some special cases of the Prabhakar fractional integral.

- For $\gamma = 0$, Prabhakar fractional integral (28.117) coincides with (see equation (1.16) in [199]) the Riemann–Liouville fractional integral

$$(\mathcal{E}^{0}_{\rho,\mu,\omega,a+} f)(t) = (I^{\mu}_{\mathrm{RL};a+} f)(t), \tag{28.118}$$

  where $I^{\mu}_{\mathrm{RL};a+}$ is the left-sided Riemann–Liouville fractional integral of the order α ($\mathrm{Re}(\alpha) > 0$) that is defined (see equation (2.1.1) of [200, p. 69]) by the equation

$$(I^{\mu}_{\mathrm{RL};a+} f)(t) = \frac{1}{\Gamma(\mu)} \int_a^t (t - \tau)^{\alpha - 1} f(\tau)\, d\tau. \tag{28.119}$$

- For $\gamma = 1$, we have the equality

$$E^{1}_{\rho,\mu}[\omega(t - \tau)^{\rho}] = E_{\rho,\mu}[\omega(t - \tau)^{\rho}], \tag{28.120}$$

and the Prabhakar fractional integral (28.117) takes the form

$$(\mathcal{E}^{1}_{\rho,\mu,\omega,a+}f)(t) = \int_{a}^{t} (t-\tau)^{\mu-1} E_{\rho,\mu}[\omega(t-\tau)^{\rho}] f(\tau)\, d\tau, \tag{28.121}$$

where $E_{\rho,\mu}[z]$ is the two-parameter Mittag–Leffler function [143].

- For $\gamma = 1$ and $\mu = 1$, we have the equality

$$E^{1}_{\rho,1}[\omega(t-\tau)^{\rho}] = E_{\rho}[\omega(t-\tau)^{\rho}], \tag{28.122}$$

and the Prabhakar fractional integral (28.117) can be represented by the equation

$$(\mathcal{E}^{1}_{\rho,1,\omega,a+}f)(t) = \int_{a}^{t} E_{\rho}[\omega(t-\tau)^{\rho}] f(\tau)\, d\tau, \tag{28.123}$$

where $E_{\rho}[z]$ is the classical Mittag–Leffler function [143].

- For $\gamma = 1$, $\mu = 1$, and $\rho = 1$, we get (see equation (1.8.3) in [200, p. 40]) the expression

$$E^{1}_{1,1}[\omega(t-\tau)^{\rho}] = E_{1}[\omega(t-\tau)] = \exp\{\omega(t-\tau)\}, \tag{28.124}$$

and the Prabhakar fractional integral (28.117) can be represented by the equation

$$(\mathcal{E}^{1}_{\rho,1,\omega,a+}f)(t) = \int_{a}^{t} \exp\{\omega(t-\tau)\} f(\tau)\, d\tau. \tag{28.125}$$

Note that operator (28.125) cannot describe the fading memory [384, 400, 382]. It can be interpreted as an integral operator of the first order with exponentially distributed lag [400].

- For $\rho = 1$, we have (see equation (5.1.18) in [143, p. 99]) the equality

$$E^{\gamma}_{1,\mu}[z] = \frac{1}{\Gamma(\mu)}\, {}_1F_1(\gamma,\mu,z), \tag{28.126}$$

where ${}_1F_1(\gamma,\mu,z)$ is the confluent hypergeometric Kummer function, which is defined by the equation

$${}_1F_1(\gamma,\mu,z) = \sum_{k=0}^{\infty} \frac{(\gamma)_k}{(\mu)_k} \frac{z^k}{k!}, \tag{28.127}$$

where $\gamma,\mu,z \in \mathbb{C}$ and $\mathrm{Re}(\mu) \neq 0,-1,-2,\ldots$, and $(\gamma)_k$ is the Pochhammer symbol. In this case, the memory function (28.114) is the kernel of the Abel-type fractional integral (see equation (37.1) in [335, p. 731]) that has the form

$$M_{AT}(t-\tau) = \frac{(t-\tau)^{\mu-1}}{\Gamma(\mu)}\, {}_1F_1(\gamma,\mu,\omega(t-\tau)). \tag{28.128}$$

Note that the Abel-type fractional integrals and derivatives describe processes with the simultaneous action of the power-law fading memory and the gamma distributed lag [400, 404, 402, 407, 405].

The fractional derivatives, which are left-inverse operators for the Prabhakar fractional integrals, were first proposed in 2004 by Anatoly A. Kilbas, Megumi Saigo and Ram K. Saxena in work [199].

The operator, which is left inverse for the Prabhakar fractional integral, is defined (see equation (6.5) and Theorem 9 in [199, p. 47]) in the form

$$(\mathbf{D}^{\gamma}_{\rho,\mu,\omega,a+}f)(t) = (D^{\mu+\nu}_{\mathrm{RL};a+}\mathcal{E}^{-\gamma}_{\rho,\nu,\omega,a+}f)(t) = D^{\mu+\nu}_{\mathrm{RL};a+}\int_a^t (t-\tau)^{\nu-1}E^{-\gamma}_{\rho,\nu}[\omega(t-\tau)^{\rho}]f(\tau)\,d\tau, \tag{28.129}$$

where $\rho, \mu, \gamma, \nu \in \mathbb{C}$, $\mathrm{Re}(\mu) > 0$, $\mathrm{Re}(\nu) > 0$, $\mathrm{Re}(\rho) > 0$, and $D^{\mu+\nu}_{\mathrm{RL};a+}$ is the Riemann–Liouville derivative of the order $\mu+\nu$. The function $f(\tau)$ belongs to the space $L(a,b)$ of Lebesgue measurable functions on a finite interval $[a,b]$ of the real line. This statement was proved in [199] (see proof of Theorem 9 in [199, p. 47]). Equation (28.129) is equation (6.5) in [199, p. 47].

The fractional derivative (28.129) can be represented in the following equivalent form:

$$(\mathbf{D}^{\gamma}_{\rho,\mu,\omega,a+}f)(t) = \frac{d^n}{dt^n}\int_a^t (t-\tau)^{n-\mu-1}E^{-\gamma}_{\rho,n-\mu}[\omega(t-\tau)^{\rho}]f(\tau)\,d\tau, \tag{28.130}$$

where $n > \mu$ and $\rho, \mu, \gamma, \omega \in \mathbb{C}$, with $\mathrm{Re}(\mu) > 0$, $\mathrm{Re}(\rho) > 0$. Note that $\gamma \in \mathbb{C}$, and Therefore, we can consider operators with positive and negative values of γ.

Remark 28.1. In article [199], operator (28.130), which is left inverse for the Prabhakar fractional integral, is called the Prabhakar fractional derivative of Riemann–Liouville type. However, this operator was not proposed in the works of Prabhakar. The first time this operator was proposed by Anatoly A. Kilbas, Megumi Saigo and Ram K. Saxena in article [199] in 2004. Therefore, this operator (28.130) can be called as the Kilbas–Saigo–Saxena (KSS) fractional derivative [406, p. 17]. These operators can be considered as generalized fractional derivatives involving generalized Mittag–Leffler function in the kernels [199]. Note that the fractional differential and integral operators with the Prabhakar function in the kernel are also considered in [198, 352, 310, 132, 138].

We should note that the kernel of the Kilbas–Saigo–Saxena fractional derivative (28.130) can demonstrate three types of behavior at zero [406, p. 17]. This behavior significantly distinguishes this operator from other fractional derivatives, which usually have a singularity at zero [406].

For the fractional derivative (28.130), we can consider negative values of the parameter γ ($-\gamma > 0$), and we can use the following special cases.

- For $\gamma = 0$, the fractional derivative (28.130) coincides with the Riemann–Liouville fractional derivative.
- For $\gamma = -1$, we get the fractional derivative (28.130) with the two-parameter Mittag–Leffler function in the kernel.
- For $\gamma = -1$ and $n - \mu = 1$, we get the fractional derivative (28.130), the kernel of which is the classical Mittag–Leffler function.
- For $\gamma = -1$, $n - \mu = 1$, and $\rho = 1$, we get the integer-order derivatives with exponentially distributed lag [400].
- For $\rho = 1$, we get the fractional derivative (28.130) with the confluent hypergeometric Kummer function in the kernel, which can be used to describe processes with the simultaneous action of power-law memory and distributed lag [400, 404, 402, 407, 405].

We should emphasize that the Kilbas–Saigo–Saxena fractional derivative (28.130) is a left-inverse operator to the Prabhakar fractional integral (28.117), such that

$$(\mathbf{D}^{\gamma}_{\rho,\mu,\omega,a+}\mathcal{E}^{\gamma}_{\rho,\mu,\omega,a+}f)(t) = f(t). \tag{28.131}$$

The proof of this equality is given as Theorem 9 in [199, p. 47].

In fractional economic dynamics, we can also use the Prabhakar fractional derivative of Caputo type (the regularized Prabhakar fractional derivative), which was proposed by Mirko D'Ovidio, Federico Polito in [96] in 2013 (see also [94, 95, 137]).

28.9 Model of learning with memory of Prabhakar type

In order to describe the possible connection between evidence of learning on individual product lines and productivity growth in an economy, Robert E. Lucas considers [239, 238] the labor-only technology.

Let us consider a generalization of the Lucas model for learning with memory function (28.114) that was proposed in article [391, 389]. To describe the possible connection between evidence of learning on individual product lines and productivity growth with memory, we consider [391, 389] the equation of the labor-only technology with memory in the form

$$\frac{dZ(t)}{dt} = \int_{t_0}^{t} (t-\tau)^{\mu-1} E^{\gamma}_{\rho,\mu}[\omega(t-\tau)^{\rho}]\, n(\tau) Z^{a}(\tau)\, d\tau \tag{28.132}$$

that can be rewritten as

$$\frac{dZ(t)}{dt} = (\mathcal{E}^{\gamma}_{\rho,\mu,\omega,t_0+} n(\tau) Z^{a}(\tau))(t). \tag{28.133}$$

Equation (28.133) is a generalization of the equation (28.106) of the standard Lucas learning model without memory.

Using property (28.131), the action of the Kilbas–Saigo–Saxena fractional derivative (28.130) on equation (28.133) gives

$$(\mathbf{D}^{\gamma}_{\rho,\mu,\omega,t_0+}Z^{(1)})(t) = n(t)Z^{a}(t). \tag{28.134}$$

Equation (28.134) is nonlinear fractional differential equation with the Kilbas–Saigo–Saxena fractional derivative (28.130).

Let us consider the model of learning with Prabhakar memory function is described by the nonlinear fractional differential equation (28.134), where the employment $n(t)$ is described by equation

$$n(t) = n_0\Gamma(\beta-\mu)(t-t_0)^{\delta}E^{-\gamma}_{\rho,\beta-\mu}[\omega(t-t_0)^{\rho}]. \tag{28.135}$$

We use the coefficient $\Gamma(\beta-\mu)$ in order to have $n(t) = n_0(t-t_0)^{\beta}$ for the case $\gamma = 0$.

To obtain an analytical expression of the exact solutions of nonlinear equation (28.134) with the employment (28.135), we will search solution in the power-law form

$$Z(t) = Z_0(t-t_0)^{\beta}. \tag{28.136}$$

To obtain solution of equation (28.134), we will use the equation for the Kilbas–Saigo–Saxena (KSS) fractional derivative (28.130) of the power-law functions that is given by the expression

$$(\mathbf{D}^{\gamma}_{\rho,\mu,\omega,a+}(\tau-a)^{\beta-1})(t) = \Gamma(\beta)(t-a)^{\beta-\mu-1}E^{-\gamma}_{\rho,\beta-\mu}[\omega(t-a)^{\rho}], \tag{28.137}$$

if $\mathrm{Re}(\beta) > \mathrm{Re}(\mu) > 0$, where $\mathrm{Re}(\rho) > 0$, and $\mathrm{Re}(\mu) < n \in \mathbb{N}$. Equation (28.137) is proved in [391, 389].

The first-order derivative for equation (28.136) gives

$$Z^{(1)}(t) = \beta Z_0(t-t_0)^{\beta-1}. \tag{28.138}$$

Using equation (28.137), we get

$$(\mathbf{D}^{\gamma}_{\rho,\mu,\omega,t_0+}Z^{(1)})(t) = \beta Z_0(\mathbf{D}^{\gamma}_{\rho,\mu,\omega,t_0+}(\tau-t_0)^{\beta-1})(t) = \beta Z_0\Gamma(\beta)(t-t_0)^{\beta-\mu-1}E^{-\gamma}_{\rho,\beta-\mu}[\omega(t-t_0)^{\rho}], \tag{28.139}$$

if $\mathrm{Re}(\beta) > \mathrm{Re}(\mu) > 0$, where $\mathrm{Re}(\rho) > 0$, and $\mathrm{Re}(\mu) > n \in \mathbb{N}$.

In case (28.135) and (28.136), the right side of equation (28.134) is given by the expression

$$n(t)Z^{a}(t) = n_0Z_0^{a}\Gamma(\beta-\mu)(t-t_0)^{a\beta+\delta}E^{-\gamma}_{\rho,\beta-\mu}[\omega(t-t_0)^{\rho}]. \tag{28.140}$$

We can see that equation (28.139) and (28.140), give the same expression, if the following conditions are satisfied

$$\beta Z_0 \Gamma(\beta) = n_0 Z_0^a \Gamma(\beta - \mu), \tag{28.141}$$

$$\beta - \mu - 1 = a\beta + \delta. \tag{28.142}$$

Using the equality $\Gamma(\beta + 1) = \beta\Gamma(\beta)$, and

$$\beta = \frac{\delta + \mu + 1}{1 - a}, \tag{28.143}$$

we get

$$Z_0 = \left(\frac{n_0 \Gamma(\beta - \mu)}{\Gamma(\beta)} \right)^{1/(1-a)}. \tag{28.144}$$

As a result, the cumulative experience in the production of a good is described [391, 389] by the equation

$$Z(t) = Z_0 (t - t_0)^{\beta} = \left(\frac{n_0 \Gamma(\beta - \mu)}{\Gamma(\beta)} \right)^{1/(1-a)} (t - t_0)^{(\delta+\mu+1)/(1-a)}, \tag{28.145}$$

which is the solution of equation (28.134), where the dynamics of the employment $n(t)$ is described by (28.135).

Using equation (28.145), the rate of production of a good is given by the equation

$$\begin{aligned} X(t) &= kn(t) Z^a(t) = kn_0 Z_0^a (t - t_0)^{\beta-\mu-1} E^{-\gamma}_{\rho,\beta-\mu}[\omega(t - t_0)^{\rho}] \\ &= kn_0 \left(\frac{n_0 \Gamma(\beta - \mu)}{\Gamma(\beta)} \right)^{a/(1-a)} (t - t_0)^{(\delta+a(\mu+1))/(1-a)} E^{-\gamma}_{\rho,\beta-\mu}[\omega(t - t_0)^{\rho}], \end{aligned} \tag{28.146}$$

where we used (28.142), and the parameter β is defined by equation (28.143). Let us use equation (5.1.15) of [143, p. 99] that has the form

$$\left(\frac{d}{dt} \right)^n (z^{\beta-1} E^{\gamma}_{\alpha,\beta}[\omega z^{\alpha}]) = z^{\beta-n-1} E^{\gamma}_{\alpha,\beta-n}[\omega z^{\alpha}], \tag{28.147}$$

where $\mathrm{Re}(\beta) > n$, $n \in \mathbb{N}$, $\alpha, \beta, \gamma, z, \omega \in \mathbb{C}$. Using equation (28.147), we get

$$X^{(1)}(t) = kn_0 Z_0^a (t - t_0)^{\beta-\mu-2} E^{-\gamma}_{\rho,\beta-\mu-1}[\omega(t - t_0)^{\rho}], \tag{28.148}$$

if $\beta - \mu > 1$.

Using equations (28.146) and (28.148), the rate of productivity growth with the memory, which is described by the memory function (28.114), is approximately equal to

$$R_X(\gamma, \rho, \mu, \beta) = \frac{E^{-\gamma}_{\rho,\beta-\mu-1}[\omega(t - t_0)^{\rho}]}{(t - t_0) E^{-\gamma}_{\rho,\beta-\mu}[\omega(t - t_0)^{\rho}]}, \tag{28.149}$$

where $t - t_0$ is the time after the start of production at t_0.

For $\gamma = -1$, $n - \mu = 1$, $\beta - \mu = 1$ and $\rho = 1$, we get the rate of production of a good in the form

$$X(t) = kn_0 Z_0^a \exp\{\omega(t - t_0)\}, \tag{28.150}$$

and the rate of productivity growth has the form

$$R_X(-1, 1, n - 1, n) = \omega. \tag{28.151}$$

Note that in this case, there is no memory, since the Prabhakar integral operator with $\gamma = -1$, $n - \mu = 1$, $\beta - \mu = 1$ and $\rho = 1$, describes the exponential delay, and cannot describe fading memory [384].

For the case $\gamma = 0$, we get that the rate (28.149) takes the form

$$R_X(0, \rho, \mu, \beta) = \frac{\Gamma(\beta - \mu)}{(t - t_0)\Gamma(\beta - \mu - 1)} = \frac{\beta - \mu - 1}{t - t_0}, \tag{28.152}$$

where we used $\Gamma(z + 1) = z\Gamma(z)$ and the equality

$$E^0_{\rho,\beta-\mu}[\omega(t - t_0)^\rho] = \frac{1}{\Gamma(\beta - \mu)}. \tag{28.153}$$

As a result, using equation (28.143), the rate of productivity growth at $t > t_0$ is approximately equal to

$$R_X(0, \rho, \mu, \beta) = \frac{(\mu + 1)a + \delta}{(1 - a)(t - t_0)} \tag{28.154}$$

instead of (28.105). For $\mu = 0$, expression (28.154) takes the form

$$R_X(0, \rho, 0, \beta) = \frac{a + \delta}{(1 - a)(t - t_0)} \tag{28.155}$$

Expression (28.155) describes the rate of growth without memory ($mu = 0$ or $\alpha = 1$). As a result, we see that equation (28.154) with $\mu = 0$ ($\alpha = 1$) gives expression (28.105).

Let us compare expressions (28.154) and (28.155). The ratio of $R(\alpha)$ and $R(1)$ is expressed by the equation

$$\frac{R_X(0, \rho, \mu, \beta)}{R_X(0, \rho, 0, \beta)} = \frac{(\mu + 1)a + \delta}{a + \delta} = 1 + \frac{a\mu}{a + \delta}, \tag{28.156}$$

where $a \in (0, 1)$. For example $\delta = 0$, relation (28.156) gives

$$\frac{R_X(0, \rho, \mu, \beta)}{R_X(0, \rho, 0, \beta)} = 1 + \mu. \tag{28.157}$$

Using equation (17.124), we can formulate the following principle.

Principle 28.3 (Principle of changing growth rates by memory). *In the case $\mu \in (0,1)$, the rate of growth $R_X(0,\rho,\mu,\beta)$ with memory can be greater than the rate of growth $R_X(0,\rho,0,\beta)$ without memory for $t > t_0$, if the parameter of memory fading $\alpha = \mu + 1$ is more than one ($\alpha > 1$) and $\delta \geq -a$.*

As a result, we can conclude that in learning models, memory effects can significantly change growth rates and we should not neglect the memory in economic models.

28.10 Conclusion

In this chapter, an approach to describe processes with memory of the general form by using the fractional calculus is suggested. This approach is based on the generalized Taylor series that has been proposed by J. J. Trujillo, M. Rivero and B. Bonilla. It has been proved that equation of the generalized accelerator with the memory of TRB type can be represented by as a composition of actions of the accelerator with power-law memory and the parallel action of multipliers with power-law memory. This proposed approach has been applied to generalize the Harrod–Domar and Lucas models with memory. The proposed approach makes it possible to expand the possibilities of fractional calculus for describing economic processes with wide type of memory functions.

Let us give some other examples of possible generalizations of the economic models with memory.

(1) We can use the fractional integration and differentiation of variable order. This allows us to describe economic processes with memory, in which the fading parameter changes with time.

(2) In the general case, the parameter of memory fading can be distributed over a certain interval. This distribution is described by a weight function that characterizes distribution of the memory fading on a set of economic agents. The influence of this distribution is important for the economics, since various types of economic agents may have different parameters of memory fading. In this case, we should use the fractional derivatives and integrals of distributed orders.

(3) To consider economic models with general memory function, we can use the fractional derivatives and integrals with the Mittag–Leffler, Prabhakar, Kummer (confluent hypergeometric) and other functions in the operator kernels.

In the general case, it is important to understand how a change in the type of memory function can affect the behavior of economic processes with memory.

Part VIII: **Instead of conclusion**

29 Fractional calculus in economics and finance

In this chapter, we note some published works on the application of fractional calculus in economics and finance. Not all of these works are directly connected with the economics with memory. Most of these works are not related to attempts to describe the phenomena of memory in finance and economics. Most part of these works is devoted to fractional generalizations of equations and models that are used in finance and not economics. Another part of works is devoted to equations and models with nonlocality in space, and not in time, due to this it is not directly related to the memory effects. Almost all of these works do not propose generalizations of economic concepts. Not many of them offer detailed economic conclusions (and interpretations) from obtained mathematical results. However, these works may be of interest for the further development of the economics with memory. These works can be conditionally divided into the following approaches: ARFIMA; Fractional Brownian Motion; Econophysics; Deterministic Chaos; Economics with Memory. Note that these approaches can be considered, in a sense, as stages of formation of economics with memory. This chapter is based on article [385] (see also book [254, pp. 5–32]).

29.1 ARFIMA approach

This approach is characterized by economic models with discrete time and application of the Grunwald–Letnikov fractional differences.

For the first time the importance of long-range time dependence in economic data was recognized by Clive W. J. Granger in his preprint [147] and article [148] (see also [148, 152, 155, 153]). To describe economic processes with long memory Granger and Joyeux [155] in 1980 (see also [180]) proposed the fractional ARIMA models, which are also called ARFIMA(p, d, q). Fractional ARIMA models allow use of statistical methods for describing processes with memory [29, 295, 30]. In 2003, C. W. J. Granger received the Nobel memorial prize in economic sciences "for methods of analyzing economic time series with common trends (cointegration)" [479].

To describe economic processes with memory Granger, and Joyeux [155, 153, 180] proposed the difference operators that were called the fractional differencing and integrating. Then these operators began to be used in various economic models with discrete time (e.g" see books [29, 295, 30, 481, 485] and reviews [151, 21, 303, 23, 23]).

These fractional operators were proposed in 1980–1981 [155, 180], and began to be used up to the present time without any connection with the fractional calculus and the well-known fractional differences of noninteger orders. In fact, the fractional differencing and integrating of Granger and Joyeux are the well-known Grunwald–Letnikov fractional differences, which were suggested in 1867 and 1868 in works [157, 222]. These fractional differences are actively used in the fractional calculus [335, 308, 200]. It is known that in the continuous limit these fractional differences of positive

https://doi.org/10.1515/9783110627459-029

orders give the fractional derivatives, which are called the Grunwald–Letnikov derivatives [335, 308, 200].

Among economists, the approach proposed by Granger and Joyeux (and the discrete operators proposed by them) is the most common and is used without an explicit connection with the development of modern mathematics and fractional calculus. It is obvious that the restriction of mathematical tools only to the Grunwald–Letnikov fractional differences significantly reduces the possibilities for studying processes with memory and nonlocality in time. Moreover, in the framework of the ARFIMA approach, new economic concepts were not proposed (such as marginal value with memory, elasticity with memory, multiplier with memory, accelerator with memory, and others) to describe processes with memory in the modern economics. Unfortunately, in the ARFIMA approach, economics with memory was not formed as a separate branch of economics that is based on new concepts, notions, principles, rules and models for economic processes with memory.

29.2 Fractional Brownian motion approach

This approach is characterized by financial models and application of stochastic calculus methods and stochastic differential equations.

Starting with the article by L. C. G. Rogers [324] in 1997, various authors have begun to use fractional Brownian motion (fBm) to describe different financial processes. The fractional Brownian motion is not a semimartingale and the stochastic integral with respect to it is not well-defined in the classical Ito's sense. Therefore, this approach is connected with the development of fractional stochastic calculus [97, 486, 274].

At present time, this approach, which can be also called as a "fractional mathematical finance" [385], is connected with the development of fractional stochastic calculus, the theory fractional stochastic differential equations and their application in finance. The fractional mathematical finance is a branch of applied mathematics, concerned with mathematical modeling of financial markets by using the fractional stochastic differential equations.

Here, we can note the fractional generalization of the Black–Scholes pricing model. For the first time a fractional generalization of the Black–Scholes equation was proposed by Walter Wyss [506] in 2000. Wyss [506] considered the pricing of option derivatives by using the time-fractional Black–Scholes equation. The Black–Scholes equation was generalized by replacing the first derivative in time by a fractional derivative in time of the order $\alpha \in (0, 1)$. The solution of this fractional Black–Scholes equation was considered in paper [506]. However, there are no financial reasons in the Wyss paper to explain why the time-fractional derivative should be used, and memory effects are not discussed.

The works of Cartea and Meyer-Brandis [61, 59] proposed a stock price model that uses information about the waiting time between trades. In this model, the arrival of trades is driven by a counting process, in which the waiting-time between trades possesses is described by the Mittag–Leffler survival function (see also [5]). In the paper [59], Cartea proposed that the value of derivatives satisfies the fractional Black–Scholes equation that contains the Caputo fractional derivative with respect to time. It should be noted that, in general, the presence of a waiting time and a delay time does not mean the presence of memory in the process.

In the framework of fractional Brownian motion approach, a lot of papers [276, 277, 206, 473, 60, 228, 33, 262], [215, 251, 34, 468, 183, 98, 512, 332, 190, 205, 213, 4, 3] and books [331, 118] can be used to describe financial processes with memory and nonlocality.

As a rule, in the fractional mathematical finance, dynamic models are created without establishing links with economic theory and without formulating new economic or financial concepts, taking only observable market prices as input data. In the fractional mathematical finance approach, the compatibility with economic theory is not the key point.

29.3 Econophysics approach

This approach is characterized by financial models and application of physical methods and equations.

At present time, this approach can be considered as new branch of the econophysics, which is connected with application of fractional calculus. In fact, this branch, which can be called fractional econophysics, was born in 2000 and it can be primarily associated with the works of Francesco Mainardi, Rudolf Gorenflo, Enrico Scalas and Marco Raberto [341, 248, 141] on the continuous-time finance. In fractional econophysics, the fractional diffusion models [341, 248, 141] are used in finance, where price jumps replace the particle jumps in the physical diffusion model. The corresponding stochastic models are called continuous time random walks (CTRWs), where random walks are also incorporate a random waiting time between jumps. In finance, the waiting times describes delay between transactions. These two random variables (price change and waiting time) are used to describe the long-time behavior in financial markets. The diffusion limit, which is used in physics, is considered for continuous time random walks [341, 248, 141]. It was shown that the probability density function for the limit process obeys a fractional diffusion equation [341, 248, 141].

After the pioneering works [341, 248, 141] that laid the foundation for the new direction of econophysics (fractional econophysics), various papers were written on the application of fractional dynamics methods and physical models to describe processes in finance and economics (e. g., see [142, 319, 342, 340, 261, 218, 504, 307, 22]). The history and achievements of the econophysics approach in the first 5 years are described by Enrico Scalas in the article [339] in 2006, and in the article of Francesco Mainardi [246], [254, pp. 33–42].

The fractional econophysics, as a branch of econophysics, can be defined as a new direction of research applying methods developed in physical sciences, to describe processes in economics and finance, in which the effects of fading memory, nonlocality in time and spatial nonlocality are manifested. For example, the study of continuous time finance by using methods and results of fractional kinetics and anomalous diffusion. Another example, which is not related to finance, is the time-dependent fractional dynamics with memory in quantum and economic physics [398].

In this approach, the fractional calculus was applied mainly to financial processes. In the papers on fractional econophysics, generalization of basic economic concepts and principles for economic processes with memory are not suggested.

29.4 Deterministic chaos approach

This approach is characterized by financial (and economic) models and application of methods of nonlinear dynamics. Strictly speaking, this approach can be attributed to the econophysics approach.

Nonlinear dynamics models are useful to explain irregular and chaotic behavior of complex economic and financial processes. The complex behavior of nonlinear economic processes does not allow in many cases to use analytical methods to study nonlinear economic models.

In 2008, for the first time, Wei-Ching Chen proposed in [64] a generalization of a financial model with deterministic chaos. Chen [64] studied the fractional-dynamic behaviors and describes fixed points, periodic motions, chaotic motions, period doubling and intermittency routes to chaos in the financial process that is described by system of three equations with the Caputo fractional derivatives. He demonstrates by numerical simulations that chaos exists when orders of derivatives are less than 3 and that the lowest order at which chaos exists was 2.35.

In the framework of the deterministic chaos approach, many papers [1, 65, 500, 296, 243, 84, 508, 510, 182, 158, 358, 502, 344] can be used to describe financial processes with memory, in spite of the fact that such an interpretation is not considered in most of these works. In some of the papers [83, 501, 25, 173, 86, 87, 509], economic models were considered.

We should note that the various approaches did not develop in complete isolation from each other. For example, for the fractional Chen model of dynamic chaos in the economy, Tomas Skovranek, Igor Podlubny and Ivo Petras [349] applied the concept of the state space (the configuration space, the phase space) that arose in physics more than 100 years ago. As state variables authors consider the gross domestic product, inflation and unemployment rate. The dynamics of the modeled economy in time, which is represented by the values of these three variables, is described as a trajectory in state-space. The system of three fractional order differential equations is used to describe the dynamics of the economy by fitting the available economic data. Then Jose

A. Tenreiro Machado, Maria E. Mata and Antonio M. Lopes developed the concept of the state space in the papers [465, 466, 471, 472]. The economic growth is described by using the multidimensional scaling (MDS) method for visualizing information in data. The state space is used to represent the sequence of points (the fractional state space portrait (fSSP), and pseudo phase plane (PPP)) corresponding to the states over time.

29.5 Economics with memory

This approach is characterized by economic models with continuous time and generalization of basic economic concepts and notions.

From our subjective point of view, the formation of new basic economic concepts and notions of 'economics with memory" began with a proposal of generalizations of the basic economic concepts and notions at the beginning of 2016, when the concept of elasticity for economic processes with memory was proposed [418] and [430, 419]. Then in 2016, we proposed the concepts of the marginal values with memory [424, 416, 425], the concepts of accelerator and multiplier with memory [412, 417], the nonlocal measures of risk aversion [428] and nonlocal deterministic factor analysis [414]. In 2016, these concepts are used in generalizations of some standard economic models [420, 421, 426, 423, 427, 422] that allow us to describe the dynamics of economic processes with memory.

It should be noted that formal replacements of derivatives of integer order by fractional derivatives in standard differential equations, which describe economic processes, and solutions of the obtained fractional differential equations were considered in papers published before 2016. However, these papers were purely mathematical works, in which generalizations of economic concepts and notions were not proposed. In these works, fractional differential equations have not been derived, since a formal replacement of integer-order derivatives by fractional derivatives cannot be recognized as a derivation of the equations. It is not enough to generalize the differential equations describing the dynamic model. It is necessary to generalize the whole scheme of obtaining (all steps of derivation) these equations from the basic principles, concepts and assumptions. In this sequential derivation of the equations, we should take into account the nonstandard characteristic properties of fractional derivatives and integrals. If necessary, generalizations of the notions, concepts and methods, which are used in this derivation, should also be obtained. Formulations of economic conclusions and interpretations from the obtained solutions are not usually suggested in these papers. Examples of incorrectness and errors in such generalizations are given in review article [386] (see also book [254, pp. 43–92]).

The most important purpose of the modern stage of development of economics with memory is the inclusion of memory into the economic theory, into the basic economic concepts and methods. The economics should be extended and generalized such that it takes into account the memory and nonlocality in time. Fractional generalizations of standard economic models should be constructed only on this conceptual basis. The most important purpose of studying such generalizations, which take into account fading memory, is the search and formulation of qualitatively new effects and phenomena caused by memory in the behavior of economic processes. In this case, these results of economics with memory can be further used in computer simulations of real economic processes and in econometric studies.

During 2016–2020, we defined the basic concepts of economics with memory and described the dynamics of economic processes with memory in the framework of generalizations of the basic economic models. Let us give a list of some generalizations of economic concepts and generalizations of economic models that have already been proposed in these years.

The list of economic concepts, which were proposed in 2016–2020, primarily includes the following concepts:

1. the marginal value of noninteger order [424, 416, 425, 439, 446] with memory and nonlocality;
2. the economic multiplier with memory [412, 397];
3. the economic accelerator with memory [412, 397];
4. the exact discretization of economic accelerators and multipliers [441, 442, 431, 432] based on exact fractional differences;
5. the accelerator with memory and periodic sharp bursts [417, 436, 444];
6. the duality of the multiplier with memory and the accelerator with memory [412, 397];
7. the accelerators and multipliers with memory and distributed lag [405, 404, 402];
8. the elasticity of fractional order [418, 430, 429, 419] for processes with memory and nonlocality;
9. the measures of risk aversion with nonlocality and with memory [428, 449];
10. the warranted (technological) rate of growth with memory [450, 447, 445, 453, 393, 387];
11. the nonlocal methods of deterministic factor analysis [414, 452];
12. the productivity with fatigue and memory [454];
13. the chronological memory ordering [433, 398];
14. and some others.

The use of these notions and concepts makes it possible for us to generalize some classical economic models.

The list of economic models, which were proposed in 2016–2020, primarily includes the following models with memory:

1. the natural growth model with memory [420, 440];
2. the growth model with constant pace and memory [447, 438];
3. the Harrod–Domar model with memory [421, 426] and [450, 445, 453, 393];
4. the Keynes model with memory [423, 427, 422] and [402, 407];

5. the dynamic Leontief (intersectoral) model with memory [437, 451, 433, 398];
6. the model of dynamics of fixed assets (or capital stock) with memory [447, 438];
7. the model of price dynamics with memory [438];
8. the logistic growth model with memory [386] (see also [403]);
9. the model of logistic growth with memory and periodic sharp splashes (kicks) [444];
10. the time-dependent dynamic intersectoral model with memory [433, 398];
11. the Phillips model with memory and distributed lag [405];
12. the Harrod–Domar growth model with memory and distributed lag [404];
13. the dynamic Keynesian model with memory and distributed lag [402, 407];
14. the model of productivity with fatigue and memory [454];
15. the Solow–Swan model with memory [386, 383];
16. the Kaldor-type model of business cycles with memory [386];
17. the Evans model with memory [390];
18. the Cagan model with memory [388];
19. the Lucas models with memory [391, 389, 381];
20. and some other economic models.

In our works, which are published in 2016–2020, new concepts, principles and model are proposed for economic dynamics with memory. Qualitatively new effects due to the presence of memory in the economic process are described. Most of the introduced new concepts, principles and models of economics with memory are described in this book.

Let us also note works, in which generalizations of economic models, were proposed without introducing new economic notions and concepts.

1. Michele Caputo proposed some fractional dynamic economic models in works [47, 48, 51, 49, 52, 50, 53, 54, 58, 55]: A generalization of the standard equation of relaxation to equilibrium, where the memory was introduced in the reactivity term; the continuous-time IS-LM model with memory; the tax version of the Fisher model with memory for stock prices and inflation rates. Some of proposed models can be considered as econophysics approach, which are based on fractional generalization of the standard damped harmonic oscillator equation, where the memory has been introduced in the frictional term by using fractional derivative instead of first-order derivative.
2. Mathematical description of some fractional generalization of economic models was proposed by the Kabardino–Balkarian group: Adam M. Nakhushev [280] (and Section 4.6 of [281, pp. 145–153]), Khamidbi Kh. Kalazhokov [186], Zarema A. Nakhusheva [282].
3. Mathematical description of some fractional generalization of economic models were proposed by the Kamchatka group: Viktoriya V. Samuta, Viktoriya A. Strelova and Roman I. Parovik [336], Yana E. Shpilko, Anastasiya E. Solomko, Roman I. Parovik [346], Danil M. Makarov [249].

4. The one-parameter Mittag–Leffler function is proposed by Tomas Skovranek [348] to describe the relation between the unemployment rate and the inflation rate, also known as the Phillips curve. Shiou-Yen Chu and Christopher Shane proposed the hybrid Phillips curve model with memory to describe dynamic process of inflation with memory in work [74].
5. Rituparna Pakhira, Uttam Ghosh and Susmita Sarkar derived [290, 288, 289, 291, 292, 293, 294] some inventory models with memory.
6. Computer simulation for modeling the national economies in the framework of the fractional generalizations of the Gross domestic product (GDP) model was proposed by Ines Tejado, Duarte Valerio, Nuno Valerio, Emiliano Perez [456, 457, 458, 459, 460, 461, 462, 463] and by Dahui Luo, JinRong Wang, Michal Feckan [242], and Hao Ming, JinRong Wang, Michal Feckan [273].
7. We also note the economic models that were proposed in works [83, 501, 25, 173, 86, 87, 509, 507] that are related to the "deterministic chaos" approach.

The most important elements in the construction of the economics with memory as a branch of economics are the emergence and the formation of new notions, concepts, effects, phenomena, principles and methods, which are specific only to this branch.

Let us note that the problems and difficulties arising in the construction of fractional generalizations of standard economic models by using the fractional calculus to take into account a memory and nonlocality in time are described in review [386] (see also the book [254, pp. 43–91]).

29.6 Conclusion

In our opinion, this stage of the development of economics with memory actually includes (absorbs) the approach based on ARFIMA model using only the Grunwald–Letnikov fractional differences. This opinion is based on the obvious fact that Economics with Memory allows the AFRIMA approach to go beyond the restrictions of the Grunwald–Letnikov operators, and use different types of fractional finite differences and fractional derivatives of noninteger orders.

Economics with memory can also include (absorbs) approaches based on the fractional econophysics and deterministic chaos. For the econophysics approach, new opportunities are opening up on the way to formulating economic analogues of physical concepts and notions that will be more understandable to economists. This will significantly simplify the implementation of the concepts and methods of fractional econophysics in economic theory and application.

Moreover, the standard economics, which does not take into account the fading memory, can be considered as a special case of the economics with memory, when the parameter of memory fading takes integer values. This statement is based on the fact that fractional differential equations of economic models with memory include differential equations of integer orders that describe standard models. The situation is similar for generalized economic concepts. For example, a marginal value with a power-law memory contains the standard marginal and average values as special cases corresponding to the fading parameter equal to zero and one.

The most important element in the construction of the economics with memory as a new branch of economics is the emergence and the formation of new notions, concepts, effects, phenomena, principles and methods, which are specific only to this theory. Economics with memory is a new science because it has something of its own that is not in other economic sciences.

30 Future directions of economics with memory

In place of the Conclusion, which summarizes what was written, we decided to speculate a bit about open questions and future tasks in the construction of economics with memory.

In this chapter, we propose some assumptions about possible direction of development of the economics with memory. The proposed areas of development are primarily associated with the underdevelopment of the mathematical apparatus for describing economic processes with memory, poor development of the observational base and analysis methods for such processes. Because of this, the various areas of the economics and the possibilities of their modification to take into account the effects of memory and nonlocality in time are not discussed. In our opinion, in almost all areas of the modern economics, to one degree or another, the effects of memory should be taken into account. This chapter is based on the works [385], [254, pp. 5–32].

30.1 Self-organization in economics with memory

An important direction in the development of theory of economic processes with memory is the construction of models, in which the economic agents are represented as a self-organizing set of interacting objects with memory [387]. For example, this approach can be used to generalize models of self-organization in economics [513, 207, 353] by taking into account a fading memory.

30.2 Simultaneous action of distributed lag and fading memory

The continuously distributed lag has been considered in economics starting with the works of Michal A. Kalecki [189] and Alban W. H. Phillips [305, 306]. The continuous uniform distribution of delay time is considered by M. A. Kalecki [189] (see also Sections 8.4 of [10, pp. 251–254]) for dynamic models of business cycles in 1935. The continuously distributed lag with the exponential distribution of delay time is considered by A. W. H. Phillips [305, 306] in 1954. The operators with continuously distributed lag were considered by Roy G. D. Allen [8, pp. 23–29] in 1956. Currently, economic models with delay are actively used to describe the processes in the economy. The time delay is caused by finite speeds or inertia of processes and therefore it cannot be considered as processes with memory.

To take into account the memory and lag in economic and physical models, the fractional differential and integral operators with continuously distributed lag (time delay) were proposed in [400]. The distributed lag fractional operators are compositions of fractional differentiation or integration and continuously distributed translation (shift). The kernels of these operators are Laplace convolution of probability

https://doi.org/10.1515/9783110627459-030

density function and the kernels of fractional derivatives or integrals. The random variable is delay time that is distributed by probability law (distribution) on a positive semi-axis. These operators allow us to describe simultaneous action of distributed lag and fading memory. Examples of economic application of the lag-distributed fractional operators have been suggested in works [405, 404, 402, 407], where the economic concepts of accelerator and multipliers with distributed lag and memory were proposed.

30.3 Memory with distributed fading

The order α of fractional derivative or integral, which is interpreted as parameter of memory fading, can be distributed on interval $[\alpha_1, \alpha_2]$, where the distribution is described by a weight function $\rho(\alpha)$. The functions $\rho(\alpha)$ describes distribution of the parameter of the memory fading on a set of economic agents. Different types of economic agents may have different parameters of memory fading. In this case, we should use the fractional operators, which depends on the weight function $\rho(\alpha)$ and the interval $[\alpha_1, \alpha_2]$.

The concept of the integrals and derivatives with distributed orders was first proposed by Michele Caputo in [45] in 1995, and then these operators are applied and developed in different works (e. g., see [19, 20, 46, 234, 234]).

Let us note the following three cases:

1. The simplest distribution of the order of fractional derivatives and integrals is the continuous uniform distribution. The fractional operators with uniform distribution were proposed in [278, 279], and were called as the Nakhushev operators. Adam M. Nakhushev [278, 279] proposed the continual fractional derivatives and integrals in 1998. The fractional operators, which are inversed to the continual fractional integrals and derivatives, have been proposed by Arsen V. Pskhu [317, 318]. In the articles [450, 448], we proved that the fractional integrals and derivatives of the uniform distributed order can be expressed (up to a numerical factor) thought the continual fractional integrals and derivatives, which were suggested by A. M. Nakhushev. The proposed fractional integral and derivatives of uniform distributed order were called as the Nakhushev fractional integrals and derivatives in article [450]. The corresponding inverse operators, which contains the two parameter Mittag–Leffler functions in the kernel, were called as the Pskhu fractional integrals and derivatives
2. In articles [450, 448], we proposed concept of "weak" memory and the distributed order fractional operators with the truncated normal distribution of the order. The truncated normal distribution with integer mean and small variance can be used to describe economic processes with memory, in which the fading parameter is distributed around the integer value. In our opinion, the truncated normal distribution can be of great interest for economic models. In application, the truncated

normal distribution with integer mean values (e. g., equal to 1 or 2) and small variance can be used to describe wide class of economic processes with "weak" memory. The processes with "weak" memory, which is interpreted as small deviation from classical case of amnesia, can be represented by the Gaussian distribution functions with integer mean and infinitesimally small variance. It is known that the delta-function can be considered as a limit of a family of the Gaussian functions, when the variance becomes smaller (tends to zero). The processes without memory are described by the Dirac delta-function $\delta(\alpha - n)$. Therefore, the Gaussian distribution functions with integer mean and infinitesimally small variance can be used to describe economic processes with memory, in which the fading parameter is distributed around the integer value.

3. As a special case of the general fractional operators, which were proposed by Anatoly N. Kochubei, the fractional derivatives and integrals of distributed order are investigated in works [209, 210, 208, 211].

Fractional differential equations are actively used to describe physical processes. However, at present time, equations with distributed order operators have not yet been used to describe economic processes. We hope that new interesting effects in Economics with memory can be described by using fractional operators with distributed orders.

30.4 Generalized fractional calculus in economics

Generalized fractional calculus was proposed by Virginia Kiryakova and described in detail in book [202] in 1994. The brief history of the generalized fractional calculus is given in paper [203]. Operators of generalized fractional calculus [202, 203] can be used to describe complex processes with memory and nonlocality in real economy. In application of the generalized fractional operators, an important question arises about the correct economic interpretation of these operators. It is important to emphasize that not all fractional operators can describe the processes with memory [406]. It is important to clearly understand what type of phenomena can be described by a given operator. For example, among these types of phenomena, in addition to memory, we can specify the time delay (lag) and the scaling. Let us give two examples to clarify this problem:

1. The Abel-type fractional integral (and differential) operator with Kummer function in the kernel, which is described in the book [335] (see equation (37.1) in [335, p. 731]), can be interpreted as the Riemann–Liouville fractional integral (and derivatives) with gamma distribution of delay time [400]. Some Prabhakar fractional operators with three-parameter Mittag–Leffler functions in the kernel can also be interpreted as a Laplace convolution of the Riemann–Liouville (or Caputo) fractional operators with continuously distributed lag (time delay) [400].

2. We can state that the Kober fractional integration of noninteger order [335, 202, 200] can be interpreted as an expected value of a random variable up to a constant factor [401] (see also Section 9 in [400], and [404]). In this interpretation, the random variable describes dilation (scaling), which has the gamma distribution. The Erdelyi–Kober fractional integration also has a probabilistic interpretation. Fractional differential operators of Kober and Erdelyi–Kober type have analogous probabilistic interpretation, i. e., these operators cannot describe the memory. These operators describe integer-order operator with continuously distributed dilation (scaling). The fractional generalizations of the Kober and Erdelyi–Kober operators, which can be used to describe memory and distributed dilation (scaling), were proposed in [400, 402]. In articles [400, 406], we proposed fractional integral and differential operators that describe simultaneously action of distributed scaling and fading memory. Therefore, an important part of the application of generalized fractional calculus is a clear understanding of what types of processes and phenomena can describe fractional operators of noninteger order [406].

30.5 General fractional calculus and memory

The concept of general fractional calculus was suggested by Anatoly N. Kochubei [209, 210, 208, 211] by using the differential-convolution operators. In works [209, 210], Kochubei describe the conditions under which the general operator has a right inverse (a kind of a fractional integral) and produce, as a kind of fractional derivative, equations of evolution type. A solution of the relaxation equations with Kochubei general fractional derivative with respect to the time variable is described in works [209, 210].

In the economics, various growth models are used to describe real processes in economy. Therefore it is very important to describe conditions, for which the Cauchy problem for the growth equation has a solution. Solution of this Cauchy problem in the general case will allow us to accurately describe the conditions on the operator kernels (the memory functions), under which equations for models of economic growth with memory have solutions. An article dedicated to solving this mathematical problem was written by Anatoly N. Kochubei and Yuri Kondratiev [211] in 2019. The application of these mathematical results in economics and their economic interpretation is an open question at the moment.

To understand the warranted growth rate of the economy, it is important to obtain the asymptotic behavior for solutions of the general growth equation.

In application of the general fractional operators, it is also important to have correct economic interpretations that will connect the types of operator kernels with the types of phenomena. For example, it is obvious that the kernels of general fractional operators satisfying the normalization condition will describe distributed delays in time (lag), and not memory.

30.6 Fractional variational calculus in economics with memory

In economics, the existence of extreme values of certain parameters, the properties of equilibrium points and equilibrium growth trajectories are actively studied [217, 66, 80]. The existence of optimal solutions for fractional differential equations should be considered for economic processes with memory and nonlocality.

Methods of the fractional calculus of variations are actively developing [250, 14]. However, at present time, none of the variational problems that are described for processes with memory.

In the variation approach, there are some problems that restrict the possibilities of its application. One of the problems associated with the property of integration in parts, which actually turns the left-sided fractional derivative into a right-sided derivative. As a result, we will obtain equations in which, in addition to being dependent on the past, there is a dependence on the future, that is, the principle of causality is violated.

We assume that this problem cannot be solved within the framework of using the principle of stationarity of the holonomic functional (action). It is necessary to use nonholonomic functionals. We can also consider nonholonomic constraints with fractional derivatives of noninteger orders [409]. We can also consider variations of noninteger order [394] and fractional variational derivatives [369].

Another problem is the mathematical interpretation and the economic interpretations of extreme values.

However, we assume that the variational calculus is important for the economics with memory.

30.7 Fractional differential games in economics with memory

Models of differential games in which derivatives of noninteger order are used and, thereby, the power-fading memory is taken into account were proposed in the works of Arkadiy A. Chikriy (Arkadii Chikrii), Ivan I. Matychyn and Alexander G. Chentsov [69, 70, 73, 72, 71] (see also [259, 258, 514]), which are clearly not related to economy. Note that the construction of models of economic behavior, using differential games with fading memory, currently remains an open question. The construction of such models requires further research on economic behavior within the framework of game theory. The basis for such constructions can be the methods and results described in [69, 70, 73, 72, 71, 259, 258, 514].

30.8 Economic data and modeling of economics with memory

We should note the importance of computer simulations and modeling of real economic data, including data related to both macroeconomics and microeconomics.

The first works that can be attributed to the mathematical economics stage/approach are works published in 2014–2015 by Ines Tejado, Duarte Valerio, Nuno Valerio [456, 457]. An application of fractional calculus for modeling the national economies in the framework of the fractional generalizations of the Gross domestic product (GDP) model, which are described by the fractional differential equations are used, were considered by Ines Tejado, Duarte Valerio, Nuno Valerio and Emiliano Perez [456, 457, 458, 459, 460, 461, 462, 463] in 2014–2019, and by Dahui Luo, JinRong Wang, Michal Feckan and Hao Ming in 2018–2019 in the articles [242, 273]. The fractional differential equation used in the fractional GDP model was obtained by replacing first-order derivatives with fractional derivatives. Therefore, this model requires theoretical justification and consistent derivation.

30.9 Big data and memory

It is obvious that Big Data that describes behavior of peoples and other economic agents should contain information about various manifestations of memory. It would be strange if these Big Data neglected memory, since people have memory if they do not suffer from amnesia. We can assume that economic modeling in the "Era of Big Data" will describe the memory effects in microeconomics and macroeconomics. The Big Data will give us a possibility to take into account the effects of memory and nonlocality in time for those economic and financial processes, in which they were not even suspected.

30.10 Numerical methods for economics with memory

Fractional derivatives have many nonstandard properties [367, 377, 376, 382, 81, 225, 386] and, therefore, the analytical solutions of fractional differential and difference equations are much smaller in comparison with differential and difference equations of the integer order. Because of this, computer modeling will play an important role in describing economic processes with memory. To implement computer modeling of real economic processes with memory, it is necessary to develop numerical methods of fractional calculus. Computer simulation of economic processes with memory and nonlocality in time should use such numerical methods for equations with fractional derivatives of noninteger order, which take into account memory and nonlocality in time. Numerical approximation should not use only local information. Numerical scheme should contain a term of the memory (or nonlocality). Numerical methods that neglect nonlocality and memory are not reliable and often lead to incorrect results, since the nonlocal nature of fractional differential operators of noninteger orders cannot be neglected [133, 91]. Numerical approaches to fractional integrals and derivatives are important for economics with memory. Note that these methods are actively developing in modern fractional calculus (e. g., see the review articles [224, 44] and the books [227, 223, 166]). However, their use for modeling economic processes with memory is currently minimal. We hope that numerical methods will be actively applied and developed for modeling and describing economic processes with memory.

30.11 Econometrics for processes with memory

Economic theory is a branch of economics that employs models and abstractions of economic processes and objects to rationalize, explain and predict economic phenomena.

One of the main goals of economic theory is to explain the processes and phenomena in economy and make predictions. To achieve this goal, within the framework of economic theory, new notions, concepts, tools and methods should be developed in economics with memory.

Obviously, it is impossible to explain processes with memory without having adequate concepts.

To explain and predict the processes and phenomena with memory in economy, we must have a good instrument for conducting observations and their adequate description. This instrument is econometrics, or rather econometrics with memory. Econometrics is a link that connects economic theory with the phenomena and processes in real economy.

Econometrics mainly based on statistics for formulating and testing models and hypotheses about economic processes or estimating parameters for them. Theoretical econometrics considers the statistical properties of assessments and tests, while applied econometrics deals with the use of econometric methods for evaluating economic models. Theoretical econometrics develop tools and methods, and also study the properties of econometric methods. Applied econometrics develops quantitative economic models and the application of econometric methods to these models using economic data. Applied econometrics use theoretical econometrics and real-world data for assessing economic theories, developing econometric models, analyzing economic dynamics and forecasting.

Econometrics with memory can reach new opportunities in the development of new econometric methods and their use in describing economic reality by applying methods of modern fractional calculus, various types of fractional finite differences, differential and integral operators of noninteger order.

As a result, we can state that the main goals of economics with memory, which include explaining the economy and predicting for processes with memory, and correctly describing economic events, data, processes, and building adequate forecasts, cannot be realized without the econometrics with memory.

30.12 Development concept of memory

One of the main goals of economics with memory is to explain the processes and phenomena with memory in economy and make predictions. However, to explain the economic processes with memory, it is necessary to understand what memory is and how to describe it.

Further development of the concept of memory for economic processes is of great importance. It should be emphasized that not all types of fractional derivatives and integrals of noninteger order can describe processes with memory.

For example, within the framework of fractional calculus, it is necessary to distinguish between fractional operators that describe distributed time delay and distributed scaling from operators describing memory, and the combination of memory with these phenomena. However, there are open questions about what types of memory we can describe by using fractional calculus (e. g., see [399, 380, 406]), and in what directions the concept of memory for economic processes should be developed.

30.13 Conclusion

We can hope that the further development of economics with memory to describe economic phenomena and processes will take an important place with modern economics. Generally speaking, it is strange to neglect memory in economics, since the most important actors are people with memory.

Bibliography

[1] Abd-Elouahab, M. S.; Hamri, N.-E.; Wang, J. Chaos control of a fractional-order financial system. *Mathematical Problems in Engineering* **2010**. Vol. 2010. Article ID: 270646. 18 pages. DOI: 10.1155/2010/270646.

[2] Advances in the Applications of Nonstandard Finite Difference Schemes. Edited by R. E. Mickens. Singapore: World Scientific, 2005. 664 pages. DOI: 10.1142/5884.

[3] Aguilar, J. P.; Coste, C. G. Non-Gaussian analytic option pricing: A closed formula for the Levy-stable model. *SSRN Electronic Journal* **2016**. Article ID: 2828673. DOI: 10.2139/ssrn.2828673. URL: https://ssrn.com/abstract=2828673 or http://dx.doi.org/10.2139/ssrn.2828673 (arXiv:1609.00987).

[4] Aguilar, J.-P.; Korbel, J.; Luchko, Yu. Applications of the fractional diffusion equation to option pricing and risk calculations. *Mathematics* **2019**. Vol. 7. No. 9. Article ID: 796. 23 pages. DOI: 10.3390/math7090796.

[5] Akrami, M. H.; Erjaee, G. H. Examples of analytical solutions by means of Mittag-Leffler function of fractional Black-Scholes option pricing equation. *Fractional Calculus and Applied Analysis* **2015**. Vol. 18. No. 1. P. 38–47. DOI: 10.1515/fca-2015-0004.

[6] Al-Refai, M.; Pal, K. New aspects of Caputo-Fabrizio fractional derivative. *Progress in Fractional Differentiation and Applications* **2019**. Vol. 5. No. 2. P. 157–166.

[7] Alber, H. D. *Materials with Memory. Initial Boundary Value Problems for Constitutive Equations with Internal Variables*. Berlin: Springer-Verlag, 1998. 171 pages.

[8] Allen, R. G. D. *Mathematical Economics*. London: Macmillan, 1956. 768 pages.

[9] Allen, R. G. D. *Mathematical Economics*. Second Edition. London: Macmillan, 1960. 812 pages. ISBN: 978-1-349-81547-0. DOI: 10.1007/978-1-349-81547-0.

[10] Allen, R. G. D. *Mathematical Economics*. Second Edition. London: Macmillan, 1963. 812 pages. ISBN: 978-1-349-81547-0. DOI: 10.1007/978-1-349-81547-0.

[11] Allen, R. G. D. *Macro-Economic Theory. A Mathematical Treatment*. London, UK: Palgrave Macmillan, 1967. 420 pages. ISBN: 978-1-349-81541-8. DOI: 10.1007/978-1-349-81541-8.

[12] Allen, R. G. D. *Macro-Economic Theory. A Mathematical Treatment*. London: Macmillan, 1968. 420 pages. ISBN: 978-033304112.

[13] Almeida, R. A Caputo fractional derivative of a function with respect to another function. *Communications in Nonlinear Science and Numerical Simulation* **2017**. Vol. 44. P. 460–481. DOI: 10.1016/j.cnsns.2016.09.006 (arXiv:1609.04775).

[14] Almeida, R.; Tavares, D.; Torres, D. F. M. *The Variable-Order Fractional Calculus of Variations*. Heidelberg, New York: Springer, 2019. 124 pages. DOI: 10.1007/978-3-319-94006-9.

[15] Amendola, G.; Fabrizio M.; Golden, J. M. *Thermodynamics of Materials with Memory Theory and Applications*. New York: Springer Science+Business Media, LLC, 2012. 576 pages. ISBN: 978-1-4614-1691-3. DOI: 10.1007/978-1-4614-1692-0.

[16] Applications of Nonstandard Finite Difference Schemes. Edited by R. E. Mickens. Singapore: World Scientific, 2000. 264 pages. DOI: 10.1142/4272.

[17] Area, I.; Losada, J.; Nieto, J. J. A note on the fractional logistic equation. *Physica A: Statistical Mechanics and its Applications* **2016**. Vol. 444. P. 182–187. DOI: 10.1016/j.physa.2015.10.037. (arXiv:1508.04419).

[18] Ashby, W. R. Principles of the self-organizing dynamic system. *Journal of General Psychology* **1947**. Vol. 37. No. 2. P. 125–128. DOI: 10.1080/00221309.1947.9918144.

[19] Bagley, R. L.; Torvik, P. J. On the existence of the order domain and the solution of distributed order equations - Part I. *International Journal of Applied Mathematics* **2000**. Vol. 2. P. 865–882.

https://doi.org/10.1515/9783110627459-031

[20] Bagley, R. L.; Torvik, P. J. On the existence of the order domain and the solution of distributed order equations - Part II. *International Journal of Applied Mathematics* **2000**. Vol. 2. P. 965–987.

[21] Baillie, R. N. Long memory processes and fractional integration in econometrics. *Journal of Econometrics* **1996**. Vol. 73. P. 5–59. DOI: 10.1016/0304-4076(95)01732-1.

[22] Baleanu D.; Diethelm, K.; Scalas, E.; Trujillo, J. J. *Fractional Calculus: Models and Numerical Methods*. Second Edition. Singapore: World Scientific, 2016. 478 pages. ISBN: 978-9813140035, 9813140038.

[23] Banerjee, A.; Urga, G. Modelling structural breaks, long memory and stock market volatility: an overview. *Journal of Econometrics* **2005**. Vol. 129. No. 1-2. P. 1–34. DOI: 10.1016/j.jeconom.2004.09.001.

[24] Barro, R. J.; Sala-i-Martin, X. I. *Economic Growth*. Second Edition. Cambridge, Massachusetts, London: The MIT Press, 2003. 654 pages. ISBN-13: 978-0262025539.

[25] Baskonus, H. M.; Mekkaoui, T.; Hammouch, Z.; Bulut, H. Active control of a chaotic fractional order economic system. *Entropy* **2015**. Vol. 17. P. 5771–5783. DOI: 10.3390/e17085771.

[26] Bateman, H. *Tables of Integral Transforms*. Volume I. New York: McGraw-Hill, 1954. 391 pages. ISBN: 07-019549-8.

[27] Baumol, W. J.; Benhabib, J. Chaos: Significance, mechanism, and economic applications. *Journal of Economic Perspectives* **1989**. Vol. 3. No. 1. P. 77–105. DOI: 10.1257/jep.3.1.77.

[28] Bellman, R. *Introduction to Matrix Analysis*. Second Edition. New York: McGraw-Hill, 1997. 399 pages. DOI: 10.1137/1.9781611971170.

[29] Beran, J. *Statistics for Long-Memory Processes*. New York: Capman and Hall, 1994. 315 pages. ISBN: 0-412-04901-5.

[30] Beran, J.; Feng, Y.; Ghosh, S.; Kulik, R. *Long-Memory Processes: Probabilistic Properties and Statistical Methods*. Berlin, Heidelberg, New York: Springer-Verlag, 2013. 884 pages. ISBN: 978-3-642-35511-0. DOI: 10.1007/978-3-642-35512-7.

[31] Bernanke, B.; Gertler, M.; Gilchrist, S. The financial accelerator and the flight to quality. *Review of Economics and Statistics* **1996**. Vol. 78. No. 1. P. 1–15.

[32] Bischi, G. I.; Chiarella, C.; Sushko, I. *Global Analysis of Dynamic Models in Economics and Finance: Essays in Honour of Laura Gardini*. Berlin: Springer-Verlag, 2013. 444 pages. ISBN: 978-3-642-29502-7. DOI: 10.1007/978-3-642-29503-4.

[33] Blackledge, J. Application of the fractal market hypothesis for modelling macroeconomic time series. *ISAST Trans Electron Signal Process* **2008**. Vol. 2. No. 1. P. 89–110. DOI: 10.21427/D7091P.

[34] Blackledge, J. Application of the fractional diffusion equation for predicting market behavior. *International Journal of Applied Mathematics* **2010**. Vol. 40. No. 3. P. 130–158. DOI: 10.21427/D7HK8R.

[35] Boltzmann, L. Theory of elastic aftereffect [Zur Theorie der elastischen Nachwirkung]. *Wien Akad. Sitzungsber* **1874**. Band 70 (2). S. 275–306 [in German].

[36] Boltzmann, L. Theory of elastic aftereffect [Zur Theorie der elastischen Nachwirkung]. *Annalen der Physik und Chemie: Erganzungsband VII* **1876**. Band 7. Stuck 4. S. 624–654 [in German]. URL: http://gallica.bnf.fr/ark:/12148/bpt6k15009g/f637.image.langDE.

[37] Boltzmann, L. On some problems of the theory of elastic aftereffect and on a new method to observe vibrations by means of mirror reading, without burdening the vibrating body with a mirror of considerable mass [Uber einige Probleme der Theorie der elastischen Nachwirkung und uber eine neue Methode, Schwingungen mittels Spiegelablesung zu beobachten, ohne den schwingenden Korper mit einem Spiegel von erheblicher Masse zu belasten]. In *Wissenschaftliche Abhandlungen*. Volume 2. Edited by F. Hasenohrl. Cambridge: Cambridge University Press, 2012. pp. 224–249 [in German]. DOI: 10.1017/CBO9781139381437.012.

[38] Boltzmann, L. Theory of elastic aftereffect [Zur Theorie der elastischen Nachwirkung]. In *Wissenschaftliche Abhandlungen*. Volume 1. Edited by F. Hasenohrl. Cambridge: Cambridge University Press, 2012. pp.616–644 [in German]. DOI: 10.1017/CBO9781139381420.031.

[39] Boltzmann, L. Theory of elastic aftereffect [Zur Theorie der elastischen Nachwirkung]. In *Wissenschaftliche Abhandlungen*. Volume 2. Edited by F. Hasenohrl. Cambridge: Cambridge University Press, 2012. pp.318–320 [in German]. DOI: 10.1017/CBO9781139381437.015.

[40] Boole, G. *A Treatise on the Calculus of Finite Differences*. Cambridge: Cambridge University Press, 2009. DOI: 10.1017/CBO9780511693014.

[41] Bruno, M.; Fischer, S. Seigniorage, operating rules, and the high inflation trap. *The Quarterly Journal of Economics* **1990**. Vol. 105. No. 2. P. 353–374. DOI: 10.2307/2937791.

[42] Bruno, M.; Fischer, S., Seigniorage operating rules, and the high inflation trap. *NBERWorking Paper* No. 2413. October. **1987**. URL: https://www.nber.org/papers/w2413.pdf.

[43] Cagan, P. The Monetary Dynamics of Hyperinflation. In *Studies in the Quantity Theory of Money*. Edited by M. Friedman. Chicago, London: University of Chicago Press, 1956. pp. 25–117. ISBN: 0-226-26406-8.

[44] Cai, M.; Li, C. P. Numerical Approaches to Fractional Integrals and Derivatives: A Review. *Mathematics* **2020**. Vol. 8. No. 1. Article ID: 43. 53 pages. DOI: 10.3390/math8010043. URL: https://www.mdpi.com/2227-7390/8/1/43.

[45] Caputo, M. Mean fractional-order-derivatives differential equations and filters. *Annali dell'Universita di Ferrara* **1995**. Vol. 41. No. 1. P. 73–84. DOI: 10.1007/BF02826009.

[46] Caputo, M. Distributed order differential equations modelling dielectric induction and diffusion. *Fractional Calculus and Applied Analysis* **2001**. Vol. 4. No. 4. P. 421–442.

[47] Caputo, M. The memory function of the reactivity of investments caused by step of the prime rate. *Economia Politica* **2005**. Vol. 22. No. 3. P. 365–379. DOI: 10.1428/21175. URL: https://www.rivisteweb.it/doi/10.1428/21175.

[48] Caputo, M. Planning output dynamics with memory. *Economia Politica* **2009**. Vol. 26. No. 1. P. 79–97. DOI: 10.1428/29093. URL: https://www.rivisteweb.it/doi/10.1428/29093.

[49] Caputo, M. The memory response of the populations to extreme events. In *Atti dell'Accademia delle Scienze di Ferrara. Volume 88. Anno Accademico 188*. 2010–2011. pp. 225–240. 433 pages. ISBN: 88-7400-041-3.

[50] Caputo, M. The convergence of economic developments. *Studies in Nonlinear Dynamics and Econometrics* **2012**. Vol. 16. No. 2. Article 7. 23 pages. DOI: 10.1515/1558-3708.1986.

[51] Caputo, M. The memory response of populations and markets to extreme events. *Economia Politica* **2012**. Vol. 29. No. 2. P. 261–282. DOI: 10.1428/37632. URL: https://www.rivisteweb.it/doi/10.1428/37632.

[52] Caputo, M. Memory oscillator for modeling socio-economic oscillatory responses to extreme events. *Economia Politica* **2013**. Vol. 30. No. 2. P. 221–239. DOI: 10.1428/74241. URL: https://www.rivisteweb.it/doi/10.1428/74241.

[53] Caputo, M. The evolution and homogeneity of EU economies (with an econometric approach). *Meccanica* **2014**. Vol. 49. No. 9. P. 2237–2246. DOI: 10.1007/s11012-014-9966-1.

[54] Caputo, M. The role of memory in modeling social and economic cycles of extreme events. Chapter 11 In *A Handbook of Alternative Theories of Public Economics*. Edited by F. Forte, R. Mudambi, P. M. Navarra. Cheltenham: Edward Elgar Publishing, 2014. pp. 245–259. 576 pages. ISBN: 9781781004715. URL: https://books.google.ru/books?hl=ru&lr=&id=bekyAwAAQBAJ&oi=fnd&pg=PA245&ots=4pzWJ8won7&sig=BKAM1zBH1dezS9CD78TOKK8wK8o&redir_esc=y#v=onepage&q&f=false.

[55] Caputo, M.; Di Giorgio, G. Monetary Policy, Memory and Output Dynamics. In *Monetary Policy and Institutions*. Edited by G. Di Giorgio, F. Neri. Rome: LUISS University Press, 2006. pp. 165–176.

[56] Caputo, M.; Fabrizio, M. A new definition of fractional derivative without singular kernel. *Progress in Fractional Differentiation and Applications* **2015**. Vol. 1. No. 2. P. 73–85. DOI: 10.12785/pfda/010201.

[57] Caputo, M.; Fabrizio, M. Applications of new time and spatial fractional derivatives with exponential kernels. *Progress in Fractional Differentiation and Applications* **2016**. Vol. 2. No. 1. P. 1–11. DOI: 10.18576/pfda/020101.

[58] Caputo, M.; Kolari, J. An analytical model of the Fisher equation with memory functions. *Alternative Perspectives on Finance and Accounting* **2001**. Vol. 1. P. 1–14.

[59] Cartea, A. Derivatives pricing with marked point processes using tick-by-tick data. *Quantitative Finance* **2013**. Vol. 13. No. 1. P. 111–123. DOI: 10.1080/14697688.2012.661447.

[60] Cartea, A.; Del-Castillo-Negrete, D. Fractional diffusion models of option prices in markets with jumps. *Physica A: Statistical Mechanics and its Applications* **2007**. Vol. 374. No. 2. P. 749–763. DOI: 10.2139/ssrn.934809.

[61] Cartea, A.; Meyer-Brandis, T. How duration between trades of underlying securities affects option prices. *Review of Finance* **2010**. Vol. 14. No. 4. P. 749–785. DOI: 10.1093/rof/rfp013.

[62] Chang, W. W.; Smyth, D. J. The existence and persistence of cycles in a nonlinear model: Kaldor's 1940 model re-examined. *The Review of Economic Studies* **1971**. Vol. 38. No. 1. P. 37–44. DOI: 10.2307/2296620.

[63] Chaos and Nonlinear Dynamics: Application to Financial Markets. Edited by R. A. Meyers. New York: Springer-Verlag 2011. 406 pages.

[64] Chen, W. C. Nonlinear dynamics and chaos in a fractional-order financial system. *Chaos, Solitons and Fractals* **2008**. Vol. 36. No. 5. P. 1305–1314. DOI: 10.1016/j.chaos.2006.07.051.

[65] Chen, L.; Chai, Y.; Wu, R. Control and synchronization of fractional-order financial system based on linear control. *Discrete Dynamics in Nature and Society* **2011**. Vol. 2011. Article ID: 958393. 21 pages. DOI: 10.1155/2011/958393.

[66] Cheremnykh, Yu. N. *Microeconomics. Advanced Level*. Moscow: INFRA-M, 2015. 844 pages. ISBN: 978-5-16-002059-4.

[67] Chian, A. C.-L. *Complex Systems Approach to Economic Dynamics*. Berlin, Heidelberg, New York: Springer-Verlag, 2007. 101 pages. ISBN: 978-3-540-39752-6.

[68] Chian, A. C.-L.; Rempel, E. L.; Rogers, C. Complex economic dynamics: Chaotic saddle, crisis and intermittency. *Chaos, Solitons and Fractals* **2006**. Vol. 29. No. 5. P. 1194–1218. DOI: 10.1016/j.chaos.2005.08.218.

[69] Chikrii, A. A. Game dynamic problems for systems with fractional derivatives. In *Pareto Optimality, Game Theory and Equilibria*. Vol. 17. Edited by A. Chinchuluun, P. M. Pardalos, A. Migdalas, L. Pitsoulis. New York: Springer, 2008. pp. 349–386. DOI: 10.1007/978-0-387-77247-9_13.

[70] Chikrii, A. A.; Matychyn, I. I. Game problems for fractional-order linear systems. *Proceedings of the Steklov Institute of Mathematics. Supplementary issues 1* **2010**. Vol. 268. No. 1. P. 54–70. DOI: 10.1134/S0081543810050056. (*Works of the Institute of Mathematics and Mechanics. Ural Branch of the Russian Academy of Sciences*. **2009**. Vol. 15. No. 3. P. 262–278 [in Russian]. URL: http://mi.mathnet.ru/eng/timm419).

[71] Chikrii, A. A.; Matychyn, I. I. On linear conflict-controlled processes with fractional derivatives. *Works of the Institute of Mathematics and Mechanics. Ural Branch of the Russian Academy of Sciences* **2011**. Vol. 17. No. 2. P. 256–270 [in Russian]. URL: http://mi.mathnet.ru/eng/timm711.

[72] Chikrii, A. A.; Matychyn, I. I. Riemann–Liouville, Caputo, and sequential fractional derivatives in differential games. In *Advances in Dynamic Games*. Edited by M. Breton, K. Szajowski. Boston: Birkhauser, Springer Science+Business Media, 2011. pp. 61–81. DOI: 10.1007/978-0-8176-8089-3_4.

[73] Chikriy, A. A.; Chentsov, A. G.; Matychyn, I. I. Differential games of the fractional order with separated dynamics. *Journal of Automation and Information Sciences* **2009**. Vol. 41. No. 11. P. 17–27. DOI: 10.1615/JAutomatInfScien.v41.i11.20.

[74] Chu, Sh. Y.; Shane, Ch. Using the hybrid Phillips curve with memory to forecast US inflation. *Studies in Nonlinear Dynamics and Econometrics* **2017**. Vol. 21. No. 4. Article ID: 20160088. DOI: 10.1515/snde-2016-0088.

[75] Clark, J. M. Business acceleration and the law of demand: A technikal factor in economic cycles. *Journal of Political Economy* **2017**. Vol. 25. P. 217–235. DOI: 10.1086/252958. URL: https://www.journals.uchicago.edu/doi/pdfplus/10.1086/252958.

[76] Cobb, C. W.; Douglas, P. H. A theory of production. *American Economic Review* **1928**. Vol. 18. No. 1. Supplement. P. 139–165. URL: https://www.aeaweb.org/aer/top20/18.1.139-165.pdf.

[77] Coleman, B. D.; Mizel, V. J. Norms and semi-groups in the theory of fading memory. *Archive for Rational Mechanics and Analysis* **1966**. Vol. 23. No. 2. P. 87–123. DOI: 10.1007/BF00251727.

[78] Coleman, B. D.; Mizel V. J. A general theory of dissipation in materials with memory. *Archive for Rational Mechanics and Analysis* **1967**. Vol. 27. No. 4. P. 255–274. DOI: 10.1007/BF00281714.

[79] Coleman, B. D.; Mizel, V. J. On the general theory of fading memory. *Archive for Rational Mechanics and Analysis* **1968**. Vol. 29. No. 1. P. 18–31. DOI: 10.1007/BF00256456.

[80] Cowell, F. A. *Microeconomics. Principles and Analysis*. Oxford: Oxford University Press, 2006. 678 pages. ISBN: 0199267774 978-0199267774.

[81] Cresson, J.; Szafranska, A. Comments on various extensions of the Riemann-Liouville fractional derivatives: About the Leibniz and chain rule properties. *Communications in Nonlinear Science and Numerical Simulation* **2020**. Vol. 82. Article ID: 104903. 9 pages. DOI: 10.1016/j.cnsns.2019.104903.

[82] Cyclical Growth in Market and Planned Economies. Edited by C. Hillinger. Oxford: Clarendon Press, 1992. 224 pages. ISBN: 0198283148.

[83] Dadras, S.; Momeni, H. R. Control of a fractional-order economical system via sliding mode. *Physica A: Statistical Mechanics and its Applications* **2010**. Vol. 389. No. 12. P. 2434–2442. DOI: 10.1016/j.physa.2010.02.025.

[84] Danca, M. F.; Garrappa, R.; Tang, W. K. S.; Chen, G. Sustaining stable dynamics of a fractional-order chaotic financial system by parameter switching. *Computers and Mathematics with Applications* **2013**. Vol. 66. No. 5. P. 702–716. DOI: 10.1016/j.camwa.2013.01.028.

[85] Das, S.; Gupta, K. A mathematical model on fractional Lotka–Volterra equations. *Journal of Theoretical Biology* **2011**. Vol. 277. No. 1. P. 1–6. DOI: 10.1016/j.jtbi.2011.01.034.

[86] David, S. A.; Tenreiro Machado, J. A.; Quintino, D. D.; Balthazar, J. M. Partial chaos suppression in a fractional order macroeconomic model. *Mathematics and Computers in Simulation* **2016**. Vol. 122. P. 55–68. DOI: 10.1016/j.matcom.2015.11.004.

[87] David, S. A.; Fischer, C.; Tenreiro Machado, J. A. Fractional electronic circuit simulation of a nonlinear macroeconomic model. *AEU - International Journal of Electronics and Communications* **2018**. Vol. 84. P. 210–220. DOI: 10.1016/j.aeue.2017.11.019.

[88] Day, W. A. *The Thermodynamics of Simple Materials with Fading Memory*. Berlin: Springer-Verlag, 1972. 134 pages.

[89] De Vroey, M. *A History of macroeconomics from Keynes to Lucas and Beyond*. Cambridge: Cambridge University Press, 2016. 450 pages. ISBN: 978-0-521-89843-0. DOI: 10.1017/CBO9780511843617.

[90] Diethelm, K. *The Analysis of Fractional Differential Equations: An Application-Oriented Exposition Using Differential Operators of Caputo Type*. Berlin: Springer-Verlag, 2010. 247 pages. DOI: 10.1007/978-3-642-14574-2.

[91] Diethelm, K.; Garrappa, R.; Stynes, M. Good (and not so good) practices in computational methods for fractional calculus. *Mathematics* **2020**. Vol. 8. No. 3. Article ID: 324. 21 pages. DOI: 10.3390/math8030324.

[92] Domar, E. D. Capital expansion, rate of growth and employment. *Econometrica* **1946**. Vol. 14. No. 2. P. 137–147. DOI: 10.2307/1905364.

[93] Domar, E. D. Expansion and employment. *The American Economic Review* **1947**. Vol. 37. No. 1. P. 34–55.

[94] D'Ovidio, M.; Polito, F. Fractional diffusion-telegraph equations and their associated stochastic solutions. *Theory of Probability and Its Applications*. (ISSN: 0040-361X (print), 2305-3151 (online)) **2017**. Vol. 62. No. 4. P. 692–718. DOI: 10.4213/tvp5150.

[95] D'Ovidio, M.; Polito, F. Fractional diffusion-telegraph equations and their associated stochastic solutions. *Theory of Probability and Its Applications*. (ISSN: Electronic: 1095-7219 Print: 0040-585X) **2018**. Vol. 62. No. 4. P. 552–574. DOI: 10.1137/S0040585X97T988812.

[96] D'Ovidio, M.; Polito, F. Fractional diffusion-telegraph equations and their associated stochastic solutions. Cornell University arXiv. **2013**. arXiv:1307.1696. 23 pages. https://arxiv.org/abs/1307.1696. URL: https://arxiv.org/pdf/1307.1696v1.pdf.

[97] Duncan, T. E.; Hu, Y.; Pasik-Duncan, B. Stochastic calculus for fractional Brownian motion. I. Theory. *SIAM Journal on Control and Optimization* **2000**. Vol. 38. No. 2. P. 582–612. DOI: 10.1137/S036301299834171X.

[98] Dung, N. T. Fractional stochastic differential equations with applications to finance. *Journal of Mathematical Analysis and Applications* **2013**. Vol. 397. No. 1. P. 334–348. DOI: 10.1016/j.jmaa.2012.07.062.

[99] Dynamical Inverse Problems: Theory and Application. Edited by G. M. L. Gladwell, A. Morassi. Wien: Springer-Verlag, 2011. 226 pages. DOI: 10.1007/978-3-7091-0696-9.

[100] Dzherbashian, M. M. *Integral Transforms and Representation of Functions in the Complex Domain*. Moscow: Nauka, 1966. 672 pages [in Russian].

[101] Edelman, M. Fractional standard map: Riemann-Liouville vs. Caputo. *Communications in Nonlinear Science and Numerical Simulation* **2011**. Vol. 16. No. 12. P. 4573–4580. DOI: 10.1016/j.cnsns.2011.02.007.

[102] Edelman, M. Fractional maps and fractional attractors. Part I: alpha-families of maps. *Discontinuity, Nonlinearity, and Complexity* **2013**. Vol. 1. No. 4. P. 305–324. DOI: 10.5890/DNC.2012.07.003 (arXiv:1209.5681).

[103] Edelman, M. Universal fractional map and cascade of bifurcations type attractors. *Chaos: An Interdisciplinary Journal of Nonlinear Science* **2013**. Vol. 23. No. 3. Article ID: 033127. DOI: 10.1063/1.4819165.

[104] Edelman, M. Caputo standard alpha-family of maps: fractional difference vs. fractional. *Chaos: An Interdisciplinary Journal of Nonlinear Science* **2014**. Vol. 24. No. 2. Article ID: 023137. DOI: 10.1063/1.4885536 (arxiv:1406.4059).

[105] Edelman, M. *Fractional maps as maps with power-law memory*. Nonlinear Dynamics and Complexity. Volume 8. Edited by A. Afraimovich, A. C. J. Luo, X. Fu. New York: Springer, 2014. pp. 79–120. DOI: 10.1007/978-3-319-02353-3_3.

[106] Edelman, M. Universality in fractional dynamics. In *International Conference on Fractional Differentiation and Its Applications (ICFDA)*. 2014. 6 pages. DOI: 10.1109/ICFDA.2014.6967376 (arXiv:1401.0048).

[107] Edelman, M. Fractional maps and fractional attractors. Part II: Fractional difference α-families of maps. *Discontinuity, Nonlinearity, and Complexity* **2015**. Vol. 4. P. 391–402. DOI: 10.5890/DNC.2015.11.003 (arXiv:1404.4906).

[108] Edelman, M. On nonlinear fractional maps: Nonlinear maps with power-law memory. In *Proceedings of the International Conference CCT15 – Chaos, Complexity and Transport 2015, June 1–5, 2015, Marseilles, France*. Edited by X. Leoncini, Ch. Eloy, G. Boedec. Singapore: World Scientific, 2017. pp. 119–130. ISBN: 978-981-3202-74-0. DOI: 10.1142/10332. (arXiv:1612.01174).

[109] Edelman, M. Dynamics of nonlinear systems with power-law memory. In *Handbook of Fractional Calculus with Applications. Volume 4. Application in Physics. Part A*. Edited by V. E. Tarasov. Berlin, Boston: Walter de Gruyter GmbH, 2019. pp. 103–132. 306 pages. ISBN: 978-3-11-057088-5. DOI: 10.1515/9783110571707-005.

[110] Edelman, M.; Tarasov, V. E. Fractional standard map. *Physics Letters A* **2009**. Vol. 374. No. 2. P. 279–285. DOI: 10.1016/j.physleta.2009.11.008.

[111] Elagan, S. K. On the invalidity of semigroup property for the Mittag-Leffler function with two parameters. *Journal of the Egyptian Mathematical Society* **2016**. Vol. 24. No. 2. P. 200–203. DOI: 10.1016/j.joems.2015.05.003.

[112] Elsgolts, L. E.; Norkin, S. B. *Introduction to the Theory and Application of Differential Equations with Deviating Arguments*. New York: Academic Press, 1973. 356 pages.

[113] Erdelyi, A. Hypergeometric functions of two variables. *Acta Mathematica* **1950**. Vol. 83. P. 131–164. DOI: 10.1007/BF02392635. URL: https://projecteuclid.org/euclid.acta/1485888583.

[114] Erdelyi, A.; Magnus, W.; Oberhettinger, F.; Tricomi, F. G. *Higher Transcendental Functions*. Volume I. (Bateman Manuscript Project). New York: McGraw-Hill, 1953.

[115] Eriksson, C. *Economic Growth and the Environment. An Introduction to the Theory*. Oxford: Oxford University Press, 2013. 217 pages. ISBN: 978-0-19-966389-7.

[116] Evans, G. C. The dynamics of monopoly. *The American Mathematical Monthly* **1924**. Vol. 31. No. 2. P. 77–83. DOI: 10.1080/00029890.1924.11986301. URL: https://www.tandfonline.com/doi/abs/10.1080/00029890.1924.11986301. https://docuri.com/download/the-dynamics-of-monopoly_59c1e3b2f581710b286ad64e_pdf.

[117] Evans, G. C. *Mathematical Introduction to Economics*. New York, London: McGraw-Hill Book Company, 1930. 177 pages. URL: https://archive.org/details/in.ernet.dli.2015.263264.

[118] Fallahgoul, H. A.; Focardi, S. M.; Fabozzi, F. J. *Fractional Calculus and Fractional Processes with Applications to Financial Economics, Theory and Application*. London: Academic Press, 2016. 118 pages. ISBN: 9780128042489.

[119] Feckan, M.; Wang, J. R.; Pospisil, M. *Fractional-Order Equations and Inclusions*. De Gruyter, 2017. 367 pages. ISBN: 978-3-11-052138-2, 978-3-11-052155-9.

[120] Ferrara, M.; Guerrini, L. The neoclassical model of Solow and Swan with logistic population growth. In *Proceedings of the 2nd International Conference of IMBIC on Mathematical Sciences for Advancement of Science and Technology (MSAST), Kolkata, India, 21–23 December* 2017. pp. 119–127. URL: https://www.unirc.it/documentazione/materiale_didattico/600_2008_108_1691.pdf.

[121] Fichtenholz, G. M. *Differential and Integral Calculus. Volume 2*. 7th edition. Moscow: Nauka, 1969. 800 pages [in Russian].

[122] Fichtenholz, G. M. *Functional Series*. Pocket Mathematical Library. Gordon and Breach Science, 1970. 176 pages. ISBN: 0677209509, ISBN: 978-0677209500.

[123] Fichtenholz, G. M. *Infinite Series: Ramifications. (Pocket Mathematical Library)* Routledge, 1970. 139 pages. ISBN: 0677209401, ISBN: 978-0677209401.

[124] Forbes, C.; Evans, M.; Hastings, N.; Peacock, B. *Statistical Distributions*. Fourth Edition. Hoboken, New Jersey: John Wiley and Sons, 2011. 212 pages. ISBN: 978-0-470-39063-4.

[125] Gabaix, X. Power laws in economics and finance. *Annual Review of Economics* **2009**. Vol. 1. No. 1. P. 255–293. DOI: 1941-1383/09/0904-0255.

[126] Gabaix, X. Power laws in economics: An introduction. *Journal of Economic Perspectives* **2016**. Vol. 30. No. 1. P. 185–206. DOI: 10.1257/jep.30.1.185.

[127] Gabisch, G.; Lorenz, H. W. *Business Cycle Theory: A Survey of Methods and Concepts*. Berlin, Heidelberg, New York: Springer-Verlag, 1987. 229 pages. DOI: 10.1007/978-3-662-01178-2.

[128] Gabisch, G.; Lorenz H. W. *Business Cycle Theory: A Survey of Methods and Concepts. Second, Revised, and Enlarged Edition*. Berlin, Heidelberg: Springer-Verlag, 1989. 248 pages. DOI: 10.1007/978-3-642-74715-1.

[129] Galiullin, A. S. *Inverse Problems of Dynamics*. Moscow: Mir Publishers, 1984. 149 pages.

[130] Galiullin, A. S. *Method for Solving Inverse Problems of Dynamics*. Moscow: Nauka, 1986. 224 pages [in Russian].

[131] Gantmakher, F. R. *Theory of Matrices*. Third Edition. Moscow: Nauka, 1967. 576 pages [in Russian].

[132] Garra, R.; Garrappa, R. The Prabhakar or three parameter Mittag–Leffler function: theory and application. *Communications in Nonlinear Science and Numerical Simulation* **2018**. Vol. 56. P. 314–329. DOI: 10.1016/j.cnsns.2017.08.018 (arXiv:1708.07298v2).

[133] Garrappa, R. Neglecting nonlocality leads to unreliable numerical methods for fractional differential equations. *Communications in Nonlinear Science and Numerical Simulation* **2019**. Vol. 70. P. 302–306. DOI: 10.1016/j.cnsns.2018.11.004.

[134] Gel'fand, I. M.; Shilov, G. E. *Generalized Functions. Vol. I: Properties and Operations*. Boston: Academic Press, 1964 and American Mathematical Society, 2016. 423 pages. ISBN: 1-4704-2658-7. ISBN: 978-1-4704-2658-3.

[135] Gil-Alana, L. A.; Hualde, J. Fractional Integration and Cointegration: An Overview and an Empirical Application. In *Palgrave Handbook of Econometrics. Volume 2: Applied Econometrics*. Edited by T. C. Mills, K. Patterson. Berlin: Springer-Verlag, 2009. pp. 434–469. DOI: 10.1057/9780230244405_10.

[136] Girdzijauskas, S.; Streimikiene, D.; Mialik, A. Economic growth, capitalism and unknown economic paradoxes. *Sustainability* **2012**. Vol. 4. P. 2818–2837.

[137] Giusti, A. General fractional calculus and Prabhakar's theory. *Communications in Nonlinear Science and Numerical Simulation* **2020**. Vol. 83. Article ID: 105114. 7 pages. DOI: 10.1016/j.cnsns.2019.105114.

[138] Giusti, A.; Colombaro, I.; Garra, R.; Garrappa, R.; Polito, F.; Popolizio, M.; Mainardi, F. A practical guide to Prabhakar fractional calculus. *Fractional Calculus and Applied Analysis* **2020**. Vol. 23. No. 1. P. 9–54. DOI: 10.1515/fca-2020-0002.

[139] Goodwin, R. M. The nonlinear accelerator and the persistence of business cycles. *Econometrica* **1951**. Vol. 19. No. 1. P. 1–17. DOI: 10.2307/1907905.

[140] Gorenflo, R.; Mainardi, F.; Fractional calculus, integral and differential equations of fractional order. In *Fractals and Fractional Calculus in Continuum Mechanics*. Edited by A. Carpinteri, F. Mainardi. Wien: Springer Verlag, 1997. pp. 223–276. ISBN: 3-211-82913-X. (E-print http://arxiv.org/abs/0805.3823).

[141] Gorenflo, R.; Mainardi, F.; Raberto, M.; Scalas, E. Fractional diffusion in finance: basic theory. In *MDEF2000. Workshop "Modelli Dinamici in Economia e Finanza". Urbino (Italy), September 28–30*, 2000. Available at http://www.mdef.it/fileadmin/user_upload/mdef/meetings/MDEF2000/MainardiMDEF.pdf.

[142] Gorenflo, R.; Mainardi, F.; Scalas, E.; Raberto, M. Fractional calculus and continuous-time finance III: the diffusion limit. In *Mathematical Finance*. Edited by A. Kohlmann, S. Tang. Basel: Birkhauser, 2001. pp. 171–180. DOI: 10.1007/978-3-0348-8291-0_17.

[143] Gorenflo, R.; Kilbas, A. A.; Mainardi, F.; Rogosin, S. V. *Mittag-Leffler Functions, Related Topics and Applications*. Berlin: Springer-Verlag, 2014. 443 pages. DOI: 10.1007/978-3-662-43930-2.

[144] Gossen, H. H. *Die Entwickelung der Gesetze des menschlichen Verkehrs, und der daraus fliebenden Regeln fur menschliches Handeln*. Braunschweig: F. Vieweg, 1854. 295 pages [in German]. ISBN: 978-3878810186.

[145] Gossen, H. H. *The Laws of Human Relations and the Rules of Human Action Derived Therefrom*. Cambridge: MIT Press, 1983. 465 pages. ISBN: 9780262070904.

[146] Granberg, A. G. *Dynamic Models of National Economy*. Moscow: Ekonomika, 1985. 240 pages [in Russian].

[147] Granger, C. W. J. The typical spectral shape of an economic variable. *Technical Report No. 11. January 30, 1964. Department of Statistics. Stanford University*, **1964**. 21 P. URL: https://statistics.stanford.edu/sites/default/files/PAR%20NSF%2011.pdf. URL: https://statistics.stanford.edu/research/typical-spectral-shape-economic-variable.

[148] Granger, C. W. J. The typical spectral shape of an economic variable. *Econometrica* **1966**. Vol. 34. No. 1. P. 150–161. DOI: 10.2307/1909859.

[149] Granger, C. W. J. Investigating causal relations by econometric models and cross-spectral methods. *Econometrica* **1969**. Vol. 37. No. 3. P. 424–438.

[150] Granger, C. W. J. Testing for causality: A personal viewpoint. *Journal of Economic Dynamics and Control* **1980**. Vol. 2. P. 329–352. DOI: 10.1016/0165-1889(80)90069-X.

[151] Granger, C. W. J. Current perspectives on long memory processes. *Academia Economic Papers* **2000**. Vol. 28. No. 1. P. 1–16. DOI: 10.2307/1909859. URL: http://citeseerx.ist.psu.edu/viewdoc/download?doi=10.1.1.454.949&rep=rep1&type=pdf.

[152] Granger, C. W. J. *Essays in Econometrics: Collected Papers of Clive W. J. Granger. Volume I. Spectral Analysis, Seasonality, Nonlinearity, Methodology, and Forecasting*. Edited by E. Ghysels, N. R. Swanson, M. W. Watson. Cambridge, New York: Cambridge University Press, 2001. 523 pages.

[153] Granger, C. W. J. *Essays in Econometrics Collected Papers of Clive W.J. Granger. Volume II: Causality, Integration and Cointegration, and Long Memory*. Edited by E. Ghysels, N. R. Swanson, M. W. Watson. Cambridge: Cambridge University Press, 2001. 398 pages. ISBN: 978-0-521-79207-3.

[154] Granger, C. W. J. Testing for causality: A personal viewpoint. In *Essays in Econometrics Collected Papers of Clive W. J. Granger. Volume II: Causality, Integration and Cointegration, and Long Memory*. Edited by E. Ghysels, N. R. Swanson, M. W. Watson. Cambridge: Cambridge University Press, 2001. pp. 31–43. 398 pages. ISBN: 978-0-521-79207-3. URL: http://ecsocman.hse.ru/data/390/696/1219/sampler.pdf.

[155] Granger, C. W. J.; Joyeux R. An introduction to long memory time series models and fractional differencing. *Journal of Time Series Analysis* **1980**. Vol. 1. P. 15–39. DOI: 10.1111/j.1467-9892.1980.tb00297.x.

[156] Grigoletto, E. C.; De Oliveira E. C. Fractional versions of the fundamental theorem of calculus. *Applied Mathematics* **2013**. Vol. 4. P. 23–33. DOI: 10.4236/am.2013.47A006.

[157] Grunwald, A. K. About "limited" derivations their application [Uber "begrenzte" Derivationen und deren Anwendung]. *Zeitschrift fur angewandte Mathematik und Physik* **1867**. Vol. 12. P. 441–480 [in German]. URL: https://www.deutsche-digitale-bibliothek.de/item/7OWDE57YKUI6KP2N2ZI3SDEW2B2EZ6AP.

[158] Hajipour, A.; Hajipour, M.; Baleanu, D. On the adaptive sliding mode controller for a hyperchaotic fractional-order financial system. *Physica A: Statistical Mechanics and its Applications* **2018**. Vol. 497. P. 139–153. DOI: 10.1016/j.physa.2018.01.019.

[159] Haken, H. *Synergetics. An Introduction. Nonequilibrium Phase Transitions and Self-Organization in Physics, Chemistry and Biology*. Berlin, Heidelberg, New York: Springer-Verlag, 1977. 325 pages. ISBN: 978-3-642-96365-0. DOI: 10.1007/978-3-642-96363-6.

[160] Haken, H. *Advanced Synergetics: Instability Hierarchies of Self-Organizing Systems and Devices*. Berlin, Heidelberg, New York, Tokyo: Springer-Verlag 1983. (2nd ed. 1987) 356 pages. ISBN: 978-3-642-45555-1. DOI: 10.1007/978-3-642-45553-7.

[161] Haken, H. *Synergetics. An Introduction*. Berlin, Heidelberg: Springer-Verlag, 1983. 390 pages. DOI: 10.1007/978-3-642-88338-5.

[162] Haken, H. Slaving principle revisited. *Physica D: Nonlinear Phenomena* **1996**. Vol. 97. No. 1–3. P. 95–103. DOI: 10.1016/0167-2789(96)00080-2.

[163] Haken, H. *Synergetics. Introduction and Advanced Topics*. Berlin, Heidelberg, New York: Springer-Verlag, 2004. 758 pages. ISBN: 978-3-642-07405-9. DOI: 10.1007/978-3-662-10184-1.

[164] Handbook of Fractional Calculus with Applications. Volume 1. Basic Theory. Edited by A. Kochubei, and Yu. Luchko. Berlin, Boston: Walter de Gruyter GmbH, 2019. 481 pages. ISBN: 978-3-11-057081-6 DOI: 10.1515/9783110571622.

[165] Handbook of Fractional Calculus with Applications. Volume 2. Fractional Differential Equations. Edited by A. Kochubei, and Yu. Luchko. Berlin, Boston: Walter de Gruyter GmbH, 2019. 519 pages. ISBN: 978-3-11-057082-3 DOI: 10.1515/9783110571660.

[166] Handbook of Fractional Calculus with Applications. Volume 3. Numerical Methods. Edited by G. E. Karniadakis. Berlin, Boston: Walter de Gruyter GmbH, 2019. 350 pages. ISBN: 978-3-11-057083-0 DOI: 10.1515/9783110571684.

[167] Handbook of Fractional Calculus with Applications. Volume 4. Application in Physics. Part A. Edited by V. E. Tarasov. Berlin, Boston: Walter de Gruyter GmbH, 2019. 306 pages. ISBN: 978-3-11-057088-5 DOI: 10.1515/9783110571707.

[168] Handbook of Fractional Calculus with Applications. Volume 5. Application in Physics. Part B. Edited by V. E. Tarasov. Berlin, Boston: Walter de Gruyter GmbH, 2019. 319 pages. ISBN: 978-3-11-057089-2 DOI: 10.1515/9783110571721.

[169] Hardy, G. H. *Divergent Series. Secondd Edition*. New York: American Mathematical Society, 2000. 396 pages.

[170] Harrod, R. F. An Essay in dynamic theory. *Economic Journal* **1939**. Vol. 49 (193). P. 14–33. DOI: 10.2307/2225181. URL: http://piketty.pse.ens.fr/files/Harrod1939.pdf.

[171] Harrod, R. F. *Towards a Dynamic Economics*. London: Macmillan, 1948.

[172] Harrod, R. F. *Economic Dynamics*. London: Macmillan, 1973. ISBN: 978-0-333-14247-9.

[173] He, L.; Yi, L.; Tang, P. Numerical scheme and dynamic analysis for variable-order fractional van der Pol model of nonlinear economic cycle. *Advances in Difference Equations* **2016**. Vol. 2016. Article ID: 195. DOI: 10.1186/s13662-016-0920-5.

[174] Hillinger, C. Evidence and ideology in macroeconomics: the case of investment cycles. *Munich Discussion Paper*. No. 2005-16. Ludwig-Maximilians-Universitat Munchen, Volkswirtschaftliche Fakultat, Munchen, 2005. 54 pages. DOI: 10.5282/ubm/epub.694. Available at SSRN: https://ssrn.com/abstract=814527 or Econstor: https://www.econstor.eu/handle/10419/104163.

[175] Hillinger, C.; Reiter, M.; Weser, T. *Micro Foundations of the Second-Order Accelerator and of Cyclical Behavior. Chapter 8 in Cyclical Growth in Market and Planned Economies*. Oxford: Clarendon Press, 1992. 224 pages. ISBN: 0198283148.

[176] Hoover, K. D. *Causality in Macroeconomics*. Cambridge: Cambridge University Press, 2001. 326 pages. ISBN: 978-0521002882.

[177] Hoover, K. D. *The Methodology of Empirical Macroeconomics*. Cambridge: Cambridge University Press, 2004. 186 pages. ISBN: 0-521-80272-5.

[178] Hoover, K. D. Causality in economics and econometrics. In *The New Palgrave Dictionary of Economics. Volume 1*. Edited by S. N. Durlauf, L. E. Blume. Second Edition. New York: Palgrave Macmillan, 2008. pp. 719–728. ISBN: 978-1-349-58804-6. DOI: 10.1007 /978-1-349-58802-2.

[179] Hori, H. *Business Cycle Dynamics and Stabilization Policies. A Keynesian Approach*. Singapore: Springer, 2017. 142 pages. ISSN 2197-8859. ISSN: 2197-8859. DOI: 10.1007/978-981-10-3081-9.

[180] Hosking, J. R. M. Fractional differencing. *Biometrika* **1981**. Vol. 68. No. 1. P. 165–176. DOI: 10.1093/biomet/68.1.165.

[181] Husain, H. S.; Mariam Sultana, M. Principal parts of a vector bundle on projective line and the fractional derivative. *Turkish Journal of Mathematics* **2019**. Vol. 43. P. 2697–2706. DOI: 10.3906/mat-1901-23.

[182] Jajarmi, A.; Hajipour, M.; Baleanu, D. New aspects of the adaptive synchronization and hyperchaos suppression of a financial model. *Chaos, Solitons and Fractals* **2017**. Vol. 99. P. 285–296. DOI: 10.1016/j.chaos.2017.04.025.

[183] Jiang, Y.; Wang, X.; Wang, Y. On a stochastic heat equation with first order fractional noises and applications to finance. *Journal of Mathematical Analysis and Applications* **2012**. Vol. 396. No. 2. P. 656–669. DOI: 10.1016/j.jmaa.2012.07.003.

[184] Jiao, Z.; Chen, Y. Q.; Podlubny, I. *Distributed-Order Dynamic Systems: Stability, Simulation, Applications and Perspectives*. London: Springer, 2012. 90 pages. DOI: 10.1007/978-1-4471-2852-6.

[185] Kahn, R. F. The relation of home investment to unemployment. *The Economic Journal* **1931**. Vol. 41. No. 162. P. 173–198. DOI: 10.2307/2223697.

[186] Kalazhokov, Kh. Kh. Some classes of dynamic models of productive and economic systems with "memory". *News of the Kabardino-Balkarian Scientific Center of the Russian Academy of Sciences [Izvestiya Kabardino-Balkarskogo Nauchnogo Centra RAN]* **2012**. No. 4 (48). P. 26–34 [in Russian]. URL: https://elibrary.ru/item.asp?id=17905118.

[187] Kaldor, N. A model of the trade cycle. *The Economic Journal* **1940**. Vol. 50. No. 197. P. 78–92. DOI: 10.2307/2225740.

[188] Kaldor, N. Capital accumulation and economic growth. Chapter 10 In *The Theory of Capital*. Edited by D. C. Hague. London: Palgrave Macmillan, 1961. P. 177–222. DOI: 10.1007/978-1-349-08452-4_10. URL: http://www.masterhdfs.org/masterHDFS/wp-content/uploads/2014/05/Kaldor_1961.pdf.

[189] Kalecki, M. A macrodynamic theory of business cycles. *Econometrica* **1935**. Vol. 3. No. 3. P. 327–344. DOI: 10.2307/1905325. URL: https://archive.org/details/Kalecki1935.AMacrodynamicTheoryOfBusinessCycles/page/n1 (accessed on 9.05.2019).

[190] Kerss, A.; Leonenko, N.; Sikorskii, A. Fractional Skellam processes with applications to finance. *Fractional Calculus and Applied Analysis* **2014**. Vol. 17. No. 2. P. 532–551. DOI: 10.2478/s13540-014-0184-2.

[191] Keynes, J. M. *The General Theory of Employment, Interest and Money*. London: Macmillan 1936.

[192] Keynes, J. M. *The Collected Writings of John Maynard Keynes. Volumes 1–30*. Edited by E. Johnson, D. Moggridge, A. Robinson. Cambridge: Cambridge University Press, 2013. 15126 pages. ISBN: 978-1107677722. Reprint edition.

[193] Keynes, J. M. *The General Theory of Employment, Interest and Money: With the Economic Consequences of the Peace. Series: Classics of World Literature. Ware (Hertfordshire, UK)*: Wordsworth Editions, 2017. 576 pages. ISBN: 978-1840227475.

[194] Keynes, J. M. *The General Theory of Employment, Interest and Money: With the Economic Consequences of the Peace*. Palgrave Macmillan, 2018. 438 pages. DOI: 10.1007/978-3-319-70344-2.

[195] Kilbas, A. A.; Marzan, S. A. Cauchy problem for differential equation with Caputo derivative. *Fractional Calculus and Applied Analysis* **2004**. Vol. 7. No. 3. P. 297–321. URL: http://www.diogenes.bg/fcaa/volume7/fcaa73/73_Kilb_Marzan.pdf.

[196] Kilbas, A. A.; Marzan, S. A. Nonlinear differential equations with the Caputo fractional derivative in the space of continuously differentiable functions. *Differential Equations* **2005**. Vol. 41. No. 1. P. 84–89. DOI: 10.1007/s10625-005-0137-y.

[197] Kilbas, A. A.; Saigo, M. On solution of integral equation of Abel-Volterra type. *Differential and Integral Equations* **1995**. Vol. 8. No. 5. P. 993–1011. URL: https://projecteuclid.org/download/pdf_1/euclid.die/1369056041.

[198] Kilbas, A. A.; Saigo, M.; Saxena, R. K. Solution of Volterra integro-differential equations with generalized Mittag-Leffler function in the kernels. *Journal of Integral Equations and Applications* **2002**. Vol. 14. No. 4. P. 377–396. DOI:10.1216/jiea/1181074929.

[199] Kilbas, A. A.; Saigo, M.; Saxena, R. K. Generalized Mittag-Leffler function and generalized fractional calculus operators. *Integral Transforms and Special Functions* **2004**. Vol. 15. No. 1. P. 31–49. DOI: 10.1080/10652460310001600717.

[200] Kilbas, A. A.; Srivastava, H. M.; Trujillo J. J. *Theory and Applications of Fractional Differential Equations*. Amsterdam: Elsevier, 2006. 540 pages.

[201] Kiryakova, V. Convolutions of Erdelyi-Kober fractional integration operators. In *Proceedings in Complex Analysis and Applications (Varna, 1987)*. Sofia, 1989. pp. 273–283.

[202] Kiryakova, V. *Generalized Fractional Calculus and Applications*. New York: Longman and J. Wiley, 1994. 360 pages. ISBN: 9780582219779.

[203] Kiryakova, V. A brief story about the operators of the generalized fractional calculus. *Fractional Calculus and Applied Analysis* **2008**. Vol. 11. No. 2. P. 203–220. URL: https://pdfs.semanticscholar.org/5a45/82f7ba9d2bdb2d05f70c316b3347876ad1f3.pdf.

[204] Kiryakova, V.; Luchko, Yu. Riemann-Liouville and Caputo type multiple Erdelyi-Kober operators. *Central European Journal of Physics* **2013**. Vol. 11. No. 10. P. 1314–1336. DOI: 10.2478/s11534-013-0217-1.

[205] Kleinerta, H.; Korbel, J. Option pricing beyond Black-Scholes based on double-fractional diffusion. *Physica A: Statistical Mechanics and its Applications* **2016**. Vol. 449. P. 200–214. DOI: 10.1016/j.physa.2015.12.125 (arxiv:1503.05655).

[206] Kluppelberg, C.; Kuhn, C. Fractional Brownian motion as a weak limit of Poisson shot noise processes with applications to finance. *Stochastic Processes and their Applications* **2004**. Vol. 113. No. 2. P. 333–351. DOI: 10.1016/j.spa.2004.03.015.

[207] Koblo, R. B. *The Visible Hand: Synergetic Microfoundation of Macroeconomic Dynamics*. Berlin, Heidelberg: Springer-Verlag, 1991. 131 pages. ISBN: 978-3-540-54595-8. DOI: 10.1007/978-3-642-46753-0.

[208] Kochubei, A. N. Distributed order calculus and equations of ultraslow diffusion. *Journal of Mathematical Analysis and Applications* **2008**. Vol. 340. No. 1. P. 252–281. DOI: 10.1016/j.jmaa.2007.08.024.

[209] Kochubei A. N. General fractional calculus, evolution equations, and renewal processes. *Integral Equations and Operator Theory* **2011**. Vol. 71. No. 4. P. 583–600. DOI: 10.1007/s00020-011-1918-8 (arXiv:1105.1239).

[210] Kochubei, A. N. General fractional calculus. Chapter 5. In *Handbook of Fractional Calculus with Applications. Volume 1. Basic Theory*. Edited by A. Kochubei, Yu. Luchko. Series edited by J. A. Tenreiro Machado. Berlin, Boston: De Gruyter, 2019. pp. 111–126. DOI: 10.1515/9783110571622-005.

[211] Kochubei A. N.; Kondratiev, Yu. Growth equation of the general fractional calculus. *Mathematics* **2019**. Vol. 7. No. 7. Article ID: 615. 8 pages. DOI: 10.3390/math7070615.

[212] Kolemaev, V. A. *Mathematical economics*. Third Edition. Moscow: Unity-Dana, 2005. 399 pages [in Russian]. ISBN: 5-238-00794-9.

[213] Korbel, J.; Luchko, Yu. Modeling of financial processes with a space-time fractional diffusion equation of varying order. *Fractional Calculus and Applied Analysis* **2016**. Vol. 19. No. 6. P. 1414–1433. DOI: 10.1515/fca-2016-0073.

[214] Korolyuk, V. S.; Portenko, N. I.; Skorokhod, A. V.; Turbin, A. F. *Handbook on Probability Theory and Mathematical Statistics*. Second Edition. Moscow: Nauka, 1985. 640 pages.

[215] Kuroda, K.; Murai, J. Long memory in finance and fractional Brownian motion. *Progress of Theoretical Physics. Supplement* **2009**. Vol. 179. P. 26–37. DOI: 10.1143/PTPS.179.26.

[216] Kwasnicki, W. Logistic growth of the global economy and competitiveness of nations. *Technological Forecasting and Social Change* **2013**. Vol. 80. No. 1. P. 50–76. URL: http://kwasnicki.prawo.uni.wroc.pl/todownload/WKLogisticGrowthTF&SChFinal.pdf.

[217] Lancaster, K. *Mathematical Economics*. New York: The Macmillan Company, 1968. ISBN: 978-0023221316. (Revised edition. Dover Publications, 2011. 448 pages. ISBN: 978-0486653914).

[218] Laskin, N. Fractional market dynamics. *Physica A: Statistical Mechanics and its Applications* **2000**. Vol. 287. No. 3. P. 482–492. DOI: 10.1016/S0378-4371(00)00387-3.

[219] Leontief, W. W. *The Structure of the American Economy 1919–1939: An Empirical Application of Equilibrium Analysis*. Second Edition. New York: Oxford University Press, 1951. 282 pages.

[220] Leontief, W. W. *Input–Output Economics*. Second Edition. New York: Oxford University Press, 1986.

[221] Letnikov, A. V. On the historical development of the theory of differentiation with arbitrary index. *Mathematical Collection [Matematicheskii Sbornik]* **1868**. Vol. 3. No. 2. P. 85–112 [in Russian]. URL: http://mi.mathnet.ru/eng/msb8048.

[222] Letnikov, A. V. Theory of differentiation with arbitrary pointer [Teoriya differenchirovaniya s proizvolnym ukazatelem]. *Matematicheskii Sbornik* **1868**. Vol. 3. No. 1. P. 1–68 [in Russian]. URL: http://mi.mathnet.ru/eng/msb8039.

[223] Li, C. P.; Cai, M. *Theory and Numerical Approximations of Fractional Integrals and Derivatives*. Philadelphia: SIAM, 2019. 312 pages. ISBN: 978-1-61197-587-1. DOI: 10.1137/1.9781611975888. URL: https://epubs.siam.org/doi/book/10.1137/1.9781611975888.

[224] Li, C. P.; Chen, A. Numerical methods for fractional partial differential equations. *International Journal of Computer Mathematics* **2018**. Vol. 95. No. 6–7. P. 1048–1099. DOI: 10.1080/00207160.2017.1343941.

[225] Li, C. P.; Deng, W. H. Remarks on fractional derivatives. *Applied Mathematics and Computation* **2007**. Vol. 187. P. 777–784. DOI: 10.1016/j.amc.2006.08.163.

[226] Li, C. P.; Sarwar, S. Existence and continuation of solutions for Caputo type fractional differential equations. *Electronic Journal of Differential Equations* **2016**. Vol. 2016. No. 207. P. 1–14. ISSN: 1072-6691. URL: http://ejde.math.txstate.edu or http://ejde.math.unt.edu. URL: https://ejde.math.txstate.edu/Volumes/2016/207/li.pdf.

[227] Li, C. P.; Zeng, F. H. *Numerical Methods for Fractional Calculus*. Boca Raton: Chapman and Hall/CRC, 2015. 281 pages. ISBN: 9781482253801.

[228] Li, M.; Lim, S. C.; Hu, B. J.; Feng, H. Towards describing multi-fractality of traffic using Local Hurst function. In *International Conference on Computational Science. ICCS 2007: Computational Science – ICCS 2007*. Lecture Notes in Computer Science. Vol. 4488. Berlin, Heidelberg: Springer, 2007. pp. 1012–1020. DOI: 10.1007/978-3-540-72586-2_143.

[229] Lighthill, M. J. *An Introduction to Fourier Analysis and Generalised Functions*. Cambridge: Cambridge University Press, 1958. 79 pages. ISBN: 9781139171427. DOI: 10.1017/CBO9781139171427.

[230] Lokshin, A. A.; Suvorova, Yu. V. *Mathematical Theory of Wave Propagation in Media with Memory*. Moscow: Moscow State University, 1982. 152 pages [in Russian].

[231] Long Memory in Economics. Edited by G. Teyssiere, A. P. Kirman. Berlin, Heidelberg: Springer-Verlag, 2007. 390 pages. DOI: 10.1007/978-3-540-34625-8.

[232] Lorenz, H. W. *Nonlinear Dynamical Economics and Chaotic Motion*. Berlin, Heidelberg: Springer-Verlag, 1989. 248 pages. DOI: 10.1007/978-3-662-22233-1.

[233] Lorenz, H. W. *Nonlinear Dynamical Economics and Chaotic Motion. Second, Revised and Enlarged Edition*. Berlin, Heidelberg, New York: Springer-Verlag, 1993. 319 pages. DOI: 10.1007/978-3-642-78324-1.

[234] Lorenzo, C. F.; Hartley, T. T. Variable order and distributed order fractional operators. *Nonlinear Dynamics* **2002**. Vol. 29. No. 1. P. 57–98. DOI: 10.1023/A:1016586905654.

[235] Lucas, R. E. *Studies in Business-Cycle Theory*. Cambridge: MIT Press, 1981. 300 pages. ISBN: 978-0262620444.

[236] Lucas, R. E. *Models of Business Cycles*. Yrjo Jahnsson Lectures, Oxford: Basil Blackwell, 1987. ISBN: 9780631147916.

[237] Lucas, R. E. On the mechanics of economic development. *Journal of Monetary Economics* **1988**. Vol. 22. No. 1. P. 3–42. DOI: 10.1016/0304-3932(88)90168-7.

[238] Lucas, R. E. Jr. Making a Miracle. *Econometrica* **1993**. Vol. 61. No. 2. P. 251–272. DOI: 10.2307/2951551. URL: http://www.dklevine.com/archive/refs42101.pdf.

[239] Lucas, R. E. Jr. *Lectures on Economic Growth*. Cambridge and London: Harvard University Press, 2002. 204 pages. ISBN: 0-674-00627-5.

[240] Luchko, Yu. Fractional derivatives and the fundamental theorem of fractional calculus. *Fractional Calculus and Applied Analysis*. **2020**. Vol. 23. P. 939–966. DOI: 10.1515/fca-2020-0049 (arXiv:2006.14383)

[241] Luchko, Yu.; Trujillo, J. J. Caputo-type modification of the Erdelyi-Kober fractional derivative. *Fractional Calculus and Applied Analysis* **2007**. Vol. 10. No. 3. P. 249–268.

[242] Luo, D.; Wang, J. R.; Feckan, M. Applying fractional calculus to analyze economic growth modelling. *Journal of Applied Mathematics, Statistics and Informatics* **2018**. Vol. 14. No. 1. P. 25–36. DOI: 10.2478/jamsi-2018-0003.

[243] Ma, S.; Xu, Y.; Yue, W. Numerical solutions of a variable order fractional financial system. *Journal of Applied Mathematics*. **2012**. Vol. 2012. Article ID: 417942. 14 pages. DOI: 10.1155/2012/417942.

[244] Mainardi, F. Fractional relaxation-oscillation and fractional diffusion-wave phenomena. *Chaos, Solitons and Fractals* **1996**. Vol. 7. No. 9. P. 1461–1477. DOI: 10.1016/0960-0779(95)00125-5.

[245] Mainardi, F. *Fractional Calculus and Waves Linear Viscoelasticity: An Introduction to Mathematical Models*. London: Imperial College Press, 2010. 368 pages.

[246] Mainardi, F. On the advent of fractional calculus in econophysics via continuous-time random walk. *Mathematics* **2020**. Vol. 8. No. 4. Article ID: 641. DOI: 10.3390/math8040641.

[247] Mainardi, F.; Gorenflo, R. Fractional relaxation-oscillation phenomena. Chapter 3 In *Handbook of Fractional Calculus with Applications. Volume 4. Application in Physics. Part A*. Edited by V. E. Tarasov. Berlin, Boston: Walter de Gruyter GmbH, 2019. pp. 45–73. DOI: 10.1515/9783110571707-003.

[248] Mainardi, F.; Raberto, M.; Gorenflo, R.; Scalas, E. Fractional calculus and continuous-time finance II: The waiting-time distribution. *Physica A: Statistical Mechanics and its Applications* **2000**. Vol. 287. No. 3–4. P. 468–481. DOI: 10.1016/S0378-4371(00)00386-1.

[249] Makarov, D. V. On a dynamic hereditarity system that simulates the economic cycle. *Bulletin KRASEC. Physical and Mathematical Sciences [Vestnik KRAUNC. Fiziko-Matematicheskie Nauki, ISSN 2079-6641]* **2016**. No. 2 (13). P. 55–61. DOI: 10.18454/2079-6641-2016-13-2-55-61. URL: http://mi.mathnet.ru/eng/vkam132.

[250] Malinowska, A. B.; Odzijewicz, T.; Torres, D. F. M. *Advanced Methods in the Fractional Calculus of Variations*. Heidelberg, New York: Springer, 2015. 135 pages. DOI: 10.1007/978-3-319-14756-7.

[251] Marom, O.; Momoniat, E. A comparison of numerical solutions of fractional diffusion models in finance. *Nonlinear Analysis: Real World Applications* **2009**. Vol. 10. No. 6. P. 3435–3442. DOI: 110.1016/j.nonrwa.2008.10.066.

[252] Mathai, A. M.; Haubold H. J. Erdelyi-Kober fractional integral operators from a statistical perspective - I. *Tbilisi Mathematical Journal* **2017**. Vol. 10. No. 1. P. 145–159. DOI: 10.1515/tmj-2017-0009 (arXiv:1303.3978).

[253] Mathai, A. M.; Haubold, H. J. *An Introduction to Fractional Calculus*. New York: Nova Science Publishers, 2017. 258 pages. ISBN: 978-1536120424.

[254] Mathematical Economics: Application of Fractional Calculus. Edited by V. E. Tarasov. Basel, Beijing: MDPI, 2020. 278 pages. ISBN: 978-3-03936-118-2; 978-3-03936-119-9. DOI: 10.3390/books978-3-03936-119-9. URL: https://www.mdpi.com/books/pdfview/book/2350.

[255] Mathematical works: memories and notes [Opere matematiche: memorie e note]/Vito Volterra; pubblicate a cura dell'Accademia nazionale dei Lincei col concorso del Consiglio nazionale delle ricerche; 5: 1926–1940/Vito Volterra; corredato dall'Elenco cronologico generale delle pubblicazioni. Roma: Accademia nazionale dei Lincei, 1962. 538 pages [in Italian].

[256] Matsumoto, A.; Szidarovszky, F. Nonlinear multiplier–accelerator model with investment and consumption delays. *Structural Change and Economic Dynamics* **2015**. Vol. 33. P. 1–9. DOI: 10.1016/j.strueco.2015.01.003.

[257] Matthews, R. C. O. *The Trade Cycle*. Cambridge: Cambridge University Press, 1959. 300 pages. ISBN: 0521087635, ISBN: 978-0521087636.

[258] Matychyn, I. I.; Onyshchenko, V. V. Differential games of fractional order with impulse effect. *Journal of Automation and Information Sciences* **2015**. Vol. 47. No. 4. P. 43–53. DOI: 10.1615/JAutomatInfScien.v47.i4.50.

[259] Matychyn, I. I.; Chikriy, A. A.; Onyshchenko, V. V. Conflict-controlled processes involving fractional differential equations with impulses. *Mathematica Balkanica* **2012**. Vol. 26. No. 1-2. P. 159–168. DOI: 10.13140/2.1.1835.6167.

[260] May, R. M. Simple mathematical models with very complicated dynamics. *Nature* **1976**. Vol. 261 (5560). P. 459–467. DOI: 10.1038/261459a0.

[261] Meerschaert, M. M.; Scalas, E. Coupled continuous time random walks in finance. *Physica A: Statistical Mechanics and its Applications* **2006**. Vol. 370. No. 1. P. 114–118. DOI: 10.1016/j.physa.2006.04.034 (arXiv:physics/0608281).

[262] Mendes, R. V. A fractional calculus interpretation of the fractional volatility model. *Nonlinear Dynamics* **2009**. Vol. 55. No. 4. P. 395–399. DOI: 10.1007/s11071-008-9372-0.

[263] Menger, C. *Grundsatze der Volkswirtschaftslehre*. Austria, Wien: Braumuller, 1871. 285 pages [in German]. URL: https://oll.libertyfund.org/titles/menger-grundsatze-der-volkswirtschaftslehre.

[264] Menger, K. *Selected Works*. Moscow: Publishing House "Territoriya Budushchego", 2005. 496 pages [in Russian]. ISBN: 5-733-0175-9.

[265] Menger, C. *Principle of Economics*. Auburn, Alabama: Ludwig von Mises Institute, 2007. Translated by J. Dingwall and B.F. Hoselitz. 328 pages. ISBN: 978-1-933550-12-1.

[266] Mickens, R. E. Difference equation models of differential equations. *Mathematical and Computer Modelling* **1988**. Vol. 11. P. 528–530. DOI: 10.1016/0895-7177(88)90549-3.

[267] Mickens, R. E. *Nonstandard Finite Difference Models of Differential Equations*. Singapore: World Scientific, 1994. 262 pages. ISBN: 9810214588. ISBN: 978-9810214586.

[268] Mickens, R. E. Discretizations of nonlinear differential equations using explicit nonstandard methods. *Journal of Computational and Applied Mathematics* **1999**. Vol. 110. No. 1. P. 181–185. DOI: 10.1016/S0377-0427(99)00233-2.

[269] Mickens, R. E. Nonstandard finite difference schemes for differential equations. *Journal of Difference Equations and Applications* **2002**. Vol. 8. No. 9. P. 823–847. DOI: 10.1080/1023619021000000807.

[270] Mikhailov, A. S. *Foundations of Synergetics I: Distributed Active Systems*. Berlin, Heidelberg: Springer, 1990. 187 pages. ISBN: 978-3-642-97271-3. DOI: 10.1007/978-3-642-97269-0.

[271] Mikhailov, A. S.; Loskutov, A. Yu. *Foundations of Synergetics II: Complex Patterns*. Berlin, Heidelberg, New York: Springer-Verlag, 1991. 210 pages. ISBN: 978-3-642-97296-6. DOI: 10.1007/978-3-642-97294-2.

[272] Mikhailov, A. S.; Loskutov, A. Yu. *Foundations of Synergetics II: Chaos and Noise. Second Revised and Enlarged Edition*. Berlin, Heidelberg, New York: Springer-Verlag, 1996. 277 pages. ISBN: 978-3-642-80198-3. DOI: 10.1007/978-3-642-80196-9.

[273] Ming, H.; Wang, J. R.; Feckan, M. The application of fractional calculus in Chinese economic growth models. *Mathematics* **2019**. Vol. 7. No. 8. Article ID: 665. 6 pages. DOI: 10.3390/math7080665.

[274] Mishura, Yu. S. *Stochastic Calculus for Fractional Brownian Motion and Related Processes. Lecture Notes in Mathematics*. Berlin Heidelberg: Springer-Verlag, 2008. 393 pages. ISBN: 978-3540758723. DOI: 10.1007/978-3-540-75873-0.

[275] Miskiewicz, J.; Ausloos, M. A logistic map approach to economic cycles. (I). The best adapted companies. *Physica A: Statistical Mechanics and its Applications* **2004**. Vol. 336. No. 1–2. P. 206–214.

[276] Muniandy, S. V.; Lim, S. C. Modeling of locally self-similar processes using multifractional Brownian motion of Riemann-Liouville type. *Physical Review E* **2001**. Vol. 63. Article ID: 046104 (7 pages). DOI: 10.1103/PhysRevE.63.046104.

[277] Muniandy, S. V.; Lim, S. C.; Murugan, R. Inhomogeneous scaling behaviors in Malaysian foreign currency exchange rates. *Physica A: Statistical Mechanics and its Applications* **2001**. Vol. 301. No. 1-4. P. 407–428. DOI: 10.1016/S0378-4371(01)00387-9.

[278] Nakhushev, A. M. On the positivity of continuous and discrete differentiation and integration operators that are very important in fractional calculus and in the theory of equations of mixed type. *Differential Equations* **1998**. Vol. 34. No. 1. P. 103–112. URL: http://mi.mathnet.ru/eng/de9635.

[279] Nakhushev, A. M. *Fractional Calculus and its Application*. Moscow: Fizmatlit, 2003. 272 pages. ISBN: 5-9221-0440-3 [in Russian].

[280] Nakhushev, A. M. Loaded equations of mathematical economics. Reports of the Adyghe (Circassian). *International Academy of Sciences. [Doklady Adygskoi (Cherkesskoi) Mezhdunarodnoi Akademii Nauk, ISSN: 1726-9946]* **2010**. Vol. 12. No. 1. P. 91–97 [in Russian]. URL: https://elibrary.ru/item.asp?id=16950586.

[281] Nakhushev, A. M. Loaded equations of mathematical economics. In *Loaded equations and their application*. Edited by A. M. Nakhushev. Moscow: Nauka, 2012. pp. 145–153. 232 pages. ISBN: 978-5-02-037977-0 [in Russian].

[282] Nakhusheva, Z. A. On a one-sector macroeconomic model of long-term prognosis. *News of the Kabardino-Balkarian Scientific Center of the Russian Academy of Sciences [Izvestiya Kabardino-Balkarskogo Nauchnogo Centra RAN]* **2012**. No. 4 (48). P. 35–38 [in Russian].

[283] Nicholson, W.; Snyder, C. *Microeconomic Theory. Basic Principles and Extensions. 11th ed. Masson, Ohio: South-Western*, Cengage Learning, 2012. 758 pages. ISBN: 978-111-1-52553-8.

[284] Nicolis, G.; Prigogine, I. *Self-Organization in Nonequilibrium Systems: From Dissipative Structures to Order through Fluctuations*. Wiley, 1977. 512 pages. ISBN: 978-0471024019.

[285] Nikaido, H. *Convex Structures and Economic Theory*. New York and London: Academic Press, 1968. 405 pages.

[286] Odibat Z. M.; Shawagfeh N.T. Generalized Taylor's formula. *Applied Mathematics and Computation* **2007**. Vol. 186. No. 1. P. 286–293. DOI: 10.1016/j.amc.2006.07.102.

[287] Ortigueira, M. D.; Tenreiro Machado, J. A. What is a fractional derivative?. *Journal of Computational Physics* **2015**. Vol. 293. P. 4–13. DOI: 10.1016/j.jcp.2014.07.019.

[288] Pakhira, R.; Ghosh, U.; Sarkar, S. Application of memory effects in an inventory model with linear demand and no shortage. *International Journal of Research in Advent Technology* **2018**. Vol. 6. No. 8. P. 1853–1871. URL: http://www.ijrat.org/downloads/Vol-6/august-2018/paperID-68201804.pdf.

[289] Pakhira, R.; Ghosh, U.; Sarkar, S. Study of memory effect in an inventory model with linear demand and salvage value. *International Journal of Applied Engineering Research* **2018**. Vol. 13. No. 20. P. 14741–14751. URL: https://www.ripublication.com/ijaer18/ijaerv13n20_40.pdf.

[290] Pakhira, R.; Ghosh, U.; Sarkar, S. Study of memory effects in an inventory model using fractional calculus. *Applied Mathematical Sciences* **2018**. Vol. 12. P. 797–824. DOI: 10.12988/ams.2018.8578. URL: http://www.m-hikari.com/ams/ams-2018/ams-17-20-2018/p/ghoshAMS17-20-2018.pdf.

[291] Pakhira, R.; Ghosh, U.; Sarkar, S. Study of memory effect in an inventory model with linear demand and shortage. *International Journal of Mathematical Sciences and Computing* **2019**. Vol. 5. No. 2. P. 54–70. DOI: 10.5815/ijmsc.2019.02.05.

[292] Pakhira, R.; Ghosh, U.; Sarkar, S. Study of memory effect in an inventory model with quadratic type demand rate and salvage value. *Applied Mathematical Sciences* **2019**. Vol. 13. No. 5. P. 209–223. DOI: 10.12988/ams.2019.9111. URL: http://www.m-hikari.com/ams/ams-2019/ams-5-8-2019/p/pakhiraAMS5-8-2019.pdf.

[293] Pakhira, R.; Ghosh, U.; Sarkar, S. Study of memory effect in an inventory model with price dependent demand. *Journal of Applied Economic Sciencese Dependent Demand* **2019**. Vol. 14. No. 2(64). P. 360–367. DOI: 10.14505/jaes.v14.2(64).06.

[294] Pakhira, R.; Ghosh, U.; Sarkar, S. Study of memory effect in an economic order quantity model with quadratic type demand rate. *Computational Methods in Science and Technology* **2019**. Vol. 25. No. 2. P. 71–80. DOI: 10.12921/cmst.2019.0000004.

[295] Palma, W. *Long-Memory Time Series: Theory and Methods*. Hoboken: Wiley-InterScience, 2007. 304 pages. ISBN: 978-0-470-11402-5. DOI: 10.1002/97804701314.

[296] Pan, I.; Korre, A.; Das, S.; Durucan, S. Chaos suppression in a fractional order financial system using intelligent regrouping PSO based fractional fuzzy control policy in the presence of fractional Gaussian noise. *Nonlinear Dynamics* **2012**. Vol. 70. No. 4. P. 2445–2461. DOI: 10.1007/s11071-012-0632-7.

[297] Paris, R. B. Exponentially small expansions in the asymptotics of the Wright function. *Journal of Computational and Applied Mathematics* **2010**. Vol. 234. No. 2. P. 488–504. DOI: 10.1016/j.cam.2009.12.040.

[298] Paris, R. B. Exponentially small expansions of the Wright function on the Stokes lines. *Lithuanian Mathematical Journal* **2014**. Vol. 54. No. 1. P. 82–105. DOI: 10.1007/s10986-014-9229-9.

[299] Paris, R. B. Asymptotics of the special functions of fractional calculus. In *Handbook of Fractional Calculus with Applications. Volume 1: Basic Theory*. Berlin: De Gruyter, 2019. pp. 297–325. DOI: 10.1515/9783110571622-012.

[300] Paris, R. B. Some remarks on the theorems of Wright and Braaksma on the Wright function ${}_p\Psi_q$. **2017**. arXiv:1708.04824. URL: https://arxiv.org/abs/1708.04824.

[301] Paris, R. B.; Kaminski, D. *Asymptotics and Mellin-Barnes Integrals*. Cambridge: Cambridge University Press, 2001. 440 pages. DOI: 10.1017/CBO9780511546662.

[302] Paris, R. B.; Vinogradov, V. Asymptotic and structural properties of the Wright function arising in probability theory. *Lithuanian Mathematical Journal* **2016**. Vol. 56. No. 3. P. 377–409. DOI: 10.1007/s10986-016-9324-1 (arXiv:1508.00863).

[303] Parke, W. R. What is fractional integration?. *Review of Economics and Statistics* **1999**. Vol. 81. No. 4. P. 632–638. DOI: 10.1162/003465399558490.

[304] Peng, J.; Li, K. A note on property of the Mittag-Leffler function. *Journal of Mathematical Analysis and Applications* **2010**. Vol. 370. No. 2. P. 635–638. DOI: 10.1016/j.jmaa.2010.04.031.

[305] Phillips, A. W. Stabilisation policy in a closed economy. *Economic Journal* **1954**. Vol. 64. No. 254. P. 290–323. DOI: 10.2307/2226835.

[306] Phillips, A. W. *A. W. H. Phillips Collected Works in Contemporary Perspective*. Edited by R. Leeson. Cambridge: Cambridge University Press, 2000. 515 pages. ISBN: 9780521571357.

[307] Picozzi, S.; West, B. J. Fractional Langevin model of memory in financial markets. *Physical Review E*. **2002**. Vol. 66. Article ID: 046118. 12 pages. DOI: 10.1103/PhysRevE.66.046118.

[308] Podlubny, I. *Fractional Differential Equations*. San Diego: Academic Press, 1998. 340 pages.

[309] Pokrovskii, V. N. *Econodynamics: The Theory of Social Production*. Second Edition. Dordrecht, Heidelberg, London, New York: Springer, 2012. 200 pages. DOI: 10.1007/978-94-007-2096-1.

[310] Polito, F.; Tomovski, Z. Some properties of Prabhakar-type fractional calculus operators. *Fractional Differential Calculus* **2016**. Vol. 6. No. 1. P. 73–94. DOI: 10.7153/fdc-06-05.

[311] Polyanin, A. D.; Manzhirov, A. V. *Handbook of Integral Equations. Second Edition. Boca Raton, London*, New York: Chapman and Hall/CRC Press, 2008. 1108 pages. ISBN: 978-1584885078.

[312] Potts, R. B. Differential and difference equations. *American Mathematical Monthly* **1982**. Vol. 89. No. 6. P. 402–407.

[313] Potts, R. B. Ordinary and partial difference equations. *Journal of the Australian Mathematical Society B* **1986**. Vol. 27. No. 4. P. 488–501. DOI: 10.1017/S0334270000005099.

[314] Prabhakar, T. R. A singular integral equation with a generalized Mittag-Leffler function in the kernel. *Yokohama Mathematical Journal* **1971**. Vol. 19. P. 7–15. URL: https://irdb.nii.ac.jp/00822/0001826467.

[315] Prudnikov, A. P.; Brychkov, Yu. A.; Marichev, O. I. *Integrals and Series. Volume 1. Elementary Functions*. Moscow: Nauka, 1981. 798 pages [in Russian].

[316] Prudnikov, A. P.; Brychkov, Yu. A.; Marichev, O. I. *Integrals and Series. Volume 1. Elementary Functions*. London: Taylor & Francis, 2002. 798 pages. ISBN: 2-88124-089-5.

[317] Pskhu, A. V. On the theory of the continual integro-differentiation operator. *Differential Equations* **2004**. Vol. 40. No. 1. P. 128–136. DOI: 10.1023/B:DIEQ.0000028722.41328.21.

[318] Pskhu, A. V. *Partial Differential Equations of Fractional Order*. Moscow: Nauka, 2005. 199 pages. ISBN: 5-02-033721-8 [in Russian].

[319] Raberto, M.; Scalas, E.; Mainardi, F. Waiting-times and returns in high-frequency financial data: an empirical study. *Physica A: Statistical Mechanics and its Applications* **2002**. Vol. 314. No. 1-4. P. 749–755. DOI: 10.1016/S0378-4371(02)01048-8 (arXiv:cond-mat/0203596).

[320] Rabotnov, Yu. N. *Elements of Hereditary Solid Mechanics*. Moscow: Nauka, 1977. 384 pages [in Russian].

[321] Rabotnov, Yu. N. *Elements of Hereditary Solid Mechanics*. Moscow: Mir Publishers, 1980. 387 pages. ISBN: 978-0714714578.

[322] Razumikhin, B. S. *Stability of Hereditary Systems*. Moscow: Nauka, 1988. 108 pages. ISBN: 5-02-006601-X [in Russian].

[323] Rehman, H. U.; Darus, M.; Salah, J. A note on Caputo's derivative operator interpretation in economy. *Journal of Applied Mathematics* **2018**. Vol. 2018. Article ID: 1260240. 7 pages. DOI: 10.1155/2018/1260240. URL: https://www.hindawi.com/journals/jam/2018/1260240/.

[324] Rogers, L. C. G. Arbitrage with fractional Brownian motion. *Mathematical Finance* **1997**. Vol. 7. No. 1. P. 95–105. DOI: 10.1111/1467-9965.00025. URL: http://www.skokholm.co.uk/wp-content/uploads/2013/02/fbm.pdf.

[325] Romer, D. *Advanced Macroeconomics*. New York: McGraw-Hill Companies, 1996. 540 pages. ISBN: 0-07-053667-8.

[326] Romer, D. *Advanced Macroeconomics*. Third Edition. Boston: McGraw-Hill Companies, 2006. 678 pages. ISBN: 978-0-07-287730-8.

[327] Romer, D. *Advanced Macroeconomics*. Fourth Edition. New York: McGraw-Hill Companies, 2012. 716 pages. ISBN: 978-0-07-351137-5.

[328] Ross, B. A brief history and exposition of the fundamental theory of fractional calculus. In *Fractional Calculus and Its Applications. Proceedings of the International Conference Held at the University of New Haven, June 1974*. Series: Lecture Notes in Mathematics. Vol. 457. Berlin, Heidelberg: Springer, 1975. pp. 1–36. DOI: 10.1007/BFb0067096.

[329] Ross, B. Fractional Calculus. *Mathematics Magazine* **1977**. Vol. 50. No. 3. P. 115–122. DOI: 10.2307/2689497. URL: https://www.jstor.org/stable/2689497?seq=1#metadata_info_tab_contents.

[330] Ross B. The development of fractional calculus 1695–1900. *Historia Mathematica* **1977**. Vol. 4. No. 1. P. 75–89. DOI: 10.1016/0315-0860(77)90039-8.

[331] Rostek, S. Option Pricing in Fractional Brownian Markets. Berlin, Heidelberg: Springer-Verlag, 2009. 137 pages. DOI: 10.1007/978-3-642-00331-8.

[332] Rostek, S.; Schobel, R. A note on the use of fractional Brownian motion for financial modeling. *Economic Modelling* **2013**. Vol. 30. P. 30–35. DOI: 10.1016/j.econmod.2012.09.003.

[333] Saad, K. M.; Al-Shomrani, A. A.; Mohamed, M. S.; Yang, X. J. Solving fractional order logistic equation by approximate analytical methods. *International Journal of Open Problems in Computer Science and Mathematics* **2016**. Vol. 9. No. 2. Article ID: 3. 15 pages. URL: http://www.i-csrs.org/Volumes/ijopcm/vol.9/9.2.3.pdf.

[334] Sadeghi, A.; Cardoso, J. R. Some notes on properties of the matrix Mittag-Leffler function. *Applied Mathematics and Computation* **2018**. Vol. 338. P. 733–738. DOI: 10.1016/j.amc.2018.06.037.

[335] Samko, S. G.; Kilbas, A. A.; Marichev, O. I. *Fractional Integrals and Derivatives Theory and Applications*. New York: Gordon and Breach, 1993. 1006 pages. ISBN: 9782881248641.

[336] Samuta, V. V.; Strelova,V. A.; Parovik, R. I. Nonlocal Solow model of neoclassical economic growth. *Bulletin KRASEC. Physical and Mathematical Sciences [Vestnik KRAUNC. Fiziko-Matematicheskie Nauki, ISSN 2079-6641]* **2012**. No. 2 (5). P. 37–41. DOI: 10.18454/2079-6641-2012-5-2-37-41. URL: http://mi.mathnet.ru/eng/vkam72.

[337] Saut, J. C.; Joseph, D. D. Fading memory. *Archive for Rational Mechanics and Analysis* **1983**. Vol. 81. No. 1. P. 53–95. DOI: 10.1007/BF00283167.

[338] Sayevand, K.; Tenreiro Machado, J.; Baleanu, D. A new glance on the Leibniz rule for fractional derivatives. *Communications in Nonlinear Science and Numerical Simulation* **2018**. Vol. 62. P. 244–249. DOI: 10.1016/j.cnsns.2018.02.037.

[339] Scalas, E. Five years of continuous-time random walks in econophysics. In *The Complex Networks of Economic Interactions*. Lecture Notes in Economics and Mathematical Systems. Volume 567. Berlin: Springer-Verlag, 2006. pp. 3–16. DOI: 10.1007/3-540-28727-2_1.

[340] Scalas, E. The application of continuous-time random walks in finance and economics. *Physica A: Statistical Mechanics and its Applications* **2006**. Vol. 362. No. 2. P. 225–239. DOI: 10.1016/j.physa.2005.11.024.

[341] Scalas, E.; Gorenflo, R.; Mainardi, F. Fractional calculus and continuous-time finance. *Physica A: Statistical Mechanics and its Applications* **2000**. Vol. 284. No. 1–4. P. 376–384. DOI: 10.1016/S0378-4371(00)00255-7.

[342] Scalas, E.; Gorenflo, R.; Luckock, H.; Mainardi, F.; Mantelli, M.; Raberto, M. Anomalous waiting times in high-frequency financial data. *Quantitative Finance* **2004**. Vol. 4. No. 6. P. 695–702. DOI: 10.1080/14697680500040413.

[343] Schuster, H. G. *Deterministic Chaos*. Fourth, Revised and Enlarged Edition. Weinheim: WILEY-VCH Verlag GmbH & Co. KGaA, 2005. ISBN: 9783527604807. DOI: 10.1002/3527604804.

[344] Shaojie, W.; Shaobo, H.; Yousefpour, A.; Jahanshahi, H.; Repnik, R.; Perc, M. Chaos and complexity in a fractional-order financial system with time delays. *Chaos, Solitons and Fractals* **2020**. Vol. 131. Article ID: 109521. DOI: 10.1016/j.chaos.2019.109521.

[345] Sheremet, A. D. *Theory of Economic Analysis*. 2nd Edition. Moscow: Infra-M, 2005. 366 pages [in Russian]. ISBN: 5-16-002027-6.

[346] Shpilko, Y. E.; Solomko, A. A.; Parovik, R. I. Parametrization Samuelson equation model for Evans fixing, equilibrium price of the same product market. *Bulletin KRASEC. Physical and Mathematical Sciences [Vestnik KRAUNC. Fiziko-Matematicheskie Nauki, ISSN 2079-6641]* **2012**. No. 2 (5). P. 33–36 [in Russian]. DOI: 10.18454/2079-6641-2012-5-2-33-36. URL: http://mi.mathnet.ru/eng/vkam71.

[347] Sin, Ch.-S. Well-posedness of general Caputo-type fractional differential equations. *Fractional Calculus and Applied Analysis* **2018**. Vol. 21. No. 3. P. 819–832. DOI: 10.1515/fca-2018-0043.

[348] Skovranek, T. The Mittag-Leffler fitting of the Phillips curve. *Mathematics* **2019**. Vol. 7. No. 7. Article ID: 589. 11 pages. DOI: 10.3390/math7070589.

[349] Skovranek, T.; Podlubny, I.; Petras, I. Modeling of the national economies in state-space: A fractional calculus approach. *Economic Modelling* **2012**. Vol. 29. No. 4. P. 1322–1327. DOI: 10.1016/j.econmod.2012.03.019.

[350] Solow, R. M. A contribution to the theory of economic growth. *Quarterly Journal of Economics* **1956**. Vol. 70. No. 1. P. 65–94. DOI: 10.2307/1884513. URL: http://piketty.pse.ens.fr/files/Solow1956.pdf.

[351] Solow, R. M. Neoclassical growth theory. Chapter 9. In *Handbook of Macroeconomics*. Volume 1. Part A. Elsevier Science B.V., 1999. pp. 637–667. DOI: 10.1016/S1574-0048(99)01012-5.

[352] Srivastava, H. M.; Tomovski, Z. Fractional calculus with an integral operator containing a generalized Mittag-Leffler function in the kernel. *Applied Mathematics and Computation* **2009**. Vol. 211. No. 1. P. 198–210. DOI: 10.1016/j.amc.2009.01.055.

[353] Stanley, H. E.; Amaral, L. A. N.; Buldyrev, S. V.; Gopikrishnan, P.; Plerou, V.; Salinger, M. A. Self-organized complexity in economics and finance. *Proceedings of the National Academy of Sciences of the United States of America* **2002**. Vol. 99. No. Supplement 1. P. 2561–2565. DOI: 10.1073/pnas.022582899.

[354] Studies in the Structure of the American Economy. Theoretical and Empirical Explorations in Input-Output Analysis. Edited by W. W. Leontief. New York: Oxford University Press, 1953. 561 pages.

[355] Studies in the Structure of the American Economy. Theoretical and Empirical Explorations in Input-Output Analysis. Edited by W. W. Leontief. Moscow: Gosstatizdat, 1958. 640 pages [in Russian]. URL: https://drive.google.com/file/d/14GhmO_ar_wGfGHnXln8eyZhAlktAynqS/view.

[356] Sverdlove, R. Inverse problems for dynamical systems. *Journal of Differential Equations* **1981**. Vol. 42. No. 1. P. 72–105. DOI: 10.1016/0022-0396(81)90034-6.

[357] Swan, T. W. Economic growth and capital accumulation. *Economic Record* **1956**. Vol. 32. No. 2. P. 334–361. DOI: 10.1111/j.1475-4932.1956.tb00434.x.

[358] Tacha, O. I.; Munoz-Pacheco, J. M.; Zambrano-Serrano, E.; Stouboulos, I. N.; Pham, V.-T. Determining the chaotic behavior in a fractional-order finance system with negative parameters. *Nonlinear Dynamics* **2018**. Vol. 94. No. 2. P. 1303–1317. DOI: 10.1007/s11071-018-4425-5.

[359] Tarasov, V. E. Continuous limit of discrete systems with long-range interaction. *Journal of Physics A* **2006**. Vol. 39. No. 48. P. 14895–14910. DOI: 10.1088/0305-4470/39/48/005 (arXiv:0711.0826).

[360] Tarasov, V. E. Fractional vector calculus and fractional Maxwell's equations. *Annals of Physics* **2008**. Vol. 323. No. 11. P. 2756–2778. DOI: 10.1016/j.aop.2008.04.005 (arXiv:0907.2363).

[361] Tarasov, V. E. Differential equations with fractional derivative and universal map with memory. *Journal of Physics A*. **2009**. Vol. 42. No. 46. Article ID: 465102.

[362] Tarasov, V. E. Discrete map with memory from fractional differential equation of arbitrary positive order. *Journal of Mathematical Physics* **2009**. Vol. 50. No. 12. Article ID: 122703.

[363] Tarasov, V. E. *Fractional Dynamics: Applications of Fractional Calculus to Dynamics of Particles, Fields and Media*. New York: Springer, 2010. 505 pages. DOI: 10.1007/978-3-642-14003-7.

[364] Tarasov, V. E. Fractional Zaslavsky and Henon map. In *Long-range Interactions, Stochasticity and Fractional Dynamics*. Edited by A. C. J. Luo, V. Afraimovich. New York: Springer and HEP, 2010. pp. 1–26. 275 pages. DOI: 10.1007/978-3-642-12343-6_1 (arXiv:1107.5148).

[365] Tarasov, V. E. Quantum dissipation from power-law memory. *Annals of Physics* **2012**. Vol. 327. No. 6. P. 1719–1729. DOI: 10.1016/j.aop.2012.02.011.

[366] Tarasov, V. E. Lattice model with power-law spatial dispersion for fractional elasticity. *Central European Journal of Physics* **2013**. Vol. 11. No. 11. P. 1580–1588. DOI: 10.2478/s11534-013-0308-z (arXiv:1501.01201).

[367] Tarasov, V. E. No violation of the Leibniz rule. No fractional derivative. *Communications in Nonlinear Science and Numerical Simulation* **2013**. Vol. 18. No. 11. P. 2945–2948. DOI: 10.1016/j.cnsns.2013.04.001.

[368] Tarasov, V. E. Fractional gradient elasticity from spatial dispersion law. *ISRN Condensed Matter Physics* **2014**. Vol. 2014. Article ID: 794097. 13 pages. DOI: 10.1155/2014/794097 (arXiv:1306.2572).

[369] Tarasov, V. E. Fractional-order variational derivative. *International Journal of Applied Mathematics* **2014**. Vol. 27. No. 5. P. 491–518. DOI: 10.12732/ijam.v27i5.7 (arXiv:1502.07677).

[370] Tarasov, V. E. Toward lattice fractional vector calculus. *Journal of Physics A* **2014**. Vol. 47. No. 35. Article ID: 355204. 51 pages. DOI: 10.1088/1751-8113/47/35/355204.

[371] Tarasov, V. E. Exact discrete analogs of derivatives of integer orders: Differences as infinite series. *Journal of Mathematics* **2015**. Vol. 2015. Article ID: 134842. DOI: 10.1155/2015/134842.

[372] Tarasov, V. E. Fractional-order difference equations for physical lattices and some applications. *Journal of Mathematical Physics* **2015**. Vol. 56. No. 10. Article ID: 103506. DOI: 10.1063/1.4933028.

[373] Tarasov, V. E. Lattice fractional calculus. *Applied Mathematics and Computation* **2015**. Vol. 257. P. 12–33. DOI: 10.1016/j.amc.2014.11.033.

[374] Tarasov, V. E. Non-linear fractional field equations: weak non-linearity at power-law non-locality. *Nonlinear Dynamics* **2015**. Vol. 80. No. 4. P. 1665–1672. DOI: 10.1007/s11071-014-1342-0.

[375] Tarasov, V. E. Geometric interpretation of fractional-order derivative. *Fractional Calculus and Applied Analysis* **2016**. Vol. 19. No. 5. P. 1200–1221. DOI: 10.1515/fca-2016-0062.

[376] Tarasov, V. E. Leibniz rule and fractional derivatives of power functions. *Journal of Computational and Nonlinear Dynamics* **2016**. Vol. 11. No. 3. Article ID: 031014. DOI: 10.1115/1.4031364.

[377] Tarasov, V. E. On chain rule for fractional derivatives. *Communications in Nonlinear Science and Numerical Simulation* **2016**. Vol. 30. No. 1-3. P. 1–4. DOI: 10.1016/j.cnsns.2015.06.007.

[378] Tarasov, V. E. Exact discretization by Fourier transforms. *Communications in Nonlinear Science and Numerical Simulation* **2016**. Vol. 37. P. 31–61. DOI: 10.1016/j.cnsns.2016.01.006.

[379] Tarasov, V. E. Interpretation of fractional derivatives as reconstruction from sequence of integer derivatives. *Fundamenta Informaticae* **2017**. Vol. 151. No. 1–4. P. 431–442. DOI: 10.3233/FI-2017-1502.

[380] Tarasov, V. E. Generalized memory: fractional calculus approach. *Fractal and Fractional* **2018**. Vol. 2. No. 4. Article ID: 23. DOI: 10.3390/fractalfract2040023.

[381] Tarasov, V. E. Generalization of Lucas model that taking into account power memory. *Economy and Entrepreneurship (ISSN: 1999-2300)* **2020**. Vol. 14. No. 10 (123). P. 195–201. DOI: 10.34925/EIP.2020.123.10.034. https://www.elibrary.ru/item.asp?id=44195379.

[382] Tarasov, V. E. No nonlocality. No fractional derivative. *Communications in Nonlinear Science and Numerical Simulation* **2018**. Vol. 62. P. 157–163. DOI: 10.1016/j.cnsns.2018.02.019 (arXiv:1803.00750).

[383] Tarasov, V. E. Non-linear macroeconomic models of growth with memory. *Mathematics* **2020**. Vol. 8. No. 11. Article ID: 2078. 22 pages. DOI: 10.3390/math8112078. URL: https://www.mdpi.com/2227-7390/8/11/2078.

[384] Tarasov, V. E. Caputo-Fabrizio operator in terms of integer derivatives: memory or distributed lag?. *Computational and Applied Mathematics* **2019**. Vol. 38. Article ID: 113. 15 pages. DOI: 10.1007/s40314-019-0883-8.

[385] Tarasov, V. E. On history of mathematical economics: application of fractional calculus. *Mathematics* **2019**. Vol. 7. No. 6. Article ID: 509. 28 pages. DOI: 10.3390/math7060509.

[386] Tarasov, V. E. Rules for fractional-dynamic generalizations: difficulties of constructing fractional dynamic models. *Mathematics* **2019**. Vol. 7. No. 6. Article ID: 554. 50 pages. DOI: 10.3390/math7060554.

[387] Tarasov, V. E. Self-organization with memory. *Communications in Nonlinear Science and Numerical Simulation* **2019**. Vol. 72. P. 240–271. DOI: 10.1016/j.cnsns.2018.12.018.

[388] Tarasov, V. E. Cagan model of inflation with power-law memory effects. *Computational and Applied Mathematics* **2020**. Vol. 39. No. 3. Article ID: 207. 17 pages. DOI: 10.1007/s40314-020-01240-5.

[389] Tarasov, V. E. Corrigendum to "Fractional nonlinear dynamics of learning with memory". *Nonlinear Dynamics* **2020**. Vol. 100. P. 1231–1242. *Nonlinear Dynamics*. **2021**. Submitted 2020.06.28. To be published in 2021.

[390] Tarasov, V. E. Fractional econophysics: market price dynamics with memory effects. *Physica A: Statistical Mechanics and its Applications* **2020**. Vol. 557. Article ID: 124865. 16 pages. DOI: 10.1016/j.physa.2020.124865.

[391] Tarasov, V. E. Fractional nonlinear dynamics of learning with memory. *Nonlinear Dynamics* **2020**. Vol. 100. No. 2. P. 1231–1242. DOI: 10.1007/s11071-020-05602-w.

[392] Tarasov, V. E. Mathematical Economics: Application of Fractional Calculus. *Mathematics* **2020**. Vol. 8. No. 5. Article ID: 660. 3 pages. DOI: 10.3390/math8050660.

[393] Tarasov, V. E. Economic models with power-law memory. Chapter 1 In *Handbook of Fractional Calculus with Applications. Volume 8: Applications in Engineering, Life and Social Sciences. Part B.* Edited by D. Baleanu, A. M. Lopes. Series edited by J. A. Tenreiro Machado. Berlin, Boston: De Gruyter, 2019. pp. 1–32. ISBN: 978-3-11-057092. DOI: 10.1515/9783110571929-001.

[394] Tarasov, V. E. Fractional Stability. *Cornell University Preprint.* **2007**. arXiv:0711.2117. URL: https://arxiv.org/pdf/0711.2117.pdf.

[395] Tarasov, V. E.; Edelman, M. Fractional dissipative standard map. *Chaos: An Interdisciplinary Journal of Nonlinear Science* **2010**. Vol. 20. No. 2. Article ID: 023127. DOI: 10.1063/1.3443235.

[396] Tarasov, V. E.; Tarasova, V. V. Long and short memory in economics: fractional-order difference and differentiation. *IRA-International Journal of Management and Social Sciences* **2016**. Vol. 5. No. 2. P. 327–334. DOI: 10.21013/jmss.v5.n2.p10 (arXiv:1612.07903).

[397] Tarasov, V. E.; Tarasova, V. V. Accelerator and multiplier for macroeconomic processes with memory. *IRA-International Journal of Management and Social Sciences* **2017**. Vol. 9. No. 3. P. 86–125. DOI: 10.21013/jmss.v9.v3.p1.

[398] Tarasov, V. E.; Tarasova, V. V. Time-dependent fractional dynamics with memory in quantum and economic physics. *Annals of Physics* **2017**. Vol. 383. P. 579–599. DOI: 10.1016/j.aop.2017.05.017.

[399] Tarasov, V. E.; Tarasova, V. V. Criterion of existence of power-law memory for economic processes. *Entropy* **2018**. Vol. 20. No. 6. Article ID: 414. 24 pages. DOI: 10.3390/e20060414.

[400] Tarasov, V. E.; Tarasova, S. S. Fractional and integer derivatives with continuously distributed lag. *Communications in Nonlinear Science and Numerical Simulation* **2019**. Vol. 70. P. 125–169. DOI: 10.1016/j.cnsns.2018.10.014.

[401] Tarasov, V. E.; Tarasova, S. S. Probabilistic interpretation of Kober fractional integral of non-integer order. *Progress in Fractional Differentiation and Applications* **2019**. Vol. 5. No. 1. P. 1–5. DOI: 10.18576/pfda/050101. URL: http://www.naturalspublishing.com/files/published/a4x7w1w734sun1.pdf.

[402] Tarasov, V. E.; Tarasova, V. V. Dynamic Keynesian model of economic growth with memory and lag. *Mathematics* **2019**. Vol. 7. No. 2. Article ID: 178. DOI: 10.3390/math7020178.

[403] Tarasov, V. E.; Tarasova, V. V. Logistic equation with continuously distributed lag and application in economics. *Nonlinear Dynamics* **2019**. Vol. 97. No. 2. P. 1313–1328. DOI: 10.1007/s11071-019-05050-1.

[404] Tarasov, V. E.; Tarasova, V. V. Harrod-Domar growth model with memory and distributed lag. *Axioms* **2019**. Vol. 8. No. 1. Article ID: 9. DOI: 10.3390/axioms8010009.

[405] Tarasov, V. E.; Tarasova, V. V. Phillips model with exponentially distributed lag and power-law memory. *Computational and Applied Mathematics* **2019**. Vol. 38. No. 1. Article ID: 13. DOI: 10.1007/s40314-019-0775-y.

[406] Tarasov, V. E.; Tarasova, S. S. Fractional derivatives and integrals: What are they needed for?. *Mathematics* **2020**. Vol. 8. No. 2. Article ID: 164. DOI: 10.3390/math8020164.

[407] Tarasov, V. E.; Tarasova, V. V. Dynamic Keynesian model of economic growth with memory and lag. In *Advanced Mathematical Methods Theory and Applications*. Edited by F. Mainardi, A. Giusti. Basel, Beijing, Wuhan, Barcelona, Belgrade: MDPI, 2020. pp. 116–132. 198 pages. ISBN: 978-3-03928-247-0. DOI: 10.3390/books978-3-03928-247-0.

[408] Tarasov, V. E.; Trujillo, J. J. Fractional power-law spatial dispersion in electrodynamics. *Annals of Physics* **2013**. Vol. 334. P. 1–23. DOI: 10.1016/j.aop.2013.03.014 (arXiv:1503.04349).

[409] Tarasov, V. E.; Zaslavsky, G. M. Nonholonomic constraints with fractional derivatives. *Journal of Physics A* **2006**. Vol. 39. No. 31. P. 9797–9815. DOI: 10.1088/0305-4470/39/31/010 (arXiv:math-ph/0603067).

[410] Tarasov, V. E.; Zaslavsky, G. M. Fractional equations of kicked systems and discrete maps. *Journal of Physics A* **2008**. Vol. 41. No. 43. Article ID: 435101. 16 pages DOI: 10.1088/1751-8113/41/43/435101 (arXiv:1107.3953).

[411] Tarasova, S. S. *Theory of Probability in Problems of Aerospace Engineering*. Moscow: Max Press, 2018. 104 pages. ISBN: 978-5-317-05780-0. URL: https://b-ok.cc/book/3582033/f604ab.

[412] Tarasova, V. V.; Tarasov, V. E. A generalization of concepts of accelerator and multiplier to take into account memory effects in macroeconomics. *Journal of Economy and Entrepreneurship [Ekonomika i Predprinimatelstvo]* **2016**. No. 10-3 (75-3). P. 1121–1129. [in Russian]. URL: https://elibrary.ru/item.asp?id=27211611.

[413] Tarasova, V. V.; Tarasov, V. E. Criteria hereditarity of economic process and memory effect. *Young scientist [Molodoj Uchenyj]* **2016**. No. 14 (118). P. 396–399 [in Russian]. URL: https://elibrary.ru/item.asp?id=26389117 and https://moluch.ru/archive/118/32551/.

[414] Tarasova, V. V.; Tarasov, V. E. Deterministic factor analysis: methods of integro-differentiation of non-integral order. *Actual Problems of Economics and Law [Aktualnye Problemy Ekonomiki i Prava]* **2016**. Vol. 10. No. 4. P. 77–87 [in Russian]. DOI: 10.21202/1993-047X.10.2016.4.77-87. URL: https://www.elibrary.ru/item.asp?id=27469729.

[415] Tarasova, V. V.; Tarasov, V. E. Economic indicators: Ambiguity and memory effects. *Economics. Management. Law. [Ekonomika. Upravlenie. Pravo]* **2016**. No. 3 (66). P. 3–5. [in Russian]. URL: https://istina.msu.ru/media/publications/article/925/f43/26341086/EUP2016.pdf.

[416] Tarasova, V. V.; Tarasov, V. E. Economic indicator that generalizes average and marginal values. *Journal of Economy and Entrepreneurship [Ekonomika i Predprinimatelstvo]* **2016**. No. 11-1 (76-1). P. 817–823. [in Russian]. URL: https://elibrary.ru/item.asp?id=27340824.

[417] Tarasova, V. V.; Tarasov, V. E. Economic accelerator with memory: discrete time approach. *Problems of Modern Science and Education* **2016**. No. 36 (78). P. 37–42. DOI: 10.20861/2304-2338-2016-78-002 (arXiv:1612.07913).

[418] Tarasova, V. V.; Tarasov, V. E. Elasticity for economic processes with memory: fractional differential calculus approach. *Fractional Differential Calculus* **2016**. Vol. 6. No. 2. P. 219–232. DOI: 10.7153/fdc-06-14.

[419] Tarasova, V. V.; Tarasov, V. E. Elasticity of OTC cash turnover of currency market of Russian Federation. *Actual Problems of Humanities and Natural Sciences. [Aktualnye Problemy Gumanitarnyh i Estestvennyh Nauk]* **2016**. No. 7-1 (90). P. 207–215. [in Russian]. URL: https://www.elibrary.ru/item.asp?id=26365901 and URL: https://publikacia.net/archive/uploads/pages/2016_7_1/49.pdf.

[420] Tarasova, V. V.; Tarasov, V. E. Fractional dynamics of natural growth and memory effect in economics. *European Research* **2016**. No. 12 (23). P. 30–37. DOI: 10.20861/2410-2873-2016-23-004 (arXiv:1612.09060).

[421] Tarasova, V. V.; Tarasov, V. E. Hereditary generalization of Harrod-Domar model and memory effects. *Journal of Economy and Entrepreneurship [Ekonomika i Predprinimatelstvo]* **2016**. No. 10-2 (75-2). P. 72–78 [in Russian]. URL: https://elibrary.ru/item.asp?id=27220550.

[422] Tarasova, V. V.; Tarasov, V. E. Influence of memory effects on world economics and business. *Azimuth Scientific Research: Economics and Management [Azimut Nauchnih Issledovanii: Ekonomika i Upravlenie]* **2016**. Vol. 5. No. 4 (17). P. 369–372 [in Russian]. URL: https://elibrary.ru/item.asp?id=28383526 and URL: https://cyberleninka.ru/article/n/vliyanie-effektov-pamyati-na-mirovuyu-ekonomiku-i-biznes.

[423] Tarasova, V. V.; Tarasov, V. E. Keynesian model of economic growth with memory. *Economics and Management: Problems, Solutions [Ekonomika i Upravlenie: Problemy i Resheniya]* **2016**. No. 10-2 (58). P. 21–29. [in Russian]. URL: https://elibrary.ru/item.asp?id=27542599.

[424] Tarasova, V. V.; Tarasov, V. E. Marginal utility for economic processes with memory. *Almanac of Modern Science and Education [Almanah Sovremennoj Nauki i Obrazovaniya]* **2016**. No. 7 (109). P. 108–113. [in Russian]. URL: https://www.elibrary.ru/item.asp?id=26452585. URL: http://www.gramota.net/materials/1/2016/7/28.html.
[425] Tarasova, V. V.; Tarasov, V. E. Marginal values of non-integer order in economic analysis. *Azimuth Scientific Research: Economics and Management [Azimut Nauchnih Issledovanii: Ekonomika i Upravlenie]* **2016**. No. 3 (16). P. 197–201. [in Russian]. URL: https://elibrary.ru/item.asp?id=27198798.
[426] Tarasova, V. V.; Tarasov, V. E. Memory effects in hereditary Harrod-Domar model. *Problems of Modern Science and Education [Problemy Sovremennoj Nauki i Obrazovaniya]* **2016**. No. 32 (74). P. 38–44. DOI: 10.20861/2304-2338-2016-74-002 [in Russian]. URL: https://elibrary.ru/item.asp?id=27253247.
[427] Tarasova, V. V.; Tarasov, V. E. Memory effects in hereditary Keynes model. *Problems of Modern Science and Education [Problemy Sovremennoj Nauki i Obrazovaniya]* **2016**. No. 38 (80). P. 56–61. DOI: 10.20861/2304-2338-2016-80-001 [in Russian]. URL: https://www.elibrary.ru/item.asp?id=27537339 and URL: https://cyberleninka.ru/article/n/effekty-pamyati-v-ereditarnoy-modeli-keynsa.
[428] Tarasova, V. V.; Tarasov, V. E. Non-local measures of risk aversion in the economic process. *Economics: Theory and Practice [Ekonomika: Teoriya i Praktika]* **2016**. No. 4 (44). P. 54–58. [in Russian]. URL: https://elibrary.ru/item.asp?id=27541974 and URL: http://econ.kubsu.ru/images/4_2016.pdf.
[429] Tarasova, V. V.; Tarasov, V. E. On applicability of point elasticity of demand on price to exchange trading on US dollar. *Scientific Perspective [Nauchnaya Perspektiva]* **2016**. No. 6 (76). P. 6–11. [in Russian]. URL: https://elibrary.ru/item.asp?id=26344263 and URL: http://naupers.ru/6archive.html and URL: https://docplayer.ru/55299569-O-primenimosti-tochechnoy-elastichnosti-sprosa-po-cene-dlya-birzhevyh-torgov-po-dollaru-ssha.html.
[430] Tarasova, V. V.; Tarasov, V. E. Price elasticity of demand with memory. *Economics, Sociology and Law. [Ekonomika, Sociologiya i Pravo]* **2016**. No. 4-1. P. 98–106. [in Russian]. URL: https://elibrary.ru/item.asp?id=26093967.
[431] Tarasova, V. V.; Tarasov, V. E. Accelerators in macroeconomics: Comparison of discrete and continuous approaches. *Scientific Journal [Nauchnyy Zhurnal]* **2017**. No. 8 (21). P. 4–14. [in Russian] (arXiv:1712.09605). URL: https://elibrary.ru/item.asp?id=30561974.
[432] Tarasova, V. V.; Tarasov, V. E. Accelerators in macroeconomics: Comparison of discrete and continuous approaches. *American Journal of Economics and Business Administration* **2017**. Vol. 9. No. 3. P. 47–55. DOI: 10.3844/ajebasp.2017.47.55.
[433] Tarasova, V. V.; Tarasov, V. E. Chronological exponent for processes with memory and dynamic intersectoral economic models. *Science and Education Today [Nauka i Obrazovanie Segodnya]*. **2017**. No. 4 (15). P. 29–39. [in Russian]. URL: https://www.elibrary.ru/item.asp?id=29460893 and URL: https://cyberleninka.ru/article/n/hronologicheskaya-eksponenta-dlya-protsessov-s-pamyatyu-i-dinamicheskie-mezhotraslevye-modeli-ekonomiki.
[434] Tarasova, V. V.; Tarasov, V. E. Comments to the article "Long and short memory in economics: fractional-order difference and differentiation". *Problems of Modern Science and Education [Problemy Sovremennoj Nauki i Obrazovaniya]* **2017**. No. 31 (113). P. 26–28. DOI: 10.20861/2304-2338-2017-113-002.
[435] Tarasova, V. V.; Tarasov, V. E. Comments to the article "Long-term and short-term memory in discrete economic models". *Problems of Science [Problemy Nauki]*. **2017**. No. 8 (21). P. 12–13 [in Russian]. URL: https://elibrary.ru/item.asp?id=29862856.
[436] Tarasova, V. V.; Tarasov, V. E. Discrete accelerator with memory in macroeconomics. *Economics (ISSN: 2410-289X)*. **2017**. No. 8 (29). P. 32–40 [in Russian]. URL: https://elibrary.ru/item.asp?id=29944864.

[437] Tarasova, V. V.; Tarasov, V. E. Dynamic intersectoral models with memory that generalize Leontief model. *Journal of Economy and Entrepreneurship [Ekonomika i Predprinimatelstvo]* **2017**. No. 2-1 (79-1). P. 913–924 [in Russian]. URL: https://elibrary.ru/item.asp?id=28791089.

[438] Tarasova, V. V.; Tarasov, V. E. Economic growth model with constant pace and dynamic memory. *Problems of Modern Science and Education* **2017**. No. 2 (84). P. 40–45. DOI: 10.20861/2304-2338-2017-84-001 (arXiv:1701.06299).

[439] Tarasova, V. V.; Tarasov, V. E. Economic interpretation of fractional derivatives. *Progress in Fractional Differentiation and Applications* **2017**. Vol. 3. No. 1. P. 1–7. DOI: 10.18576/pfda/030101 (arXiv:1712.09575).

[440] Tarasova, V. V.; Tarasov, V. E. Economic model of natural growth with dynamic memory. *Actual Problems of Humanities and Natural Sciences. [Aktualnye Problemy Gumanitarnyh i Estestvennyh Nauk]* **2017**. No. 4-2. P. 51–58. [in Russian]. URL: https://elibrary.ru/item.asp?id=28855511 and URL: https://publikacia.net/archive/uploads/pages/2017_4_2/15.pdf.

[441] Tarasova, V. V.; Tarasov, V. E. Exact discretization of economic accelerator and multiplier with memory. *Fractal and Fractional.* **2017**. Vol. 1. No. 1. Article ID: 6. DOI: 10.3390/fractalfract1010006.

[442] Tarasova, V. V.; Tarasov, V. E. Exact discretization of economic accelerators and multipliers with memory. *Journal of Economy and Entrepreneurship [Ekonomika i Predprinimatelstvo]* **2017**. No. 7 (84). P. 1063–1069. [in Russian]. URL: https://elibrary.ru/item.asp?id=29774025.

[443] Tarasova, V. V.; Tarasov, V. E. Long-term and short-term memory in discrete economic models. *Competitiveness in a global world: economics, science, technology. [Konkurentosposobnost v Globalnom Mire: Ekonomika, Nauka, Tekhnologii]* **2017**. No. 7. P. 155–160. [in Russian]. URL: https://elibrary.ru/item.asp?id=29901818.

[444] Tarasova, V. V.; Tarasov, V. E. Logistic map with memory from economic model. *Chaos, Solitons and Fractals* **2017**. Vol. 95. P. 84–91. DOI: 10.1016/j.chaos.2016.12.012 (arXiv:1712.09092).

[445] Tarasova, V. V.; Tarasov, V. E. Macroeconomic models with dynamic memory. *Journal of Economy and Entrepreneurship [Ekonomika i Predprinimatelstvo]* **2017**. No. 3-2 (80-2). P. 26–35. [in Russian]. URL: https://elibrary.ru/item.asp?id=29078471.

[446] Tarasova, V. V.; Tarasov, V. E. Microeconomic meaning of derivatives of non-integer order. *Science and Education Today [Nauka i Obrazovanie Segodnya]* **2017**. No. 8 (19). P. 32–39. [in Russian]. URL: https://elibrary.ru/item.asp?id=29870013.

[447] Tarasova, V. V.; Tarasov, V. E. Model of economic growth with constant rate and dynamic memory. *Economics, Sociology and Law. [Ekonomika, Sociologiya i Pravo]* **2017**. No. 8. P. 18–24. [in Russian]. URL: https://elibrary.ru/item.asp?id=29938971.

[448] Tarasova, V. V.; Tarasov, V. E. Notion of dynamic memory in economic theory. *Journal of Economy and Entrepreneurship [Ekonomika i Predprinimatelstvo]* **2017**. No. 6 (83). P. 868–880. [in Russian]. URL: https://elibrary.ru/item.asp?id=29679697.

[449] Tarasova, V. V.; Tarasov, V. E. Risk aversion for investors with memory: Hereditary generalizations of Arrow-Pratt measure. *Financial Journal [Finansovyj Zhurnal]* **2017**. No. 2 (36). P. 46–63. [in Russian]. URL: https://elibrary.ru/item.asp?id=28966889 and URL: http://www.nifi.ru/images/FILES/Journal/Archive/2017/2/statii_2/fm_2017_2_05.pdf.

[450] Tarasova, V. V.; Tarasov, V. E. Concept of dynamic memory in economics. *Communications in Nonlinear Science and Numerical Simulation* **2018**. Vol. 55. P. 127–145. DOI: 10.1016/j.cnsns.2017.06.032 (arXiv:1712.09088).

[451] Tarasova, V. V.; Tarasov, V. E. Dynamic intersectoral models with power-law memory. *Communications in Nonlinear Science and Numerical Simulation* **2018**. Vol. 54. P. 100–117. DOI: 10.1016/j.cnsns.2017.05.015 (arXiv:1712.09087).

[452] Tarasova, V. V.; Tarasov, V. E. Fractional deterministic factor analysis of economic processes with memory and nonlocality. Chapter 9. In *Chaotic, Fractional, and Complex Dynamics: New Insights and Perspectives*. Edited by M. Edelman, E. Macau, M. A. F. Sanjuan. Springer International Publishing, 2018. pp. 173–189. DOI: 10.1007/978-3-319-68109-2_9 (Book DOI: 10.1007/978-3-319-68109-2. eBook ISBN: 978-3-319-68109-2. Hardcover ISBN: 978-3-319-68108-5).

[453] Tarasova, V. V.; Tarasov, V. E. Macroeconomic models with dynamic memory: Fractional calculus approach. *Applied Mathematics and Computation* **2018**. Vol. 338. P. 466–486. DOI: 10.1016/j.amc.2018.06.018.

[454] Tarasova, V. V.; Tarasov, V. E. Productivity with fatigue and long memory: Fractional calculus approach. *International Journal of Applied and Computational Mathematics* **2019**. Vol. 5. No. 1. Article ID: 16. DOI: 10.1007/s40819-019-0599-x.

[455] Tavares, D.; Almeida, R.; Torres, D. F. M. Caputo derivatives of fractional variable order: Numerical approximations. *Communications in Nonlinear Science and Numerical Simulation* **2016**. Vol. 35. P. 69–87. DOI: 10.1016/j.cnsns.2015.10.027.

[456] Tejado, I.; Valerio, D.; Valerio, N. Fractional calculus in economic growth modeling. The Portuguese case. In *Conference: 2014 International Conference on Fractional Differentiation and its Applications (FDA'14)*. IEEE, 2014. ISBN: 978-1-4799-2591-9. DOI: 10.1109/ICFDA.2014.6967427.

[457] Tejado, I.; Valerio, D.; Valerio, N. Fractional calculus in economic growth modelling. The Spanish case. In *CONTROLO'2014 – Proceedings of the 11th Portuguese Conference on Automatic Control*. Lecture Notes in Electrical Engineering. Volume 321. Edited by A. P. Moreira, A. Matos, G. Veiga. Cham: Springer International Publishing, 2015. P. 449–458. DOI: 10.1007/978-3-319-10380-8_43.

[458] Tejado, I.; Valerio, D.; Perez, E.; Valerio, N. Fractional calculus in economic growth modelling: The economies of France and Italy. In *Proceedings of International Conference on Fractional Differentiation and its Applications, Novi Sad, Serbia, July 18–20*. Edited by D. T. Spasic, N. Grahovac, M. Zigic, M. Rapaic, T. M. Atanackovic. Novi Sad, 2016. P. 113–123.

[459] Tejado, I.; Valerio, D.; Perez, E.; Valerio, N. Fractional calculus in economic growth modelling: the Spanish and Portuguese cases. *International Journal of Dynamics and Control* **2017**. Vol. 5. No. 1. P. 208–222. DOI: 10.1007/s40435-015-0219-5.

[460] Tejado, I.; Perez, E.; Valerio, D. Economic growth in the European Union modelled with fractional derivatives: First results. *Bulletin of the Polish Academy of Sciences, Technical Sciences* **2018**. Vol. 66. No. 4. P. 455–465. DOI: 10.24425/124262.

[461] Tejado, I.; Perez, E.; Valerio, D. Fractional calculus in economic growth modelling of the group of seven. *SSRN Electronic Journal*. **2018**. DOI: 10.2139/ssrn.3271391.

[462] Tejado, I.; Perez, E.; Valerio, D. Fractional calculus in economic growth modelling of the group of seven. *Fractional calculus and Applied Analysis* **2019**. Vol. 22. No. 1. P. 139–157. DOI: 10.1515/fca-2019-0009.

[463] Tejado, I.; Perez, E.; Valerio, D. Fractional derivatives for economic growth modelling of the Group of Twenty: Application to Prediction. *Mathematics* **2020**. Vol. 8. No. 1. Article ID: 5. 21 pages. DOI: 10.3390/math8010050.

[464] Tenreiro Machado, J. A.; Kiryakova, V. The chronicles of fractional calculus. *Fractional Calculus and Applied Analysis* **2017**. Vol. 20. No. 2. P. 307–336. DOI: 10.1515/fca-2017-0017.

[465] Tenreiro Machado, J. A.; Mata, M. E. A fractional perspective to the bond graph modelling of world economies. *Nonlinear Dynamics* **2015**. Vol. 80. No. 4. P. 1839–1852. DOI: 10.1007/s11071-014-1334-0.

[466] Tenreiro Machado, J. A.; Mata, M. E. Pseudo phase plane and fractional calculus modeling of western global economic downturn. *Communications in Nonlinear Science and Numerical Simulation* **2015**. Vol. 22. No. 1-3. P. 396–406. DOI: 10.1016/j.cnsns.2014.08.032.

[467] Tenreiro Machado, J. A.; Kiryakova, V.; Mainardi, F. Recent history of fractional calculus. *Communications in Nonlinear Science and Numerical Simulation* **2011**. Vol. 16. No. 3. P. 1140–1153. DOI: 10.1016/j.cnsns.2010.05.027.

[468] Tenreiro Machado, J. A.; Duarte, F. B.; Duarte, G. M. Fractional dynamics in financial indices. *International Journal of Bifurcation and Chaos* **2012**. Vol. 22. No. 10. Article ID: 1250249. 12 pages. DOI: 10.1142/S0218127412502495.

[469] Tenreiro Machado, J. A.; Galhano, A. M.; Trujillo, J. J., Science metrics on fractional calculus development since 1966. *Fractional Calculus and Applied Analysis* **2013**. Vol. 16. No. 2. P. 479–500. DOI: 10.2478/s13540-013-0030-y.

[470] Tenreiro Machado, J. A.; Galhano, A. M.; Trujillo, J. J., On development of fractional calculus during the last fifty years. *Scientometrics* **2014**. Vol. 98. No. 1. P. 577–582.

[471] Tenreiro Machado, J. A.; Mata, M. E.; Lopes, A. M. Fractional state space analysis of economic systems. *Entropy* **2015**. Vol. 17. No. 8. P. 5402–5421. DOI: 10.3390/e17085402.

[472] Tenreiro Machado, J. A.; Mata, M. E.; Lopes, A. M. Fractional dynamics and pseudo-phase space of country economic processes. *Mathematics* **2020**. Vol. 8. No. 1. Article ID: 81. 17 pages. DOI: 10.3390/math8010081.

[473] Thao, T. H. An approximate approach to fractional analysis for finance. *Nonlinear Analysis: Real World Applications* **2006**. Vol. 7. No. 1. P. 124–132. DOI: 10.1016/j.nonrwa.2004.08.012.

[474] The Logistic Map and the Route to Chaos: From The Beginnings to Modern Applications. Part III. Edited by M. Ausloos, M. Dirickx. Berlin: Springer 2006.

[475] The Royal Swedish Academy of Sciences (1995). The Scientific Contributions of Robert E. Lucas, Jr. URL: https://www.nobelprize.org/prizes/economic-sciences/1995/advanced-information/.

[476] The Sveriges Riksbank Prize in Economic Sciences in Memory of Alfred Nobel 1973. Press release. NobelPrize.org. Nobel Media AB 2019. Available online URL: https://www.nobelprize.org/prizes/economic-sciences/1973/press-release/.

[477] The Sveriges Riksbank Prize in Economic Sciences in Memory of Alfred Nobel 1987. Press Release. NobelPrize.org. Nobel Media AB 2019. Available online URL: https://www.nobelprize.org/prizes/economic-sciences/1987/press-release/.

[478] The Sveriges Riksbank Prize in Economic Sciences in Memory of Alfred Nobel 1995. NobelPrize.org. Nobel Media AB 2019. URL: https://www.nobelprize.org/prizes/economic-sciences/1995/summary/.

[479] The Sveriges Riksbank Prize in Economic Sciences in Memory of Alfred Nobel 2003. Press release. NobelPrize.org. Nobel Media AB 2019. URL: https://www.nobelprize.org/prizes/economic-sciences/2003/summary/.

[480] Thirlwall, A. P.; Stella, L.; Rampa, G. *Economic Dynamics, Trade and Growth: Essays on Harrodian Themes*. London, UK: Palgrave Macmillan, 1998. ISBN: 978-1-349-26933-4; 978-1-349-26931-0. DOI: 10.1007/978-1-349-26931-0.

[481] Time Series with Long Memory. Series: Advanced Texts in Econometrics. Edited by P. M. Robinson. Oxford: Oxford University Press, 2003. 392 pages. ISBN: 978-0199257300.

[482] Tricomi, F. G. *Integral Equations*. New York, London: Interscience Publishers, 1957. 238 pages.

[483] Tricomi, F. G. *Integral Equations*. Revised edition. New York, London: Dover Publications, 1985. 256 pages. ISBN: 978-0486648286.

[484] Trujillo, J. J.; Rivero, M.; Bonilla, D. On a Riemann–Liouville generalized Taylor's formula. *Journal of Mathematical Analysis and Applications* **1999**. Vol. 231. P. 255–265. DOI: jmaa.1998.6224.

[485] Tschernig, R. *Wechselkurse, Unsicherheit und Long Memory*. Heidelberg: Physica-Verlag, 1994 [in German]. 232 pages. ISBN: 978-3-7908-0753-0. DOI: 10.1007/978-3-642-95912-7.

[486] Tudor, C.; Viens, F. Statistical aspects of the fractional stochastic calculus. *The Annals of Statistics* **2007**. Vol. 35. No. 3. P. 1183–1212. DOI: 10.1214/009053606000001541. URL: https://projecteuclid.org/download/pdfview_1/euclid.aos/1185304003.

[487] Tumanova, E. A.; Shagas, N. L. *Macroeconomics. Elements of an Advanced Approach*. Moscow: INFRA-M, 2004 [in Russian]. 400 pages. ISBN: 5-16-001864-6.

[488] Vadzinsky, R. N. *Handbook of Probabilistic Distributions*. St. Petersburg: Nauka, 2001. 295 pages. ISBN: 5-02-024919X.

[489] Valerio, D.; Machado, J.; Kiryakova, V. Some pioneers of the applications of fractional calculus. *Fractional Calculus and Applied Analysis* **2014**. Vol. 17. No. 2. P. 552–578. DOI: 10.2478/s13540-014-0185-1.

[490] Varian, H. *Microeconomic Analysis*. 3rd Edition. New York: W. W. Norton, 1992. 548 pages. ISBN: 0-393-95735-7.

[491] Varian, H. R. *Intermediate Microeconomics. A Modern Approach*. 8th Edition. New York, London: W. W. Norton and Company, 2010. ISBN: 978-0-393-93424-3.

[492] Verhulst, P. F. Mathematical researches into the law of population growth increase. ["Recherches mathematiques sur la loi d'accroissement de la population"]. *Nouveaux Memoires de l'Academie Royale des Sciences et Belles-Lettres de Bruxelles* **1845**. Vol. 18. P. 1–42 [in French]. URL: https://gdz.sub.uni-goettingen.de/id/PPN129323640_0018.

[493] Volgina, O. A.; Golodnaya, N. Yu.; Odiyako, N. N.; Shuman, G. I. *Mathematical Modeling of Economic Processes and Systems*. Moscow: Knorus, 2011. 200 pages [in Russian]. ISBN: 978-5-406-00453-1.

[494] Volgina, O. A.; Golodnaya, N. Yu.; Odiyako, N. N.; Shuman, G. I. *Mathematical Modeling of Economic Processes and Systems*. Third Edition. Moscow: Knorus, 2016. 196 pages [in Russian]. ISBN: 978-5-406-04805-4.

[495] Volterra, V. Functional theory applied to hereditary phenomena [La teoria dei funzionali applicata ai fenomeni ereditari]. *Proceedings of the International Congress of Mathematicians: Bologna [Atti del Congresso internazionale dei matematici: Bologna]* **1928**. Vol. 1. P. 215–232 [in Italian]. URL: https://www.liberliber.it/mediateca/libri/v/volterra/la_teoria_dei_funzionali/pdf/volterra_la_teoria.pdf.

[496] Volterra, V. On the mathematical theory of hereditary phenomena [Sur la theorie mathematique des phenomenes hereditaires]. *Journal de Mathematiques Pures et Appliquees* **1928**. Serie 9. Tome 7. P. 249–298 [in French]. URL: http://sites.mathdoc.fr/JMPA/PDF/JMPA_1928_9_7_A14_0.pdf. URL: http://gallica.bnf.fr/ark:/12148/bpt6k107620n/f257n50.capture.

[497] Volterra, V. *Theory of Functionals and of Integral and Integro-Differential Equations*. London and Glasgow: Blackie and Son Ltd., 1930. 226 pages.

[498] Volterra, V. *Theory of Functionals and of Integral and Integro-Differential Equations*. New York: Dover, 2005. 288 pages. ISBN: 978-0486442846.

[499] Wang, C. C. The principle of fading memory. *Archive for Rational Mechanics and Analysis* **1965**. Vol. 18. No. 5. P. 343–366. DOI: 10.1007/BF00281325.

[500] Wang, Z.; Huang, X.; Shi, G. Analysis of nonlinear dynamics and chaos in a fractional order financial system with time delay. *Computers and Mathematics with Applications* **2011**. Vol. 62. No. 3. P. 1531–1539. DOI: 10.1016/j.camwa.2011.04.057.

[501] Wang, Z.; Huang, X.; Shen, H. Control of an uncertain fractional order economic system via adaptive sliding mode. *Neurocomputing* **2012**. Vol. 83. P. 83–88. DOI: 10.1016/j.neucom.2011.11.018.

[502] Wen, C. H.; Yang, J. H. Complexity evolution of chaotic financial systems based on fractional calculus. *Chaos, Solitons and Fractals* **2019**. Vol. 128. P. 242–251. DOI: 10.1016/j.chaos.2019.08.005.

[503] West, B. J. Exact solution to fractional logistic equation. *Physica A: Statistical Mechanics and its Applications* **2015**. Vol. 444. P. 182–187. DOI: 10.1016/j.physa.2015.10.037 (arxiv1508. 04419).

[504] West, B. J.; Picozzi, S. Fractional Langevin model of memory in financial time series. *Physical Review E* **2002**. Vol. 65. Article ID: 037106. DOI: 10.1103/PhysRevE.65. 037106.

[505] Wunderlin, A. On the Slaving Principle. In *Lasers and Synergetics*. Springer Proceedings in Physics. Volume 19. Edited by R. Graham, A. Wunderlin. Berlin, Heidelberg: Springer, 1987. ISBN: 978-3-642-72760-3. DOI: 10.1007/978-3-642-72758-0_10.

[506] Wyss, W. The fractional Black–Scholes equation. *Fractional Calculus and Applied Analysis* **2000**. Vol. 3. No. 1. P. 51–61.

[507] Xie, Y. K.; Wang, Z.; Meng, B. Stability and bifurcation of a delayed time-fractional order business cycle model with a general liquidity preference function and investment function. *Mathematics* **2019**. Vol. 7. No. 9. Aricle ID: 846. 10 pages. DOI: 10.3390/math7090846.

[508] Xu, Y.; He, Z. Synchronization of variable-order fractional financial system via active control method. *Central European Journal of Physics* **2013**. Vol. 11. No. 6. P. 824–835. DOI: 10.2478/s11534-013-0237-x.

[509] Yousefpour, A.; Jahanshahi, H.; Munoz-Pacheco, J. M.; Bekiros, S.; Wei, Z. A fractional-order hyper-chaotic economic system with transient chaos. *Chaos, Solitons and Fractals* **2020**. Vol. 130. Article ID: 109400. 12 pages. DOI: 10.1016/j.chaos.2019.109400.

[510] Yue, Y.; He, L.; Liu, G. Modeling and application of a new nonlinear fractional financial model. *Journal of Applied Mathematics* **2013**. Vol. 2013. Article ID: 325050. 9 pages. DOI: 10.1155/2013/325050.

[511] Zaslsvsky, G. M. *Hamiltonian Chaos and Fractional Dynamics*. Oxford: Oxford University Press, 2008. 432 pages. ISBN: 0199535485, 978-0199535484.

[512] Zeng, C.; Chen, Y. Q.; Yang, Q. Almost sure and moment stability properties of fractional order Black-Scholes model. *Fractional Calculus and Applied Analysis* **2013**. Vol. 16. No. 2. P. 317–331. DOI: 10.2478/s13540-013-0020-0.

[513] Zhang, W.-B. *Synergetic Economics. Time and Change in Nonlinear Economics*. Berlin, Heidelberg: Springer-Verlag, 1991. 246 pages. ISBN: 978-3-642-75911-6. DOI: 10.1007/978-3-642-75909-3.

[514] Zhukovskiy, V. J.; Chikriy, A. A. On discrete conflict-controlled processes described by Grunvald-Letnikov fractional systems. *Journal of Automation and Information Sciences* **2015**. Vol. 47. No. 1. P. 24–34. DOI: 10.1615/JAutomatInfScien.v47.i1.30.

Index

Fractional Calculus in Applied Sciences and Engineering

Volume 7
Manuel Duarte Ortigueira, Duarte Valério
Fractional Signals and Systems, 2020
ISBN 978-3-11-062129-7, e-ISBN (PDF) 978-3-11-062458-8,
e-ISBN (EPUB) 978-3-11-062132-7

Volume 6
Yingjie Liang, Wen Chen, Wei Cai
Hausdorff Calculus. Applications to Fractal Systems, 2018
ISBN 978-3-11-060692-8, e-ISBN (PDF) 978-3-11-060852-6,
e-ISBN (EPUB) 978-3-11-060705-5

Volume 5
JinRong Wang, Michal Fečkan
Fractional Hermite-Hadamard Inequalities, 2018
ISBN 978-3-11-052220-4, e-ISBN (PDF) 978-3-11-052362-1,
e-ISBN (EPUB) 978-3-11-052244-0

Volume 4
Kecai Cao, YangQuan Chen
Fractional Order Crowd Dynamics. Cyber-Human System Modeling and Control, 2018
ISBN 978-3-11-047281-3, e-ISBN (PDF) 978-3-11-047398-8,
e-ISBN (EPUB) 978-3-11-047283-7

Volume 3
Michal Fečkan, JinRong Wang, Michal Pospíšil
Fractional-Order Equations and Inclusions, 2017
ISBN 978-3-11-052138-2, e-ISBN (PDF) 978-3-11-052207-5,
e-ISBN (EPUB) 978-3-11-052155-9

Volume 2
Bruce J. West
Nature's Patterns and the Fractional Calculus, 2017
ISBN 978-3-11-053411-5, e-ISBN (PDF) 978-3-11-053513-6,
e-ISBN (EPUB) 978-3-11-053427-6

Bei Fragen zur Produktsicherheit wenden Sie sich bitte an:
If you have any questions regarding product safety,
please contact:

Walter de Gruyter GmbH
Genthiner Straße 13
10785 Berlin
productsafety@degruyterbrill.com